Immunochemistry of viruses II

The basis for serodiagnosis and vaccines

Immunochemistry of viruses II

The basis for serodiagnosis and vaccines

edited by

M.H.V. Van Regenmortel
Institut de Biologie Moléculaire et Cellulaire (IBMC)
15 rue René Descartes, 67084 Strasbourg, France

and

A.R. Neurath
The Lindsley F. Kimball Research Institute,
New York Blood Center, 310 East 67th Street,
New York, NY 10021, U.S.A.

1990
Elsevier
Amsterdam · New York · Oxford

ISBN 0 444 81182 6

Library of Congress Cataloging-in-Publication Data

Immunochemistry of viruses II : the basis for serodiagnosis and
vaccines / edited by M.H.V. Van Regenmortel and A.R. Neurath.
p. cm.
Includes bibliographical references.
Includes index.
ISBN 0-444-81182-6
1. Viral antigens. 2. Virus diseases--Serodiagnosis. 3. Viral
vaccines. 4. Immunochemistry. I. Van Regenmortel, M. H. V.
II. Neurath, A. Robert. III. Title: Immunochemistry of viruses 2.
[DNLM: 1. Immunochemistry. 2. Serodiagnosis. 3. Viral Vaccines.
4. Virus Diseases--immunology. 5. Viruses--immunology. QW 160
I33]
QR186.6.V57I46 1990
616'.0194--dc20
DNLM/DLC
for Library of Congress 90-3815
CIP

Published by:
Elsevier Science Publishers B.V. (Biomedical Division)
P.O. Box 1527
1000 BM Amsterdam
The Netherlands

Sole distributors for the USA and Canada:
Elsevier Science Publishing Co. Inc.
655 Avenue of the Americas
New York, NY 10010
U.S.A.

Printed in The Netherlands

Cover illustration shows the antibody NC41 epitope on the influenza virus neuraminidase, indicating the contact amino acids as seen by X-ray crystallography of the NA-Fab complex, and the positions of escape mutants.
Contributed by G.M. Air, University of Alabama, Birmingham, AL, USA.

Preface

Since the first volume of *Immunochemistry of Viruses* appeared five years ago, our knowledge of the antigenic properties of viruses has continued to expand very rapidly. In the earlier volume, 13 chapters were devoted to viral immunology, serological techniques and vaccines while the remaining chapters dealt with the antigenic properties of 13 major families of viruses. The present volume *Immunochemistry of Viruses II* contains five chapters that review some major advances in viral immunology (e.g., processing of antigens, local immunity, autoimmunity) and three chapters devoted to applications of recombinant DNA technology and synthetic peptides. The remaining 15 chapters each deal with the antigenic structure of distinct families of viruses. Some of these families (picorna, influenza viruses, etc.) had already been discussed in the first volume but so much has been learnt in the intervening years that it seemed necessary to update the earlier chapters. Additional virus families and groups have been included in volume II such as the arena-, bunya-, corona-, lenti- and poxviruses as well as bacteriophages.

The aim of this second volume is very much the same as that of the first: to bring under one cover data on the antigenic properties of a wide range of different viruses. We believe this helps to reveal the underlying unity of viral immunology and immunochemistry. Advances in techniques and analytical concepts that are made in the study of a particular group of viruses are usually applicable to most other groups. We hope therefore that the present volume will help in spreading useful information across the man-made barriers which sometimes keep the world of viruses in neat, separate compartments. An interesting example of such compartmentalization concerns the discovery made in 1963-1965 by the German virologist F.A. Anderer that synthetic peptides are potentially useful as viral vaccines. Anderer and his colleagues, working with tobacco mosaic virus, showed that short synthetic peptides corresponding to the carboxy-terminal region of the viral protein were able to induce antibodies that neutralized viral infectivity (Anderer F.A. and Schlumberger H.D., 1965, *Biochim. Biophys. Acta* 97, 503). Unfortunately, because of the host-based divisions between virologists, it took 15 years for the approach that Anderer showed was successful with a plant virus to be applied to animal viruses.

It is our belief that an understanding of viral immunochemistry and of the immune response to viral infections provides the necessary scientific base for designing better diagnostic reagents and better immunogens for prophylaxis and immunotherapy of viral diseases.

We dedicate this volume to the memory of Dr. František Link who, under difficult political circumstances, introduced one of us (ARN) to the field of Virology.

M.H.V. Van Regenmortel
A.R. Neurath

List of contributors

G.M. Air – *University of Alabama, Birmingham, AL, U.S.A.*

F. Brown – *Wellcome Biotech, Beckenham, Kent, U.K.*

M.J. Buchmeier – *Research Institute of Scripps Clinic, La Jolla, CA, U.S.A.*

T. Cabezón – *Smith Kline Biologicals, Rixensart, Belgium*

J.L. Carrascosa – *Universidad Autónoma de Madrid, Madrid, Spain*

D. Cavanagh – *AFRC Institute for Animal Health, Houghton, Huntingdon, Cambs., U.K.*

E. Celis – *Cellcor Therapies, Newton, MA, U.S.A.*

G.H. Cohen – *University of Pennsylvania, Philadelphia, PA, U.S.A.*

J.A. Cohen – *University of Pennsylvania, Philadelphia, PA, U.S.A.*

R.W. Compans – *University of Alabama, Birmingham, AL, U.S.A.*

M. De Wilde – *Smith Kline Biologicals, Rixensart, Belgium*

R.J. Eisenberg – *University of Pennsylvania, Philadelphia, PA, U.S.A.*

R.S. Fujinami – *University of California, La Jolla, CA, U.S.A.*

R. Garofalo – *State University of New York, Buffalo, NY, U.S.A.*

M.I. Greene – *University of Pennsylvania, Philadelphia, PA, U.S.A.*

F.X. Heinz – *University of Vienna, Vienna, Austria*

Y. Ichihashi – *Niigata University, Niigata, Japan*

V.J. Isola – *University of Pennsylvania, Philadelphia, PA, U.S.A.*

E. Kellenberger – *University of Basel, Basel, Switzerland*

H.-D. Klenk – *Philipps-Universität, Marburg, F.R.G.*

W.G. Laver – *John Curtin School of Medical Research, Canberra, ACT, Australia*

S. Modrow – *Max von Pettenkofer Institute, Munich, F.R.G.*

R.C. Montelaro – *Louisiana State University, Baton Rouge, LA, U.S.A.*

W.F. Mueller – *Children's Hospital, Buffalo, NY, U.S.A.*
M.I. Muggeridge – *University of Pennsylvania, Philadelphia, PA, U.S.A.*
A.R. Neurath – *Lindsley F. Kimball Research Institute, New York, NY, U.S.A.*
P.L. Ogra – *State University of New York, Buffalo, NY, U.S.A.*
E. Paoletti – *Virogenetics Corp., Troy, NY, U.S.A.*
S.D. Putney – *Repligen Corporation, Cambridge, MA, U.S.A.*
R. Ray – *Secretech, Inc., Birmingham, AL, U.S.A.*
S.R. Roberts – *University of Pennsylvania, Philadelphia, PA, U.S.A.*
J.T. Roehrig – *Centers for Disease Control, Fort Collins, CO, U.S.A.*
T. Rutgers – *Smith Kline Biologicals, Rixensart, Belgium*
A. Schmaljohn – *USAMRIID, Frederick, MD, U.S.A.*
C. Schmaljohn – *USAMRIID, Frederick, MD, U.S.A.*
J.A. Smith – *Massachusetts General Hospital, Boston, MA, U.S.A.*
W. Spaan – *Leiden University, Leiden, The Netherlands*
J. Tartaglia – *Virogenetics Corp., Troy, NY, U.S.A.*
Y. Thanavala – *Roswell Park Memorial Institute, Buffalo, NY, U.S.A.*
M.H.V. Van Regenmortel – *Institut de Biologie Moléculaire et Cellulaire du CNRS, Strasbourg, France*
D.B. Weiner – *University of Pennsylvania, Philadelphia, PA, U.S.A.*
K.P. Williams – *Harvard Medical School, Boston, MA, U.S.A.*
W.V. Williams – *University of Pennsylvania, Philadelphia, PA, U.S.A.*
H. Wolf – *Max von Pettenkofer Institut, Munich, F.R.G.*
K.E. Wright – *University of Ottawa, Ottawa, Ontario, Canada*

Contents

Eds. M.H.V. van Regenmortel and A.R. Neurath
Immunochemistry of viruses, II. The basis for serodiagnosis and vaccines.

The structure of viral epitopes 1

M.H.V. VAN REGENMORTEL

Institut de Biologie Moléculaire et Cellulaire du CNRS, 15 rue Descartes, 67084 Strasbourg, France

1. Types of epitopes

The antigenic sites or epitopes of a virus particle correspond to those parts of the capsid or envelope proteins that are specifically recognized by the binding sites or paratopes of antibody molecules. In the absence of further qualification, the term epitope refers to a B cell epitope, i.e., the site recognized by a B cell receptor and by an antibody. A second type of epitope known as T cell epitope corresponds to peptide fragments of viral proteins that have undergone metabolic processing within an antigen-presenting cell. These T cell epitopes are recognized by T cell receptors in association with proteins of the major histocompatibility complex (Berzofsky, 1988; Berzofsky et al., 1987; Livingstone and Fathman, 1987). T cell epitopes of viruses are further discussed in Chapter 3.

It has been customary to divide antigenic determinants into a number of conceptual categories that are sometimes difficult to distinguish experimentally. Sequential epitopes have been defined as random coil peptides that are recognized by an antibody independently of the conformation which the corresponding regions possess in the native, intact protein. In contrast, conformational determinants are said to require the native conformation present in the intact protein in order to be recognized by their complementary paratopes (Sela, 1969). Since a sequence of residues cannot be recognized independently of its conformation, sequential epitopes should not be equated with conformation-independent epitopes.

It is often difficult to establish if the epitope that reacts in a particular assay corresponds to the native conformation of the antigen or to a more or less denatured form of the protein. This is particularly the case when a solid-phase immunoassay is used, since proteins usually undergo some denaturation or conformational distortion when they are adsorbed to plastic (Soderquist and Walton, 1980; Friguet et al., 1984; McCullough et al., 1985; Darst et al., 1988). For instance, the fact that influenza hemagglutinin becomes denatured when adsorbed to plastic during solid-phase immunoassay is responsible for the claim, made a few years ago, that immunization with hemagglutinin peptides leads to a very high frequency of induction of antibodies able to recognize *native* hemagglutinin (Green et al., 1982; Niman et al., 1983). In these experiments, no evidence was in fact presented that the anti-peptide antibodies actually recognized the native protein. In order to demonstrate this, it would have been necessary to include controls in which the liquid-phase protein was allowed to compete with plate-bound antigen and to show that inhibition to a level approaching 100% occurred (Jemmerson, 1987). If the hemagglutinin peptides had been able to induce neutralizing antibodies, this might have been taken as indirect evidence that the native hemagglutinin conformation was recognized by some of the antibodies. However, no such neutralization has been demonstrated (Air and Laver, 1986).

It should be noted that it is not only monomeric proteins that are altered antigenically when they are adsorbed to plastic in solid-phase immunoassays. Intact virus particles also undergo changes in their antigenicity when bound to plastic and will expose epitopes that are inaccessible to antibodies when the particle is in solution (Al Moudallal et al., 1985; McCullough et al., 1985). Since the virus is usually incubated at pH 9.6 during the coating process, some dissociation of the virions occurs and monomers or oligomers of the viral protein become adsorbed. This phenomenon can be visualized by immunoelectron microscopy by performing the first coating step of the ELISA on an electron microscope grid instead of in the wells of a microtiter plate and identifying the reactive antigen with gold-labeled instead of enzyme-labeled antibodies (Dore et al., 1988). In sandwich ELISA the virus is mostly incubated at neutral pH in antibody-coated wells, which explains why the antigenicity of the intact particles tends to be preserved in this ELISA format (Al Moudallal et al., 1985; Parry et al., 1988).

1.1. Continuous and discontinuous epitopes

At present, the most common way of classifying epitopes consists in distinguishing continuous and discontinuous epitopes (Atassi and Smith, 1978). A continuous epitope is defined as a short linear stretch of residues endowed with distinctive conformational features. A discontinuous epitope is defined as a group of residues that are not contiguous in the sequence but are brought together by the folding of the polypeptide chain or by the juxtaposition of two separate peptide chains. Discontinuous epitopes are also called assembled topographic epitopes (Benjamin et al., 1984).

When peptide fragments are analyzed for antigenic reactivity, the label 'continuous epitope' is given to any linear fragment that is found to react with antibodies raised against the intact protein. It should be realized, however, that what is called a continuous epitope of the protein, in fact, usually represents only a portion of a larger discontinuous epitope. Antibodies induced by a native protein and directed to a complex discontinuous epitope may be able to cross-react weakly with an epitope fragment constituted by a linear peptide. The low cross-reactivity of such linear peptides is partly due to the fact that they rarely assume the same conformation as in the corresponding region in the protein. There is also evidence that not every residue in a so-called continuous epitope is necessarily able to make contact with the paratope (Geysen et al., 1988). Since some of the residues in a continuous epitope can be replaced by any of the other 19 amino acids without impairing the antigenicity, the so-called continuous epitope may in reality by antigenically discontinuous.

It is widely believed at present that the majority of protein epitopes are discontinuous (Leach, 1983; Berzofsky, 1985; Van Regenmortel, 1986). This belief stems from the fact that very few linear stretches of residues are found at the surface of globular proteins, as is easily ascertained from space-filling models of proteins. Furthermore, if it is assumed that all epitope-paratope interfaces in proteins involve a surface of about 700 Å as shown in the case of lysozyme epitopes (Amit et al., 1986; Sheriff et al., 1987), it follows that no protein epitope is likely to contain only atoms from a continuous stretch of residues (Barlow et al., 1986).

In a discussion of viral antigens, it is useful to distinguish four other categories of epitopes, the so-called cryptotopes, neotopes, metatopes and neutralization epitopes. Cryptotopes are epitopes that are hidden in the native form of the protein and become antigenically active only after fragmentation, depolymerization or denaturation of the antigen (Jerne, 1960). In viral capsids, cryptotopes are found for instance on the surfaces of the subunits that are turned inward and become buried during the association of the coat protein monomers (Van Regenmortel, 1966, 1982). Neotopes are epitopes that are specific for the quaternary structure of virions and are thus absent in the constituent monomeric subunits of the capsid (Van Regenmortel, 1966; Neurath and Rubin, 1971; Meloen et al., 1983; Ferguson et al., 1984). Neotopes arise either through conformational changes of the monomers induced by intersubunit bonding or through the juxtaposition of residues from neighboring subunits. Recent data obtained with neotope-specific monoclonal antibodies (mAbs) directed against tobacco mosaic virus (TMV) protein showed that these antibodies were able in some cases to induce a neotope conformation in the viral protein when the subunits were presented free in solution to antibody-coated plates (Dekker et al., 1989). In contrast, the neotope conformation could not be induced when the subunits were first immobilized on the solid phase and were then tested for their ability to react with anti-neotope mAbs.

Since the labels cryptotope and neotope were used for epitopes specific for the dissociated and polymerized forms of viral proteins, the additional label metatope was introduced for epitopes found in both states of the coat protein (Van Regenmortel,

1966). In a recent electron microscope study of 22 metatope-specific mAbs directed to TMV protein, it was found that all these antibodies reacted with only one of the two extremities of TMV particles representing less than 1% of the virion surface (Dore et al., 1988).

1.2. Neutralization epitopes

Neutralization epitopes correspond to those epitopes of the virus that are specifically recognized by antibody molecules able to neutralize the infectivity of the virus. Elucidating the structure of neutralization epitopes is important for the development of synthetic vaccines since it is this type of epitope which presumably should be mimicked by synthesis and tested for its ability to elicit protective immunity.

Neutralization refers to the inactivation of virus infectivity by antibody alone and should be distinguished from more complex inactivation phenomena that involve accessory factors such as complement and antibody-dependent cytotoxic cells. It is of course the antibody that neutralizes and which is therefore labeled a *neutralizing* antibody; the epitope itself is not neutralizing and should be referred to as a *neutralization* epitope. The concept of neutralization epitope is entirely operational since the mechanism of neutralization depends on properties intrinsic to the virus, the antibody and the host cell (Dimmock 1987). Several mechanisms appear to be implicated in the neutralization of viral infectivity (Mandel, 1985). One of them is the loss of conformational flexibility imposed upon the viral protein by neutralizing antibody and which may prevent viral uncoating. In poliovirus, cross-linking of adjacent pentamers by antibody prevents uncoating (Icenogle et al., 1983) and a single antibody molecule bivalently bound to a single virion may be sufficient for neutralization (Emini et al., 1983; Wetz et al., 1986). Cooperative effects are sometimes observed when neutralizing or non-neutralizing mAbs are mixed (Kimura-Kuroda and Yasui, 1983). Such synergy also appears to play a role during in vivo protection, as indicated by passive mouse protection tests (Mathews and Roehrig, 1984). Furthermore, non-neutralizing mAbs may also be able to induce passive protection (Gould et al., 1986; Brandriss et al., 1986).

Viral epitopes defined in terms of their binding reactivity to a mAb are often said to possess also 'functional' activity if they are able to induce neutralization, hemagglutination inhibition or hemadsorption inhibition. What is meant is that the mAb is able to inhibit a viral 'function' such as infectivity or hemagglutination. However, this does not imply that the site recognized by the antibody (the epitope) is identical to the functional receptor site involved in the recognition of a host cell or red blood cell. The antibody may simple interfere with such functional sites by steric hindrance or by inducing a conformational change.

The fact that some mAbs only neutralize the infectivity of a heterologous viral strain and not of the homologous virus (Heinz, 1986; Kimura-Kuroda and Yasui, 1986) indicates that the notion of neutralization epitope is an elusive one. The ability of an antibody to interfere with the process of viral infectivity does not depend only on its capacity to bind to a specific epitope. Whether such binding results in the loss of infectivity depends on the inherent susceptibility of the receptor binding site which may vary in different viral strains. Furthermore, the virus may only be neutralized with respect to one type of host cell and may retain full infectivity in another type of cell (Grady and Kinch, 1985).

Studies with a number of different viruses have shown that many epitopes recognized by neutralizing antibodies cannot be mimicked by linear peptides (Ferguson et al., 1984; Diamond et al., 1985; Sherry et al., 1986; Stave et al., 1986; Xie et al., 1987). There are, however, notable exceptions to this generalization (Pfaff et al., 1988; Bolwell et al., 1989). For instance, the sequence ^{145}RGDL^{148} in the VP1 protein of foot-and-mouth disease virus (FMDV) has been shown to be an important component of a neutralization epitope (Pfaff et al., 1988; Parry et al., 1989). Since the sequence ^{145}RGD^{147} corresponds to a tripeptide sequence that plays a major role in the binding of fibronectin to the surface of many cells (Ruoslahti and Pierschbacher, 1986), it seems plausible that antibodies that bind to the RGD sequence act in virus neutralization by blocking a cell binding site on FMDV (Pfaff et al., 1988; Acharya et al., 1989; Fox et al., 1989).

The regions 141–160 and 200–213 in the VP1 of FMDV have been shown to contain continuous epitopes that are able, when injected singly or together, to induce neutralizing antibodies (Strohmaier et al., 1982; Bittle et al., 1982; Pfaff et al., 1982; DiMarchi et al., 1986; Meloen et al., 1987; Francis et al., 1987a, Doel et al., 1988). Recently, it was shown that both peptide regions 141–160 and 200–213 could be recognized by the same neutralizing mAb (Parry et al., 1989). This means that these two regions that were initially considered to be two continuous neutralization epitopes probably correspond to a single discontinuous, neutralization epitope. In the case of poliovirus 1, a neutralization epitope was shown to comprise short linear stretches of residues from two different structural proteins, i.e., VP1 residues 221–226 and VP2 residues 164–170 (Page et al., 1988).

Attempts to vaccinate animals with FMDV subunits and peptides have demonstrated that the intact quaternary structure of the virus must be present in the immunogen if maximum protection is to be induced (Cartwright et al., 1982; Meloen and Barteling, 1986). Although immunization with the VP1 protein on its own is able to induce neutralizing antibody (Laporte et al., 1973; Kleid et al., 1981), the level of protection in pigs and cattle was insufficient and much lower than when animals were vaccinated with intact FMDV particles (Meloen and Barteling, 1986).

1.3. Epitopes of viral glycoproteins

Although the present discussion of viral epitopes is limited mainly to unmodified viral proteins, it should be stressed that most enveloped viruses possess glycosylated surface proteins (Feizi and Childs, 1987). Usually the carbohydrate moiety is attached to the peptide chain by *N*-glycosidic linkages but *O*-linked carbohydrate side chains are also found (see Chapter 2). Methods used for the enzymatic deglycosylation of glycoproteins have been described by Thotakura and Bahl (1987). Since the carbohydrate moieties are synthesized by the host cellular enzymes, the type of oligosaccharide side chain present on viral glycoproteins may be altered when the virus is grown in different types of host cells. For instance, the type and extent of glycosylation present on the protein is usually modified when viral proteins are produced by recombinant DNA technology in yeast or insect cells (see Chapter 7).

The presence of the same oligosaccharide side chains on the viral glycoprotein and on host components is responsible for the antigenic cross-reactivity that is sometimes observed between virus and host components (Ward et al., 1983). In general, however, the oligosaccharide side chains of viral glycoproteins do not act as epitopes per se, but only modulate the expression of neighboring epitopes constituted by residues of the underlying polypeptide chain (Alexander and Elder, 1984).

In a study of the reactivity of mAbs to tick-borne encephalitis virus, it was found that the presence of carbohydrate on the glycoprotein E preserved the conformational integrity of an epitope; in contrast, the same epitope in the deglycosylated protein lost its activity in the presence of 0.5% SDS (Guirakhoo et al., 1989). In the case of influenza virus hemagglutinin, it was found that a single substitution of position 63 which led to the attachment of an additional oligosaccharide prevented a mAb from binding to its underlying epitope (Skehel et al., 1984). The ability of carbohydrate side chains to sterically hinder the binding of antibodies to their epitopes is further discussed in Chapter 2.

2. Methods used in the localization of viral epitopes

Several reviews describing the methods used to localize epitope in proteins are available (Atassi, 1984; Benjamin et al., 1984; Berzofsky, 1985; Jemmerson and Paterson, 1986; Van Regenmortel, 1984, 1989a). The different approaches that have been used for mapping viral epitopes are summarized in Table 1.1. As discussed elsewhere (Van Regenmortel, 1989b), only method 1 (Table 1.1) is based on a structural analysis and describes the spatial relationships found in an antigen-antibody complex. In contrast, all other methods correspond to binding assays that are necessarily of an operational nature and are based on activity measurements, i.e., on a functional analysis.

2.1. X-ray crystallography

At present only one viral epitope of influenza neuraminidase has been analyzed by X-ray crystallography of antigen-Fab complexes (Colman et al., 1987a, Colman, 1988). As in the case of the two lysozyme epitopes delineated by the same method

TABLE 1.1

Methods used to localize epitopes in viruses

Method	Type of epitope recognized	Selected references
1 X-ray crystallography of antigen-Fab complexes	Discontinuous epitope reacting with homologous antibody	Colman et al., 1987a Colman, 1988
2 Study of cross-reactive binding of natural or synthetic peptide fragments with viral antibodies	Continuous epitope cross-reacting with heterologous antibody	Altschuh et al., 1983 Geysen et al., 1984 Houghten, 1985 Gnann et al., 1987
3 Study of cross-reactive binding of fusion proteins with viral antibodies	Cross-reactive continuous epitope	Van der Werf et al., 1983 Offensperger et al., 1985 Lenstra et al., 1989
4 Study of cross-reactive binding of virus with anti-peptide antibodies	Cross-reactive continuous epitope	Lerner, 1984 Walter, 1986 Dyson et al., 1988 Strynadka et al., 1988
5 Analysis of viral mutants with monoclonal antibodies	Neutralization epitopes and discontinuous epitopes	Wiley et al., 1981 McCullough., 1986 Minor et al., 1986 Sherry et al., 1986
6 Competitive binding assays with pairs of monoclonal antibodies	Only relative position of epitopes is defined	Mathews and Roehrig, 1984 Heinz, 1986 Cecilia et al., 1988

(Amit et al., 1986; Sheriff et al., 1987), the neuraminidase epitope was found to be discontinuous. It is composed of four surface loops and comprises about 15 residues. All six complementarity determining regions (CDR) of the antibody (i.e., about 16 residues), with the possible exception of light chain CDR 1, appear to make contact with the viral epitope. It should be noted that the notion of what constitutes a contact residue contributing to the structure of the epitope is less straightforward than might have been thought. As pointed out by Getzoff et al.(1988), various crystallographic groups use different definitions of contact residues, for instance: residues within van der Waals contact distance, residues buried to a certain radius probe sphere or residues with side chains interacting directly. Although the crystallographic map (Colman et al., 1987a) indicates that about 15 residues of neuraminidase are involved in the interface with the paratope, it is uncertain how many of these residues actually contribute to the binding energy of interaction. It is likely, for instance, that not all of the 12 hydrogen bonds identified in the lysozyme-antibody complex by Amit et al.(1986) contribute to the free energy of the system which is rather small since the affinity constant of the antibody is about 10^8. In general, the loss of one or two hydrogen bonds would lead to a considerable reduction in the binding constant of several orders of magnitude, which should effectively abolish the binding interaction (Colman et al., 1987b). Such considerations provide a rationale for explaining the abolition of antibody binding that is frequently observed when a reactive antigen is replaced by a mutant with a single residue exchange.

Since the three-dimensional structure of several plant and animal viruses and viral proteins has now been determined by X-ray crystallography (Bloomer et al., 1978; Harrison et al., 1978; Abad-Zapatero et al., 1980; Wilson et al., 1981; Hogle et al., 1985, 1986; Luo et al., 1987; Wilson et al., 1981; Rossmann et al., 1985; Acharya et al., 1989), it is likely that the structure of additional viral antigen-antibody complexes will be solved in the near future. Furthermore, crystallographic data of peptide-antibody complexes may help to resolve the question of how anti-peptide antibodies are able to recognize both a small peptide antigen and the same sequence in the intact protein (Schulze-Gahmen et al., 1988). Modeling studies based on sequence analysis and computer graphics may also help to identify likely epitopes in viruses that are related to one of those for which the three-dimensional structure has been elucidated (Argos and Fuller, 1988; Luo et al., 1988; Pevear et al., 1988; Page et al., 1988).

2.2. Studies with synthetic peptides as antigenic probes

The most common method used for localizing epitopes in proteins consists in measuring the ability of antibodies raised against the intact molecule to cross-react with natural or synthetic peptide fragments. If a peptide is able to bind to the protein antibodies, it is assumed that it contains a continuous epitope of the protein. A more precise localization of the epitope is achieved by testing overlapping peptides of decreasing size and determining which is the smallest peptide that still possesses a significant level of antigenic reactivity (Benjamini, 1977). Usually, the degree of antigenic cross-reactivity observed with small peptides is very low, the major exception to this rule being peptides that correspond to chain termini. In a majority of proteins, the terminal segments are surface-oriented (Thornton and Sibanda, 1983) and are more mobile than internal sections of the polypeptide chain. This may explain why the cross-reactivity observed with short terminal peptides is usually much higher than with internal peptides (Milton and Van Regenmortel, 1979; Walter et al., 1980; Muller et al., 1982; Westhof et al., 1984; Tainer et al., 1985a; Walter, 1986).

Instead of testing systematically a very large number of peptides representing the entire sequence of a viral protein (Markoff et al., 1988), another approach consists in synthesizing a limited number of peptides predicted to correspond to antigenic regions on the basis of various algorithms (Gnann et al., 1987; Trudel et al., 1987; Strynadka et al., 1988; Mäkelä et al., 1989; Portetelle et al., 1989; Roehrig et al., 1989). By testing such peptides with neutralizing mAbs, it is possible to identify putative neutralization epitopes. Immunization with the corresponding peptides will establish if the region is indeed capable of inducing neutralizing antibodies.

The use of synthetic peptides in viral diagnosis is discussed in Chapter 6 and applications of the synthetic peptide methodology to the study of the human immunodeficiency virus are described in two recent reviews (Kennedy et al., 1988; Wang, 1988).

2.2.1. Free peptides or peptides conjugated to a carrier

The most common method for preparing synthetic peptides is solid-phase synthesis on a polymeric resin. This can be carried out by means of various, commercially available automated instruments (Kent and Clark-Lewis, 1985). After synthesis, the peptides are cleaved from the resin and are then used in their free form or after conjugation to a carrier molecule. Methods for synthesizing peptides and for testing them in a variety of immunoassays have been described in detail in a recent text (Van Regenmortel et al., 1988). In order to localize epitopes in a protein, peptide fragments are tested for their capacity to inhibit the reaction of the protein with its homologous antibodies. When the peptide is tested in its free form in solution, the likelihood that it will have a suitable conformation for binding to the antibody is enhanced (Altschuh and Van Regenmortel, 1982; Nestorowicz et al., 1985). Sometimes, however, better results are obtained when the capacity of the peptide to be recognized by anti-protein antibodies is tested with a peptide-carrier conjugate adsorbed to a solid-phase (Al Moudallal et al., 1985). It is possible that the microenvironment at the surface of the carrier protein induces the peptide to assume a more appropriate conformation for antibody recognition. Cyclization of the peptide is another procedure that has been used successfully for inducing a conformation that approaches the one found in the intact protein (Dreesman et al., 1982; Dorow et al., 1985; Schulze-Gahmen et al., 1986).

Another commonly used procedure consists in simply adsorbing the peptide to the plastic solid-phase and testing its ability to be recognized by anti-protein antibodies. If during the adsorption process the solution of peptide is not allowed to dry up in the wells, increased antigenic activity in viral peptides may be observed (Norrby et al., 1987). If the peptide becomes adsorbed only with difficulty, a variety of buffers in the pH range 4–8 can be tried (Geerligs et al., 1988). It should be stressed that the antigenic activity of peptides may differ when they are tested as free peptides in solution, conjugated to a carrier or adsorbed to a solid-phase (Muller et al., 1986). If the first approach that is used for testing the peptide is unsuccessful, it is obviously important to try additional methods. Recognition of peptides by antibodies may depend on subtle conformational requirements which are not necessarily fulfilled in all types of assay. In studies with FMDV type O for example, the majority of neutralizing mAbs failed to recognize peptides from the region 140–160 of VP1 when they were bound directly to ELISA plates; in contrast mAbs to type A virus did recognize the corresponding immobilized peptides (Bolwell et al., 1989). The choice of conjugation procedure used for coupling the peptide to the carrier may also be important, since the antigenic or immunogenic activity of the peptide may be drastically affected when different coupling reagents are used (Briand et al., 1985; Dyrberg and Oldstone, 1986; Van Regenmortel et al., 1988).

When large numbers of peptides need to be synthesized, the simultaneous multiple peptide synthesis method of Houghten (1985) is very useful. In this so-called teabag method, the individual resins used for solid-phase synthesis of different peptides are contained in solvent-permeable packets. This allows the numerous repetitive steps in the synthesis to be performed simultaneously for different peptides. This method has been used for obtaining large numbers of peptides of influenza virus (Houghten 1985), respiratory syncytial virus (Norrby et al., 1987) and FMDV (Bolwell et al., 1989).

2.2.2. Peptides still attached to the support used for synthesis

Peptides can be tested without necessarily being cleaved from the polystyrene support used during the synthesis (Hurrell et al., 1977; Shi et al., 1984). Resin-bound synthetic peptides have been used for detecting viral antibodies in human sera as well as for raising antibodies in animals (Kennedy et al., 1987; Modrow et al., 1989).

Another method for producing very large numbers of synthetic peptides was developed by Geysen et al.(1984, 1987a). This method known as the pepscan method allows the concurrent synthesis of hundreds of peptides on polyethylene rods. The rods are assembled into a polyethylene holder with the format and spacing of a microtiter plate. This allows the peptides to be tested easily by ELISA while they are still attached to the rod support. After each ELISA, the rods can be freed of bound antibody by sonication and can be tested again with different antisera as many as 30 times.

The method is ideally suited for testing all possible overlapping peptides of a viral protein, for instance all hexa, hepta, octa, nona and decapeptides starting for the N-terminal residue down to the C-terminal one (Meloen et al., 1987; Middeldorp and Meloen, 1988; Aaskov et al., 1989). The role of individual amino acids in the binding interaction between peptide and antibody can also be determined by pepscan analysis by systematically testing sets of peptides in which each of the residues has been replaced by the other, 19 amino acids (Geysen et al., 1985, 1988; Schoofs et al., 1988).

Clearly, pepscan analysis will only detect so-called continuous epitopes and will fail to detect antigenic activity in peptides that represent too small a fragment of a complex discontinuous epitope. Attachment of the peptide by its C-terminal residue as well as the chemical environment of the solid support may impair antigenic reactivity in peptides that could be active when tested in a different assay format. There is also evidence that some peptide sequences may give rise to nonspecific binding, as shown by the spurious ELISA results obtained with certain control Sera. In view of these limitations and because it is impractical to chemically verify the sequence of peptides attached to the rods, results obtained by the pepscan method should be confirmed by testing peptides obtained by classical methods. The main value of the method lies in its ability to effect an initial screening of very large numbers of peptides including peptide analogues.

2.3. Studies with fusion proteins and peptides

A number of viral polypeptides have been expressed by recombinant DNA techniques in a variety of microbial and mammalian cell systems (Kleid et al., 1981; Murray et al., 1984; Lasky et al., 1986; Cane and Gould, 1988; Emini et al., 1988; Putnak et al., 1988; Salfeld et al., 1989). When used to immunize animals, such recombinant proteins are able to induce neutralizing antibodies (see Chapters 7 and 8). An epitope of FMDV (residues of 142–160 of VP1) inserted into hepatitis B virus core particles was found to be 500 times more immunogenic than the corresponding free peptide (Clarke et al., 1987). Recombinant baculoviruses containing sequences of animal viruses (Matsuura et al., 1987; Luckow and Summers, 1988) have also been shown to induce protective immunity in rats and rabbits (Van Wyke Coelingh et al., 1987; Zhang et al., 1988; Ray et al., 1989).

A sequence corresponding to a DNA fragment encoding a poliovirus neutralization epitope (residues 93–103 of VP1) has been inserted into the S gene encoding the major envelope protein of hepatitis B virus (HBV). The inserted poliovirus VP1 sequence expressed in hybrid particles was immunogenic and induced antibodies that recognized the synthetic peptide 93–103 (Delpeyroux et al., 1986). A poliovirus neutralization epitope has also been inserted at the C-terminal end of tobacco mosaic virus (TMV) coat protein. The hybrid TMV protein produced in *Escherichia coli* was able to assemble into virus-like rods and induced the formation of poliovirus neutralizing antibodies (Haynes et al., 1986). Several other virus recombinant systems expressing foreign viral epitopes have been described in recent years (Valenzuela et al., 1985; Andrew et al., 1987; Michel et al., 1988). A genetically engineered vaccinia virus has been described which expressed at the same time epitopes of HBV, herpes simplex virus and influenza hemagglutinin (Perkus et al., 1985).

Epitopes can also be localized by expressing parts of the viral protein in a prokaryotic expression system and measuring the antigenic reactivity of the expression products. In the pEX expression system, the product of an inserted DNA fragment becomes the C-terminal portion of a hybrid-galactosidase protein which precipitates in the cell. In order to assay its antigenic reactivity, the hybrid protein is solubilized in SDS and transferred to a nitrocellulose filter (Stanley, 1983). Another commonly used system is the bacteriophage λgt11 expression vector (Mehra et al., 1986).

A number of continuous epitopes have been identified by means of such prokaryotic expression vectors, for instance in poliovirus (Van der Werf et al., 1983), feline leukemia virus (Nunberg et al., 1984), HBV (Offensperger et al., 1985), Japanese encephalitis virus (Mason et al., 1987) and infectious bronchitis virus (Lenstra et al., 1989). By its very nature, this approach is only able to detect continuous epitopes. It cannot identify discontinuous epitopes or any epitopes for which the conformation needed for antigenicity is not retained in the fusion product.

2.4. Studies with anti-peptide antibodies

This method consists in using peptide fragments for immunization and then ascertaining if the resulting anti-peptide antibodies react with the intact viral protein. A positive cross-reaction is taken as an indication that the peptide approximates to an epitope of the protein. The claim that immunization with peptides always leads to a very high frequency of induction of antibodies capable of recognizing the native protein (Lerner 1984) has remained controversial since no evidence was presented that the observed cross-reactions were not due to anti-peptide antibodies simply cross-reacting with denatured protein molecules. The contention that antibodies against a highly disordered state (the peptide) are able to recognize the highly ordered state (the folded protein in its native conformation), while the reverse is not necessarily the case, has been called the order-disorder paradox (Dyson et al., 1988). It has been suggested that this type of result is obtained because peptides show a marked conformational preference in solution. It is conceivable that such a preferred conformation becomes stabilized on the surface of the carrier protein or when the peptide binds to the B cell receptor during immune stimulation (Dyson et al., 1988). However, if a process of induced fit is able to influence the conformation of the immunogenic form of the peptide, a similar induction of conformation should also occur when the peptide interacts with anti-protein antibodies. This means that antibodies against the ordered state should also be able to recognize the disordered peptide, which was in fact not observed in the experiments with hemagglutinin peptides (Green et al., 1982). As discussed above (section 1), these results can be explained by assuming that the protein was denatured on the solid-phase and that induction of a conformation complementary to the antibody cannot occur when the antigen is adsorbed to plastic (Dekker et al., 1989).

Methods used for producing anti-peptide antisera have been reviewed by Palfreyman et al.(1984), Walter (1986) and Van Regenmortel et al.(1988). Peptides of less than 10–15 amino acids are usually coupled to a carrier protein for immunization purposes. The method of conjugation used may strongly influence the type of anti-peptide antibodies that are obtained (Dyrberg and Oldstone, 1986; Schaaper et al., 1989). Furthermore, non-responsiveness to a peptide in a particular immunized animal may be overcome by attaching the peptide to a second one that is capable of inducing helper T cell activity (Francis et al., 1987b). Anti-peptide mAbs and antisera have been successfully used for identifying viral epitopes (Wilson et al., 1984; Kennedy et al., 1986; Tang et al., 1988) as well as for locating neutralization epitopes that could play a role as potential vaccine candidates, for instance for hepatitis A virus (Emini et al., 1985), rhinovirus (McCray and Werner, 1987), feline leukemia virus (Elder et al., 1987), herpes simplex virus (Strynadka et al., 1988), and bovine leukemia virus (Portetelle et al., 1989).

2.5. Analysis of viral mutants with monoclonal antibodies

The use of mAbs for elucidating the structure of viral epitopes has been discussed in several reviews (Yewdell and Gerhard, 1981; Carter and ter Meulen, 1984; Van Regenmortel, 1984; McCullough, 1986). Viral variants possessing a small number of amino acid substitutions in one of their structural proteins can be analyzed with a panel of mAbs. If it is found that the substition leads to a change in antibody binding, it is mostly assumed that the mutated residue is directly involved in the structure of an epitope. Although this assumption may often be valid (Hornbeck and Wilson, 1984; Knossow et al., 1984; Varghese et al., 1988), it is known that amino acid substitutions that affect antibody binding are sometimes located at a distance from the known epitope. This suggests that certain mutations may alter the antigenicity of viral proteins by an indirect conformational effect (Al Moudallal et al., 1982; Diamond et al., 1985; Blondel et al., 1986; Nicolaisen-Strouss et al., 1987; Barnett et al., 1989). In general, however, it is likely that single substitutions at the surface of viral proteins only cause local structural adjustments with no long-range conformational effects.

Instead of inferring the position of epitopes in a viral protein from the ability of mAbs to distinguish between naturally occurring mutants, the same can be done by using mAbs for the selection of non-neutralizable virus mutants (Pollock et al., 1984). By growing the virus in the presence of neutralizing mAbs, escape mutants can be selected which are then sequenced to identify the residue substitution that rendered the mutant neutralization-resistant. This approach was first applied to the study of the antigenic structure of influenza virus hemagglutinin (Gerhard and Webster, 1978; Wiley et al., 1981; Air and Laver, 1986; Wiley and Skehel, 1987) and has been used to map the epitopes of many different viral antigens (Diamond et al., 1985; Minor et al., 1986; Sherry et al., 1986; Webster et al., 1987).

As pointed out by Page et al.(1988), the characterization of neutralization-resistant mutants cannot identify every amino acid in an epitope. For instance, any mutation in the epitope which would interfere with capsid stability, virus assembly, receptor recognition or any other requirement of the virus replication cycle would not be viable and would thus not be detected.

An additional method for distinguishing among mAb-selected variants which has been used with herpes simplex and measles viruses consists in analyzing the peptides produced by proteolysis of mAb-bound antigen (Eisenberg et al., 1982; Sheshberadaran and Payne, 1988).

2.6. Topographic epitope mapping by competitive binding assays

Competitive binding assay with pairs of mAbs can be used to determine the relative position of epitopes as well as the degree of physical overlap between individual epitopes (Breschkin et al., 1981; Berzofsky 1984). mAbs that do not block each other define separate epitopes. Usually a number of such epitopes can be linked by other overlapping epitopes to form a series of separate antigenic domains (Heinz et al., 1983; Schlesinger et al., 1984; Cammack and Gould, 1986; Adam et al., 1987; Cecilia et al., 1988; Guirakhoo et al., 1989). When the number of mAbs used in epitope mapping is large, it is not uncommon to observe a continuum of epitopes that can no longer be subdivided into discrete, separate antigenic domains (Underwood 1982; Mathews and Roehrig, 1984; Berton and Webster, 1985).

It should be noted that when two mAbs completely block each other in a competitive binding assay, this does not mean that the corresponding epitopes are necessarily the same. Because of the bulkiness of antibody molecules, considerable steric hindrance always occurs and this may prevent mAbs directed against neighboring epitopes from binding simultaneously to the antigen surface. This phenomenon is clearly illustrated by the fact that the antigenic valence of a virus (i.e., the maximum number of antibody molecules that can be bound simultaneously to the virus particle) is often much smaller than the number of protein subunits that make up the capsid (Van Regenmortel, 1988).

In the case of complex antigens such as viruses that possess a large number of identical subunits, the interpretation of competition assays is rarely straightforward. Although individual viral subunits always possess a number of different epitopes, when incorporated into a capsid they are usually unable, because of steric hindrance, to bind more than one antibody molecule. In principle, therefore, any specific mAb preparation, if used at a sufficiently high concentration, should be able to saturate the viral surface. When a mixture of two mAbs is allowed to bind to a virus particle, the one with the highest avidity will tend to compete successfully for the single binding space that is available on each subunit, irrespective of whether it recognizes the same or a different epitope.

In competitive binding assays, the competing antibody instead of blocking binding may actually enhance the binding of the second antibody (Lubeck and Gerhard, 1982; Heinz et al., 1984; Cecilia et al., 1988). The enhancement of binding may be unidirectional (i.e., caused by only one of the two antibodies) or bidirectional (Heinz, 1986). On the basis of the results of Scatchard analysis, it has been shown in the tick-borne encephalitis virus system that one enhancing antibody caused an increase in avidity of the second antibody (Heinz et al., 1984). Such an increase in avidity due to cooperative antibody interactions is probably the result of an allosteric change brought about by the binding of the first antibody and illustrates the heterospecific binding capacity of antibodies (Underwood, 1985). Heterospecificity (also called heteroclitic binding) refers to the ability of an antibody to bind more strongly to a number of structurally related epitopes than to the one against which it was raised, and appears to be a universal feature of antibodies (Al Moudallal et al., 1982; Harper et al., 1987).

3. Prediction of viral epitopes

The accumulated knowledge concerning the antigenic structure of a small number of well-studied proteins has stimulated attempts to predict the location of continuous epitopes in proteins from certain features of the primary structure (Van Regenmortel et al., 1988; Van Regenmortel, 1989b). A number of studies have shown that the location of continuous epitopes in proteins tends to correlate with structural features of segments of the polypeptide chain such as their hydrophilicity (Hopp and Woods, 1981; Kyte and Doolittle, 1982), atomic mobility (Westhof et al., 1984; Tainer et al., 1984, 1985a), static accessibility (Thornton et al., 1986; Novotny and Haber 1986) and sequence variability (Bittle et al., 1982; Modrow et al., 1987).

There is considerable disagreement among different workers about which structural parameter correlates best with antigenicity (Jemmerson and Paterson, 1985; Van Regenmortel and Altschuh, 1985; Tainer et al., 1985b, Novotny et al., 1987a,b; Geysen et al., 1987b). It is clear, however, that the three parameters of hydrophilicity, static accessibility and segmental mobility are not independent variables but represent interconnected aspects of the folding pattern of globular proteins. Loops and turns, for example, usually correspond to surface projections and they also tend to contain hydrophilic residues and to possess higher than average mobility.

Several relative scales describing the structural propensity of each of the 20 amino acids have been derived and these are used for constructing structural profiles and for predicting the location of continuous epitopes in proteins (Hopp and Woods, 1983; Hopp, 1986; Karplus and Schulz, 1985; Welling et al., 1985; Parker et al., 1986; Sette et al., 1986).

Plots that describe a structural propensity, for example hydrophilicity, are constructed by averaging the values of this parameter over segments of 6 or 7 residues at a time, along the entire length of the polypeptide chain. The position of peaks and valleys in such hydrophilicity plots tends to occur in the same places as in plots that predict the location of helices and β-strands (Chou and Fasman, 1978; Garnier et al., 1978). This correlation arises because the centers of the largest helices and β-strands usually correspond to the tightly packed hydrophobic cores of proteins. When the helix and β-strands prediction plots are inverted, the resulting plots show peaks in the same positions as in hydrophilicity plots (Hopp, 1986). An algorithm that computes the so-called antigenic index along the protein sequence has been developed which combines the predicted secondary structure plots with hydrophilicity, accessibility and mobility plots (Jameson and Wolf, 1988; Wolf et al., 1988). A combination of various parameters was also used by Parker et al. (1986) who concluded that a composite profile based on the maximum values for individual residue positions obtained from a hydrophilicity scale, an accessibility scale and a mobility scale gave the best prediction results.

Although antigenicity prediction methods are extremely popular for predicting the location of viral epitopes (Pauletti et al., 1985; Modrow et al., 1987; Gnann et al., 1987; Strynadka et al., 1988; Portetelle et al., 1989; Roehrig et al., 1989), there is evidence that none of the methods achieves a high level of correct prediction (Van Regenmortel and Daney de Marcillac 1988; Getzoff et al., 1988; Fasman, 1989). Several methods for predicting antigenicity were compared by Hopp (1986) on the basis of known epitopes in 12 proteins. When only the top three peaks in the plots were considered, the percentage of correct predictions was found to be 75%, 68% and 65% using the hydrophilicity scales of Hopp and Woods (1981), Rose et al (1985) and Kyte and Dootlittle (1982), respectively.

In another study, eight of the commonly used prediction methods were compared with respect to their ability to correctly identify 29 known continuous epitopes in four well-studied proteins (Van Regenmortel and Daney de Marcillac, 1988). In the case of myoglobin, the method of Hopp and Woods (1981) correctly identified only 39% of the antigenic residues (using as threshold of correct prediction the value of the average hydrophilicity over the entire plot +0.7 standard deviation). Of the eight prediction methods that were compared, the scales of Karplus and Schulz (1985) and Parker et al.(1986) gave slightly better results than the others (Van Regenmortel and Daney de Marcillac, 1988).

Another comparative study has been done by Getzoff et al.(1988) who assessed the value of three predictive algorithms at two peptide lengths (7 and 15 residues, respectively). It was found that none of the three algorithms predicted better than random at a peptide length of 7 residues, but that the predictions were somewhat better for 15-residue peptides. Since the studied proteins were relatively small, it was to be expected that randomly chosen 15-residue peptides would have a fair probability of binding to the protein antibodies. In conclusion, it seems that the belief that prediction methods are fairly efficient stems from the fact that most workers use fairly long peptides to search for continuous epitopes in proteins which are replete with antigenically active regions.

4. Operational aspects of viral antigenicity

An operational definition gives meaning to a term by making it possible for the experimentalist to decide when the term applies by observing a response under specified test conditions. From an operationalist perspective, many scientific objects of study can be properly defined only through the description of the experimental conditions that allow these objects to be observed.

As discussed elsewhere (Van Regenmortel, 1986, 1989b), antigen-antibody binding measurements are always of an operational nature. When different types of probes are used for assessing immunochemical binding (for example, free peptide versus conjugated peptide, short peptide versus a more constrained long peptide, liquid versus solid-phase immunoassay), it is unavoidable that the same residues of a peptide will not always be scored as contributing to the binding. The experimental conditions of an assay may influence the conformation of the peptide and determine whether binding activity is observed or not.

The literature on protein antigenicity abounds with examples of the operational limitations of our perception of what constitutes a separate epitope. For instance, two viral antigens that appear to be identical when tested with a very large number of mAbs may eventually be recognized as different when a single discriminating mAb is discovered. The ability of the mAb to detect an antigenic difference may also depend on the test system used (Kendal et al., 1981).

Usually the degree of antigenic relationship found between individual members of a virus family is strongly dependent on the type of serological assay used. For instance, hemagglutination inhibition (H) assays tend to reveal extensive cross-reactivities whereas neutralization tests are more specific for individual virus species. When mAbs instead of polyclonal sera are used for subdividing the members of a virus family according to antigenic cross-reactivity, the resulting classification does not necessarily identify the same serogroups (Heinz, 1986).

Although some authors maintain that operational definitions of epitopes are inconsistent and lead to confusion, there is no way to avoid the operational bias of a functional binding assay (Van Regenmortel, 1989b). If epitopes are defined by their ability to bind paratopes, the binding capacity of an epitope is the only criterion which allows it to be recognized. An interesting example of the difficulty of deciding whether two epitopes are functionally equivalent is given by the phenomenon of molecular mimicry observed in virus infections (Fujinami et al., 1983; Srinivasappa et al., 1986; Oldstone et al., 1986) and in viral anti-idiotypes (Kennedy and Dreesman, 1985; Bruck et al., 1986; Gaulton and Greene, 1986).

5. References

Aaskov, J.G., Geysen, H.M. and Mason, T.J. (1989) *Arch. Virol.* 105, 209–221.

Abad-Zapatero, C., Abdel-Meguid, S.S., Johnson, J.E., Leslie, A.G.E., Rayment, I., Rossmann, M.G., Suck, D. and Sukihara, T. (1980) *Nature* 286, 33–39.

Acharya, R., Fry, E., Stuart, D., Fox, G., Rowlands, D. and Brown, F. (1989) *Nature* 337, 709–716.

Adam, E., Nasz, I., Lengyel, A., Erdei, J. and Fachet, J. (1987) *Arch. Virol.* 93, 261–271.

Air, G.M. and Laver, W.G. (1986) *Adv. Vir. Res.* 31, 53–102.

Al Moudallal, Z., Briand, J.P. and Van Regenmortel, M.H.V. (1982) *EMBO J.* 1, 1005–1010.

Al Moudallal, Z., Briand, J.P. and Van Regenmortel, M.H.V. (1985) *EMBO J.* 4, 1231–1235

Alexander, S. and Elder, J.H. (1984) *Science* 226, 1328–1330.

Altschuh, D. and Van Regenmortel, M.H.V. (1982) *J. Immunol. Methods* 50, 99–108.

Altschuh, D., Hartmann, D., Reinbolt, J. and Van Regenmortel, M.H.V. (1983) *Mol. Immunol.* 20, 271–278

Amit, A.G., Mariuzza, R.A., Phillips, S.E.V. and Poljak, R.J. (1986) *Science* 233, 747–753.

Andrew, M.E., Boyle, D.B., Coupar, B.E.H., Whitfeld, P.L., Both, G.W. and Bellamy, A.R. (1987) *J. Virol.* 61, 1054–1060.

Argos, P. and Fuller, S.D. (1988) *EMBO J.* 7, 819–824.

Atassi, M.Z. (1984) *Eur. J. Biochem.* 145, 1–20.

Atassi, M.Z. and Smith, J.A. (1978) *Immunochemistry* 15, 609–610.

Barlow, D.J., Edwards, M.S. and Thornton, J.M. (1986) *Nature* 322, 747–748.

Barnett, P.V., Ouldridge, E.J., Rowlands, D.J., Brown, F. and Parry, N.R. (1989) *J. Gen. Virol.* 70, 1483–1491.

Benjamin, D.C., Berzofsky, J.A., East, I.J., Gurd, F.R.N., Hannum, C., Leach, S.J., Margoliash, E., Michael, J.G., Miller, A., Prager, E.M., Reichlin, M., Sercaz, E.E., Smith-Gill, S.J., Todd, P.E. and Wilson, A.C. (1984) *Annu.Rev.Immunol.* 2, 67–101.

Benjamini, E. (1977) In *Immunochemistry of Proteins* (M.Z. Atassi, Ed.) Plenum Press, New York. pp. 2–265.

Berton, M.T. and Webster, R.G. (1985) *Virology* 143, 583–594.

Berzofsky, J.A. (1984) In *Monoclonal and Anti-idiotypic Antibodies: Probes for Receptor Structure and Function* (J.C. Venter, C.M. Fraser, and J. Lindstrom, Eds.) Alan R. Liss, New York. pp. 1–19.

Berzofsky, J.A. (1985) *Science* 229, 932–940.

Berzofsky, J.A. (1988) *J. Clin. Invest.* 82, 1811–1817.

Berzofsky, J.A., Cease, K.B., Cornette, J.L., Margalit, H., Berkower, I.J., Good, M.F., Miller, L.H. and DeLisi, C. (1987) *Immunol. Rev.* 98, 9–52.

Bittle, J.L., Houghten, R.A., Alexander, H., Shinnick, T., Sutcliffe, J.G., Lerner, R.A., Rowlands, D.J. and Brown, F. (1982) *Nature* 298, 30–33.

Blondel, B., Crainic, R., Fichot, O., Dufraisse, G., Cardea, A., Girard, M. and Horaud, F. (1986) *J. Virol.* 57, 81–90.

Bloomer, A.C., Champness, J.N., Bricogne, G., Staden, R. and Klug, A. (1978) *Nature* 276, 362–368.

Bolwell, C., Clarke, B.E., Parry, N.R., Ouldridge, E.J., Brown, F. and Rowlands, D.J. (1989) *J. Gen. Virol.* 70, 59–68.

Brandriss, M.W., Schlesinger, J.J., Walsh, E.E. and Briselli, M. (1986) *J. Gen. Virol.* 67, 229–234.

Briand, J.P., Muller, S. and Van Regenmortel, M.H.V. (1985) *J. Immunol. Methods* 78, 59–69.

Breschkin, A.M., Ahern, J. and White, D.O. (1981) *Virology* 113, 130–140.

Bruck, C., Co, M.S., Slaoui, M., Gaulton, G.N., Smith, T., Fields, B.N., Mullins, J.I. and Greene, M.I. (1986) *Proc. Natl. Acad. Sci. USA* 83, 6578–6582.

Cammack, N. and Gould, E.A. (1986) *Virology* 150, 333–341.

Cane, P.A. and Gould, E.A. (1988) *J. Gen. Virol.* 69, 1241–1246.

Carter, M.J. and Ter Meulen, V. (1984) *Adv. Virus Res.* 29, 95–130.

Cartwright, B., Morrell, D.J. and Brown, F. (1982) *J. Gen. Virol.* 63, 375–381.

Caton, A.J., Brownlee, G.G., Yewdell, J.W. and Gerhard, W. (1982) *Cell* 31, 417–427.

Cecilia, D., Gadkari, D.A., Kedarnath, N. and Ghosh, S.N. (1988) *J. Gen. Virol.* 69, 2741–2747.

Chou, P.Y. and Fasman, G.D. (1978) *Adv. Enzymol.* 47, 45–148.

Clarke, B.E., Newton, S.E., Carroll, A.R., Francis, M.J., Appleyard, G., Syred, A.D., Highfield, P.E., Rowlands, D.J. and Brown, F. (1987) *Nature* 330, 381–384.

Colman, P.M. (1988) *Adv. Immunol.* 43, 99–132.

Colman, P.M., Laver, W.G., Varghese, J.N., Baker, A.T., Tulloch, P.A., Air, G.M. and Webster, R.G. (1987a) *Nature* 326, 358–363.

Colman, P.M., Air, G.M., Webster, R.G., Varghese, J.N., Baker, A.T., Lentz, M.R., Tulloch, P.A. and Laver, W.G. (1987b) *Immunol. Today* 8, 323–326.

Darst, S.A., Robertson, C.R. and Berzofsky, J.A. (1988) *Biophys. J.* 53, 533–539.

Dekker, E.L., Porta, C. and Van Regenmortel, M.H.V. (1989) *Arch. Virol.* 105, 269–286.

Delpeyroux, F., Chenciner, N., Lim, A., Malpiece, Y., Blondel, B., Crainic, R., Van der Werf, S. and Streeck, R.E. (1986) *Science* 233, 472–475.

DiMarchi, R., Brooke, G., Gale, C., Cracknell, V., Doel, T. and Mowat, N. (1986) *Science* 232, 639–641.
Diamond, D.C., Jameson, B.A., Bonin, J., Kohara, M., Abe, S., Itoh, H., Komatsu, T., Arita, M., Kuge, S., Nomoto, A., Osterhaus, A.D.M.E., Crainic, R. and Wimmer, E. (1985) *Science* 229, 1090–1093.
Dimmock, N.J. (1987) *Trends Biochem. Sci.* 12, 70–75.
Doel, T.R., Gale, C., Brooke, G. and DiMarchi, R. (1988) *J. Gen. Virol.* 69, 2403–2406.
Dore, I., Weiss, E., Altschuh, D. and Van Regenmortel, M.H.V. (1988) *Virology* 162, 279–289.
Dorow, D.S., Shi, P.T., Carbone, F.R., Minasian, E., Todd, P.E.E. and Leach, S.J. (1985) *Mol. Immunol.* 22, 1255–1264.
Dressman, G.R., Sanchez, Y., Ionescu-Martiu, I., Sparrow, J., Six, H.R., Peterson, D.L., Hollinger, F.B. and Melnick, J.L. (1982) *Nature* 295, 158–160.
Dyrberg, T. and Oldstone M.B.A. (1986) *J. Exp. Med.* 164, 1344–1349.
Dyson, H.J., Lerner, R.A. and Wright, P.E. (1988) *Annu. Rev. Biophys. Biophys. Chem.* 17, 305–324.
Eisenberg, R.J., Long, D., Pereira, L., Hampar, B., Zweig, M. and Cohen, G.H. (1982) *J. Virol.* 41, 478–488.
Elder, J.H., McGee, J.S., Munson, M., Houghten, R.A., Kloetzer, W., Bittle, J.L. and Grant, C.K. (1987) *J. Virol.* 61, 8–15.
Emini, E.A., Ostapchuk, P. and Wimmer, E. (1983) *J. Virol.* 48, 547–550.
Emini, E.A., Hughes, J.V., Perlow, D.S. and Boger, J. (1985) *J. Virol.* 55, 836–839.
Emini, E.A., Schleif, W.A., Armstrong, M.E., Silberklang, M., Schultz, L.D., Lehman, D., Maigetter, R.Z., Qualtiere, L.F., Pearson, G.R. and Ellis, R.W. (1988) *Virology* 166, 387–393.
Fasman, G.D. (1989) *Trends Biochem. Sci.* 14, 295–299.
Feizi, T. and Childs, R.A. (1987) *Biochem. J.* 245, 1–11.
Ferguson, M., Minor, P.D., Magrath, D.I., Yi-Hua, Q., Spitz, M. and Schild, G.C. (1984) *J. Gen. Virol.* 65, 197–201.
Fox, G., Parry, N.R., Barnett, P.V., McGinn, B., Rowlands, D.J. and Brown, F. (1989) *J. Gen. Virol.* 70, 625–637.
Francis, M.J., Fry, C.M., Rowlands, D.J., Bittle, J.L., Houghten, R.A., Lerner, R.A. and Brown, F. (1987a) *Immunology* 61, 1–6.
Francis, M.J., Hastings, G.Z., Syred, A.D., McGinn, B., Brown, F. and Rowlands, D.J. (1987b) *Nature* 330, 168–170.
Friguet, B., Djavadi-Ohaniance, L. and Goldberg, M.E. (1984) *Mol. Immunol.* 21, 673–677.
Fujinami, R.S., Oldstone, M.B.A., Wroblewska, Z., Frankel, M.E. and Koprowski, H. (1983) *Proc. Natl. Acad. Sci. USA* 80, 2346–2350.
Garnier, J., Osguthorpe, D.J. and Robson, B. (1978) *J. Mol. Biol.* 120, 97–120.
Gaulton, G.N. and Greene, M.I. (1986) *Annu. Rev. Immunol.* 4, 253–280.
Geerligs, H.J., Weijer, W.J., Bloemhoff, W., Welling, G.W. and Welling-Wester, S. (1988) *J. Immunol. Methods* 106, 239–244.
Gerhard, W. and Webster, R.G. (1978) *J. Exp. Med.* 148, 383–392.
Getzoff, E.D., Tainer, J.A., Lerner, R.A. and Geysen, H.M. (1988) *Adv. Immunol.* 43, 1–98.
Geysen, H.M., Meloen, R.H. and Barteling, S.J. (1984) *Proc. Natl. Acad. Sci. USA* 81, 3998–4002.
Geysen, H.M., Barteling, S.J. and Meloen, R.H. (1985) *Proc. Natl. Acad. Sci. USA* 82, 178–182.
Geysen, H.M., Rodda, S.J., Mason, T.J., Tribbick, G. and Schoofs, P.G. (1987a) *J. Immunol. Methods* 102, 259–274.
Geysen, H.M., Rodda, S.J., Mason, T.J., Tainer, J.A., Alexander, H., Getzoff, E.D. and Lerner, R.A. (1987b) *Science* 238, 1585–1586.
Geysen, H.M., Mason, T.J. and Rodda, S.J. (1988) *J. Mol. Recognition* 1, 32–41.
Gnann, J.W., Nelson, J.A. and Oldstone, M.B.A. (1987) *J. Virol.* 61, 2639–2641.
Gould, E.A., Buckley, A., Barrett, A.D.T. and Cammack, N. (1986) *J. Gen. Virol.* 67, 591–595.
Grady, L.J. and Kinch, W. (1985) *J. Gen. Virol.* 66, 2773–2776.
Green, N., Alexander, H., Olson, A., Alexander, S., Shinnick, T.M., Sutcliffe, J.G. and Lerner, R.A. (1982) *Cell* 28, 477–487.

Guirakhoo, F., Heinz, F.X. and Kunz, C. (1989) *Virology* 169, 90–99.

Harper, M., Lema, F., Boulot, G. and Poljak, R.J. (1987) *Mol. Immunol.* 24, 97–108.

Harrison, S.C., Olson, A.J., Schutt, C.E., Winkler, F.K. and Bricogne, G. (1978) *Nature* 276, 368–373.

Haynes, J.R., Cunningham, J., von Seefried, A., Lennick, M., Garvin R.T. and Shen, S.-H. (1986) *Biotechnology* 4, 637–640.

Heinz, F.X. (1986) *Adv. Vir. Res.* 31, 103–168.

Heinz, F.X., Berger, R., Tuma, W. and Kunz, C. (1983) *Virology* 126, 525–537.

Heinz, F.X., Mandl, C., Berger, R., Tuma, W. and Kunz, C. (1984) *Virology* 133, 25–34.

Hogle, J.M., Chow, M. and Filman, D.J. (1985) *Science* 229, 1358–1365.

Hogle, J.M., Maeda, A. and Harrison, S.C. (1986) *J. Mol. Biol.* 191, 625–638.

Hopp, T.P. (1986) *J. Immunol. Methods* 88, 1–18.

Hopp, T.P. and Woods, K.R. (1981) *Proc. Natl. Acad. Sci. USA* 78, 3824–3828.

Hopp, T.P. and Woods, K.R. (1983) *Mol. Immunol.* 20, 483–489.

Hornbeck, P.V. and Wilson, A.C. (1984) *Biochemistry* 23, 998–1002.

Houghten, R.A. (1985) *Proc. Natl. Acad. Sci. USA* 82, 5131–5135.

Hurrell, J.G.R., Smith, J.A., Todd, P.E. and Leach, S.J. (1977) *Immunochemistry* 14, 283–288.

Icenogle, J., Shiwen, H., Duke, G., Gilbert, S., Rueckert, R. and Anderegg, J. (1983) *Virology* 127, 412–425.

Jameson, B.A. and Wolf, H. (1988) *CABIOS* 4, 181–186.

Jemmerson, R. (1987) *Proc. Natl. Acad. Sci. USA* 84, 9180–9184.

Jemmerson, R. and Paterson, Y. (1985) *Nature* 317, 89–90.

Jemmerson, R. and Paterson, Y. (1986) *BioTechniques* 4, 18–31.

Jerne, N.K. (1960) *Annu. Rev. Microbiol.* 14, 341–358.

Karplus, P.A. and Schulz, G.E. (1985) *Naturwissenschaften* 72, 212–213.

Kendal, A.P., Phillips, D.J., Webster, R.G., Galland, G.G. and Reimer, C.B. (1981) *J. Gen. Virol.* 54, 253–261.

Kennedy, R.C. and Dreesman, G.R. (1985) *Prog. Med. Virol.* 31, 168–182.

Kennedy, R.C., Sparrow, J.T., Sanchez, Y., Melnick, J.L. and Dreesman, G.R. (1984) *Virology* 136, 247–252.

Kennedy, R.C., Henkel, R.D., Pauletti, D., Allan, J.S., Lee, T.H., Essex, M. and Dreesman, G.R. (1986) *Science* 231, 1556–1559.

Kennedy, R.C., Dreesman, G.R., Chanh, T.C., Boswell, R.N., Allan, J.S., Lee, T.-H., Essex, M., Sparrow, J.T., Ho, D.D. and Kanda, P. (1987) *J. Biol. Chem.* 262, 5769–5774.

Kennedy, R.C., Chanh, T.C., Allan, J.S., Dreesman, G.R., Eichberg, J.W., Cauda, R. and Kanda, P. (1988) In *Synthetic Peptides in Biotechnology* (A. Mizrahi, ed.), Alan R. Liss, New York, pp. 149–172.

Kent, S. and Clark-Lewis, I. (1985) In *Synthetic Peptides in Biology and Medicine* (K. Alitalo, P. Partanen and A. Vaheri, eds), pp. 29–57, Elsevier Science Publishers, Amsterdam.

Kimura-Kuroda, J. and Yasui, K. (1983) *J. Virol.* 45, 124–132.

Kimura-Kuroda, J. and Yasui, K. (1986) *J. Gen. Virol.* 67, 2663–2672.

Kleid, D.G., Yansura, D., Small, B., Dowbenko, D., Moore, D.M., Grubman, M.J., McKercher, P.D., Morgan, D.O., Robertson, B.H. and Bachrach, H.L. (1981) *Science* 214, 1125–1129.

Knossow, M., Daniels, R.S., Douglas, A.R., Skehel, J.J. and Wiley, D.C. (1984) *Nature* 311, 678–680.

Kyte, J. and Doolittle, R.F. (1982) *J. Mol. Biol.* 157, 105–132.

Laporte, J., Grosclaude, J., Wantyghem, J., Bernard, S. and Rouze, P. (1973) *C.R. Acad. Sci.* (Paris) (Sér. D), 276, 3399–3401.

Lasky, L.A., Groopman, J.E., Fennie, C.W., Benz, P.M., Capon, D.J., Dowbenko, D.J., Nakamura, G.R., Nunes, W.M., Renz, M.E. and Berman, P.W. (1986) *Science* 233, 209–212.

Leach, S.J. (1983) *Biopolymers* 22, 425–440.

Lenstra, J.A., Kusters, J.G., Koch, G. and Van Der Zeijst, B.A.M. (1989) *Mol. Immunol.* 26, 7–15.

Lerner, R.A. (1984) *Adv. Immunol.* 36, 1–44.

Livingstone, A.M. and Fathman, C.G. (1987) *Annu. Rev. Immunol.* 5, 477–501.

Lubeck, M. and Gerhard, W. (1982) *Virology* 118, 1–7.

Luckow, V.A. and Summers, M.D. (1988) *Biotechnology* 6, 47–55.

Luo, M., Vriend, G., Kamer, G., Minor, I., Arnold, E., Rossmann, M.G., Boege, U., Scraba, D.G., Duke, G.M. and Palmenberg, A.C. (1987) *Science* 235, 182–191.

Luo, M., Rossmann, M.G. and Palmenberg, A.C. (1988) *Virology* 166, 503–514.

Mäkelä, M.J., Lund, G.A. and Salmi, A.A. (1989) *J. Gen. Virol.* 70, 603–614.

Mandel, B. (1985) in Immunochemistry of viruses. *The basis for serodiagnosis and vaccines* (M.H.V. Van Regenmortel and A.R. Neurath, Eds.) pp. 53–70. Elsevier, Amsterdam.

Mandl, C.W., Guirakhoo, F., Holzmann, H., Heinz, F.X. and Kunz, C. (1989) *J. Virol.* 63, 564–571.

Markoff, L.J., Bray, M., Lai, C.J., Chanock, R.M., Eckels, K., Summers, P., Gentry, M.R., Houghten, R.A. and Lerner, R.A. (1988) In *Vaccines 88*. Cold Spring Harbor Laboratory, Cold Spring Harbor NY. pp. 161–165.

Mason, P.W., McAda, P.C., Dalrymple, J.M., Fournier, M.J. and Mason, T.L. (1987) *Virology* 158, 361–372.

Mathews, J.H. and Roehrig, J.T. (1984) *J. Immunol.* 132, 1533–1537.

Matsuura, Y., Possee, R.D., Overton, H.A. and Bishop, D.H.L. (1988) *J. Gen. Virol.* 68, 1233–1250.

McCray, J. and Werner, G. (1987) *Nature* 329, 736–738.

McCullough, K.C. (1986) *Arch. Virol.* 87, 1–36.

McCullough, K.C., Crowther, J.R. and Butcher, R.N. (1985) *J. Immunol. Methods* 82, 91–100.

Mehra, V., Sweetser, D. and Young, R.A. (1986) *Proc. Natl. Acad. Sci. USA* 83, 7013–7017.

Meloen, R.H. and Barteling, S.J. (1986) *J. Gen. Virol.* 67, 289–294.

Meleon, R.H., Briaire, J., Woortmeyer, R.J. and Van Zaane, D. (1983) *J. Gen. Virol.* 64, 1193–1198.

Meloen, R.H., Puyk, W.C., Meijer, D.J.A., Lankhof, H., Posthumus, W.P.A. and Schaaper, W.M.M. (1987) *J. Gen. Virol.* 68, 305–314.

Michel, M-L., Mancini, M., Sobczak, E., Favier, V., Guetard, D., Bahraoui, E.M., Tiollais, P. (1988) *Proc. Natl. Acad. Sci. USA* 85, 7957–7961.

Middeldorp, J.M. and Meloen, R.H. (1988) *J. Virol. Methods* 21, 147–159.

Milton, R.C. de L. and Van Regenmortel, M.H.V. (1979) *Mol. Immunol.* 16, 179–184.

Minor, P.D., Ferguson, M. Evans, D.M.A., Almond, J.W. and Icenogle, J.P. (1986) *J. Gen. Virol.* 67, 1283–1291.

Modrow, S., Hahn, B.H., Shaw, G.M., Gallo, R.C., Wong-Staal, F. and Wolf, H. (1987) *J. Virol.* 61, 570–578.

Modrow, S., Höflacher, B., Mertz, R. and Wolf, H. (1989) *J. Immunol. Methods* 118, 1–7.

Muller, S., Himmelspach, K. and Van Regenmortel, M.H.V. (1982) *EMBO J.* 1, 421–425.

Muller, S., Plaue, S., Couppez, M. and Van Regenmortel, M.H.V. (1986) *Mol. Immunol.* 23, 593–601.

Murray, K., Bruce, S.A., Hinnen, A., Wingfield, P., Van Erd, P.M.C.A., De Reus, A. and Schellekens, H. (1984) *EMBO J.* 3, 645–650.

Nestorowicz, A., Tregear, G.W., Southwell, C.N., Martyn, J., Murray, J.M., White, D.O. and Jackson, D.J. (1985) *Mol. Immunol.* 22, 145–154.

Neurath, A.R. and Rubin, B.A. (1971) *Viral Structural Components as Immunogens of Prophylactic Value.* Karger, Basel. pp. 1–87.

Neurath, A.R., Kent, S.B.H. and Strick, N. (1986) *J. Gen. Virol.* 67, 453–461.

Nicolaisen-Strouss, K., Kumar, H.P.M., Fitting, T., Grant, C.K. and Elder, J.H. (1987) *J. Virol.* 61, 3410–3415.

Niman, H.L., Houghten, R.A., Walker, L.A., Reisfeld, R.A., Wilson, I.A., Hogle, J.M. and Lerner, R.A. (1983) *Proc. Natl Acad. Sci. USA* 80, 4949–4953.

Norrby, E., Mufson, M.A., Alexander, H., Houghten, R.A. and Lerner, R.A. (1987) *Proc. Natl. Acad. Sci. USA* 84, 6572–6576.

Novotny, J. and Haber, E. (1986) *Biochemistry* 25, 6748–6754.

Novotny, J., Bruccoleri, R.E., Carlson, W.D., Handschumacher, M. and Haber, E. (1987a) *Science* 238, 1584–1584.

Novotny, J., Handschumacher, M. and Bruccoleri, R.E. (1987b) *Immunol. Today* 8, 26–31.

Nowak, T. and Wengler, G. (1987) *Virology* 156, 127–137.

Nunberg, J.H., Rodgers, G., Gilbert, J.H. and Snead, R.M. (1984) *Proc. Natl. Acad. Sci. USA* 81, 3675–3679.

Offensperger, W., Wahl, S., Neurath, A.R., Price, P., Strick, N., Kent, S.B.H., Christman, J.K. and Acs, G. (1985) *Proc. Natl. Acad. Sci. USA* 82, 7540–7544.

Oldstone, M.B.A., Schwimmbeck, P., Dyrberg, T. and Fujinami, R. (1986) In *Progress in Immunology* VI (B. Cinader and R.G. Miller, Eds.) Academic Press, New York, pp. 787–795.

Page, G.S., Mosser, A.G., Hogle, J.M., Filman, D.J., Rueckert, R.R. and Chow, M. (1988) *J. Virol.* 62, 1781–1794.

Palfreyman, J.W., Aitcheson, T.C. and Taylor, P. (1984) *J. Immunol. Methods* 75, 383–393.

Parker, J.M.R., Guo, D. and Hodges, R.S. (1986) *Biochemistry* 25, 5425–5432.

Parry, N.R., Syred, A., Rowlands, D.J. and Brown, F. (1988) *Immunology* 64, 567–572.

Parry, N.R., Barnett, P.V., Ouldridge, E.J., Rowlands, D.J. and Brown, F. (1989) *J. Gen. Virol.* 70, 1493–1503.

Pauletti, D., Simmonds, R., Dreesman, G.R. and Kennedy, R.C. (1985) *Anal. Biochem.* 151, 540–546.

Perkus, M.E., Piccini, A., Lipinskas, B.R. and Paoletti, E. (1985) *Science* 229, 981–984.

Pevear, D.C., Luo, M. and Lipton, H.L. (1988) *Proc. Natl. Acad. Sci. USA* 85, 4496–4500.

Pfaff, E., Mussgay, M., Bohm, H.O., Schulz, G.E. and Schaller, H. (1982) *EMBO J.* 1, 869–874.

Pfaff, E., Thiel, H.J., Beck, E., Strohmaier, K. and Schaller, H. (1988) *J. Virol.* 62, 2033–2040.

Pollock, R.R., Teillaud, J-L. and Scharff, M.D. (1984) *Annu. Rev. Microbiol.* 38, 389–417.

Portetelle, D., Dandoy, C., Burny, A., Zavada, J., Siakkou, H., Gras-Masse, H., Drobecq, H. and Tartar, A. (1989) *Virology* 169, 34–41.

Putnak, J.R., Charles, P.C., Padmanabhan, R., Irie, K., Hoke, C.H. and Burke, D.S. (1988) *Virology* 163, 93–103.

Ray, R., Galinski, M.S. and Compans, R.W. (1989) *Virus Res.* 12, 169–180.

Roehrig, J.T., Hunt, A.R., Johnson, A.J. and Hawkes, R.A. (1989) *Virology* 171, 49–60.

Rose, G.D., Geselowitz, A.R., Lesser, G.L., Lee, R.H. and Zehfus, M.H. (1985) *Science* 229, 834–838.

Rossmann, M.G., Arnold, E., Erickson, J.W., Frankenberger, A.E., Griffith, J.P., Hecht, H.J., Johnson, J.E., Kamer, G., Luo, M., Mosser, A.G., Rueckert, R.R., Sherry, B. and Vriend, G. (1985) *Nature* 317, 145–153.

Rothbard, J.B. and Taylor, W.R. (1988) *EMBO J.* 7, 93–100.

Ruoslahti, E. and Pierschbacher, M.D. (1986) *Cell* 44, 517–518.

Salfeld, J., Pfaff, E., Noah, M. and Schaller, H. (1989) *J. Virol.* 63, 798–808.

Schaaper, M.M., Lankhof, H., Pujik, W.C. and Meloen, R.H. (1989) *Mol. Immunol.* 26, 81–85.

Schlesinger, J.J., Walsh, E.E. and Brandriss, M.W. (1989) *Virology* 65, 1637–1644.

Schoofs, P.G., Geysen, H.M., Jackson, D., Brown, L.E., Tang, X.L. and White, D.O. (1988) *J. Immunol.* 140, 611–616.

Schulze-Gahmen, U., Klenk, H.D. and Beyreuther, K. (1986) *Eur. J. Biochem.* 159, 283–289.

Schulze-Gahmen, U., Rini, J.M., Arevalo, J., Stura, E.A., Kenten, J.H. and Wilson, I.A. (1988) *J. Biol. Chem.* 263, 17100–17105.

Sela, M. (1969) *Science* 166, 1365–1374.

Sette, A., Doria, G. and Adorini, L. (1986) *Mol. Immunol.* 23, 807–810.

Sheriff, S., Silverton, E.W., Padlan, E.A., Cohen, G.H., Smith-Gill, S., Finzel, B.C. and Davies, D.R. (1987) *Proc. Natl. Acad. Sci. USA* 84, 8075–8079.

Sherry, B., Mosser, A.G., Colonno, R.J. and Rueckert, R.R. (1986) *J. Virol.* 57, 246–257.

Sheshberadaran, H. and Payne, L.G. (1988) *Proc. Natl. Acad. Sci. USA* 85, 1–5.

Shi, P.T., Riehm, J.P., Todd, P.E.E. and Leach, S.J. (1984) *Mol. Immunol.* 21, 489–496.

Skehel, J.J., Stevens, D.J., Daniels, R.S., Douglas, A.R., Knossow, M., Wilson, I.A. and Wiley, D.C. (1984) *Proc. Natl. Acad. Sci USA* 81, 1779–1783.

Soderquist, M.E. and Walton, A.G. (1980) *J. Colloid Interface Sci.* 75, 386–397.

Srinivasappa, J., Saegusa, J., Prabhakar, B.S., Gentry, M.K., Buchmeier, M.J., Wiktor, T.J., Koprowski, H., Oldstone, M.B.A. and Notkins, A.L. (1986) *J. Virol.* 57, 397–401.

Stanley, K.K. (1983) *Nucleic Acids Res.* 11, 4077–4092.

Stave, J.W., Card, J.L. and Morgan, D.O. (1986) *J. Gen. Virol.* 67, 2083–2092.

Strohmaier, K., Franze, R. and Adam, K.H. (1982) *J. Gen. Virol.* 59, 295–306.

Strynadka, N.C.J., Redmond, M.J., Parker, J.M.R., Scraba, D.G. and Hodges, R.S. (1988) *J. Virol.* 62, 3474–3483.

Tainer, J.A., Getzoff, E.D., Alexander, H., Houghten, R.A., Olson, A.J., Lerner, R.A. and Hendrickson, W.A. (1984) *Nature* 312, 127–133.

Tainer, J.A., Getzoff, E.D., Paterson, Y., Olson, A.J. and Lerner, R.A. (1985a) *Annu. Rev. Immunol.* 3, 501–535.

Tainer, J.A., Getzoff, E.D., Alexander, H., Houghten, R.A., Olson, A.J. and Lerner, R.A. (1985b) *Nature* 317, 90.

Tang, X.L., Tregear, G.W., White, D.O. and Jackson, D.C. (1988) *J. Virol.* 62, 4745–4751.

Thornton, J.M., Edwards, M.S., Taylor, W.R. and Barlow, D.J. (1986) *EMBO J.* 5, 409–413.

Thornton, J.M. and Sibanda, B.L. (1983) *J. Mol. Biol.* 167, 443–460.

Thotakura, N.R. and Bahl, O.M. (1987) *Methods Enzymol.* 138, 350–359.

Trudel, M., Nadon, F., Seguin, C., Dionne, G. and Lacroix, M. (1987) *J. Gen. Virol.* 68, 2273–2280.

Underwood, P.A. (1982) *J. Gen. Virol.* 62, 153–169.

Underwood, P.A. (1985) *J. Immunol. Methods* 85, 295–307.

Valenzuela, P., Coit, D., Medina-Selby, A., Kuo, C., Van Nest, G., Burke, R.L., Bull, P., Urdea M.S. and Graves, R.V. (1985) *Biotechnology* 3, 323–326.

Van der Werf, S., Wychowski, C., Bruneau, P., Blondel, B., Crainic, R., Horodniceanu, F. and Girard, M. (1983) *Proc. Natl. Acad. Sci. USA* 80, 5080–5084.

Van Regenmortel, M.H.V. (1966) *Adv. Virus Res.* 12, 207–271.

Van Regenmortel, M.H.V. (1982) *Serology and Immunochemistry of Plant Viruses.* Academic Press, New York. pp. 1–302.

Van Regenmortel, M.H.V. (1984) In *Hybridoma Technology in Agricultural and Veterinary Research* (N.J. Stern and H.R. Gamble, Eds) Rowman and Allanheld, Totowa NJ. pp. 43–82.

Van Regenmortel, M.H.V. (1986) *Trends Biochem. Sci.* 11, 36–39.

Van Regenmortel, M.H.V. (1988) *Mol. Immunol.* 25, 565–567.

Van Regenmortel, M.H.V. (1989a) *Phil. Trans. R. Soc. Lond. B* 323, 451–466.

Van Regenmortel, M.H.V. (1989b) *Immunol. Today* 10, 266–272.

Van Regenmortel, M.H.V. and Altschuh, D. (1985) *Nature* 317, 90.

Van Regenmortel, M.H.V. and Daney de Marcillac, G. (1988) *Immunol. Lett.* 17, 95–108.

Van Regenmortel, M.H.V., Briand, J.P., Muller, S. and Plaue, S. (1988) *Synthetic Polypeptides as Antigens.* Elsevier, Amsterdam. pp. 1–227.

Van Wyke Coelingh, K.L., Murphy, B.R., Collins, P.L., Lebacq-Verheyden, A.M. and Battey, J.F. (1987) *Virology* 160, 465–472.

Varghese, J.N., Webster, R.G., Laver, W.G. and Colman, P.M. (1988) *J. Mol. Biol.* 200, 201–203.

Walter, G. (1986) *J. Immunol. Methods* 88, 149–161.

Walter, G., Scheidtmann, K.H., Carbone, A., Laudano, A. and Doolittle, R. (1980) *Proc. Natl. Acad. Sci.USA* 77, 5197–5200.

Wang, C.Y. (1988) In *Synthetic Peptides in Biotechnology* (A. Mizrahi, Ed.) Alan R. Liss, New York. pp. 131–148.

Ward, C.W., Murray, J.M., Roxburgh, C.M. and Jackson, D.C. (1983) *Virology* 126, 370–375.

Webster, R.G., Air, G.M., Metzger, D.W., Colman, P.M., Varghese, J.N., Baker, A.T. and Laver, W.G. (1987) *J. Virol.* 61, 2910–2916.

Welling, G.W., Weijer, W.J., Van der Zee, R. and Welling-Wester, S. (1985) *FEBS Lett.* 188, 215–218.

Westhof, E., Altschuh, D., Moras, D., Bloomer, A.C., Mondragon, A., Klug, A. and Van Regenmortel, M.H.V. (1984) *Nature* 311, 123–126.

Wetz, K., Willingmann, P., Zeichhardt, H. and Habermehl, K.O. (1986) *Arch. Virol.* 91, 207–220.

Wiley, D.C. and Skehel, J.J. (1987) *Annu. Rev. Biochem.* 56, 365–394.

Wiley, D.C., Wilson, I.A. and Skehel, J.J. (1981) *Nature* 289, 373–378.

Wilson, I.A., Skehel, J.J. and Wiley, D.C. (1981) *Nature* 289, 366–378.

Wilson, I.A., Niman, H.L., Houghten, R.A., Cherenson, A.R., Connolly, M.L. and Lerner, R.A. (1984) *Cell* 37, 767–778.

Wolf, H., Modrow, S., Motz, M., Jameson, B.A., Hermann, G. and Förtsch, B. (1988) *CABIOS* 4, 187–191.

Xie, Q.C., McCahon, D., Crowther, J.R., Belsham, G.J. and McCullough, K.C. (1987) *J. Gen. Virol.* 68, 1637–1647.

Yewdell, J.W. and Gerhard, W. (1981) *Annu. Rev. Microbiol.* 35, 185–206.

Zhang, Y.M., Hayes, E.P., McCarty, T., Dubois, D.R., Summers, P.L., Eckels, K.H., Chanock, R.M. and Lai, C.J. (1988) *J. Virol.* 62, 3027–3031.

Eds. M.H.V. van Regenmortel and A.R. Neurath
Immunochemistry of viruses, II. The basis for serodiagnosis and vaccines.

Influence of glycosylation on antigenicity of viral proteins

2

H.-D. KLENK

Institut für Virologie, Philipps-Universität, Robert-Koch-Strasse 17, D-3550 Marburg, F.R.G.

Glycosylation is a common asset of most eukaryotic membrane and secretory proteins. Evidence has been obtained that oligosaccharide side chains may have a variety of biological functions including maintenance of structure and stability of glycoproteins, targeting of proteins to different cellular compartments, and mediating recognition in cell adhesion and tissue organization (for reviews see Gibson et al., 1980; Elbein, 1981, 1987; Olden et al., 1985). Furthermore, oligosaccharides have been shown to be important antigenic determinants of cellular glycoproteins involved in embryonic development, cell differentiation and oncogenesis (for review see Feizi and Childs, 1987). Most lipid-containing and some nonenveloped viruses, such as rotaviruses, also have glycosylated surface proteins, and there is increasing evidence that their carbohydrate side chains may affect antigenicity and immunogenicity, too.

1. Carbohydrate moiety of viral glycoproteins

1.1. Structure and biosynthesis of *N*-glycans

The majority of the viral glycoproteins analyzed to date have their carbohydrates linked by *N*-glycosidic bonds (*N*-glycans). Such glycoproteins are present in alphaviruses (Simons et al., 1980), flaviviruses (Winkler et al., 1987), retroviruses (Schlüter et al., 1985), orthomyxoviruses (Keil et al., 1979, 1985), paramyxoviruses (Diabaté et al., 1984), rhabdoviruses (Hunt et al., 1983), bunyaviruses (Vorndam and Trent, 1979), coronaviruses (Siddell et al., 1982), arenaviruses (Boersma et al., 1982), rotaviruses (Estes et al., 1983), herpes viruses (Hope and Marsden, 1983), and hepadnaviruses (Heermann et al., 1984; Tiollais et al., 1987).

The carbohydrates attached by *N*-glycosidic linkages to viral glycoproteins are synthesized by cellular enzymes and their biosynthesis follows the general pathway established for this class of glycans (for review see Kornfeld and Kornfeld, 1985). Two stages can be distinguished in this pathway: firstly, assembly of a precursor oligosaccharide consisting of 3 glucose, 9 mannose, and 2 *N*-acetylglucosamine residues ($Glc_3Man_9GlcNAc_2$) on a dolichol pyrophosphate (Dol-PP) carrier and, secondly, processing of the precursor after its transfer to the polypeptide.

Assembly of the precursor is initiated by the transfer of GlcNAc-1-P from UDP-GlcNAc to dolichol phosphate (P-Dol) giving GlcNAc-PP-Dol, to which a second *N*-acetylglucosamine residue is added from UDP-GlcNAc, forming $GlcNAc_2$-PP-Dol. This molecule is elongated by the transfer of five mannose residues directly from GDP-Man, yielding the Man_5-$GlcNAc_2$-PP-Dol intermediate. Addition of four more mannose and three glucose residues from Man-P-Dol and Glc-P-Dol, respectively, completes the assembly of the lipid-linked oligosaccharide Glc_3Man_9Glc-NAc_2-PP-Dol which is the common precursor of most *N*-linked oligosaccharides. Transfer of the completed oligosaccharide to asparagine residues in the consensus sequence Asn-X-Thr (Ser) is a cotranslational event, occurring at the luminal face of the rough endoplasmic reticulum.

Once the oligosaccharide is protein bound, it undergoes a sequence of trimming reactions. The three glucose residues are rapidly removed by glucosidases in the endoplasmic reticulum. Subsequent removal of mannose residues by Golgi α-mannosidase I results in the formation of a $Man_5GlcNAc_2$ intermediate. Incomplete removal of mannose at this point yields oligomannosidic oligosaccharides containing 5–8 mannose residues. The Man_5-intermediate is a substrate for *N*-acetylglucosamine transferase I which adds *N*-acetylglucosamine to the free mannose residue linked α-1.3 to the β-mannose residue. Mannosidase II removes two more mannose residues. *N*-Acetylglucosamine transferase I and mannosidase II are located in the medial compartment of the Golgi. Addition of further *N*-acetylglucosamine residues and residues of galactose, fucose and sialic acid, catalyzed by specific glycosyl-transferases in the trans-Golgi, completes the assembly of the complex-type oligosaccharides (for literature see McDowell and Schwarz, 1988).

Depending on the extent of processing, a large variety of oligomannosidic and *N*-acetyllactosaminic oligosaccharides have been observed on viral glycoproteins. Table 2.1 shows some representative structures that have been found in a study on the fowl plague virus hemagglutinin employing NMR spectroscopy and methylation analysis. In the same study, analysis of tryptic and thermolytic glycopeptides of the hemagglutinin allowed the allocation of these oligosaccharides to the individual glycosylation sites of the hemagglutinin. Despite some micro-heterogeneity, a high degree of structural specificity was found in each position (Fig. 2.1). Localization of

TABLE 2.1

N-glycosidic oligosaccharides of the hemagglutinin of influenza virus A/FPV/Rostock/34 (H7N1) (fowl plague virus)

(a) Galβ4GlcNAcβ2Manα6\
Manβ4GlcNAcβ4GlcNAcOH
Galβ4GlcNAcβ2Manα3/

(b) Glβ4GlcNAcβ6\
Galβ4GlcNAcβ2Manα6\
Manβ4GlcNAcβ4GlcNAcOH
Galβ4GlcNAcβ2Manα3/

(c) Galβ4GlcNAcβ2Manα6\
GlcNAcβ4 – Manβ4GlcNAcβ4GlcNAcOH
Galβ4GlcNAcβ2Manα3/

(d) Manα2Manα6\
Manα6\
Manα2Manα3/ Manβ4GlcNAcβ4GlcNAcOH
Manα2Manα2 – Manα3/

Biantennary oligosaccharides without (a) and with (c) bisecting *N*-acetylglucosamine and triantennary oligosaccharide (b) of the *N*-acetyllactosamine type as well as oligomannosidic oligosaccharide (d) are shown. The data are from Keil et al., 1985.

the oligosaccharides on the three-dimensional structure of the hemagglutinin revealed that the oligomannosidic glycans are attached to glycosylation sites at which the enzymes responsible for carbohydrate processing do not have proper access. These observations demonstrated that an important structural determinant for the oligosaccharide side chains is the structure of the glycoprotein itself (Keil et al., 1985).

In addition to the ordinary side chains of the oligomannosidic and *N*-acetyllactosaminic type, occasionally other structures, only rarely found among *N*-glycans, have been reported for viral glycoproteins. Thus, the influenza virus neuraminidase of the N2 serotype has the unusual constituent *N*-acetylgalactosamine in one of its oligosaccharides (Ward et al., 1983). The NB protein of influenza B virus has been reported to contain polylactosaminoglycan which is characterized by a variable number of repeating units of β 1-4 *N*-acetylglucosamine attached to the $Man_3GlcNAc_2$ core (Williams and Lamb, 1988).

An important consequence of the cellular origin of the enzymes responsible for carbohydrate synthesis are host-specific variations in the structure of the oligosaccharides. For instance, some of the side chains of Sindbis virus glycoproteins differed in the extent of processing depending on virus replication in avian or mammalian cells (Hsieh et al., 1984). Such differences may become more dramatic, when viral glycoproteins are produced in foreign hosts using recombinant DNA techniques. Thus, in yeast glycoproteins are decorated with mannose polymers in *N*-glycosidic and *O*-glycosidic linkages (Tanner, 1987). When glycoproteins are expressed in in-

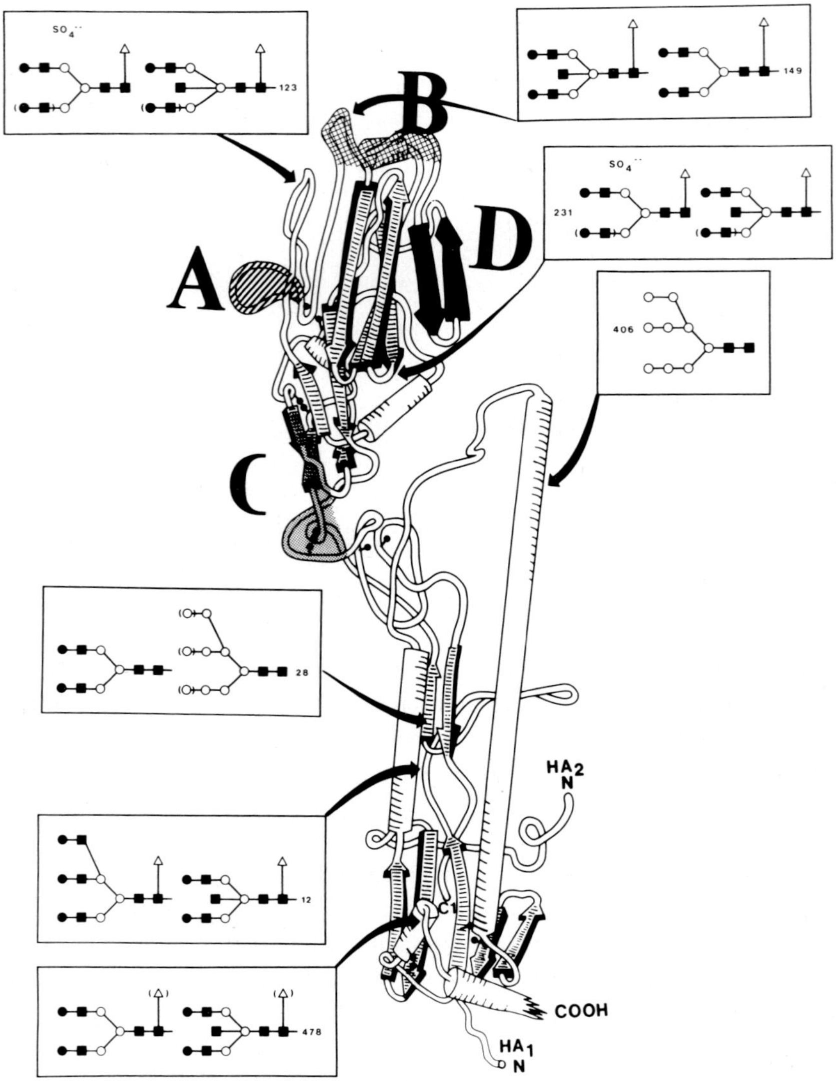

Fig. 2.2. The position of the oligosaccharide side chains on the influenza virus hemagglutinin. The three-dimensional model has been established on the hemagglutinin of strain A/Aichi/68 (H3N2) (Wilson et al., 1981). The glycosylation sites and the structure of the individual carbohydrate side chains as determined for the hemagglutinin of A/FPV/Rostock/34 (H7N1) (Keil et. al., 1985) are shown. ■, *N*-Acetylglycosamine, ○, mannose, ●, galactose, △, fucose. The shaded areas indicate antigenic sites A, B, C, and D.

sect cells using baculovirus vectors, oligosaccharides of the *N*-acetyllactosamine type are replaced by truncated side chains, because this host system lacks most of the enzymes required for elongation of the trimmed oligosaccharides (Kuroda et al., 1989).

1.2. Structure and biosynthesis of *O*-glycans

O-linked carbohydrate side chains (*O*-glycans) have been detected so far only in a few viral glycoproteins. The first viral protein in which this was discovered was the envelope protein E1 of bovine and murine coronaviruses (Niemann and Klenk, 1981; Holmes et al., 1981; Rottier et al., 1981). Whereas E1 contains only *O*-linked oligosaccharides, this type has been found together with *N*-glycans on glycoproteins of herpes simplex virus (Olofsson et al., 1981; Johnson and Spear, 1983), on the G-protein of respiratory syncytial virus (Gruber and Levine, 1985; Olmsted et al., 1989) and on the nonstructural hemagglutinin of vaccinia virus (Shida and Dales, 1981).

Structural analysis revealed two size classes of *O*-linked oligosaccharides on glycoprotein E1 of murine coronavirus A59 (Table 2.2). Oligosaccharide A (Neu5Ac α2,3 Gal β1,3 GalNAc) comprised 35% of the total carbohydrate side chains, while the remaining 65% of the oligosaccharides of E1 had the branched structure B: Neu5Ac α2,3 Gal β1,3 (Neu5Ac α 2,6) GalNAc. Both oligosaccharides were linked to the E1 polypeptide via *N*-acetylgalactosamine, and 20% of the sialic acids present in the E1 glycopeptides were found to consist of *N*-acetyl-9-*O*-acetylneuraminic acid. Evidence had been obtained that these oligosaccharides were attached to a cluster of four hydroxyamino acids at the amino-terminus of the polypeptide (Niemann et al., 1984). As believed to be generally the case with *O*-glycosidic oligosaccharides, glycosylation of E1 is a post-translational event, presumably without lipid-linked precursors being involved (Niemann et al., 1982).

TABLE 2.2

O-glycosidic oligosaccharides of glycoprotein E1 of murine coronavirus A 59

A	Neu5Ac(Neu5,9Ac2)α2,3Galβ1,3GalNAcOH
B	Neu5Ac(Neu5,9Ac2)α2,6 │ Neu5Ac(Neu5,9Ac2)α2,3Galβ1,3GalNAcOH

The data are from Niemann et al., 1984.

1.3. Biological role of glycans

In addition to their role as modulators of immunogenicity and antigenicity, which is the principle subject of this chapter and will be discussed in detail below, oligosaccharides have been found to affect several other biological properties of viral glycoproteins. Viruses are particularly useful for the analysis of carbohydrate functions, because there is easy access to host range and temperature-sensitive mutants and to host-dependent variations in oligosaccharide structure. Useful information has also been obtained from studies using drugs that are specific inhibitors of the addition or modification of *N*-linked oligosaccharide side chains (Elbein, 1987; McDowell and Schwarz, 1988). Finally, the advent of recombinant DNA technology has allowed studies involving the removal of individual oligosaccharides or the addition of new side chains.

There is increasing evidence with influenza viruses that glycosylation at specific sites of the hemagglutinin may influence receptor specificity and thus host range. It has been shown, for instance, that adaptation of influenza B virus to growth in eggs selects variants which differ from virus grown from the same source in mammalian cells by the loss of an oligosaccharide at Asn 196. This carbohydrate appears to inhibit attachment of the mammalian variant to allantois cells (Robertson et al., 1985). A similar phenomenon has been observed with influenza A (H1) isolates, where the differential attachment to avian and mammalian cells also depends on the presence of an oligosaccharide adjacent to the receptor binding site (I. Günther, G. Döller, R. Rott, W. Garten and H.-D. Klenk, unpublished results). Variations in the structure of a specific oligosaccharide may also influence host range. This is the outcome of a study on a variant of the WSN strain of influenza A virus which was interpreted to demonstrate that the oligosaccharide attached in MDBK cells to Asn 125 of the hemagglutinin reduced its affinity for these cells, whereas the oligosaccharide attached to the same site in chick embryo cells left the affinity for MDBK cells unaltered (Deom et al., 1985, 1986).

A direct interaction between a carbohydrate side chain and a functional site has also been found to be an important control mechanism for proteolytic activation of the influenza hemagglutinin. This oligosaccharide which is attached to Asn 28 is located in the vicinity to the cleavage site and interferes with access of activating proteases. The loss of this carbohydrate facilitated proteolytic activation and as a result spread of infection and pathogenicity of an avian influenza virus of the H5 serotype (Deshpande et al., 1988). It was also found that the masking effect of the oligosaccharide was overcome by an increase in the number of the basic amino acids at the cleavage site (Ohuchi et al., 1989).

Many studies have suggested that oligosaccharides play an important role in the structure and stability of viral glycoproteins. Specifically, it has been shown that supernumerary oligosaccharides may interfere with the trimer assembly and intracellular transport of the influenza virus hemagglutinin (Schuy et al., 1986; Gething et al. 1986; Gallagher et al., 1988).

2. Interference of oligosaccharides with polypeptide epitopes

2.1. Masking of polypeptide epitopes by oligosaccharides

Because viruses use the cellular glycosylation machinery, the oligosaccharides of the viral glycoproteins are often not recognized by the host's immune system as foreign structures. In fact, the carbohydrate side chains may shield antigenic determinants of the virus-specific polypeptide from immune recognition.

This concept has been derived mainly from observations made on influenza virus hemagglutinin. The hemagglutinin comprises a large family of highly pleomorphic glycoproteins which may vary up to 60% in amino acid sequence while still retaining very similar structural and biological properties (Wilson et al., 1981; Wiley et al., 1981). This extensive sequence variation, which allows newly evolved viruses to escape immunity acquired by the host during previous infection, may either occur gradually by the accumulation of single amino acid changes (antigenic drift), or abruptly after the introduction of a novel hemagglutinin gene from an animal reservoir (antigenic shift) (for review see Air and Laver, 1986). There is also considerable variation in the position and number of carbohydrate attachment sites on the different hemagglutinin molecules (Klenk, 1980; Wilson et al., 1983; Klenk et al., 1983; Keil et al., 1984; Feldmann et al., 1988). It has therefore been postulated several times that acquisition of a new oligosaccharide masking an epitope is one of the mechanisms underlying antigenic drift (Wiley et al., 1981: Keil et al., 1985; Raymond et al., 1986). Direct evidence has been obtained by Skehel and co-workers (1984). A single amino acid substitution, Asp 63 to Asn, was found in a variant of influenza virus A/Aichi/68 (H3N2) hemagglutinin selected with a monoclonal antibody. The mutation generates a potential oligosaccharide attachment site, (located in antigenic site C shown in Fig. 2.1), and it was shown that Asn 63 has carbohydrate attached. Immunoprecipitation experiments with extracts from cells infected with variant virus, prepared in the presence or absence of the glycosylation inhibitor tunicamycin, demonstrated that the addition of carbohydrate is required to prevent binding of monoclonal antibody. Similar experiments with strains A/England/878/69 (Air and Laver, 1986) and A/Victoria/3/75 (Wiley and Skehel, 1987), which have also a glycosylation site at position 63, support this conclusion and provide evidence for the epidemiological significance of carbohydrate-mediated modulation of hemagglutinin antigenicity. Similar modifications of antigenicity may occur in other regions of the hemagglutinin. From the three-dimensional coordinates of the oligosaccharide attached to residue 165 in the structure of the A/Aichi/68 hemagglutinin structure, it is clear that access of antibodies to the surface of the molecule would be blocked between residues 166 and 168. No variation in this region has been seen either in strains isolated from patients between 1968 and 1982 or in variants of H3 hemagglutinin selected with monoclonal antibodies. In contrast, the H1 subtype hemagglutinin of strain A/PR/8/34 has no glycosylation sites in this region, and an analysis of monoclonal antibody-selected variants (Caton et al., 1982) has shown that amino acids 165, 166 and 167 can be recognized by such antibodies. These observations made on virus variants are complemented to some extent by a study employing site-directed mutagenesis of the hemagglutinin. When a supernumerary carbohydrate was introduced at Asn 188, which is located in antigenic site B (see Fig. 2.2), recognition by a B-specific monoclonal antibody was significantly reduced (Gallagher et al., 1988).

Considerably less information on carbohydrate masking of antigenic epitopes is available for other viral glycoproteins. Experiments with monoclonal antibodies raised against Sindbis virus suggested the existence of a strain-specific neutralizing antigenic site (E2-b) on the E2 protein. Analysis of neutralization escape mutants selected with E2-b-specific antibodies indicated that amino acid 216 is a major determinant of the E2-b antigenic site, since eight of nine mutants showed a coding change in this position. One mutant had a new glycosylation site at position 213 which according to an altered migration rate in SDS-PAGE appeared to be glycosylated (Davis et al., 1987). Although not providing definite proof, these results are compatible with carbohydrate shielding. Similar observations have been made on mutants of rotavirus SA 11 selected by a monoclonal antibody against the VP 7 protein. The mutations found produced two new glycosylation sites which were utilized. The mutations lead to gross antigenic changes, and these were found to be reversible upon removal of the attached carbohydrate. These results were also interpreted to show an important role of carbohydrate in influencing the exposure of antigenic determinants (Caust et al., 1988).

Evidence for a protective effect of carbohydrate side chains against the immune response has also been obtained with retroviruses. Antisera raised against Rauscher leukemia virus contained a preponderance of antibodies against glycoprotein gp70 that were dependent on the presence of carbohydrate side chains for reactivity, as judged by immunoprecipitation or Western blotting. However, the majority of neutralizing antibodies were not dependent on the presence of carbohydrate, as indicated by (1) the ability of the deglycosylated virus to adsorb neutralizing antibody from sera as efficiently as glycosylated virus and (2) the ability of glycosylated virus to induce neutralizing antibody responses when injected into rabbits. In fact, a faster response was obtained with deglycosylated than with untreated virus in the latter experiments. The results are compatible with the view that carbohydrates diminish the reactivity of glycoprotein epitopes responsible for neutralization (Elder et al., 1986).

2.2. Stabilization of polypeptide epitopes by oligosaccharides

Comparative analysis of glycoproteins with their corresponding unglycosylated polypeptides have suggested that oligosaccharides may also influence antigenic properties by maintaining the integrity of the respective epitopes, which usually appear to be conformation-dependent. In a study of this type, the antigenicity of the envelope glycoproteins of Semliki forest virus has been analyzed at different stages of maturation using polyclonal antisera. The glycosylated intracellular proteins were antigenically similar to the unglycosylated polypeptides but differed in antigenicity from the mature glycoproteins present in virions. This conversion was interpreted as a change in conformation of the glycoproteins causing exposure of different antigenic determinants not identical with the sugar moiety (Kaluza et al., 1980).

The glycoprotein of influenza C virus has been analyzed using a panel of monoclonal antibodies that could be divided into those that had neutralizing activity (group A) and those that did not (group B). Operational analysis with a panel of antigenic variants selected with each of the group A antibodies identified two non-overlapping antigenic sites, A_1 and A_2. Competitive binding analysis with group B antibodies identified two additional non-overlapping antigenic sites, sites B_1 and B_2. It was found in radioimmunoprecipitation experiments that antibodies to sites B_1 and B_2 were reactive not only with the mature glycoprotein, but also with its non-glycosylated form synthesized in the presence of tunicamycin. In contrast, antibodies to sites A_1 and A_2 immunoprecipitated the unglycosylated polypeptide only in trace amounts or not at all. Additionally, Western blot analysis showed that the denatured glycoprotein blotted on nitrocellulose was reactive with antibodies to sites B_1 and B_2, but not with those to sites A_1 and A_2. These observations suggested that glycosylation of the influenza C glycoprotein selectively influences the integrity of antigenic sites A_1 and A_2 which are composed of conformation-dependent epitopes (Sugawara et al., 1988).

A similar phenomenon was observed when the antigenic structure of the glycoprotein of tick-borne encephalitis virus, a flavivirus, was analyzed by a panel of monoclonal antibodies. Most of the 19 different epitopes identified clustered to form three non-overlapping domains termed A, B, and C. The structural properties of the epitopes were assessed by analyzing, among other treatments, the effect of SDS denaturation on the antigenic reactivities. Deglycosylation by *N*-glycanase abolished the SDS resistance of domain C, providing evidence for the role of the single carbohydrate side chain of this glycoprotein in stabilizing the epitopes of this domain. The observation that deglycosylation of the native protein does not abolish antigenic reactivity of these epitopes suggested that the oligosaccharide is essential for shielding these sites from the action of SDS or allowing renaturation to occur during the dot immunoassay employed (Guirakhoo et al., 1989).

3. Carbohydrate epitopes of viral glycoproteins

With some viral glycoproteins, oligosaccharide side chains themselves have been found to act as antigenic determinants. Because of their host specificity, these carbohydrate epitopes are responsible, by a mechanism of molecular mimicry, for cross-reactivities with host components or with other viral glycoproteins carrying similar oligosaccharides.

The presence in purified preparations of influenza virus of an antigen that reacts with antibodies raised against uninfected chick chorioallantoic membranes or allantoic fluid has been recognized for many years (Knight, 1944; Schäfer et al., 1952; Harboe, 1963). The host antigen present in uninfected chick embryos has been shown to be a sulfated mucopolysaccharide (Haukenes et al., 1965; Howe et al., 1967). Laver and Webster (1966) showed that about 50% of the host antigen associated with influenza virions was covalently bound to the hemagglutinin. Structural analysis of the A/Memphis/102/72 (H3) hemagglutinin revealed that two of the six oligosaccharides of the HA_1 subunit are antigenically related to the host antigen. These are the *N*-acetyllactosamine type carbohydrate units attached to asparagine residues 8 and 22 (Ward et al., 1981).

Host cell antigens are also present on the neuraminidase of influenza virus. Chemical analysis of the neuraminidase heads of strain A/Tokyo/3/67 (N2) revealed two oligomannosidic side chains attached to asparagine residues 86 and 200 and two *N*-acetyllactosamine side chains at positions 146 and 234. Antigenic analysis of glycopeptides derived from these glycosylation sites suggested that only the oligosaccharide at asparagine 146 was capable of binding antibodies raised against uninfected chick chorioallantoic membranes (Ward et al., 1983).

It is remarkable that, on both the hemagglutinin and the neuraminidase, only one or two of several structurally similar carbohydrate side chains proved to be antigenically related to host antigen. Whereas the oligosaccharide at asparagine 146 of the neuraminidase contained *N*-acetylgalactosamine which, as pointed out above, is a sugar residue only rarely found in *N*-glycosidically linked carbohydrates, nothing unusual was found with the complex side chains attached to asparagine residues 8 and 22 of the H3 hemagglutinin. It is also puzzling that, according to the available evidence, these *N*-glycosidically linked viral carbohydrate units differ greatly from the chick chorioallantoic membrane carbohydrate in composition and general structure, since the latter has been reported to be an *O*-glycosidically linked glycan-containing *N*-acetylgalactosamine, *N*-acetylglycosamine, galactose, neuraminic acid and fucose, but lacking mannose residues (Meyer et al., 1967). Thus, the exact nature of the cross-reactivity between *N*-linked viral carbohydrate side chains and host antigens is still obscure and requires further structural analysis.

N-linked carbohydrates have also been suspected to mediate cross-reactivity between the human immunodeficiency virus (HIV) and equine infectious anemia virus (EIAV). The reported serological relatedness between the major glycoproteins of these viruses (HIV gp 120 and EIAV gp90) was examined using purified antigens and reference sera from AIDS patients, and anti-gp120 goat serum and an EIAV infected horse serum. To assess the contributions of glycoprotein oligosaccharide and peptide components, antigens treated with endoglycosidase F to remove carbohydrate were assayed in parallel with the intact glycoprotein. Deglycosylation of HIV gp 120 increased its reactivity with the AIDS patient serum, had no significant effect on its reactivity with the goat antiserum, and essentially abolished its reactivity with the EIAV reference serum. Thus, it appeared that the serological cross-reactivity observed between HIV gp120 and sera from EIAV-infected horses can be attributed to the oligosaccharide rather than the peptide components of the viral glycoproteins (Montelaro et al., 1988).

Tentative evidence for molecular mimicry by *O*-linked carbohydrates has been obtained with the murine coronavirus A59 (Niemann et al., 1984). The E1 glycoprotein of this virus contains *O*-linked oligosaccharides similar to these found on glycophorin A, the major glycoprotein of the human erythrocyte membrane (Thomas and Winzler, 1969). In glycophorin A (Tomita et al., 1978) as well as the viral glycoprotein, these oligosaccharides are attached to the tetrapeptide sequence Ser-Ser-Thr-Thr. The oligosaccharides constitute, in conjunction with the amino acid sequence, the blood group M determinant of glycophorin A (Lisowska and Kordowicz, 1977); this blood group activity has also been observed in purified E1 glycoprotein.

4. Conclusions

Viral glycoproteins contain oligosaccharides in *N*-glycosidic or *O*-glycosidic linkage. None of the carbohydrate structures hitherto identified is unique to viral glycoproteins; they are also present in a large variety of other membrane and secretory glycoproteins. By molecular mimicry, oligosaccharides on viral glycoproteins may therefore be responsible for serological cross-reactivities between different viruses or between viral and host antigens. It is, thus, tempting to speculate that carbohydrate side chains may be an important factor for autoimmune reactions developing during virus infections. The evidence available so far indicates, however, that the role of the oligosaccharides as carbohydrate antigens is secondary to their significance as modulators of antigenic polypeptide epitopes of viral glycoproteins. Oligosaccharides may affect these epitopes by two different mechanisms. On the one hand, they may stabilize the epitopes by maintaining the correct folding of the polypeptides. On the other hand, evidence has been obtained that carbohydrates shield antigenic sites from immune recognition by steric hindrance. The acquisition of oligosaccharides antigenically indifferent to the host may therefore be one of the mechanisms that help a virus to escape the immune defense of the host organism.

5. Acknowledgements

I thank W. Garten and R. T. Schwarz for critical discussions and S. Fischbach for expert secretarial help.

6. References

Air, G.M., and Laver, W.G. (1986) *Adv. Virus Res.* 31, 53–102.

Boersma, D.P., Saleh, F., Nakamura, K., and Compans, R.W. (1982) *Virology* 123, 452–456.

Caton, A.J., Brownlee, G.G., Yewdell, J.W. and Gerhard, W. (1982) *Cell* 31, 417–427.

Caust, J., Dyall-Smith, M.L., Lazdins, I. and Holmes, I.H. (1987) *Arch. Virol.* 96, 123–134.

Davis, N.C., Pence, D.F., Meyer, W.J., Schmaljohn, A.C., and Johnston, R.E. (1987) *Virology* 161, 101–108.

Deom, C.M. and Schulze, I.T. (1985) *J. Biol. Chem.* 260, 14771–14774.

Deom, C.M., Caton, A.J. and Schulze, I.T. (1986) Proc. Natl. Acad. Sci. USA 83, 3771–3775.

Deshpande, K.L., Fried, V.A., Ando, M. and Webster, R.G. (1987) *Proc. Natl. Acad. Sci. USA* 84, 36–40.

Diabaté, S., Geyer, R., and Stirm, S. (1984) *Eur. J. Biochem.* 139, 329–336.

Elbein, A.D. (1981) *Trends Biochem. Sci.* 6, 219–221.

Elbein, A.D. (1987) *Annu. Rev. Biochem.* 56, 497–534.

Elder, J.H., McGee, J.S. and Alexander, S. (1986) *J. Virol* 57, 340–342.

Estes, M.K., Palmer, E.L. and Obijeski, J.F. (1983) *Curr. Top. Microbiol. Immunol.* 105, 123–184.

Feizi, T. and Childs, R.A. (1987) *Biochem. J.* 245, 1–11.

Feldmann, H., Kretzschmar, E., Klingeborn, B., Rott, R., Klenk, H.-D. and Garten, W. (1988) *Virology* 165, 428–437.

Gallagher, P., Henneberry, J., Wilson, I.A., Sambrook. J. and Gething, M.-J. (1988) *J. Cell. Biol.* 107, 2059–2073.

Gething, M.-J., McCammon, K. and Sambrook, J. (1986) *Cell* 46, 939–950.

Gibson, R., Kornfeld, S. and Schlesinger, S. (1980) *Trends Biochem. Sci.* 5, 290–293.

Gruber, C. and Levine, S. (1985) *J. Gen. Virol.* 66, 417–432.

Guirakhoo, F., Heinz, F.X. and Kunz, C., (1989) *Virology* 169, 90–99.

Haukenes, G., Harböe, A., and Mortensson-Egmund, K. (1965) *Acta Pathol. Microbiol. Scand.* 64, 534–542.

Harböe, A. (1963) *Acta Pathol. Microbiol. Scand.* 57, 488–492.

Heermann, K.H., Goldmann, U., Schwartz, U., Seyffarth, T., Baumgarten, H. and Gerlich, W.H. (1984) *J. Virol.* 52, 396–402.

Holmes, K.V., Doller, E.W. and Sturman, L.S. (1981) *Virology* 115, 334–344.

Hope, R.G. and Marsden, H.S. (1983) *J. Gen. Virol.* 64, 1943–1953.

Howe, C., Lee, L.T., Harböe, A., and Haukenes, G. (1967) *J. Immunol.* 98, 543–557.

Hsieh, P., Rosner, M.R. and Robbins, P.W. (1984) *J. Biol. Chem.* 258, 2548–54.

Hunt, L.A., Davidson, S.K. and Golemboski, D.B. (1983) *Arch. Biochem. Biophys.* 226, 347–356.

Johnson, D.C. and Spear, P.G. (1983) *Cell* 32, 987–997.

Kaluza, G., Rott, R. and Schwarz, R.T. (1980) *Virology* 102, 286–299.

Keil, W., Klenk, H.-D. and Schwarz, R.T. (1979) *J. Virol.* 31, 253–256.

Keil, W., Niemann, H., Schwarz, R.T. and Klenk, H.-D. (1984) *Virology* 133, 77–91.

Keil, W., Geyer, R., Dabrowski, J., Dabrowski, U., Niemann, H., Stirm, S. and Klenk, H.-D. (1985) *EMBO J.* 4, 2711–2720.

Klenk, H.-D. (1980) In *Structure and Variation in Influenza Virus* (W.G. Laver and G.M. Air, Eds.) Elsevier, New York. pp. 213–222.

Klenk, H.-D., Keil, W., Niemann, H., Geyer, R. and Schwarz, R.T. (1983) *Curr. Topics Microbiol. Immunol.* 104, 247–257.

Knight, C.A. (1944) *J. Exp. Med.* 80, 83–99.

Kornfeld, R., and Kornfeld, S. (1985) *Annu. Rev. Biochem.* 54, 631–664.

Kuroda, K., Gröner, A., Frese, K., Drenckhahn, D., Hauser, C., Rott, R., Doerfler, W. and Klenk, H.-D. (1989) *J. Virol.* 63, 1677–1685.

Laver, W.G. and Webster, R.G. (1966) *Virology*30, 104–114.

Lisowska, E. and Kordowicz, M. (1977) In *Human Blood Groups* (J.F. Mohn, Ed.) Karger, Basel. p. 188.

McDowell, W. and Schwarz, R.T. (1988) *Biochimie* 70, 1535–1549.

Meyer, K., Bhavanandan, V.P., Yung, D., Lee, L.T. and Howe, C. (1967) *Proc. Natl. Acad. Sci.* USA 58, 1655–1659.

Montelaro, R.C., Robey, W.G., West, M.D., Issel, C.J. and Fischinger, P.J. (1988) *J. Gen. Virol.* 69, 1711–1717.

Niemann, H., and Klenk, H.-D. (1981) *J. Mol. Biol.* 153, 993–1010.

Niemann, H., Boschek, B., Evans, D., Rosing, M., Tamura, T. and Klenk, H.-D. (1982) *EMBO J.* 1, 1499–1505.

Niemann, H., Geyer, R., Klenk, H.-D., Stirm, S. and Wirth, M. (1984) *EMBO J.* 3, 665–670.

Ohuchi, M., Orlich, M., Ohuchi, R., Simpson, B.E.J., Garten, W., Klenk, H.-D. and Rott, R., (1989) *Virology* 168, 274–280.

Olden, K., Bernard, B.A., Humphries, M.J., Yeo, T.-K., Yeo, K.-T., White, S.L., Newton, S.A., Bauer, H.C. and Parent, J.P. (1985) *Trends. Biochem. Sci.* 10, 78–82.

Olmsted, R.A., Murphy, B.R., Lawrence, L.A., Elango, N., Moss, B. and Collins, P.L. (1989) *J. Virol.* 63, 411–420.

Olofsson, S., Jeansson, S. and Lycke, E. (1981). *J. Virol.* 38, 564–570.

Raymond, F.L., Caton, A.J., Cox, N.J., Kendal, A.P. and Brownlee, G.G. (1986) *Virology* 148, 275–287.

Robertson, J.S., Naeve, C.W., Webster, R.G., Bootman, J.S., Newman, R. and Schild, G.C. (1985) *Virology* 143, 166–174.

Rottier, P.J.M., Horzinek, M.C. and van der Zeijst, B.A.M. (1981) *J. Virol.* 40, 350–357.

Schäfer, W., Munk, K. and Armbruster, O. (1952) *Z. Naturforsch.* 76, 29–33.

Schlüter, M., Linder, D. and Geyer, R. (1985) *Carbohydrate Res.* 138, 305–314.

Schuy, W., Will, C., Kuroda, C., Scholtissek, C., Garten, W. and Klenk, H.-D. (1986) *EMBO J.* 5, 2831–2836.

Shida, H. and Dales, S. (1981) *Virology* 111, 56–72.

Siddell, S.G., Wege, H. and ter Meulen, V. (1982) *Curr. Top. Microbiol. Immunol.* 99, 131–163.

Simons, K., Garoff, H. and Helenius, A. (1980) In *The Togaviruses* (R.W. Schlesinger, Ed.) Academic Press, New York. pp. 317–333.

Skehel, J.J., Stevens, D.J., Daniels, R.S., Douglas, A.R., Knossow, D. M., Wilson, I.A. and Wiley, D.C. (1984a) *Proc. Natl. Acad. Sci. USA* 81, 1779–1783.

Sugawara, K., Kitame, F., Nishimura, H. and Nakamura, K. (1988) *J. Gen. Virol* 69, 537–547.

Tanner, W. (1987) *Biochim. Biophys. Acta* 906, 81–99.

Thomas, D.B. and Winzler, R.J. (1969) *J. Biol. Chem.* 244, 5943–5946.

Tiollais, P., Buendia, M.-A., Brechot, C., Dejean, A., Michel, M.-L. and Pourcel, C. (1988) In *Viral Hepatitis and Liver Disease* (A.J. Zuckerman, Ed.) Alan R. Liss, New York. pp. 295–300.

Tomita, M., Furthmeyr, H. and Marchesi, V.T. (1978) *Biochemistry* 17, 4756–4770.

Vorndam, A.V. and Trent, D.W. (1979) *Virology* 95, 1–7.

Ward, C.W., Brown, L.E., Downie, J.C. and Jackson, D.C. (1981) *Virology* 108, 71–79.

Ward, C.W., Murray, J.M., Roxburgh, C.M. and Jackson, D.C. (1983) *Virology* 126, 370–375.

Wiley, D.C. and Skehel, J.J. (1987) *Annu. Rev. Biochem.* 56, 365–394.

Wiley, D.C., Wilson, I.A. and Skehel, J.J. (1981) *Nature* 289, 373–378.

Williams, M.A. and Lamb, R.A. (1988) *Mol. Cell. Biol.* 8, 1186–1196.

Wilson, I.A., Skehel, J.J. and Wiley, D.C. (1981) *Nature* 289, 366–373.

Wilson, I.A., Ladner, P.C., Skehel, J.J. and Wiley, D.C. (1983) *Biochem. Soc. Trans.* 11, 145–147.

Winkler, G., Heinz, F.X. and Kunz, Ch. (1987) *J. Gen. Virol.* 68, 2239–2244.

Eds. M.H.V. van Regenmortel and A.R. Neurath
Immunochemistry of viruses, II. The basis for serodiagnosis and vaccines.

Processing and presentation of viral antigens 3

K.P. WILLIAMS[1] and J.A. SMITH[2]

Departments of [1,2] Molecular Biology and [2]Pathology, Massachusetts General Hospital, and Departments of [1]Genetics and [2]Pathology, Harvard Medical School, Boston, MA 02114, U.S.A.

1. Introduction

The immune response to many viral infections is characterized by the induction of a response from both the cytolytic (CTL) and helper (T_h) T lymphocyte subsets. Classically described antiviral CTL are restricted by class I major histocompatibility (MHC) gene products in their recognition of viral and other foreign antigens, while T_h are restricted by MHC class II gene products (Zinkernagel and Doherty, 1979; Schwartz, 1984). Expression of MHC class II molecules is restricted to certain cells with specialized antigen-presenting functions (antigen-presenting cells (APC) e.g., dendritic cells, macrophages, B cells), whereas MHC class I molecules are expressed on most cells.

For most viruses, de novo viral protein synthesis in the infected cell is needed for target cell sensitization (Braciale and Yap 1978). Evidence accumulated in the past 5 years suggests that T cells recognize degraded forms of viral antigens bound to class I or class II MHC molecules in a manner analogous to that first proposed for the presentation of soluble exogenous protein antigens. However, the processing mechanisms for soluble exogenous antigens and endogenously derived viral antigens are quite distinct. It is only now becoming apparent that a number of alternative routes exist in the cell for processing of viral antigens and which pathway is followed depends on the mode of entry of the virus into the cell and whether the antigen is exogenously or endogenously derived. This review concerns itself with the problem of what occurs between viral infection of a cell and the exposure of epitopes on the plasma membrane associated with either class I or II MHC molecules.

2. Processing and presentation of endogenous viral antigen for recognition by CTL

2.1. Modes of virus entry

Enveloped viruses enter the host cell by membrane fusion. Two classes of virus can be distinguished based on the pH at which fusion occurs. One class is represented by viruses such as influenza virus that can fuse with host membranes only at acidic pH. Fusion occurs after receptor-mediated endocytosis in the acidic environment of endosomes. The second class of enveloped viruses, such as paramyxoviruses, do not show pH dependence in their fusion activity. For these viruses, fusion with the host cell membranes occurs at neutral pH, and endocytosis is not a requirement for virus entry (e.g., human immunodeficiency virus).

2.2. Influenza A virus as a model

Influenza A virus has been used extensively as a model to study the recognition of viral antigens by CTL (reviewed in Townsend and Bodmer, 1989). The viral genome consists of eight RNA segments, each coding for at least one viral polypeptide. The virion is composed of four major virion polypeptides; two external envelope-associated glycoproteins, hemagglutinin (HA) and neuraminidase (NA), and two internal nonglycosylated virion polypeptides, the matrix (M) and nucleocapsid (NP) proteins.

2.3. Requirement for de novo synthesis of viral proteins

For influenza virus, fusion of viral and cellular membranes is an essential step in the cellular processing of virion proteins for CTL recognition. Class I MHC-restricted CTL efficiently lyse infected target cells but fail to lyse target cells treated with noninfectious (heat-inactivated virus which retains fusion activity but is no longer competent to replicate or induce new viral protein synthesis) virion preparations (Braciale and Yap, 1978). In contrast, class II MHC-restricted CTL lyse target cells treated with either infectious or noninfectious influenza virion preparations. These studies suggested that de novo synthesis of viral proteins was required for class I restricted CTL-mediated lysis of influenza virus-infected target cells. This requirement for de novo viral protein synthesis was confirmed by the finding that emetine, a protein synthesis inhibitor, completely abolished target cell recognition by class I-restricted CTL clones directed to the influenza HA or NP when the inhibitor is added simultaneously with the virus. Emetine failed to inhibit target cell recognition by class I-restricted CTL if added 4–6 h after infection when substantial de novo protein synthesis had already occurred (Braciale et al., 1988). Inhibition of viral protein synthesis had no effect on recognition of virions by class II-restricted CTL (Morrison et al. 1986). These findings have been complicated by the observation that target cells treated with intact noninfectious influenza virus with inactivated neuraminidase activity can be recognized by class I-restricted anti-influenza CTL (Yewdell et al., 1988). This was explained by the observation that viral polypeptides were reaching the cytoplasm. Two mechanisms were proposed to account for this phenomenon (Yewdell et al., 1988). Inactivation of the receptor-destroying activity of the virus might alter the manner in which cells handle the virus. Alternatively, NA inactivation might increase the amount of virus available for processing by preventing the virus from leaving receptor molecules. Early evidence strongly favored the view that class I MHC-restricted T lymphocytes recognized foreign antigens expressed in their native form on cell surfaces (i.e., CTL recognition is mainly directed at viral glycoproteins which are transported to and inserted in the target cell plasma membrane) (Sugamura et al., 1977; Gething et al., 1978; Rott and Scholtissek, 1982). However, target cells expressing a mutant HA that was prevented from leaving the cytoplasm by deletion of the N-terminal hydrophobic signal sequence (Gething and Sambrook, 1982) were recognized efficiently by class I-restricted CTL originally induced by influenza infection (Townsend et al. 1986a). Such evidence suggested that the mutant HA was retained in the cytoplasm and rapidly degraded. Moore et al. (1988) demonstrated that the requirement for de novo synthesis of viral antigen could be circumvented by introducing the protein into the target cell cytoplasm by osmotic lysis of pinosomes that have taken up the antigen.

2.4. Class I-restricted CTL recognition is primarily directed against internal viral polypeptides

Although HA, the major glycoprotein of the virus, could be shown to be a target antigen (Braciale et al. 1984; Townsend et al. 1984; Bennink et al. 1984), a much greater proportion of the class I-restricted activity in polyclonal CTL cultures was directed at NP and other internal proteins. Transfected L cells expressing NP were recognized efficiently by class I-restricted CTL, despite the lack of detectable NP at the cell surface (Townsend et al. 1984). In contrast, target cells exposed to purified undenatured NP were not sensitized for lysis by class I-restricted CTL.

2.5. Viral epitopes can be defined by short synthetic peptides

More recently, it was found that the requirement for de novo synthesis of viral proteins could be circumvented by incubating cells with fragments of viral proteins, including synthetic peptides consisting of as few as 9 amino acid residues (Townsend et al. 1985; Townsend et al. 1986b). Prior work had also shown that CTL could be induced in vitro with a fragment of HA (Wabuke-Bunoti et al. 1981, 1984). Four NP epitopes were identified that resulted in target cell lysis (Townsend et al. 1986a; Bastin et al. 1987), and each was recognized by characterized $CD8^+$ T cells in association with a different class I molecule. Similar results have been reported for a number of other class I-restricted T cell responses to proteins expressed by transfection or recombinant vaccinia viruses (Braciale et al. 1987a, Takahashi et al. 1988; Nixon et al. 1988; Parham et al. 1987). These findings have led to the idea that internal viral proteins are degraded in the cytoplasm of the infected cell, producing short denatured peptides that then move to the cell surface and become available for recognition by CTL in association with class I MHC molecules in a way similar to that in which T_h cells recognize denatured or degraded soluble proteins in association with class II molecules.

2.6. Site of endogenous antigen processing

The primary requirement for presentation by class I molecules appears to be the cytoplasmic localization of antigen. Soluble viral proteins synthesized on free ribosomes in the cytoplasm of the target cell are also proposed to be degraded in this same cellular location. The identity of the proteolytic system involved in the presentation of cytoplasmic proteins remains unknown, although the more effective presentation of a ubiquitin-NP fusion protein (Townsend et al. 1988) and the lack of inhibitory effect of lysosomotropic agents (Townsend et al., 1986b; Morrison et al., 1986) raises the possibility that ubiquitin-dependent degradation may play a role (Hershko and Ciechanover, 1986; Mayer, 1987).

It is unclear how peptides generated in the cytoplasm cross a lipid membrane and gain access to class I molecules in the endoplasmic reticulum (ER) or Golgi. However, labeled proteins microinjected into the cytosol of a cell have been detected in a system of vesicles soon after injection (Stacey and Allfrey, 1977), although the nature of this mechanism has not been elucidated.

Recent studies have provided new clues to the fate of cytoplasmic antigens destined for class I presentation (Yewdell and Bennink, 1989; Nuchtern et al. 1989). The antibiotic brefeldin A (BFA), which blocks the transport of proteins out of the ER and causes disassembly of the *cis*/medial Golgi complex (Fujiwara et al. 1988; Lippincott-Schwartz et al. 1989), totally abolished the recognition of influenza infected target cells by CTL specific for the matrix protein (Nuchtern et al. 1989). This inhibition was reversible and did not affect the recognition of target cells pulsed with a synthetic influenza matrix-derived peptide. In another study (Yewdell and Bennink, 1989), presentation of several viral antigens by infected cells was completely blocked by BFA. The temperature dependence of recovery from inhibition was the same for the reappearance at the cell surface of newly synthesized class I molecules and for CTL recognition. These two studies strongly suggest that processed cytoplasmic antigens associate with class I molecules prior to their egress from the ER.

3. Processing and presentation of exogenous antigens

Sources of exogenous viral antigen comprising inactive viral particles and also those intact viruses that enter the host via endocytosis and pH-dependent fusion become available for association with class II MHC in a manner analogous to that detailed below for the processing of soluble antigens.

3.1. Requirement for antigen processing

Much of our current understanding of antigen processing and presentation was initially worked out using $CD4^+$ T cells specific for soluble protein antigens in combination with class II MHC molecules. These studies (Harding et al. 1988; Werdelin et al. 1988) have demonstrated that in general, native antigens do not associate directly with MHC molecules; instead, most must be unfolded, if not fragmented, to reveal specific amino acid residues that bind to the MHC molecule (Braciale et al. 1987b, Werdelin et al. 1988). A role for antigen processing prior to presentation was postulated from a number of studies (reviewed by Grey and Chesnut, 1985) observing that an approximately 1 h delay occurs between initial antigen binding to APC and effective antigen presentation to T cells (Ziegler and Unanue, 1981). Shimonkevitz et al. (1983) elegantly demonstrated that whatever in vivo APC processing events occurred, they could be bypassed if the native antigen was proteolysed in a test tube and then added to gluteraldehyde-fixed APC. The same fixed APC could not present intact native antigen. Peptides in the digest were subsequently identified and synthesized (Shimonkevitz et al. 1984), and these were shown to be able to substitute for native antigen. Similar results have now been obtained in a number of other antigen systems (Allen and Unanue, 1984; Kovac and Schwartz, 1985), establishing that 'processed' forms of the antigen can be presented by fixed APC. Competition studies, using peptides derived from two model proteins, staphylococcal nuclease and λ-repressor cI, initially performed at the cellular level (Guillet et al. 1986) and subsequently with purified class II proteins (Guillet et al., 1987), suggested that both the class I and class II molecules have a single antigen-combining site. Further, the peptides could be subdivided experimentally into two components (Allen et al. 1987; Sette et al. 1987), the amino-acid residues that interact with the T cell receptor (epitope) and the amino-acid residues that interact with the MHC molecule (agretope) (Schwartz, 1984).

The recent solution of the three-dimensional structure of HLA-A2, a class I MHC molecule, has revealed details of the MHC-peptide interaction (Bjorkman et al. 1987a, 1987b). It has been proposed (Ajitkumar et al. 1988) that the T cell receptor simultaneously contacts residues on two α-helices of the MHC molecule whilst also interfacing with the foreign or self peptide in the antigen binding site. MHC class II molecules can be modeled by sequence homology to adopt a similar structure (Brown et al. 1988). Furthermore, peptidyl material has been eluted from purified

I-A^d MHC molecules (Buus et al. 1988) and shown to bind specifically to I-A^d but not I-E^d MHC molecules.

3.2. Mechanism of processing of exogenous antigens

Ziegler and Unanue (1982) found that processing of *Listeria monocytogenes* by an APC was inhibited by reagents that elevate intracellular pH (e.g. lysosomotrophic drugs). Subsequently, the same effect on processing of soluble protein antigens by these lysosomotrophic drugs (i.e., chloroquine and monensin) was also demonstrated (Chesnut et al. 1982; Allen et al. 1984). Chloroquine was also found to inhibit the process of class II-restricted CTL recognition of target cells when either infectious or noninfectious virus preparations were used to sensitize target cells (Braciale et al. 1988). In the same study chloroquine did not inhibit de novo viral protein synthesis in infected target cells, and the drug did not affect recognition of virally infected cells by class I-restricted CTL. Highly specific protease inhibitors, such as leupeptin, can also block exogenous antigen processing (Buus and Werdelin, 1986).

3.3. Site of exogenous antigen processing

A recent study by Lin et al. (1988), using colloidal gold conjugates of an antigen, supports the notion that a protein antigen is handled intracellularly by an APC and then recycled to the surface before presentation. Using biotinylated antigen and an antigen-specific B cell line, Watts and Davidson (1988) have demonstrated the presence of a rapid and efficient pathway for the endocytosis and recycling of antigen in B cells. Antigen-presenting cells with a temperature-sensitive defect in endosomal acidification were unable to process native antigen, although their ability to present immunogenic peptides was unaffected (McCoy and Schwartz, 1988; McCoy et al. 1989). These results confirmed without any pharmacologic manipulation that foreign antigen moves into an acidic compartment, likely to be acidified early endosomes, during some phase of its processing. Acidic proteases capable of proteolytic processing have been localized in macrophage endosomes (Diment et al. 1988) and APC plasma membrane fractions (Chain et al. 1989).

3.4. Site of association of immunogenic fragments and class II MHC molecules

How degradation products reach the site of their association with class II molecules and how immunogenic moieties are translocated to this site of association remains to be elucidated. However, endocytic vesicles have been suggested to be the site of association of the immunogenic protein fragments with class II molecules, as well as the site of antigen processing. Further, antigenic peptides may encounter recycling class II molecules in the endosome or newly synthesized class II in the *trans* cisternae of the Golgi. Theoretically, binding could occur at any time between the time of folding of the MHC molecule (i.e., after formation of its binding site) and the appearance of the complete molecule at the cell surface. An intracellular site for the association of peptide with MHC molecules has been suggested, as only in this case would the immune activity of the antigen-specific T cells be localized to the appropriate APC (Pernis, 1988).

These findings have led to a currently favored hypothesis for class II processing (Germain, 1988; Gerlier and Rabourdin-Combe, 1989). Exogenous protein antigens are taken up by APC via nonspecific fluid phase or adsorptive endocytosis. These early endosomes fuse with Golgi-derived endosomes containing acidic proteases. The fusion is energy-dependent and requires cytosolic protein factors (Mayorga et al. 1989). Acidification of the fused endosomes activates the proteases. Proteolysis by the acid proteases generates peptides capable of associating with class II MHC molecules at some intracellular site, and these peptide:MHC complexes are then transported to the cell surface for recognition by T cells (Germain, 1986; Allen, 1987; Unanue and Allen, 1987; Berzofsky et al., 1988).

4. Compartmentation in antigen processing and presentation

4.1. Mode of antigen entry determines mechanism of processing

Morrison et al. (1986) demonstrated using different CTL specific for HA associated with either class I or class II molecules, that the pathway by which the HA entered the target cell played the major role in determining whether effective antigen presentation occurred in the context of class I or class II molecules. Thus, antigen entering the cell in endocytic vesicles was selectively seen in the context of class II, whereas antigen synthesized in the cytoplasm of the cell was seen in the context of class I. It was also shown (Morrison et al. 1986) that treatment of the target cells with chloroquine abolished recognition of infected cells by class II-restricted CTL without diminishing class I-restricted recognition of infected target cells. Similarly, class I CTL specific for ovalbumin could recognize target cells synthesizing ovalbumin endogenously, target cells incubated with synthetic peptides, or target cells in which ovalbumin was introduced into the cytoplasm by osmotic lysis of pinocytic vesicles but could not recognize targets incubated with intact ovalbumin without a means of cytoplasm delivery (Moore et al. 1988). This result indicates that it is not where the protein is originally made, but whether the protein has access to the proper processing compartment that determines how it is processed by the cell.

These data suggest that at least two distinct cellular pathways exist for the generation of protein fragments suitable for binding to MHC molecules, one an endocytic pathway for the processing of exogenously derived antigens and the other a cytosolic degradative system for the processing of endogenously derived viral polypeptides (Germain, 1986). It is likely that both of these pathways produce denatured and/or degraded forms of the antigen. For transmembrane proteins such as HA, synthesized in the ER, processing may occur in this compartment to generate fragments for presentation (Lippincott-Schwartz et al. 1988).

4.2. Factors determining the preferential recognition by class I and class II MHC-restricted T lymphocytes of antigenic moieties generated by the endogenous and exogenous pathways

Mechanisms must exist for the intracellular separation of the reactions by which class I and class II molecules bind peptide. The supply of peptides to these molecules is presumably dependent upon the pathways of unfolding and degradation for all cellular and internalized proteins and upon intracellular transport of the peptide products. If processed transmembrane or cytoplasmic antigens bind to class I molecules in the ER, why is it that they cannot associate with class II molecules for presentation? This inability could easily be explained by postulating that newly synthesized class I molecules on route to the cell surface do not traverse the endocytic pathway. In contrast, class II molecules on their way to the cell surface do intersect the endocytic pathway as demonstrated by the observation that internalization of transferrin-neuraminidase conjugates led to the desialyation of intracellular class II molecules (Cresswell, 1985).

4.3. Putative role of the invariant chain in preventing the association of endogenously derived antigenic fragments with class II MHC molecules

It has been suggested that the mechanism that would prevent class II molecules from associating with peptides in the ER or *cis*-Golgi involves the invariant (Ii) chain which complexes with class II molecules shortly after biosynthesis. The Ii subunit could protect the peptide-binding site until it was removed by proteolysis in an acidic compartment allowing endocytically generated peptides to bind (Blum and Cresswell, 1988). Alternatively, the role of the invariant chain may be to act as a signal to target class II molecules to the endosomes.

4.4. Association of class I heavy chain with β_2-microglobulin is induced by viral peptides

Further insight into the association of peptides with class I molecules has come from a recent article (Townsend et al. 1989) using a mutant mouse cell line (RMA-S) that has a reduced expression of class I molecules. The defect in RMA-S is not in the structure or expression of the genes encoding β_2-microglobulin or the class I heavy chain but rather in the association of their products. Influenza virus infection of normal target cells (RMA) resulted in lysis by CTL directed against NP. In contrast, RMA-S cells infected with influenza virus are not lysed by CTL, but both cell types are lysed by incubation with a synthetic peptide containing the epitope against which the CTL are directed. For RMA-S this susceptibility to lysis could not be explained simply by the binding of peptide to the class I molecules already present on the cell surface. It was found that incubation of RMA-S with peptide increased the number of class I at the cell surface, suggesting that the peptide may be correcting the defective intracellular association of class I MHC heavy chain with β_2-microglobulin. It was found that a peptide only increases expression of the MHC molecule to which it is known to bind. Townsend et al. (1989) suggest that the lesion may lie in proteins that serve to transport peptides across the ER membrane. Uptake of peptide presumably corrects the defect by providing an alternative pathway for transporting the peptides into the ER. An interpretation of the data is that before the heavy chain interacts with β_2-microglobulin, it folds in a manner dependent upon interactions with processed peptide. Heavy chains that fold around appropriately restricted peptides reach a conformation that can interact with β_2-microglobulin to form a stable complex that is transported to the cell surface, while those that fold in the absence of peptide or that bind an inappropriate peptide would not fold correctly nor be transported to the cell surface (Parham, 1989). This concept does not rule out a role for peptide exchange once the assembled class I molecule is on the cell surface (Townsend et al. 1989). The results of Townsend et al. (1989) may explain the earlier findings of Ploegh et al. (1979) who reported that translation of mRNA for class I heavy chain and β_2-microglobulin lead to both polypeptides being made and transferred into the membrane but without the expected association between the two polypeptide chains. Townsend's results may also offer an explanation for the difficulties reported in demonstrating the direct binding of peptides to purified class I molecules. Chen and Parham (1989) estimated that less than 0.3% of a purified HLA preparation could bind added peptide of the correct restriction.

5. Summary

It is now apparent that cytotoxic T cells recognize degraded forms of viral antigens bound to class I or class II MHC molecules in a manner analogous to that first proposed for the presentation of soluble exogenous protein antigens. The idea that recognition by class I and class II restricted T cells is mechanistically similar is consistent with the related structures of class I and class II molecules (Brown et al. 1988) and the evidence that T cells of both types select receptors from a common V gene pool (Kronenberg et al. 1986; Davis and Bjorkman, 1988).

Although some form of intracellular processing of endogenously derived, as well as exogenously derived proteins occurs prior to their effective association with MHC molecules, the processing pathways utilized are mechanistically distinct (Germain, 1988; Long and Jacobson, 1989) (Fig. 1). In general, fragments generated for one class of MHC molecules are unable under most circumstances to associate with the

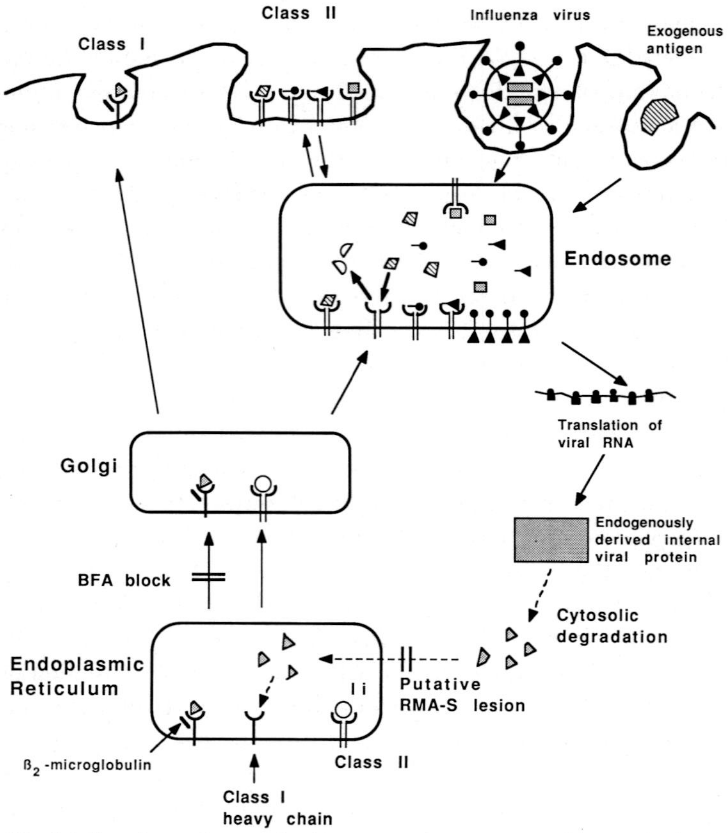

Fig. 3.1. Putative pathways of endogenous and exogenous antigen processing. *Proposed endogenous pathway for influenza virus.* The virus enters the target cell via receptor-mediated endocytosis and fuses with host cell membranes in an endosomal compartment. For internal polypeptides, viral RNA is translated on free ribosomes in the target cell cytoplasm. The putative pathways for the processing and class I association of endogenously derived antigen are shown with broken arrows. Peptides are generated from the endogenous antigen by an undefined cytosolic proteolytic system and the resulting peptides then pass into the endoplasmic reticulum, perhaps via a peptide transporter (Townsend and Bodmer, 1989). This same transporter may also be the site of the lesion in RMA-S cells for which reduced assembly of class I molecules occurs (Townsend et al. 1989). In the endoplasmic reticulum (ER), appropriately restricted peptides first bind to class I heavy chains, and this followed by an association with β_2-microglobulin. It is proposed (Blum and Cresswell, 1988; Long, 1989) that peptides derived from endogenous antigens are unable to associate with newly synthesized class II molecules in the ER because their peptide binding site is occupied by the invariant chain (Ii). Assembled class I molecules are transported via the Golgi to the cell surface for recognition by $CD8^+$ T cells. Transport of class I MHC molecules out of the ER can be blocked by brefeldin A (Nuchtern et al. 1989; Yewdell and Bennink, 1989).
Exogenous processing pathway. The transmembrane and internal polypeptides present in the input virion are degraded in the low-pH evironment of the endosome by acidic proteases. Soluble exogenous antigens entering the cell by fluid phase endocytosis would also be partially degraded in the endosome. In the endosomal compartment, the resulting peptides either combine with recycling class II molecules or with newly synthesized class II molecules after the invariant chain has been proteolytically removed (shown with bold arrows). The resulting complexes of peptide and class II molecules are then transported to the cell surface for recognition by $CD4^+$ T cells.

other class of MHC molecules. Exogenous protein antigens and polypeptides present in a virus (if entry is by endocytosis), are degraded in the acidic environment of the endosome and the resulting processed fragments associate with class II molecules in the *trans*-Golgi or an endosomal compartment. Endogenously derived viral antigens are synthesized and degraded into fragments in the cytoplasm and subsequently transported into the endoplasmic reticulum for association with class I MHC molecules. The cytoplasmic proteolytic mechanism involved may be no different than that involved in normal protein turnover and the rapid degradation of abnormal proteins. The inhibition of class I-restricted CTL recognition of influenza infected target cells by the antibiotic brefeldin A (preventing egress from the ER) suggested that fragments of newly synthesized viral antigen associate with class I molecules in the ER. It is proposed that peptides derived from endogeous antigens do not associate with class II MHC molecules, because they can not displace the invariant chain associated with class II molecules, which may protect the peptide binding site. The invariant chain is thought to be proteolytically removed in a *trans*- or post-Golgi compartment to allow peptides derived from exogenous antigens to bind.

Studies using a mutant cell line possessing decreased expression of class I MHC molecules indicated that the correct folding of class I MHC molecule is critically dependent upon a supply of appropriately processed peptides (Townsend et al. 1989). The demonstration that peptides can prevent their own presentation by inhibiting exocytosis has implications for tolerance, autoimmunity and the evolutionary strategies adopted by pathogens to evade the immune system (Parham, 1989).

6. Acknowledgements

The authors were supported by a grant from Hoechst Aktiengesellschaft (F.R.G.).

7. References

Ajitkumar, P., Geier, S.S., Kesari, K.V., Borriello, F., Nakagwawa, M., Bluestone, J.A., Saper, M.A., Wiley, D.C. and Nathenson, S.G. (1988) *Cell* 54, 47–56.

Allen, P.M. and Unanue, E.R. (1984) *J. Immunol.* 132, 1077–1079.

Allen, P.M. (1987) *Immunol. Today* 8, 270–273.

Allen, P.M., Matsueda, G.R., Evans, R.J., Dunbar, J.B., Marshall, G.R., and Unanue, E.R. (1987) *Nature* 327, 713–715.

Bastin, J., Rothbard, J., Davey, J., Jones, I. and Townsend, A. (1987) *J. Exp. Med.* 165, 1508–1523.

Bennink, J.R., Yewdell, J.W., Smith, G.L., Moller, C. and Moss, B. (1984) *Nature* 311, 578–579.

Berzofsky, J.A., Brett, S.J., Streicher, H.Z. and Takahashi, H. (1988) *Immunol. Rev.* 106, 5–31.

Bjorkman, P.J., Saper, M.A., Samraoui, B., Bennett, S.W., Strominger, J.L and Wiley, D.C. (1987a) *Nature* 329, 506–512.

Bjorkman, P.J., Saper, M.A., Samraoui, B., Bennett, S.W., Strominger, J.L and Wiley, D.C. (1987b) *Nature* 329, 512–518.

Blum, J.S. and Cresswell, P. (1988) *Proc. Natl. Acad. Sci. USA* 85, 3975–3979.

Braciale, T.J. and Yap, K.L. (1978) *J. Exp. Med.* 147, 1236–1252.

Braciale, T.J., Braciale, V.L., Henkel, T.J., Sambrook, J. and Gething, M.J. (1984) *J. Exp. Med.* 159, 341–354.

Braciale, T.J., Braciale, V.L., Winkler, M., Stroynowski, I., Hood, L., Sambrook, J. and Gething, M.-J. (1987a) *J. Exp. Med.* 166, 678–692.

Braciale, T.J., Morrison, L.A., Sweetser, M.T., Sambrook, J., Gething, M.-J. and Braciale, V.L. (1987b) *Immunol. Rev.* 98, 95–114.

Braciale, T.J., Morrison, L.A., Henkel, T.J. and Braciale, V.L. (1988) In *Processing and Presentation of Antigens* (B. Pernis, S.C. Silverstein and H.J. Vogel, Eds.) Academic Press, San Diego CA. pp. 69–79.

Brown, J.H., Jardetzky, T., Saper, M.A., Samraoui, B., Bjorkman, P.J. and Wiley, D.C. (1988) *Nature* 332, 845–850.

Buus, S. and Werdelin, O. (1986) *J. Immunol.* 136, 452–465.

Buus, S. Sette, A., Colon, S.M. and Grey, H.M. (1988) *Science* 242, 1045–1047.

Chain, B.M., Bou-Gharios, G. and Olsen, I. (1989) *Clin. Exp. Immunol.* 75, 87–92.

Chesnut, R.W., Colon, S.M. and Grey, H.M. (1982) *J. Immunol.* 129, 2382–2388.

Cresswell, P. (1985) *Proc. Natl. Acad. Sci. USA* 82, 8188–8192.

Davis, M.M. and Bjorkman, P.J. (1988) *Nature* 334, 395–402.

Diment, S., Leech, M.S. and Stahl, P. (1988) *J. Biol. Chem.* 263, 6901–6907.

Fujiwara, T., Oda, K., Yokota, S., Takatsuki, A. and Ikehara, Y. (1988) *J. Biol. Chem.* 263, 18545–18552.

Gerlier, D. and Rabourdin-Combe, C. (1989) *Immunol. Today* 10, 3–5.

Germain, R.N. (1986) *Nature* 322, 687–689.

Germain, R.N. (1988) *Cell* 54, 441–444.

Gething, M.-J., Koszinowski, U. and Waterfield, M. (1978) *Nature* 274, 689–691.

Gething, M.-J. and Sambrook, J. (1982) *Nature* 300, 598–603.

Grey, H.M. and Chesnut, R. (1985) *Immunol. Today* 6, 101–106.

Guillet, J.-G., Lai, M.-Z., Briner, T.J., Smith, J.A. and Gefter, M.L. (1986) *Nature* 324, 260–262.

Guillet, J.-G., Lai, M.Z., Briner, T.J., Buus, S., Sette, A., Grey, H.M., Smith, J.A. and Gefter, M.L. (1987) *Science* 235, 865–870.

Harding, C.V., Leyva-Cobian, F. and Unanue, E.R. (1988) *Immunol. Rev.* 106, 77–92.

Hershko, A. and Ciechanover, A. (1986) *Prog. Nucleic Acid Res. Mol. Biol.* 33, 19–56.

Kovac, Z. and Schwartz, R.H. (1985) *J. Immunol.* 134, 3233–3240.

Lin, J., Berzofsky, J.A. and Delovitch, T.L. (1988) *J. Mol. Cell Immunol.* 3, 321–343.

Lippincott-Schwartz, J., Bonifacino, J.S., Yuan, L.C. and Klausner, R.D. (1988) *Cell* 54, 209–220.

Lippincott-Schwartz, J., Yuan, L.C., Bonifacino, J.S. and Klausner, R.D. (1989) *Cell* 56, 801–813.

Long, E.O. (1989) *Immunol. Today* 10, 232–234.

Long, E.O. and Jacobson, S. (1989) *Immunol. Today* 10, 45–48.

Kronenberg, M., Sui, G., Hood, L.E. and Shastri, N. (1986) *Annu. Rev. Immunol.* 4, 529–591.

Mayer, R.J. (1987) *Biochem. Soc. Trans.* 15, 820–823.

Mayorga, L.S., Diaz, R. and Stahl, P.D. (1989) *J. Biol. Chem.* 264, 5392–5399.

McCoy, K.L. and Schwartz, R.H. (1988) *Immunol. Rev.* 106, 129–147.

McCoy, K.L., Miller, J., Jenkins, M., Ronchese, F., Germain, R.N. and Schwartz, R.H. (1989) *J. Immunol.* 143, 29–38.

Moore, M.W., Carbone, F.R. and Bevan, M.J. (1988) *Cell* 54, 777–785.

Morrison, L.A., Lukacher, A.E., Braciale, V.L., Fan, D.P. and Braciale, T.J. (1986) *J. Exp. Med.* 163, 903–921.

Nixon, D.F., Townsend, A.R.M., Elvin, J.G., Rizza, C.R., Gallwey, J. and McMichael, A.J. (1988) *Nature* 336, 484–487.

Nuchtern, J.G., Bonifacino, J.S., Biddison, W.E. and Klausner, R.D. (1989) *Nature* 339, 223–226.

Parham, P. (1989) *Nature* 340, 426–428.
Parham, P., Clayberger, C., Zorn, S.L., Ludwig, D.S., Schoolnik, G.K. and Krensky, A.M. (1987) *Nature* 325, 625–628.
Pernis, B. (1988) In *Processing and Presentation of Antigens* (B. Pernis, S.C. Silverstein and H.J. Vogel, Eds.) Academic Press, San Diego CA. pp. 247–259.
Ploegh, H.L., Cannon, L.E. and Strominger, J.L. (1979) *Proc. Natl. Acad. Sci. USA* 76, 2273–2277.
Rott, R. and Scholtissek, C. (1982) In *Basic and Applied Influenza Research,* CRC Press, Boca Raton FL. p. 211.
Schwartz, R.N. (1984) In *Fundamental Immunology* (W.E. Paul, Ed.) Raven Press, New York. pp. 385–392.
Sette, A., Buus, S., Colon, S., Smith, J.A., Miles, C. and Grey, H.M. (1987) *Nature* 328, 395–399.
Shimonkevitz, R., Kappler, J.W., Marrack, P. and Grey, H.M. (1983) *J. Exp. Med.* 158, 303–316.
Shimonkevitz, R., Colon, S., Kappler, J.W., Marrack, P. and Grey, H.M. (1984) *J. Immunol.* 133, 2067–2074.
Stacey, D.W. and Allfrey, V.G. (1977) *J. Cell Biol.* 75, 807–817.
Sugamura, K., Shemezu, R., Zarling, J. and Bach, F.H. (1977) *Nature* 270, 251–253.
Takahashi, H., Cohen, J., Hosmalin, A., Cease, K., Houghten, R., Cornette, J., DeLisi, C., Moss, B., Germain, R. and Berzofsky, J.A. (1988) *Proc. Natl. Acad. Sci. USA* 85, 3105–3109.
Townsend, A.R.M., McMichael, A.J., Carter, N.P., Huddleston, J.A., and Brownlee, G.G. (1984) *Cell* 39, 13–25.
Townsend, A.R.M., Gotch, F.M. and Davey, J. (1985) *Cell* 42, 457–467.
Townsend, A.R.M., Bastin, J., Gould, K. and Brownlee, G.G. (1986a) *Nature* 234, 575–577.
Townsend, A.R.M., Rothbard, J., Gotch, F.M., Bahadur, G., Wraith, D. and McMichael, A.J. (1986b) *Cell* 44, 959–968.
Townsend, A., Bastin, J., Gould, K., Brownlee, G., Andrew, M., Coupar, B., Boyle, D., Chan, S. and Smith, G. (1988) *J. Exp. Med.* 168, 1211–1224.
Townsend, A. and Bodmer, H. (1989) *Annu. Rev. Immunol.* 7, 601–624.
Townsend, A., Ohlen, C., Bastin, J., Ljunggren, H.G., Foster, L. and Karre, K. (1989) *Nature* 340, 443–448.
Unanue, E.R. and Allen, P.M. (1987) *Science* 236, 551–557.
Wabuke-Bunoti, M.A.N., Fan, D. and Braciale, T.J. (1981) *J. Immunol.* 127, 1122–1125.
Wabuke-Bunoti, M.A.N., Taku, A., Fan, D., Kent, S., Webster, R.G. (1984) *J. Immunol.* 133, 2194–2201.
Watts, C. and Davidson, H.W. (1988) *EMBO J.* 7, 1937–1945.
Werdelin, O., Mouritsen, S., Peterson, B.L., Sette, A. and Buus, S. (1988) *Immunol. Rev.* 106, 181–193.
Yewdell, J.W., Bennink, J.R. and Hosaka, Y. (1988) *Science* 239, 637–640.
Yewdell, J.W. and Bennink, J.R. (1989) *Science* 244, 1072–1075.
Ziegler, H.K. and Unanue, E.R. (1981) *J. Immunol.* 127, 1869–1875.
Ziegler, H.K. and Unanue, E.R. (1982) *Proc. Natl. Acad. Sci. USA* 79, 175–178.
Zinkernagel, R.M. and Doherty, P.C. (1979) *Adv. Immunol.* 27, 51–177.

Eds. M.H.V. van Regenmortel and A.R. Neurath
Immunochemistry of viruses, II. The basis for serodiagnosis and vaccines.

Local immune response to viruses 4

W.F. MUELLER, R. GAROFALO and P.L. OGRA

Departments of Pediatrics and Microbiology, State University of New York at Buffalo School of Medicine, and Division of Infectious Diseases, The Children's Hospital, 219 Bryant Street, Buffalo, NY 14222, U.S.A.

1. Introduction

The skin and mucosal sites of the vertebrate organism are continuosly exposed to the surrounding environment and have to face changes in these conditions directly, whether these are potentially harmful or beneficial. During the last three decades, most of the scientific work concerned with the events happening around the large surface area of mucosa-associated lymphoid tissue (MALT) of the lungs and gut was performed to better define the structure and properties of this first line of host defense (Chodirker and Tomasi, 1963; Tomasi and Bienenstock, 1968; Ogra, 1969). As a result, this work led to a somewhat arbitrary definition of a common local immune system that, apart from lungs and gut, included mucosal elements of salivary glands, conjunctivae, urogenital tract, skin and mammae (Bienenstock et al., 1979). Functionally, however, local immune reactions are closely linked to reactions of systemic immune organs like thymus, spleen, bone marrow, histiocytic cells and peripheral lymph nodes (Zinkernagel et al., 1985; Butcher, 1988; Tonegawa, 1988). Moreover, the deleterious effects for a given host, such as undue physicochemical conditions (i.e., pollution), nonreplicable molecules and macromolecules (i.e., food proteins) and replicable microorganisms like bacteria, fungi, parasites and viruses, have been studied more extensively up to now than the effects of interaction between pathogens and host in terms of evolutionary benefits for the host arising from this interaction (Breuel, 1969; Zinkernagel et al., 1985; Freihorst et al., 1988; Bergmann and Waldman, 1988; Mueller, 1988; Ogra, 1989). Only under certain circumstances are pathogens able to attach to mucosal surfaces, overcome these defending mucosal sites, invade and spread (infection) and finally cause disease.

Focusing on the events during this possibly harmful interaction with respect to host and virus properties, this chapter will discuss the functional structures of the MALT and the spectrum of local immune reactions that are involved in fighting viral infectious agents. Eventually, a better understanding of virus-mucosa interaction may help contribute to the development of new and better vaccines, that under physiological routes of local application might provide immunity in terms of protection (Chretien et al., 1984; Flexner et al., 1988; Ogra and Garofalo, 1989).

2. Anatomic elements of the mucosa-associated lymphoid tissue (MALT)

2.1. Definition of a common MALT

Historically, much work had been done to study locally induced immune reactions after Paul Ehrlich had successfully induced local and circulating anti-toxin by feeding mice with ricin (Ehrlich, 1892). All these attempts, however, were difficult to interpret since the measurement of secretory, humoral and cellular immune responses lacked sensitive and specific assays. Morphologic investigations at that time were mainly performed to describe anatomic elements of gut and lung tissue and properties of parenchymal cells.

Interest in local immune reactions arose again when in the early sixties Tomasi in the U.S.A. (Tomasi and Tisdale, 1964) and Hanson in Europe (Hanson, 1961) independently reported that the distribution of immunoglobulins in certain secretions was different from that in serum. This polymeric IgA, connected with a J-chain, receives its secretory component and is released as secretory IgA (SIgA) into the mucosal lumen (McGhee et al., 1989). Physicochemically and functionally, polymeric SIgA behaves differently from monomeric serum IgA and represents the cornerstone in the characterization of local immune phenomena (Mestecky and McGhee, 1987).

The common MALT includes the skin and the mucosal sites of the orogastrointestinal tract, upper and lower respiratory tracts, salivary glands, conjunctivae, genitourinary tract and the mammae (Scicchitano et al., 1988). Luminal secretions, including a variety of nonspecific and specific immune factors, together with mechanical propulsive (gut) or ciliary movement (lung) contribute to the clearing mechanisms necessary to maintain anatomical integrity (Pabst, 1985; Parrott, 1987). Some properties of mucosal defense are listed in Table 4.1. Only recently have factors with anti-viral neutralizing activity other than of an immunoglobulin nature been described in human milk (Hanson et al., 1988).

The importance of nonspecific mucosal defense mechanisms is not well established. They might play a more important role in defense against bacterial and parasitic infections (Speer et al., 1986; Hanson et al., 1988; Ogra and Ogra, 1988). It must be mentioned, however, that most of the proteins listed in Table 4.1, like lyso-

TABLE 4.1

Non-specific and specific factors of mucosal defense

Non-specific	Specific
Mechanical movement	SIgA
Mucus	SIgM
Milieu (pH)	IgG
Fibronectin	IgE
Lactoferrin	Cytotoxic cells
Lysozyme	Epithelial cell membranes
Complement factors	
Proteolytic enzymes	
Bacterial flora (gut)	
Macrophages	
Polymorphonuclear cells	
NK cells (IELs?)	
(Antiviral activity of human milk of non-immunoglobulin nature)	

(Hanson et al., 1988; Pabst, 1985; Brandtzaeg, 1984; Parrott, 1987)

zyme, lactoferrin, as well as anti-proteases like α_1-antitrypsin, and cells like macrophages, are present in breast milk in high amounts and, when delivered to the offspring, may offer protection against a variety of gastrointestinal and respiratory diseases in the young (Hanson et al., 1988; Cairo, 1989). Specific SIgA of mother's milk and actively produced SIgA at other mucosal surfaces, however, are the most important immunoglobulin providing local immunity (Mestecky and Russell, 1986; Hanson et al., 1988). This is highlighted not only by the fact that IgA is produced locally in higher amounts than all the other Ig isotypes, but also by the fact that selective IgA deficiency exposes patients unable to compensate this defect with an efficient SIgM production to recurrent infections of the airways and the gut (Piedra and Ogra, 1986; Savilahti et al., 1986).

Underneath the mucus layer, epithelial cells face the lumen. Lymphocytes are found between epithelial cells (intraepithelial lymphocytes, IEL) and in the lamina propria adjacent to the epithelial cell layer. In the gut, distinct areas of accumulation of lymphoid tissue can be detected macroscopically as whitish lymph follicles opposite mesenteric insertions. These Peyer's patches contain abundant naive B cells, surrounded by a B cell corona and T cells in an interfollicular area. Towards the lumen, microfold cells (M cells) lay between epithelial cells and seem to be involved in the uptake of particulate and soluble antigenic material (Bienenstock, 1984; Parrott, 1987). It is still a matter of discussion as to whether M cells contribute mainly to the uptake of antigen or pathogens like viruses, or whether part of the antigen uptake is mediated via attachment to glycoproteins present on epithelial cells or via paracellular uptake across the zonula occludens (tight junctions) (Madara, 1989).

Depending on the nature, amount and other properties of antigenic material, naive B cells, when provided with antigenic information, leave Peyer's patches via the lymph for remote sites of mucosal surfaces to initiate mucosal SIgA production or to return to the gut-associated lymphoid tissue (homing). The mechanism and the information necessary to return to the mucosal site of origin or to other mucosal tissues are still unclear but seem to be mediated via specific receptor molecules of endothelial cells present on certain post-capillary venules (see below) (Parrott, 1987; Butcher, 1988).

As a result of the antigenic stimulation along or across mucous membranes and of the cell traffic, specific SIgA antibody usually appears at mucosal sites together with little but mostly detectable IgA, IgG and IgM in the circulation. Long-lived B cell memory (or B and T cell memory) is responsible for the secondary events following priming or for long-lived specific immunity that usually arises after acute viral infections like measles, mumps or vaccination using attenuated live vaccine (Lamm, 1976; Ogra et al., 1984; Mestecky and McGhee, 1987; McGhee et al., 1989).

2.2. Soluble elements of the MALT

Shortly after the description of an IgA-system present mainly at mucosal surfaces, it was found that immune responses after local (oral or inhalational) antigenic stimulation differ highly from responses elicited via subcutaneous or intravenous routes. As these latter routes give mainly rise to an IgM, IgG and monomeric IgA response in the circulation, local immunization along the mucosa give rise to $SIgA_1$ ($SIgA_1$ > $SIgA_2$) antibodies, sometimes without detectable antibody-forming cells in the spleen or in peripheral lymph nodes. Nevertheless, SIgA antibody appears at remote mucosal sites, with a distribution of $SIgA_1/SIgA_2$ that may differ between distinct mucosal sites (Table 4.2). In patients with IgA deficiency, it has been demonstrated that SIgM (SIgG?) can replace SIgA partly (Savilahti et al., 1986). IgE antibody does often arise at mucosal sites along with SIgA, although most of the specific IgE appears in the upper (nasal, tonsillar) respiratory tract and is bound to mast cells and basophils with high-affinity IgE receptors (Brandtzaeg, 1984; Blank et al., 1989). Abu-Ghazaleh (1989) has recently shown that SIgA is able to stimulate the degranulation of eosinophils and might play a role in the pathogenesis of bronchial hyperreactivity, especially in asthma attacks precipitated by viral infections.

Following specific antigenic stimulation in primed individuals, mucosal macrophages (low-affinity IgE receptors) and mucosal mast cells (high-affinity IgE receptors) are able to release a large variety of pro-inflammatory substances and factors leading to early and/or late hypersensitivity reactions (Nsouli et al., 1988) (see also Table 4.4). Although there is no direct evidence in humans that viral hypersensiti-

TABLE 4.2

Soluble elements of the MALT

Source	Soluble products	
B cells	SIgA	(IgA_1:IgA_2 %)
	Mammae	60:40
	Salivary glands	50:50
	Bronchus	70:30
	Serum	80:20
	IgM	
	IgE (mainly cell bound)	
	Lymphokines:	
T cells	IL-1, IL-2, IL-4, IL-5, IL-6, IFN-gamma	
Macrophages	Chemotactic factors, IL-1, IL-6	
PMN, eosinophils	Lysosomal enzymes, etc.	
Mucosal mast cell (MMC)	Histamine, leukotrienes, etc.	

(Brandtzaeg, 1984; Pabst, 1985; Mestecky and Russell, 1986; Clark and Karmen, 1987; Mestecky and McGhee, 1987; Nsouli et al., 1988; McGhee et al., 1989; Ogra, 1989; Rankin, 1989)

vity may actually account for substantive disease, there are speculations that in certain immunologic conditions, like the first trimenon of infancy, viral hypersensitivity may be also responsible for some cases of the sudden infant death syndrome (SIDS) (Guntheroth et al., 1980; Halstead, 1983; Huang, 1983; Zink et al., 1987).

With respect to other Ig isotypes and subclasses, these have been mainly investigated in sera. IgE antibodies to viral determinants have been found in sera and secretions after RSV infections, and the amount of specific IgE seems to correlate directly to the severity of disease and to the genetic disposition for atopy (Welliver et al., 1986; Bui et al., 1987; Freihorst et al., 1988). IgE production has been demonstrated during Epstein-Barr virus infections as well (Heiner and Rose, 1970). The anti-viral antibodies in sera of the IgG isotype have been mainly demonstrated in the IgG_1 and IgG_3 subclasses (Smith et al., 1984; Julkunen et al., 1985; Linde, 1985; Linde et al., 1987). Subclass determination of HIV-1 antibody could be helpful in detection of early active infection in infants born to HIV-1-positive mothers, with respect to the rise in IgG_1- and IgG_3-anti-HIV-1 in the infected infant at two to three months of age when total IgG-anti-HIV-1 of passive origin from the mother is still declining (Pyun et al., 1987).

2.3. Cellular elements of the MALT

PMN, eosinophils and macrophages are present at the luminal sides of the mucosa and seem to be mainly involved in antigen processing, antigen clearing and promotion of inflammatory signals as well as in the repair of post-inflammatory tissue injury. These cells can also be found scattered in mucosal structures during inflammation and, more rarely, in a normal mucosal tissue. Some of the mediators released from leukocytes and macrophages serve as afferent signals as well, transmitting information to nerve endings and receptors present in the submucosa. Thus, stimulation of mucus secretions, permeability of capillary vessels, smooth-muscle contraction and triggering of pain-sensitive nerve endings can all be mediated via mediators release by these cells (Taussig et al., 1988; Moore et al., 1989; Nsouli et al., 1988).

More specific cellular events with immune regulatory effect come from T and B cells present at mucosal sites. As mentioned earlier, uncommitted B-precursors after interaction with antigen, $CD4^+$ T cells and/or other activating signals, leave the mucosal site to form B lymphoblasts (Lamm, 1976; Pabst, 1985; Butcher, 1988; Forrest, 1988; Phillips-Quaglita and Lamm, 1988). Stimulated T cells also leave the mucosa and, probably upon proliferation, they have to undergo clonal selection in the thymus before they are committed to become specific cytotoxic $CD8^+$ T cells. $CD8^+$ suppressor T cells eventually home to the spleen where they seem to be involved in the suppression of further proliferation towards the offending pathogen (Ritter et al., 1988; Smith et al., 1989). T cells require expression of MHC-class I and/or class II antigen on cell surfaces together with the antigen in order to elicit

TABLE 4.3

Cellular elements of the MALT

Cell type	Localization	Cluster of designation	Function
Epithelial cells	Epithelium	–	Express class II antigen
M cells	Epithelium, dome	–	Antigen uptake
$CD3^+$ T cells	Lamina propria	CD3 pan T	Mature T cell
$CD3^+$ T cells $CD4^+$ T cells	Lamina propria	CD4 helper/ inducer	B-cell help
$CD3^+$ T cells $CD8^+$ T cells	Interfollicular regions, intraepithelial	CD8 suppressor/ cytotoxic	NK?, ADCC?
B cells	PP, follicular regions	CD19	Antibody source
$CD3^+$ T cells	Intraepithelial	$CD3^+$, $CD4^-$, $CD8^-$ $CD3^+$, $CD4^-$, $CD8^+$ NK?	Antigen processing? NK activity? Homing?
MM cells	Lamina propria intraepithelial, submucosa	–	Release of mediators, interaction with nerve terminals?

cytotoxic effects (Zinkernagel and Doherty, 1979; Tonegawa, 1988; Smith et al., 1989). $CD4^+$ T cell interaction with B cells is also depending on class II antigen expression of antigen-presenting cells for the proliferation of B cells as is highlighted by MHC-class II-deficient infants with chronic diarrhea and failure to thrive who lack the ability to mount a specific antibody response (Durandy et al., 1989).

Intraepithelial lymphocytes (IEL) lie scattered between parenchymal epithelial cells and remain a matter of scientific discussion with respect to their designation and function (Scicchitano et al., 1988; Tonegawa, 1988). Only recently it has been shown that apart from skin dendritic cells, IEL seem to represent an appreciable number of $CD3^+$ T cells bearing the $\gamma\delta$-T cell receptor. It has been speculated that expression of this unusual type of T cell receptor might reflect characteristics needed for homing mechanisms (Tonegawa, 1988). Earlier evidence, however, preferred to refer IEL to the NK-lineage on the basis of morphological and developmental criteria. Moreover, it was shown that some of the IEL population expresses the CD8 suppressor/cytotoxic surface marker (Pabst, 1985) (Table 4.3).

Mast cells in the mucosal tissue have been found to differ in some respects from connective tissue mast cells. Mucosal mast cells (MMC) have different staining characteristics, seem to be bone-marrow derived and are T cell-dependent with respect to T cell derived IL-4 (Nsouli et al., 1988). The presence of high-affinity IgE receptor on MMCs and the ability to release a variety of preformed or newly generated mediators has gained wide interest in this type of cell of allergic immediate (and late phase) type reactions. However, most studies of MCs have used rat peritoneal MCs, so that the relevance of the observations in human respiratory tract pathology and in chronic inflammatory diseases of the gut must still be proven. Moreover, more work seems to be needed which combines the interaction of viruses and MMCs. Clinical studies have shown that in asthmatics, viral infections of the upper respiratory tract are the main cause of asthmatic attacks, rather than specific allergen contacts (Carlsen et al., 1984).

Some of the events occurring in response to mediators of MMC release are listed in Table 4.4. Usually, in type I hypersensitivity no infiltration of mucosal tissue with PMNs with long-standing inflammatory processes and cell damage are demonstrated. Interestingly, however, bronchiolitis obliterans in humans, which is associated with viral respiratory tract infection, is an example of chronic cell destruction and proliferation which might involve the whole cascade of chronic inflammatory processes (Tristram et al., 1988; King, 1989; Mueller, 1989). Mucosal mast cells have a structural relationship with neuropeptide-containing axons. Several neuropeptides and endorphins are potent mast cell secretagogues. Substance P (SP) especially causes histamine release from mucosal mast cells in the intestinal lamina propria of rats (Stead et al., 1987). Neuropeptides like SP modulate the lymphocyte response both in Peyer's patches and in the spleen. In vitro SP enhances DNA synthesis of murine lymphocytes and increases immunoglobulin synthesis of either the IgM or the IgA isotype. Animals that received a constant infusion of SP in vivo showed in-

TABLE 4.4

Mediators released by MMC and biologic effects

Preformed mediators (rapidly released)	Effects
Histamine	Bronchospasm
Chemotactic factors	Accumulation of inflammatory cells
NCF	
ECF-A	
Lysosomal enzymes (acid hydrolases)	Inactivation of leukotrienes C_4, D_4, E_4
Serotonin[a]	Bronchoconstriction – increased G1 motility – stimulation of sensory nerves
Bradykinin	
Newly generated mediators	
Superoxide anions	Cell damage and collagen deposition
Leukotrienes, C_4, D_4 and E_4	Bronchospasm – mucus secretion – edema
Prostaglandins and tromboxane	
PAF	
LTB_4 (small amount)	Chemotaxis of PMN
Preformed granule-associated mediators	
Neutral proteases	
tryptase	Cleaves C3 – C3a – cleaves kininogen
chymase	Convert angiotensin I to angiotensin II
Heparin	Increases the inhibitory capability of antitrombin towards trombin

[a] Not demonstrated in humans.

creased proliferation and immunoglobulin synthesis by both Peyer's patch and spleen lymphocytes (Moore et al., 1989). Parainfluenza 3 virus has also been shown to increase airway smooth muscle contractibility to SP in vitro (Stead et al., 1987).

2.4. Cell traffic of MALT-derived cells

Conflicting results have been obtained in both animals and humans in attempting to elicit immune responses at remote mucosal sites after local antigen administration (Turner-Warwick, 1978; Jakab, 1984; Bice and Shopp, 1988; Taussig et al., 1988). The differences might be due to the nature, route and dose of antigens used (Dhar and Ogra, 1985; Bergmann and Waldmann, 1988; Friedmann et al., 1989). Pre-existing immunity (passive immunity), cross-reactivity of antibody and specificity and sensitivity of the assay systems used may further explain the differences found (Aizpurua and Russel-Jones, 1988). In summary, however, it is now accepted that the induction of secretory and long-lasting immune responses with or without a measurable systemic response is possible in several animal models and in humans. Induction of immune responses of remote mucosal sites is depending on extensive cell traffic of T and B cells and unique properties of these naive, primed precursors or mature cells to migrate and home to MALT structures (Sanders et al., 1988). From an evolutionary point of view, probably the most important part of cell traffic occurs during pregnancy and lactation in the mammary glands (Hanson et al., 1988; Ogra and Ogra, 1988). There are several studies implicating active, de novo synthesis of SIgA antibody in the lactating mammal, specific for the viral antigen that the mother has encountered via gut or lung mucosal surfaces. As colostrum and mother's milk are the only source of passive antibody protection for several species of animals, including rodents, pigs and calves, it is important that antiviral antibody of the IgA and other isotypes is produced in the mammary gland and transferred to the offspring. This can make use of IgA antibodies, 'bathing' its own mucosal surfaces, and can transport IgG antibodies via the gut mucosa to be passively supplied with circulating IgG until weaning (Parkam, 1989; Simister and Mostov, 1989). As protection of an offspring is not only dependent on passively acquired antibody, it is important to state that mother's milk does also show unspecific anti-inflammatory properties, containing specialized milk macrophages and other immunocompetent cells that are, for example, able to elicit NK and ADCC against HSV-infected cells (Hayward et al., 1987). Thus, the local immune system and its mobile cells are designated to supply species' offspring with a complete passive immune system adapted to the needs at birth but not preventing upcoming active immune responses of the growing offspring (Ogra et al., 1977; Ogra, 1979; Hanson et al., 1988; Ogra and Ogra, 1988).

Recently, more information has been uncovered to explain the unique features of cell trafficking and homing mechanisms of T and B cells (Czinn and Lamm, 1986; Butcher, 1988; Berg et al., 1989). Cell migration may be initiated and maintained upon antigenic stimulation, upon humoral and chemotactic systemic or local influences, and is mediated by means of receptor expression on the cell surface of migrating cells so to enable them to adhere to tissue-specific ligands on high endothelial venules (HEV) (Butcher, 1988; Nakache et al., 1989). Having attached to these 'vascular addressins', cells leave the circulation through the endothelial cell layer to seed or home into different compartments. As a result, a nonrandom distribution of lymphocytes occurs. So, surface IgA^+ B cells segregate mainly into the mucosal tissue and spleen, whereas IgM^+ or IgG^+ B cells are represented in peripheral lymph nodes (Berg et al., 1989). Whereas antigenic stimulation seems to be needed to induce cell traffic leading to primary immune responses, mature unprimed (naive) B cells (and $CD4^+$ T cells) prefer to migrate to mucosal sites, without antigenic stimulation. Upon antigen stimulation and after a short transient nonmigratory stage during which dividing blast cells adhere to each other by expression of the leukocyte function antigen-1 (LFA-1), B cells acquire HEV-receptors that enable them to migrate and home, though it is not yet clear whether this is only possible for primed B-memory cells, as circulating B cells are not staining for the major HEV adhesion receptor with mAb CD44 (Butcher, 1988). Dividing migratory blast cells, however, could also find their way through different ligand sites than the specificities known up to now, as B as well as T immunoblasts can migrate specifically to the mucosal tissues. It is of great interest in this context that a nondialyzable chemotactic activity has been described that is present in the milk and is able to attract mature IgA B cells in vitro (Czinn and Lamm, 1986). T cells may, upon antigenic stimulation, behave differently from B cells. Although it has been shown that mature unprimed $CD4^+$ T cells, like B cells, mainly migrate to mucosal sites in the immune system, T cells leave mucosal sites soon after local antigenic stimulation (Butcher, 1988). They are found in the major thoracic duct not yet expressing HEV receptors but still have to undergo clonal selection in the thymus as they eventually express high-density HEV receptors on re-entering the peripheral blood. In a macaque model this post-thymic differentiative step includes expression of the IL-2 receptor, cell division and loss of CD45 receptor, another marker of unprimed mature T cells (Willerford et al., 1989). Primed CD45 T cells then express the UCHL1-marker of the leukocyte common antigen (LCA) family as a memory T cell marker and home to the submucosal lymphocyte aggregates, as has been shown recently by immunofluorescence studies in human gut (Janossy et al., 1989).

Besides cell traffic via specialized ligand-receptor interactions specific for the tissue source and/or type and subtype of cells, chemotactic and humoral influences seem to regulate the nonrandom distribution of immunocompetent cells as well (Czinn and Lamm, 1986; Berg et al., 1989). Although granulocytes and monocytes have also been shown to express homing receptors for HEV but do not bind to HEV normally, other inflammatory stimuli seem to be needed to mediate signals onto these cells. Besides IL-1 and interferon-α, which seem to be involved in regulating the architecture of HEV itself, glucocorticoids are also able to influence lymphocyte traffic in a direction towards the bone marrow with decreased binding of lymphocytes to HEV (Butcher, 1988).

3. Local immune response to viruses

3.1. Determinants of virus-specific responses

Though most studies to prove a common MALT have been done in experimental animals, it is now generally accepted that MALT is also functionally present in man (Forrest, 1988; Scicchitano et al., 1988). Under in vivo conditions, however, local immune reactions may very well be connected to some systemic events, especially with invasive spread of locally acquired viral infection and the type of viral tissue tropism (Uhnoo et al., 1989). The outcome of any viral infection may thus be determined by properties residing in the virus itself and on the other hand by factors residing in the host (or its present condition). The outcome of this interaction may vary (health, recovery, immunity, chronic disease, death). Some of the prerequisites of virus-host interaction are listed in Table 4.5. With respect to the properties elicited by viruses, most of them are discussed in detail in Chapters 1 and 2. Besides species and strain-specific differences to be infected by a given strain of virus, which seem to be mediated by glycoprotein receptors of the virus that find ligands on susceptible host cells, the mechanisms of virus-host cell fusion and virus replication may vary with the pathogen in question (Lotan, 1975; Willoughby et al., 1989). So, the primary part of viral entry on mucosal sites (oral or inhalatory) has to provide cell ligands for viral receptors, thus facilitating and determining the localization of primary viral replication (Notkins, 1985; Holt et al., 1988). For example, in intestinal infection with MRV it has been shown recently that virus binds to asialo GM1-carbohydrate via its VP7 to epithelial cells of the intestine (Willoughby et al., 1989). Infected epithelial cells that replicate the virus are irreversibly damaged. It is not known, however, whether these infected cells or the uptake of virus through M cells regulates the induced local immune response (Madara, 1989). The early expression of MHC-class II (Ia) antigens in virus-infected cells would favor the view that besides cytotoxic effector cell responses, epithelial infected cells are a major source of antigen recognition and are important for the induction of immune responses (Askonas and Taylor, 1987). In vivo, this viral cytopathogenicity and natural or anti-

TABLE 4.5

Properties of pathogenesis and defense in viruses and mammalian hosts

Condition	Virus-related	Host-related
Genetic	Pathogenicity for a host species Cytopathogenicity (virulence) Virulence factors (+toxins) Antigenic shift (mutation) Tissue tropism	Genetic disposition Genetic resistance Immune competence
Developmental and recent	Ability to attack Infectious dose Adequate route of exposure Ability to persist Concomitant pathogen	Developmental (mal)formation Passively acquired immunity (+breast feeding) Natural immunity Cross-protective immunity Prior exposure (immunity)
Environmental	Climate, season, contact, vectors of transmission	Health (somatic, psychologic, social) Nutrition Adequate child care

body-dependent cytotoxicity may act together. RSV and influenza viruses are known for early induction of cell alteration as measured by inhibition of ciliary movement of the respiratory epithelium, for example. Recently, in RSV infected cotton rats the electrophysiologic properties of rat trachea were found altered, at a time (72 h post-infection) when light microscopy of the mucosa was still unchanged. Namely, a significant decrease in short-circuit current and a marked increase in tissue resistance were noted (Cloutier et al., 1989). These early functional cell alterations are supposedly due to inhibition of chloride secretion (short circuit current) and restriction of NaCl movement through the paracellular pathway (resistance). In sequence, changes in active and passive ion fluxes should influence secondary water transport across the airway epithelium and increase cell injury.

The unique ability of viruses to replicate in the host in huge numbers makes it necessary for the host to defend itself by a number of molecules and cells that may equal the rapid increase of replicated viruses (Zinkernagel and Doherty, 1979). This can only be accomplished by an immune system able to proliferate on antigenic stimulation by means of effector T cells, and B cells proliferating to antibody secreting plasma cells. Evolving antibody specificities should be neutralizing the infectivity of the offending virus. The neutralization of influenza virus, for example, is mediated by antibodies directed against the heamagglutinin (Ada and Jones, 1987; Ada and Jones, 1988). Among the variety of immunogenic viral proteins, virus- and viral protein-specific cytotoxic T cells do also arise, as has been shown for several viral agents up to now (Askonas and Taylor, 1987). As these viral proteins are now being studied in detail by use of recombinant molecular techniques, these studies may eventually help to develop viral vaccines that include responses against neutralization and cytolytic epitopes of the offending pathogen (Ogra and Garofalo, 1989).

3.2. Determinants of host-specific responses

A number of host conditions have been studied in experimental animal models in recent years (Duffy et al., 1986; Dharakul et al., 1988). These include the viruses important for acute infections of intestinal and respiratory tracts, namely RSV, RV and influenza virus (see Table 6). RSV and RV produce clinical disease mainly in young infants (up to 2 years of age) (MacDonald et al., 1982; Fosarelli et al., 1985; Welliver et al., 1986; Pickering et al., 1988; Ruuskanen et al., 1988; Uhnoo et al., 1988; Riepenhoff-Talty et al., 1989). Similarly, murine rotavirus (MRV) produces diarrhea in an age-dependent way, in the suckling mouse up to the age of 3 weeks. The course of the disease closely resembles human disease with complete recovery after several weeks (Uhnoo et al., 1988). Defined immune deficiency states in mice (nude, T cell-deficient mice; and a CB-17 strain with nonfunctioning T and B cells, SCID mice) furthermore allowed studies to determine the contribution of different compartments of the immune system on the course of MRV infection. Only suckling SCID mice developed severe disease with persistent mucosal infection and virus

TABLE 4.6

Experimental animal models of interactions of MALT and viral infections

Animal model	Virus	Results	Reference
Suckling mouse	MRV	Infection and disease is age-dependent (up to 3 weeks of age)	Uhnoo et al., 1988a
Suckling nude mouse	MRV	As in normal suckling mice	Uhnoo et al., 1989
SCID mouse	MRV	Chronic shedding of virus, moderate disease	Uhnoo et al., 1989
Malnourished mouse	MRV	Higher incidence of infection (36% vs. 18% in normal)	Riepenhoff-Talty et al., 1989
SCID mouse	MRV vaccine strains	Hepatitis with different candidate vaccine strains, altered tissue tropism?	Uhnoo et al., 1989
Rabbit	RSV + food antigen	IgG and SIgA formation and transfer to different sites of MALT (only IgG against food protein)	Peri et al., 1982
Mouse	RSV + food antigen	Specific IgE to food protein appears with RSV-infected animals	Freihorst et al., 1988
Rat	RSV	NK + ADCC against RSV *only* after intranasal exposure, but *not* after subcutaneous exposure	Kumagai, et al., 1985
Rat	RSV	Early tissue injury (72 h p.i.) as measured by electrophysiologic changes of decreased short circuit current and increased resistance of rat trachea	Cloutier et al., 1989
Nude mouse	Influenza	Fatality depends on non-specific response and/or IgM- antibody, no complete recovery	Yap and Ada, 1979
Mx gene$^+$ mouse	Influenza	Enhanced production of interferons depending on genetic background (increased resistance)	Haller, 1981

shedding, whereas in athymic nude mice the course was very similar to that in normal mice, underlining the point that T cell-mediated effects may contribute little to the course of the disease. Due to rarity of primary immunodeficiency syndromes (PIDS) in humans, limited information exists about the outcome of viral mucosal infections. The evidence gathered up to now from small series of patients, however, is very similar to the mouse model. So, children with thymic dysplasia (DiGeorge Sequence) are able to live into adulthood with very little morbidity from infections, whereas SCID infants usually acquire several pathogens (including viruses) before they are diagnosed and undergo bone marrow transplantation (Jarvis et al., 1983; Milner et al., 1985; Dressler et al., 1989; Mueller et al., 1989). Interestingly, SCID patients after BMT with reconstitution of immune competence develop more severe clinical disease (bronchitis, pneumonia, diarrhea) than that due to the mere presence of the pathogen itself (infection) (Dressler et al., 1989). Again, this points to the fact that host-related defense mechanisms are not only needed to overcome and clear the pathogen, but may be related to the tissue damage as well (Piedra et al., 1989). Influenza virus pneumonia has also been studied in athymic nude mice (Eiden et al., 1986). Contrary to the observations in MRV infection, nude mice did not recover from viral pneumonia and exhibited pneumonia in a dose-dependent manner with protection that was attributed to nonspecific or IgM antibody dependent mechanisms (Eiden et al., 1986; Uhnoo et al., 1989). Among the nonspecific protective effects, enhanced production of interferons should be mentioned, as shown for the genetically determined increased resistance (high interferon α and β in Mx gene-bearing mice) in experimental influenza virus infection (Haller, 1981). In B cell disorders like X-linked hypogammaglobulinemia, it has long been known that viral infections with enteroviruses may lead to chronic disease and even viral encephalomyelitis (Ogra, 1969; Walzer et al., 1976; Wright et al., 1977; Dressler et al., 1989; Sotomayer et al., 1989). Shedding of oral live poliovaccine (OPV) in these patients is prolonged and viral respiratory tract infections remain the main causes of morbidity in these boys, even under proper Ig-substitution therapy (Fishaut et al., 1980; Dressler et al., 1989; Sotomayer et al., 1989). So, both in animal models and in humans, specific mucosal antibody formation and the development of mucosal immunity are necessary to successfully overcome mucosal viral infections.

Besides genetic and developmental factors like age, several environmental conditions of the host may influence the outcome of viral infections. It has long been known that malnutrition, especially in underdeveloped countries, contributes to morbidity and mortality seen with diarrheal diseases or respiratory tract infections of bacterial and viral origin (Winick and Noble, 1966; South et al., 1975; Glaser et al., 1984; Wang et al., 1984; Ogra et al., 1988; Uhnoo et al., 1988b, Riepenhoff-Talty et al., 1989). In particular, measles in humans with malnutrition remains a considerable mortality factor (Ellison, 1931; Friedman et al., 1989; Mueller, 1989). In the murine rotavirus model, it has been shown recently that malnourished mice exhibit disease patterns different from normal and SCID mice. Malnourished mice need a lower minimal infectious dose, have a shortened incubation period until diarrhea develops and increased severity of disease, compared to normal mice. Infectious complications, like RV-associated hepatitis, were found in 36% of malnourished mice as compared to 18% in normal mice and 80% in SCID mice when candidate heterologous vaccine strains were used (Riepenhoff-Talty et al., 1989). Furthermore, the concomitant uptake of an unrelated nutritional antigen like ovalbumin was increased in malnutrition, as was the number of epithelial cells replicating the infectious agent (RV). So, proper nutritional conditions may be crucial for mucosal clinical immunity.

Different from the effects of specific SIgA formation at the mucosal site of the gut in response to gastrointestinal viral infection, the presence or formation of antibody against respiratory viral pathogens like RSV may provide little protection against the infection (Piedra et al., 1989). Both in rabbits and in lactating human mothers it has been shown, however, that specific SIgA against RSV is induced or boosted in colostrum and milk (Peri et al., 1982; Theodore et al., 1982). Successful viral elimination may be dependent upon cellular events like cytotoxicity of NK or ADCC-dependent mechanisms, as has been shown in cotton rats (Kumagai et al., 1985). In this model, no cytotoxicity was observed after subcutaneous RSV inoculation but was easily detected after intranasal inoculation of live RSV. Prolonged viral shedding has also been reported in T cell deficient humans for RSV, parainfluenza virus and other respiratory viruses (Hall et al., 1978; Hall et al., 1986). As T cell associated regulatory disturbances have been implicated in the development of atopic diseases, it is interesting to note that RSV infection regularly gives rise to virus-specific IgE in human infants (Welliver et al., 1986; Bui et al., 1987). The IgE response usually is transient and declines within a few days. However, patients developing more severe lower respiratory tract diseases, like bronchiolitis and long-lasting bronchial obstruction, do continue to produce specific IgE and IgG_4 (Bui et al., 1987). As IgE had also been found in secretions of these patients, increased histamine content in nasopharyngeal secretions has been reported (Welliver et al., 1986).

Since these studies suggested that viral mucosal infection could trigger the release of preformed and newly generated mediators of mast cells, it was important to show that in relation to the appearance of RSV-IgE, measurable leukotriene C_4 could be detected in such patients (Garofalo et al., 1989). The further release of bronchoconstrictive mediators from immigrating PMNs and eosinophils and the activation of the oxidative metabolic pathway may then contribute to tissue injury and damage.

3.3. Function of virus-specific local immunity

Among the most important issues for the development and function of mucosal immunity to any kind of pathogen is the question of passively acquired protection, namely in the perinatal period (Losonsky et al., 1982; Okamoto et al., 1989). Systemic viral infection of a mother during pregnancy is one of the well-known conditions for miscarriages, embryonic malformation and congenital infections (South et al., 1975). The importance of invasive viral disease during pregnancy has been also highlighted recently with the upcoming HIV-1 infection epidemic. Besides HIV-1, varicella-zoster virus, herpes simplex virus and rubella virus are all able to induce immuno-regulatory disturbances of the T and B cell system in the offspring (Pyun et al., 1987; Namikawa et al., 1988). Despite specific antibody formation in congenitally infected infants, viral shedding is prolonged for the excretion of rubella virus or cytomegalovirus (CMV) for many years. Very early acquisition of some viruses is, however, associated with decreased cellular effects against the offending viruses, as has been shown for HSV and CMV infections (Hayward et al., 1987). This may be related to decreased expression of MHC-class II antigen on virus-infected cells, when infection has to be encountered in utero (Rognum et al., 1988).

After birth, feeding the offspring with its own mother's milk provides protection against a variety of infectious diseases (Hanson et al., 1988). This seems to be mediated through transfer of a complete secretory passive immune system, containing unspecific and specific soluble and cellular immune elements. For example, the composition of milk carbohydrate (lactose mainly) can interfere with the attachment of pathogens, as has been shown for bacterial attachment, but may be important for glycoprotein attachment of viral origin and lectins as well (Sharon and Lis, 1972; Hanson et al., 1988; Willoughby et al., 1989).

Passively acquired SIgA via mother's milk and passively acquired IgG via placenta are efficient means of protection to early infection in infancy as outlined by many epidemiological studies towards the effects of breast feeding (Duffy et al., 1986; Hanson et al., 1988; Krause et al., 1989). Other factors might also contribute to this observed passive protection, though most of the components of mother's milk still await careful scientific studies with respect to functional antiviral properties. Milk macrophages, when tested for cytotoxicity against HSV, exhibited less cytotoxicity than blood monocytes, for example (Speer et al., 1986). The conditions that govern cytotoxicity of milk cells in an infant's gut may, however, differ completely from other in vitro systems or the blood. This has been elegantly demonstrated by functional in vitro measurement of oxidative metabolic pathway products of milk macrophages at different pH levels (Speer et al., 1986). Milk macrophages exhibited production of oxygen radicals even after incubation at pH for 30 min, when activity of blood monocytes of the same individuals was abolished.

Transplacentally acquired IgG has been studied eventually with respect to active antibody formation in the offspring. In the rat model, there are several studies suggesting that passive IgG prevents IgE formation in rat pups (Jarrett et al., 1977; Hall et al., 1988). In human infants, passive immunity against poliovirus does not interfere with OPV vaccination in infants (Ogra, 1969; Ogra and Karzon, 1971). Otherwise, little information exists regarding the protective effects of passive IgG towards viral infections. In hepatitis B virus infection of the mother, the risk of the infant being infected is very high, thus passive antibody in timely conjunction with the virus obviously does not provide protection, but early active/passive immunization after birth is able to provide protection. Dengue hemorrhagic fever is another example of a viral disease that may not be protected by passive antibody and has a high fatality rate (Halstead, 1983). Many similarities of the course of disease and the peak incidence at two months of age have led to the hypothesis that the suddent infant death syndrome (SIDS) may be related to some course of viral infection, or infection related viral hypersensitivity. Up to now, however, no clear-cut studies, either epidemiological or serological/pathological, have conclusively shown anykind of serum sickness-like connection between SIDS and viral mucosal infection. This important point, however, urgently needs further investigation (Huang, 1983; Zink et al., 1987).

Evaluation of functional effects between virus-host interaction is most important clinically with respect to possible protective effects. These events, therefore, have to be studied carefully, also with respect to advanced vaccine development (Ogra and Garofalo, 1989; Welliver et al., 1989). The contribution of mucosal immune reactions to subsequent antiviral immunity has hitherto been generally underestimated. Thus, live OPV vaccine is still the only oral viral vaccine used in most countries of the world. More than 20 years ago, however, it was shown that viral replication gives rise to specific immunity in the gut and nasopharynx and provides protection against subsequent mucosal reinfection (Ogra et al., 1968). On the other hand, nonreplicating Salk poliovaccine results in little or no mucosal immune response. When, however, a high-potency inactivated poliovaccine (IPV) was used parenterally for priming, followed by oral OPV, significant amounts of SIgA were found in nasopharyngeal secretions (Dhar and Ogra, 1985). Since parenteral/mucosal scheme of vaccination induce a local immune response with no signs of adverse reactions, it is interesting to note that several studies carried out to achieve protection of the airways to viral disease, actually did result in increased severity of disease after natural virus exposure or on challenge (Piedra et al., 1989). Using different routes of primary RSV inoculation (intratracheal, intragastric and subcutaneous) in mice, it has recently been shown that upon intranasal challenge 21–28 days after primary immunization with live RSV, considerable lung pathology occurred in subcutaneously primed mice (Piedra et al., 1989). This also held true for sham-primed mice, which had been primed with cell culture medium (used for RSV cell culture) and challenged with RSV. Thus, parenteral priming and local challenge may be associated with increased pathology, at least as far as the lung is concerned (King, 1989). Multiple intragastric doses of killed Sendai virus together with cholera toxin protected mice from Sendai virus infection, at least when oral killed virus was combined with nasal killed virus application (Nedrud et al., 1987). In mice and in humans, it has been repeatedly shown that repeated oral application of enteric-coated influenza hemagglutinin is able to induce secretory antibodies in tears, saliva and nasal secretions, without detectable serum antibody response (Bergmann et al., 1986; Waldman et al., 1986; Waldman et al., 1987). No side effects were noted. Longer epidemiological studies have to be carried out, however, for an increasing number of candidate locally administered vaccines, to ensure protection and safety.

4. Concluding remarks

The description and definition of a common MALT has renewed scientific interest in eliciting immune reactions via local application of both nonreplicable and replicable antigens. Although up to now live-attenuated oral poliovaccine (OPV) is still the only oral viral vaccine against a disabling disease in most countries of the world, knowledge of the distribution and function of mucosal immune reactions will eventually be helpful in finding more and better vaccines against other major pathogens.

Respiratory tract infections caused by viruses like RSV and influenza virus, viral gastroenteritis and hepatitis, as well as HIV infections, remain challenging disease entities to be elaborated towards local disease and immune mechanisms. Advances in mucosal immunology like new insights in cell homing mechanisms may be helpful to better address antigens and induced immune reactions at the mucosal site to be protected.

Animal models and primary immune deficiency states in humans are still incompletely used to study the importance of different arms of defense against infectious diseases.

The influence of passive immunity in terms of protection as naturally supplied by breast feeding is another important part of the overall protective local immune system, functioning in conjunction with systemic immune mechanisms. Allergy and other states of hypersensitivity are now seen as entities involving the common MALT and should be studied with a view to solving the clinical problem of allergy triggered and modulated by viral infections.

5. References

Abu-Ghazaleh, R.I., Fujisawa, T., Mestecky, J., Kyle, R.A. and Gleich, G.J. (1989) *J. Immunol.* 142, 2393–2400.

Ada, G.L. and Jones, P.D. (1987) In *Recent Advances in Mucosal Immunology* Vol. 216B (J.R. McGhee, J. Mestecky, J. Bienenstock and P.L. Ogra, Eds.) Plenum Press, New York. pp. 1033–1046.

Ada, G.L. and Jones, P.D. (1988) *Curr. Top. Microbiol. Immunol.* 128, 1–54.

Aizpurua, H.J. and Russell-Jones, G.J. (1988) *J. Exp. Med.* 167, 440–451.

Askonas, B.A. and Taylor, P.M. (1987) *Immunol. Lett.* 16, 337–346.

Berg, E.L., Goldstein, L.A., Juhila, M.A., Nakache, M., Picker, L.J., Streeter, P.R. Wu, N.W. Zhou, D. and Butcher, E.C. (1989) *Immunol. Rev.* 108, 6–18.

Bergmann, K.C., Waldman, R.H., Tischner, H. and Pohl, W.D. (1986) *Int. Arch. Allergy Appl. Immun.* 80, 107–109.

Bergmann, K.C. and Waldman, R.H. (1988) *Rev. Infect. Dis.* 10, 939–950.

Bice, E.D. and Shopp, G.M. (1988) *Exp. Lung Res.* 14, 133–155.

Bienenstock, J., McDermott, M. and Befus, D. (1979) In *Immunology of Breast Milk* (P.L. Ogra and D.H. Dayton, Eds.) Raven Press, New York. pp. 91–104.

Blank, K., Ra, C., Miller, L., White, K., Metzger, H. and Kinet, J.P. (1989) *Nature* 337, 187–189.

Brandtzaeg, P. (1984) In *Immunology of the Lung and Upper Respiratory Tract* (J. Bienenstock, Ed.) McGraw-Hill, New York. pp. 28–95.

Breuel, H.P. (1969) *Local Immunization.* Gustav Fischer Verlag, Stuttgart. p. 52.

Bui, R.H.D., Molinaro, G.A., Kettering, J.D., Heiner, D.C., Imagawa, D.T. and St. Geme, J.W., Jr. (1987) *J. Pediatr.* 110, 87–90.

Butcher, E.C. (1988) *Curr. Top. Microbiol. Immunol.* 128, 85–122.

Cairo, M.S. (1989) *Am. J. Dis. Child.* 143, 40–46.

Carlsen, K.H., Orstavik, J., Leegaard, J. and Hoeg, H. (1984) *Arch. Dis. Child.* 59, 310–315.

Chretien, J., Holland, W., Macklem, P., Murray, J. and Woolcock, A. (1984) *New Engl. J. Med.* 310, 982–984.

Chodirker, W.B. and Tomasi, T.B. (1963) *Science* 142, 1080.

Clark, S.C. and Karmen, R. (1987) *Science* 236, 1229–1237.

Cloutier, M.M., Wong, D. and Ogra, P.L. (1989) *Pediatr. Pulmonol.* 6, 164–168.

Czinn, S.J. and Lamm, M.E. (1986) *J. Immunol.* 136, 3607–3611.

Dhar, R. and Ogra, P.L. (1985) *Br. Med. Bull.* 41, 28–33.

Dharakul, T., Riepenhoff-Talty, M., Albini, B. and Ogra, P.L. (1988) *Clin. Exp. Immunol.* 74, 14–19.

Dressler, F., Mueller, W., Franz, A., Freihorst, J., Schmidt, R.E., Welte, K. and von der Hardt, H. (1990) *Prax. Klin. Pneumol.,* in press.

Duffy, L.C., Riepenhoff-Talty, M., Byers, T.E., La Scolea, L.J., Jr., Zielezny, M.A., Dryja, D.M. and Ogra, P.L. (1986) *Am. J. Dis. Child.* 140, 1164–1168.

Durandy, A., Mangeney, M., Griscelli, C., Forveille, M., LeDeist, F. and Fischer, A. (1989) *J. Clin. Microbiol.* 9, 125–131.

Ehrlich, P. (1892) *Zbl. J. Hyg.* 12, 183–203.

Eiden, J., Lederman, H.M., Vonderfecht, S. and Yolken, R. (1986) *J. Virol.* 57, 706–708.

Ellison, J.B. (1931) *Arch. Dis. Child.* 6, 37–52.

Ennis, F.A. (1983) In *Regulation of the Immune Response* (P.L. Ogra and D.M. Jacobs, Eds.) Karger, Basel. pp. 259–266.

Fishaut, M., Tubergen, D. and McIntosh, K. (1980) *J. Pediatr.* 96, 179–186.

Flexner, C., Murphy, B.R., Rooney, J.F., Wohlenberg, C., Yuferov, V., Notkins, A.L. and Moss, B. (1988) *Nature* 335, 259–262.

Forrest, B.C. (1988) *Lancet* i, 81–83.

Fosarelli, P.D., DeAngelis, C., Winkelstein, J. and Mellits, E.D. (1985) *Pediatr. Infect. Dis.* 4, 153–159.

Freihorst, J., Piedra, P.A., Okamoto, Y. and Ogra, P.L. (1988) *Proc. Soc. Exp. Biol. Med.* 188, 191–197.

Friedman, M.G., Phillip, M. and Dagan, R. (1989) *Clin. Exp. Immunol.* 75, 58–63.

Garofalo, R., Welliver, R.C., and Ogra, P.L. (1989) *Pediatr. Res.* 25, 163A.

Glaser, J.H., Balistreri, W.F. and Morecki, R. (1984) *J. Pediatr.* 105:912–915.

Guntheroth, W.G., Kawaboni, J., Breazeale, D.G., Garlinghouse, L.E. and Van Hoosier, G.L. (1980) *Am. J. Dis. Child.* 134, 364–366.

Hall, C.B., Douglas, R.G., Jr. and Simons, R.L. (1978) *J. Pediatr.* 93, 28–32.

Hall, C.B., Powell, K.R., MacDonald, N.E., Gala, C.L., Menegus, M.E., Suffin, S.C. and Cohen, H.J. (1986) *New Engl. J. Med.* 315, 77–81.

Haller, O. (1981) *Curr. Top. Microbiol. Immunol.* 92, 25–52.

Halstead, S.B. (1983) In *Regulation of the Immune Response* (P.L. Ogra and D.M. Jacobs, Eds.) Karger, Basel. pp. 249–258.

Hanson, L.A. (1961) *Int. Arch. Allergy Appl. Immunol.* 18, 241–253.

Hanson, L.A., Carlsson, B., Jalit, F., Halin-Zovic, M., Hernodson, S., Kaulberg, J., Mellander, L., Khan, S.R., Lindblad, B., Thiringer, K. and Zaman, S. (1988) In *Biology of Human Milk* (L.A. Hanson, Ed.) Raven Press, New York. pp. 141–158.

Hayward, A., Leibson, P. and Arvin, A. (1987) In *Immunology of the Neonate* (G.R. Burgio, L.A. Hanson and A.G. Ugazio, Eds.) Springer-Verlag, New York. pp. 112–119.

Heiner, D.C. and Rose, B. (1970) *J. Allergy* 45, 30–42.

Holmes, M.J. and Ramsay, A.J. (1988) *J. Med. Virol.* 25, 271–280.

Holt, P.G., Vires, J. and Bilyk, N. (1988) *Int. Arch. Allergy Appl. Immunol.* 86, 121–123.

Huang, S.W. (1983) In *Sudden Infant Death Syndrome* (Tyson, T.J., Ed.) Academic Press, New York. pp. 593–606.

Jakab, G.C. (1984) In *Immunology of the Lung and Upper Respiratory Tract* (Bienenstock, J., Ed.) McGraw-Hill, New York. pp. 298–310.

Janossy, G., Bofill, M., Rowve, D., Muir, J. and Beverley, P.C.L. (1989) *Immunology* 66,517–525.

Jarrett, E. (1977) *Lancet* ii, 223–225.

Jarvis, W.R., Middleton, P.J. and Gelfand, E.W. (1983) *Pediatr. Infect. Dis.* 2, 187–192.

Julkunen, I., Hovi, T., Seppala, J. and Makela, O. (1985) *Clin. Exp. Immunol.* 60, 130–138.

King, T.E., Jr. (1989) *Lung* 167, 69–93.

Krause, P.J. Henson, V.C., Eisenfeld, L. and Johnson, G.M. (1989) *Pediatr. Infect. Dis. J.* 8, 382–389.

Kumagai, T., Wong, D.T. and Ogra, P.L. (1985) *Clin. Exp. Immunol.* 61, 351–359.

Lamm, M.E. (1976) *Adv. Immunol.* 22, 223–290.

Leibovitz, E., Freihorst, J., Piedra, P.A. and Ogra, P.L. (1988) *Int. Arch. Allergy Appl. Immunol.* 86, 112–116.

Linde, A. (1985) *J. Clin. Microbiol.* 21, 117–121.

Linde, G.A., Granstrom, M., and Oervell, C. (1987) *J. Clin. Microbiol.* 25, 653–1658.

Losonsky, G.G., Strussenberg, J., Fishaut, J.M. and Ogra, P.L. (1982) *J. Infect. Dis.* 145, 654–660.

Lotan, R., Skutelski, E., Danon, D. and Sharon, N. (1975) *J. Biol. Chem.* 250, 8518–8523.

Lotan R. (1986) In *The Lectins: Properties, Functions and Application in Biology and Medicines* (I.E. Liener, N. Sharon and I.J. Goldstein, Eds.) pp. 266–285. Academic Press, London and Florida.

McDonald, N.E., Hall, C.E., Suffin, S.C., Alexson, C., Harris, P.J. and Manning, J.A. (1982) *New Engl. J. Med.* 307, 397–400.

Madara, J.L. (1989) *J. Clin. Invest.* 83, 1089–1094.

McGhee, J.R., Mestecky, J., Elson, C.O. and Kiyono, H. (1989) *J. Clin. Immunol.* 9, 175–199.

Mestecky, J. and Russell, M.W. (1986) *Monogr. Allergy* 19, 277–301.

Mestecky, J. and McGhee, J. (1987) *Adv. Immunol.* 40, 153–245.

Milner, M.E., de la Monte, S.M. and Hutchins, G.M. (1985) *Am. J. Dis. Child.* 139, 1111–1114.

Moore, T.C., Lami, J.C. and Spruck, C.H. (1989) *Immunology* 67, 109–114.

Mueller, W. (1987) In *Paediatrische Allergologie und Immunologie* (U. Wahn, R. Seger and V. Wahn, Eds.) Gustav Fischer, Stuttgart. pp. 127–136.

Mueller, W. (1988) *Milchantiküorper im Kindesalter.* Thieme, Stuttgart.

Mueller, W. (1989) *Atemwegs Lungenkrankh.*, 15, 285–287.

Mueller, W., Peter, H.H., Franz, A., Kallfelz, H.C. and Rieger, C.H.L. (1989) *Eur. J. Pediatr.*, 149, 96–103.

Murray, J.C. (1988) *Thorax* 43, 9–11.

Nakache, M., Berg, E.L., Streeter, P.R. and Butcher, E.C. (1989) *Nature* 337, 179–181.

Namikawa, R., Kaneshima, H., Lieberman, M., Weissman, I.L. and McCune, J.M. (1988) *Science* 242, 1684–1686.

Nedrud, J.G., Liang, X., Hague, N. and Lamm, M.E. (1987) *J. Infect. Dis.* 150, 3484–3492.

Notkins, A.L. (1985) *New Engl. J. Med.* 312, 507–509.

Nsouli, T.M., Nsouli, S.M. and Bellanti, J.A. (1988) *Allergy* 60, 379–389.

Ogra, P.L., Karzon, D.T., Righthand, F. and MacGillivray, M. (1968) *New Engl. J. Med.* 279, 893.

Ogra, P.L. (1969) In *The Secretory Immunologic System* (D.H. Dayton, Jr., P.A. Small, Jr., R.M. Chanock, H.E. Kaufman and T.B. Tomasi, Eds.) U.S. Department of Health, Education and Welfare, Washington, DC. pp. 259–279.

Ogra, P.L. and Karzon, D.T. (1971) *Prog. Med. Virol.* 13,156–193.

Ogra, P.L., Cumella, J.C. and Welliver, R.C. (1984) In *Immunology of the Lung and Upper Respiratory Tract* (J. Bienenstock, Ed.) McGraw-Hill, New York. pp. 242–263.

Ogra, P.L. and Ogra, S.S. (1988) In *Biology of Human Milk* (L.A. Hanson, Ed.) Raven Press, New York. pp. 185–196.

Ogra, P.L., Volovitz, B., Riepenhoff-Talty, M. and Freihorst, J. (1988) *Monogr. Allergy* 24, 266–273.

Ogra, P.L. (1989) *Pediatr. Infect. Dis. J.* 8, 341–352.

Ogra, P.L. and Garofalo, R. (1989) *Prog. Med. Virol.*, in press.

Ogra, S.S., Weintraub, D.J. and Ogra, P.L. (1977) *Adv. Exp. Med. Biol.* 107, 95–112.

Okamoto, Y., Tsutsumi, H., Kumar, N.S. and Ogra, P.L. (1989) *J. Immunol.* 142, 2507–2512.

Pabst, R. (1985) In *Nahrungsmittelallergie* (W. Schmutzler and H. Reimann, Eds.) Dustri, Munich. pp. 5–13.

Parkam, P. (1989) *Nature* 337, 118–120.

Parrott, D.M.V. (1987) In *Food Allergy and Intolerance* (J. Brostoff and S. Challacombe, eds.) Bailliere Tindall, London. pp. 3–26.

Peri, B.A., Theodore, C.M., Losonsky, G.A., Fishaut, J.M., Rothberg, R.M. and Ogra, P.L. (1982) *Clin. Exp. Immunol.* 48, 91–101.

Phillips-Quaglita, J.M. and Lamm, M.E. (1988) In *Migration and Homing of Lymphoid Cells* (A.J. Husband, Ed.) Vol. II. CRC Press, Boca Raton FL. pp. 53–76.

Pickering, L.K., Bartlett, A.V., Reves, R.R. and Morrow, A. (1988) *J. Pediatr.* 112, 361–365.

Piedra, P.A., Faden, H.S., Camussi, G., Wong, D.T. and Ogra, P.L. (1989) *Vaccine* 7, 34–38.

Piedra, P. and Ogra, P.L. (1986) *J. Pediatr.* 108, 817–823.

Pyun, K.H., Ochs, H.D., Dufford, M.T.W. and Wedgewood, R.J. (1987) *New Engl. J. Med.* 317, 611–614.

Rankin, J.A. (1989) *J. Allergy Clin. Immunol.* 83, 722–729.

Riepenhoff-Talty, M., Uhnoo, I., Chegas, P. and Ogra, P.L. (1989) *Immun. Invest.* 18, 127–139.

Ritter, M.A., Rozing, J. and Schuurman, H.J. (1988) *Immunol. Today* 9, 189–193.

Rognum, T.O., Throne, P.S. and Brandtzaeg, P. (1988) *International Congress of Gastroenterology and Digestive Endoscopy, Rome* (Abstract).

Rooney, J.F., Wohlenberg, C., Cremer, K.J. and Notkins, A.L. (1989) *J. Infect. Dis.* 159, 974–976.

Ruuskanen, O., Mertsola, J. and Meurman, O. (1988) *Arch. Dis. Child.* 63, 1250–1253.

Sanders, M.E., Makgoba, M.W. and Shaw, S. (1988) *Immunol. Today* 9, 195–203.

Savilahti, E., Klemola, T., Hovi T., Carlson, B., Stenvik, M., Mellander, L. and Tiainen, P. (1986) In *Progress in Immunodeficiency Research and Therapy II* (J. Vossen and C. Griscelli, Eds.) Elsevier, Amsterdam. pp. 261–267.

Schimke, R.N. Bolano, C. and Kirkpatrick, C.H. (1969) *Am. J. Dis. Child.* 118, 626–630.

Scicchitano, R., Stanisz, A., Ernst, P. and Bienenstock, J. (1988) In *Migration and Homing of Lymphoid Cells* (A.J. Husband, Ed.) Vol. II. CRC Press, Boca Raton, FL. pp. 1–34,

Sharon, N. and Lis, H. (1972) *Science* 177, 949–959.

Sharpe, A.H. and Fields, B.N. (1985) *New Engl. J. Med.* 312, 486–497.

Simister, N.E. and Mostov, K.E. (1989)*Nature* 337, 184–187.

Smith, C.A., Williams, G.T., Kingston, R., Jenkinson, E.J. and Owen, J.J.T. (1989) *Nature* 337, 181–184.

Smith, T.F., Mornis, E.C. and Bain, R.P. (1984) *J. Pediatr.* 105: 896–900.

Sotomayer, J.L., Douglas, S.D. and Wilmott, R.W. (1989) *Pediatr. Pulmonol.* 6, 275–292.

South, M.A., Montgomery, J.R. and Rawls, W.E. (1975) *Birth Defects 11,* 234–238.

Speer, C.H.P., Galin, M. and Pabst, M.H. (1986) *Acta Paediatr. Scand.* 75, 444–451.

Stead, R.H., Bienenstock, J. and Stavist, A.M. (1987) *Immunol. Rev.* 100, 333–359.

Taussig, L.M., Busse, W.W., Lemen, R.J. and Sri Ram J. (1988) *Am. Rev. Respir. Dis.* 137, 979–984.

Theodore, C.M., Losonsky, G., Peri, B., Fishaut, M., Rothberg, R.M. and Ogra, P.L. (1982) In *Recent Advances in Mucosal Immunity* (W. Strober, L.A. Hanson and K.W. Sell, Eds.) Raven Press, New York. pp. 393–403.

Tomasi, T.B. and Bienenstock, J. (1968) Adv. Immunol. 9, 2–96.

Tomasi, T.B., Jr. (1976) *The Immune System of Secretions.* Prentice-Hall, Englewood Cliffs, NJ.

Tonegawa, S. (1988) *J. Am. Med. Assoc.* 259, 1845–1847.

Turner-Warwick, M. (1978) *Immunology of the Lung.* Edward Arnold, London. pp. 118–147.

Tristram, D.A., Miller, R.W., McMillian, J.A. and Weiner, L.B. (1988) *Am. J. Dis. Child.* 142, 834–836.

Uhnoo, I., Dharakul, T., Riepenhoff-Talty, M. and Ogra, P.L. (1988a), *Immunol. Cell Biol.* 66, 135–145.

Uhnoo, I., Olding-Stenkvist, E. and Wadell, G. (1988b) *Dig. Dis.* 6, 102–116.

Uhnoo, I., Riepenhoff-Talty, M., Chegas, P., Fisher, J., Greenberg, H.B. and Ogra, P.L. (1989) *Pediatr. Res.* 25, 192A.

Waldman, R.H., Stone, J., Bergmann, K.C., Khakoo, R., Lazzell, V., Jacknowitz, A., Waldmann, E.R. and Howard, S. (1986) *Am. J. Med. Sci.* 292, 367–371.

Waldman, R.H., Bergmann, K.C., Stone, J., Howard, S., Chiodo, V., Jacknowitz, A., Waldman, E.R. and Khakoo, R. (1987) *J. Clin. Immunol.* 7, 327–332.

Walzer, P.D., Schultz, M.G., Western, K.A. and Robbins, J.B. (1976) *Natl. Cancer Inst. Monogr.* 43, 65–75.

Wang, E.E.L., Prober, C.G., Manson, B., Corey, M. and Levison, H. (1984) *New Engl. J. Med.* 311, 1653–1658.

Welliver, R.C., Sun, M., Rinaldo, D. and Ogra, P.L. (1986) *J. Pediatr.* 109, 776–780.

Welliver, R.C., Sun, M., Hildreth, S.W., Arummighan, R. and Ogra, P.L. (1989) *J. Clin. Microbiol.* 27, 295–299.

Willerford, D.M., Hoffman, P.A. and Gallatin, W.M. (1989) *J. Immunol.* 142, 3416–3422.

Williams, A.L. Cuven, E. and Brethuton, L. (1984) *Br. Med. J.* 288, 1491–1493.

Willoughby, R.E., Schnaar, R. and Yolken, R.H. (1989) *Pediatr. Res.* 25, 193A.

Winick, M. and Noble, A. (1969) *J. Nutr.* 89, 300–306.

Wright, P.F., Hatch, M.H., Kasselberg, A.G., Lowry, S.P., Wadington, W.B. and Karzon, D.T. (1977) *J. Pediatr.* 91, 408–412.

Yap, K.L. and Ada, G.L. (1979) *Scand. J. Immunol.* 10, 325–332.

Zhaori, G., Sun, M., Faden, H.S. and Ogra, P.L. (1989) *J. Infect. Dis.* 159, 1018–1024.

Zink, P., Drescher, J., Verhagen, W., Flik, J. and Milbradt, H. (1987) *Arch. Virol.* 93, 223–232.

Zinkernagel, R.M. and Doherty, P.C. (1979) *Adv. Immunol.* 27, 51–177.

Zinkernagel, R.M., Hengartner, H. and Stitz, L. (1985) *Br. Med. Bull.* 41, 92–97.

Eds. M.H.V. van Regenmortel and A.R. Neurath
Immunochemistry of viruses, II. The basis for serodiagnosis and vaccines.

Viruses and autoimmunity 5

R.S. FUJINAMI

Department of Pathology, M-012, University of California, San Diego, La Jolla, CA 92093, U.S.A.

1. Historical background

Viruses have been associated, for many decades, with autoimmune-like diseases, particularly of the central nervous system (CNS) (Waksman, 1962). An example of this is post-infectious encephalopathies. Virus is seldom isolated and immune responses to myelin components can be found in afflicted individuals. These post-viral encephalopathies have common features with an experimental autoimmune disease known as experimental allergic encephalomyelitis (EAE) (Johnson, 1982). This experimental autoimmmune disease can be induced by the injection of CNS antigens in adjuvant into experimental animals (Freund et al., 1947). The advent of this experimental autoimmune model, EAE, has its roots deeply associated with virus infection. In the late 1880s the Pasteur vaccine for rabies virus was widely utilized and its use was spread to India and Africa by the French (Remlinger, 1928). This virus vaccine was prepared in rabbits and dried infected rabbit CNS was used as the immunizing agent. It became clear that a complication of vaccination was the development of an acute inflammatory process of the CNS. The histopathology associated with this acute encephalomyelitis was very different from that observed with actual rabies virus infection of CNS. At that time, acute encephalomyelitis was suspected to be initiated by an aberrant rabies virus infection. To overcome the putative effects of live virus infection, Semple in the late 1910s developed a phenol-treated preparation in which the rabies virus was inactivated. However, even with this killed vaccine preparation the incidence of the acute encephalomyelitis continued. Furthermore, in 1928, Witebsky and Steinfeld (1928), determined that brain components could induce antibodies to the CNS in contrast to Ehrlich's concepts of self-tolerance presented earlier. Five years later, Rivers and colleagues (Rivers et al., 1933; Rivers and Schwentker, 1935) documented that an acute encephalomyelitis could be induced in monkeys with multiple injections of CNS tissue. This disease resembled the post-rabies acute encephalomyelitis induced by immunization with rabies virus vaccine containing CNS material. In the 1940s Freund developed an immunopotentiator known as Freund's adjuvant. When CNS antigens were emulsified in the adjuvant and injected into animals, these animals developed within a few weeks an acute encephalomyelitis now known as EAE (Freund et al., 1947; Morgan, 1947; Kabat et al., 1947; Morrison, 1947).

Several viruses have been associated with post-infectious encephalomyelitis. These include measles, mumps and vaccinia viruses (Johnson et al., 1984; Johnson, 1982). The histopathology associated with post-infectious encephalomyelitis is similar to that in EAE. In patients with post-infectious encephalomyelitis induced by measles virus, peripheral blood mononuclear cells were found to proliferate when cultured in the presence of myelin basic protein (MBP), a major component of myelin, indicating sensitivity to the CNS antigen (Johnson et al., 1984; Lisak et al., 1974) or to the P2 protein of peripheral nerves (Geczy et al., 1985; Hemachudha et al., 1987). Measles virus infection is associated with encephalomyelitis in about 1/1000 cases. Why or how this is initiated is still unclear.

2. Possible mechanisms

Several mechanisms have been proposed for virus induction of autoimmune disease. These mechanisms will be discussed using measles virus as an example.

Measles virus is a paramyxovirus which initially establishes a respiratory tract infection. From here the virus enters the draining lymph nodes and replicates in lymphoid cells. After 5–6 days post-infection, a viremia ensues and virus is seeded back to the respiratory tract, skin, oropharynx, conjunctiva, alimentary canal and on occasion the CNS (Mims, 1987), resulting in local cytopathic effects in this organ. Infection enhancing class II HLA antigen expression on astrocytes due to interferon induction and cytopathic effects leading to release of CNS antigens are a potential means for the initiation of an autoimmune response.

Under these circumstances, the virus could further potentiate an autoimmune reaction by deregulation of the immune response. Measles virus has the ability to infect T cells, B cells and monocytes (Joseph et al. 1975). Measles virus can cause immunoderegulation of lymphocytes, although little or no viral antigens or infectious virus are expressed in these cells (Casali et al., 1984). Virus could conceivable infect a subpopulation of T cells, i.e., the suppressor cells which modulate the immune response (Huddlestone et al., 1980) and alter the ensuing immune response. During such a situation, the generated immune response would not be properly regulated.

A local immune response in the CNS to CNS and/or viral antigens would not be restrained or limited. Under these circumstances the immune response initiated by virus infection would not be coordinated and could result in disease rather than in the generation of a limited self-regulated response.

3. Molecular mimicry

Another mechanism for the generation of autoimmunity is molecular mimicry (Fujinami and Oldstone, 1985). Viruses may have antigenic determinants in common with host components. Upon infection these common epitopes could produce cross-reacting immune responses leading to autoimmune disease. It has been established for example, that measles virus has a common determinant with one of the cytokeratin proteins found in uninfected cells (Fujinami et al., 1983). This was demonstrated by immunofluorescent staining of infected and uninfected cells with a monoclonal antibody to measles virus. Using this technique, it was found that in infected cells viral cytoplasmic inclusion bodies were stained. In uninfected cells a network-like pattern consistent with intermediate filament proteins was detected. To establish which viral protein was recognized by the anti-measles virus monoclonal antibody, Western blotting experiments were performed. The monoclonal antibody was found to react with the phosphoprotein of measles virus. Similar experiments were conducted with intermediate filament proteins from uninfected cells. The antiviral monoclonal antibody reacted with one of the cytokeratin proteins, thus confirming the presence of a shared determinant.

Many have reports have suggested a cross-reaction between measles virus and CNS components. Panith and McFarlin (1977) described a cross-reaction between measles virus and MBP using antibodies generated during virus infection. More recently, Matossian-Rogers et al. (1987) reported the generation from humans who were immune to measles virus of T cells which were specific for measles virus antigens. The cytotoxicity of the effector T cells could be enhanced by adding to the measles virus target cells MBP. Control experiments with allogeneic target cells using allospecific killers and MBP produced no enhanced killing. The measles virus-specific cytotoxic T lymphocytes (CTL) were able to kill measles virus-infected target cells as well as uninfected MBP-coated target cells. These investigators suggested that the MBP-coated target cells were recognized and lysed by measles virus-specific CTL. Since the control cells displayed no enhanced lysis, these authors speculated that the reaction was specific for measles virus and MBP.

Johnson et al. (1984) studied 19 patients with post-infectious encephalomyelitis arising after measles virus infection. Most of the patients had peripheral blood mononuclear cells which responded to MBP as measured by proliferation. Similar observations were made with individual patients with encephalomyelitis following other viral infections, i.e., varicella, rubella and rabies viruses. These data support the contention that this disease is similar to the experimental autoimmune disease, EAE.

Several viruses have been used to potentiate EAE. Hochberg et al. (1977) reported that half of rats given two intracerebral injections of herpes simplex virus-1 (HSV-1), one day prior to and one day after the induction of EAE showed an recrudescence of disease. Fifty per cent of rats infected with HSV-1 by the subcutaneous route and immunized to induce EAE developed recrudescence of EAE. One immunization by the intracerebral or subcutaneous route did not result in disease. In another animal model using hamsters, Massanari et al., (1979) could enhance the incidence of EAE in these resistant animals. This was accomplished by infecting weanling hamsters with measles virus. These persistently infected hamsters had an increased susceptibility to EAE when sensitized with neuroantigen and adjuvant.

Previously, we described a monoclonal antibody reacting with measles virus phosphoprotein and with one of the intermediate filament proteins, cytokeratin (Fujinami et al., 1983). To explain these findings, computer searches of existing databases were performed. This identified a common region between the encephalitogenic regions of MBP for the rabbit and hepatitis B virus polymerase (Fujinami and Oldstone, 1985). The viral peptide in common was synthesized and used to immunize rabbits. Most of the animals developed autoantibodies to MBP. Fifty per cent of the animals developed cellular reactivity to the CNS antigen. Lesions similar to those found in EAE were observed in some of the animals (Fujinami and Oldstone, 1985). This was the first demonstration that viral determinants could induce autoimmune disease in an animal model system.

More recently, we produced a monoclonal antibody which reacts with Theiler's murine encephalomyelitis virus (Fujinami et al., 1988). This monoclonal antibody reacted with VP-1 of the viral capsid in Western blots. The monoclonal antibody has the ability to neutralize the infectivity of the virus, thus defining a neutralizing site. By ELISA, the monoclonal antibody reacts with galactocerebroside, a major component of myelin. This monoclonal antibody can bind to oligodendrocyte-like cells in newborn mouse brain cultures. The disease induced by Theiler's murine encephalomyelitis virus infection is an EAE-like disease in the overall pathology. Some of the contributing factors associated with the demyelination are immune-mediated. Antibodies like the one described above could enhance the demyelinating attributes of the disease. For example, in a model of Brosnan et al. (1977) demyelination only occurred in the presence of anti-CNS antibodies. Inflammation alone at the target site was not sufficient for demyelination to develop.

Webb and colleagues (Fazakerley et al., 1988) presented data indicating a common determinant between Semliki Forest virus derived from brain and CNS glycolipids. Monoclonal antibodies to the brain glycolipids could inhibit the neutralizing activity of monoclonal antibodies directed to other viral components. Induction of such antibodies to CNS components could allow virus to persist within the confines of the CNS.

Bartholomaeus et al. (1988) exploited a model of murine cytomegalovirus. Infection of BALB/c, C43BL/10 and C3H mice with cytomegalovirus led to the production of antibodies with multiple organ, tissue and cell specificities. The distribution of the antibodies varied among the three mouse strains, suggesting the involvement of a genetic component in the production and specificity of the autoantibodies. Interestingly, in BALB/c mice, myocarditis and inflammation of the salivary gland were evident two months post-infection. Murine cytomegalovirus could not be demonstrated in these tissues at this time, although antibodies to cardiac muscle and salivary duct epithelium were evident. These data suggest a relationship between virus infection and the induction of autoimmune disease.

Viruses can also possess epitopes that mimic regions from major histocompatibility antigens. Interestingly, human cytomegalovirus codes for an immediate early protein which shares a common stretch of five acids with the HLA DR β chain (Fujinami et al., 1988). ELISA and Western blotting experiments show that antibodies made to the viral peptide also bound to the HLA DR β chain from B lymphocytes. The binding of the antibody to the HLA DR β chain could be inhibited with excess peptide, demonstrating the specificity of the interaction. The peptide is from an interesting region in the molecule. Within the common region, amino acid 57 is important for the development of autoimmune diabetes in humans (Todd et al., 1988; see below). In addition, an immune response against the viral protein, which has a common determinant with HLA DR β chain, could lead to immunomodulation. Alternatively, antibodies to the viral determinants may also react with the DR molecule and possibly enhance graft rejection. These two phenomena, immunomodulation and graft rejection, have been associated with human cytomegalovirus infection.

Golding and colleagues (1988) have found homologous regions between the human immunodeficiency virus HIV-1 gp41 and human class II HLA β chain. The common regions are highly conserved between the various DR and DQ alleles and among the different HIV-1 isolates. Monoclonal antibodies against the HIV-1 gp41 region bound to a class II derived peptide. In addition, these antibodies were able to bind to native MHC class II β chain on human lymphoblastoid cells. Lastly, these authors were able to determine that more than one-third of AIDS patients had antibodies which reacted with the HLA class II peptide and with HLA class II enriched preparations. They suggested that autoimmune mechanisms can contribute to the development of AIDS.

Thivolet and Trepo (1988) have summarized data suggesting an association between viral infections and autoimmune type I diabetes. Some of the candidate viruses are cytomegalovirus, Coxsackie virus and hepatitis B virus. These viruses have a tropism for β cells. These authors suggest that molecular mimicry is one mechanism for the development of autoimmune disease against β cells. Gerling et al. (1988) presented evidence that Coxsackie virus B4 infection in mice induced the expression of a 64-kDa autoantigen. They suggested that the appearance of this protein in infected mice was associated with the development of autoimmune diabetes. Similarly, Miller et al. (1987) suggested that polymyositis could be induced by viral infection as the result of molecular mimicry.

In the same vein, Walker and Jeffrey (1986) identified two proteins involved as autoantigens in polymyositis. Using sequence alignment programs, *Escherichia coli* histidyl-tRNA synthetase and alanyl-tRNA synthetase showed similar sequences with viral and muscle proteins. These authors suggest that these data lend strong support for the mechanism of autoimmunity involving molecular mimicry between viral and host muscle proteins.

Vento et al. (1988) investigated the development of autoreactivity to liver components following viral hepatitis. They studied 144 sera from 35 patients with hepatitis virus A, B and non-A non-B infections. Reactivity to two antigens was followed, namely liver-derived lipoprotein complex and asialoglycoprotein receptor. Most of the individuals infected with A or B virus but not the non-A non-B virus patients had antibodies against these two antigens. In addition, cellular reactivity was detected in four of the six hepatitis B virus patients tested with the lipoprotein complex antigen. They concluded that liver-specific autoreactions have the potential to contribute to liver damage in hepatitis virus A and B infections but no in non-A non-B hepatitis infections. Zauli et al. (1986) described studies demonstrating the presence of autoantibodies during hepatitis δ virus infection. Higher titers of antibodies to basal cell layer, intermediate filaments and microtubules were found when infection with hepatitis δ virus was present versus infection with just hepatitis B virus. The mechanism(s) of induction of these antibodies were not investigated, however, and secondary events may have contributed to the generation of these autoantibodies.

Cafruny and Hovinen (1988) studied lactate dehydrogenase virus-induced autoimmunity. After infection of BALB/c mice and C58 mice, autoantibodies to CNS, liver, kidney and spleen cell antigens could be detected. The patterns of reactivity varied between the two inbred strains. In addition, in outbred Swiss mice infected with lactate dehydrogenase virus, autoantibodies to heart tissues were also observed. These investigators suggested that this form of autoimmunity occurs following the deregulation of the immune system after lactate dehydrogenase virus infection. However, the presence of these autoantibodies was not correlated with disease induction or with inflammation within the target organs.

Query and Keene (1987) observed an interesting relationship between retrovirus infection and the development of connective tissue disease. Patients with connective tissue disease have autoantibodies against a 70-kDa protein (70K) associated with U1, a small nuclear ribonucleoprotein. This has a sequence similarity with the p30 gag protein of mammalian type C retroviruses. Antibodies to p30 gag protein bound to a 70K-LacZ fusion protein as well as to U1 small nuclear ribonucleoprotein. Thus, antibodies to the U1 small nuclear ribonucleoproteins could be induced by immunization with p30 gag protein. The authors suggested that immune responses to this region could initiate autoimmunity to the U1 small nuclear ribonucleoprotein and result in connective tissue disorders, Similarly, Wick et al. (1987) and Ziemiecki et al. (1988) suggested that endogenous viruses may play a role in the disease process of spontaneous autoimmune thyroiditis in chickens.

Thus, there is a considerable amount of evidence that viruses and self determinants have common epitopes. Determining the biological relevance of the peptide sharing may prove more difficult. For example, Schwimmbeck et al. (1987a) uncovered a six-amino-acid region common to both an Epstein-Barr virus (EBV) protein and HLA B27. The HLA B27 allele is linked with ankylosing spondylitis (AS) and Reiter's syndrome (RS). Ninety five percent of AS and over 80% of RS patients have this allele. The region on HLA B27 is from a common region domain on the molecule. None of over 50 patients with AS or RA had detectable antibodies in their serum to the corresponding EBV peptide (Schwimmbeck et al., 1987a). In contrast, other studies (Schwimmbeck et al., 1987b) identified similar sequences in another region of HLA B27 and in a segment from the *Klebsiella pneumoniae* nitrogenase. Interestingly, this region is on the hypervariable domain of the HLA B27 corresponding to the HLA B27.1 haplotype. This area is in common with the*Klebsiella* nitrogenase and relates to the specific haplotype. A high proportion of RA and AS patients' sera reacted with the *K. pneumoniae* nitrogenase-derived peptide. In experiments approaching the question from a different angle, Chen et al., (1987) found that antibody to the hypervariable region of HLA B27.1 bound to *Yersinia,* another microorganism which has been implicated in AS and RS.

A common or shared region between a virus and a host protein does not necessarily indicate that a cross-reacting immune response can be mounted. In experiments using rabbits, Dyrberg and Oldstone (1986) found that a six- or eight-amino-acid identity between the polyoma virus middle T antigen and the acetylcholine receptor α chain was not sufficient to generate a cross-reacting immune response, even when the peptide was coupled to a carrier. However, with another viral peptide from the Herpes simplex D protein, sharing four or eight amino acids with the acetylcholine receptor peptide, a highly cross-reacting antibody could be generated. Thus, sequence alignment studies are not sufficient to predict relevance to autoimmune phenomena.

4. Acknowledgements

I thank Jan Richards for her excellent manuscript preparation. This work was supported by Public Health Service grants NS23162 from the National Institutes of Health, by the National Multiple Sclerosis Society grant GR 1780 and University of California Task Force on AIDS grant R8882D027.

5. References

Bartholomaeus, W.N., O'Donoghue, H., Foti, D., Lawson, C.M., Shellam, G.R. and Reed, W.D.(1988) *Immunology* 64, 397–405.

Brosnan, C.F., Stoner, G.L., Bloom, B.R. and Wisniewski, H.M. (1977) *J. Immunol.* 118, 2103–2110.

Cafruny, W.A. and Hovinen, D.E. (1988) *J. Gen. Virol.* 3, 723–729.

Casali, P., Rice, G.P.A. and Oldstone, M.B.A. (1984) *J. Exp. Med.* 159, 1322–1337.

Chen, J.H., et al. (1987) *J. Immunol.* 139, 3003–3011.
Dyrberg, T. and Oldstone, M.B.A. (1986) *Immunology* 130, 25–27.
Fazakerley, J.K., Khalili-Shirazi, A. and Webb, H.E. (1988) *Ann. NY Acad. Sci.* 540, 672–673.
Freund, J., Stern, E.R. and Pisani, T.M. (1947) *J. Immunol.* 51, 179–194.
Fujinami, R.S. and Oldstone, M.B.A. (1985) *Science* 230, 1043–1045.
Fujinami, R.S., Nelson, J.A., Walker, L. and Oldstone, M.B.A. (1988) *J. Virol.* 62, 1001–1005.
Fujinami, R.S., Oldstone, M.B.A., Wroblewska, Z., Frankel, M.E. and Koprowski, H. (1983) *Proc. Natl. Acad. Sci. USA* 80, 2346–2350.
Fujinami, R.S., Zurbriggen, A. and Powel, H.C. (1988) *J. Neuroimmunol.* 20, 25–32.
Geczy, C., Raper, R., Roberts, I.M., Meyer, P. and Bernard, C.C.A. (1985) *J. Neuroimmunol.* 9, 179–191.
Gerling, I., Nejman, C. and Chatterjee, N.K. (1988) *Diabetes* 37, 1419–1425.
Golding, H., Robey, F.A., Gates, F.T., Linder, W., Beining, P.R., Hoffman, T. and Golding, B. (1988) *J. Exp. Med.* 167, 914–923.
Hemachudha, T., Griffin, D.E., Giffels, J.J., Johnson, R.T., Moser, A.B. and Phanuphak, P. (1987) *New Engl. J. Med.* 316, 369–374.
Hochberg, F.H., Lehrich, J.R. and Arnason, B.G.W. (1977) *Neurology* 27, 584–587.
Huddlestone, J.R., Lampert, P.W. and Oldstone, M.B.A. (1980) *Clin. Immunol. Immunopathol.* 15, 502–509.
Johnson, R.T. (1982) *Viral Infections of the Central Nervous System.* Raven Press, New York.
Johnson, R.T., Griffin, D.E., Hirsch, R.L., Wolinsky, J.S., Roedenbeck, S., Lindo deSoriano, I. and Vaisberg, A. (1984) *New Engl. J. Med.* 310, 137–141.
Joseph, B.S., Lampert, P.W. and Oldstone, M.B.A. (1975) *J. Virol.* 16, 1638–1649.
Kabat, E.A., Wolf, A. and Beze, A.E. (1947) *J. Exp. Med.* 85, 117–129.
Lisak, R.P., Behan, P.O., Zweiman, B. and Shetty, T. (1974) *Neurology* 24, 560–564.
Massanari, R.M., Paterson, P.Y. and Lipton, H.L. (1979) *J. Infect. Dis.* 139, 297–303.
Matossian-Rogers, A., Dos Santos, A. and Festenstein, H. (1987) *Int. Arch. Allergy Appl. Immunol.* 84, 159–164.
Miller, F.W., Love, L.A., Biswas, T., McClintock, P.R., Notkins, A.L. and Plotz, P.H. (1987) *Arthritis Rheum.* 30, 549–556.
Mims, C. (1987) *The Pathogenesis of Infectious Disease.* Academic Press, London.
Morgan, I.M. (1947) *J. Exp. Med.* 85, 131–140.
Morrison, L.R. (1947) *Arch. Neuro-Psychiatry* 58, 391–416.
Panitch, H.S. and McFarlin, D.E. (1977) *Fed. Proc* 36, 1207.
Query, C.C. and Keene, J.D. (1987) *Cell* 51, 211–220.
Remlinger, P. (1928) *Ann. Inst. Pasteur* 42 (Suppl) 71–132.
Rivers, T.M. and Schwentker, F.F. (1935) *J. Exp. Med* 61, 689–702.
Rivers, T.M., Sprunt, D.H. and Berry, G.P. (1933) *J Exp Med* 61, 39–54.
Schwimmbeck, P. and Oldstone, M.B.A. (1987a) *Am. Assoc. Phys. Transact* C, 28–39.
Schwimmbeck, P., Yu, D.T.Y. and Oldstone, M.B.A. (1987b) *J. Exp. Med.* 166, 173–181.
Thivolet, C. and Trepo, C. (1988) *Pathol. Biol.* (Paris) 36, 182–186.
Todd, J.A., Acha-Orbea, H., Bell, J.L., Chao, N., Fronek, Z., Jacob, C.O., McDermott, M., Sinha, A.A., Timmerman, L., Steinman, L. and McDevitt, H.O. (1988) *Science* 240, 1003–1009.
Vento, S., McFarlane, B.M., McSorley, C.G., Ranieri, S., Guiliani-Piccari, G., Del Monte, P.R. and Verucchi, G. (1988) *Immunology* 25, 1–7.
Waksman, B.H. (1962) *Medicine* 41, 93–141.
Walker, E.J. and Jeffrey, P.D. (1986) *Lancet* ii, 605–607.
Wick, G., Dromer, G., Neu, N., Fassier, R., Ziemiecki, A., Muller, R.G., Ginzel, M., Beladi, I., Kuhr, T. and Hala, K. (1987) *Immunol. Lett* 16, 249–257.
Witebsky, E. and Steinfeld, J. (1928) *Z. Immun. Forsch. Exp. Ther.* 58, 271–296.
Zauli, D., Crespi, C., Bianchi, F.B., Craxi, A. and Pisi, E. (1986) *J. Clin. Pathol.* 39, 897–899.
Ziemiecki, A., Kromer, G., Mueller, R.G., Hala, K. and Wick, G. (1988) *Arch. Virol* 100, 267–271.

Eds. M.H.V. van Regenmortel and A.R. Neurath
Immunochemistry of viruses, II. The basis for serodiagnosis and vaccines.

Use of synthetic peptides as diagnostic reagents in virology

6

S. MODROW and H. WOLF

Max von Pettenkofer Institute, Ludwig Maximilians University, Pettenkoferstrasse 9a, D-8000 Munich 2, F.R.G.

1. Introduction

Since solid phase peptide synthesis was introduced by B. Merrifield (1963), chemical methods for synthesis have been considerably improved by the use of milder conditions (Atherton et al., 1978) and the methods were adapted to automated systems which allow synthesis of peptide sequences in a relatively short time with high yields. As peptides became readily available, they found widespread use as reagents in natural sciences and medicine. In particular, synthetic peptides were introduced as specific antigens in diagnostic ELISA test systems for the detection of antibodies in patients' pathological sera (Neurath et al., 1982; Gerin et al., 1983; Modrow and Wolf, 1986; Gnann et al., 1987a,b,c; Goudsmit, 1988). One of the major advantages of peptide tests was shown to be an easy and fast distinction between virus subtype infections, e.g., diagnosis of HIV-1/HIV-2 infections (Norrby et al., 1987b; Gnann et al., 1987). Monoclonal or monospecific antibodies with predefined specifity against selected peptide sequences have been used in the identification of native polypeptides in cellular and molecular systems (Lerner et al., 1981; Lerner, 1982; Pfaff et al., 1982; Dietzschold et al., 1984) and have proven to be of value in immunological test systems for detection in serum samples of viral proteins synthesized during the course of infection (Brahm et al., 1988). Since oligopeptides represent mostly sequential epitopes of limited size, their use in diagnostic test systems requires methods for the definition of those dominant antigenic sites that are recognized by the immune response and accessible to antibody binding. Some knowledge about the function of the proteins selected as diagnostic marker in the particular virus system is also of interest. The present chapter will discuss the use of peptides for viral diagnosis and the problems encountered in this field.

2. Selection of epitopes relevant for diagnosis of infection

Until now, synthetic oligopeptides have mainly been used as antigens for studying the humoral immune response developing during the course of infection. In most viral systems, (e.g., polio-, influenza-, herpesvirus, hepatitis B virus, foot-and-mouth disease virus),the humoral immune reaction, i.e.,the production of antibody molecules, will be directed against viral proteins which are part of the viral core or envelope structures. In special cases, virus-encoded enzymes or regulatory proteins prove to be a useful tool for antibody screening, especially during early or latent phases of viral infections (Bayliss and Wolf, 1981). This class of proteins is also recognized by the cellular immune system (influenza virus, cytomegalovirus) and may even play a dominant role in immunological control of disease. Identification and characterization of antigenic sites is a prerequisite, when peptides are to be used either as antigen for antibody detection in sera or for raising specific antipeptide sera in animal systems. The smallest epitopes for antibody recognition have been described as four to six amino acids (Geysen et al., 1985; Tang et al., 1988); generally, peptides of about 12–20 amino acids are used. These sequences mainly represent sequential epitopes, since structural epitopes formed by the folding of the amino acid chain into a secondary or a tertiary structure cannot be simulated by short peptides in the same way as they are formed in the native viral protein. During the past years two main procedures have been developed, which are frequently used in the identification of sequential epitopes: (1) analysis of the amino acid sequence and prediction of epitopes by computer programs based on theoretical and empirical data linking amino acid sequences to structural conformation parameters, and (2) synthesis of overlapping peptides spanning the entire protein sequence or the use of peptide fragments obtained by proteolytic cleavage of the purified polypeptide and testing of these fragments for reactivity with sera derived from infected individuals.

Despite consistent improvements of these methods for identification and characterization of antigenic and immunogenic sites in proteins, when an identified region has been synthesized as an oligopeptide, without the surrounding regions naturally occurring in the native protein, it may show all possible reactions from nonrecognition to specific binding to nonspecific positive reaction. Possible reasons for nonrecognition of peptides are manifold: some flanking sequences stabilizing a defined structure of the epitope may be missing or the selected region presented in the form of an oligopeptide may mimic epitopes in host proteins, even in the absence of any sequence homology. Such instances have been described (Geysen, 1985; Nestorowicz et al. 1985; van Regenmortel, 1987) and in general the suitability and usefulness of a synthetic peptide can only be established experimentally.

2.1. Computer analysis for identification of antigenic sites

Although precise information concerning protein structure and surface exposed areas can only be obtained by methods such as X-ray diffraction (Liljas and Rosman, 1974) and NMR spectroscopy (Wagner and Wüthrich, 1982), certain features have been found to be common for antigenic sites. Based on these observations, programs for computer-assisted identification of sequential epitopes from the primary amino acid sequence have been developed (Van Regenmortel and Daney de Marcillac, 1988). For the prediction of continuous antigenic sites and the immunodominant determinants in a protein it appeared important not only to analyze individual parameters reported to be common to epitopes, such as hydrophilicity (Hopp and Woods, 1981; Westhof et al., 1984; Tainer et al., 1984; Karplus and Schulz, 1985), surface accessibility (Janin et al., 1978; Emini et al., 1985) or probabilities for secondary structures as developed by Chou and Fasman (1978) and Garnier et al. (1978), but also to combine those values in a weighted manner, resulting in an algorithm for the prediction of sequential antigenic sites (Jameson and Wolf, 1988; Wolf et al., 1988). These predictions of antigenic determinants have been successfully applied to various viral proteins and the identified regions have been synthesized as oligopeptides (Cohen et al., 1984; Eisenberg et al., 1985; Modrow et al., 1987; Middledorp and Meloen, 1988; Modrow et al., 1989a) or produced by recombinant gene technology (Motz et al., 1986a,b; Wolf et al., 1987). In Fig. 6.1 the amino acid sequence analysis of gp41 of HIV-1 (Ratner et al., 1985), which is a glycosylated, membrane-anchored protein, is shown as an example. It is important to mention that an antigenic peptide for use in diagnostic test systems, especially when derived from an envelope protein, should preferably be located on the external side of particular structures or membranes and should not contain potential *N*-glycosylation sites, since sequential epitopes may be highly altered by the addition of carbohydrate groups (Rademacher et al., 1988). Antibodies in patients' sera elicited by modified epitopes are often directed against the glycosyl groups or structural determinants induced by the modification. Similarly, antipeptide sera induced in animals will not recognize the corresponding epitope in a native, modified polypeptide (Emini et al., 1988). Phosphorylation of peptide regions has also been shown to lead to an altered backbone structure resulting in a changed antigenic determinant (Otvos et al., 1988).

In the case of the transmembrane protein of HIV-1 (human immunodeficiency virus type 1), only one diagnostically relevant protein region located in the β-turn rich region preceding the glycosylation sites could be predicted to be positioned on the membrane without additional modification. In a series of recent studies (Wang et al., 1986; Gnann et al., 1987a,b; Motz et al., 1987; Modrow et al., 1989c), this fairly constant region was found to be generally recognized by patients' sera and was therefore an important component of diagnostic test systems for HIV infection.

2.2 Epitope mapping using overlapping peptides

Independent of the development of more reliable prediction methods for antigenic epitopes, another approach to analyzing immunodominant epitopes in protein sequences has been developed by Geysen et al. (1984, 1987). Here, short overlapping peptides are synthesized onto chemically activated polystyrene pins. After deprotection of side groups these fixed peptides are tested for their ability to react with sera derived from individuals infected with the corresponding virus. This approach has been successfully applied to map epitope clusters in the amino acid sequence derived from structural proteins of foot-and-mouth disease virus (Meloen and Barteling, 1986; Schaaper et al., 1988), influenza virus (Schoofs et al., 1988), respiratory syncytial virus (Norrby et al., 1987a) and Epstein-Barr virus (Middeldorp and Meloen,

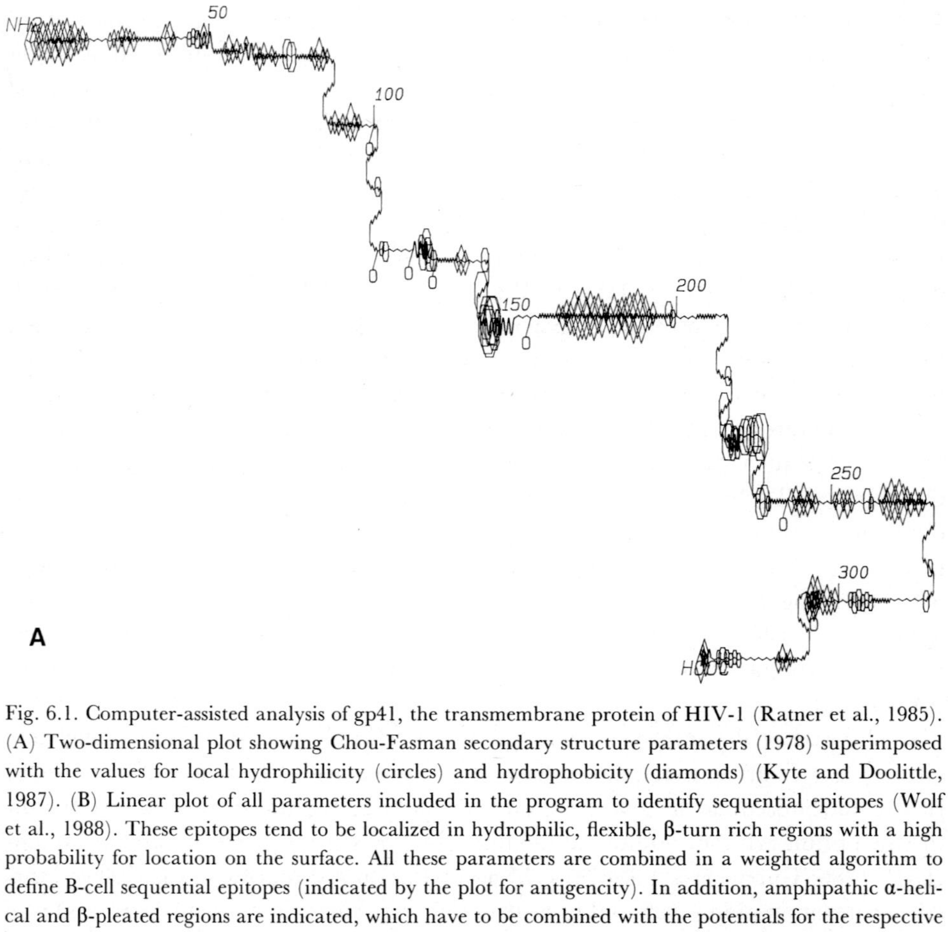

Fig. 6.1. Computer-assisted analysis of gp41, the transmembrane protein of HIV-1 (Ratner et al., 1985). (A) Two-dimensional plot showing Chou-Fasman secondary structure parameters (1978) superimposed with the values for local hydrophilicity (circles) and hydrophobicity (diamonds) (Kyte and Doolittle, 1987). (B) Linear plot of all parameters included in the program to identify sequential epitopes (Wolf et al., 1988). These epitopes tend to be localized in hydrophilic, flexible, β-turn rich regions with a high probability for location on the surface. All these parameters are combined in a weighted algorithm to define B-cell sequential epitopes (indicated by the plot for antigencity). In addition, amphipathic α-helical and β-pleated regions are indicated, which have to be combined with the potentials for the respective structures according to Chou-Fasman or Garnier et al. In both plots, the transmembrane region can be identified by a stretch of hydrophobic residues; potential *N*-glycosylation sites are indicated—this region is supposed to be localized outside the membrane. The fusogenic amino-terminal end is rather hydro- →

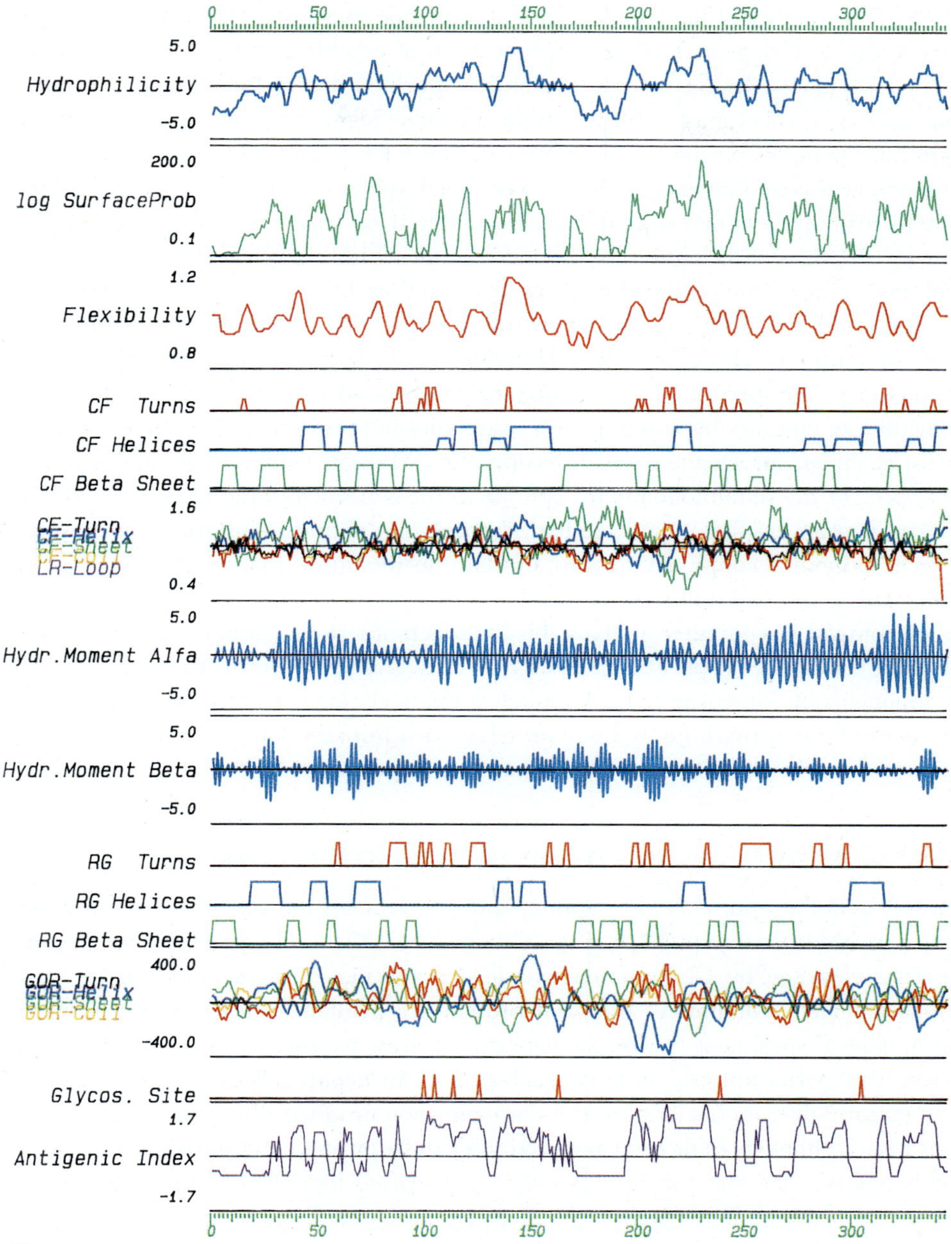

B

phobic. In conclusion, only one antigenic region between the hydrophobic amino terminus and the glycosylated region remains to be represented by synthetic peptides. The two-dimensional plots are useful in obtaining a quick first impression of the nature of the protein. The combination of linear parameters, in contrast, allows a defined analysis of the diverse functions localized in the individual regions of the protein.

1988). The quality of the identification of antigenic sites using this method of 'peptide walking' is dependent on the optimization of peptide synthesis, since the individual sequences cannot be characterized or purified due to their fixation to pins. Incorrect sequences may result in a nonreactive or highly cross-reactive peptide. In addition, peptides synthesized onto pins should have a length of at least nine residues, since shorter sequences, especially of rather apolar amino acids may form quite stable structures, either with the plastic material or with themselves resulting in 'sticky' peptides with highly nonspecific reactions with many types of proteins, including antibody molecules. Hence, positive reactions have to be confirmed with purified peptides. Unconjugated free peptides, which are synthesized using multiple peptide synthesis (Houghten, 1985; Houghten et al., 1986) allowing the fast production of large numbers of peptide analogues, can be used in ELISA tests for the identification of epitopes by reaction with antibodies in sera derived from infected persons as shown for Dengue virus glycoprotein (Markoff et al., 1988) and for fine mapping of the immunodominant epitope in the transmembrane protein of HIV-1 (Norrby et al., in press). This method also does not permit monitoring the completion of individual coupling reactions, but has the advantage that the individual peptides can be purified and characterized.

In contrast to computer analysis this approach of testing a series of peptides with positive sera makes it possible to clearly identify antigenic determinants relevant as components of diagnostic tests. Although a relatively expensive synthesis of a large number of oligopeptides may be required at least initially, for products with a vast potential market this argument is of secondary importance.

2.3. Epitope mapping by the use of proteolytic cleavage products

Another similar approach for mapping epitopes does not use chemically synthesized peptides but utilizes peptide products obtained from the native protein through proteolytic cleavage by CNBr or enzymatic degradation. This method can be used when sufficient amounts of the viral protein are available, either by purification from clinical specimens (serum for hepatitis B virus, parvovirus; stool for rotavirus, hepatitis A virus; autopsy material, such as liver for hepatitis B virus), from infected cell cultures or from bacterial or cellular expression products obtained by recombinant gene technology. The cleavage products can be purified and tested for reactivity with positive sera and the epitope presenting regions identified or fine-mapped by further processing. The reactive oligopeptides are either sequenced by Edman degradation or, provided the sequence of the protein is known, classified with the amino acid sequence of the native polypeptide according to their values for size, ionic charge or hydropathy and tested for reactivity with positive sera. This approach to epitope mapping has been successfully applied to a number of viral and nonviral proteins (Baur et al., 1982; Berzofsky et al., 1982; Eisenberg et al., 1982; Smith-Gill et al., 1982; Jemmerson and Paterson, 1986a,b). Using CNBr cleavage of a bacterial fusion protein of β-galactosidase with the major part of the structural protein of parvovirus B19 (Shade et al., 1986) the causative agent of erythema infectiosum (Anderson et al., 1984) was analyzed (Fig. 6.2). The HPLC purified fragment showing reactivity with antibodies of parvovirus B19-positive individuals could be used in diagnostic ELISA tests, while fine mapping of the antigenic determinants could be done with chemically synthesized peptides (manuscript in preparation).

3. Use of synthetic peptides as antigen for antibody identification in serum samples

Serodiagnosis of most viral infections is done routinely in ELISA (enzyme-linked immunosorbent assay) or RIA (radioimmunoassay) test systems using purified viral preparations as antigen for antibody detection in the samples. In the past these antigen preparations were obtained mainly from infected cell cultures (adeno-, poliovirus, Tickborne encephalitis virus etc), stools (hepatitis A virus, rotavirus) or serum samples of infected individuals (hepatitis B virus, parvovirus B19). In many cases, these methods required the manipulation of highly infectious agents and were—due to the use of cell culture and consecutive purification steps—very expensive and time-consuming. Advances in gene technology and chemistry led to the stepwise introduction of bacterially expressed proteins (Motz et al., 1986b) or synthetic peptides as a cheap source of antigen material usable in diagnostic tests. Both methods have their specific advantages and disadvantages. Proteins produced by recombinant DNA technology must be highly purified when used in ELISA to avoid false positive results due to contamination with cellular or bacterial cross-reacting proteins. A high level of purity is often difficult to achieve and requires extensive research and laborious methods. Confirmation by Western blot or radioimmunoprecipitation experiments may be necessary if purity problems cannot be solved. However, antigen can be produced in almost unlimited quantities and production costs are minimal. Since the expressed regions may span relatively large parts of the protein, several epitopes, also those defined by tertiary structure parameters, can be presented by only one product. Alternatively, automated systems now allow the chemical synthesis of peptides in a rather short time with high yields; the products are free of contaminating proteins and can often be used after a one-step column purification in specific ELISA test systems. Due to their limited size of about 25–30 amino acid residues, they usually represent small protein regions spanning only one sequential epitope. The synthesis of the longer peptides is, despite progress in chemical methods during the last years, still rather difficult and expensive. As a consequence peptides are preferentially used for special diagnostic problems (e.g., differentiation between two similar virus types) and in order to minimize false negative results they should be applied as a set of peptides representing several epitopes when used for serum screening. It can be expected that due to the genetic basis of the im-

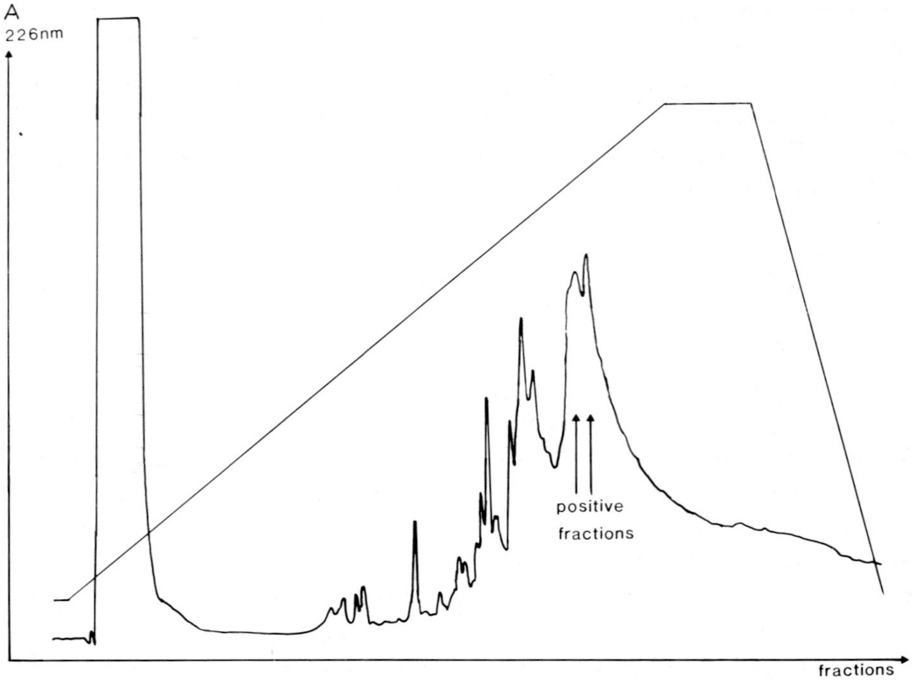

Fig. 6.2. Separation of proteolytic peptides derived from parvovirus B19 structural protein β-galactosidase fusion protein. After CNBr cleavage in 70% formic acid the lyophylized fragments were suspended in 6 M guanidine hydrochloride and purified by reversed phase HPLC (high performance liquid chromatography) using a C18 column (TSK ODS 12OT, Pharmacia/LKB) and a gradient of 0–70% acetonitrile:isopropanol (2:1) in 0.1% TFA. The individual fractions were dried, suspended in 0.2 M sodium carbonate buffer (pH 9.5), coupled to polystyrene wells of ELISA plates and tested for their reactivity with a panel of sera from parvovirus B19 positive and negative individuals. Only two neighboring fractions at the end of the gradient showed a specific reaction with positive sera. The fractions were rechromatographed on C4 reversed phase column resulting in a single peak. According to the selected running condition and the charge and hydrophilicity of the expected cleavage products the reactive fragment spans a region near the amino-terminus of the protein and includes approximately three potentially antigenic sites according to computer analysis.

mune system of each individual, antibodies in sera derived from different persons may not be directed to the same epitope and may differ in quality and quantity.

The demands on quality of synthetic peptides are often underestimated. Every incomplete cycle of amino acid condensation opens a complete tree of false products which may be undetectable and barely removable if longer products are assayed. The latest developments in synthesizers have provided economic and efficient ways to obtain optimal products by emphasizing end-group protection of unreacted groups and the monitoring of each individual step of amino acid coupling (Horn and Novak, 1987).

3.1. Peptide conjugation and presentation

In the majority of cases, use of synthetic peptides as antigen in ELISA assays as well as for eliciting specific antibodies to a predefined sequence requires the attachment or coupling of the peptide to a carrier system. This coupling procedure of the purified antigen can have a major influence on the reactivity of antibodies and on the potential of antipeptide sera to recognize the native protein sequence (Briand et al., 1985). There is general agreement that conjugation procedures should be selected which allow the presentation of the peptide in a way similar to that in the native protein, whatever the function of the respective protein may be. Dyrberg and Oldstone (1986, 1987) have noticed major differences in reactivity of sera to the same peptide when the antigen is coupled via the carboxy- or amino-terminal residue. Covalent linkage to the carrier should preferably be done via the terminal residues. One possibility is coupling via the SH group of an additionally included cysteine (Liu et al., 1979), which does away with the need to use the carboxy or amino functions of internal amino acids (Glu, Asp, Lys); these residues are often important for antigenicity and the corresponding antigenic sites may be destroyed when coupling agents such as carbodiimides (Arnon et al., 1980) or glutaraldehyde (Reichlin, 1980) are used. To induce antipeptide sera alternate protocols have been developed and applied successfully. Two important protocols involve covalent coupling of palmitic acid molecules to a terminal lysine residue (Hopp, 1984) and linkage of the peptides to liposomes (Neurath et al., 1984).

When used as diagnostic antigen in ELISA, the peptide needs to be attached to the solid phase. Two major routes are commonly used: (i) the coupling of the free purified peptides either by nonspecific adsorption to the plastic support (Cantarero et al., 1980) mostly at high pH values or covalent linkage of the peptides by functional NH_2 groups after treatment of polystyrene wells with glutaraldehyde; and (ii) the application of peptides covalently coupled to a carrier protein (e.g., BSA or KLH) and the use of those complexes for adsorption to the plastic surface. All methods may lead to the alteration or destruction of antigenic determinants, but they have been successfully applied in different test systems according to the peptide used (Modrow and Wolf, 1986; Neurath et al., 1986b; Wang et al., 1986; Gnann et al., 1987a,b; Norrby et al., 1987b; Goudsmit et al., 1988; Närvänen et al., 1988; Modrow et al., 1989a).

Since neither of the described methods fulfills optimal conditions for peptide presentation, we developed a system where the same polystyrene carrier used during peptide synthesis is also used as support for antibody reaction in an ELISA test system (Modrow et al., 1989b). In this method the peptides are still covalently bound via their carboxy-terminal ends (Fig. 6.3). This simple and cheap method uses polystyrene resin with the first amino acid complex in ester linkage to the chloromethylated benzyl groups (generally used in N-t-Boc peptide chemistry) and the utilization of Fmoc-protected residues. After each cycle, Fmoc groups are removed by

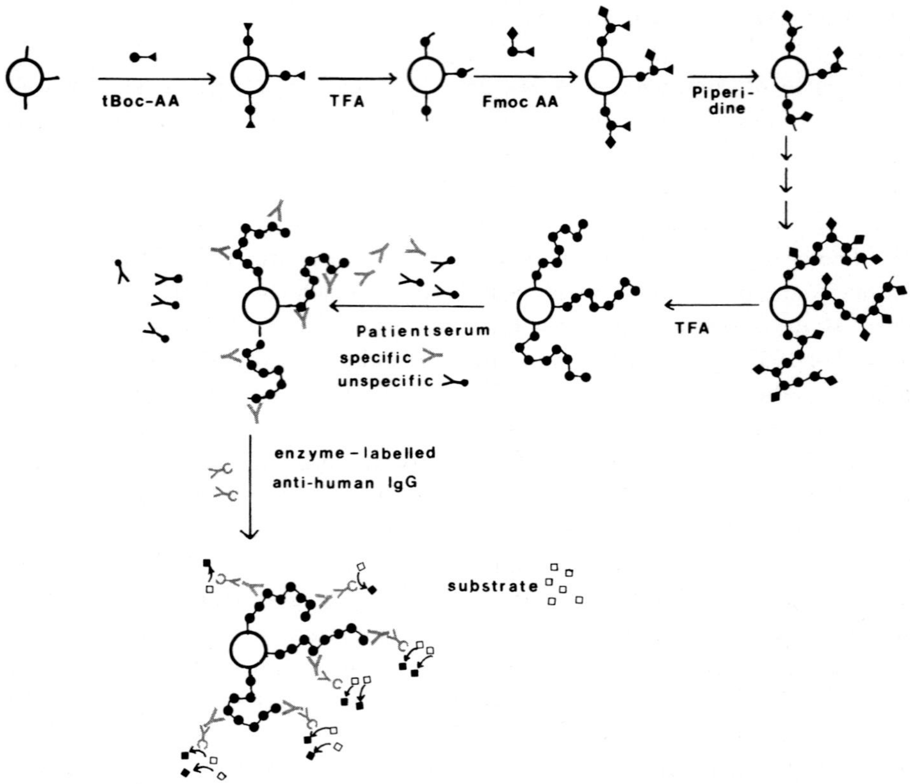

Fig. 6.3. Schematic presentation of the synthesis and use of synthetic oligopetides in solid-phase immunoassays without cleavage of the peptide from the resin. Due to the high substitution with peptide antigen (0.4–0.6 meg/g), a specific and easy peptide test system may be established in a rather short time. Due to their coupling to a polystyrene carrier, the beads may be directly used after cleavage of side chain protection groups to characterize the peptides by gas phase sequencing and to react with patients' antisera.

treatment with secondary amine. When the synthesis is completed, permanent side-chain blocking groups are deprotected using diluted trifluoroacetic acid, a relatively mild deprotection method not associated with extensive peptide destruction as may occur with HF or TFMSA cleavage. The peptide-resin linkage is resistant to both reactions. The amino acid sequence of these carrier-bound peptides can easily be confirmed by gas phase sequencing. Using only the carboxy-terminal group for covalent coupling to the polystyrene matrix, the peptide antigens can be used without further purification and have a highly reactive conformation for reaction with antibodies during the following steps of an ELISA assay 'in solution'. We have applied these 'peptide beads', which are highly substituted with antigen, in a simple and cheap system for antibody detection in sera obtained from HIV-infected individuals (Modrow et al., 1989b,c). In a similar approach Kennedy et al. (1987) and Fischer et al. (1989) have used polyamide based solid phase support with an acid stable *p*-

(oxymethyl)benzoic acid linker for peptide synthesis and further reactions, in these cases for antibody elicitation.

3.2. Peptides used for antibody detection in sera

Test systems using synthetic oligopeptides as antigens in diagnostic test systems are especially useful for antibody detection of viral infections, when it is difficult to grow the infectious agent in cell culture systems. The first test systems based on peptide antigens were developed to demonstrate the presence of antibody directed against hepatitis B virus surface antigen (HBsAg) during the course of infection or during vaccination; these studies were mainly carried out to answer scientific and experimental questions, such as the location of immunodominant epitopes and the time course of antibody production. For these initial experiments hepatitis B virus proved to be an excellent choice, since the viral genome encodes only two structural proteins (HBsAg, HBcAg), in addition to the DNA polymerase and X-gene product. Sequencing data of several virus isolates revealed no regions of major variability. Thus, a fairly uniform pattern of antibodies directed mainly to conserved epitopes (Lerner et al., 1981; Bhatnager et al., 1982) was found in patients. Synthetic peptides derived from the preS1 and preS2 sequences of the large and middle form of HBsAg, which are underrepresented in infectious viral particles, have shown the presence and the functional importance of antibodies directed against these sequences during the early stages of viral infection. They have also been used for analyzing functional parameters of the preS regions involved in virus neutralization and cell adsorption (Neurath et al., 1985; Neurath et al., 1986a,b; Neurath et al., 1987a,b; Steward et al., 1988).

Other viral systems are considerably more complex than hepatitis B virus with respect to genomic organization, expression of viral antigens during the course of the lytic and latent stages of infection and sequence variation. One example is Epstein-Barr virus (EBV), a member of the human herpesvirus family. This virus encodes more than 80 viral proteins and several polypeptides constitute structural components forming the viral core, tegument and envelope. Different viral proteins are expressed during the lytic infection leading to infectious mononucleosis and chronic EBV infection and during the latent stage of the disease. Latent EBV genomes are associated with nasopharyngeal carcinoma and African Burkitt's lymphoma. Due to different viral gene expression associated with the individual diseases, characteristic antibody patterns are observed and utilized for individual diagnosis. Before diagnostic test systems based on peptide antigens can be developed, basic knowledge on biological features of the virus, sequence and antigen variation must be available. In initial experiments, antigens of importance in diagnosing disease were identified and mapped to specific reading frames of the viral genome (Seibl and Wolf, 1985; Wolf et al., 1985). From these data recombinant antigens were developed for diagnostic test systems (Motz et al., 1986a,b; Wolf et al., 1987). Peptide antigens have been derived from regions of EBNA 1 (Epstein-Barr virus nuclear antigen), one member of the protein family representing the latent viral infection with anti-EBNA 1 antibodies as an immune reaction signalling past EBV-infection. These were synthesized as oligopeptides and used for antibody screening in sera of EBV-positive individuals (Dillner et al., 1985; Rhodes et al., 1985; Rumpold et al., 1987; Linde et al., 1989, in press). Recently a commercial anti-EBNA 1 test system based on synthetic peptide antigen has been developed (Johnson and Johnson, La Jolla, CA, U.S.A.). IgG and IgA antibodies directed to an amino terminal epitope of another latent viral protein, the membrane antigen LMP also associated with cellular immune response to EBV-infected B-lymphocytes, was found to be elevated in patients with infectious mononucleosis or nasopharyngeal carcinoma (Modrow and Wolf, 1986). In similar test systems Fox et al. (1987) showed, predominantly in infectious mononucleosis patients, the presence of immunoglobulins to epitopes localized in a nuclear protein expressed early. These rather promising first results represent, however, only a few aspects of the complicated system of EBV infections and their diagnosis.

Recently, the demand for an easy and safe test system for HIV infections led to the first commercial ELISA test systems based totally on synthetic peptide antigens (Biochrom KG, Berlin, FRG; Clonatec, Paris, France; Labsystems, Helsinki, Finland; Johnson and Johnson, La Jolla, CA, USA). Major efforts were made to identify and localize relevant diagnostic epitopes in the sequence of the structural core proteins and the envelope protein complex. Antibodies directed to epitopes localized in an external loop region of gp41 proved to be most useful in diagnosing HIV infections (Wang et al., 1986; Gnann et al., 1987b,c; Smith et al., 1987; Modrow et al., 1989c). In the case of HIV, another problem arose which was associated with test systems based on synthetic peptides, caused by the extreme variability found mainly in antigenic regions of the protein (Alizon et al., 1986; Starcich et al., 1986; Modrow et al., 1987). This high degree of variation exists in antigenic epitopes located mostly in the envelope protein complex but is also present in the sequences of the core proteins. In addition, it must be anticipated that due to the differences in the genetic basis of the immune system of each individual, antibodies in sera derived from different persons may not be directed to the same epitope and may differ in quality and quantity. These reasons require the use of a set of oligopeptides if false negative results are to be minimized. Furthermore, the final antibody concentration has to be determined especially in problematic sera, e.g.,with high background and cross-reaction. The use of a set of peptide antigens and the combination of the titration curves will ultimately lead to reliable determination of the specific antibody concentration allowing a clear diagnosis (Modrow et al., 1989c).

3.3. Use of peptides for special diagnostic problems

One of the major advantages using peptides in diagnostic systems is the possibility of measuring antibody concentrations to one specific epitope. Recently, it has been shown that the prognostic situation in HIV-infected individuals can be predicted by the observation of antibody titers against defined regions of the envelope proteins gp120 (Modrow et al., 1989a) and gp41 (Klasse et al., 1988). When testing consecutive sera of the same individuals it was observed that the humoral immune reaction in AIDS patients was mainly directed to a peptide located in residues 315–326 of gp120. In HIV-positive individuals in an early stage of disease infection not yet showing signs of the progressing AIDS disease, reaction was observed to peptides derived from various other epitopes of gp120, despite major differences in antibody titer and reactivity in the individual samples. The reduced antibody formation to certain epitopes in AIDS patients can be used as a diagnostic marker when the individual patient enters the final stage of infection. For human papilloma virus infections, Dillner et al. (1989, in press) could show the presence of antibodies directed against a synthetic peptide, predominantly in patients with cervical intraepithelial neoplasia.

In a similar way, testing for antibodies against specific epitopes allows the distinction between virus infection caused by related virus strains or subtypes of the same virus types. Applying this sort of site-directed serology Dietzschold et al. (1984) could distinguish HSV-1 from HSV-2. Gnann et al. (1987a) and Norrby et al. (1987b, 1989) could show that it is possible to discriminate between HIV-1, HIV-2 and SIV infections by using peptides derived from a region of the transmembrane part of the envelope proteins gp36/gp41, which is diverse in the protein sequences of the related virus strains. This sort of single peptide-antigen test, albeit useful for definition of certain virus strains, may, however, be of limited value in geographic regions where a large degree of sequence variation exists among the individual virus isolates. Diagnosis of HIV infections in Europe is associated with the problem that virus isolates originating in different parts of the world (e.g.,U.S.A., Caribbean, various parts of Africa) show extreme sequence and antigen variation; this results in a considerably elevated degree of cross-reaction in diagnostic test systems, since the sequences of individual isolates are generally not known and tests are done using one representative peptide. This high diversity in combinations with the different immune reaction of each individual based on the genetic HLA constellation makes it necessary to use a set of peptides representing epitopes from the same or various structural proteins encoded by the virus. We used a set of HIV-1 and HIV-2 specific peptides derived from core and envelope protein sequences, which allow diagnosis of HIV infections in general and distinguishing between the two virus types, even when sera failed to react with individual epitopes or were cross-reactive to one of the peptides (Modrow et al., 1989,c). Protein regions 12–20 residues in length, without cross-reacting or surrounding sequences homologous between the two virus types, can only be represented by chemically synthesized peptides. Such test systems will have a wide application to a large number of viral, bacterial and immunological diseases for strain differentiation or for identifying different phases of disease.

4. Use of antipeptide antibodies for diagnosis

The major application of monospecific or monoclonal antipeptide antibodies is the identification and characterization of the respective complete native protein, mostly in experimental systems where other specifically reactive test systems are not available. The use of such antibodies in diagnostic serology of viral infections is limited because large amounts of viral protein in serum, stool or other samples are present only during a short period in a few viral infections (hepatitis B virus, parvovirus B19, rotavirus, hepatitis A virus). The recently developed methods for amplification of short regions of viral nucleic acids by polymerase chain reaction (PCR) (Saiki et al., 1985; Mullis and Faloona, 1987) allow an extremely sensitive determination of even very small amounts of viral DNA in biological materials (sera, urine etc.), which cannot be achieved by antipeptide sera at the protein level.

One possible exception is hepatitis B virus infection. During the acute phase large amounts of the surface antigen (HBsAg) including the preS sequences are synthesized and released into the serum. The HBsAg can be determined and screened by the use of monoclonal antipeptide antibodies, as has been shown by Budkowska et al. (1986) and Brahm et al. (1988). In addition to the detection of hepatitis B virus infection these authors showed the presence of preS2 sequences on the Dane particle during the acute phase of hepatitis B and the relevance of these antibodies for predicting an individual's risk of commencing a chronic or nonchronic stage of infection. This type of predictive diagnosis cannot be accomplished by mere detection of viral DNA.

4.1. Identification of viral antigens in tissue biopsies

The development of highly sensitive techniques for antigen detection in ultrathin sections of frozen tissue using immunofluorescence or APAAP methodology requires a highly specific detection system without cross-reactivities to cellular material. Using antipeptide sera to epitopes derived from immediate early or early proteins produced during the lytic EBV infection we could demonstrate the presence of these viral proteins in regions of hairy leukoplakia of the tongue of AIDS patients (Becker et al., 1989; Wolf et al., 1989). In these experiments using frozen tissue sections of hairy leukoplakias from AIDS patients, it could be shown that local activation of the productive EBV cycle occurs in these lesions dependent on the differentiation of the cells from basic epithelial cells to keratinocytes. Due to the immune suppression in AIDS, cellular immune control does not take place and infectious virus is released into the saliva. Similar leukoplakias in the tongues of smokers or tobacco chewers show no virus production; here only immediate early protein synthesis is observed in the basal layers together with simultaneous immigration of T lymphocytes which eliminate EBV protein-producing cells in this early stage of the lytic cycle (Fig. 6.4). Monospecific antipeptide sera were used rather than sera raised against the

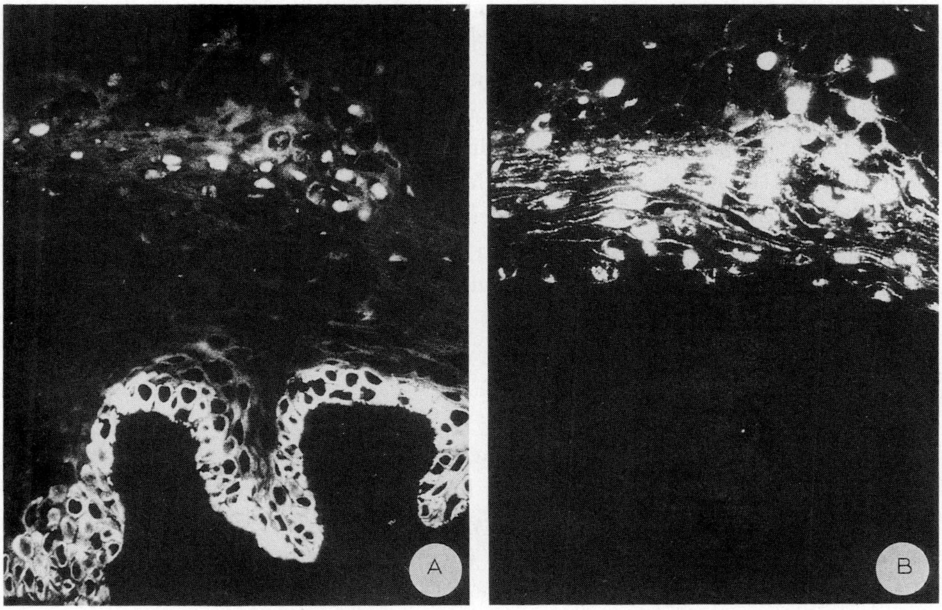

Fig. 6.4. Ultrathin frozen section of a biopsy derived from hairy leukoplakia on the tongue of an AIDS patient. The section in 4A has been stained with an anti-peptide serum directed against transinducing immediate early protein BZLF1 of Epstein-Barr virus. This protein is known to have regions homologous to the cellular protein c-fos. By selecting a nonhomologous region for peptide synthesis, a highly specific rabbit anti-peptide serum could be obtained without cross-reaction to cellular proteins. In A., the induction of viral protein synthesis, dependent on cell differentiation, is observed in the basal cell layer, where the product remains in a cytoplasmatic, inactive localization. In a layer of differentiating cells above this product is activated and transported to the nucleus, its functional localization, where early viral protein synthesis is induced by BZLF1. On the surface of the tongue, late viral structural proteins are synthesized and infectious virus is released into the saliva (4B).

intact protein. The specificity of these tests was demonstrated by the fact that segments of homology to cellular regulatory genes such as c-*fos* did not interfere. Not only the presence of the individual EBV proteins but also the intracellular localization could be studied and a shift of the immediate early regulatory transactivation protein from a cytoplasmic to a nuclear localization according to the stage of cell differentiation could be shown.

5. Conclusion

Synthetic peptides have been demonstrated to be extremely useful as antigens for early and fast, highly specific diagnostic test systems, as well as for the elicitation of monospecific and monoclonal antipeptide antibodies. Progress has been rapid during the last 8–10 years mainly because of the adaptation of improved chemistry to automated systems. Synthetic peptides will be introduced as antigens in latex agglutination or similar tests increasing the sensitivity and resulting in a system which allows easy and rapid serology without the need for expensive detection systems or instruments. Beside their application in diagnostic systems, synthetic peptides may prove useful as a vaccine without side effects, especially if a specific T cell response can be elicitated. The use of such vaccines is highly dependent on our growing knowledge of the mechanisms involved in cellular and humoral immune response to viral infections and on an adequate knowledge of the respective viral systems.

6. Acknowledgements

The authors thank Wolfgang Jilg, Michael Roggendorf and Lutz Gürtler, Max von Pettenkofer-Institute, Munich, for many helpful discussions and critical reading of the manuscript and Tanya Polonsky for typing it. Work quoted from our own group was supported in part by the Fonds der Chemischen Industrie and the Bundesministerium für Forschung und Technologie (01 ZR 8506, FVP/BGA 188).

7. References

Alizon, M., Wain-Hobson, S., Montagnier, L. and Sonigo, P. (1986) *Cell* 46, 63–74.

Anderson, M.J., Lewis, E., Kidd, J.M., Hall, S.M. and Cohen, B.J. (1984) *J. Hyg.* 93, 85–93.

Arnon, R., Sela, M., Parant, M. and Chedid, L. (1980) *Proc. Natl. Acad. Sci. USA* 77, 6769–6772.

Atherton, E., Fox, H., Logan, C.J., Harkiss, D. Sheppard, R.C. and Williams, B.J. (1978) *J. Chem. Commun.* 13, 537–539.

Baur, X., Dewair, M., Fruhmann, G., Aschauer, H., Pfletschinger, J. and Braunitzer, G. (1982) *J. Allergy. Clin. Immun.* 69, 66–76.

Becker, J., Leser, U., Langford, A., Jilg, W., Reichart, P., Gelderblom, H. and Wolf, H. (1989) In (*Epstein-Barr Virus and Human Disease II* D.V. Ablashi et al., Eds.) Humana Press, New Jersey, pp. 267–270.

Bayliss, G.J. and Wolf, H. (1981) *J. Gen. Virol.* 56, 105–118.

Berzofsky, J.A., Buckenmeyer, G.K., Hicks, G., Gurd, F.R.N., Feldman, R.J., and Minna, J. (1982) *J. Biol. Chem.* 257, 3189–3195.

Bhatnager, P.K., Papes, E., Blum, H.E., Milich, D.R., Nitecki, D., Karels, M.J. and Vyas, G.N. (1982) *Proc. Natl. Acad. Sci. USA.* 79, 4400–4404.

Brahm, J., Vento, S., Rondanelli, E.G., Ranieri, S., Fagan, E.A., Williams, R. and Eddleston, A.L.W. (1988) *J. Med. Virol.* 24, 205–209.

Briand, J.P., Muller, S. and Van Regenmortel, M.H.V. (1985) *J. Immunol Methods* 78, 59–69.

Budkowska, A., Riotto, M.M., Dubrenil, P., Lazizi, Y., Brechot, C., Sobezak, E., Petit, M.A. and Pillot, J. (1986) *J. Med. Virol.* 22, 111–125.

Canterero, L.A., Butler, J.E. and Osborne, J.W. (1980) *Anal. Biochem.* 105, 375–382.

Chou, P.Y. and Fasman, G.D. (1978) *Adv. Enzymol.* 47, 45–148.

Cohen, G.H., Dietzschold, B., Ponce de Leon, M., Long, D., Golub, E., Varrichio, A., Pereira, L. and Eisenberg, R.J. (1984) *J. Virol.* 49, 102–108.

Dietzschold, B., Eisenberg, R.J., Ponce de Leon, M., Golub, E., Hudecz, F., Varrichio, A. and Cohen, G.H. (1984) *J. Virol.* 52, 431–435.

Dillner, J., Eliasson, L., Sternas, L., Kallin, B., Klein, G. and Lerner, R.A. (1985) In *Epstein-Barr Virus and Associated Disease* P.H. Levine, D.V. Ablashi, G.R. Pearson, and S.D. Kottaridis, Eds.) Martinus Nijhoff, Boston. pp. 446–455.

Dillner, J., Dillner, L., Robb, J., Willems, J., Jones, I., Lancester, W., Smith, R. and Lerner, R. (1989) *Proc. Natl. Acad. Sci. USA.* 86, 3838–3841.

Dyrberg, T. and Oldstone, M.B.A. (1986) *J. Exp. Med.* 164, 1344–1349.

Dyrberg, T. and Oldstone, M.B.A. (1987) *Vaccines* 87, 38–43.

Eisenberg, R.J., Long, D., Pereira, L., Hampar, B., Zweig, M. and Cohen, G.H. (1982) *J. Virol.* 41, 478–488.

Eisenberg, R.J., Long, D., Ponce de Leon, M., Matthews, J.T., Speer, P.G., Gibson, M.G., Lasky, L.A., Berman, P., Golub, E. and Cohen, G.H. (1985) *J. Virol.* 53, 634–644.

Emini, E.A., Hughes, J.V., Perlon, D.S. and Boger, J. (1985) *J. Virol.* 55, 836–839.

Emini, E.A., Schleif, W.A., Armstrong, M.E., Silberklang, M., Schultz, L.D., Lemann, D., Maigatter, R.Z., Qualtiere, L.F., Perason, G.R. and Ellis, R.W. (1988) *Virology* 166, 387–394.

Fischer, P.M., Comis, A. and Howden, M.E.H. (1989) *J. Immunol. Methods* 118, 119–123.

Fox, R.J., Houghten, R., Whalley, A., Geltofsky, J., Vaughan, J. and Smith, R. (1987) *J. Clin. Lab. Anal.* 1, 140–145.

Garnier, J., Osguthorpe, O.J. and Robson, B. (1978) *J. Mol. Biol.* 120, 97–120.

Gerin, J.L., Alexander, H., Shih, J.W.K., Purcell, R.H., Dapolit, G., Engle, R., Green, N., Sutcliffe, J.G., Shinnick, T.M. and Lerner, R.A. (1983) *Proc. Natl. Acad. Sci. USA.* 80, 2365–2369.

Geysen, H.M., Meloen, R.H. and Barteling, S.J. (1984) *Proc. Natl. Acad. Sci. USA.* 81, 3998–4002.

Geysen, H.M. (1985) *Immunol. Today* 6, 364–369.

Geysen, H.M., Barteling, S.J. and Meloen, R.H. (1985) *Proc. Natl. Acad. Sci. USA.* 82, 178–182.

Geysen, H.M., Rodda, S.J., Mason, T.J., Tribbick, G. and Schoofs, P.J. (1987) *J. Immunol. Methods* 102. 259–274.

Gnann, J.W., McCormick, J.B., Mitchell, S., Nelson, J.A. and Oldstone, M.B.A. (1987a) *Science* 237, 1346–1349.

Gnann, J.W., Nelson, J.A. and Oldstone, M.B.A. (1987b) *J. Virol.* 61, 2639–2641.

Gnann Jr., J.W., Schwimmbeck, P.L., Nelson, J.A., Truax, B. and Oldstone, M.B.A. (1987c) *J. Infect. Dis.* 159, 261–267.

Goudsmit, J., Boucher, C.A.B., Meloen, R.H., Epstein, L.G., Smit, L., van der Hoek, L. and Bakker, M. (1988) *AIDS* 2, 157–164.

Hopp, T.P. (1984) *Mol. Immunol.* 21, 13–16.

Hopp, T.P. and Woods, K.R. (1981) *Proc. Natl. Acad. Sci. USA.* 78, 3824–3828.

Horn, M. and Novak, C. (1987) *Am. Biotech. Lab.* 5, 12–17.

Houghten, R.A. (1985) *Proc. Natl. Acad. Sci.* 78, 3824–3828.

Houghten, R.A., DeGraw, S.T., Bray, M.K., Hoffman, S.R. and Frizzell, N.D. (1986) *Biotechniques* 4, 522–528.

Jameson, B.A. and Wolf, H. (1988) *CABIOS* 4, 181–186.

Janin, J., Wodak, S., Levitt, M. and Maigret, M. (1978) *J. Mol. Biol.* 125, 357–386.

Jemmerson, R. and Paterson, Y. (1986a) *Science* 232, 1001–1004.

Jemmerson, R. and Paterson, Y. (1986b) *Biotechniques* 4, 18–31.

Karplus, P.A. and Schulz, G.E. (1985) *Naturwissenchaften* 72, 212–213.

Kennedy, R.C., Dreesman, G.R., Chanti, T.C., Boswell, R.W., Allan, J.S., Lee, T.H., Essex, M., Sparrow, J.T., Ho, D.D. and Kanda, P. (1987) *J. Biol. Chem.* 262, 5769–5774.

Klasse, P.J., Pipkorn, R. and Blomberger, J. (1988) *Proc. Natl. Acad. Sci. USA* 85, 5225–5229.

Kyte, J. and Doolittle, R.F. (1983) *J. Mol. Biol.* 157, 105–132.

Lerner, R.A. (1982) *Nature* 299, 592–596.

Lerner, R.A., Green, N., Alexander, H., Liu, F.T., Sutcliffe, J.G. and Shinnick, T.M. (1981) *Proc. Natl. Acad. Sci. USA.* 78, 3403–3407.

Liljas, A. and Rosman, M.G. (1974) *Annu. Rev. Biochem.* 43, 475–507.

Linde, A., Kallin, B., Dillner, J., Anderrson, J., Jägdahl, L., Lindvall, A. and Wahren, B. (1989) *J. Infect. Dis.*, in press.

Liu, F.T., Zinnecker, M., Hamaloka, T. and Katz, D.H. (1979) *Biochemistry* 18, 690–697.

Markoff, L.J., Bray, M., Lai, C.J., Chanock, R.M., Eckels, K., Summers, P., Gentry, M.K., Houghten, R. and Lerner, R.A. (1988) *Vaccines* 88, 161–165.

Meloen, R.H. and Barteling, S.J. (1986) *Virology* 149, 55–63.

Merrifield, R.B. (1963) *J. Am. Chem. Soc.* 85, 2149–2154.

Middledorp, J.M. and Meleon, R.H. (1988) *J. Virol. Methods* 21, 147–159.

Modrow, S. and Wolf, H. (1986) *Proc. Natl. Acad. Sci. USA.* 83, 5703–5707.

Modrow, S., Hahn, B.H., Shaw, G.M., Gallo, R.C., Wong-Staal, F. and Wolf, H. (1987) *J. Virol.* 61, 570–578.

Modrow, S., Höflacher, B., Mellert, W., Erfle, V., Wahren, B. and Wolf, H. (1989a) *J. AIDS,* 2, 21–27.

Modrow, S., Höflacher, B., Mertz, R. and Wolf, H. (1989b) *J. Immunol. Methods,* 118, 1–7.

Modrow, S., Höflacher, B., Gürtler, L., Deinhardt, F., Wolf, H. (1989c) *J. AIDS,* 2, 141–148.

Motz, M., Deby, G., Jilg, W. and Wolf, H. (1986a) *Gene* 44, 353–359.

Motz, M., Fan, J., Seibl, R., Jilg, W. and Wolf, H. (1986b) *Gene* 42, 303–312.

Motz, M., Soutschek-Bauer, E., Frösner, G.G., Gürtler, L., Schall, M. and Wolf, H. (1987) *Lancet ii,* 1093.

Mullis, K. and Faloona, F.A. (1987) *Methods Enzymol.* 155, 335–350.

Närvänen, A., Korkolainen, M., Kontio, S., Suni, Y., Korpela, J., Partnanen, P., Soos, J., Vahen, A. and Huhtala, M.L. (1988) *AIDS* 2, 119–123.

Nestorowicz, A., Tregear, G.W., Southwell, C.N., Martya, J., Murray, J.M., White, D.O. and Jackson, D.C. (1985) *Mol. Immunol.* 22, 145–154.

Neurath, A.R., Kent, S.B.H. and Strick, N. (1982) *Proc. Natl. Acad. Sci. USA.* 79, 7871–7875.

Neurath, R.A., Kent, S.B.H. and Strick, N. (1984) *J. Gen. Virol.* 65, 1009–1014.

Neurath, A.R., Kent, S.B.H., Strick, N., Taylor, P. and Stevens, L.E. (1985) *Nature* 315, 154–156.

Neurath, A.R., Kent, S.B.H., Parker, K., Prince, A.M., Strick, N., Brotman, B. and Sproul, P. (1986a) *Vaccine* 4, 35–40.

Neurath, A.R., Kent, S.B.H. and Strick, N., (1986b) *J. Gen. Virol.* 67, 453–461.

Neurath, A.R., Kent, S.B.H., Adamowicz, P., Riotto, M.M., Price, P., Strick, N., Parker, K., Petit, M.A., Budkowska, A., Girard, M. and Pillot, J. (1987b) *Mol. Immunol.* 24, 561–568.

Neurath, A.R., Kent, S.B.H., Strick, N., Parker, K., Conronce, A.M., Riotto, M.M., Petit, M.A., Budkowska, A., Girard, M. and Pillot, J. (1987a) *Mol. Immunol.* 24, 975–980.

Norrby, E., Mufson, M.A., Alexander, H., Haughten, R.A. and Lerner, R.A. (1987a) *Proc. Natl. Acad. Sci. USA* 84, 6572–6576.

Norrby, E., Biberfeld, G., Chiodi, F., von Gegerfeld, A., Naucler, A., Parks, E. and Lerner, R.A. (1987b) *Nature* 329, 248–250.

Norrby, E., Chiodi, F., Whalley, A., Parks, E., Nauclér, A., Costa, C.M., Torstensson, R. and Biberfeld, G. (1989) *AIDS* 3, 17–20.

Norrby, E., Biberfeld, G., Johnson, P.R., Parks, D.E., Houghten, R.A. and Lerner, R.A. (1989) *AIDS Res. Human Retroviruses* 5, 487–493.

Otvos, L., Hollosi, M., Perczel, A., Lee, V.M.Y., Dietzschold, B. and Fasman, G.A. (1988) *J. Protein Chem.* 7, 365–376.

Pfaff, E., Mussgay, M., Böhm, H.O., Schulz, G.E. and Schaller, H. (1982) *EMBO J.* 1, 869–874.

Rademacher, T.W., Parekh, R.B. and Dwak, R.A. (1988) *Annu. Rev. Biochem.* 57, 785–838.

Ratner, L., Haseltine, W., Patarca, R., Livak, K.J., Starcich, B., Josephs, S.F., Doran, E.R., Rafelske, J.A., Whitehorn, E.A., Baumeister, K., Ivanoff, L., Petteway, S.R., Pearson, M.L., Lautenberger,

J.A., Papas, T.A., Ghyrayeb, J., Chang, N.T., Gallo, R.C. and Wong-Staal, F. (1985) *Nature* 313, 227–233.

Reichlin, M. (1980) *Methods Enzymol.* 70, 159–165.

Rhodes, G., Carson, D.A., Valbracht, J., Houghten, R. and Vaughan, J.H. (1985) *J. Immunol.* 134, 211–217.

Rumpold, H., Rhodes, G.H., Bloch, P.L., Carson, D.A. and Vaughan, J.H. (1987) *J. Immunol.* 138, 593–599.

Saiki, R., Scharf, S., Faloona, F., Mullis, K., Horn, G., Erlich, H.A. and Arnheim, N. (1985) *Science* 230, 1350–1353.

Schaaper, W.M.H., Puijk, W.C., Lankhof, H., Thomas, A. and Meloen, R.H. (1988) In *Peptides 1988* (E. Bayer and G. Jung, Eds.) Walther de Gruyter Verlag, pp. 226–228.

Schade, R.O., Blundell, M.C., Cotmora, S.F., Tatersall, P. and Astell, C.R. (1986) *J. Virol.* 58 921–936.

Schoofs, P.G., Geysen, H.M., Jackson, D.C., Brown, L.E., Tang, X.L. and White, D.O. (1988) *J. Immunol.* 140, 611–616.

Seibl, R. and Wolf, H. (1985) *Virology* 141, 1–13.

Smith, R.S., Naso, R.B., Rosen, J., Whalley, A., Hom, Y.L., Hoey, K., Kennedy, C.J., McCutchan, J.A., Spector, S.A. and Richman, D.D. (1987) *J. Clin. Microbiol.* 25, 1498–1504.

Smith-Gill, S.J., Wilson, A.C., Potter, M., Prager, E.M., Feldmann, R.J. and Mainhart, C.R. (1982) *J. Immunol.* 128, 314–322.

Starcich, B.R., Hahn, B.H., Shaw, G.M., McNeeley, P.D., Modrow, S., Wolf, H., Parks, E.S., Parks, W.P., Josephs, S.F., Gallo, R.C. and Wong-Staal, F. (1986) *Cell* 45, 637–648.

Steward, M.W., Sisley, B.M., Stanley, C., Brown, S.E. and Howard, C.R. (1988) *Clin. Exp. Immunol.* 71, 19–25.

Tainer, J.A., Getzhoff, E.D., Alexander, H., Houghten, R.A., Olson, A.J., Lerner, R.A. and Hendrickson, W.A. (1984) *Nature* 312, 127–134.

Tang, X.L., Tregear, G.W., White, D.O. and Jackson, D.C. (1988) *J. Virol.* 62, 4745–4751.

Van Regenmortel, M.H.V. (1987) *Trends Biochem. Sci.* 12, 237–240.

Van Regenmortel, M.H.V. and Daney de Marcillac, G. (1988) *Immunol Lett.* 17, 95–108.

Wagner, G. and Wüthrich, K. (1982) *J. Mol. Biol.* 155, 347–366.

Wang, J.J.G., Steel, S., Wisniewolski, R. and Wang, C.Y. (1986) *Proc. Natl. Acad. Sci. USA* 83, 6159–6163.

Westhof, E., Altschuh, D., Moras, D., Bloomer, A.C., Mondragon, A., Klug, A. and Van Regenmortel, M.H.V. (1984) *Nature* 311, 123–126.

Wolf, H., Motz, M., Seibl, R., Bayliss, G.J., Modrow, S. and Fan, J. (1985) in *Epstein-Barr Virus and Associated Diseases* (P.H. Levine, D.V. Ablashi, G.R. Pearson and S.D. Kottaridis, Eds.) Martinus Nijhoff, Boston. pp. 485–494.

Wolf, H., Motz, M., Kühbeck, R., Jilg, W., Fan, J., Pi, G.H. and Zeng, Y. (1987) In *EBV and its Diseases* (Nonoyama et al., Eds.) Humana Press, New Jersey. pp. 179–182.

Wolf, H., Modrow, S., Motz, M., Jameson, B., Herrmann, G. and Förtsch, B. (1988) *CABIOS* 4, 187–191.

Wolf, H., Jilg, W., Markert, C., Modrow, S., Zietz, C. and Wilmes, E. (1989) In *Epstein-Barr Virus and its Disease II.* (D.V. Ablashi et al., eds.) Humana Press, New Jersey, pp. 99–103.

Eds. M.H.V. van Regenmortel and A.R. Neurath
Immunochemistry of viruses, II. The basis for serodiagnosis and vaccines.

Production of viral antigens in prokaryotic and eukaryotic cells

7

T. RUTGERS, T. CABEZÓN and M. DE WILDE

Department of Molecular and Cellular Biology, Smith Kline Biologicals, 89 rue de l'Institut, B-1330 Rixensart, Belgium

Over the last decade a wide variety of molecular cloning vectors have been developed that allow expression of heterologous genes, or gene fragments, in prokaryotic and eukaryotic cells. Cloning and expression of viral genes have enabled numerous studies to identify and characterize the antigenic and immunogenic properties of viral proteins, examples of which are found throughout this book. Ultimately, these studies will lead to the design and development of new vaccines. Approaches involve introduction of genetically engineered mutations in the viral genome leading to well-defined, stable, attenuated viral strains, chimeric viruses, recombinant live viral or bacterial vectors expressing foreign genes, and purified viral antigens produced by genetically engineered cells (bacteria, yeasts, insect or mammalian cells).

Purified viral antigens produced by recombinant DNA technology offer the possibility to limit the number of antigens in a vaccine to those which are protective, to eliminate sequences that could trigger immunosuppression mechanisms or autoimmune reactions, and to avoid the inclusion of genetic material. However, few vaccines based on this technology have been developed yet. One major obstacle has been an inability to render the purified antigens sufficiently immunogenic. An important exception is the yeast-derived recombinant DNA vaccine against hepatitis B, which was licensed for human use in 1986. The reasons why the hepatitis B vaccine became an early success in the recombinant DNA field will be examined in this chapter. Observations made in the case of HBV will be related to other attempts to develop subunit vaccines, without the intention of reviewing all data concerning viral antigen production by recombinant DNA technology. The identification of antigens able to induce a protective immune response, their production in heterologous cell systems, and options to improve their immunogenicity and protective efficacy will be addressed.

1. Identification of protective antigens

Hepatitis B virus (HBV) consists of an envelope and a nucleocapsid containing a circular DNA molecule (Fig. 7.1). The envelope carries the hepatitis B surface antigen (HBsAg) and the capsid the hepatitis B core antigen (HBcAg). Besides virions, large quantities of empty viral envelopes (HBsAg) in the form of 22-nm particles circulate in the bloodstream of persons carrying an HBV infection. HBsAg has a complex macromolecular structure containing viral protein and host-derived lipid. The most abundant protein is the S protein encoded by the S gene (Fig. 1). Two minor proteins are present, the middle (M) protein encoded by the preS2 region and the S gene, and the large (L) protein encoded by the preS1 and preS2 regions and the S gene. These proteins are each present in two differently glycosylated forms (Tiollais et al., 1985).

HBsAg is a conformational antigen. A dimer of two S protein molecules linked by disulfide bridges bears HBsAg antigenicity and dimers are 20 to 1000-fold more immunogenic than monomers (Mishiro et al., 1980). Delipidation of 22-nm particles and formation of protein micelles generates an excellent immunogen (Skelly et al., 1981; Howard et al., 1985; Hollinger et al., 1986).

An effective vaccine against HBV has been available for human application since 1981 (Heptavax-B, Merck, Sharp and Dohme; Hevac-B, Institut Pasteur Production). The 22-nm particles are isolated from the plasma of chronic HBV carriers and subsequently inactivated. The vaccine is safe, but high production costs due to the fact that human blood is used as raw material (requiring elaborate purification and inactivation procedures to avoid possible contamination with other infectious agents) was a major impetus to seek an alternative source of antigen.

These data obtained with plasma-derived HBsAg had defined the antigen re-

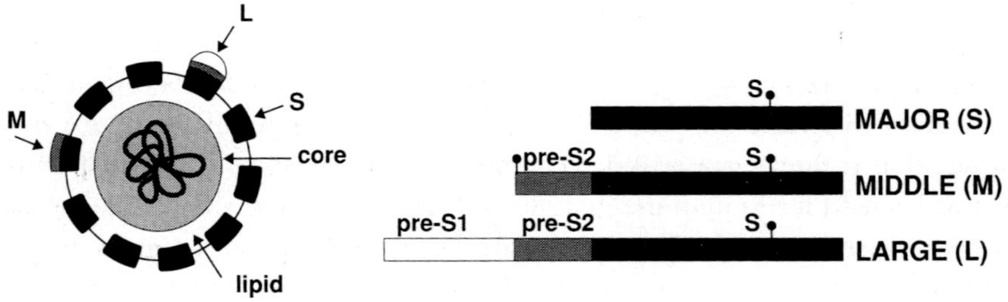

Fig. 7.1. Schematic representation of the HBV virion (left) and its envelope proteins (right). ● indicates attached *N*-linked glycans. Each envelope protein exists in two forms, one form has the structure as indicated in the figure, the other form lacks the *N*-linked glycan in the S part of the molecules.

quired for vaccine development. In the absence of a cell culture system to propagate HBV, the alternative was to produce S protein by recombinant DNA technology and if necessary, render it immunogenic by dimerization, micelle formation or mimicry of 22-nm particles.

The identification of M and L proteins in the envelope of HBV (Stibbe and Gerlich, 1982, 1983; Heermann et al., 1984) led to an analysis of the contribution of the preS1 and preS2 regions of these proteins to protective immunity. Utilizing synthetic peptides and recombinant envelope proteins in a congenic mouse model, additional neutralizing and protective epitopes showing independent H-2 restriction patterns have been identified in the preS region as well as preS-localized T-helper epitopes which can provide functional help for S-specific antibody production (Itoh et al., 1986a; Neurath et al., 1986, 1988; Milich, 1988). If the same mechanisms of immune response are operative in humans, addition of preS determinants to the present vaccines based on S protein should be taken into consideration. The present vaccines have great efficacy but a certain percentage of vaccinees, especially immunocompromised persons like hemodialysis patients, are non- or slow responders (Hadler et al., 1986; Bruguera et al., 1987). At the present time, human trials are under way to establish if such nonresponders to the S protein will respond to preS determinants and if the addition of preS can circumvent the nonresponsiveness to the S protein.

As was the case for the analysis of the antigenic and immunogenic properties of the preS region of the HBV surface proteins, a combination of research tools such as recombinant DNA-derived viral proteins, synthetic peptides, and monoclonal antibodies have contributed to the efforts to define the immunological properties of the proteins of other viruses. In general, in vaccination induction of the highest possible titer of neutralizing antibodies is important. However, induction of cellular immunity is also a critical component. Not only is potent T cell help usually needed to regulate most effector components of the immune system, but an active cytotoxic T cell function is also often needed to destroy infected cells and eradicate infection. The element of the immune response (neutralizing antibodies, cytotoxic T lymphocytes, antibody-dependent cell killing) which plays the major role in protection against a given pathogen largely determines which viral antigen(s) will be the best vaccine candidate.

For viruses other than HBV, the antigens suitable for vaccine development have been defined to varying extents. For instance, induction of neutralizing and protective immunity to foot-and-mouth disease virus (FMDV) has been localized to specific sequences in one of the viral proteins. For respiratory syncytial virus (RSV) and other paramyxoviruses the viral proteins, but not the epitopes, presumably important for induction of a protective immune response have been identified while for human immunodeficiency virus (HIV) in vitro neutralizing activities are just beginning to be defined.

The amino acid sequence 141–160 of protein VP1 of FMDV has been identified as an immunodominant region and synthetic peptides bearing this sequence can elicit neutralizing antibodies that protect guinea pigs against experimental infection. However, the immunizing activity of such a peptide is only 1% or less of that of inactivated virus particles on an equal weight basis (Bittle et al., 1982). The peptide contains both B and T-helper epitopes (Francis et al., 1985). Due to its small size it can contain only a few T_H-cell epitopes which could give rise to problems in obtaining good immunity within an outbred population. The objective in the development of an FMDV vaccine is therefore to present the B cell epitopes in an immunogenic form to the immune system and to add sufficient T cell epitopes to avoid MHC restriction of the immune response.

Studies in which monoclonal antibodies and vaccinia-RSV recombinants were employed have shown that RSV F and G glycoproteins individually stimulate neutralizing antibodies and protective immunity (Taylor et al., 1984; Stott et al., 1987; Wertz et al., 1987; Olmsted et al., 1988). Cytotoxic T cell (CTL) immune responses are directed primarily toward the F glycoprotein and the nucleoprotein (Bangham et al., 1986; Pemberton et al., 1987). Development of subunit vaccines (Murphy et al., 1988a) against RSV is complicated by the possibility of disease potentiation following virus infection as occurred previously with formalin-inactivated RSV vaccines (Prince et al., 1986). Furthermore, the vaccine should be able to elicit a protective immune response in infants before the second month of life and overcome the immunological immaturity of the infant, and the immune suppression by maternally derived antibodies (Murphy et al., 1986b,c; 1988b). Although the RSV proteins that could serve as immunogens are defined, their presentation to the immune system may be a critical parameter to obtain protection.

Purified gp120 of HIV has been shown to elicit in immunized animals antibodies able to neutralize HIV infectivity in vitro (Robey et al., 1986). HIV is characterized by a high rate of mutation (Hahn et al., 1986) occurring mostly in the *env* gene. The antigenic variation of HIV probably represents a major obstacle for vaccine development. Indeed, the neutralizing response of animals to in vitro expressed *env* gene products has appeared to remain mostly strain-specific. Neutralizing antibodies to sequences of gp41 (Chanh et al., 1986; Dalgleish et al., 1988) and gag p17 protein (Papsidero et al., 1989) have also been reported. CTL-specific responses to the envelope (Walker et al., 1987), reverse transcriptase (Walker et al., 1988) and gag p24 protein (Nixon et al., 1988) have been demonstrated in HIV-seropositive individuals. However, the immune mechanisms that may play a role in preventing AIDS are not clear and protective immunity to HIV in humans has yet to be demonstrated. There is a clinical progression in virus-infected individuals to AIDS in the presence of both humoral and cellular immunity. The critical question in HIV research is the definition of a protective immune response.

2. Production of viral antigens in prokaryotic cells

Expression in *Escherichia coli* of the S protein of HBV has not been successful. Only very low expression levels have been obtained and cells expressing S protein showed a growth inhibition. The S protein is extremely hydrophobic and it has been speculated that the growth inhibition might be due to embedding of the hydrophobic protein into the *E. coli* membrane causing interference with some membrane function (Fujisawa et al., 1983). Growth inhibition of *E. coli* was also observed upon expression of the vesicular stomatitis virus (VSV) G protein and the NH_2-terminal hydrophobic domain of the G protein was shown to be responsible (Rose and Shafferman, 1981). Removal of the sequences coding for the NH_2-terminal hydrophobic domain of the S protein resulted in normal growth of the cells expressing this modified S protein but expression levels were extremely low (Fujisawa et al., 1983). The hemagglutinin (HA) surface glycoprotein of influenza virus was expressed as a fusion protein in *E. coli* as direct expression was only marginal (Davis et al., 1983). Although the fusion protein is made in abundance, the antibody response to this polypeptide appears to be qualitatively different from the response to the native viral HA. Lack of glycosylation and incorrect folding are presumably the reason for the altered immunogenicity of the protein. This demonstrates the general advantages and disadvantages of the *E. coli* system. Heterologous proteins, often in the form of fusion proteins, can routinely be expressed at high levels ($> 15\%$ of total cell protein). The recombinant proteins are usually found in an insoluble form in the bacterial cells. Aggregates in the form of inclusion bodies are visible by phase contrast microscopy. Although the existence of these inclusion bodies is an advantage for purification, the protein in these aggregates has no biological activity. Recovery of product requires the inclusion body to be dissolved and the proteins denatured in a chaotropic agent such as urea or guanidinium hydrochloride. Refolding and oxidizing to form native-folded proteins with the correct pairing of disulfide bridges is not always effective. The immunogenic properties of the isolated recombinant proteins are usually not equivalent to those of the native form of the protein integrated in the virions. The *E. coli* system can be useful for 'simple' antigens where activity does not depend on conformation or post-translational modifications like glycosylation or disulfide bonds.

Expression of protein VP1 of FDMV in *E. coli* (Kleid et al. 1981) is another example of recovery of a protein without biological activity. The immunogenicity of VP1 isolated from *E. coli* was very low ($< 0.1\%$) compared to that of the intact virus particle. Broekhuijsen et al. (1986) used the *E. coli* system to express the immunodominant region of VP1 as a fusion protein with β-galactosidase. 1, 2, 4 or 8 repeats of amino acids 137–162 of VP1 were fused N-terminally to β-galactosidase. The fusion proteins were found in soluble form in the bacterial cells and showed β-galactosidase activity. As only the tetramer of β-galactosidase has enzymatic activity, the fusion proteins are apparently folded in a comparable oligomeric structure.

Immunization of guinea pigs with the purified fusion proteins containing two or four copies of the VP1 peptide sequence resulted in a superior immune response as compared to the fusion protein containing only one copy of the peptide sequence or the chemically synthesized peptide (Broekhuijsen et al., 1987). The reason for this enhanced response may be that the repetition of the antigenic determinant provides an epitope density required for good immune recognition.

Recombinant proteins that contain portions of gp120 of HIV have been produced in *E. coli* (Putney et al., 1986), and have served as important research tools. A fragment covering the carboxy-terminal half of gp120 was expressed as a fusion protein (PB1) at levels between 10 and 20% of total cellular protein. The protein was solubilized using protein denaturants and purified. Antibodies from AIDS patients recognized the fusion protein on Western blots. Immunization of either goats or rabbits with PB1 induced a strong immune response with type-specific neutralizing antibody titers, as measured by inhibition of viral replication in an in vitro assay, comparable to those elicited by native gp120. This demonstrated that, in this case, glycosylation is not required to elicit neutralizing antibodies and helped to localize the major neutralizing epitope in gp120 (Rusche et al., 1988).

Although the immunogenic properties of *E. coli*-derived viral proteins in general are disappointing, their antigenic properties can be very useful for the development of diagnostic test systems. For instance, screening and confirmatory tests to detect antibodies to HIV have been developed that use recombinant DNA-derived protein fragments produced in *E. coli* (Chang et al. 1985; Crowl et al., 1985; Dowbenko et al., 1985; Dawson et al., 1988 Cabradilla et al. 1986). In contrast to assays based on inactivated disrupted whole virus, recombinant DNA-derived protein fragments provide homogeneous, noninfectious, reproducible target antigens and can give specific information about the in vivo targets of the immune response. A potential disadvantage is that recombinant proteins are usually derived from selected portions of the HIV genome and may not detect all of the individuals who are truly seropositive for HIV. The sensitivity and specificity of the tests have to be rigorously analyzed.

A remarkable exception to the general rule that the *E. coli* system is unable to produce proteins of viruses with eukaryotic host range, in a correct conformational form, is the core antigen of HBV (HBcAg). HBcAg produced in *E. coli* forms particles that resemble the 27-nm core particles isolated from human liver. The particles which contain nucleic acid as well as the core protein are stable and can be purified (Stahl et al., 1982). They are immunogenic in laboratory animals and are a valuable diagnostic reagent (Stahl et al., 1982; Cohen et al., 1984). Murray et al. (1984, 1987) immunized chimpanzees with the *E. coli*-derived HBcAg particles and observed a partial protection against HBV challenge. This may be explained by the ability of HBcAg primed T cells to provide help for antibody production to HBsAg epitopes (Milich et al., 1987; Milich, 1988).

3. Production of viral antigens in eukaryotic cells

3.1. Expression, post-translational modifications, and transport of viral antigens

Post-translational modifications such as processing, disulfide bond formation, glycosylation, phosphorylation, myristoylation, palmitoylation can influence the conformation of a protein and consequently profoundly affect its antigenicity and immunogenicity profile. Most viral envelope proteins are glycoproteins containing intra- or intermolecular disulfide bonds and are present in the virion envelope as homo- or hetero-oligomers. Core proteins which are in contact with the viral nucleic acid are often phosphorylated. Over the last few years a variety of viral proteins have been shown to be *N*-myristoylated, others are palmitoylated (Schultz et al., 1988; Towler et al., 1988). These modifications should be taken into consideration when an expression system for a particular protein is chosen.

Available evidence indicates that membrane glycoproteins of enveloped viruses that bud from the surface of the infected cell (for example: influenza HA and NA, RSV G and F, VSV G, HIV gp160), utilize the same biosynthetic, processing and transport pathways as those employed by the cellular membrane proteins (Pferrer and Rothman, 1987). They are synthesized on membrane-bound polyribosomes and are cotranslationally translocated across the endoplasmic reticulum (ER). Translocation is directed by a signal sequence present at the N terminus of most of these membrane proteins and which is in most instances cotranslationally cleaved from the protein. In the lumen of the ER, *N*-linked high-mannose type oligosaccharides are added to the polypeptide chain, and oligomerization and disulfide bond formation take place. Subsequently, the viral glycoproteins move in a rate-determing step from the ER to the Golgi complex, where the *N*-linked oligosaccharides can be processed and modified into complex glycans, and *O*-linked oligosaccharides added. These proteins are then further transported to the plasma membrane where they remain anchored by a stop-transfer sequence usually present at the C terminus of viral membrane proteins. These very complex processes of synthesis, modification and transport are interrelated and by producing viral glycoproteins in heterologous cell systems, there is a risk of incorrect execution of one or several of the steps involved, with dramatic consequences for the overall process.

3.2. Expression in yeast

As an eukaryotic organism that is easy to manipulate, the yeast *Saccharomyces cerevisiae* is a valuable candidate for expression of foreign genes. Like bacteria, yeast can be cultivated relatively easily in large amounts. On the other hand, yeast is a typical eukaryote able to execute many post-translational modifications required for biological activity of a foreign protein.

Protein glycosylation in *S. cerevisiae* (Kukuruzinska et al. 1987; Tanner and Lehle, 1987) shares many features with that occurring in animal cells. *N*-linked glycosylation of asparagine residues is similar in yeast and mammalian cells up to the first stages of glycan processing. However, *N*-linked chains in *S. cerevisiae* are uniformly of the high-mannose type. Processing and conversion to the complex type does not take place. Instead, outer-chain mannose residues can be added resulting in a highly branched outer-chain consisting of 50–150 mannose residues. *O*-linked glycosylation of serine or threonine residues proceeds in an unusual way in *S. cerevisiae*. *O*-linked glycans consist of mannose residues (up to four) only. The first mannose transfer reaction takes place in the ER, extension to oligosaccharides most likely takes place in the Golgi. It is to be expected, therefore, that when viral glycoproteins are expressed in *S. cerevisiae*, their glycosylation pattern will be different from that in the virion.

The S protein of HBV does not contain a typical N-terminal signal peptide which is cleaved upon translocation. Instead, the S protein possesses two complex signal sequences, one N-terminal and one internal, whose conjoint action determines its final membrane topology (Eble et al., 1986, 1987). Expression of the S gene of HBV in *S. cerevisiae* (Valenzuela et al., 1982; Harford et al., 1983, 1987; Hitzeman et al., 1983; De Wilde et al., 1985; Bitter et al., 1988) results in the synthesis of unglycosylated S protein. The *N*-glycan which is present on a proportion of the S molecules synthesized in human liver cells (Fig. 7.1), is not added in *S. cerevisiae*. The glycosylation site is apparently not accessible to, or not recognized by, the yeast oligosaccharyltransferase. In contrast, both the M and L protein of HBV are found in glycosylated form upon synthesis in *S. cerevisiae*, however the glycosylation is both quantitatively and qualitatively different from the natural situation (Fig. 7.1). The natural *N*-glycosylation in the S region of these molecules is not present but the majority of M molecules and about 50% of the L molecules carry a high-mannose type glycan at the *N*-glycosylation site in the preS2 region. Furthermore, these *N*-glycosylated forms of the M and L proteins also carry several *O*-linked mannose chains (Itoh et al., 1986b; Kniskern et al., 1988; Kobayashi et al., 1988; Langley et al., 1988; Rutgers et al., 1988). *O*-linked chains are also present, although less abundantly, on the molecules that do not carry the *N*-glycan (S. Carr, personal communication).

The L protein is also modified by an *N*-myristoyl group, the *N*-glycosylated form to a lesser extent than the form not carrying the *N*-glycan (R. Biemans, personal communication). *N*-myristoylation has also been demonstrated for L protein expressed in COS cells (Persing et al., 1987). It is not yet known whether the L protein in human plasma-derived virions is also *N*-myristoylated.

HA of influenza virus is a typical viral membrane protein containing an N-terminal signal sequence and a C-terminal transmembrane anchor. Expression of the influenza HA protein in *S. cerevisiae* under constitutive promoter control resulted in signal-processing, *N*-glycosylation, and membrane-bound location of the protein indicating that the signal sequence of influenza HA functions in this yeast. Indeed, elimination of the signal sequences resulted in the synthesis of nonglycosylated HA and a cytoplasmic location. Expression of complete hemagglutinin, but not the signal-minus form, slowed cell growth and expression levels of the proteins were rather low (Jabbar et al., 1985). Expression of the HA1 sequences with the signal sequence (thus eliminating the anchor sequences and the hydrophobic fusion peptide contained in the HA2 moiety of HA) resulted in a protein that was inefficiently translocated across the ER membrane, hyperglycosylated by outer-chain mannose addition and secreted into the medium (Jabbar and Nayak, 1987). Wen and Schlesinger (1986) used an inducible promoter system to express the G protein gene of VSV. Relatively high expression levels were obtained but the cells almost stopped growing after induction. The G protein was *N*-glycosylated, no outer-chain mannose addition was observed. Palmitoylation of the G protein was not detected although *S. cerevisiae* has the enzymatic capability for palmitoylation (Wen and Schlesinger, 1984; Fujiyama and Tamanoi, 1986). The protein made in *S. cerevisiae* was isolated as multimers covalently linked through disulfide bridges (Wen et al., 1986).

The general picture arising from these results is that *S. cerevisiae* is not a suitable host for the production of viral glycoproteins. Toxic effects on the host and aberrant glycosylation limit the usefulness of the yeast system for this kind of proteins.

However, by eliminating signal sequences and expression in the form of fusion proteins, sequences of these glycoproteins can be expressed in nonglycosylated form. For instance, Barr et al. (1987) expressed sequences from the HIV *env* gene representing the majority of the polypeptide moiety of gp120 minus the signal peptide. The protein is found in nonglycosylated form in the insoluble fraction of a cell lysate and has been purified after solubilization in SDS in the presence of reducing agents. When injected into experimental animals, antibodies are elicited that react with fully glycosylated native gp120 of HIV virions. Certain, but not all, sera showed cross-neutralization with certain, but not all, HIV isolates (Steimer et al., 1988). A phase I clinical trial has been started in 1988 to establish safety and immune responses (both humoral and cellular) in humans.

Internal viral proteins which do not enter the secretory pathway can be successfully produced in yeast. The core protein of HBV has been expressed at a level of about 40% of soluble yeast protein (Kniskern et al., 1986; Imamura et al., 1988). The HIV gag precursor p55 has been expressed together with the gag-specific protease (Kramer et al., 1986). The p55 is processed efficiently into its p17, p24 and p16 components, and the p24 and p16 proteins are phosphorylated. Jacobs et al. (1989a) showed that the gag p55 precursor is *N*-myristoylated in yeast and that localization of the p55 protein at the plasma membrane requires *N*-myristoylation.

3.3. Expression in insect cells

Production of foreign proteins in insect cells utilizing an expression vector system based on baculovirus has lately gained great popularity (Luckow and Summers,

1988a). High expression levels of the foreign protein (theoretically up to 40% of the cellular protein) can be attained by placing the foreign sequences under the regulation of the polyhedrin promoter of the *Autographa californica* nuclear polyhedrosis virus (AcNPV). The recombinant viruses are used to infect cell cultures of *Spodoptera frugiperda* cells. This system has the advantage of an inherent product safety as no transformed cells or transforming elements are employed.

Insect cells appear to provide proper post-translational modifications (Luckow and Summers, 1988a), but much remains to be learned about the molecular details of these processes in insect cells. Hsieh and Robbins (1984) have demonstrated that insect cells have the capacity to process glycoproteins of the high-mannose type by partially trimming their mannose residues just before export, but the conversion to the complex type has not been demonstrated in insect cells.

Expression of the S protein of HBV by recombinant baculovirus-infected insect cells resulted in variable expression levels (Kang et al., 1987; Takehara et al., 1988; Lanford et al., 1989). Expression levels of foreign genes by recombinant AcNPV appear in some cases to be related to the length and nature of the leader sequence preceding the foreign gene (Matsuura et al., 1987; Luckow and Summers, 1988b; Lanford et al., 1989).

A proportion of the S molecules are glycosylated with unprocessed *N*-linked high-mannose oligosaccharides. The L protein is both glycosylated and myristoylated (Lanford et al., 1989). Glycosylation was also observed upon expression of influenza HA (Kuroda et al., 1986), HIV gp160 (Cochran et al., 1987; Rusche et al., 1987) or parainfluenza virus (PIV) F protein (Ray et al., 1989). These proteins are located at the cell surface of the infected insect cells. Neither HA, gp160 nor F protein is cleaved to any reasonable extent into its respective subunits (HA1 and HA2, gp120 and gp41, F1 and F2).

Immunization of experimental animals with insect cell-derived glycoproteins elicited antibodies with in vitro neutralization and fusion-blocking activities (HA, gp160, F) and the animals were partially protected against viral challenge (HA, F). Purified gp160 (Cochran et al., 1987) has been inoculated into chimpanzees. HIV neutralizing and a cell-mediated response with no toxic side effects were observed. Currently, a phase I human trial is being carried out (Megil et al., 1988).

3.4. Expression in mammalian cells

Expression of viral proteins in a cell line of the same histological type as that in which the protein is expressed during normal infection would give the best assurance for its correct folding and post-translational modification. However, as only a limited number of continuous mammalian cell lines have been established this ideal situation can rarely be attained in practice. Depending on the host cell, differences in post-translational modifications of viral proteins can be observed. For instance, in certain cases, there are differences in the sialic acid content of viral glycoproteins or the host cell proteases do not process viral proteins correctly. Furthermore, during the normal replication cycle of a virus, gene expression (splicing events) and post-translational modification of viral protein (processing by virus-encoded protease or conformation changes induced by association of different viral proteins) can be stage-specific. These events should be taken into account in preparing expression systems for a particular viral protein.

Continuous cell lines are in general transformed by oncogenic elements and may also carry endogenous retroviruses which gives reason for concern in using these cell lines for production of substances destined for human application. Current biotechnology production and analysis methods, however, should be able to control and diminish the potential risk to an acceptable level.

In addition to expression systems based on stable multiple integration of the foreign DNA in the cellular genome or a stable replicating multicopy episomal expression vector, the vaccinia vector replicating in the cytoplasm of the infected cell can be used as an antigen production system.

The S, M and L proteins of HBV have been expressed in a variety of cell lines under the control of different promoter-enhancer elements (Dubois et al., 1980; Michel et al., 1984; Patzer et al., 1984; Caputo et al., 1988; Yoneyama et al., 1988; Samanta and Youn, 1989). The secreted proteins are glycosylated in a manner similar to the plasma-derived HBV proteins as judged by their molecular weight and endoglycosidase sensitivity. No detailed structural analysis of the glycosyl chains has been reported. However Mizuochi et al. (1988) have analyzed the structure of the *N*-linked oligosaccharides present on HIV gp120, expressed and secreted as a fusion protein from chinese-hamster ovary (CHO) cells (Lasky et al., 1986). An enormous diversity of oligosaccharide structures was found including high-mannose type, hybrid type and four categories of complex chains: mono-, bi-, tri- and tetra-antennary, with or without *N*-acetyl-lactosamine repeats, and with or without a core region fucose residue. This illustrates nicely the complexity of glycosylation and the impossibility (at least at present) of predicting or controlling the events.

The gp120 used in these studies was expressed as a truncated form of the gp160 precursor protein terminating 20 amino acids downstream of the gp160 processing site and was fused to the signal sequence of glycoprotein D (gD) of herpes simplex virus in order to obtain efficient production and secretion of the protein (Lasky et al., 1986). Full length gp160 or truncated forms in which C-terminal hydrophobic domains were deleted, showed inefficient expression in CHO cells. Synthesis of full length gp160 fused to the gD signal sequence resulted in a protein that was highly glycosylated with high-mannose type chains. Although expressed abundantly, it was inefficiently exported to the cell surface and retained in an intracellular compartment (Berman et al., 1988a). Expression of gp160 by recombinant human adenovirus (Dewar et al., 1989) or vaccinia virus vectors (Chakrabarti et al., 1986; Hu et al., 1986; Kieny et al., 1986) produced plasma membrane localized, processed gp160 in several human cell lines. The extent of processing and transport is cell type dependent.

Purified gp120 fusion protein (Lasky et al., 1986) has been injected into chimpanzees. Despite repeated boosting, only low titers of neutralizing antibody and low T cell proliferation following stimulation with gp120 were detected. The immune response elicited in chimpanzees was not effective in preventing viral infection after intravenous challenge with HIV (Berman et al., 1988b).

3.5. Secretion and assembly

Expression of the S protein of HBV in mammalian cells leads to secretion into the medium of S protein containing particles which closely resemble the 22-nm particles present in human plasma from HBV patients (Dubois et al., 1980; Michel et al., 1984; Patzer et al., 1984). Selective expression of the M or L protein by recombinant vaccinia virus vectors (Cheng et al., 1986; Cheng and Moss, 1987) results in secretion of particles composed of M protein but the L protein is not secreted into the medium. Co-expression of the M and S proteins in a variety of mammalian expression systems (Michel et al., 1984; Yoneyama et al., 1988) leads to secretion of particles. However, additional expression of L protein inhibits particle secretion (Persing et al., 1986; McLachlan et al., 1987; Ou and Rutter, 1987; Molnar-Kimber et al., 1988), although particles containing S, M as well as L protein have been isolated from the culture medium of a mammalian expressing system (Youn and Samanta, 1989). Whether the blockage is at the level of assembly or secretion is not clearly established.

Delpeyroux et al. (1987, 1988a) have created in-phase insertions into different regions of the S gene. Secretion of particles is influenced by the length of the insert, its primary structure, and the site of insertion into the S molecule. Hybrid proteins which are not secreted may, nevertheless, be isolated as particles from cell extracts, indicating that assembly is not sufficient to assure secretion.

Analysis by electron microscopic methods of mammalian cells expressing the S protein demonstrated the presence of particles in the ER. Transfer from ER to Golgi seems to be the rate-limiting step in secretion. Particles isolated from cell extracts carry ER-like (high-mannose type) oligosaccharide chains while particles isolated from the medium carry the complex-type glycan (Patzer et al., 1984, 1986). In vitro translation-translocation studies of the S protein have led to a model in which the S protein is inserted into the ER membrane as an integral membrane protein. Budding into the ER lumen leads to formation of particles which are subsequently transported through the secretory organelles (Eble et al., 1986, 1987).

Expression of the S protein in insect cells (Cochran et al., 1987; Kang et al., 1987; Takehara et al., 1988; Lanford et al., 1989) does not lead to active secretion of particles (Lanford et al., 1989). Particles are found in the medium after prolonged culture (Kang et al., 1987) when the infected insect cells have become very fragile and cell rupture is frequent. Particles similar to the 22-nm particles in human plasma can be isolated from cell lysates. However, a proportion of the S protein is apparently membrane associated (Lanford et al., 1989). The distinction between association of particles to the membrane or the presence of unassembled S protein has not been made. It seems that in insect cells either budding is inefficient or particle detachment is incomplete and that transport of particles is blocked.

No secretion of particles takes place from yeast cells expressing HBV envelope proteins. Analysis of cell extracts from *S. cerevisiae* expressing the S protein showed the presence of lipoprotein particles (Valenzuela et al., 1982; Harford et al., 1983; Hitzeman et al., 1983; Miyanohara et al., 1983). M protein of HBV expressed in yeast was also found assembled into particles (Valenzuela et al., 1985; Itoh et al., 1986b; Kobayashi et al., 1988; Langley et al., 1988; Rutgers et al., 1988). Conflicting reports exist concerning the assembly of L protein. The L protein is associated with lipid (membrane) but particles similar to those formed by S and M protein are either absent or formed only to a minor extent. (Dehoux et al., 1986; Imamura et al., 1987; Kniskern et al., 1988; Rutgers et al., 1988).

The observation that the S protein expressed in yeast is not glycosylated has led to the speculation that the S protein does not enter the yeast secretory pathway and has raised questions about time and place of particle assembly (Hitzeman et al., 1983; Bitter et al., 1988). The M protein clearly enters the yeast secretory pathway and possibly reaches the Golgi apparatus in view of the presence of *O*-linked mannose chains of more than one mannose residue length on the protein (Tanner and Lehle, 1987). Co-expression of S and M proteins in the same yeast cell results in particles of mixed (S, M) subunit composition (Jacobs et al., 1989b) implying that the S protein follows the same pathway as the M protein, at least when these proteins are synthesized in the same cell.

As in the insect cell system, a substantial quantity of the S protein (and M protein) is membrane associated. Extraction with nonionic detergents at detergent/protein ratios too low for membrane solubilization (Helenius and Simons, 1975) releases particles from the membrane fraction. Apparently particle formation takes place but the particles are not completely released into the lumen. Possibly, transport of particles occurs in the lipid phase. The S protein-containing particles extracted in the presence of low detergent concentrations have biophysical and biochemical characteristics similar to the 22-nm particles derived from human plasma. Extensive and detailed antigenic and immunogenic analyses demonstrated that purified yeast-derived particles are also immunologically similar to plasma-derived particles (McAleer et al., 1984; Hauser et al., 1987; Zahradnik et al., 1987; Bitter et al., 1988).

In 1986, two HBV vaccines based on purified yeast-derived particles containing the S protein (Engerix-B, Smith Kline Biologicals; Recombivax-HB, Merck, Sharp and Dohme) were licensed for human application. Their protective efficacy is qualitatively and quantitatively similar to that obtained with plasma-derived vaccines (Zajac et al., 1986; André and Safary, 1987).

In retrospect, it appears that the main reason that these hepatitis B vaccines have become the first human vaccines developed by recombinant DNA technology, is the self-assembly of the S protein into immunogenic and otherwise innocuous particles. The observation that lack of glycosylation of the S protein does not affect the protective immunogenicity of the particles, made yeast with its easily scaled up cell growth, the most cost-effective production system (Stephenne, 1988).

The self-assembly of the S protein is a rather unique process. Other examples of assembly are the hepatitis core antigen and the HIV gag precursor p55. As mentioned before, HBcAg assembles into 27-nm particles in *E. coli* and in other host cells such as yeast (Kniskern et al. 1986; Imamura et al. 1988), insect (Takehara et al. 1988) and mammalian cells (Will et al., 1984; Roossinck et al., 1986). The particles are not secreted, they remain in the cytoplasm. The gag precursor protein p55 of HIV expressed by recombinant baculovirus-infected insect cells, assembles into budding 100–120-nm particles. Particles can be found in the medium but release seems to be inefficient. The budding at the plasma membrane requires *N*-myristoylation of the protein as a nonmyristoylated variant forms particulate structures in the cytoplasm and the nucleus (Gheysen et al., 1989 and personal communication).

Assembly of viral glycoproteins during their transport to the plasma membrane has been demonstrated for influenza HA and VSV G protein in infected cell cultures and recombinant mammalian expression systems. Several data suggest that trimerization is required for transport from the ER to the Golgi complex. However, trimerization alone is not sufficient for transport and conformation of the individual subunits seems to be a critical factor (Pferrer and Rothman, 1987; Rose and Doms, 1988). Analysis of oligomerization of other viral glycoproteins expressed in heterologous systems is generally lacking or inconclusive.

4. Antigen-presenting systems

Synthesis of monomeric protein by recombinant DNA technology seldom leads to an immunogen with good efficacy. Improvements are usually required in presentation of these antigens to render them more effective in inducing both humoral and cellular immunity. Ideally, the antigen should have a conformation that mimics the native situation in the virion where a high concentration of conformational antigen is exposed on a high-molecular-weight carrier. One approach is the incorporation of the isolated proteins in a complex with phospholipid (virosome) or the saponin Quil A (ISCOM) (Morein et al., 1987). Another approach is the use of recombinant DNA technology to solve the presentation problem at the level of antigen production.

Valenzuela et al. (1985) were the first to investigate whether the self-assembly of the hepatitis B surface protein into a complex multimeric lipoprotein particle could be exploited for the presentation of heterologous antigenic sequences. A 300-amino-acid coding region of the herpes simplex-1 viral glycoprotein D was inserted into the preS2 region of the M protein. Expression of this hybrid protein in yeast resulted in particles similar to the 22-nm particles produced by the S protein alone. Both HBsAg and gD sequences were accessible to their respective antibodies in ELISA assays but their immunogenicity was not tested. This showed that significant amounts of heterologous sequences can be fused to the S protein without interference with the self-assembly process of the S protein. Michel et al. (1988) inserted HIV gp120 sequences within the preS2 part of the M protein and showed that particles carrying a 84 amino acid sequence encoding part of a domain involved in binding to the CD4 receptor, could induce an immune response which had in vitro neutralizing activity.

Delpeyroux et al. (1986) inserted a neutralization epitope of poliovirus type 1 (amino acid residues 93–103 of VP1) in the S protein (at residue 113). Expression in mouse L cells resulted in secretion of hybrid particles which reacted with a poliovirus-neutralizing monoclonal antibody (McAbC3). Mice immunized with these hybrid particles showed a poor immune response to HBsAg and low (probably due to antigen restriction) but significant titers of polio-neutralizing antibodies. In rabbits, higher titers of polio-neutralizing antibodies were obtained which showed an increase after a booster injection (Delpeyroux et al. 1988b). However, the antiserum obtained was specific for the Mahoney strain. The Sabin type 1 strain, the Mahoney-derived attenuated strain which differs in 2 of the 11 amino acids of the VP1 peptide, was not neutralized by the rabbit antisera. In contrast, the monoclonal antibody (McAbC3) which was isolated from a mouse immunized with heated Mahoney virions neutralizes both poliovirus strains (Horaud et al., 1987). This suggests that different paratopes were preferentially induced by the hybrid polio-HBsAg particles, as compared with the natural poliovirus epitope. The low response in mice and rabbits to HBsAg is probably due to distortion of the antigenic region as a result of the insertion of the polio sequence. By co-expressing the hybrid polio-HBsAg with unmodified HBsAg, mixed particles containing both proteins were obtained which showed normal levels of HBsAg immunogenicity (Delpeyroux et al., 1988b).

In the search for possible particulate antigen carriers other systems have been developed including one based on the hepatitis B core antigen. As mentioned before the antigen self-assembles into 27-nm particles when produced in *E. coli*, yeast or animal cells (Stahl et al. 1982; Will et al., 1984; Kniskern et al. 1986; Roossinck et al., 1986; Imamura et al. 1988). Sequences encoding the neutralizing epitope of FMDV (amino acids 142–160 of VP1) were fused to the core gene, six codons upstream in the precore region, and the fusion gene was placed in a vaccinia virus vector (Clarke et al., 1987). Cell lysates from recombinant vaccinia virus infected cells contained 27-nm particles demonstrating that the FMDV sequence does not interfere with the assembly of core particles. On a weight for weight basis, 142–160 peptide on core particles is 30 to 40 times more immunogenic than peptide-β-galactosidase fusion protein containing tandem copies of the FDMV epitope (Broekhuijsen et al., 1986; 1987), 500 times more immunogenic than free peptide (Francis et al., 1987), and almost as immunogenic as inactivated virus particles. The remarkably good response might be related to the observation that the core particle itself is capable of acting as a T cell-independent as well as a T cell-dependent antigen (Milich and McLachlan, 1986; Milich et al., 1987).

Another approach is based on the yeast retrotransposon Tyl. The replication or transposition cycle involves the packaging of Ty RNA into virus-like particles (VLPs). The Ty element contains two genes called TYA and TYB. TYA is analogous to the retroviral gag gene. Expression of the TYA gene alone results in synthesis of the precursor protein pl which in the absence of the TYB products remains unprocessed but is still assembled into precursor VLPs of about 60-nm diameter (Kingsman and Kingsman, 1988). Adams et al. (1987) demonstrated that pl fusion proteins, comprising most of pl and part of HIV gp120 protein, form hybrid HIV: Ty-VLPs.

A drawback of the latter two systems (HBcAg and Ty) is that particle assembly might be dependent on interaction with nucleic acid. Nucleic acid has been detected in bacterial or yeast-derived HBcAg particles (Stahl et al., 1982; Imamura et al., 1988). Elimination of the nucleic acid binding domain might resolve this problem (Imamura et al., 1988). However, particles formed by such deleted core protein are less stable and less immunogenic, presumably because of the T cell dependence of antibody production while complete HBcAg particles function as a T cell-independent as well as T cell-dependent antigen (Milich et al., 1988).

These approaches to antigen presentation can be extended to sophisticated combinations of B and T cell epitopes by co-expression and co-assembly of different hybrid proteins in well-defined relative amounts.

5. Conclusion

From a general viewpoint, progress in the production of viral antigens by recombinant DNA technology is proceeding at a dazzling pace. Virtually any protein can be produced in large quantities, purified and studied in detail. However, to the individual researcher trying to decide which expression system to use for production of a particular protein, the choice is not obvious. As exemplified in this chapter, expression of viral glycoproteins in a structure and conformation identical to that in the native virion, is to a large extent a process of trial and error. Signal sequences are not functionally equivalent in all host cells, conformational variations or anchor sequences can provoke blockage of the protein in intermediate secretory organelles, glycosylation can be aberrant or lacking, processing can be absent.... At present, production still has to be tried in several expression systems, gaining on the way precious information about the protein to be expressed and the cell system employed.

Once antigens are produced, one is confronted with problems of immunogenicity and protective efficacy with respect to vaccine development. Several antigen formulation and presenting systems may have to be tried to obtain information about the epitopes involved in protective immunity and the mechanisms and regulation of the immune response.

The vaccine against HBV is an exception in the sense that the S protein, self-assembles into a particle exposing protective conformational epitopes at a high density on the surface of a high-molecular-weight structure. The application of this self-assembly mechanism to expose other well-defined heterologous protective epitopes to the immune system, is an exciting possibility for future vaccine development.

6. Acknowledgements

We wish to thank our colleagues C. Bruck, D. Gheysen, N. Harford, E. Jacobs and C. Locht for critical reading of the manuscript. We are especially grateful to Catherine Van Roy and Hélène Monnig for expert preparation of the manuscript. Research at Smith Kline Biologicals has been sponsored in part by grants from the Institut pour l'Encouragement de la Recherche Scientifique dans l'Industrie et l'Agriculture (IRSIA) and the Walloon region (STN-1363).

7. References

Adams, S.E., Dawson, K.M., Gull, K., Kingsman, S.M. and Kingsman, A.J. (1987) *Nature* 329, 68–70.

André, F.E. and Safary, A. (1987) *Postgrad. Med. J.* 63 (Suppl. 2), 169–177.

Bangham, C.R.M., Openshaw, P.J.M., Ball, L.A., King, A.M.O., Wertz, G.W. and Askonas, B.A. (1986) *J. Immunol.* 137, 3973–3977.

Barr, P.J., Steimer, K.S., Sabin, E.A., Parkes, D., George-Nascimento, C., Stephans, J.C., Powers, M.A., Gyenes, A., Van Nest, G.A., Miller, E.T., Higgins, K.W. and Luciw, P.A. (1987) *Vaccine* 5, 90–101.

Berman, P.W., Nunes, W.M. and Haffar, O.K. (1988a) *J. Virol.* 62, 3135–3142.

Berman, P.W., Groopman, J.E., Gregroy, T., Clapham, P.R., Weiss, R.A., Ferriani, R., Riddle, L., Shimasaki, C., Lucas, C., Lasky, L.A. and Eichberg J.W. (1988b) *Proc. Natl. Acad. Sci. USA* 85, 5200–5204.

Bitter, G.A., Egan, K.M., Burnette, W.N., Samal, B., Fieschko, J.C., Peterson, D.L., Downing, M.R., Wypych, J. and Langley, K.E. (1988) *J. Med. Virol.* 25, 123–140.

Bittle, J.L., Houghten, R.A., Alexander, H., Shinnick, T.M., Sutcliffe, J.G., Lerner, R.A., Rowlands, D.J. and Brown, F. (1982) *Nature* 298, 30–33.

Broekhuijsen, M.P., Blom, T., van Rijn, J., Pouwels, P.H., Klasen, E.A., Fasbender, M.J. and Enger-Valk, B.E. (1986) *Gene* 49, 189–197.

Broekhuijsen, M.P., van Rijn, J.M.M., Blom, A.J.M., Pouwels, P.H., Enger-Valk, B.E., Brown, F. and Francis, M.J. (1987) *J. Gen. Virol.* 68, 3137–3143.

Bruguera, M., Cremades, M., Mayor, A., Sanchez Tapia, J.M. and Rodès, J. (1987) *Postgrad. Med. J.* 63 (Suppl. 2), 155–158.

Cabradilla, C.D., Groopman, J.E., Lanigan, J., Renz, M., Lasky, L.A. and Capon, D.J. (1986) *Bio/Technology* 4, 128–133.

Caputo, A., Barbanti-Brodano, G., Reschiglian, P., Gianni, M., Mottes, M., Miranda, P., Milanesi, G., Knowles, B.B. and Ricciardi, R.P. (1988) *J. Gen. Virol.* 69, 459–466.

Chang, T.W., Kato, I., McKinney, S., Chanda, P., Barone, A.D., Wong-Staal, F., Gallo, R.C. and Chang, N.T. (1985) *Bio/Technology* 3, 905–909.

Chanh, T.C., Dreesman, G.R., Kanda, P., Linette, G.P., Sparrow, J.T., Ho, D.D. and Kennedy, R.C. (1986) *EMBO J.* 5, 3065–3071.

Chakrabarti, S., Robert-Guroff, M., Wong-Staal, F., Gallo, R.C. and Moss, B. (1986) *Nature* 320, 535–537.

Cheng, K.C. and Moss, B. (1987) *J. Virol.* 61, 1286–1290.

Cheng, K.C., Smith, G.L., and Moss, B. (1986) *J. Virol.* 60, 337–344.

Clarke, B.E., Newton, S.E., Carroll, A.R., Francis, M.J., Appleyard, G., Syred, A.D., Highfield, P.E., Rowlands, D.J. and Brown, F. (1987) *Nature* 330, 381–384.

Cochran, M.A., Ericson, B.L., Knell, J.D. and Smith, G.E. (1987) In *Vaccines 87: Modern Approaches to New Vaccines* (R.M. Chanock, R.A. Lerner, F. Brown and H. Ginsberg, Eds.) Cold Spring Harbor Laboratory, Cold Spring Harbor NY, pp. 384–388.

Cohen, B.J., Litton, P.A., Mortimer, P. and Simmons, P. (1984) *J. Hyg.* 93, 225–232.

Crowl, R., Ganguly, K., Gordon, M., Conroy, R., Schaber, M., Kramer, R., Shaw, G., Wong-Staal, F. and Reddy, E.P. (1985) *Cell* 41, 979–986.

Dalgleish, A.G., Chanh, T.C., Kennedy, R.C., Kanda, P., Clapham, P.R. and Weiss, R.A. (1988) *Virology* 165, 209–215.

Davis, A.R., Bos, T., Ueda, M., Nayak, D.P., Dowbenko, D. and Compans, R.W. (1983) *Gene* 21, 273–284.

Dawson, G.J., Heller, J.S., Wood, C.A., Gutierrez, R.A., Webber, J.S., Hunt, J.C., Hojvat, S.A., Senn, D., Devare, S.G. and Decker, R.H. (1988) *J. Infect. Dis.* 157, 149–155.

Dehoux, P., Ribes, V., Sobczak, E. and Streeck, R.E. (1986) *Gene* 48, 155–163.

Delpeyroux, F., Chenciner, N., Lim, A., Malpièce, Y., Blondel, B., Crainic, R., van der Werf, S. and Streeck, R.E. (1986) *Science* 233, 472–475.

Delpeyroux, F., Chenciner, N., Lim, A., Lambert, M., Malpièce, Y. and Streeck, R.E. (1987) *J. Mol. Biol.* 195, 343–350.

Delpeyroux, F., Crainic, R., Blondel, B., Horaud, F., Van der Werf, S., Girard, M., Lagarde, D., Mazert, M.C. and Streeck, R.E. (1988a) *Biochimie* 70, 1065–1073.

Delpeyroux, F., Peillon, N., Blondel, B., Crainic, R. and Streeck, R.E. (1988b) *J. Virol.* 62, 1836–1839.

Dewar, R.L., Natarajan, V., Vasudevachari, M.B. and Salzman, N.P. (1989) *J. Virol.* 63, 129–136.

De Wilde, M., Cabezon, T., Harford, N., Rutgers, T., Simoen, E. and Van Wijnendaele (1985) *Dev. Biol. Stand.* 59, 99–107,

Dowbenko, D.J., Bell, J.R., Benton, C.V., Groopman, J.E., Nguyen, H., Vetterlein, D., Capon, D.J. and Lasky, L.A. (1985) *Proc. Natl. Acad. Sci. USA* 82, 7748–7752.

Dubois, M.F., Pourcel, C., Rousset, S., Chany, C. and Tiollais, P. (1980) *Proc. Natl. Acad. Sci. USA* 77, 4549–4553.

Eble, B.E., Lingappa, V.R. and Ganem, D. (1986) *Mol. Cell. Biol.* 1454–1463.

Eble, B.E., MacRae, D.R., Lingappa, V.R. and Ganem, D. (1987) *Mol. Cell. Biol.* 3591–3601.

Francis, M.J., Fry, C.M., Rowlands, D.J., Brown, F., Bittle, J.L., Houghten, R.A. and Lerner, R.A. (1985) *J. Gen. Virol.* 66, 2347–2354.

Francis, M.J., Fry, C.M., Rowlands, D.J., Bittle, J.L., Houghten, R.A., Lerner, R.A. and Brown, F. (1987) *Immunology* 61, 1–6.

Fujisawa, Y., Ito, Y., Sasada, R., Ono, Y., Igarashi, K., Marumoto, R., Kikuchi, M. and Sugino, Y. (1983) *Nucleic Acids Res.* 11, 3581–3591.

Fujiyama, A. and Tamanoi, F. (1986) *Proc. Natl. Acad. Sci. USA* 83, 1266–1270.

Gheysen, D., de Foresta, F., Jacobs, E., Thines, D., Francotte, M., Van Wijnendaele, F., Bruck, C., Pêtre, J. and De Wilde, M. (1989) In *Advances in Animal Cell Biology and Biotechnology for Bioprocesses* (J. Bryant, Ed.) Butterworth Scientific, London. pp. 501–507.

Hadler, S.C., Francis, D.P., Maynard, J.E., Thompson, S.E., Judson, F.N., Echenberg, D.F., Ostrow, D.G., O'Malley, P.M., Penley, K.A., Altman, N.L., Braff, E., Shipman, G.F., Coleman, P.J. and Mandel, E.J. (1986) *New Engl. J. Med.* 315, 209–215.

Hahn, B.H., Shaw, G.M., Taylor, M.A., Redfield, R.R., Markham, P.D., Salahuddin, S.Z., Wong-Staal, F., Gallo, R.C., Parks, E.S. and Parks, W.P. (1986) *Science* 232, 1548–1553.

Harford, N., Cabezon, T., Crabeel, M., Simoen, E., Rutgers, A. and De Wilde, M. (1983) *Dev. Biol. Stand.* 54, 125–130.

Harford, N., Cabezon, T., Colau, B., Delisse, A.M., Rutgers, T. and De Wilde, M. (1987) *Postgrad. Med. J.* 63 (Suppl. 2), 65–70.

Hauser, P., Voet, P., Simoen, E. Thomas, H.C., Pêtre, J., De Wilde, M. and Stephenne, J. (1987) *Postgrad. Med. J.* 63 (Suppl. 2), 83–91.

Heermann, K.H., Goldmann, U., Schwartz, W., Seyffarth, T., Baumgarten, H. and Gerlich, W.H. (1984) *J. Virol.* 52, 396–402.

Helenius, A. and Simons, K. (1975) *Biochim. Biophys. Acta* 415, 29–79.

Hitzeman, R.A., Chen, C.Y., Hagie, F.E., Patzer, E.J., Liu, C., Estell, D.A., Miller, J.V., Yaffe, A., Kleid, D.G., Levinson, A.D. and Oppermann, H. (1983) *Nucleic Acids Res.* 11, 2745–2763.

Hollinger, F.B., Troisi, C., Heiberg, D., Sanchez, Y., Dreesman, G.R. and Melnick, J.L. (1986) *J. Med. Virol.* 19, 229–240.

Horaud, F., Crainic, R., Van der Werf, S., Blondel, B., Wichowski, C., Akacem, O., Bruneau, P., Couillin, P., Siffert, O. and Girard, M. (1987) *Prog. Med. Virol.* 34, 129–155.

Howard, C.R., Young, P., Steward, M.W. and Zuckerman, A.J. (1985) *Dev. Biol. Stand.* 59, 89–98.

Hsieh, P. and Robbins, P.W. (1984) *J. Biol. Chem.* 259, 2375–2382.

Hu, S.L., Kosowski, S.G., Dalrymple, M. (1986) *Nature* 320, 537–540.

Imamura, T., Araki, M., Miyanohara, A., Nakao, J., Yonemura, H., Ohtomo, N. and Matsubara, K. (1987) *J. Virol.* 61, 3543–3549.

Imamura, T., Sugahara, K., Adachi, S., Miyatsu, Y., Mizokami, H. and Matsusaka, T. (1988) *J. Biotechno.* 8, 149–162.

Itoh, Y., Takai, E., Ohnuma, H., Kitajima, K., Tsuda, F., Machida, A., Mishiro, S., Nakamura, T. Miyakawa, Y. and Mayumi, M. (1986a) *Proc. Natl. Acad. Sci. USA* 83, 9174–9178.

Itoh, Y., Hayakawa, T. and Fujisawa, Y. (1986b) *Biochem. Biophys. Res. Commun.* 138, 268–274.

Jabbar, M.A. and Nayak, D.P. (1987) *Mol. Cell. Biol.* 7, 1476–1485.

Jabbar, M.A., Sivasubramanian, N. and Nayak, D.P. (1985) *Proc. Natl. Acad. Sci. USA* 82, 2019–2023.

Jacobs, E., Gheysen, D., Thines, D., Francotte, M. and De Wilde, M. (1989a) *Gene* 79, 71–81.

Jacobs, E., Rutgers, T., Voet, P., Dewerchin, M., Cabezon, T. and De Wilde, M. (1989b) *Gene* 80, 279–291.

Kang, C.Y., Bishop, D.H.L., Seo, J-S., Matsuura, Y. and Choe, M. (1987) *J. Gen. Virol.* 68, 2607–2613.

Kieny, M.P., Rautmann, G., Schmitt, D., Dott, K., Wain-Hobson, S., Alizon, M., Girard, M., Chamaret, S., Laurent, A., Montagnier, L. and Lecocq, J.P. (1986) *Biotechnology* 4, 790–795.

Kingsman, S.M. and Kingsman, A.V. (1988) *Vaccine* 6, 304–306.

Kleid, D.G., Yansura, D.G., Small, B., Dowbenko, D., Moore, D.M., Grubman, M.J., McKercher, P.D., Morgan, D.O., Robertson, B.H. and Bachrach, H.L. (1981) *Science* 214, 1125–1129.

Kniskern, P.J., Hagopian, A., Montgomery, D.L., Burke, P., Dunn, N.R., Hofmann, K.J., Miller, W.J. and Ellis, R.W. (1986) *Gene* 46, 135–141.

Kniskern, P.J., Hagopian, A., Burke, P., Dunn, N., Emini, E.A., Miller, W.J. Yamazaki, S. and Ellis, R.W. (1988) *Hepatology* 8, 82–87.

Kobayashi, M., Asano, T., Utsunomiya, M., Itoh, Y., Fujisawa, Y., Nishimura, O., Kato, K. and Kakinuma, A. (1988) *J. Biotechnol.* 8, 1–22.

Kramer, R.A., Schaber, M.D., Skalka, A.M., Ganguly, K., Wong-Staal, F. and Reddy, E.P. (1986) *Science* 231, 1580–1584.

Kukuruzinska, M.A., Bergh, M.L.E. and Jackson, B.J. (1987) *Annu. Rev. Biochem.* 56, 915–944.

Kuroda, K., Hauser, C., Roff, R. Klenk, H.D. and Doerfler, W. (1986) *EMBO J.* 5, 1359–1365.

Lanford, R.E., Luckow, V., Kennedy, R.C., Dreesman, G.R., Notvall, L. and Summers, M.D. (1989) *J. Virol.* 63, 1549–1557.

Langley, K.E., Egan, K.M., Barendt, J.M., Parker, C.G. and Bitter, G.A. (1988) *Gene* 67, 229–245.

Lasky, L.A., Groopman, J.E., Fennie, C.W., Benz, P.M., Capon, D.J., Dowbenko, D.J., Nakamura, G.R., Nunes, W.M., Renz, M.E. and Berman, P.W. (1986) *Science* 233, 209–212.

Luckow, V.A. and Summers, M.D. (1988a) *Bio/Technology* 6, 47–55.

Luckow, V.A. and Summers, M.D. (1988b) *Virology* 167, 56–71.

McAleer, W.J., Buynak, E.B., Maigetter, R.Z., Wampler, D.E., Miller, W.J. and Hilleman, M.R. (1984) *Nature* 307, 178–180.

McLachlan, A., Milich, D.R., Raney, A.K., Riggs, M.G., Hughes, J.L., Sorge, J. and Chisari, F.V. (1987) *J. Virol.* 61, 683–692.

Matsuura, Y., Posse, R.D., Overton, H.A. and Bishop, D.H.L. (1987) *J. Gen. Virol.* 68, 1233–1250.

Megil, M.E., Deyton, L., Metcalf, J.A., Salzman, N., Vasudevachari, M., Baseler, M., Martin, M., Folks, T.M., Smith, G., Volvovitz, F., Masur, H. Fauci, A.S. and Lane, H.C. (1988) *IV International Conference on AIDS. Stockholm* (Abstract 6569).

Michel, M.L., Pontisso, P., Sobczak, E., Malpièce, Y., Streeck, R.E. and Tiollais, P. (1984) *Proc. Natl. Acad. Sci. USA* 81, 7708–7712.

Michel, M.L., Mancini, M., Sobczak, E., Favier, V., Guetard, D., Bahraoui, E.M. and Tiollais, P. (1988) *Proc. Natl. Acad. Sci. USA* 85, 7957–7961.

Milich, D.R. (1988) *Immunol. Today* 9, 380–386.

Milich, D.R. and McLachlan, A. (1986) *Science* 234, 1398–1401.

Milich, D.R., McLachlan, A., Thornton, G.B. and Hughes, J.L. (1987) *Nature* 329, 547–549.

Milich, D.R., McLachlan, A., Stahl, S., Wingfield, P., Thornton, G.B., Hughes, J.L. and Jones, J.E. (1988) *J. Immunol.* 141, 3617–3624.

Mishiro, S., Imai, M., Takahashi, K., Machida, A., Gotanda, T., Miyakawa, Y. and Mayumi, M. (1980) *J. Immunol.* 124, 1589–1593.

Miyanohara, A., Toh-e, A., Nozaki, C., Hamada, F., Ohtomo, N. and Matsubara, K. (1983) *Proc. Natl. Acad. Sci. USA* 80, 1–5.

Mizuochi, T., Spellman, M.W., Larkin, M., Solomon, J., Basa, L.J. and Feisi, T. (1988) *Biochem. J.* 254, 599–603.

Molnar-Kimber, K.L., Jarocki-Witek, V., Dheer, S.K., Vernon, S.K., Conley, A.J., Davis, A.R. and Hung, P.P. (1988) *J. Virol.* 62, 407–416.

Morein, B., Lövgren, K., Höglund, S. and Sundquist, B. (1987) *Immunol. Today* 8, 333–338.

Murphy, B.R., Alling, D.W., Snyder, M.H., Walsh, E.E., Prince, G.A., Chanock, R.M., Hemming, V.G., Rodriguez, W.J., Kim, H.W., Graham, B.S. and Wright, P.F. (1986a) *J. Clin. Microbiol.* 24, 894–898.

Murphy, B.R., Graham, B.S., Prince, G.A., Walsh, E.E., Chanock, R.M. Karzon, D.T. and Wright, P.F. (1986b) *J. Clin. Microbiol.* 23, 1009–1014.

Murphy, B.R., Prince, G.A., Walsh, E.E., Kim, H.W., Parrott, R.H., Hemming, V.G., Rodriguez, W.J. and Chanock, R.M. (1986c) *J. Clin. Microbiol.* 24, 197–202.

Murphy, B.R., Prince, G.A., Collins, P.L., Van Wyke Coelingh, K., Olmsted, R.A., Spriggs, M.K., Parrott, R.H., Kim, H.W., Brandt, C.D. and Chanock, R.M. (1988a) *Virus Res.* 11, 1–15.

Murphy, B.R., Olmsted, R.A., Collins, P.L., Chanock, R.M. and Prince, G.A. (1988b) *J. Virol.* 62, 3907–3910.

Murray, K., Bruce, S.A., Hinnen, A., Wingfield, P., van Erd, P.M.C.A., de Reus, A. and Schellekens, H. (1984) *EMBO J.* 3, 645–650.

Murray, K., Bruce, S.A., Wingfield, P., van Eerd, P., de Reus, A. and Schellekens, H. (1987) *J. Med. Virol.* 23, 101–107.

Neurath, A.R., Kent, S.B.H., Parker, K., Prince, A.M., Strick, N., Brotman, B. and Sproul, P. (1986) *Vaccine* 4, 35–37.

Neurath, A.R., Strick, N., Kent, S.B.H., Parker, K., Seto, B. and Girard, M. (1988) In *Vaccines 88: New Chemical and Genetic Approaches to Vaccination* (H. Ginsberg, F. Brown, R.A. Lerner and R.M. Channock eds.) Cold Spring Harbor Laboratory, Cold Spring Harbor NY. pp. 229–234.

Nixon, D.F., Townsend, A.R.M., Elvin, J.G., Rizza, C.R., Gallwey, J. and McMichael, A.J. (1988) *Nature* 336, 484–487.

Olmsted, R.A., Buller, R.M.L., Collins, P.L., London, W.T., Beeler, J.A., Prince, G.A., Chanock, R.M. and Murphy, B.R. (1988) *Vaccine* 6, 519–524.

Ou, J.H. and Rutter, W.J. (1987) *J. Virol.* 61, 782–786.

Papsidero, L.D., Sheu, M. and Ruscetti, F.W. (1989) *J. Virol.* 63, 267–272.

Patzer, E.J., Nakamura, G.R. and Yaffe, A. (1984) *J. Virol.* 51, 346–353.

Patzer, E.J., Nakamura, G.R., Simonsen, C.C., Levinson, A.D. and Brans, R. (1986) *J. Virol.* 58, 884–892.

Pemberton, R., Cannon, M., Openshaw, P., Ball, L.A., Wertz, G.W. and Askonas, B. (1987) *J. Gen. Virol.* 68, 2177–2182.

Persing, D.H., Varmus, H.E. and Ganem, D. (1986) *Science* 234, 1388–1391.

Persing, D.H., Varmus, H.E. and Ganem, D. (1987) *J. Virol.* 61, 1672–1677.

Pfeffer, S.R. and Rothman, J.E. (1987) *Annu. Rev. Biochem.* 56, 829–852.

Prince, G.A., Jenson, A.B., Hemming, V.G., Murphy, B.R., Walsh, E.E., Horswood, R.L. and Chanock, R.M. (1986) *J. Virol.* 57, 721–728.

Putney, S.D., Matthews, T.J., Robey, W.G., Lynn, D.L., Robert-Guroff, M., Mueller, W.T., Langlois, A.J., Ghrayeb, J., Petteway, S.R., Jr., Weinhold, K.J., Fischinger, P.J., Wong-Staal, F., Gallo, R.C. and Bolognesi, D.P. (1986) *Science* 234, 1392–1395.

Ray, R., Galinski, M.S. and Compans, R.W. (1989) *Virus Res.* 12, 169–180.

Robey, W.G., Arthur, L.O., Matthews, T.J., Langlois, A., Copeland, T.D., Lerche, N.W., Oraszian, S., Bolognesi, D.P., Gilden, R.V., Fischinger, P.J. (1986) *Proc. Natl. Acad. Sci. USA* 83, 7023–7027.

Roossinck, M., J., Jameel, S., Loukin, S.H. and Siddiqui, A. (1986) *Mol. Cell. Biol.* 6, 1393–1400.

Rose, J.K. and Shafferman, A. (1981) *Proc. Natl. Acad. Sci. USA* 78, 6670–6674.

Rose, J.K. and Doms, R.W. (1988) *Annu. Rev. Cell. Biol.* 4, 257–288.

Rusche, J.R., Lynn, D.L., Robert-Guroff, M., Langlois, A.J., Lyerly, H.K., Carson, H., Krohn, K., Ranki, A., Gallo, R.C., Bolognesi, D.P., Putney, S.D. and Matthews, T.J. (1987) *Proc. Natl. Acad. Sci. USA* 84, 6924–6928.

Rusche, J.R., Javaherian, K., McDanal, C., Petro, J., Lynn, D.L., Grimaila, R., Langlois, A., Gallo, R.C., Arthur, L.O., Fischinger, P.J., Bolognesi, D.P., Putney, S.D., Mattheus, T.J. (1988) *Proc. Natl. Acad. Sci. USA* 85, 3198–3202.

Rutgers, T., Cabezon, T., Harford, N., Vanderbrugge, D., Descurieux, M., Van Opstal, O., Van Wijnendaele, F., Hauser, P., Voet, P. and De Wilde, M. (1988) In *Viral Hepatitis and Liver Disease* (A.J. Zuckerman, Ed.) Alan R. Liss, New York, pp. 304–308.

Samanta, H. and Youn B.W. (1989) *Vaccine* 7, 69–76.

Schultz, A.M., Henderson, L.E. land Oroszlan, S. (1988) *Annu. Rev. Cell. Biol.* 4, 611–647.

Skelly, J., Howard, C.R. and Zuckerman, A.J. (1981) *Nature* 290, 51–54.

Stahl, S., MacKay, P., Magazin, M., Bruce, S.A. and Murray, K. (1982) *Proc. Natl. Acad. Sci. USA* 79, 1606–1610.

Steimer, K.S., Van Nest, G.A., Haigwood, N.L., Tillson, E.M., George-Nascimento, C., Barr, P.J. and Dina, D. (1988) In *Vaccines 88: New Chemical and Genetic Approaches to Vaccination* (H. Ginsberg, F. Brown, R.A. Lerner and R.M. Channock, Eds.) Cold Spring Harbor Laboratory, Cold Spring Harbor, NY, pp. 347–355.

Stéphenne, J. (1988) *Vaccine* 6, 299–303.

Stibbe, W. and Gerlich, W.H. (1982) *Virology* 123, 436–442.

Stibbe, W. and Gerlich, W.H. (1983) *J. Virol.* 46, 626–628.

Stott, E.J., Taylor, G., Ball, L.A., Anderson, K., King, A. and Wertz, G. (1987) *J. Virol.* 61, 3855–3861.

Takehara, K., Ireland, D. and Bishop, D.H.L. (1988) *J. Gen. Virol.* 69, 2763–2777.

Tanner, W. and Lehle, L. (1987) *Biochem. Biophys. Acta* 906, 81–99.

Taylor, G., Stott, E.J., Bew, M., Fernie, B.F., Cote, P.J., Collins, A.P., Hughes, M. and Jebbett, J. (1984) *Immunology* 52, 137–142.

Tiollais, P., Pourcel, C. and Dejean, A. (1985) *Nature* 317, 489–495.
Towler, D.A., Gordon, J.I., Adams, S.P. and Glaser, L. (1988) *Annu. Rev. Biochem.* 57, 69–99.
Valenzuela, P., Medina, A., Rutter, W.J., Ammerer, G. and Hall, B.D. (1982) *Nature* 298, 347–350.
Valenzuela, P., Coit, D., Medina-Selby, M.A., Kuo, C.H., Van Nest, G., Burke, R.L., Bull, P., Urdea, M.S. and Graves, P.V. (1985) *Biotechnology* 3, 323–326.
Walker, B.D., Chakrabarti, S., Moss, B., Paradis, T.J., Flynn, T., Durno, A.G., Blumberg, R.S., Kaplan, J.C., Hirsch, M.S. and Schooley, R.T. (1987) *Nature* 328, 345–348.
Walker, B.D., Flexner, C., Paradis, T.J., Fuller, T.C., Hirsch, M.S., Schooley, R.T. and Moss, B. (1988) *Science* 240, 64–66.
Wen, D. and Schlesinger, M.J. (1984) *Mol. Cell. Biol.* 4, 688–694.
Wen, D. and Schlesinger, M.J. (1986) *Proc. Natl. Acad. Sci. USA* 83, 3639–3643.
Wen, D., Ding, M. and Schlesinger, M.J. (1986) *Virology* 153, 150–154.
Wertz, G.W., Stott, E.J., Young, K.K.Y., Anderson, K. and Ball, L.A. (1987) *J. Virol.* 61, 293–301.
Will, H., Cattaneo, R., Pfaff, E., Kuhn, C., Roggendorf, M. and Schaller, H. (1984) *J. Virol.* 50, 335–342.
Yoneyama, T., Akatsuka, T. and Miyamura, T. (1988) *J. Gen. Virol.* 69, 1931–1939.
Youn, B.W. and Samanta, H. (1989) *Vaccine* 7, 60–68.
Zabraknik, J.M., Couch, R.B. and Gerin, J.L. (1987) *J. Infect. Dis.* 155, 903–908.
Zajac, B.A., West, D.J., McAleer, W.J. and Scolnick, E.M. (1986) *J. Infect.* 13 (Suppl. A) 39.

Eds. M.H.V. van Regenmortel and A.R. Neurath
Immunochemistry of viruses, II. The basis for serodiagnosis and vaccines.

Live recombinant viral vaccines 8

J. TARTAGLIA and E. PAOLETTI

Virogenetics Corporation, 465 Jordan Road, Rensselaer Technology Park, Troy, NY 12180, U.S.A.

1. Introduction

New vaccine types are greatly needed in both the human and veterinary fields to either complement or replace existing conventional vaccines. This is due either to the inability to generate safe and effective vaccines by conventional means or to the inherent limitations of conventional vaccines. Advances in molecular biology and biochemical procedures have brought forth a new era in the development of live and inactivated vaccines, which hold promise for overcoming the limitations of conventional vaccines. Table 8.1 summarizes these new potential vaccine types and provides recent reviews as sources for information pertaining to these approaches.

Animal viruses such as adenoviruses, herpesviruses and poxviruses have been engineered as vaccine candidates. These vectors may provide advantages over other approaches since upon viral replication the foreign gene product is amplified, in-

TABLE 8.1

New potential vaccine types

Synthetic peptides (Arnon, 1987)
Subunit vaccines based on expression in bacteria, yeast, and mammalian cells (Zanetti et al., 1987)
Anti-idiotype immunogen (Dalgleish and Kennedy, 1988)
Immunostimulating complexes (Morein et al., 1987)
Live recombinant vectors such as bacteria and viruses (Nakayama et al., 1988)

creasing the antigen load. Additionally, foreign antigens should be processed and presented to immune effector cells in a manner similar to natural infection. This chapter discusses the engineering of adenoviruses, herpesviruses (cytomegalovirus, herpes simplex virus, pseudorabies virus, varicella-zoster virus), and poxviruses (vaccinia and fowlpox virus) as expression vectors and examines their potential use as live recombinant vaccine candidates for both human and veterinary use. A brief description of the biology of these virus families is also provided for an understanding of their molecular biology and potential limitations as live vectors.

2. Adenovirus vectors

2.1. Biology of adenovirus

Adenoviruses have been isolated from mammalian and avian species (reviewed by Horwitz, 1985a,b). Approximately 40 human serotypes have been identified. Although a number cause acute respiratory illness such as pharyngitis, pneumonia, conjunctivitis, epidemic keratoconjunctivitis, hemorrhagic cystitis, and acute gastroenteritis in susceptible populations, in general, the adenoviruses have not been shown to cause severe human disease.

The human adenovirus serotypes 2 and 5 (Ad2 and Ad5) are the most well analyzed members of this viral family. Adenoviruses contain a linear double-stranded DNA genome of approximately 36 kb, replicate within the cytoplasm of infected cells and utilize many host functions to synthesize virus-specified products. Viral gene products are designated as early or late on the basis of their sensitivity to DNA synthesis inhibitors (Logan and Shenk, 1982). Temporal regulation of adenovirus gene expression is mediated by both transcriptional and post-transcriptional mechanisms. At very early times post-infection the E1 region (Fig. 8.1) is transcribed. Mature E1 mRNAs resulting from differential mRNA splicing schemes encode the EIA and EIB polypeptides. The EIA early gene products act as regulatory proteins and induce the expression of the other early transcription units E2, E3, and E4 (Fig. 8.1).

The progression to the late stage of the adenovirus lytic replicative cycle occurs upon the onset of DNA replication. Early gene products are required for the switch to late gene expression. E2-specified gene products which are generated from alternate splicing patterns are involved in the replication of the adenovirus genome. E2A encodes a 72-kDa DNA binding protein (Van der Vliet et al., 1975; Van der Vliet and Levine, 1973) while E2B encodes an 87-kDa polypeptide which becomes covalently linked to the 5′-termini of the viral genome (Rekosh et al., 1977, Robinson et al., 1973). The E2 region also encodes the 140-kDa viral DNA polymerase (Lichy et al., 1982). Similarly, the E1B and E4 specified products contribute to the switch from early to late adenovirus gene expression. Both transcription units encode prod-

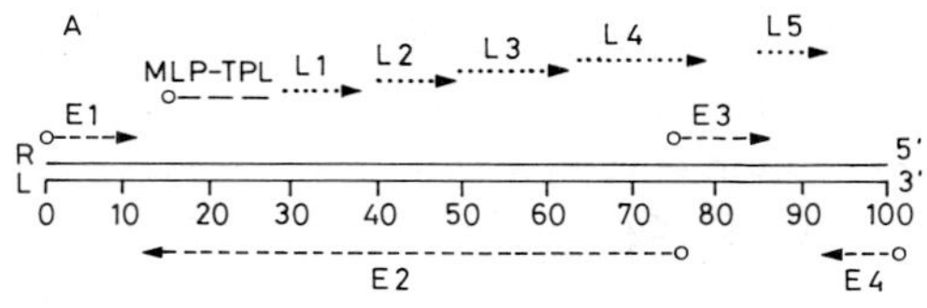

Fig. 8.1. Schematic representation of the adenovirus transcription map. The 36-kb adenovirus genome is divided into 100 map units. Early transcription units (E1, E2, E3, and E4) are represented by short dashed lines with arrowheads indicating the direction of transcription. Open circles represent sites of transcription initiation. Each of these units give rise to families of mRNAs derived by differential mRNA splicing events. Late transcription units (L1 to L5) are represented by dotted lines with arrowheads again indicating the direction of transcription. The L1 to L5 mRNAs are generated by differential splicing and polyadenylation site utilization from a major late transcription unit which extends from map units 16.4 to map unit >99 (Logan and Shenk, 1982). The major late promoter (MLP) and tripartite leader (TPL) are represented by long dashed lines. Although the VA RNAs are not shown in this figure, they are transcribed in a rightward direction between map units 30 and 35. Used by permission of Hung et al. (1988).

ucts responsible for the stabilization and transport of late mRNAs and the shutoff of host macromolecular synthesis (Sandler and Ketner, 1989; Bridge and Ketner, 1989; Babiss et al., 1985).

The E3 region appears as the only early gene product dispensable for viral replication in cell culture. The 19-kDa gene product may, however, affect growth *in vivo,* since it has been shown to directly abrogate the expression of the major histocompatibility locus (MHC) class I molecules on the surface of infected cells (Andersson et al., 1985; Burgert and Kvist, 1985; Severinsson and Peterson, 1983). It has also been demonstrated that deletion of the E3 region increases the pathogenicity of the virus, possibly by enhancing the responsiveness of cytotoxic T cells (CTL) and facilitating the spread of virus within the respiratory tract (Ginsberg et al., 1987).

Upon the shift to late adenovirus gene expression, transcription is no longer initiated from the early promoter regions. Late transcription is initiated at the major late promoter (MLP) and is extended towards the right end of the genome (Fig. 8.1). Five individual gene products are subsequently generated by differential mRNA splicing events which join the common 5′-untranslated tripartite leader (TPL) to individual genes encoding the virion structural components.

Translation of the late adenovirus mRNAs is extremely efficient and outcompetes translation of early mRNAs and host cell messages. This translational control can be attributed, at least in part, to the 200-nucleotide TPL found on all late transcripts (Kaufman, 1985; Berkner and Sharp, 1985; Berget et al., 1977). Possible mechanisms for this property of more efficient translation of the TPL-containing mRNAs may stem from a decreased requirement for the cap-binding protein complex (Dolph et al., 1988), an increase in mRNA transport, and/or an increase in the half-life of these mRNAs (Moore and Shenk, 1988). It should be noted that sequences within the TPL have been found to enhance transcriptional levels of adenovirus mRNAs (Alonso-Caplen et al., 1988). Efficient translation of adenoviral late mRNAs also requires the expression of virus-associated (VA) RNAs which map at approximately position 30 (Fig. 8.1). Transcription of these RNAs is initiated by the host cell RNA polymerase III (Weinmann et al., 1974; Reich et al., 1966). Adenovirus deletion mutants lacking these RNAs produce wild-type levels of viral mRNAs but show a 10-fold decrease in translation (Thimmappaya et al., 1982) due to a defect in initiation complex formation (Siekierka et al., 1985).

2.2. Generation of adenovirus recombinants

The generation of adenovirus recombinants is facilitated by the infectious nature of isolated viral DNA. Conventional techniques to introduce isolated exogenous genetic material into mammalian cells (i.e., Ca^{2+}-phosphate precipitation) have allowed the generation of novel adenovirus recombinants by two basic schemes. The first approach is represented schematically in Fig. 8.2. Insertion into the E1 region of the Ad5 genome is shown in Fig. 8.2A, where DNA isolated from a deletion mutant is digested with the restriction endonucleases *Xba*I and *Cla*I. The digested genomic DNA is subsequently mixed with plasmid DNA containing the left end of the Ad5 genome with a foreign genetic element (hatched segment) inserted within the viral DNA sequences. Upon cotransfection of mammalian cells with these DNA species,

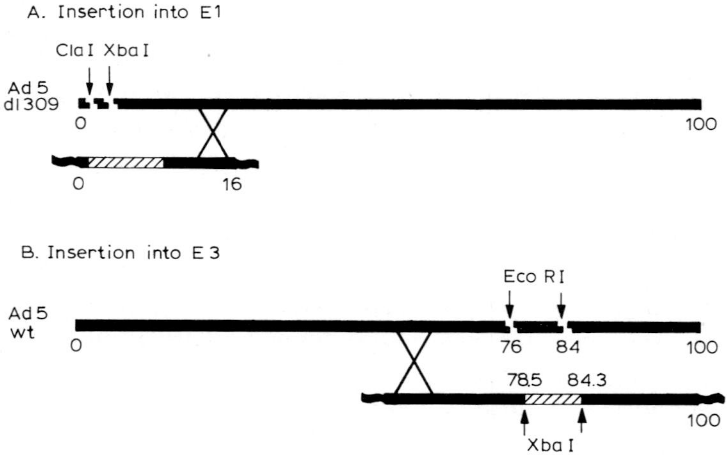

Fig. 8.2. Conventional approaches for generating adenovirus recombinants. (A) Insertion into the essential Ad5 E1 region. Ad5 DNA (darkened segment) from a deletion mutant (d1309), lacking a portion of the E1 region, was digested with restriction endonucleases *Cla*I and *Xba*I. This DNA was used to cotransfect 293 cells with plasmid DNA containing the left end of the Ad5 genome flanking a foreign genetic element (hatched segment). In vivo recombination results in the recovery of the foreign DNA into infectious progeny virus. The 36 kb Ad5 genome is represented by map units 0–100. (B) Insertion of foreign DNA into the nonessential E3 region in a similar manner as depicted in (A). Used by permission of Graham et al. (1988).

recombinants are generated via an intermolecular recombination event. Fig. 8.2B illustrates insertion of a foreign gene (hatched segment) into the E3 region of the Ad5 genome by this process. Recombinant progeny are then identified by hybridization and/or immunological assays.

The second scheme takes advantage of the fact that adenovirus DNA circularizes in infected cells (Graham, 1984). Based upon this observation, bacterial plasmids have been constructed which contain the entire adenovirus genome and are capable of initiating viral replication upon introduction into human cells (Graham et al., 1988). These plasmids are infectious, but they are unable to be packaged because they exceed the packaging capacity of the Ad5 virion due to an insertion of plasmid sequences within the region that a foreign gene is to be targeted to. However, upon recombination with a plasmid containing the foreign gene flanked by adenovirus DNA sequences homologous to the insertion site, bacterial sequences are removed from the infectious plasmid and recombinant genomes of appropriate size are formed and packaged. Infectious plasmids of this type have been engineered for insertions into both the E1 and E3 regions (Graham et al., 1988; Ghosh-Choudhury et al., 1986).

Foreign genetic elements have been inserted into three distinct adenovirus genetic loci. These are the E1 region (Graham et al., 1988; Zhu et al., 1988; Davis et al., 1985; Ballay et al., 1985), the E3 region (Dewar et al., 1989; Prevec et al., 1989; Graham et al., 1988; Hung et al., 1988), and a region between the right terminus of the genome and the E4 promoter sequences (Hung et al., 1988; Saito et al., 1985). Insertion into E3 and the intergenic region near E4 result in the generation of replication-competent adenovirus recombinant, while disruption of the E1 region by a foreign gene results in a replication-defective virus. Recombinant viruses containing an interruption of the E1 region, however, are easily propagated in the 293 cell line, which is an Ad5 transformed human cell line that provides the E1 gene products (Graham et al., 1977).

Insertion of exogenous DNA into the adenovirus genome is quite limited due to the stringent packaging constraints (Prevec et al., 1989; Graham et al., 1988; Hung et al., 1988). Approximately 105% of the total genomic size can be efficiently packaged to produce infectious progeny. The restriction imposed upon packaging coupled to the essential nature of most of the viral DNA permits, at best, the incorporation of 5.3 kb of foreign DNA into the adenovirus genome (Prevec et al., 1989; Hung et al., 1988).

Expression of foreign genes by adenovirus recombinants has been achieved at early times post-infection, at late times post-infection, and throughout the infectious cycle. Adequate early expression levels were demonstrated using either endogenous E1 or E3 promoter sequences (Prevec et al., 1989; Graham et al., 1988; Morin et al., 1987; Ballay et al., 1985). Attempts have been made, with limited success, to couple the SV40 early promoter to foreign genes (Johnson et al., 1988; Ghosh-Choudhury et al., 1986). This may stem from repression of the SV40 enhancer by the E1A gene products (Velcich and Ziff, 1985; Borrelli et al., 1984) and the occurrence of a cryptic splice acceptor situated just downstream from the TATA sequence motif (Johnson et al., 1988). Since the early promoters are not recognized upon the onset of DNA replication and the resultant mRNAs lack the TPL, they are not expressed at late times post-infection even if they persist.

The MLP has been employed to express heterologous gene products at late times post-infection (Davis et al., 1985). Efficient translation of the transcripts initiated at late times post-infection from the MLP required the TPL sequence (Kaufman, 1985; Logan and Shenk, 1984). The proximity of the TPL to the initiation codon of the exogenous gene appears important for optimal translational efficiency. Recombinant viruses which have the TPL positioned closest to the initiation codon produce higher levels of the foreign gene product (Berkner et al., 1987).

Expression of a foreign gene product has also been achieved at early and late times post-infection. Morin et al., (1987) utilized the **Y** leader to express HBsAg constitutively throughout the infectious cycle in adenovirus-infected cells. This was accomplished by the deletion of specific sequences within the E3 region which altered the splicing pattern and resulted in enhanced utilization of the **Y** leader (Morin et al., 1987; Bhat and Wold, 1986). Significant levels of HBsAg expression were observed by placing the foreign gene proximal to the **Y** leader sequence (Morin et al., 1987).

Successful expression of a foreign gene product in the adenovirus system requires a careful consideration of the complex RNA splicing patterns which occur during viral gene expression. The potential for altering these RNA splicing patterns when engineering these viral genomes must be investigated to ensure that the synthesis of undesirable fusion products does not occur.

Adenovirus vectors have been used to express abundant levels of heterologous proteins from various sources, which appear antigenically and functionally similar to the gene products as they exist in their native system. Included in this list are the HBsAg (Hung et al., 1988; Morin et al., 1987; Ballay et al., 1985; Davis et al., 1985; Saito et al., 1985), the HSV-1 gB and gC glycoproteins (McDermott et al., 1989; Johnson et al., 1988; Graham et al., 1988), the HIV *env* gene (Dewar et al., 1989; Hung et al., 1988), the VSV G protein (Prevec et al., 1989; Schneider et al., 1989), the rabies virus glycoprotein (Graham et al., 1988), dihydrofolate reductase (Berkner et al., 1987), the T-antigens of polyoma virus (Davidson and Hassell, 1987; Berkner et al., 1987; Mansour et al., 1986; Massie et al., 1986; Thummel et al., 1983), and the HSV-1 immediate early gene ICPO (Zhu et al., 1988). Analysis of the expressed foreign gene products have demonstrated that correct post-translational modifications (i.e., glycosylation, proteolysis) and cellular localization occurred. For example, the HIV *env* polypeptide expressed by adenovirus vectors was shown to produce a gp120 and gp41 species and to be present on the surface of infected cells (Dewer et al., 1989; Hung et al., 1988).

2.3. Adenovirus vectors as vaccine candidates

Certain characteristics of adenoviruses have provided the impetus to engineer viral recombinants as live vaccine candidates. This virus family has been studied extensively, with respect to genome structure and gene expression and, therefore, the genome can be manipulated with some insight. Ad4 and Ad7 serotypes have been used for the effective immunoprophylaxis against acute respiratory diseases in military recruits for the past 20–30 years with no apparent complications (Top et al., 1971). These vaccines are administered orally as enteric-coated tablets (Hung et al., 1988; Johnson et al., 1988). The viruses replicate asymptomatically in the intestinal tract where they induce a strong mucosal response (Horwitz, 1985a,b). The sites of adenovirus replication (upper respiratory and gastrointestinal tracts) and the ability to induce a strong mucosal response may make adenovirus vectors useful as vaccines for certain viruses, such as influenza and HSV (McDermott et al., 1989; Johnson et al., 1988). Additionally, adenoviruses are stable and are easily propagated in culture (Johnson et al., 1988).

The feasibility of using adenovirus as infectious vectored vaccines was described using human Ad5/HBsAg recombinants (Morin et al., 1987; Ballay et al., 1985). These recombinant viruses were found to elicit a humoral response to both adenovirus and HBsAg upon inoculation into rabbits (Ballay et al., 1985) and hamsters (Morin et al., 1987). The adenovirus vector technology has recently been extended to the human adenovirus vaccine strains, Ad4 and Ad7, by engineering these viruses to express the HBsAg and the HIV *env* gene products (Hung et al., 1988). An Ad7/HBsAg recombinant was shown to evoke an immune response to both Ad7 and HBsAg in dogs (Hung et al., 1988). Other adenovirus recombinants which have shown promise include an Ad5/VSV-G (Schneider et al., 1989; Prevec et al., 1989; Graham et al., 1988), an Ad5/HIV *env* (Dewar et al., 1989), and an Ad5/HSVgB recombinant virus (McDermott et al., 1989; Johnson et al., 1988).

An interesting observation from these animal studies was that immune responses, including protection, were apparent in animal species (i.e., mice and dogs) in which human adenoviruses either replicate poorly or not at all in cell culture (McDermott et al., 1989; Schneider et al., 1989; Prevec et al., 1989; Hung et al., 1988). Additionally, a replication-defective adenovirus recombinant resulting from an insertion into the essential E1 region provided an immune response in rabbits following inoculation via an unconventional intravenous route (Ballay et al., 1985). The ability of these potentially nonreplicating vectors to induce a significant immune response may provide safety advantages. However, it has not yet been firmly established whether these viruses totally lack the potential to replicate in the species in question (Prevec et al., 1989).

Wide acceptance of the use of adenovirus-based vaccine candidates in both the human and veterinary fields requires further evaluation with regard to safety. Limited information exists pertaining to the pathogenic properties of adenoviruses and to the relative roles of individual gene products in the pathology of disease. This is, in part, due to the absence of appropriate animal model systems for adenovirus infections. Recently, however, Ad5 was shown to induce a pneumonia in the cotton rat, *Sigmadon hispidus,* that is pathologically similar to that observed in humans (Pacini et al., 1984). Using this model system, Ginsberg et al. (1987) established that the E1B 55-kDa gene product has an important role in the production of the pneumonia. Additionally, these investigators found that the E3 19-kDa protein, which is nonessential for viral growth in cell culture, affects viral pathogenesis, since deletion of this region results in higher levels of virus infiltration in the respiratory tissue (Ginsberg et al., 1987). This finding is especially significant since many foreign genes have been substituted for E3 in generating recombinant adenoviruses.

Other concerns relating to the use of adenoviruses as live recombinant vaccines stem from the oncogenic potential of these viruses (reviewed by Horwitz, 1985a, b; Bos and van der Eb, 1985; Bernards and van der Eb, 1984). All human adenoviruses have the ability to transform cells in tissue culture at a low frequency (a few foci per 10^6cells), although they vary considerably in their oncogenic potential in rodents (Horwitz, 1985a, b; Bernards and van der Eb, 1984). The Ad4 and Ad7 serotypes have a low to moderate oncogenic potential in rodents.

The E1 region is involved in the oncogenic potential of adenoviruses, since both E1A and E1B gene products function as adenovirus transforming genes (Bos and van der Eb, 1985; Bernards and van der Eb, 1984). The oncogenic potential, however, appears not to be linked solely to the E1 region (Bernards et al., 1984). The role of EIA in cellular transformation probably reflects the ability of the E1A phosphoproteins localized in nuclei to act as *trans*-activating factors. These proteins have the ability to activate and repress the transcription of both viral and cellular genes (Bos and van der Eb, 1985; Velcich and Ziff, 1985; Svonsson and Akusjari, 1984) and to influence the differentiation and growth of cells (Webster et al., 1988; Hen et al., 1986). Mutational analyses of the E1A region have demonstrated the existence of multiple functional domains within this gene product (reviewed by Moran and Matthews, 1987). The potential exists, therefore, to engineer this region such that deleterious effects caused by this promiscuous *trans*-acting factor are eliminated while retaining the domains required for viral replication. Complete removal of the E1 region may provide another means to ensure the safety of adenovirus vector-based vaccines.

In summary, adenovirus vectors provide valuable reagents for investigating the structure and function of proteins and the immunobiology of pathogens. Their acceptance as vaccine candidates, however, requires a more in-depth evaluation of their pathogenic properties in both human and animals. Further analyses are also required with regard to revaccination protocols for boosting existing immunity and for induction of immunity to novel immunogens despite pre-existing immunity to adenoviruses.

3. General biology of herpesviruses and their potential use as vaccine vectors

Members of the herpesvirus family are further divided into three subfamilies (α-, β-, and γ-herpesviruses) based upon biological properties such as host range, growth characteristics, and the induction of disease (reviewed by Roizman and Batterson, 1985). With respect to the specific herpesviruses mentioned in this chapter, human cytomegalovirus (HCMV) belongs to the betaherpesvirus subfamily while herpes simplex virus type 1 (HSV-1), pseudorabies virus (PRV), and varicella-zoster virus (VZV) are categorized as alphaherpesviruses.

Herpesviruses contain linear double-stranded DNA genomes which vary among family members with respect to size, base composition, and sequence arrangement. The five different sequence arrangements known to exist among herpesviruses are depicted in Fig. 8.3. Due to the presence of reiterated sequences flanking unique sequences, some genomes exist in multiple isomeric forms (Fig. 8.3).

The replication, expression, and packaging of herpesvirus genomes are highly regulated processes which occur within the nucleus of infected cells. Temporal regulation of the herpesvirus-encoded gene products result in the synthesis of five distinct

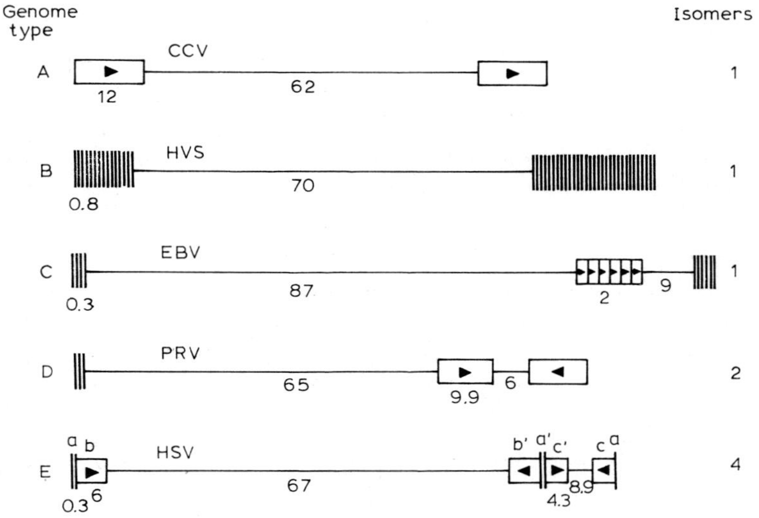

Fig. 8.3. Diagram of the five sequence arrangements observed among the herpesviruses. The genome types A through E are exemplified by channel catfish herpesvirus (CCV), herpesvirus saimiri (HVS), Epstein-Barr virus (EBV), Pseudorabies virus (PRV), and herpes simplex virus (HSV), respectively. Varicella-zoster virus (VZV) contains a type D genome while human cytomegalovirus (HCMV) has a type E sequence arrangement. Horizontal lines represent unique sequences whereas reiterated sequences larger than 1 kb are shown as rectangles. Arrowheads indicate the direction of the repeats relative to each other. The vertical lines indicate terminal sequences reported in genome types B–E. The number of isomeric forms that each genome arrangement can exist in is indicated on the right. The numbers represent molecular weights in megadaltons. Used by permission of Roizman and Batterson (1985).

kinetic classes of gene products (α, β_1, β_2, γ_1, and γ_2), which are expressed in a cascade fashion (Honess and Roizman, 1975; 1974). Transcriptional regulation of the various herpesvirus-encoded gene products is afforded by the transcriptional activity of the host RNA polymerase II in conjunction with specific temporally regulated virus-encoded *trans*-acting factors (Everett, 1987; Roizman and Batterson, 1985; Everett, 1984). Post-transcriptional modes of regulation, including translational regulation and perhaps to a limited extent mRNA splicing, may also have a role in the temporal regulation of herpesvirus gene products (Johnson and Spear, 1984; McLaughlin and Clements, 1982; Jones and Roizman, 1979; Kozak and Roizman, 1974).

At the present time, limited knowledge exists pertaining to the regulatory mechanisms involved in the complex replication cycle of herpesviruses and most of the available data concerns the replication of HSV. The paucity of information regarding specific replicative features of other herpesviruses stems from the inherent difficulty of propagating large quantities of certain herpesviruses (i.e., HCMV and VZV) in cell culture. Recent application of new technologies has now made molecular analysis of these viruses possible.

Genetic and molecular biological analyses of herpesviruses, such as HCMV, HSV, PRV, and VZV, have provided a means of engineering these viruses as eukaryotic vectors. Herpesviruses have been manipulated using isolated viral DNA and molecularly cloned subgenomic fragments in cotransfection experiments (Van Zijl et al., 1988; Post and Roizman, 1981; Mocarski, 1980). Certain characteristics of these viruses which make them attractive as live recombinant vaccine candidates are the availability of specific herpesvirus vaccine strains which appear safe and effective (sections 3.1–3.4) and their ability to induce significant cellular and humoral immune responses. The value of using herpesvirus recombinants as a general vaccine vector cannot be fully assessed at the present time, since not many immunologically pertinent foreign genes have been expressed by herpesvirus recombinants. Further studies which address the versatility of these systems with respect to: (a) expressing genes derived from a variety of sources; (b) the ability of these recombinants to induce protective immunity to the corresponding pathogens; and (c) the relative safety of the herpesvirus recombinants are required to evaluate the benefits associated with using them as live vaccine candidates. The potential use of herpesvirus vaccine candidates needs to be evaluated carefully with respect to their ability to affect cell growth and cell function. This is due in part to the indiscriminant function of certain virally-encoded *trans*-acting factors on cellular gene expression (Everett, 1987) and on the potential transactivation of latent viral infections such as HIV (Rando et al., 1987). Additionally, due to the prevalence of herpesviruses in various populations and their ability to induce a long-lasting immunity, revaccination analyses are necessary to determine the efficacy of herpesvirus recombinants in humans or animals previously exposed to these viruses.

The remainder of this section describes the progress which has been made in developing the specific herpesviruses, noted above, as mammalian virus vaccine vectors.

3.1. Human cytomegalovirus-based vectors

HCMV is a highly species-specific virus which infects a large proportion of the human population, resulting in CMV mononucleosis in healthy adults (Alford and Britt, 1985). Infection of neonates or immunodeficient individuals, however, can lead to a more severe disease state. As with other herpesvirus infections, initial exposure to CMV can result in a latent infection which has the potential to become reactivated at later times.

The prevalence of HCMV in the human population, the severity of the disease in immunosuppressed persons (i.e., transplant patients), and the risk of congenitally acquired CMV infection has prompted the development of a CMV vaccine. The Towne strain is an attenuated CMV strain which resulted from the serial passage of a wild-type HCMV isolate in human fibroblasts (Plotkin et al., 1976; Plotkin et al., 1975). Clinical testing of this attenuated vaccine candidate has demonstrated its safety and efficacy in preventing primary and latent CMV infections (Plotkin and Huang, 1985; Plotkin et al., 1984).

The availability of the Towne strain of CMV and the ability to manipulate herpesvirus genomes by inserting foreign genetic elements into defined genetic loci via homologous recombination (Post and Roizman, 1981; Mocarski et al., 1980) suggests that CMV may provide a useful recombinant vaccine vector. Spaete and Mocarski (1987) have developed methods for generating site-specific insertions and deletions of the CMV (Towne strain) genome. In these studies the *E. coli lacZ* gene was inserted into the major beta-gene within the left L-component repeat of the genome. The expression of beta-galactosidase was under the control of a beta-gene promoter and induced the accumulation of substantial levels of the protein in infected human fibroblasts (approx. 1% of total cell protein).

3.2. Live recombinant herpes simplex virus vaccine candidates

HSV-1 infections commonly involve the skin, mucous membranes, eyes, and central nervous system. Most primary infections are subclinical although they can result in gingivostomatitis, vulvovaginitis, keratoconjunctivitis, and a life-threatening encephalitis (Rawls, 1985). In most, if not all, cases the virus becomes latent and can be reactivated even in the presence of significant levels of HSV-specific antibody.

In recent years attention has focused on defining virulence-associated genes of HSV-1. Mice have served as an important model system, since intracranial inoculation of low doses of wild-type HSV-1 are lethal to these animals. Taking advantage of this animal model system it was determined that the viral thymidine kinase (*tk*) gene (Roizman and Sears, 1987; Field and Wildy, 1978) and the viral DNA polymerase (Field and Coen, 1986) are associated with HSV-1 neurovirulence. In addition, analysis of HSV-1 and HSV-2 intertypic recombinants has established that other HSV-1-encoded gene products are involved in virulence (Javier et al., 1988; 1987). The multigenic nature of the HSV-1 virulence phenotype is also supported by analysis of specific HSV-1 strains (Ben-Hur et al., 1987; Sedarti and Stevens, 1987; Centifanto-Fitzgerald et al., 1982) and genetically engineered HSV (Meignier et al., 1988a,b).

Localization of genetic loci associated with HSV-1 virulence has enabled the genetic engineering of strains R7017 and R7020 as prototypes of live attenuated HSV-1 vaccines (Meignier et al., 1988a,b). These strains were constructed by removing the *tk* locus and a 14.5-kb segment consisting of the right end of the unique (L) segment, the entire internal inverted repeat sequence, and the leftmost portion of the small unique (S) component (Meignier et al., 1988a,b) (Fig. 8.3). The internal repeat was replaced with the HSV-2 genes encoding the glycoproteins D, G, I, and a portion of E. R7020 differs from R7017 in that the HSV-1 *tk* gene was re-inserted into the R7020 genome at a novel genetic locus under the control of an α gene promoter rather than its normal β gene promoter. Interestingly, the presence of the *tk* gene in R7020 did not enhance the neurovirulence of this virus compared to strain R7017.

To evaluate the safety and efficacy of the attenuated, modified HSV-1 strains, R7017 and R7020, animal studies have been done in several species, which are models for various types of HSV-induced disease. Both recombinant viruses were capable of protecting guinea pigs, rabbits, *Aotus* monkeys, and rodents from severe or fatal disease caused by live virus challenge (Meignier et al., 1988a,b). Challenge virus replication did occur, however, at the site of inoculation resulting in the establishment of latent infection in sensory ganglia which innervate this site. Of significance, these studies demonstrated that both attenuated strains caused localized lesions which healed rapidly as compared to those induced by wild-type HSV infection, and that R7017 and R7020 have a decreased ability to establish latent infections (Meignier et al., 1988a,b). The development of HSV strains totally devoid of the ability to establish latent infections may require more in-depth genetic and molecular analyses of the specific HSV latency-associated genes, which appear to be regulated differently than the alpha, beta, and gamma HSV-induced gene products (Spivack and Fraser, 1988; Deatly et al., 1988, 1987; Fraser et al., 1986).

In addition to genetic manipulation of the HSV-1 genome to generate safe vaccine strains, HSV-1 has been engineered to express heterologous proteins to analyze its potential use as an immunization vehicle against other infectious agents. The R7017 and R7020 HSV-1 strains provide an example of such a multivalent vaccine since these viruses contain the HSV-2 genes encoding the D, G, and I glycoproteins and a portion of the E glycoprotein gene (Meignier et al., 1988a,b). It is evident from animal studies, noted above, that these recombinant viruses were capable of eliciting protective immunity against severe disease states caused by HSV-2 challenge although protection against latent infections was not achieved. Other HSV-1 recombinants which have been generated to express foreign genes include a HSV/HBsAg recombinant (Shih et al., 1984) and HSV-1 recombinant viruses that express *Plasmodium falciparum* antigens (Edwards et al., 1988). To date, no animal studies have been reported regarding the efficacy of the HSV/HBsAg recombinant while expression of the *Plasmodium* antigens by HSV-1 was shown to be inefficient in both its level of synthesis and its ability to induce a detectable anti-malarial immune response in mice.

3.3. Recombinant pseudorabies virus vaccines

PRV causes a natural infection in swine similar to that of HSV in man (reviewed by Gustafson, 1981). The clinical features of PRV infection, termed Aujeszky's disease, depend upon the age of the animal, with a more severe disease-state occurring in newborn pigs. Adult pigs are rarely fatally infected and usually suffer a subclinical or mild infection. PRV persists in a latent form in swine which have recovered from the disease and this plays a major role in virus transmission to susceptible animals.

A commercially available PRV vaccine strain (Omnivac-PRV) has been developed from the Bucharest vaccine strain by deletion of the *tk* gene (Kit et al., 1988, 1987). Removal of the *tk* gene was based upon previous findings in PRV and other herpesviruses that Tk activity is associated with viral virulence and latency (Kit et al., 1985; Field and Wildy, 1978; Tartarov and Gergov, 1969; Tartarov, 1968). Testing of the Omnivac-PRV vaccine strain in animals demonstrated the safety and efficacy of these viruses over conventional PRV vaccines, although it has not been determined whether vaccination can develop into latent infections in swine (Kit, 1988; Kit et al., 1987; Kit et al., 1985).

The *tk* locus is not the sole PRV virulence determinant. Analysis of various conventional vaccine strains and engineered PRV deletion mutants clearly demonstrate that, as with other herpesviruses, virulence of PRV appears to be multifactorial with several genetic loci responsible (Kit, 1988; Kit et al., 1987; Thomsen et al., 1987a; Lomniczi et al., 1987; McFarland et al., 1987; Mettenleiter, 1987, 1985; Marchioli et al., 1987; Van Oirschot and Gielkens, 1987; Rea et al., 1985). Included in these loci are certain genes which encode the PRV glycoproteins gpI, gpIII, and gX which have been shown to be nonessential for viral growth in cell culture (Kit, 1988; Kit et al., 1987, Thomsen et al., 1987a; Marchioli et al., 1987; Mettenleiter et al., 1987, 1985; Rea, 1985) and unnecessary for induction of protective immunity (Kit et al., 1988; Kit, 1987). Excision of these genes has provided second generation genetically engineered PRV vaccines, some of which have been extensively tested for their safety and efficacy and were found to be good vaccine candidates with no significant adverse reactions (Kit, 1988; Kit et al., 1987), although deletion of the PRV gX gene did result in a virus strain that was more virulent in certain animal species

(Marchioli et al., 1987). Deletion of these genes not only may provide less virulent PRV vaccine strains but also enables vaccinated pigs to be serologically distinguishable from those naturally infected with PRV field strains.

The wide host range of PRV, which excludes humans, and the ability to manipulate the PRV genome to generate safe vaccine strains have brought forth the possibility of employing this virus as a vector to immunize against heterologous pathogens of veterinary importance (Thomsen et al., 1987b). The use of PRV as a eukaryotic expression vector has been demonstrated by insertion of the human tissue plasminogen activator (tPA) gene into the PRV gX locus (Thomsen et al., 1987b). Resultant recombinant progeny efficiently expressed the tPA gene product utilizing the gX promoter element. To date, the tPA gene is the only reported foreign gene that has been expressed by PRV.

3.4. Varicella-zoster virus recombinants as vaccine candidates

VZV is the causal agent of chickenpox (reviewed by Grose, 1987). The virus pervades the human population such that most individuals are infected by age 20 and develop the typical benign exanthemous disease. Severe complications can occur, however, in neonates, adults and immunocompromised individuals. VZV infection also causes zoster (shingles) upon reactivation of latent virus in specific dermatomes. Clinical reactivation results most often in immunodeficient adults, although zoster has also been observed in healthy adults with a waning immunity to VZV.

Vaccination against chickenpox has been achieved using the attenuated VZV (Oka) strain. This vaccine strain, derived by serial passage of VZV in guinea pig embryonic fibroblast cells, has been extensively tested in Japan, the United States, and Europe (Takahashi, 1987; Takahashi et al., 1974). Clinical testing in both healthy and immunocompromised individuals has demonstrated the safety and efficacy of this live viral vaccine with no apparent increase in the development of latent infections. The VZV (Oka) strain is the only live viral vaccine approved for use in immunodeficient individuals. As with wild-type VZV, molecular analysis of the Oka strain has been hampered by the inability to propagate extracellular virus in culture at high titers. Therefore, the genetic basis of attenuation remains unknown, although it has been demonstrated that DNA cleavage patterns generated upon digestion with restriction endonucleases differ between wild-type and Oka strain derived genomic DNA (Takahashi, 1987).

The safety and efficacy of the VZV (Oka) strain has prompted the development of this virus strain as a live recombinant vaccine candidate. Lowe et al. (1987) have engineered a VZV (Oka) recombinant which expresses the EBV gp350/220 glycoproteins. Expression of the EBV gp350/220 gene product was achieved using the promoter element from the highly expressed VZV gI gene (Keller et al., 1984). The actual expression cassette used to generate the VZV recombinant virus also contained the gI hydrophobic leader sequence fused to the EBV gp350/220 gene so that a fusion product was synthesized in infected cells. The EBV gp350/220 gene product expressed by the VZV recombinant appeared authentic and was localized to the cellular plasma membrane. Significantly, both the gp350 and gp220 proteins were expressed demonstrating that mRNA splicing events similar to those noted in EBV-infected cells occurred with the VZV/EBV-derived gp350/220 RNA transcript.

Subsequent animal studies with the VZV/EBV recombinant were done in the common marmoset, *Callithrix jacchus,* to test the immunogenicity of recombinant virus (Ellis et al., 1988). In these studies, *C. jacchus,* which represents a reliable system for analyzing VZV replication (Provost et al., 1987) inoculated with the recombinant virus mounted a humoral response to both VZV proteins and to the heterologous gp350/220 proteins. A booster effect was also noted in the animals upon the administration of a second dose of the recombinant virus or upon challenge with live EBV virus.

4. Poxvirus vectors

4.1. Biology of poxviruses

Vaccinia virus (VV), the prototypic orthopoxvirus, has been utilized extensively as a substrate for generating live recombinant vaccine candidates (reviewed by Moss and Flexner, 1987; Mackett and Smith, 1986). This virus is attractive as a vector-based vaccine candidate because of its biological properties and its successful history as an immunizing agent (Behbehami, 1983). Recent reviews discuss in detail the biological properties of this virus (Fenner, 1985; Moss, 1985).

VV is a large double-stranded DNA-containing virus which replicates within the cytoplasm of infected cells. Virus replication is independent of host nuclear functions, although there exists evidence that a certain component of the host RNA polymerase II is involved in the virus replicative cycle (Wilton and Dales, 1989; Morrison and Moyer, 1986). The expression, replication, and packaging of this complex genome (approximately 187 kb) involves a series of sequentially regulated events. This is evident as the infectious cycle progresses through distinct stages from uncoating to assembly. The VV replicative program is dependent upon various virus-encoded enzymatic activities, including a DNA polymerase, a DNA-dependent RNA polymerase, a thymidine kinase, and various enzymatic activities necessary for the capping and polyadenylation of RNA transcripts. Enzymes involved in the initiation of the replication cycle upon virus entry into the host cell are specifically encapsidated by VV virions.

Distinct kinetic classes of virus-encoded gene products are synthesized during VV infection (Vos and Stunnenberg, 1988; Moss, 1985). These sequentially controlled gene products are apparently regulated by both transcriptional and post-transcriptional mechanisms (Hruby, 1985; Vassef et al., 1982; Paoletti and Grady, 1977; Oda and Joklik, 1967). Presently, little is known about these regulatory mechanisms or the genomic organization of VV-specified products of known function. Further investigation of the VV genome, including gene mapping and nucleotide sequence analysis, is required for a more complete understanding of the molecular organization and genetic regulation of the VV system. Definition of these molecular characteristics will be advantageous for optimizing VV as an expression vector.

4.2. Insertion and expression of foreign genes

An extension of the marker rescue technique has been used to insert exogenous genetic elements into nonessential regions of the VV genome. Unlike with other DNA-containing viruses, however, simple cotransfections experiments using purified viral DNA or infectious plasmids are not possible in the VV system due to the noninfec-

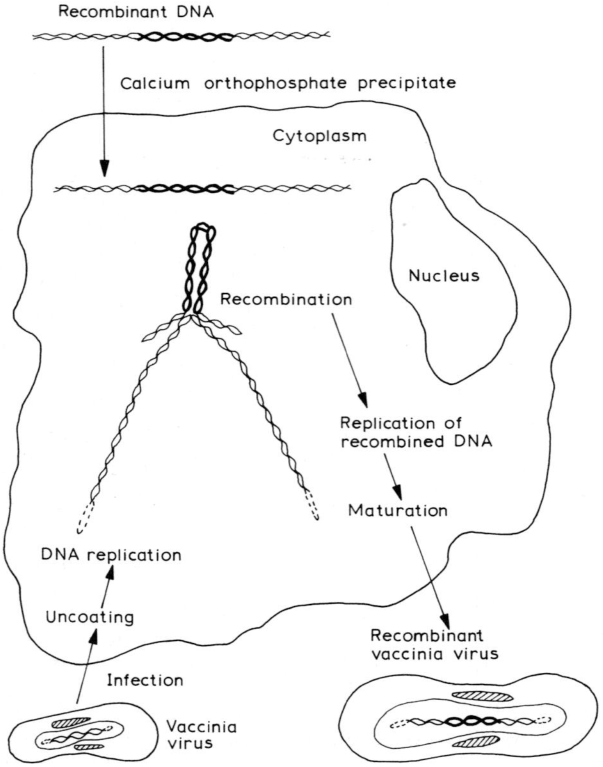

Fig. 8.4. The general procedure used for foreign gene insertion into the VV genome. The foreign gene of interest (thick double helical segment) flanked by VV DNA sequences (thin double helical segment) is inserted into a plasmid cloning vector. This recombinant plasmid vector harboring the foreign gene is transfected into cells already infected with VV. During the virus replication cycle, the foreign gene becomes recombined into the VV genome via the homologous VV flanking arms. Recombinant genomes are replicated and packaged.

tious nature and large size of VV DNA. This stems from the obligatory requirement for virion-associated enzymatic functions during the initial stages of the replicative cycle. Therefore transfected DNA can only by rescued by infectious VV particles.

The general methodology used to generate VV recombinants is illustrated in Fig. 8.4 and described in detail in recent reviews (Piccini et al., 1987; Mackett et al., 1985). The foreign gene of interest flanked by VV DNA sequences is inserted into a plasmid cloning vehicle. The flanking VV sequences, ranging from between 100–10 000 base pairs in length, serve as 'homologous arms' to direct the foreign gene to a desired location within the VV genome via homologous recombination. The relevant recombinant plasmid vector harboring the foreign gene is transfected into cells additionally infected with viable VV. The chimeric virus genomes formed by homologous recombination are replicated and packaged into infectious progeny virus and identified by standard procedures (reviewed by Piccini et al., 1987; Mackett et al., 1985).

The recovery of infectious VV recombinant progeny which expresses a foreign gene is dependent upon several factors. Insertion of exogenous genetic material cannot interfere with the expression of essential viral functions. Due to the apparently nonessential nature of much of the terminal region (Thompson and Condit, 1986; Ensinger et al., 1985), viable terminal region deletion mutants have been isolated (Perkus et al., 1986; Paez et al., 1985; Panicali et al., 1981; Moss et al., 1981), and seemingly provide targets for insertion of foreign genes. The variability observed within poxvirus terminal sequences (McCauley et al., 1987; Pickup et al., 1986a; Archard et al., 1984; Esposito et al., 1981; Moyer et al., 1980), however, may somewhat limit their use with respect to generating stable recombinant progeny. Further knowledge regarding the mechanisms involved in the replication and recombination of the VV genome may provide a means to overcome these limitations. Several appropriate stable sites have been defined by the physical mapping of nonessential genes such as the thymidine kinase gene (Hruby et al., 1983; Weir et al., 1982) and the hemagglutinin gene (Shida, 1986) and were also generated by insertional mutagenesis (Perkus et al., 1986).

Unlike with other DNA-containing viruses, splicing of exons to generate mature mRNAs has not been demonstrated in the VV system. Therefore, expression of foreign genes requires the insertion of a cDNA species into the VV genome.

Since the VV-encoded DNA-dependent RNA polymerase fails to recognize typical type II eukaryotic promoters, appropriate VV regulatory sequences must be juxtaposed at the 5′ end of the foreign gene. The initial approach was taken empirically by inserting a foreign gene into a nonessential site (Panicali and Paoletti, 1982). Expression in this case was dependent upon a VV promoter being adjacent to the site of insertion within the VV genome. Expression of foreign genes has been achieved by directly engineering VV promoters such as the 7.5-kDa promoter (Cochran et al., 1985; Mackett et al., 1984), the hemagglutinin promoter (Shida et al., 1987), the H6 promoter (Taylor et al., 1988a) and the 28-kDa promoter

(Cremer et al., 1985). VV promoters differ with respect to their programmed regulation and promoter strengths and therefore provide the potential to vary the kinetics and levels of foreign gene expression.

A variety of heterologous proteins have been expressed by VV to assess their structural and functional properties. Most of these foreign genes encode immunologically pertinent antigens from a variety of infectious agents, particularly viruses (reviewed by Moss and Flexner, 1987; Mackett and Smith, 1986). With proper DNA engineering the proteins are generally of the predicted size and undergo appropriate post-translational modifications including glycosylation and proteolytic cleavage. Foreign gene products also appear to be transported properly in VV infected cells. In a few cases, however, post-translational modifications of foreign gene products expressed by VV recombinants have been reported to vary depending upon the infected cell type. Human preproenkephalin was secreted by several cell lines infected with a recombinant encoding this product. Only a mouse pituitary cell line, however, was capable of processing proenkephalin to mature enkephalin peptides (Thomas et al., 1986). Proteolytic processing of the HIV envelope and group-associated antigens (gag) was also observed to vary in different cell lines infected with recombinant VV expressing these genes (Gowda et al., 1989; Flexner et al., 1988a; Zarling et al., 1987; Chakrabarti et al., 1986; Hu et al., 1986). Despite these apparent abnormalities, animals vaccinated with these recombinants produced antibodies that could react with the authentic antigen.

4.3. Recombinant vaccinia virus as a vaccine candidate

VV recombinants have induced immunological responses to various foreign proteins in experimental animals (reviewed by Moss and Flexner, 1987; Mackett and Smith, 1986). Significantly, stimulation of both humoral and cell-mediated responses by VV recombinants have been noted. This is apparently crucial, since both arms of the immune system are pertinent for the development of complete immunity to many pathogenic organisms, particularly viruses.

Factors influencing these immune responses included the route of vaccination and the manner in which the foreign antigen is presented to the immune system. It has been demonstrated that secretory IgA against influenza virus hemagglutinin (Inf HA) was produced only following nasal inoculation with the VV recombinant and that this correlated with prevention of local infection with Inf (Small et al., 1985). Langford et al., (1986) showed that the immunogenicity of the *Plasmodium falciparum* S-antigen was greatly enhanced when it was presented as an anchored membrane protein rather than as a secreted protein. Furthermore, the immune response to a VV expressed foreign gene product was shown to be affected by the temporal regulation of its expression. Coupar et al. (1986) showed that Inf HA expression under the control of an early VV promoter was capable of priming a CTL response. However, late expression of the HA product was unable to induce such a response. A similar finding was also reported for the induction of a T cell response to the HSVgD expressed by a VV recombinant (Wachsman et al., 1988).

Some factors which may contribute to the defective responsiveness of T cells, particularly at late times post-infection are the cessation of host cell specific protein synthesis upon VV infection, the induction of virus-encoded serine protease inhibitors (Kotwal and Moss, 1989; Boursnell et al., 1988; Pickup et al., 1986b), and/or the induction of suppressor factors (Martin and Rouse, 1987; Martin et al., 1987). Both of the former two processes could drastically affect the induction of T cells by altering the proper presentation of antigen in conjunction with class I MHC molecules on the surface of infected cells. The concept of defective antigen presentation causing a block in CTL responsiveness is supported by results reported by Townsend et al. (1988) analyzing the immune response to VV-encoded Inf antigens. Results from these studies demonstrated that enhanced degradation of antigen by either removing the NH_2-terminal signal sequence, engineering a ubiquitin-antigen fusion product, or altering the N-terminal amino acid can overcome the defective presentation of antigen to CTLs even when expressed at late times post-infection.

A large number of examples have now been gathered showing that animals immunized with VV recombinants have been protected against subsequent challenge with the corresponding pathogen. Significantly, protection was achieved against pathogens with differing abilities to produce local or systemic infections and acute or latent infections (reviewed by Moss and Flexner, 1987; Mackett and Smith, 1986). Additionally, it was observed that the absence of a humoral response does not eliminate the possibility that protection was afforded by a particular immunizing agent. For instance, even though a significant antibody response was not observed in experimental animals following inoculation with VV recombinants expressing either the HBsAg (Moss et al., 1984), HSVgD (Wachsman et al., 1987), or the Lassa virus glycoprotein (Auperin et al., 1988), the animals were nevertheless protected upon challenge with the corresponding pathogen. From this it becomes evident that VV as a vector for heterologous genes provides valuable reagents for investigating the immunobiology of particular pathogens, their component immunogens, and the basis of protective immunity.

The versatility of VV as a live recombinant vaccine is illustrated by its ability to express a variety of immunologically pertinent proteins (reviewed by Moss and Flexner, 1987; Mackett and Smith, 1986). Additionally, the flexibility in its packaging potential permits large quantities of foreign DNA to be inserted (Smith and Moss, 1983) and/or deleted (Perkus et al., 1986; Paez et al., 1985; Panicali et al., 1981; Moss et al., 1981) so that one can envisage the engineering of multivalent vaccines which could elicit immunity to a number of infectious agents with a single inoculation. Towards this end, a single VV recombinant has been generated which expresses three foreign genes: the Inf HA, HBsAg, and the HSV-1gD (Perkus et al., 1985). Significantly, animals inoculated with this triple recombinant VV produced antibodies against all three antigens. More recently, Flexner et al. (1988b) reported the generation of a VV double recombinant which expressed the Inf HA and the HSV-1gD glycoproteins. This double recombinant was able to induce an immune response to both foreign gene products and to protect mice from subsequent challenges with both HSV-1 and Inf virus.

A concern pertaining to the use of live vectors as vaccine candidates is whether previous exposure to the immunizing vehicle or to any of the expressed heterologous antigens will prevent the induction of an optimal immune response towards a novel antigen expressed by the vector. Results have been reported by both Perkus et al. (1985) and Flexner et al. (1988b) which suggest that recombinant VV can be used in revaccination protocols either to give a booster effect or to elicit an immunological response to a novel antigen. Pre-existing immunity to VV, however, has been shown to have a negative effect on the immune response to a novel antigen expressed by a VV recombinant (Rooney et al., 1988). These conflicting results noted with the various revaccination studies may well stem from differences in the parameters analyzed in their investigations and/or the vaccination protocols. In any case, it is apparent that VV recombinants can boost pre-existing immunity to a particular antigen (Rooney et al., 1988; Gillespie et al., 1986; Perkus et al., 1985) and if engineered properly and employed in an optimal immunization protocol may induce an immune response to novel immunogens despite pre-existing immunity.

Some objections have been raised concerning the acceptance of live VV recombinants as human and veterinary vaccine candidates due to the rare complications which have been observed from immunization with VV, particularly in immunocompromised individuals (Behbehami, 1983; Lane et al., 1969). Recently, however, great strides have been taken in understanding viral virulence factors with the hope of modifying virus strains for use as immunization vehicles.

The VV thymidine kinase gene (Williamson et al., 1988; Buller et al., 1985), the growth factor gene (Buller et al., 1988), and the hemagglutinin gene (Shida et al., 1988) apparently influence the pathogenic properties of VV. Additionally, deletions of the left terminal region of the genome are known to reduce viral virulence since removal of these genetic loci decreases the pathogenic properties of VV in mice (Rodriguez et al., 1989; Buller et al., 1985). Other genes which may have relevance to the pathogenesis of poxviruses include the A-type inclusion protein of cowpox virus (Joklik et al., 1988; Patel et al., 1986), the hemorrhagic gene (Pickup et al., 1986b), other genes which share sequence homology with serine protease inhibitors (Kotwal and Moss, 1989; Boursnell et al., 1988), and a VV secretory protein which shares sequence and functional homology with complement control proteins (Kotwal and Moss, 1988). Genetic manipulation of these genes and the potential to coexpress immunoregulatory products, such as IL-2 (Flexner et al., 1987; Ramshaw et al., 1987), may provide the ability to engineer live vaccine strains which are safe and effective.

4.4. Cancer prevention and immunotherapy

Specifically engineered viral recombinants may provide additional vehicles for vaccination against virus-induced cancers. Antigens derived from a variety of potentially carcinogenic viruses including EBV MAgp340 (Morgan et al., 1988; Mackett and Arrand, 1985), the HBsAg (Cheng and Moss, 1987; Moss et al., 1984; Paoletti et al., 1984; Smith et al., 1983), the polyoma virus T-antigens (Lathe et al., 1987), the feline leukemia virus *env* protein (Gilbert et al., 1987), the Friend leukemia virus (FLV) *env* protein (Earl et al., 1986), and the HTLV-I *env* protein (Shida et al., 1987) have been expressed by VV recombinants. Significantly, protection was achieved by several of these recombinants. Morgan et al. (1988) have provided evidence that a VV recombinant which expresses the EBV gp340 gene protected cottontop tamarins against EBV-induced lymphomas. Mice inoculated with VV recombinant expressing the FLV *env* protein were protected against virus-induced leukemia (Earl et al., 1986), while tumor prevention and immunotherapy of polyoma virus-induced tumors in rats was afforded by a VV recombinant which expressed the polyoma large and middle T-antigen (Lathe et al., 1987).

The use of poxvirus recombinants in cancer immunotherapy may be extended to cancers of nonviral origin. Results obtained within the human melanoma and rat *neu* experimental systems support the use of such recombinants to control tumor growth. Mice immunized with a VV recombinant expressing the human melanoma antigen, p97, rejected tumor transplants (Estin et all, 1988; Hu et al., 1988). Furthermore, Bernards et al. (1987) constructed a VV recombinant which expressed the extracellular domain of the rat *neu*-encoded transmembrane glycoprotein, p185. Mice immunized with this recombinant virus developed a strong humoral response against the *neu* gene product and were fully protected against subsequent tumor challenge.

4.5. Other poxvirus vectors

The extension of the VV technology to other members of the poxvirus family provides the potential for generating recombinant viruses which are more species-specific and which may provide vaccine candidates for defined usage in the human and veterinary fields. Recently, a raccoonpox virus recombinant was successfully engineered to express the rabies G glycoprotein (Esposito et al., 1988). This recombinant was shown to protect raccoons from a lethal rabies virus challenge.

Fowlpox virus, which has a host range limited to avian species, has also been engineered as a species-specific recombinant vaccine candidate (Taylor et al., 1988b; Boyle and Coupar, 1988). Interestingly, a fowlpox vector encoding the rabies G glycoprotein was found to protect several animal species from a lethal challenge with rabies virus (Taylor et al., 1988a). This protective immunity in nonavian systems was achieved even though fowlpox virus is incapable of a sustained productive infection in these species.

5. Summary

The advent of recombinant DNA techniques and advances in immunology have provided a means for dissecting the immunobiology of pathogens. Identification and expression of individual genes from the pathogens in heterologous systems, such as adenoviruses, herpesviruses, and poxviruses has yielded valuable information regarding the structural properties of the gene products and their role in eliciting protective immunity. Targets of both humoral and/or cellular immunity for many disease causing agents have been identified or confirmed using these expression systems.

Adenoviruses, herpesviruses, and poxviruses have certain features, including the ability to induce pertinent humoral and cellular immune responses, the potential to develop multivalent vaccines, and the accessibility of attenuated viral strains, which have provided the impetus for engineering these viruses as live recombinant vaccine candidates. Each virus type has perceived and real advantages or disadvantages as vaccine candidates. Critical to the refinement of these viruses as recombinant immunizing agents is a more in-depth knowledge of the pathogenic mechanisms of these viruses, since an ideal vaccine vector clearly should be as safe as possible. Safety issues pertaining to the oncogenic potential of adenoviruses, the development of latent infections by herpesviruses, and the occurrence of disseminated vaccinial disease should be addressed. Although significant advances have been made in these areas over the past several years, a greater understanding of viral virulence factors should enable the development of safe and effective live recombinant virus vaccine candidates for defined use in the human and veterinary fields.

6. References

Alford, C.A. Jr. and Britt, W.J. (1985) In *Virology* (B. Fields, Ed.) Raven Press, New York. pp. 629–660.

Alonso-Caplen, F.V., Katzee, M.G. and Krug, R.M. (1988) *J. Virol.* 62, 1606–1616.

Andersson, M., Paabo, S., Nilsson, T. and Peterson, P.A. (1985) *Cell* 43, 215–222.

Archard, L.C., Mackett, M., Barnes, D.E. and Dumbell, K.R. (1984) *J. Gen. Virol.* 65, 875–886.

Arnon, R. (1987) In *Synthetic Vaccines* (R. Arnon, Ed.) CRC Press, Boca Raton FL. pp. 1–5.

Auperin, D.D., Esposito, J.J., Lange, J.V., Bauer, S.P., Knight, J., Sasso, D.R. and McCormick, J.B. (1988) *Virus Res.* 9, 233–248.

Babiss, L.E., Ginsberg, H.S. and Darnell, J.E., Jr. (1985) *Mol. Cell. Biol.* 5, 2552–2558.

Ballay, A., Levrero, M., Buendia, M.A., Tiollais, P. and Perricaudet, M. (1985) *EMBO J.* 4, 3861–3865.

Behbehami, A.M. (1983) *Microbiol. Rev.* 47, 455–509.

Ben-Hur, T. Asher, Y., Tabor, E., Darai, G. and Becker, Y. (1987) *Arch. Virol.* 96, 117–122.

Berget, S.M., Moore, C. and Sharp, P.A. (1977) *Proc. Natl. Acad. Sci. USA* 74, 3171–3175.

Berkner, K.L. and Sharp, P.A. (1985) *Nucleic Acids Res.* 13, 841–857.

Berkner, K.L., Schaffhausen, B.S., Roberts, T.M. and Sharp, P.A. (1987) *J. Virol.* 61, 1213–1220.

Bernards, R. and van der Eb, A.J. (1984) *Biochim. Biophys. Acta* 783, 187–204.

Bernards, R., Destree, A., McKenzie, S., Gordon, E., Weinberg, R.A. and Panicali, D. (1987) *Proc. Natl. Acad. Sci. USA*, 84, 6854–6858.

Bhat, B.B. and Wold, W.S.M. (1986) *J. Virol.* 60, 54–63.

Borrelli, E., Hen, R. and Chambon, P. (1984) *Nature* 312, 608–612.

Bos, J.L. and van der Eb., A.J. (1985) *Trends Biochem. Sci.* 10, 310–312.

Boursnell, M.E.G., Foulds, I.J., Campbell, J.I. and Binns, M.M. (1988) *J. Gen Virol.* 69, 2995–3003.

Boyle, D.B. and Coupar, B.E.H. (1988) *Virus Res.* 10, 343–356.

Bridge, E. and Ketner, G. (1989) *J. Virol.* 63, 631–638.

Buller, R.M.L., Smith, G.L., Cremer, K., Notkins, A.L. and Moss, B. (1985) *Nature* 317, 813–815.

Buller, R.M.L., Chakrabarti, C., Cooper, J.A., Twardzik, D.R. and Moss, B. (1988) *J. Virol.* 62, 866–874.

Burgert, H.-G. and Kvist, S. (1985) *Cell 41,* 987–997.

Centifanto-Fitzgerald, Y.M., Yamaguchi, T., Kaufman, H.E., Tognon, M. and Roizman, B. (1982) *J. Exp. Med.* 155, 475–489.

Chakrabarti, S., Robert-Guroff, M., Wong-Staal, F., Gallo, R.C. and Moss, B. (1986) *Nature* 320, 535–537.

Cheng, K.C. and Moss, B. (1987) *J. Virol.* 61, 1286–1290.

Cochran, M.A., Puckett, C. and Moss, B. (1985) *J. Virol.* 54, 30–37.

Coupar, B.E.H., Andrew, M.E., Both, G.W. and Boyle, D.B. (1986) *Eur. J. Immunol.* 16, 1479–1487.

Cremer, K.J., Mackett, M., Wohlenberg, C., Notkins, A.L. and Moss, B. (1985) *Science* 228, 737–740.

Dalgleish, A.G. and Kennedy, R.C. (1988) *Vaccine,* 6, 215–220.

Davidson, D. and Hassell, J.A. (1987) *J. Virol.* 61, 1226–1239.

Davis, A.R., Kostek, B., Mason, B., Hasiao, C.L., Morin, J., Dheer, S.K. and Hung, P.P. (1985) *Proc. Natl. Acad. Sci. USA* 82, 7560–7564.

Deatly, A.M., Spivack, J.G., Lavi, E. and Fraser, N.W. (1987) *Proc. Natl. Acad. Sci. USA* 84, 3204–3208.

Deatly, A.M., Spivack, J.G., Lavi, E., O'Boyle II, D.R. and Fraser, N.W. (1988) *J. Virol.* 62, 749–756.

Dewar, R.L., Natarajan, V., Vasudevachari, M.B. and Salzman, N.P. (1989) *J. Virol.* 63, 129–136.

Dolph, P.J., Racaniello, V., Villamarin, A., Palladino, F. and Schneider, R.J. (1988) *J. Virol.* 62, 2059–2066.

Earl, P.L., Moss, B., Morrison, R.P., Wehrly, K., Nishio, J. and Chesebro, B. (1986) *Science* 234, 728–731.

Edwards, S.J., Triglia, T., Sheppard, M. and Langford, C.J. (1988) In *Technological Science in Vaccine Development* (L. Lasky, Ed) Alan R. Liss, New York. pp. 223–234.

Ellis, R.W., Keller, P.M., Provost, P.J., Han, J., Emini, E.A. and Lowe, R.S. (1988) In *Technological Advances in Vaccine Development* (L. Lasky, Ed.), pp. 235–241. Alan R. Liss, New York.

Ensinger, M.J., Weir, J.P. and Moss, B. (1985) *J. Virol.* 56, 1027–1029.

Esposito, J.J., Cabradilla, C.D., Nakano, J.H. and Obijeski, J.F. (1981) *Virology* 109, 231–243.

Esposito, J.J., Knight, J.C., Shaddock, J.H., Novembre, F.J. and Baer, G.M. (1988) *Virology* 165, 313–316.

Estin, C.D., Stevenson, U.S., Plowman, G.D., Hu, S.-L., Sridhar, P., Hellstrom, I., Brown, J.P. and Hellstrom, K.E. (1988) *Proc. Natl. Acad. Sci. USA* 85, 1052–1056.

Everett, R.D. (1984) *EMBO J.* 3, 3135–3141.

Everett, R.D. (1987) *Anticancer Res.* 7, 589–604.

Fenner, F. (1985) In *Virology* (B.N. Fields, Ed.) Raven Press, New York. pp. 661–684.

Field, H.J. and Coen, D.M. (1986) *J. Virol.* 60, 286–289.

Field, H.J. and Wildy, P. (1978) *J. Hyg.* 81, 267–277.

Flexner, C., Hugin, A. and Moss, B. (1987) *Nature* 330, 259–262.

Flexner, C., Broyles, S.S., Earl, P., Chakrabarti, S. and Moss, B. (1988a) Virology 166, 339–349.

Flexner, C., Murphy, B.R., Rooney, J.F., Wohlenberg, C., Yuferov, V., Notkins, A.L. and Moss, B. (1988b) *Nature* 335, 259–262.

Fraser, N.W., Deatly, A.M., Mellerick, D.M., Muggeridge, M.I. and Spivack, J.G. (1986) In *Human Herpesvirus Infections: Pathogenesis, Diagnosis and Treatment* (C. Lopez and B. Roizman, Eds.) Raven Press, New York. pp. 39–54.

Gilbert, J.H., Pedersen, N.C. and Nunberg, J.H. (1987) *Virus Res.* 7, 49–67.

Gillespie, J.H., Geissinger, C., Scott, F.W., Higgins, W.P., Holmes, D.F., Perkus, M., Mercer, S. and Paoletti, E. (1986) *J. Clin. Microbiol.* 23, 283–288.

Ginsberg, H.S., Valdesuso, J., Horswood, R., Chanock, R.M. and Prince, G. (1987) In *Vaccines '87: Modern Approaches to New Vaccines.* (R.M. Chanock, R.A. Lerner, F. Brown and H. Ginsberg, Eds.) Cold Spring Harbor Laboratory, Cold Spring Harbor NY. pp. 322–326.

Ghosh-Choudhury, G., Haj-Ahmad, Y., Brinkley, P., Rudy, J. and Graham, F.L. (1986) *Gene* 50, 161–171.

Gowda, S.D., Stein, B.S., Steimer, K.S. and Engelman, E.G. (1989) *J. Virol.* 63, 1451–1454.

Graham, F.L. (1984) *EMBO J.* 3, 2917–2922.

Graham, F.L., Smiley, J., Russell, W.C. and Nairn, R. (1977) *J. Gen. Virol.* 36, 59–72.

Graham, F.L., Prevec, L.A., Schneider, M., Ghosh-Choudhury, G., McDermott, M. and Johnson, D.C. (1988) In *Technological Advances in Vaccine Development* (L. Lasky, Ed.) Alan R. Liss, New York. pp. 243–253.

Grose, C. (1987) In *Natural History of Varicella-Zoster Virus* (R.W. Hyman, Ed.) CRC Press, Boca Raton, FL. pp. 1–65.

Gustafson, D.P. (1981) In *Comparative Diagnosis of Viral Diseases* Vol. III (E. Kurstak and C. Kurstak, Eds.) Academic Press, New York. pp. 205–263.

Hen, R., Borrelli, E., Fromental, C., Sassone-Corsi, P. and Chambon, P. (1986) *Nature* 321, 249–251.

Honess, R.W. and Roizman, B. (1974) *J. Virol.* 16, 1308–1326.

Honess, R.W. and Roizman, B. (1975) *Proc. Natl. Acad. Sci. USA* 72, 1276–1280.

Horwitz, M.S. (1985a) In *Virology* (B. Field, Ed.), pp. 433–476. Raven Press, New York.

Horwitz, M.S. (1985b) in *Virology* (B. Field, Ed.), pp. 477–496. Raven Press, New York.

Hruby, D.E. (1985) *Virus Res.* 2, 151–156.

Hruby, D.E., Maki, R.A., Miller, D.B. and Ball, L.A. (1983) *Proc. Natl. Acad. Sci. USA* 80, 3411–3415.

Hu, S.L., Kosowski, S.G. and Dalrymple, J.M. (1986) *Nature* 320, 537–540.

Hu, S.-L., Plowman, G.D., Sridhar, P., Stevenson, U.S., Brown, J.P. and Estin, C.D. (1988) *J. Virol.* 62, 176–180.

Hung, P.P., Bhat, B.M., Chanda, P.K., Chengalvala, M., Dheer, S.K., Lubeck, M.D., Mason, B.B., Molnar-Kimber, K.L., Morin, J.E., Natuk, R.J. and Davis, A.R. (1988) In *Technological Advances in Vaccine Development* (L. Lasky, Ed.), Alan R. Liss, New York. pp. 267–277.

Javier, R.T., Thompson, R.L. and Stevens, J.G. (1987) *J. Virol.* 61, 1978–1984.

Javier, R.T., Izumi, K.M., and Stevens, J.G. (1988) *J. Virol.* 62, 1381–1387.

Johnson, D.C. and Spear, P.G. (1984) *J. Virol.* 51, 389–394.

Johnson, D.C., Ghosh-Choudhury, G., Smiley, J.R., Fallis, L. and Graham, F.L. (1988) *Virology* 164, 1–14.

Joklik, W.K., Pickup, D.J., Patel, D.D. and Moody, M.D. (1988) *Vaccine* 6, 123–128.

Jones, P.C. and Roizman, B. (1979) *J. Virol.* 31, 299–314.

Kaufman, R.J. (1985) *Proc. Natl. Acad. Sci. USA* 82, 689–693.

Keller, P.M., Neff, B.J. and Ellis, R.W. (1984) *J. Virol.* 52, 293–297.

Kit, S. (1988) In *Technological Advances in Vaccine Development* (L. Lasky, Ed.) Alan R. Liss, New York. pp. 183–195.

Kit, S., Kit, M. and Pirtle, E.C. (1985) *Am. J. Vet. Res.* 46, 1359–1367.

Kit, S., Kit, M., Bartkoski, M.J. and Dees, C. (1987) In *Vaccines '88: Modern Approaches to New Vaccines* (R.M. Chanock, R.A. Lerner, F. Brown and H.S. Ginsberg, Eds.) Cold Spring Harbor Laboratory, Cold Spring Harbor, NY. pp. 345–349.

Kotwal, G.J. and Moss, B. (1988) *Nature* 335, 176–178.

Kotwal, G.J. and Moss, B. (1989) *J. Virol.* 63, 600–606.

Kozak, M. and Roizman, B. (1974) *Proc. Natl. Acad. Sci. USA* 71, 4322–4326.

Lane, J.M., Ruben, F.L., Neff, J.M. and Millar, J.D. (1969) *New Engl. J. Med.* 281, 1201–1208.

Langford, C.J., Edwards, S.J., Smith, G.L., Mitchell, G.F., Moss, B., Kemp, D.J. and Anders, R.F. (1986) *Mol. Cell Biol.* 6, 3191–3199.

Lathe, R.S., Kieny, M.P., Gerlinger, P., Clertant, P., Guizani, I., Cuzin, F. and Chambon, P. (1987) *Nature* 326, 878–880.

Lichy, J.H., Field, J., Horwitz, M.S. and Hurwitz, J. (1982) *Proc. Natl. Acad. Sci. USA* 79, 5225–5229.

Logan, J.S. and Shenk, T. (1982) *Microbiol. Rev.* 46, 377–383.

Logan, J.S. and Shenk, T. (1984) *Proc. Natl. Acad. Sci. USA* 81, 3655–3659.

Lomniczi, B., Kaplan, A.S. and Ben-Parat, T. (1987) *Virology* 161, 181–189.

Lowe, R.S., Keller, P.M., Keech, B.J., Davison, A.J., Whang, Y., Morgan, A.J., Kieff, E. and Ellis, R.W. (1987) *Proc. Natl. Acad. Sci. USA* 84, 3896–3900.

Macauley, C., Upton, C. and McFadden, G. (1987) *Virology* 158, 381–393.

Mackett, M. and Arrand, J.R. (1985) *EMBO J.* 4, 3229–3234.

Mackett, M. and Smith, G.L. (1986) *J. Gen. Virol.* 67, 2067–2082.

Mackett, M., Smith, G.L. and Moss, B. (1984) *J. Virol.* 49, 857–864.

Mackett, M., Smith, G.L. and Moss, B. (1985) In *DNA Cloning* Vol. II (D.M. Glover, Ed.) IRL Press, Oxford. pp. 191–212.

Mansour, S.L., Grodzicker, T. and Tijan, R. (1986) *Mol. Cell. Biol.* 6, 2684–2694.

Marchioli, C.C., Yancey, Jr., R.J., Wardley, R.C., Thomsen, D.R. and Post, L.E. (1987) *Am. J. Vet. Res.* 48, 1577–1583.

Martin, S. and Rouse, B.T. (1987) *J. Immunol.* 138, 3431–3437.

Martin, S., Moss, B., Berman, P.W., Lasky, L.A. and Rouse, B.T. (1987) *J. Virol.* 61, 726–734.

Massie, B., Gluzman, Y. and Hassell, J.A. (1986) *Mol. Cell. Biol.* 6, 2872–2883.

McDermott, M.R., Graham, F.L., Hanke, T. and Johnson, D.C. (1989) *Virology* 189, 244–247.

McFarland, M.D., Hill, H.T., Tabatabai, L.B. (1987) *Can. J. Vet. Res.* 51, 334–339.

McLaughlan, J. and Clements, J.B. (1982) *Nucleic Acids Res.* 10, 501–512.

Meignier, B., Longnecker, R. and Roizman, B. (1988a) *J. Infect. Dis.* 158, 602–614.

Meignier, B., Longnecker, R. and Roizman, B. (1988b) In *Vaccines '88* (Chanock, R.M., Lerner, R.A., Brown, F. and Ginsberg, H.S., Eds.) Cold Spring Harbor Laboratory, Cold Spring Harbor NY. pp. 193–196.

Mettenleiter, T.C., Lukacs, N. and Rziha, H.J. (1985) *J. Virol.* 56, 307–311.

Mettenleiter, T.C., Zsak, L., Kaplan, A.S., Ben-Porat, T. and Lomniczi, B. (1987) *J. Virol.* 61, 4030–4032.

Mocarski, E.S., Post, L.E. and Roizman, B. (1980) *Cell* 22, 243–255.

Moore, M.A. and Shenk, T. (1988) *Nucleic Acids. Res.* 16, 2247–2262.

Morein, B., Lovgren, K., Hoglund, S. and Sundquist, B. (1987) *Immunol. Today* 8, 333–338.

Moran, E. and Matthews, M.B. (1987) *Cell* 48, 177–178.

Morgan, A.J., Mackett, M., Finerty, S., Arrand, J.R., Scullion, F.T. and Epstein, M.A. (1988) *J. Med. Virol.* 25, 189–195.

Morin, J.E., Lubeck, M.D., Barton, J.E., Conley, A.J., Davis, A.R. and Hung, P.P. (1987) *Proc. Natl. Acad. Sci. USA* 84, 4626–4630.

Morrison, D.K. and Moyer, R.W. (1986) *Cell* 44, 587–596.

Moss, B. (1985) in *Virology* (B.N. Fields, Ed.), pp. 685–704. Raven Press, New York.

Moss, B. and Flexner, C. (1987) *Annu. Rev. Immunol.* 5, 305–324.

Moss, B., Winters, E. and Cooper, J.A. (1981) *J. Virol.* 40, 387–395.

Moss, B., Smith, G.L., Gerin, J.L. and Purcell, R.H. (1984) *Nature* 311, 67–69.

Moyer, R.W., Graves, R.L. and Rothe, C.T. (1980) *Cell* 22, 545–553.

Nakayama, K., Kelly, S.M. and Curtis III, R. (1988) *Bio/Technology* 6, 693–697.

Oda, K.I. and Joklik, W.K. (1967) *J. Mol. Biol.* 27, 395–419.

Pacini, D.L., Dubovi, E.J. and Clyde, Jr., W.A. (1984) *J. Infect. Dis.* 150, 92–97.

Paez, E., Dallo, S. and Esteban, M. (1985) *Proc. Natl. Acad. Sci. USA,* 82, 3365–3369.

Panicali, D. and Paoletti, E. (1982) *Proc. Natl. Acad. Sci. USA* 79, 4927–4931.

Panicali, D., Davis, S.W., Mercer, S.R. and Paoletti, E. (1981) *J. Virol.* 37, 1000–1010.

Paoletti, E. and Grady, L.J. (1977) *J. Virol.* 23, 608–615.

Paoletti, E., Lipinskas, B.R., Samsonoff, C., Mercer, S. and Panicali, D. (1984) *Proc. Natl. Acad. Sci. USA* 81, 193–197.

Patel, D.D., Pickup, D.J. and Joklik, W.K. (1986) *Virology* 149, 174–189.

Perkus, M.E., Piccini, A., Lipinskas, B.R. and Paoletti, E. (1985) *Science* 229, 981–984.

Perkus, M.E., Panicali, D., Mercer, S. and Paoletti, E. (1986) *Virology* 152, 285–297.

Piccini, A., Perkus, M.E. and Paoletti, E. (1987) *Methods Enzymol*, 153, 545–563.

Pickup, D.J., Ink. B.S., Hu, W., Ray, C.A. and Joklik, W.K. (1986b) *Proc. Natl. Acad. Sci. USA* 83, 7698–7702.

Pickup, D.J., Ink, B.S., Parsons, B.L., Hu, W. and Joklik, W.K. (1986a) *Proc. Natl. Acad. Sci. USA* 81, 6817–6821.

Plotkin, S.A. and Huang, E.-S. (1985) *J. Infect. Dis.* 152, 395–397.

Plotkin, S.A., Furukawa, T., Zygraich, N. and Huygelen, C. (1975) *Infect. Immun.* 2, 521–527.

Plotkin, S.A., Farquhar, J. and Hornberger, E. (1976) *J. Infect. Dis.* 134, 470–475.

Plotkin, S.A., Friedman, H.M., Fleisher, G.R., Dafoe, D.C., Grossman, R.A., Smiley, M.L., Wlodaver, C., Friedman, A.D. and Barker, C.F. (1984) *Lancet i*, 528–530.

Post, L.E. and Roizman, B. (1981) Cell 25, 227–232.

Prevec, L., Schneider, M., Rosenthal, K.L., Belbeck, L.W., Derbyshire, J.B., and Graham, F.L. (1989) *J. Gen. Virol.* 70, 429–434.

Provost, P.J., Keller, P.M., Banker, F.S., Keech, B.J., Klein, H.J., Lowe, R.S., Morton, D.H., Phelps, A.H., McAlear, W.J. and Ellis, R.W. (1987) *J. Virol.* 61, 2951–2955.

Ramshaw, I.A., Andrew, M.E., Phillips, S.M., Boyle, D.B. and Coupar, B.E.H. (1987) *Nature* 329, 545–546.

Rando, R.F., Pellett, P.E., Luciw, P.A., Bohan, C.A. and Srinivasan, A. (1987) *Oncogene* 1, 13–18.

Rawls, W.E. (1985) In *Virology* (B. Fields, Ed.) Raven Press, New York pp. 527–561.

Rea, T.J., Timmins, J.G., Long, G.W. and Post, L.E. (1985) J. Virol. 54, 21–29.

Reich, P.R., Rose, J., Forget, B. and Weissmann, S.M. (1966) *J. Mol. Biol.* 17, 357–405.

Rekosh, D.M.K., Russell, W.C., Bellet, A.J.D. and Robinson, A.J. (1977) *Cell* 11, 283–295.

Robinson, A.J., Younghusband, H.B. and Bellet, A.J.D. (1973) *Virology* 56, 54–69.

Rodriguez, D., Rodriguez, J.-R., Rodriguez, J.F., Trauber, D. and Esteban, M. (1989) *Proc. Natl. Acad. Sci. USA* 86, 1287–1291.

Roizman, B. and Batterson, W. (1985) In *Virology* (B. Fields, Ed.), Raven Press, New York. pp. 497–526.

Roizman, B. and Sears, A.E. (1987) *Annu. Rev. Microbiol.* 41, 543–571.

Rooney, J.F., Wohlenberg, C., Cremer, K.J., Moss, B. and Notkins, A.L. (1988) *J. Virol.* 62, 1530–1534.

Saito, I., Oya, Y., Yamamoto, K., Yuasa, T. and Shimojo, H. (1985) *J. Virol.* 54, 711–719.

Sandler, A.B. and Ketner, G. (1989) *J. Virol.* 63, 624–630.

Schneider, M., Graham, F.L. and Prevec, L. (1989) *J. Gen. Virol.* 70, 417–427.

Sedarti, F. and Stevens, J.G. (1987) *J. Gen. Virol.* 68, 2389–2395.

Severinsson, L. and Peterson, P.A. (1983) *J. Cell. Biol.* 99, 226–232.

Shida, H. (1986) *Virology* 150, 451–462.

Shida, H., Tochikura, T., Sato, T., Konno, T., Hirayoshi, K., Seki, M., Ito, Y., Hatanaka, M., Hinuma, Y., Sugimoto, M., Takahashi-Nishimaki, F., Maruyama, T., Miki, K., Suzuki, K., Morita, M., Sashiyama, H. and Hayami, M. (1987) *EMBO J.* 6, 3379–3384.

Shida, H., Hinuma, Y., Hatanaka, M., Morita, M., Kidokoro, M., Suzuki, K., Maruyama, T., Takahashi-Nishimaki, F., Sugimoto, M., Kitamura, R., Miyazawa, T. and Hayami, M. (1988) *J. Virol.* 62, 4474–4480.

Shih, M.F., Arsenakis, M., Tiollais, P. and Roizman, B. (1984) *Proc. Natl. Acad. Sci. USA* 81, 5867–5870.

Siekierka, J., Mariano, T.M., Reichel, P.A. and Mathews, M.B. (1985) *Proc. Natl. Acad. Sci. USA* 82 1959–1963.

Small, P.A., Jr., Smith, G.L. and Moss, B. (1985) In *Vaccinia Viruses as Vectors for Vaccine Antigens* (G.V. Quinnan, Jr., Ed.) Elsevier, Amsterdam. pp. 175–178.

Smith, G.L. and Moss, B. (1983) *Gene* 25, 21–28.

Smith, G.L., Mackett, M. and Moss, B. (1983) *Nature* 302, 490–495.

Spaete, R.R. and Mocarski, E.S. (1987) *Proc. Natl. Acad. Sci. USA* 84, 7213–7217.

Spivack, J.G. and Fraser, N.W. (1988) *J. Virol.* 62, 3281–3287.

Svonsson, C. and Akusjari, G. (1984) *EMBO J.* 3, 789–794.

Takahashi, M. (1987) In *Natural History of Varicella-Zoster Virus* (R.W. Hyman, Ed.) CRC Press, Boca Raton, FL. pp. 179–209.

Takahashi, M., Otsuka, T., Okuno, Y., Asano, Y., Yazaki, T. and Isomura, S. (1974) *Lancet ii,* 1288–1290.

Taylor, J., Weinberg, R., Languet, B., Desmettre, P. and Paoletti, E. (1988a) *Vaccine 6,* 497–503.

Taylor, J., Weinberg, R., Kawaoka, Y., Webster, R.G. and Paoletti, E. (1988b) *Vaccine* 6, 504–508.

Tartarov, G. (1968) *Zentralbl. Veterinärmed.* 15, 847–853.

Tartarov, G. and Gergov, P. (1969) *Vet. Med. Nauki* 6, 49–54.

Thimmappaya, B., Weinberger, C., Schneider, P.J. and Shenk, T. (1982) *Cell* 31, 543–551.

Thomas, G., Herbert, E. and Hruby, D.E. (1986) *Science* 232, 1641–1645.

Thompson, C.L. and Condit, R.C. (1986) *Virology* 150, 10–20.

Thomsen, D.R., Marchioli, C.C., Yancey, Jr., R.J. and Post, L.E. (1987a) *J. Virol.* 61, 229–232.

Thomsen, D.R., Marotti, K.R., Palermo, D.P. and Post, L.E. (1987b) *Gene* 57, 261–265.

Thummel, C., Tijan, R., Hu, S.-H. and Grodzicker, T. (1983) *Cell* 33, 455–464.

Top, F.H., Dudding, B.A., Russell, P.K. and Buescher, E.L. (1971) *Am. J. Epidemiol.* 94, 142–146.

Townsend, A., Bastin, J., Gould, K., Brownlee, G., Andrew, M., Coupar, B., Boyle, D., Chan, S. and Smith, G. (1988) *J. Exp. Med.* 168, 1211–1224.

Van der Vliet, P.C. and Levine, A.J. (1973) *Nature* 246, 170–174.

Van der Vliet, P.C., and Levine, A.J., Ensinger, M.J. and Ginsberg, H.S. (1975) *J. Virol.* 15, 348–354.

Van Oirschot, J.T. and Gielkens, A.L. (1987) *Vet. Q.* 9, 375–495.

Van Zijl, M., Quint, W., Briaire, J., de Rover, T., Gielkens, A. and Berns, A. Berns, A. (1988) *J. Virol.* 62, 2191–2195.

Vassef, A., Ben-Hamida, F., Dru, A. and Beaud, G.A. (1982) *Virology* 118, 45–53.

Velcich, A. and Ziff, E. (1985) *Cell* 40, 705–716.

Vos, J.C. and Stunnenberg, H.G. (1988) *EMBO J.* 7, 3487–3492.

Wachsman, M., Aurelian, L., Smith, C.C., Lipinskas, B.R., Perkus, M.E. and Paoletti, E. (1987) *J. Infect. Dis.* 155, 1188–1197.

Wachsman, M., Aurelian, L., Hunter, J.C.R., Perkus, M.E. and Paoletti, E. (1988) *Biosci. Rep.* 8, 323–334.

Webster, K.A., Muscat, G.E.O. and Kedes, L. (1988) *Nature* 332, 553–556.

Weinmann, R., Raskas, H.J. and Roeder, R.G. (1974) *Proc. Natl. Acad. Sci. USA* 71, 3426–3430.

Weir, J.P., Bajszar, G. and Moss, B. (1982) *Proc. Natl. Acad. Sci. USA* 79, 1210–1214.

Williamson, J.D., Reith, R.W., Arrand, J.R. and Mackett, M. (1988) In *Technological Advances in Vaccine Development* (L. Lasky, Ed.) Alan R. Liss, New York. pp. 205–214.

Wilton, S. and Dales, S. (1989) *J. Virol.* 63, 1540–1548.

Zanetti, M., Sercarz, E. and Salk, J. (1987) *Immunol. Today* 8, 18–25.

Zarling, J.M., Morton, W., Moran, P.A., McClure, J., Kosowski, S.G. and Hu, S.L. (1987) *Nature* 323, 344–346.

Zhu, X., Young, C.S.H. and Silverstein, S. (1988) *J. Virol.* 62, 4544–4553.

Eds. M.H.V. van Regenmortel and A.R. Neurath
Immunochemistry of viruses, II. The basis for serodiagnosis and vaccines.

Picornaviruses 9

FRED BROWN

Department of Virology, Wellcome Biotech, Langley Court, Beckenham, Kent BR3 3BS, U.K.

Picornaviruses cause a wide range of diseases in a similarly wide range of host species (Table 9.1). The best known diseases caused by these viruses are poliomyelitis, hepatitis and the common cold in man and foot-and-mouth disease in domestic animals. In addition there are many diseases which have less serious consequences but the study of their causal agents has added considerably to our knowledge of picornaviruses. In particular, the relatively simple structure of this family of viruses has allowed detailed comparisons to be made at the atomic level. Moreover, the importance of the diseases caused by several of the members has led to the development of vaccines which in turn has encouraged a study of their immunochemistry.

Poliovirus was a devastating disease of man until the Salk and Sabin vaccines were produced and applied on a global scale in the 1950s and 1960s. Unfortunately, for reasons which are not clear, a considerable number of cases of the disease still occur each year in under-developed countries, sufficient for the World Health Organization to have made it a target for eradication by the year 2000. By comparison, the common cold is trivial; nevertheless its nuisance value has meant that it has received considerable attention for several years, even to the extent of its three-dimensional structure having been determined (Rossmann et al., 1985). Hepatitis A virus also causes a devastating disease. Less is known about this virus because of the considerable difficulties in growing it in the amounts required for detailed biochemical analysis but the biological differences in terms of its growth characteristics make it an intriguing target for research in fundamental as well as in practical terms. Other diseases of man caused by picornaviruses include the wide range of symptoms caused by the Coxsackie and echoviruses, and conjunctivitis caused by enterovirus 70. In

TABLE 9.1

Diseases caused by picornaviruses

Genus	Disease	Agent	Species affected
Entero	Poliomyelitis	Polio	Man
	Hepatitis	Hepatitis A	Man
	Aseptic meningitis	Coxsackie A and B	Man
	Herpangina	ECHO	
	Pleurodynia		
	Myalgia		
	Swine vesicular disease	Coxsackie B5	Pigs
	Conjunctivitis	Enterovirus 70	Man
	Hepatitis	Duck hepatitis virus	Ducks
Cardio	Encephalomyocarditis		Mice
	Myocarditis	EMC	Pigs Chimpanzees Monkeys
	Fever		Man
Rhino	Common cold	Human rhinovirus	Man
	Rhinitis, pharyngitis	Equine rhinovirus	Horses
Aphtho	Foot-and-mouth disease	FMDV	Cattle sheep, pigs goats
Unclassified	Encephalitis	Theiler's disease virus	Mice

animals, the best known disease caused by a picornavirus is foot-and-mouth disease. In addition, however, and not generally recognized, some strains of encephalomyocarditis virus cause a devastating and frequently fatal disease in pigs (Murnane et al., 1960). This disease has occurred in several countries but no work has been done to identify those factors which determine its virulence for pigs.

There are at least 70 viruses which are included in the family Picornaviridae. These viruses were first divided into four genera on the basis of their stability at different acid pHs (Melnick et al., 1963). Subsequently, detailed examination of the particles by more sophisticated methods of analysis has confirmed this division and four genera are now recognised: entero, cardio, rhino and aphtho. Although the properties of each of the genera differ from each other in certain significant features (Table 9.2), their three-dimensional structures are very similar (Rossmann et al., 1985; Hogle et al., 1985; Luo et al., 1987; Acharya et al., 1989). Now that these

TABLE 9.2

Physico-chemical properties of picornaviruses

Genus	pH of disruption	Buoyant density in CsCl (g/cc)	Susceptibility to photodynamically active dye
Entero	<3	1.34	No
Cardio	3 But unstable at pH 6.2 in 0.1M chloride	1.34	
Rhino	5	1.40	
Aphtho	6.8	1.43	Yes

structures are available for one member of each of the genera it will be interesting to see whether the different physicochemical properties such as differential lability at pH values below 7 and density in CsCl can be explained at this level of sophistication.

1. Structure of picornaviruses

In addition to the infectious virus particle, two other well recognized virus-specified particles are found in harvests produced in a variety of cells. These are the so called empty particles and the capsid protein subunit. The intact virus particle consists of one molecule of single-stranded RNA and 60 copies of each of four proteins VP1–VP4. The empty particles, which are similar to the full particles in size and shape, lack the RNA, and the proteins VP2 and VP4 are in an uncleaved form, VP0. With the exception of the cardioviruses these particles appear to be present in all the picornaviruses.

With the viruses causing encephalomyocarditis and foot-and-mouth disease, there are also naturally occurring viral subunits composed of pentamers of the proteins VP1–VP3. These particles can also be prepared from the intact virus particles by lowering the pH of the medium marginally below 7. For EMCV there is a requirement for chloride ions for this disruption to occur (see section 1.2).

In addition to these particles an additional protein is found in large amounts in harvests of FMDV. This protein has been analysed in some detail and appears to be an inactive form of the viral polymerase (Cowan and Graves, 1966; Polatnick et al., 1967; Newman et al., 1979). It differs from the other virus specified particles in not being type-specific in serological tests.

1.1. Morphology

The intact viruses are roughly spherical nucleocapsids, 30 nm in diameter which have icosahedral symmetry with a triangulation number of 1. There is no envelope or core and the surface is almost featureless. However, complexes of the virus particles with antibody, particularly FMDV (Brown and Smale, 1970) reveal structural features which have provided important information about their antigenicity (see section 2).

The empty particles can be readily distinguished from the intact particles because they are penetrated by negative stains. However, they are the same size and shape as the intact particles.

1.2. Physicochemical properties of the virus particles

The intact viruses have a molecular weight of 8–9 $\times$ 10^6 and their sedimentation coefficients range from 140 to 165S. This difference in sedimentation behavior is probably related to their different densities. It is clear from their buoyant densities in CsCl, which vary between 1.34 g/cc for the entero and cardioviruses and 1.43 g/cc for FMDV (Table 9.2), that there are important differences in the way in which the proteins are assembled in the virus particles. In addition there are also differences in the stability of the particles below pH 7 and their susceptibility to photodynamically active dyes. Thus those viruses with a low density in CsCl (i.e.,1.34 g/cc) are stable between pH 7 and pH 3 whereas FMDV is unstable just below pH 7, disintegrating into the genomic RNA and a pentameric protein subunit consisting of the three surface proteins VP1–VP3. The internal protein VP4 forms an aggregate when released from the particles. Intriguingly, EMC virus (density 1.34 g/cc) also disrupts into similar subunits at pH 6.2 in the presence of 0.1 M chloride ions.

Photodynamic dyes such as proflavine penetrate the FMDV particle, intercalate with the RNA and disrupt the capsid. In contrast, the dyes do not penetrate enterovirus particles. The recently described three-dimensional structures of the picornaviruses provide a possible explanation for these differences (see section 1.5).

The empty particles sediment at 70S and their density in CsCl is consistent with the absence of RNA. The pentamers of VP1–VP3 sediment at 12–14S.

1.3. Properties of the genomic RNA

The virus genome consists of one molecule of ssRNA, of approximately 2.5 $\times$ 10^6 Da. In all instances where the RNA has been studied, a covalently linked protein of approximately 2.6 kDa has been found at the 5′ end of the molecule and a polyadenylic acid tract of variable length at the 3′ end. Despite considerable effort, the role of these structures is still not clear. A distinctive feature is found in the cardio and aphthovirus genomes, namely a polycytidylic acid tract near the 5′ end of the molecule. The length of the tract varies considerably between different isolates but the position in the genome is constant for the aphthoviruses (approximately 400 bases from the 5′ end) and cardioviruses (approximately 150 bases from the 5′ end). The role of the polycytidylic acid tract in the structural and biochemical organization of the particles has not been determined.

The nucleotide sequences of several members of each genus have been determined during the last decade and taxonomic relationships based on these sequences have been described (Palmenberg, 1989). It is gratifying that the earlier grouping of the picornaviruses into the four genera on the basis of acid stability and density in CsCl has been confirmed by the relationships between the nucleotide sequences although it is significant that one of the more thoroughly analysed human rhinoviruses, HRV14, fits more comfortably into the enterovirus genus on the basis of its sequence.

1.4. Properties of the virus proteins

The virus particle contains 60 copies of each of four proteins VP1–VP4. Their molecular weights and in many instances their sequences are known with certainty, having been derived from the sequences of the genes coding for them. Protein VP1 appears to be the most exposed on the surface followed by VP2 and VP3 whereas VP4 is internal. The number of amino acids in VP1 ranges from 213–306, VP2 from 218–272, VP3 from 220–238 and VP4 from 69–85. Protein VP4 also has an N-terminal myristic acid whose function is not known with certainty but is likely to be concerned with interaction with cell membranes during the early stages of infection and/or with assembly of the capsid (Chow et al., 1987).

1.5. Three-dimensional structure of the viruses

With the publication of the three-dimensional structure of FMDV early in 1989 (Acharya et al., 1989), the structure of examples of viruses belonging to each of the genera has now been determined (Rossmann et al., 1985; Hogle et al., 1985; Luo et al., 1987). The information has not only made it possible to start to reach an understanding of the conformation of antigenic and immunogenic sites and the relationships between individual epitopes but has also provided possible explanations for other properties of the viruses.

The overall shape of the well-ordered electron density is that of a spherical shell starting at a radius of 100 Å from the particle centre and extending to an outer radius of 150 Å. Whereas polio, rhino and Mengo viruses have canyons or pits on their surface, FMDV has a smooth surface and, as would be expected from the considerably smaller size of proteins VP1 and VP2, the protein shell is thinner. However, the most significant difference, particularly with regard to the antigenicity of the particle, is the presence of a prominent disordered loop encompassing residues 133–158 of VP1.

The three-dimensional structure has provided a possible explanation for the acid lability of FMDV, its relatively high density in CsCl and penetration with intercalating dyes. As mentioned above, reducing the pH below 7 disrupts the virus particle into a 12S subunit comprising a pentamer of VP1–VP3. In FMDV there is a high density of histidine residues lining the pentamer interfaces. As the pK value of this amino acid correlates with the pH at which disassembly occurs, this may provide an explanation for the observed lability. The relatively high density in CsCl and penetrability with intercalating dyes can be explained by the presence of a hole at the 5-fold axis of the particle which is large enough (11 Å average diameter) to allow these molecules to penetrate and interact with the virus RNA.

The disordered loop encompassing residues 133–158 of VP1 of FMDV contains the major immunogenic site of the virus and the flexibility may account for the high level of neutralizing antibody elicited by peptides corresponding to this region of the virus. The other region of VP1 at the C terminus which has been clearly identified as possessing immunogenic activity is located in close proximity to the loop on the surface of the particle and it is considered likely that both these regions form part of a single epitope (Parry et al., 1985; DiMarchi et al., 1986). The situation with the other picornaviruses appears to be more complex and is discussed below.

2. Antigenic properties

The most important aspect of the immunochemistry of picornaviruses relates to the protection of the host species against infection. With this family of viruses, protection can be correlated with the level of neutralizing antibody in the serum, whether this was elicited by a previous infection or by vaccination. One of the more intriguing features which has emerged is the antigenic diversity which exists with some viruses and the lack of it with others. This difference is interesting taxonomically and is also of considerable practical importance in the prevention of disease by vaccination. It is also valuable in allowing us to speculate and even to draw some tentative conclusions about the structural basis for antigenic variation. Indeed, now that nucleotide sequences are available for many of the picornaviruses (Palmenberg, 1989), enlightened comparisons can be made between the different serotypes of individual viruses. Together with the three-dimensional structures which have become available recently, this information should lead to a much clearer understanding of the immunochemistry of the viruses.

2.1. Antigenic variability

There are remarkable differences in the antigenic variability of the picornaviruses, even for those members belonging to the same genus (Table 9.3). In most instances variability is measured by virus neutralization tests since it is generally accepted that

TABLE 9.3

Antigenic variability in the picornaviruses

Genus	Virus	Number of serotypes
Entero	Polio	3
	Coxsackie A	23
	Coxsackie B	6
	Echo	32
	Hepatitis A	1
Cardio	Encephalomyocarditis	1
Rhino	Human rhino	>100
	Equine rhino	2
Aphtho	Foot-and-mouth disease	7 Multiple sub-types

protection is correlated with the level of neutralizing antibody in infections caused by picornaviruses. Because protection studies cannot be made with most of the diseases due to the lack of suitable experimental models and the impossibility in most cases of using the natural host, namely man, data from work with FMDV have proved important in establishing guidelines for the group. With FMD, the level of neutralizing antibody in the serum correlates well with the degree of protection, indicating that neutralization tests provide a relevant basis for measuring antigenic variability.

Members of the enterovirus genus show remarkable differences in antigenic variability. There are 24 serotypes of Coxsackie A virus, 6 serotypes of Coxsackie B virus and more than 30 serotypes of ECHO virus. In contrast to this wide range of variability, poliovirus occurs only as 3 serotypes and hepatitis A virus as one serotype.

The cardioviruses are regarded as an antigenic unity because they cannot be distinguished by neutralization tests. There are clear biological differences in isolates of the virus but little precise differential serology has been reported.

The rhinoviruses provide the extreme example of antigenic variation with more than 100 different serotypes. Because of the lack of an animal model there is little soundly based information on the relationship between the level of neutralizing antibody and protection. However, in necessarily limited studies, Barclay and Al-Nakib (1987) have shown that, in human volunteers, those with high levels of neutralizing antibody in the serum escape illness on intranasal challenge. It seems reasonable to conclude, therefore, that neutralizing antibody is a valuable indicator of antigenic variation and is of value in assessing immunity.

With FMDV the relationship between the level of neutralizing antibody in the serum and protection is much more firmly established because of the requirements which are imposed on the vaccines to be used in the field. To be accepted for use, vaccines must protect cattle against a severe challenge infection 28 days following a single injection of the product. Normally the neutralizing antibody titer of the serum immediately before challenge is measured. Consequently many thousands of observations have been made which have allowed the relationship between neutralization data and protection to be established.

The seven major serotypes were first clearly defined by protection studies and neutralizing antibody assays confirmed this division. In addition many sub-types have been identified. These were recognized when a vaccine which was expected to protect against viruses within the same serotype failed to do so. This is clearly of great practical importance and manufacturers are careful to check whether their available vaccines are sufficiently closely related antigenically to the viruses causing the outbreaks at any time. With the constant emergence of new strains of the viruses, the problems confronting the vaccine manufacturers can be readily appreciated.

2.2. Vaccines

One of the main reasons for studying the immunochemistry of viruses is to gain a better understanding of the structural features which induce protection against infection. It has been generally recognized for several years that with picornaviruses the intact virus particle is required to effect protection. This is particularly clear-cut for FMDV where it has been shown that the 12S protein subunit, consisting of proteins VP1, VP2 and VP3, elicits levels of neutralizing antibody which are only about 1% of those produced by the intact virus particle. A similar low response has been obtained with the 14S subunit of encephalomyocarditis virus (F. Brown, unpublished results.

With some strains of FMDV, treatment with trypsin or other proteolytic enzymes leads to the loss of immunogenic activity. The loss is caused by cleavage of the capsid protein VP1 at surface Arg and Lys residues. With virus of serotype O, which has been studied extensively, this results in the loss of fragments of the protein from the C terminal region and from between residues 138 and 154. This evidence has led to the conclusion that the major immunogenic site encompasses these regions of the protein and a synthetic peptide corresponding to 141–160 region alone elicits neutralizing antibody sufficient to protect experimental animals against infection (Bittle et al., 1982; Di Marchi et al., 1986). Since the first report with virus of serotype O, peptides corresponding to the same region of all other serotypes of the virus have been shown to elicit similar responses (Francis et al., 1990). Of considerable interest and potential importance is the observation that the antibodies elicited by the peptides are more cross-reactive than those produced by the classical inactivated vaccines. A detailed analysis of this region of the virus is made in section 3.

While the lack of immunogenicity is clear-cut with regard to the protein subunit, the role of the empty particles in eliciting protective antibody is less clear. There is substantial evidence that the empty particles of FMDV are as efficient as the intact particles in absorbing neutralizing antibody from convalescent serum (Rowlands et al., 1972; Rweyemamu et al., 1974). However, it is sometimes necessary for the empty particles to be stabilized (e.g.,by incubating with formaldehyde) for them to be effective as immunogens. With poliovirus there seems to be some discrepancy between the reports of different authors. Early work had suggested that the empty or C particles were ineffective in eliciting neutralizing antibody but there are also reports that empty particles are in fact effective. The three-dimensional structure of the empty particles, currently being studied by Hogle and his colleagues (personal communication) should help to resolve this dilemma.

Much less information is available for other picornaviruses. The considerable antigenic diversity of the common cold viruses seemed to put the possibility of effective vaccines beyond reach. This diversity, together with the lack of a suitable animal model for the testing of ideas has severely restricted research in this area and it is only fairly recently, as the sequences and three-dimensional structure of the virus have become available that there has been a re-awakening of interest in the possibility of producing effective vaccines which may be protective against more than one serotype.

3. Immunogenic activity of isolated proteins and peptides

There is considerable evidence from work with enveloped viruses that individual proteins will elicit neutralizing antibody in sufficient amounts to protect the host against infection. In all these instances, the viruses were disrupted into their constituent proteins by treatment with mild detergents. Picornaviruses require more vigorous treatment to disrupt them and the consequent loss of conformation of the individual proteins was expected to lead to inactive products. Indeed, there is much evidence that, even with the mild acid disruption of FMDV and EMC viruses, there is a considerable loss of immunogenic activity. Nevertheless, significant information has emerged from these studies to provide a framework of understanding of the immunochemistry of picornaviruses.

3.1. Isolated proteins

Early work with FMDV provided substantial evidence that the immunogenic activity of the particle was largely associated with only one of the capsid proteins, namely VP1. There were several pieces of evidence to support this contention. (a) Treatment of some strains of the virus with trypsin markedly reduced their immunogenic activity (Wild et al., 1969). The only change appeared to be the cleavage of VP1

(Burroughs et al., 1971). (b) Isolated VP1 elicited neutralizing antibody and afforded protection against challenge whereas the other proteins had no activity (Laporte et al., 1973; Bachrach et al., 1975). However the level of neutralizing antibody produced by VP1 was very low compared with that elicited by the virus particle.

It could be argued that the other proteins also contain immunogenic sites when present on the virus but that these are lost when the particle is disrupted. Nevertheless, the marked loss of immunogenic activity with virus of serotype O when it is treated with trypsin, when VP2, VP3 and VP4 are apparently unaffected, is highly suggestive that the major activity is associated with VP1.

Chow and Baltimore (1982) showed that VP1 of poliovirus will also elicit neutralizing antibody but, as with the corresponding protein in FMDV, the response was very low compared with that elicited by the intact virus particle. With poliovirus, however, there is some evidence that proteins VP2 and VP3 will also elicit low levels of neutralizing antibody (Dernick et al., 1983). There is general agreement that VP4 does not stimulate the production of neutralizing antibody.

There are no published reports on similar experiments with other picornaviruses. However, M.J. Francis (unpublished observations) has found that although disrupted rhinovirus particles elicited high levels of antibody which reacted with particles disrupted by a variety of methods, there was no measurable reaction with the intact virus particle.

The clear message that emerges from these studies is that the configurations of the individual proteins when they form part of the capsid structure are lost when the virus particle is disrupted. Consequently it is unlikely that even if the immunodominant site(s) is confined to one of the proteins, the activity will be as high as when it forms part of the particle.

3.2. Peptides

The substantial evidence that VP1 of FMDV possesses significant immunogenic activity has led to considerable work on the active sites of this protein. With the emphasis on protection and consequently on neutralization tests, most of the work has focused on those sites which are concerned with the stimulation of neutralizing antibody. Several approaches have been used to identify the antigenic sites on the protein. In pioneering work, Strohmaier and his colleagues (1982) analyzed fragments of the protein obtained by proteolytic enzyme or cyanogen bromide cleavage and concluded that the immunogenic activity would be confined to two regions encompassing amino acids 146–154 and 200–213.

In a highly variable virus such as FMDV, it would be expected that antigenic differences would correlate with amino acid sequence differences. The occurrence of seven serotypes of the virus and the multiplicity of antigenic variants within these serotypes meant that ample material was available to test the correctness of this postulated correlation. Studies of the growing number of sequences of VP1 which have been available have shown that there is about 98% conservation of the amino acid sequences of VP4, 90% of VP3 and VP2 and 80% of VP1. More particularly, sequences of variants within a serotype are more conserved. Of particular importance was the observation that sequence variability occurs mainly at three sites in VP1, namely in the regions 42–61, 131–160 and 193–204. Taken together with the evidence that proteolytic enzymes cleave intact virus particles at the latter two regions, it seemed likely that the immunogenic major site was located in those two regions.

In another approach the 'Pepscan' method described by Geysen et al. (1984) identified amino acid sequence 146–152 as a potential immunogenic site. In this method hexapeptides corresponding to overlapping sequences of the entire amino acid sequence of VP1 were allowed to react with the type-specific antibody and the reactive regions of the protein detected with anti-species antiserum. Taken together with the other two approaches, amino acid sequence 141–160 appeared to contain the most immunogenic site and in general work has concentrated on this region of VP1.

In addition, antigenic sites have been identified by growing the virus in the presence of neutralizing monoclonal antibodies. This approach has revealed the presence of sites on proteins VP2 and VP3 in addition to those on VP1 (Xie et al., 1987; Thomas et al., 1988; Stave et al., 1988; McCahon et al., 1989). However, it is likely that the immunodominant site is contained in the loop region of VP1 because peptides corresponding to this sequence absorb greatly in excess of 50% of the neutralizing antibody of polyclonal sera from both convalescent and vaccinated animals (S.M. Russell and M.J. Francis, unpublished data). Indeed, the immunogenic activity is sufficiently high to have received serious consideration as a potential vaccine (Brown, 1989). The advantages of such a vaccine are summarized in Table 9.4. In addition there is now important evidence that the antibody produced is more cross-

TABLE 9.4

Advantages of a peptide vaccine

The product would be chemically defined.
The peptide would be stable indefinitely.
No infectious agent would be used in its production.
No large scale production plant or downstream processing would be required.
The product could be designed to stimulate appropriate immune responses and omit those sequences which elicit undesirable side reactions.
An opportunity would be provided to use delayed release mechanisms.

reactive than that elicited by intact virus particles (Ouldridge et al., 1986). There is also preliminary evidence that anti-peptide antibody will neutralize viruses belonging to other serotypes and in one experiment described by Parry et al. (1989) it was shown that substitution of the Leu residue at position 148 of the type O peptide by Ser gave a peptide which elicited a protective response against both type O and type A viruses. A comparative analysis of these peptides by nuclear magnetic resonance and circular dichroism should yield useful information for the design of molecules which will elicit responses with even greater cross-reactivity (see section 3.3).

3.3. Effect of amino acid substitutions on antigenic activity of peptides

The influence of single amino acid changes at positions 148 and 153 of the loop region on the antigenic character of the virus and the corresponding peptide have been described for FMDV of serotype A, sub-type 12 (Rowlands et al., 1983; Mascagni et al., submitted for publication). Four viruses whose entire capsid protein regions differed only at positions 148 and 153 of VP1 could be distinguished by neutralization and immunoprecipitation tests with polyclonal sera against the intact virus or the anti 141–160 peptide. The results in Table 9.5 which illustrate these differences also show that the viruses can be grouped A with C and B with D.

Predictions of the structure of the peptides by the Chou-Fasman (1978) and Ni-

TABLE 9.5

Effect of single amino acid substitutions on the antigenicity of FMDV particles and peptides corresponding to the loop region 141–160

Antiserum	Amino acid at 148 and 153		Log$_{10}$ virus neutralized by 0.015 ml serum			
			A	B	C	D
		148	Ser	Leu	Ser	Phe
		153	Leu	Pro	Ser	Pro
Virus	Ser	Leu	3.3	1.3	1.7	0.7
Peptide			2.5	1.1	1.5	0.9
Virus	Leu	Pro	0.9	3.7	0.9	2.1
Peptide			0.0	3.2	0.5	2.2
Virus	Ser	Ser	3.5	2.9	4.1	2.5
Peptide			1.5	0.1	2.6	1.5
Virus	Phe	Pro	1.3	2.5	1.1	3.7
Peptide			0.0	2.0	0.0	2.7

A

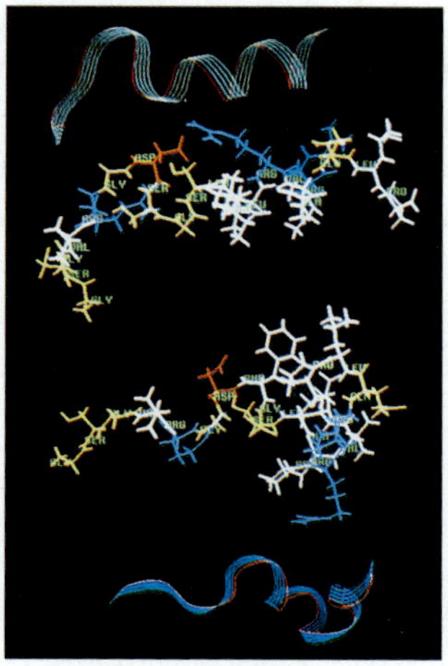

D

a

b

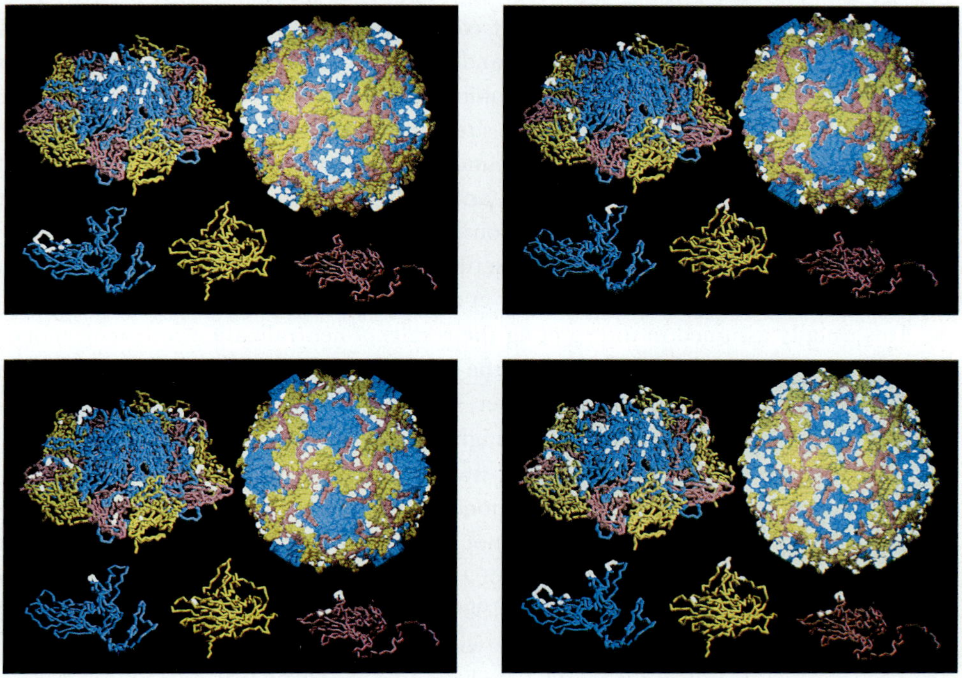

c

d

Fig. 9.1. Computer molecular models of peptide A and D based on NMR data (Mascagni et al., 1989). The ribbon-like structures are added for comparative purposes.

Fig. 9.2. The antigenic sites of poliovirus, type 1. The photograph shows, in addition to the virus particle, isolated copies of VP1 (blue), VP2 (yellow) and VP3 (red). The mutation sites are indicated in white. (a) Site 1; (b) site 2; (c) site 3; (d) all sites (see Hogle and Filman, 1989).

cholson (1982) methods had supported the division into two groups. Analysis of the peptides by circular dichroism and two-dimensional-nuclear magnetic resonance has now shown that the substitution of the leucine residue at position 153 in peptide A by proline (peptide D) resulted in a dramatic change in its shape (Fig. 9.2). Extension of this analysis to peptides B and C has confirmed the grouping of A with C and B with D established by the serological experiments. These results suggest that this type of analysis will provide useful information for the design of peptides with predictable antigenic properties.

4. Location of antigenic sites on the virus particle

The positions of the antigenic sites on the virus particle have been studied most thoroughly with polio and rhinoviruses (see reviews by Hogle and Filman, 1989; Rossmann and Rueckert, 1987; Mosser et al., 1989). All three serotypes of poliovirus have been studied. The mutations that confer resistance to monoclonal antibodies map to discrete regions of VP1, VP2 and VP3 and these have been grouped into three major sites by X-ray crystallographic studies. Site 1 is located at the top of the large protrusion near the five-fold axis, site 2 at the top of the large protrusion near the two-fold axis and site 3 in and around the smaller protrusion at the three-fold axis (Fig. 9.1). The grouping of the antigenic sites by using cross-neutralization studies with monoclonal antibodies is consistent with the structural analyses. A very similar distribution of sites has been described for rhinovirus serotype 14 (Rossmann et al., 1985; Sherry et al., 1986; Mosser et al., 1989).

It is usually assumed in interpreting the results of neutralization tests with monoclonal antibody escape mutants that the mutated site is part of the physical site which reacts with the antibody. However, there has been some discussion that a mutation located outside the binding site might allow escape from neutralization, either by altering the structure of the binding site or by preventing some structural alteration required for neutralization (Diamond et al., 1985; Blondel et al., 1986). Although Hogle and his colleagues have not observed with poliovirus a mutation that would require long-range propagation of a structural alteration to allow escape from neutralization (see Hogle and Filman, 1989) recent work with FMDV has indicated that mutation at a distance can occur with this virus (Parry et al., submitted for publication). This difference may be explained by the distinctive location and flexibility of the FMDV loop but the studies with this virus are at a much earlier stage than those with polio and rhinoviruses and further data are required before firm conclusions can be drawn.

5. Hybrid viruses

The demonstration by Racaniello and Baltimore (1981) that the DNA complementary to the genomic RNA of poliovirus was infectious opened up the possibility that hybrid viruses could be produced by appropriate manipulation of the cDNA. This avenue is now being explored intensively by several groups and it has been shown that replacement of the loop region encompassing amino acid residues 94–102 of VP1 with oligonucleotides coding for epitopes belonging to other serotypes of the virus gives cDNA which is still infectious and thus allows the production of new hybrid virus particles. Moreover, epitopes belonging to other infectious agents have also been incorporated into hybrid virus particles (Burke et al., 1989). Of particular interest has been the way in which this technology has provided information about the cell attachment site of poliovirus. In these experiments a hybrid virus was constructed from the Mahoney strain of serotype 1 (which does not infect mice) and the loop region of the Lansing strain of the virus (which belongs to serotype 2 and does infect mice). This hybrid virus infected mice and gave progeny which possessed the characteristics of the Mahoney and Lansing strains (Martin et al., 1988; Murray et al., 1988).

This approach provides a new dimension in the study of picornavirus immunochemistry because it will allow the ready production of viruses containing antigenic sites with altered amino acid sequences. Consequently the effect of single changes on the serological characteristics of the viruses will be readily determined.

6. Conclusions

The determination of the three-dimensional structures of at least one member of each of the four genera of the family Picornaviridae, the isolation of monoclonal antibody mutants and the sequencing of the RNAs coding for the capsid proteins of these mutants has provided evidence for the location and shape of the antigenic regions of these viruses. This information has also provided the opportunity to analyze the difference between antigenic variants and to perhaps generate vaccines, whether based on peptides or hybrid viruses, which will be more effective than those which are available at present.

7. References

Acharya, R., Fry, E., Stuart, D., Fox, G., Rowlands, D. and Brown, F. (1989) *Nature* 337, 709–716.

Bachrach, H.L., Moore, D.M., McKercher, P.D. and Polatnick, J. (1975) *J. Immunol.* 115, 1636–1641.

Barclay, W.S. and Al-Nakib, W. (1987) *J. Virol. Methods* 16, 53–64.

Bittle, J.L, Houghten, R.A., Alexander, H., Shinnick, T.M., Sutcliffe, J.G., Lerner, R.A., Rowlands, D.J. and Brown, F. (1982) *Nature* 298, 30–33.

Blondel, B., Crainic, R., Fichot, O., Dufraisse, G., Cardea, A., Girard, M. and Horaud, F. (1986). *J. Virol.* 57, 81–90.

Brown, F. (1989) In *Molecular Aspects of Picornavirus Infection and Detection* (B.L. Semler and E. Ehrenfeld, eds.) American Society for Microbiology, Washington DC. pp. 179–191.

Brown, F. and Smale, C.J. (1970). *J. Gen. Virol.* 7, 115–127.

Burke, K.L., Evans, D.J., Jenkins, O., Meredith, J., D'Souza, E.D.A. and Almond, J.W. (1989). *J. Gen Virol.* 70, 2475–2479.

Burroughs, J.N., Rowlands, D.J., Sangar, D.V., Talbot, P. and Brown, F. (1971). *J. Gen. Virol.* 13, 73–84.

Chou, P.Y. and Fasman, G.D. (1978). *Adv. Enzymol.* 47, 45–148.

Chow, M., Newman, J.F.E., Filman, D., Hogle, J.M., Rowlands, D.J. and Brown, F. (1987). *Nature* 237, 482–486.

Cowan, K.M. and Graves, J.H. (1966). *Virology* 30, 528–540.

Dernick, R., Heukeshoven, J. and Hilbrig, M. (1973). *Virology* 130, 243–246.

Diamond, D.C., Jameson, B.A., Bonin, J., Kohara, M., Abe, S., Itoh, H., Komatou, T., Arita, M., Kuge, S., Nomoto, A., Osterhaus, A.D.M.E., Crainic, R. and Wimmer, E. (1985). *Science* 229, 1090–1093.

DiMarchi, R., Brooke, G., Gale, C., Cracknell, V., Doel, T. and Mowat, N. (1986). *Science* 232, 639–641.

Francis, M.J., Hastings, G.Z., Clarke, B.E., Brown, A.L., Beddell, C.R., Rowlands, D.J. and Brown, F. (1990). *Immunology* 69, 171–176.

Geysen, H.M., Meloen, R.H. and Barteling, S.J. (1984). *Proc. Natl. Acad. Sci.* USA 81, 3998–4002.

Hogle, J.M., Chow, M. and Filman, D.J. (1985) *Science* 229, 1358–1365.

Hogle, J.M. and Filman, D.J. (1989) *Phil. Trans. R. Soc. Lond* 467–478.

Laporte, J., Grosclaude, J., Wantyghem, J., Bernard, S. and Rouze, P. (1973). *C.R. Acad. Sci.* 276, 3399–3401.

Luo, M., Vriend, G., Kamer, G., Minor, I., Arnold, E., Rossmann, M.G., Boege, U., Scraba, D.G., Duke, G.M. and Palmenberg, A. (1987). *Science* 235, 182–191.

Martin, A., Wychowski, C., Benichou, D., Crainic, R. and Girard, M. (1988). *Ann. Inst. Pasteur Virol.* 138, 79–88.

Mascagni, P., Siligardi, G., Brown, F., Drake, A.F., Nicolai, N., James, C.J., Rowlands, D. and Gibbons, W.A. (1989). Submitted for publication.

McCahon, D., Crowther, J.R., Belsham, G.J., Kitson, J.D.A., Duchesne, M., Hare, P., Meloen, R.H., Morgan, D.O. and de Simone, F. (1989). *J. Gen. Virol.* 70, 639–645.

Melnick, J.L. and 20 others (1974). *Intervirology* 4, 303–316.

Murnane, T.G., Craighead, J.E., Mondragon, H. and Shelokov, A. (1960). *Science* 131, 498–500.

Mosser, A.G., Leippe, D.M. and Rueckert, R.R. (1989) In *Molecular Aspects of Picornavirus Infection and Detection* (B.L. Semler and E. Ehrenfeld, Eds.) American Society for Microbiology, Washington DC. pp. 155–167.

Murray, M.G., Bradley, I., Yang, X-F, Wimmer, E., Moss, E.G. and Racaniello, V.R. (1988) *Science* 241, 213–214.

Newman, J.F.E., Cartwright, B., Doel, T.R. and Brown, F. (1979) *J. Gen. Virol.* 45, 497–507.

Nicholson, B.H. (1982) *Biochem. Soc. Trans.* 10, 387–388.

Ouldridge, E.J., Parry, N.R., Barnett, P.V., Bolwell, C., Rowlands, D.J., Brown, F., Bittle, J.L., Houghten, R.A. and Lerner, R.A. (1986). In *Vaccines 86,* (R.M. Chanock, R.A. Lerner and F. Brown, eds.) Cold Spring Harbor Laboratory, Cold Spring Harbor NY. pp. 45–49.

Palmenberg, A. (1989). In *Molecular Aspects of Picornavirus Infection and Detection.* ed. B.L. Semler and E. Ehrenfeld. American Society for Microbiology, Chapter 13, p. 211–241

Parry, N.R., Ouldridge, E.J., Barnett, P.V., Clarke, B.E., Francis, M.J., Fox, J.D., Rowlands, D.J. and Brown, F. (1989). *J. Gen Virol.* 70, 2919–2930.

Parry, N.R., Ouldridge, E.J., Barnett, P.V., Rowlands, D.J. Brown, F., Bittle, J.L., Houghten, R.A. and Lerner, R.A. (1985). In: *Vaccines 85* (R.M. Chanock, R.A. Lerner and F. Brown, eds.) Cold Spring Harbor Laboratory, Cold Spring Harbor NY. pp. 211–216.

Polatnick, J., Arlinghaus, R.B., Graves, J.H. and Cowan, K.M. (1967). *Virology* 31, 609–615.

Racaniello, V. and Baltimore, D. (1981). *Science* 214, 916–919.

Rossmann, M.G., Arnold, E., Erickson, J.W., Frankenberger, E.A., Griffith, J.P., Hecht, H.J., Johnson, J.E., Kamer, G., Luo, M., Mosser, A.G., Rueckert, R.R., Sherry, B. and Vriend, G. (1985). *Nature* 317, 145–153.

Rossmann, M.G. and Rueckert, R.R. (1987). *Microbiol. Sci,* 4, 206–214.

Rowlands, D.J., Clarke, B.E., Carroll, A.R., Brown, F., Nicholson, B.H., Bittle, J.L., Houghten, R.A. and Lerner, R.A. (1983). *Nature* 306, 694–697.

Rowlands, D.J., Sangar, D.V. and Brown, F. (1972). *Arch. Virusforsch.* 39, 274–283.

Rweweymamu, M.M., Terry, G. and Pay, T.W.F. (1979). *Arch. Virol.* 59, 69–79.

Sherry, B., Mosser, A.G., Colonno, R.J. and Rueckert, R.R. (1986). *J. Virol.* 57, 246–256.

Strohmaier, K., Franze, R. and Adam, K-H., (1982). *J. Gen. Virol.* 59, 295–306.

Thomas, A.A.M., Woortmeijer, R.K., Puijk, W. and Barteling, S.J. (1988). *J. Virol.* 62, 2782–2789.

Wild, T.F., Burroughs, J.N. and Brown, F. (1969). *J. Gen. Virol.* 4, 313–320.

Xie, Q-C., McCahon, D., Crowther, J.R., Belsham, G.J. and McCullought, K.C. (1987). *J. Gen. Virol.* 68, 1637–1647.

Eds. M.H.V. van Regenmortel and A.R. Neurath
Immunochemistry of viruses, II. The basis for serodiagnosis and vaccines.

Influenza viruses 10

G.M. AIR[1] and W.G. LAVER[2]

[1]Department of Microbiology, University of Alabama at Birmingham, Birmingham, AL 35294, U.S.A. and [2]Influenza Research Unit, John Curtin School of Medical Research, Australian National University, P. O. Box 334, Canberra, ACT 2601, Australia

Early in the 1940s, at the Rockefeller Foundation in New York City, George Hirst noticed that allantoic fluid from embryonated chicken eggs inoculated with influenza virus caused red blood cells to heavily agglutinate (Hirst, 1941). It is now known, of course, that this agglutination was caused by one of the surface 'spikes' on the virus. When the red cells, agglutinated by virus particles, were warmed to 37°C they dispersed as the virus eluted and were not re-agglutinated by fresh infectious allantoic fluid. On the other hand, virus which had eluted from the red cells at 37°C could agglutinate fresh red blood cells. Hirst interpreted these phenomena as resulting from some enzymic activity of the virus, where the substrate (virus receptor) was located on the surface of the red cell. In the 1950s, Alfred Gottschalk, working at the Walter and Eliza Hall Institute in Melbourne, realized that if the receptor-destroying activity was an enzyme, this might remove something from the surface of the red cells. He therefore set about looking for the 'split product'. This was eventually identified as *N*-acetyl neuraminic acid (sialic acid) and the enzyme was called sialidase or neuraminidase (Gottschalk, 1957), now designated acylneuraminyl hydrolase, EC 3.2.1.18, which is located in a second surface 'spike' on the virus particle. These two surface spikes are the major antigens of influenza virus, and it is variation in these which allows escape of influenza from antibodies and thus causes the annual epidemics of 'flu in the human population.

Influenza viruses are classified into three types—A, B and C—on the basis of their internal type-specific nucleoprotein and matrix protein antigens. Type A influenza viruses are further classified into subtypes based on the antigenic characteristics of their surface antigens, hemagglutinin (HA) and neuraminidase (NA). Thirteen distinct HA subtypes and nine NA subtypes are now recognized in the nomenclature system for influenza A viruses recommended by the World Health Organization (WHO, 1980). This review will concern those structures recognized by B cells (antibodies). T-cell recognition will not be discussed here.

A comprehensive review on the hemagglutinin by Don Wiley and John Skehel has supplied most of the discussion on the HA for this chapter and we are grateful to

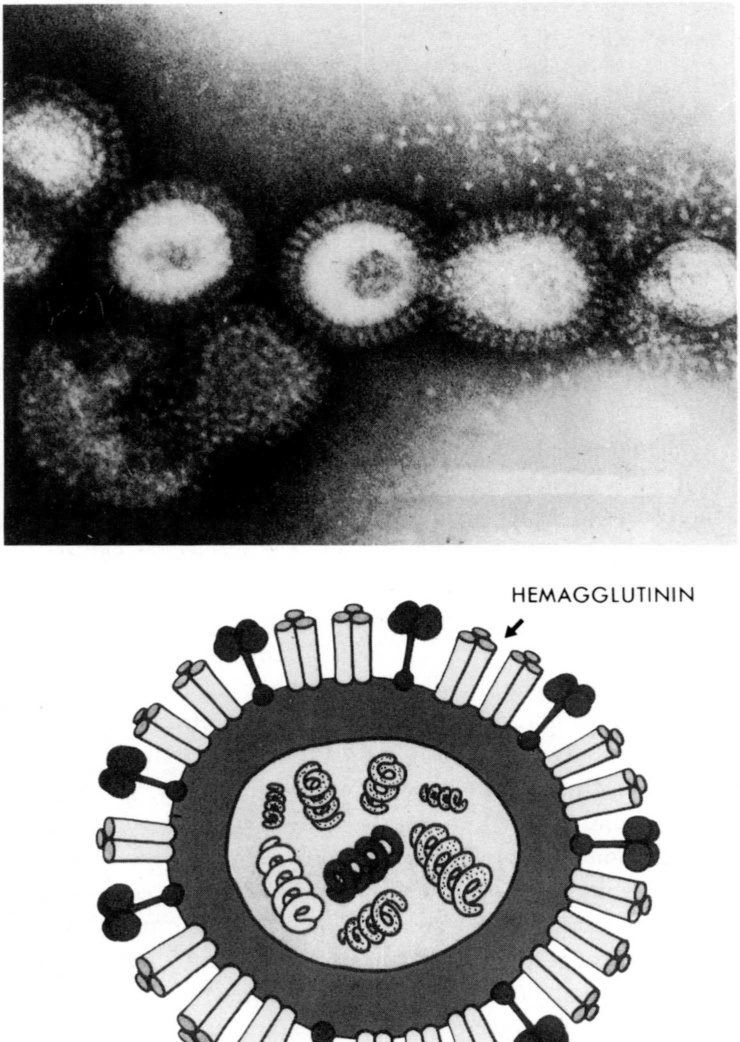

Fig. 10.1. Electron micrograph and diagram showing the surface structure of particles of influenza type A virus. There are eight RNA segments in the virus, each of which codes for one (in some cases for two, using overlapping reading frames) of the virus proteins.

these authors and Annual Reviews Inc. for allowing us to use this material (Wiley and Skehel, 1987).

1. Properties of influenza A, B and C viruses

The highly pleomorphic particles of type A influenza virus (Fig. 1) are enclosed by a lipid envelope, derived from the plasma membrane of the host cell, to which the HA and NA antigens are attached by sequences of hydrophobic amino acids at one end of the molecules. Both surface antigens are glycosylated and some of the carbohydrate side chains possess antigenic activity characteristic of the host cell in which the virus grew.

Within the lipid envelope lies the matrix protein, which is believed to be structural in function. Within the matrix shell are eight single-stranded RNA molecules of negative sense (that is, the virion RNA is complementary to the messenger RNA) associated with a nucleocapsid protein and three large proteins, PA, PB1 and PB2, which are responsible for RNA replication and transcription (Krug, 1983). At least three virus-encoded nonstructural proteins are also found in infected cells. Table 10.1 summarizes current information on the eight genes of the influenza A virus and the molecular weights of the proteins encoded by them.

The properties of type B influenza virus particles are similar to those of type A (except that an additional nonstructural protein, NB, is present) (Lamb, 1983); but type C influenza virus differs considerably from the other two types.

Although the type C virus is much less well characterized than influenza A and B viruses, it differs most significantly in possessing a unique surface glycoprotein which resembles the HA of type A and B viruses in many of its biochemical properties.

The glycoproteins of the type C virus form characteristic hexagonal arrays on the surface of both spherical and filamentous forms of the virus (Fig. 10.2). Such arrays have not been observed on influenza A or B viruses, and it has been suggested (Herrler et al., 1981) that the presence of morphologically distinct trimeric HA and tetrameric NA of different symmetry may prevent lattice formation in these strains. Herrler et al., (1981) reported that the influenza C glycoprotein spikes viewed end on in preparations negatively stained with sodium phosphotungstate (pH 7.2) have a rounded end, in contrast with the triangular shape observed for influenza A HA (Wrigley, 1979).

However, the influenza C glycoprotein is clearly seen to be a trimer in specimens prepared with uranyl stains. Three-dimensional reconstructions from naturally occurring hexagonal arrays show that at low resolution (3 nm) the influenza C glycoprotein exhibits similar features to the HA glycoprotein of influenza A (Hewat et al., 1984). Both have a triangular stalk near the membrane. Further from the membrane, the stalk becomes broader and the monomers more separated, leaving an

TABLE 10.1

Type A influenza virus genome RNA segments and coding assignments

Segment	Length[a] (nucleotides)	Encoded polypeptide[b]	Nascent polypeptide length[c] (amino acids)	Mature polypeptides	Approx. no. molecules per virion[d]	Remarks
1	2341	PB2	759		30–60	Host cell RNA cap binding; component of RNA transcriptase[e]
2	2341	PB1	757		30–60	Initiation of transcription; possible endonuclease activity; component of RNA transcriptase[e]
3	2233	PA	716		30–60	Elongation of mRNA chains; component of RNA transcriptase[e]
4	1778	HA	566	HA1 326 HA2 222	500	Surface glycoprotein; trimer; major antigenic determinant
5	1565	NP	498		1000	Associated with RNA segments to form ribonucleoprotein; structural component of RNA transcriptase; type-specific antigen
6	1413	NA	454	454	100	Surface glycoprotein; neuraminidase activity; tetramer; surface antigen
7	1027	M_1	252	252	3000	Major protein component of virus; underlies lipid bilayer; type-specific antigen
		M_2	96			Spliced mRNA; non-structural protein; function unknown[f]
		?	?9			Spliced mRNA; peptide predicted by nucleotide sequence only[f]
8	890	NS_1	230			Nonstructural protein; function unknown
		NS_2	121			Spliced mRNA; nonstructural protein; function unknown[f]

open center. The molecule narrows at the top. The regions of greatest contact between adjacent trimers in the arrays are situated nearer the distal end of the molecule. These contact zones can be related to equivalent zones on the influenza A HA. Differences between the structure of the influenza A HA glycoprotein determined by X-ray analysis and reconstructions of the influenza C glycoprotein are greatest at either end of the molecule where the reconstructions are least reliable. Ordered glycoprotein arrays have not been observed on influenza C virions incubated at low pH. The staining patterns of glycoproteins on intact virions are essentially determined by the pH at which the virus is incubated and the stain type, but not the pH of the stain.

Influenza C binds to sialic acid receptors as do influenza A and B viruses, and the influenza C glycoprotein carries both the receptor-binding and receptor destroying activities. In influenza A and B viruses the activity is a neuraminidase, but the enzymatic activity of the influenza C glycoprotein is a sialoesterase (Herrler et al., 1988).

2. Antigenic variation

Two distinct kinds of antigenic variation have been demonstrated in influenza viruses, antigenic drift and major antigenic shifts. The first consists of relatively minor changes in the surface glycoproteins that occur gradually within a family of strains, all of which are clearly related to each other with respect to both internal and surface antigens. Among influenza A strains infecting man, each successive variant replaces pre-existing ones. This may be due to a selective advantage possessed by antigenic variants in overcoming immunological host resistance. Although antigenic variation occurs in influenza B viruses, it does not follow a direct sequential pattern as in influenza A viruses (Lu et al., 1983).

There have been only very minor antigenic changes detected in influenza C viruses since the first isolation of these agents in 1947. This antigenic stability resembles that of other respiratory agents, such as the parainfluenza viruses and respiratory syncytial virus, rather than influenza A or B viruses. Nucleotide sequences of the glycoprotein (hemagglutinin-esterase) genes of several strains of influenza C have been determined (Pfeifer and Compans, 1984; Nakada et al., 1984; Buonagurio et

---

←

[a] For A/PR/8/34 strain; Winter et al. (1981); see Lamb (1983) and Lamb and Choppin (1983) for full references and references to other strains.

[b] Determined by biochemical and genetic approaches.

[c] Determined by nucleotide sequence analysis and protein sequencing.

[d] Adapted from Compans and Choppin (1975).

[e] From Krug (1983).

[f] From Lamb (1983).

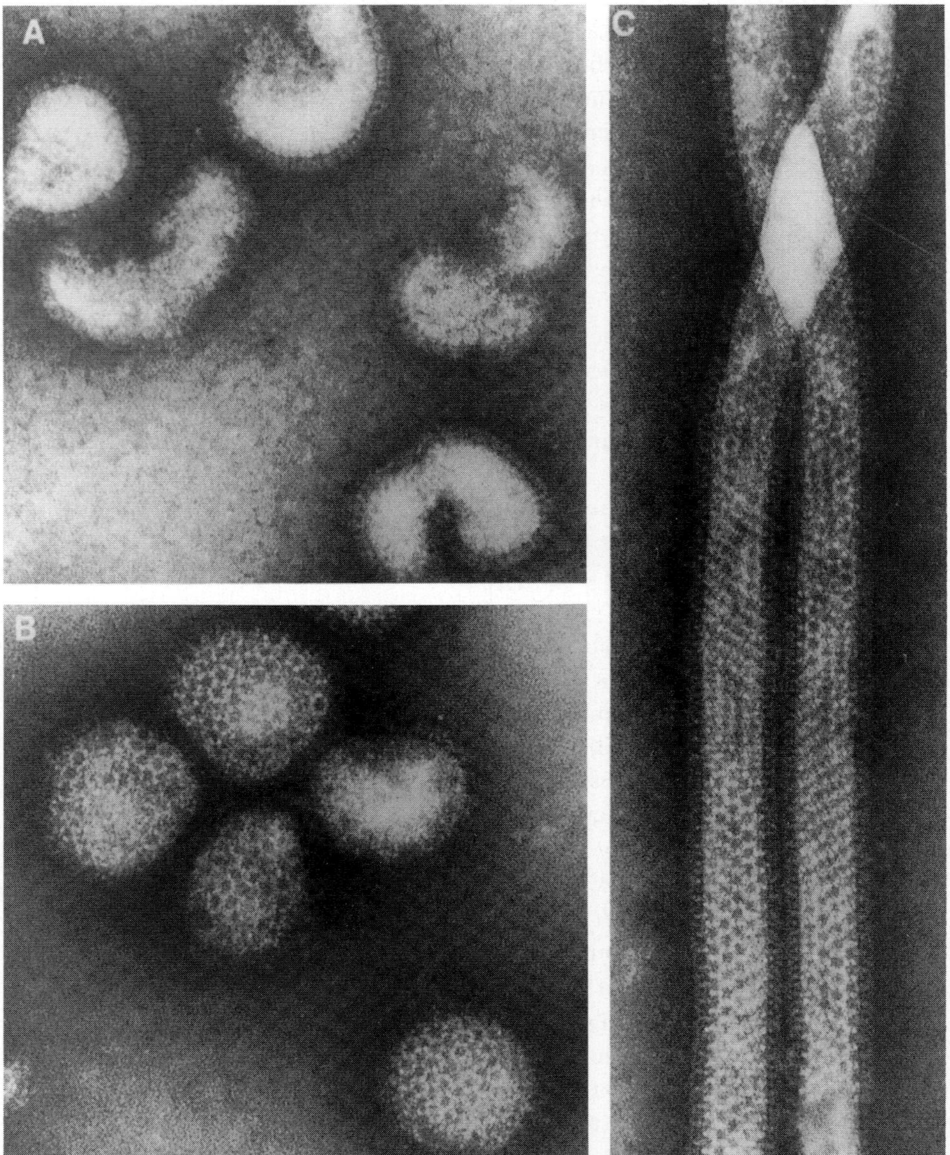

Fig. 10.2. Electron micrographs of influenza B and C virions, A, B/HK/8/73 virions; B, C/JHB/1/66 virions grown in embryonated eggs; C, Filamentous influenza C/JHB/1/66 particles grown in chick embryo fibroblast cells. Negative staining with sodium phosphotungstate, pH 7.0. Micrographs courtesy of Dr. R.W. Compans.

al., 1985), and show 25 differences in amino acid sequence of the proteins. This finding is consistent with the observation that influenza C viruses have not undergone significant antigenic variation. The reason for this difference in stability of the surface antigen is not understood, particularly since influenza C virus appears to share with the influenza A and B viruses most features of virus structure as well as virus replication.

The second kind of antigenic variation, which has been described only for influenza type A, involves much more dramatic antigenic changes. These are referred to as major antigenic shifts. These major shifts, which occur at intervals of 10–15 years, are marked by the sudden appearance of 'new' subtypes in the population which cause the major pandemics of influenza.

3. Antigenic shift

An antigenic shift in type A influenza virus occurs when a virus suddenly appears in the human populations with a 'new' HA and sometimes also a 'new' NA. Since the first human influenza virus was isolated in 1933, antigenic shifts have occurred in 1957 when the H2N2 subtype (Asian influenza) replaced the H1N1 subtype, in 1968 when the Hong Kong (H3N2) virus appeared and in 1977 when the H1N1 virus reappeared. All these major antigenic shifts in the virus occurred in China and anecdotal records suggest that previous epidemics (probably caused by the sudden emergence of 'new' viruses) also had their origin in China. Although we are still uncertain how these new subtypes appear or reappear in the human population, some clues are emerging from the accumulating sequence data.

Sequences up to 350 nucleotides from the 3′ end of the HA gene and the predicted amino acid sequences from 32 type A influenza viruses, including representatives of each of the 13 known HA subtypes, have been determined (Air, 1981; Hinshaw et al., 1982). Cysteine residues and certain other amino acids are conserved in all sequences, indicating that the 13 HA subtypes evolved from a common ancestor and share a common basic structure. When the partial amino acid sequences are compared pairwise, the most distant subtypes are H1 and H3 (25% homology). Other subtypes are more similar to each other, the highest homology (80%) being between H2 and H5. Relationships between the partial sequences can be illustrated as a dendrogram (Fig. 10.3). Where complete sequences have become available, the relationships are found to be in good agreement with those shown in Fig. 10.3 from the partial sequences (e.g., Kawaoka et al., 1984).

Within strains of single subtype, the differences in deduced amino acid sequence are generally less than 10%, but differences in the N-terminal region of HA1 from two strains each of H7 and H10 viruses are close to 20%, almost equal to the difference between the amino acid sequences of the two closest subtypes (H2 and H5).

Whereas it is possible to imagine that the H2 viruses could eventually mutate into

H5, the change to H3 must involve a more drastic mechanism. How, then, could the H3 hemagglutinin have replaced the H2 strain in the human population in 1968? First, the 'new' virus may have caused an epidemic in man many years previously and have remained hidden and unchanged in some unknown place ever since. Evidence for this kind of event has been obtained. The strain of 'Russian influenza' (H1N1) which reappeared in Anshan in northern China in May 1977 and subsequently spread to the rest of the world, seems to be identical, in all genes, to

Fig. 10.3. Dendrogram showing the relationships between the HA1 N-terminal amino acids sequences deduced from genomic sequences of viruses representing the 13 known subtypes (Air 1981). The sequences were aligned using amino acids that are invariant. The percentage sequence differences of all pairwise comparisons of the aligned sequences starting from the N-terminal Asp or corresponding amino acid were used to calculate the dendrogram. Thus, the positions of each bifurcation in the dendrogram indicate the mean sequence difference of the sequences connected through that point. Dendrogram by Adrian Gibbs.

the virus which caused an influenza epidemic in 1950 (Nakajima et al., 1978; Scholtissek et al., 1978). Where was this virus for 27 years? As yet we have no answer.

Second, some of the 'new' viruses may be derived from animal or avian viruses. One strain of human influenza has been shown to be such a recombinant (reassortant). The Hong Kong (H3N2) virus contains the NA (and other) genes from an Asian (H2N2) strain of human influenza and a HA which is antigenically related to that of A/duck/Ukraine/63 (H3N8) and A/equine/2/Miami/63 (H3N8) viruses. The amino acid sequence homology between the HAs of A/duck/Ukraine/63 and A/Aichi/2/68 viruses (both of subtype H3) is 96% (Fang et al., 1981; Ward and Dopheide, 1981a). We do not know if it was an animal or bird virus which donated the HA gene during the recombination event that led to the formation of the Hong Kong strain. The donating virus could have been one which contained Hong Kong HA that had persisted from a much earlier human influenza epidemic and was maintained unchanged in the same way as the Russian influenza virus. Antibodies in the sera of people who were born around 1900 suggest that a virus with a HA similar to that of the Hong Kong virus caused influenza at the time, but the NA of this virus was N8, not N2. It is possible that viruses such as equine (H3N8) and duck/Ukraine have evolved from the 1900 epidemic strain.

The third way in which new viruses could appear in the human population is by an animal or bird virus acquiring infectivity for man. This could be brought about by mutation, but the number of mutations required and the genes in which these would need to occur are as yet unknown.

It is possible that each of these mechanisms has operated at one time or another to produce a 'major shift' in influenza viruses infecting man. We do not know how the different subtypes of type A influenza arose, but it is reasonable to suppose that 'shuffling' of genes between influenza A strains is occurring all the time, most frequently in birds, but also in mammals, including man.

4. Structure of the hemagglutinin

In type A influenza virus, the HA accounts for about 25% of the virus protein. It is responsible for the attachment of virus to cells and for the penetration of the virus into the cell during the initial stages of infection. Antibodies to the HA neutralize the infectivity of the virus. Variation in the HA glycoprotein is mainly responsible for the continually occurring outbreaks of influenza and for our inability to control these by vaccination.

The HA of the 1968 influenza virus strain A/Aichi/2/68 (H3N2) is synthesized as a single polypeptide chain of 550 amino acids (Fig. 10.4), which is shortened by removal of a signal peptide and subsequently cleaved by removal of arginine 329 into two chains, HA_1 (36 334 daltons) and HA_2 (25 750 daltons). These chains are

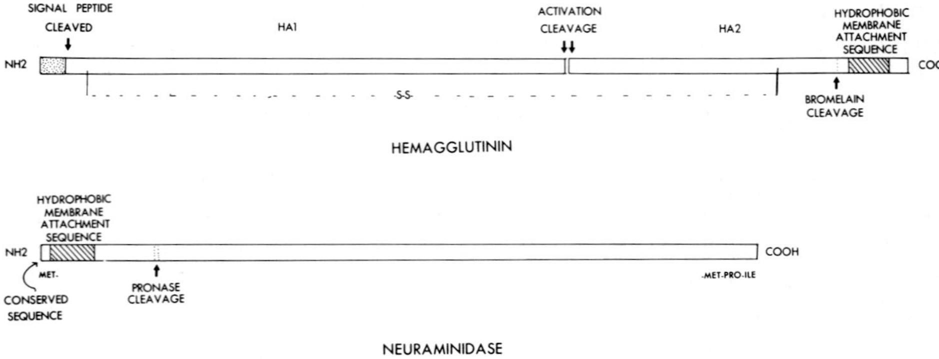

Fig. 10.4. Diagrammatic representation of certain features of the hemagglutinin and neuraminidase polypeptides. *Hemagglutinin.* The HA is synthesized as a single polypeptide. Following its synthesis, an N-terminal signal peptide is cleaved off and the molecule is cleaved further into HA1 and HA2 with the removal of one or more intervening amino acids. The latter cleavage is necessary for the virus to be infectious. HA1 and HA2 remain linked by a single disulfide bond and each HA spike contains three of these dimers. A sequence of hydrophobic amino acids near the C terminus of HA2 serves to anchor the HA in the lipid of the virus membrane. Treatment with bromelain removes this hydrophobic region without damaging the rest of the molecule, which, in some cases, can be crystallized. *Neuraminidase.* The neuraminidase is oriented in the virus membrane in the opposite way to the hemagglutinin. No post-translational cleavage of the NA polypeptide occurs, no signal peptide is split off, and even the initiating methionine is retained. No processing at the C terminus takes place; the C-terminal sequence Met-Pro-Ile predicted from the gene sequence is found in intact NA molecules of the N2 subtype isolated from virus and in the pronase-released NA heads. A sequence of six polar amino acids at the N terminus of the NA polypeptide, which are totally conserved in the nine different NA subtypes, is followed by a sequence of hydrophobic amino acids which must represent the transmembrane region of the NA stalk. This sequence is not conserved at all between subtypes (apart from conservation of hydrophobicity). Pronase cleaves the polypeptide in the positions shown, removing the stalk and releasing the enzymically and antigenically active head of the NA, which, in some cases, can be crystallized.

covalently attached by a disulfide bond from HA_1 position 14 to HA_2 position 137 (Ward, 1981), and the two-chain monomers are associated noncovalently to form trimers on the surface of membranes (Wiley et al., 1977). Bromelain treatment of virus yields a soluble trimer, BHA, of the extracellular regions, containing all of HA_1 and the first 175 of 221 amino acids of HA_2, but missing the hydrophobic membrane-anchoring peptide (Brand and Skehel, 1972). The three-dimensional structure of this trimer has been determined from X-ray studies (Wilson et al., 1981; Wiley, 1985).

4.1. Three-dimensional structure

Fig. 10.5 shows a schematic drawing of a monomer of the HA. The membrane end of the molecule is at the bottom of the drawing, and a vertical line shows the location of the threefold symmetry axis that would relate two more monomers in the

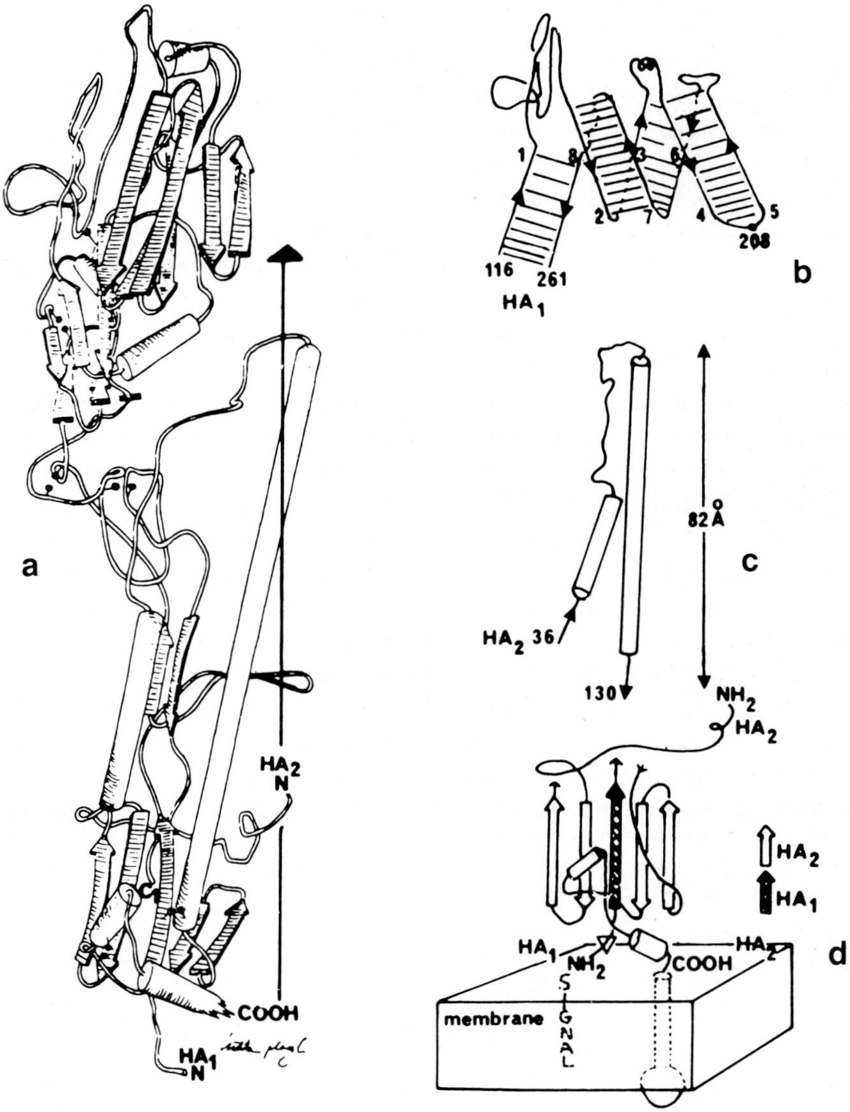

Fig. 10.5. (a) Drawing of the three-dimensional structure of the influenza virus hemagglutinin. The terminal residues of HA_1 and HA_2 are labeled. The threefold symmetry axis relating molecules in the trimer is shown: the trimer is stabilized principally by packing of the long α-helices (cylinders). Note that the amino-terminus of HA_2, the 'fusion peptide', tucks into the threefold contact. (b) The eight-stranded β-sheet structure and looped-out regions of the globular domain. (c) HA_2 residues 36–130 comprise an α-helical hairpin in the stem region of the monomer. Three of the long helices, one from each monomer, pack together as a triple-stranded coiled-coil that stabilizes the trimer. (d) The membrane end of the molecule contains a five-stranded β-sheet. The central strand is the N terminus of HA_1, and the adjacent strands come from the C-terminal portions of the chain in HA_2. Broken lines suggest the path of the hydrophobic anchoring peptide removed by bromelain cleavage. The site of the oligosaccharide attached at HA_1 8 is shown as a triangle.

trimer. Both the amino-terminus of HA_1 and the C-terminus of BHA_2 are found at the extreme membrane end of the molecule. The structure is long, projecting 135 Å off the membrane. The molecule appears to have a globular domain on top of an elongated stem region. The globular region at the top contains only part of HA_1, while the stem contains parts of HA_1 and all of HA_2. The conformation of the chain is unusually extended. From the amino-terminus at the membrane, the first 63 amino acids of HA_1 reach in a nearly extended structure 96 Å up the molecule before the first compact folding occurs. The globular regions at the top contains an eight-strand antiparallel β-sheet. This domain contains the receptor-binding site (see section 5). The remainder of HA_1 returns to the stem, running nearly antiparallel to the initial stretch of HA_1.

The amino-terminus of HA_2 is 22 Å from the C-terminus of HA_1, indicating that a conformational change accompanied the cleavage of the two polypeptide chains. The hydrophobic amino-terminal peptide that is involved in membrane fusion is tucked into the trimer interface about 35 Å from the bottom of the molecule. The major structural feature of the stem is a hairpin loop of two α-helices (cylinders in Fig. 10.5). The second helix is 80 Å long and forms the backbone of the stem region.

Independent local contacts stabilize the trimer in the globular and stem regions. The HA_1 globular domains make pairwise contacts near the upper end of the molecule. A triple-stranded coiled-coil of alpha helices formed from the top half of the long helix from each monomer provides the major contacts in the HA_2 region. These contacts are also stabilized by contacts between the second and third residues from each of the hydrophobic N-termini of HA_2 and a network of salt bridges in the lower part of the trimer.

a

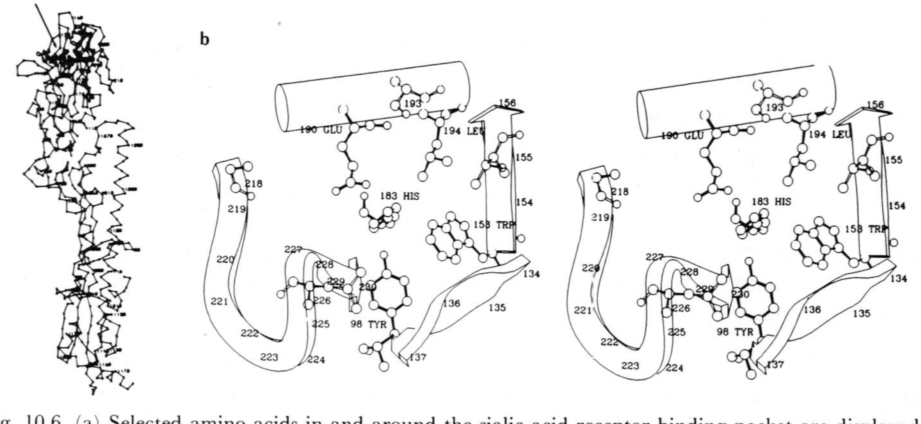

Fig. 10.6. (a) Selected amino acids in and around the sialic acid receptor binding pocket are displayed on a schematic drawing of the alpha-carbon backbone of the hemagglutinin from the 1968 influenza virus X-31. (b) A detail of the same residues shown in a view of the top of the molecule in stereo. Leucine 226 projects into the pocket near the central residue Tyr 98. The site illustrated binds NeuAcα2→6Gal linkages, while glutamine at position 226 alters the site to bind preferentially NeuAcα2→3Gal linkages.

4.2. Oligosaccharides

Six oligosaccharide chains are attached to asparagines of HA_1 (amino acid residues 8, 22, 38, 81, 165, 285), and one chain is attached to an asparagine of HA_2 (site number 154). All sites except 81 and 165 are processed, complex oligosaccharides (Ward and Dopheide, 1980). The most striking feature of the carbohydrate is its distribution (Wilson et al., 1981). All sites except 165 are on the lateral surfaces of the molecule. One site, HA_1 8, is at the extreme membrane end of the molecule. The oligosaccharide at 165 appears to stabilize the oligomeric contacts between globular units at the top of the structure. The sequence of the oligosaccharides has been determined from one strain (Keil et al.,, 1985), however, no obligatory function has been assigned to them (Wilson et al., 1981; Skehel et al., 1984; Raymond et al., 1983).

5. Receptor binding

Sialic acid (*N*-acetyl neuraminic acid) is the only known essential component of cellular receptors for influenza type A virus. Oligosaccharides terminating in sialic acid are found on many cell-surface glycoproteins and glycolipids, and the removal of sialic acid from them by neuraminidase treatment destroys receptor binding, preventing infection (Gottschalk, 1959). Furthermore, sialic-acid-containing glycolipids can serve as cellular receptors, restoring infectivity to neuraminidase-treated cells (Bergelson et al., 1982; Suzuki et al., 1985; Bukrinskaya et al., 1982).

5.1. Receptor site structure

The influenza virus hemagglutinin's receptor-binding site is a pocket located on the distal end of the molecule (Fig. 10.7) and composed of amino acid side chains (Tyr 98, His 183, Glu 190, Trp 153, Leu 194) that are largely conserved in the numerous strains of the virus (Wilson et al., 1981). Other conserved residues (Cys 97, Pro 99, Cys 139, Pro 147, Tyr 195, Arg 229) behind the pocket seem to stabilize the architecture of the site without being in a position to interact with the receptor. By contrast the perimeter of the surface of the pocket is composed of amino acid residues which have varied during the antigenic drift that accompanies recurrent epidemics (Wiley et al., 1981; Fig. 10.7).

The amino-acid side chains forming the surface of the pocket, which are positioned such that they could make direct contact with a cellular receptor, are both polar and nonpolar (hydrophobic). Histidine 183, tyrosine 98, glutamic acid 190, and serine 136 have exposed sidechain polar atoms that can act as hydrogen bond donors and acceptors. Leucine 194, tryptophan 153, and leucine 226 have exposed nonpolar side chains that have the potential to make van der Waals contacts with the hydrophobic portions of the receptor. In addition to these sidechain atoms, sections of main chain from residues 225–228 and 131–137 have carbonyl oxygens and amide nitrogens of peptide bonds exposed with potential to interact with a receptor.

Topographically, the binding site is a depression, the bottom of which is formed by the phenolic hydroxyl of tyrosine 98 and the hydrophobic surface of tryptophan

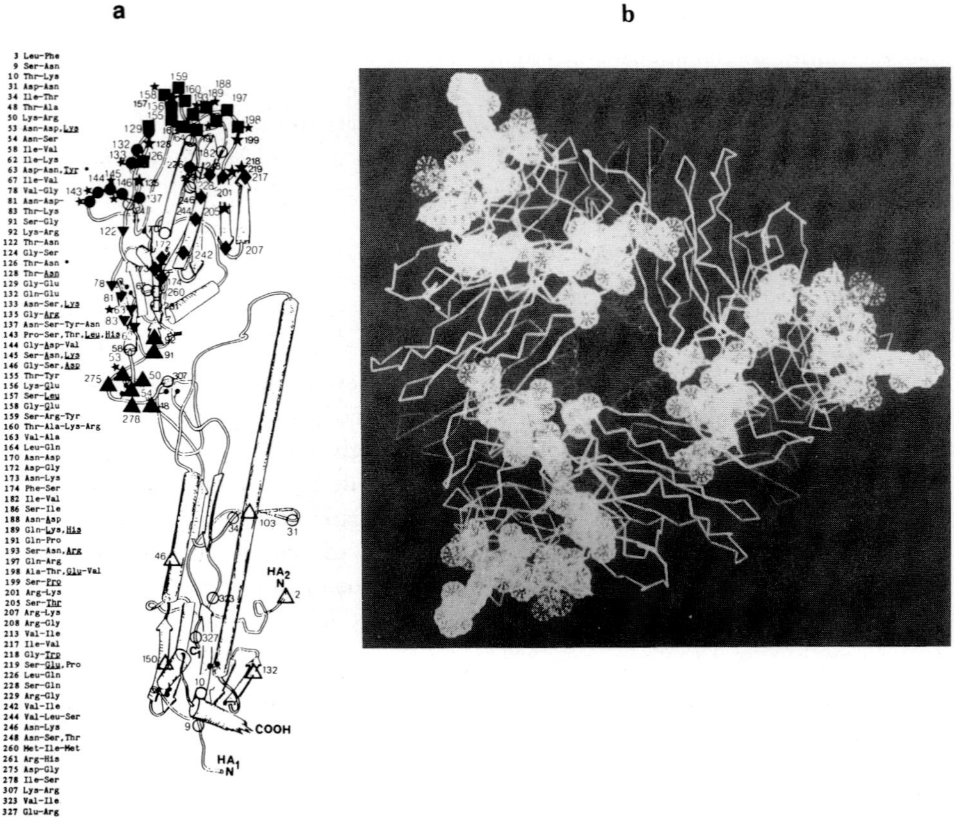

Fig. 10.7. Natural variation since 1968 and monoclonal variants suggest the antibody-binding sites on the 1968 HA. (a) ● Site A; ■ Site B; ◆ Site C; ◆ Site D; ▼ Site E. The symbols represent locations of natural sequence variation between 1968 and 1979. *, single-site monoclonally selected variant, each star represents a separate variant; * plus site symbol, represents a site of natural variation that has also been observed in a monoclonally selected variant. Underlined amino acids in the list of amino acid substitutions were observed in monoclonally selected variants only, no underline indicates in natural variants only, first letter underlined indicates substitutions found in both natural and monoclonally selected variants. An asterisk indicates addition of an *N*-glycosylation site, a minus indicates the loss of such a site. (b) Variable amino acid positions defining antibody-binding sites surround the conserved residues forming the receptor-binding pocket. The HA trimer is viewed from above, dotted spheres mark variable amino acid positions shown in Fig. 10.6a. The center of the Fig. shows the α-helices of HA_2, extending **into** the page. The receptor sites are inside the crescent of variable residues. The remaining position of HA_1, which is not in the site or variable, is covered by the oligosaccharide position 165 (not shown).

153 (Fig. 10.6). The back wall of the site has histidine 183 projecting out to hydrogen bond with tyrosine 98 and both glutamic acid 190 and leucine 194 hanging down into the site from a short α-helix along the top of the back wall. Residues 224–228 form the 'left' edge and 133–137 the 'front' edge (Fig. 10.6). The right side of the site is an extended trough terminating at theronine 155.

In crystallographically refined electron density maps at 3 Å resolution (Knossow et al., 1986), a number of water molecules are visible in the receptor-binding pocket that may mark positions where polar atoms of the receptor will bind. Such patterns of bound water molecules being displaced by substrates have been observed in a number of cases in enzyme crystallography (e.g., see James and Sielecki, 1984).

5.2. Receptor specificity mutants

Confirmation that the pocket is the receptor-binding site was provided by the selection of single-amino-acid substitution mutants of the HA with altered receptor specificity. A comparison of the binding of different strains of influenza viruses of the H3 subtype to erythrocytes derivatized with sialic acid in α2→6 and α2→3 linkages to galactose (NeuAcα2→6Gal- or NeuAcα2→3Gal-), had revealed distinct specificities. Preferential binding to sialic acid in α2→6 linkages, a trait of the 1968 HA, was found to correlate with a high sensitivity to neutralization of infection by glycoproteins (γ inhibitors) present in nonimmune horse serum. This allowed selection of receptor specificity variants by growth of virus in that serum (Rogers et al., 1983a; Rogers et al., 1983b; Rogers et al., 1985.). Five independently isolated mutants selected from two parental strains showed a reduction of affinity to α2→6 linkages and a marked increase in affinity for α2→3 linkages, so that four mutant viruses bind better to α2→3 than to α2→6 linkages, while one binds equally to both. In all five mutants, only a single-amino-acid substitution had occurred; in four cases glutamine was substituted for leucine at 226 and in one case methionine was substituted for leucine (Rogers et al., 1983b) (see Fig. 10.6), to produce the altered receptor specificity. Residue 226 is in the pocket described above (Fig. 10.6).

5.3. Structure of a complex with sialyl lactose

At high concentration, the trisaccharide sialyl lactose inhibits the binding of influenza virus to erythrocytes (F.S. Escobar and D.C. Wiley, unpublished). To visualize directly the hemagglutinin's interaction with a receptor, 0.02 M α2→6 and α2→3 sialyl lactose were diffused into crystals of the wild-type 1968 HA (226 Leu) and the specificity mutant (226 Gln), respectively.

In both cases, difference Fourier electron density maps at 3Å resolution show similar peaks in the receptor-binding site adjacent to position 226. The structure of the specificity mutant, 226 Gln, complexed with α2→3 sialyl lactose, has been determined (Weis et al., 1988).

Model building and an exhaustive six-dimensional computer search of the fit of sialic acid to the difference electron density shows both compounds in the same orientation. This orientation places the glycerol side chain of sialic acid in the vicinity of tyrosine 98, glutamic 190, and serine 228. The methyl group of the *N*-acetyl moiety is located near tryptophan 153 and leucine 194, while the polar atoms of the amide link are near main chain polar atoms at residues 134 and 135. The carboxylate of sialic acid faces into the site near Ser 136 and Asn 137.

No electron density is observed for the lactose moiety, indicating that it is spatially disordered, i.e.,flexibly linked to the sialic acid.

Some alterations in the positions of side chains (< 1Å) in the binding site are evident in the mutant HA, which may be responsible for the altered specificity.

This description of binding to a receptor fragment now allows rationalization of some relative affinity measurements of various ligands and prepares the way for the design of inhibitors to block receptor binding and, therefore, infectivity of influenza virus.

6. Affinity mutants

Variations in receptor-binding affinity have been observed in natural strains of influenza viruses, and the importance of this in the epidemiology of the disease has been discussed (Paulson, 1985; Fazekas de St. Groth, 1978; Underwood, 1982; Underwood, 1985). Relative affinities have been inferred by measuring the ability of virus to agglutinate a series of erythrocyte preparations whose receptors have been modified by partial sialidase or periodate treatment.

In one study, mutants of the 1934 virus, A/Puerto Rico/8/38, with enhanced 'affinity' for receptors were selected as inferred by the receptor gradient described above (Yewdell et al., 1986). Selection was accomplished by growth of virus in the presence of a mixture of antihemagglutinin monoclonal antibodies at subneutralizing concentrations. In the three mutants selected, a single amino acid substitution in the HA at positions 185, 231, and 244 correlated with enhanced receptor 'affinity'. In the first two cases the amino acids are in the second shell of residues around the receptor-binding site, while the third would appear to require a conformational alteration across the trimer interface adjacent to the receptor-binding site.

Twenty-two monoclonal-antibody-selected antigenic mutants of the 1968 A/Aichi (H3N2) virus with single-amino-acid substitutions in their HA, when tested on a series of erythrocytes treated with increasing amounts of periodate, exhibited altered receptor binding (Underwood et al., 1987). Some single-amino-acid substitutions at 156, 189, 193, and 135 increased the apparent 'affinity', while others at 144, 145, 158, 188, 189 decreased 'affinity', as measured by this procedure. This demonstrates that single amino acids responsible for antigenic changes can also cause altered receptor-binding characteristics. Whether variation in receptor affinities is important

TABLE 10.2

Receptor binding-membrane fusion mutants

Variant	Amino acid at position						Receptor specificity				HI[c] Eq 2M	Change in fusion pH[d]
	145	158	193	218	219	226	Hemagglutination[a]		Absorption[b]			
							SAα2,6Gal (HA titer)	SA2,3Gal	SAα2,6Gal (nmol/ml packed cells)	SAα2,3Gal		
X-31	S	G	S	G	S	L	1024	0	13	> 142	2096	–
63-E			N			P	512[e]	2[e]	30	93	0	+0.53
63-D						*	256[e]	256[e]	30	68	0	+0.20
63-3				E			1024	1024	10	54	1024	+0.43
v9A				R			256[e]	0	32	> 142	1024	+0.37
v68x			R				1024	512	24	42	0	0
X-3L HS						Q	0	1924	53	46	0	0

[a] Hemagglutination assays were performed using 12.5% (v/v) human erythrocytes enzymatically modified to contain 37 nmol NeuAc/ml packed cells for the SAα2,6Galβ1,4GlcNAc (SAα2,6Gal) sequences and 107 nmol NeuAc/ml packed cells for SAα2,3Galβ1,3 GalNAc (SAα2,3Gal) sequences. All of the viruses agglutinated native (untreated) erythrocytes but did not agglutinate sialidase-treated cells.

[b] These values represent the amount of sialic acid (in nmol/ml packed erythrocytes) required to bind 50% of the applied virus. The actual values are an average of the 50% binding point determined by assaying the amount of virus bound to derivatized cells and the amount of virus remaining in the supernatant.

[c] Hemagglutination inhibition (HI) by equine α2 macroglobulin was determined at initial Eq α_2M concentrations of 3.5 mg/ml.

[d] A resonance energy transfer assay was used to measure hemagglutinin fusion. By this assay X31 exhibited 50% fusion at pH 5.5.

[e] Indicates 'transient' agglutination where agglutination of derivatized erythrocytes was apparent after 30 min at room temperature, but no longer evident after 60 min.

* These variants have deleted amino acids 224–230 inclusive (RGLSSRI).

in generating or limiting the recurrence of epidemics is not yet clear.

In a recent study of receptor binding a series of mutant viruses selected using a single monoclonal antibody were shown to exhibit alteration in both receptor affinity and specificity as well as changes in the activation pH for membrane fusion activity by the HA (Daniels et al., 1987) (see Table 10.2). Amino acid substitutions at 218 (G to R or E) and 226 (L to P) and remarkably a deletion of seven amino acids from 224–230 cause reduction in the affinity to α2→6-linked sialic acid as measured by binding to derivatized erythrocytes, an increase in affinity for α2→3-linked sialic acid, and at the same time, substantial (0.2–0.53 pH units) changes in the pH dependence of membrane fusion activity. Residue 218 and the 224–230 loop form part of the trimeric interface between the globular domains at the top of the HA.

These observations, which indicate that alterations at the trimer interface can also influence membrane fusion, demonstrate the structural potential for information to be transmitted between the binding site and the trimer interfaces and, therefore, possible interrelationships between the process of binding to cells and the triggering of entry by membrane fusion. No evidence indicates that cell binding influences the pH optimum of membrane fusion, but whether some ligand, natural or designed, could cause such a change during normal entry or as an antiviral device to cause a premature conformational change is unexplored.

7. Antigenic drift

Antihemagglutinin antibodies neutralize virus infectivity, and as a consequence the antigenic variation for which influenza viruses are renowned involves extensive variation in the antigenic properties and structure of their hemagglutinin. For over 40 years the antigenicity of influenza isolates from outbreaks and epidemics has been monitored to ensure the inclusion in vaccines of variants closely related to the viruses in circulation (Pereira, 1982), and more recently these comparative studies have been accompanied by nucleotide sequence analyses of the genes for the hemagglutinins of representative isolates (Sleigh et al., 1981; Both and Sleigh, 1981; Both et al., 1983; Skehel et al., 1983; Daniels et al., 1985a and Laver et al., 1980).

7.1. The location of the variable antibody-binding sites

The location of antibody-binding sites on the hemagglutinins of both subtypes of influenza A currently circulating (H3 subtypes viruses since 1968 and H1 subtype since 1977) have been proposed based on the location of sequence variation in the HAs of natural isolates and on sites of single-amino-acid substitutions observed in antigenic mutants selected by growth in monoclonal antibodies (Skehel et al., 1984; Fazekas de St. Groth, 1978; Daniels et al., 1983; Gerhard and Webster, 1978; Moss et al., 1980; Laver et al., 1979; Gerhard et al., 1981; Newton et al., 1983; and Caton et al., 1982). Fig. 10.7 shows the location on a schematic diagram of the HA of the naturally occurring sequence variation of the H3 subtype HA during the last 18 years (five shaped symbols) and the single-site antigenic mutants selected from 1968 and 1972 parental strains with monoclonal antibodies (stars). (For a comparison with the H1 subtype, see Caton et al., 1982 and Figs. 22 and 23 in Wiley, 1985). The variation is predominantly in the HA_1 glycopolypeptide, involving residues covering much of the surface of the distal domain of the molecule. The substitutions have been segregated into five regions designated A to E (Wiley et al., 1981; Skehel et al., 1984), although some subdivision and overlap of these areas has been noted (Daniels et al., 1983; Caton et al., 1982).

It is currently thought that each of these regions is an antibody-binding area and that the HAs of new epidemic strains must include substitutions in each of these regions in order to escape the antisera produced during an earlier cycle of infection. The antigenic significance of the observed natural variation in regions A to E is strongly supported by the coincident mapping of monoclonal-antibody-selected variants into each of the regions (Fig. 10.7), because these variants are known to no longer bind the selecting antibody and as a consequence replicate efficiently in its presence. An additional indication that the natural variation is antigenically significant is the observation that the frequency of substitutions in HA1, about 0.8% per year since 1968, is greater than that of silent mutation, 0.3%, indicating that it probably resulted from immune selective pressure.

The assumption that the location of a single-site substitution defines the site of antibody binding has been explored in two cases by crystallizing and determining the X-ray structure of the HAs from the monoclonal-antibody-selected mutants 146 G→ D and 188N→ D (sites A and B) (Knossow et al., 1984; Weis et al., unpublished). In both cases, the structural changes between the HA of the antigenic mutant and wild type were confined to the immediate vicinity of the amino-acid substitution, demonstrating that for the antibody to recognize this mutation, it must bind directly to that region and illustrating that rather small, local variations in structure suffice to allow escape from neutralization by antibodies (Knossow et al., 1984). Evidence consistent with the conclusion that the locations of amino-acid substitutions influence antibody binding directly rather than from a distance, has also been obtained from electron microscopy of monoclonal antibody-hemagglutinin complexes using antibodies recognizing regions A, B and E (Wrigley et al., 1983). No such direct information is presently available for binding sites C and D.

Because of the proximity of antibody-binding sites and the receptor-binding pocket, certain amino-acid substitutions influence both the binding of specific monoclonal antibodies and the specificity of receptor recognition (Daniels et al., 1984). As discussed above (section 5), a number of antihemagglutinin antibodies have also been observed to select mutants with different receptor-binding specificities which are, however, not antigenic mutants, since they continue to bind the selecting antibodies (Yewdell et al., 1986; Daniels et al., 1987). The sites of binding such antibodies are not known, but they do not in these cases appear to correspond with the locations of the amino acid sequence changes.

Some of the observed variation results in the addition or loss of an *N*-linked oligosaccharide (asterisk and dash in the list of Fig. 10.7). In the case of the monoclonal antibody-selected mutant HA_1, 63 D→ N, the addition of the oligosaccharide has been proved to be the cause of the antigenic change, as mutant virus grown in the presence of tunicamycin and lacking the new oligosaccharide at 63 was not antigenically distinguishable by the antibody (Skehel et al., 1984). A similar experiment on the natural epidemic strain A/Vic/3/75, which is glycosylated at 63, indicated that it would bind antibody produced against the 1968 strain (two epidemics earlier) when grown in tunicamycin and lacking the oligosaccharide at 63. Thus, carbohydrate, which is host-specific, can mask surfaces of the hemagglutinin from the immune system. Similar masking is seen in the area of HA_1 165 when the proposed antigenic sites of H3 (glycosylated at 165) and H1 (unglycosylated at 165) are compared (Daniels et al., 1985b; Caton et al., 1982).

Fig. 10.7b shows that the proposed antibody-binding sites surround the conserved receptor binding site on the HA. A similar observation has been made on the influenza NA (Colman et al., 1983) and has been proposed for rhinovirus and poliovirus based on their recent X-ray structure determinations (Rossmann et al., 1986; Hogle et al., 1986). Thus, even if antibodies could be directed at the conserved residues in the concave binding site, because of the large footprint of an antibody (Amit et al., 1986), they would probably be dislodged by the variation that occurs readily on the rim of the binding site.

8. Mechanisms of antigenic variation: summary

Antigenic variation in influenza viruses involves two processes frequently referred to as antigenic shift and antigenic drift. The former is responsible for the introduction into the human population of viruses containing hemagglutinins antigenically similar to those of viruses circulating in animals of other species and birds. The most recent antigenic shift, which resulted in the Hong Kong influenza epidemic in 1968, was caused by a recombinant virus containing the gene for the H3 hemagglutinin and the seven other genes in the virus-genome from an H2N2 virus circulating at that time in humans (Scholtissek et al., 1978). Sequence analysis of the hemagglutinins of viruses of the H3 subtype isolated from ducks and horses before 1968 strongly suggests that the recombination involved an avian influenza virus, and occurred presumably in a bird or human simultaneously infected with a human H_2N_2 virus and an avian virus of the H_3 subtype (Laver and Webster, 1973; Ward and Dopheide, 1981b; Fang et al., 1981; Daniels et al., 1985c). The amino acid sequence homology between the Hong Kong (H_3) hemagglutinin and Asian influenza (H_2) hemagglutinins is approximately 36% HA_1 and 50% HA_2 (Gething et al., 1980; Verhoeyen et al., 1980).

Antigenic variants of the H3 subtype isolated between 1968 and 1986, however, vary in sequence by about 0.8% per year, and the process involved in this variation is termed antigenic drift. Antigenic drift appears to involve the selection under antibody pressure of antigenic mutants with the ability to reinfect at least a proportion of the population. The precise mechanism of this selective process is unknown. The frequency at which antigenic mutants occur during selection with monoclonal antibodies is between 10^{-4} and 10^{-5} (Yewdell et al., 1979), and given that antibodies against any of the five antigenic regions neutralize virus infectivity, antigenic drift to an extent that would allow reinfection of the majority of the population would occur spontaneously at a very low frequency. Changes at each of the antigenic sites may, on the other hand, occur during reinfections by antigenic mutants changed in only one or two sites in individuals who develop only partial immunity during initial infections. Analysis of the variety of antibody specificities in post-infection human sera and of the restricted ability of human sera to neutralize monoclonal-antibody-selected mutants suggest that this is the case (Natali et al., 1981; Wang et al., 1986). Information on the frequency of reinfections from more extensive serological surveys is required for assessment of this possibility.

9. Antigenic variation in influenza B virus

Sequences of the HA gene of several field isolates of influenza B virus have been determined (Krystal et al., 1983; Verhoeyen et al.,, 1983; Berton et al., 1984) and it appears that variable areas in the influenza B virus HA structure correspond to antigenic regions A, B and D of influenza A viruses. No changes were found in the influenza B virus HAs in areas corresponding to antigenic region C of influenza A viruses.

The available data for HA1 of influenza B HAs is shown in Fig. 10.8. Of 38 amino acid positions where variation is seen, 17 are conserved in all strains except B/Lee/40, and it has previously been noted that residue 90 of B/Lee/40 from another laboratory is valine, as in all other strains (Air and Compans, 1983). Otherwise substitutions tend to come and go over time, in contrast to the steady accumulation of amino acid substitutions with time found in HA1 of influenza A viruses of the H3 subtype (Both and Sleigh, 1981). Neither antigenic analyses (Lu et al., 1983) nor the sequence analyses indicate that antigenic drift occurs sequentially over time as in influenza A viruses.

The three-dimensional structure of the influenza B HA is not known but comparison of sequence data and computer-assisted model building suggested a structural similarity between influenza A and B HAs (Krystal et al., 1982).

To further investigate antigenic variation in influenza B viruses, variants have been selected with monoclonal antibodies and the sequence changes determined. Two strains have been used; B/Hong Kong/8/73 (Webster and Berton 1981; Hovan-

```
                  1                                                   50                                                    100
    [1]B/LEE/40   MKAIIVL                                                             Q        N F                 N T    V
      B/BONN/43   XXXXXXX                                                  R          Q        N F                 N      V V
   [1]B/MARY/59   MXXXXXXXXX                                                          Q        N      M                   V
   [2]B/HK/8/73   MKAIIVL                                                             Q        N
[3]B/SIN/222/79   MKAIIVL                                                                      N
  [4]B/ORE/5/80   MKAIIVL                                                                      N
[5]B/ENG/222/82   MKAIIVL                                                                      N            A
   B/USS/100/83   XXXXXXX                                                                      N
     B/VIC/3/85   XXXXXXX
     B/MEM/6/86   -------LMV VTSNADRICT GITSSNSPHV VKTATQGEVN VTGVIPLTTT PTKSHFANLK GTKTRGKLCP KCLNCTDLDV ALGRPKCMGT IPSAKASILH
                                  ↑start HA1

                  101                                                150                                                    200
       B/LEE/40       A                                  S     T    R S  K           A      N     VI  .D      I  V      Y  S
      B/BONN/43                                          S    S          KI             S   E         ...     M  V      Y  AK
      B/MARY/59                                          R               T              K   E         . K               Y   K
      B/HK/8/73                                         AR  T                                         ...               Y   K
   B/SIN/222/79                                          R       R        I                           ..D               Y   K
     B/ORE/5/80                                          R       R        I                           ..D               Y   K
   B/ENG/222/82                                      K   R       R        I                           ..D               Y   K
   B/USS/100/83                                          R       R        I                           ..D      D   I    Y   K
     B/VIC/3/85
     B/MEM/6/86   EVKPVTSGCF PIMHDRTKIR QLPNLLRGYE NIRLSTHNVI NAETAPGGPY IVGTSGSCPN VTNGNGFFAT MAWAVPKNNN NKTATNPLTV EVPFICTEGE

                  201                                                250                                                    300
       B/LEE/40               DKT  ER      N                         N T   E  K              P       V
      B/BONN/43               S T  GR      N       Y                 N T   E  K          I   P E P   V
      B/MARY/59               D TL  I                                N T   E  K          I   P       V     V
      B/HK/8/73               D T                                    N     E                 P       A     V
   B/SIN/222/79               T T                                    N T                     P       V     V
     B/ORE/5/80                 A                                    N T                     P       V     V
   B/ENG/222/82                A                                     N T                     P       V     V
   B/USS/100/83                KN                                    N T                     P   P   V     V
     B/VIC/3/85               S T
     B/MEM/6/86   DQITVWGFHS DNEIQMVKLY GDSKPQKFTS SANGVTTHYV SQIGGFPKQA EDGGLPQSGR IVVDYMVQKS GKTGTITYQR GILLPQKVWC ASGRSKVIKG

                  301                                                350                                                    400
       B/LEE/40
      B/BONN/43                                                                                           D
      B/MARY/59
      B/HK/8/73
   B/SIN/222/79
     B/ORE/5/80
   B/ENG/222/82
   B/USS/100/83
     B/VIC/3/85
     B/MEM/6/86   SLPLIGEADC LHEKYGGLNK SKPYYTGEHA KAIGNCPIWV KTPLKLANGT KYRPPAKLLK ERGFFGAIAG FLEGGWEGMI AGWHGYTSHG AHGVAVAADL
                                                                                    ↑start HA2

                  401                                                450                                                    500
       B/LEE/40                   Y                  N   D           D                                                  E
      B/BONN/43                                    E  G              D
      B/MARY/59                                                      D                                                  E
      B/HK/8/73                                                      D
   B/SIN/222/79                                                      D
     B/ORE/5/80                                                      D
   B/ENG/222/82                                                      D
   B/USS/100/83                                                      D
     B/VIC/3/85                                                      D
     B/MEM/6/86   KSTQEAINKI TKNLNSLSEL EVKNLQRLSG AMDELHNEIL ELDEKVDNLR ADTISSQIEL AVLLSNEGII NSEDEHLLAL ERKLKKMLGP SAVDIGNGCF

                  501                                                550                                  585
       B/LEE/40                           D
      B/BONN/43                    R  S
      B/MARY/59
      B/HK/8/73
   B/SIN/222/79
     B/ORE/5/80
   B/ENG/222/82
   B/USS/100/83
     B/VIC/3/85
     B/MEM/6/86   ETKHKCNQTC LDRIAAGTFN AGEFSLPTFD SLNITAASLN DDGLDNHTIL LYYSTAASSL AVTLMIAIFI VYMVSRDNVS CSICL
```

Fig. 10.8. Sequences of the HA polypeptide of influenza B strains. The amino acid sequence of B/Mem/6/86, deduced from the cDNA sequence, is shown, with differences marked for the other strains. Data from [1]B/Lee/40 (Krystal et al., 1982), [1]B/Maryland/59 (Krystal et al., 1983), [2]B/Hong Kong/8/73 (Hovanec and Air, 1984), [3]B/Singapore/222/79 (Verhoeyen et al., 1983), [4]B/Oregon/5/80 (Berton et al., 1984), [5]B/England/222/82. The sequences for B/Bonn/43, B/USSR/100/83, B/Victoria/3/85, and B/Memphis/6/86 are by Air et al., 1990. Other sequences available are only of the HA1 portion (Yamashita et al., 1988).

ec and Air, 1984) and B/Oregon/5/80 (Berton et al., 1984). For comparison, amino acid positions in the HA1 peptide can be described using the corresponding A/Aichi/2/68 residue numbers (Wiley et al., 1981 as aligned by Krystal et al., 1983).

In the B/Hong Kong/8/73 HA1 polypeptide there are two areas of antigenic variability (Fig. 10.8). These areas would correspond to antigenic regions A and B of the A/Aichi/2/68 HA molecule (Fig. 10.7). Amino acid substitutions found in these areas include residues 137 and 142 (region A) and residues 157, 158 and 193 (region B). However, two variants have a further nucleotide substitution at position 1742 which changes amino acid 211 of the HA2 peptide. This residue is located within the transmembrane region of the B/Hong Kong/8/73 HA and is therefore not even present in the bromelain-released HA structure of A/Aichi/2/68 (Wilson et al., 1981), and is unlikely to be involved in antigenicity.

A noticeable feature of the sequences of HA genes of 'monoclonal' variants of B/Hong Kong/8/73 is the relatively high incidence of multiple nucleotide changes; four of the eight 'monoclonal' variants sequenced had more than one nucleotide change.

The nucleotide sequences of the HA1 portions of the HA genes of the parent strain and of 18 laboratory-selected antigenic variants of B/Oregon/5/80 have been determined (Berton et al., 1984). Comparison of the deduced amino acid sequences of the parental and variant HA1 polypeptides revealed 16 different amino acid substitutions at nine positions. All amino acid substitutions resulted from single point mutations and no double mutants were detected. The amino acid changes cluster in four regions in the primary structure of the influenza B virus HA and appear to identify at least two immunodominant sites which correspond to proposed antigenic sites A and B on the influenza A virus H3 HA. The positions of change are indicated in Fig. 10.8. In studies of influenza A virus only single-amino-acid changes were found and monoclonal antibodies which recognized one region did not recognize any of the other antigenic sites. In both the influenza B virus studies monoclonal antibodies which belong to the same antigenic group recognize amino acid changes in regions corresponding to both antigenic sites A and B of the H3 HA (Berton et al., 1984; Hovanec and Air, 1984).

One interpretation of these results is that the loop regions of antigenic sites A and B are in closer proximity to each other in the influenza B than in the influenza A structure, although the basic three-dimensional structure of the influenza B HA may be similar to that of A/Aichi/2/68. The distance between antigenic sites A and B of the Aichi HA (Wilson et al., 1981) could possibly be spanned by an antibody-combining region, but results with H3 and H1 variants have not indicated that it is.

The sequence data of the variants suggest that there is a considerable overlap between the antigenic regions of the influenza B HA molecule. Distinct antigenic sites, as found in the HA of A/Aichi/2/68, may not exist as such in the HA of influenza B viruses. The HA1 peptide in type B viruses may be folded into a more compact structure, so that a relatively larger proportion of the sequence is recognized by a single monoclonal antibody.

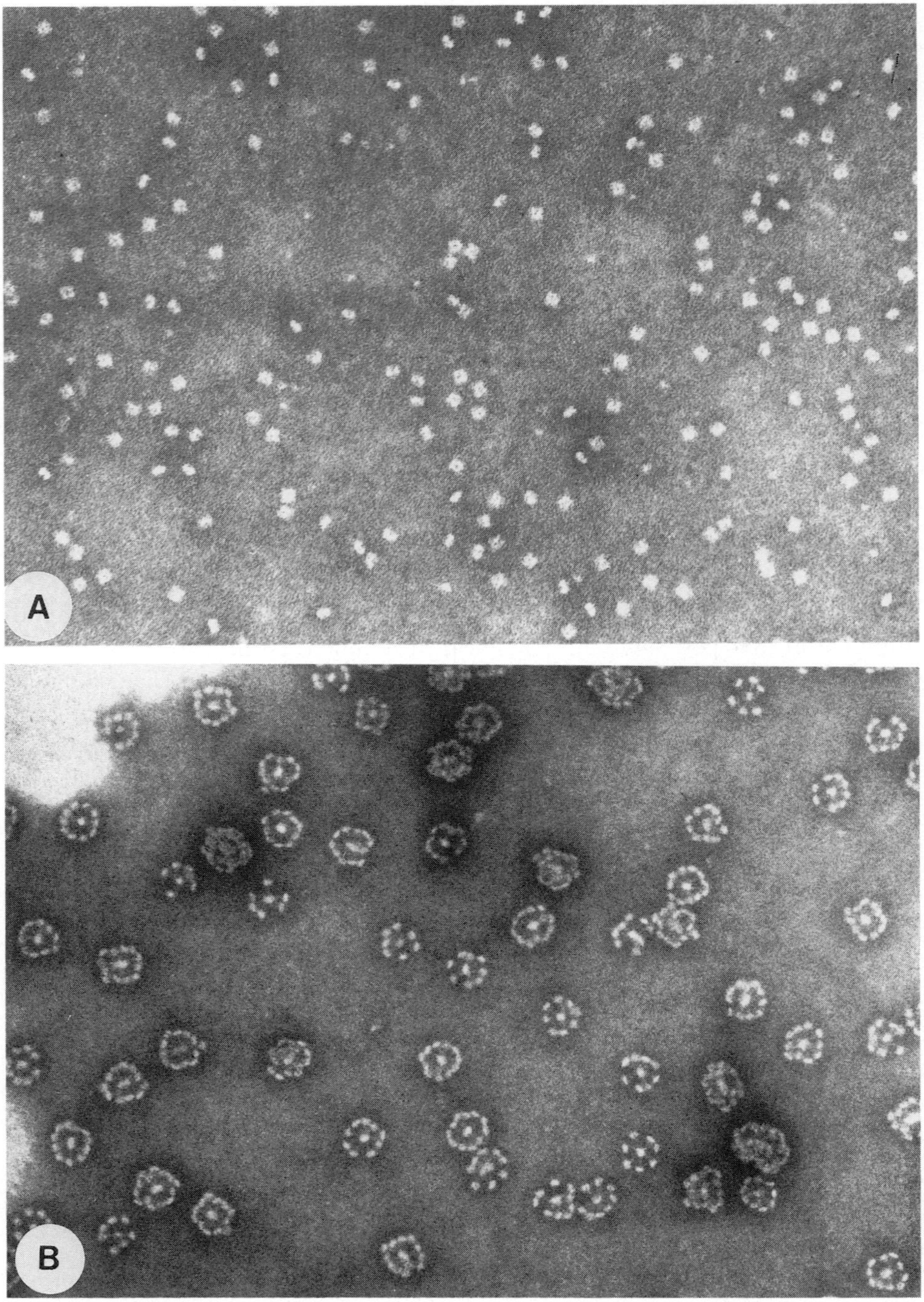

Fig. 10.9. Electron micrographs of neuraminidase molecules. (A) pronase-released NA heads (B) whole NA molecules isolated from detergent-disrupted virus. The tetrameric heads can be seen in A (both en

10. The neuraminidase of influenza virus

The neuraminidase (NA) accounts for about 5–10% of the virus protein and exists as a mushroom-shaped spike on the surface of the virion. It is a tetramer with a box-shaped head, 10.0 × 10.00 × 6.0 nm, made out of four coplanar and roughly spherical subunits and a centrally attached stalk containing a hydrophobic region by which it is embedded in the viral membrane (Fig. 10.4). The neuraminidase molecule is composed of a single polypeptide chain, coded by RNA segment 6, anchored in the virus membrane by a series of hydrophobic amino acids near the N-terminal end of the polypeptide (Fig. 10.4). This contrasts to the influenza hemagglutinin, which is anchored by a hydrophobic sequence near the C-terminus. No post-translational cleavage of the NA polypeptide occurs, no signal peptide is split off and even the initiating methionine is retained (Blok et al., 1982). Nor is there processing at the C-terminus; the C-terminal sequence, -Met-Pro-lle predicted from the gene sequence for N2 NA is found in intact NA molecules isolated from the virus. A sequence of six polar amino acids at the N-terminus of the NA polypeptide, which are

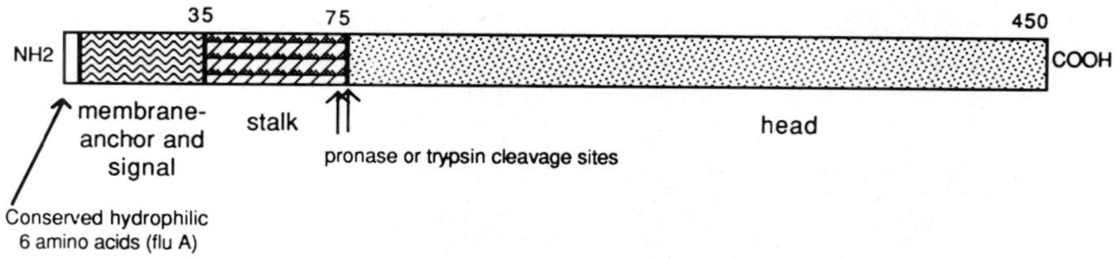

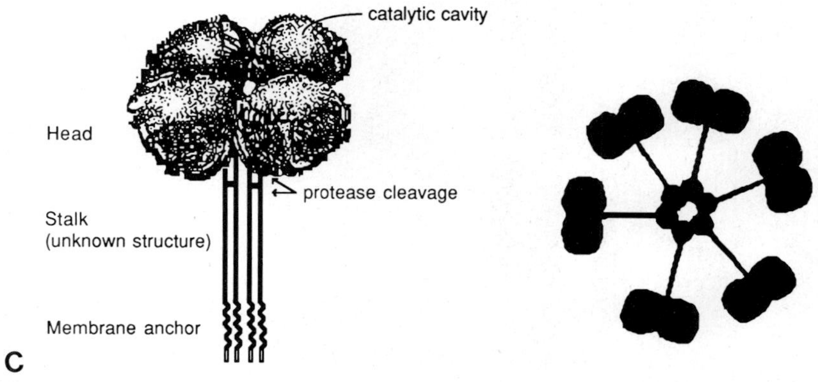

face and side views appear); in B the molecules (from which detergent has been removed) associate by the hydrophobic N-terminal sequences as diagrammed in (C). Protease digestion releases NA heads, which retain all the enzymatic and antigenic activity. Deletion in the stalk region can result in **stubby** NA molecules (Els et al., 1985). Electron micrograph courtesy of Nick Wrigley.

totally conserved in each of the nine different influenza A NA subtypes (Blok and Air, 1982; Air et al., 1985a) but not in influenza B (Shaw et al., 1982), is followed by a sequence of hydrophobic amino acids which must represent the transmembrane regions of the NA (Fields et al., 1981). This sequence is not conserved at all between subtypes (apart from conservation of hydrophobicity). Pronase cleaved the polypeptide in the positions shown (Fig. 10.4), removing the stalk and releasing the enzymically and antigenically active head of the NA (Fig. 10.9) which, in some cases, can be crystallized (Fig. 10.10).

Viruses have been obtained with 'stubby' neuraminidase molecules in which the stalk is shortened by deletions of up at 18 amino acids (Els et al., 1985).

Several roles have been suggested for NA (acylneuraminyl hydrolase, EC 3.2.1.18). The enzyme catalyzes cleavage of the α-ketosidic linkage between terminal sialic acid and an adjacent sugar residue. In mucin, removal of sialic acid lowers the viscosity and permits access of the virus to the epithelial cells. Neuraminidase also destroys the hemagglutinin receptor on the host cell, thus allowing elution of

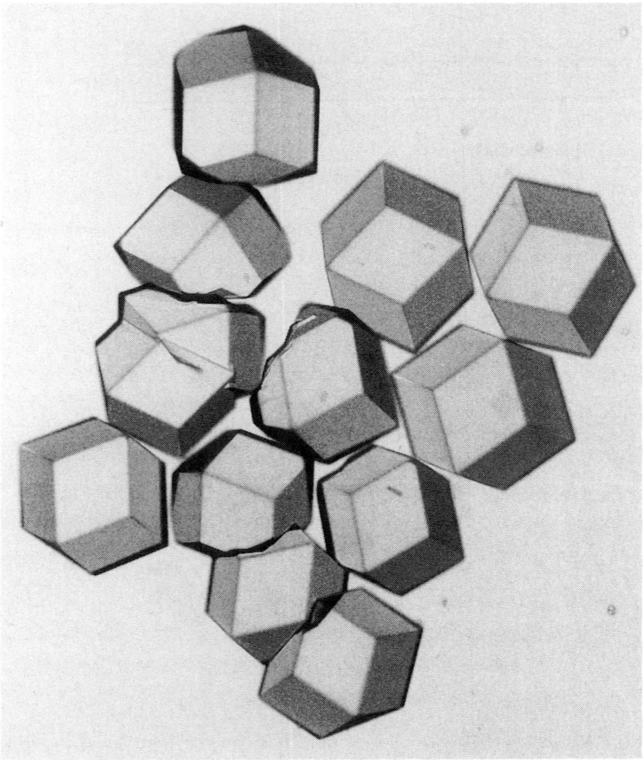

Fig. 10.10. Crystals of pronase-released NA heads of the N9 subtype used for X-ray diffraction studies. The crystals were grown from phosphate buffers by vapor diffusion.

progeny virus particles from infected cells (Palese et al., 1974). The removal of sialic acid from the carbohydrate moiety of newly synthesized HA and NA is also necessary to prevent self-aggregation of the virus. In general, then, the role of NA may be to facilitate mobility of the virus both to and from the site of infection, and although this role is not well-defined, the NA induces antibodies which protect against lethal influenza viruses (Webster et al., 1988). In type A influenza, nine subtypes of NA which lack any antigenic cross-reactivity have been classified. N1 and N2 have been found in viruses infecting man and N1–N9 in viruses infecting other animals, mainly wild birds.

10.1. Structure of N2 neuraminidase

The three-dimensional structure of N2 neuraminidase heads determined from an electron density map at 2.9 Å resolution shows that each monomer is composed of six topologically identical β-sheets arranged in a propeller formation (Varghese et al., 1983). The tetrameric enzyme has circular four-fold symmetry and is stabilized in part by metal ions bound on the symmetry axis. The catalytic sites are located in deep clefts which occur on the upper corners of the box-shaped tetramer. Sugar residues are attached to four of the five potential glycosylation sequences, and in one case contribute to the interaction between subunits in the tetramer (Fig. 10.11).

10.2. Structure of N9 neuraminidase

The three-dimensional structure of the neuraminidase antigen of subtype N9 from an avian influenza virus (A/tern/Australia/G70c/75), which shares 50% sequence identity but no antigenic cross-reactivity with the human N2 influenza virus neuraminidase, has also been determined by X-ray crystallography (Baker et al., 1987) and shown to be folded similarly to N2 neuraminidase. The result demonstrates that absence of any immunological cross-reactivity is no measure of dissimilarity of polypeptide chain folding. Small differences in the way in which the subunits are organized around the molecular fourfold axis are observed. Insertions and deletions with respect to subtype N2 neuraminidase occur in four regions, only one of which is located within the major antigenic determinants around the enzyme active site.

10.3. Influenza type B neuraminidase

Type B Influenza RNA segment 6 codes two distinct glycoproteins, NA and NB, using overlapping reading frames. The amino acid sequence of polypeptide NB deduced from the nucleotide sequence of the B/Lee/40 strain consists of 100 amino acids with a molecular weight of 11 242 (Shaw et al., 1982). The sequence contains four potential glycosylation sites, and the protein has been found to be glycosylated

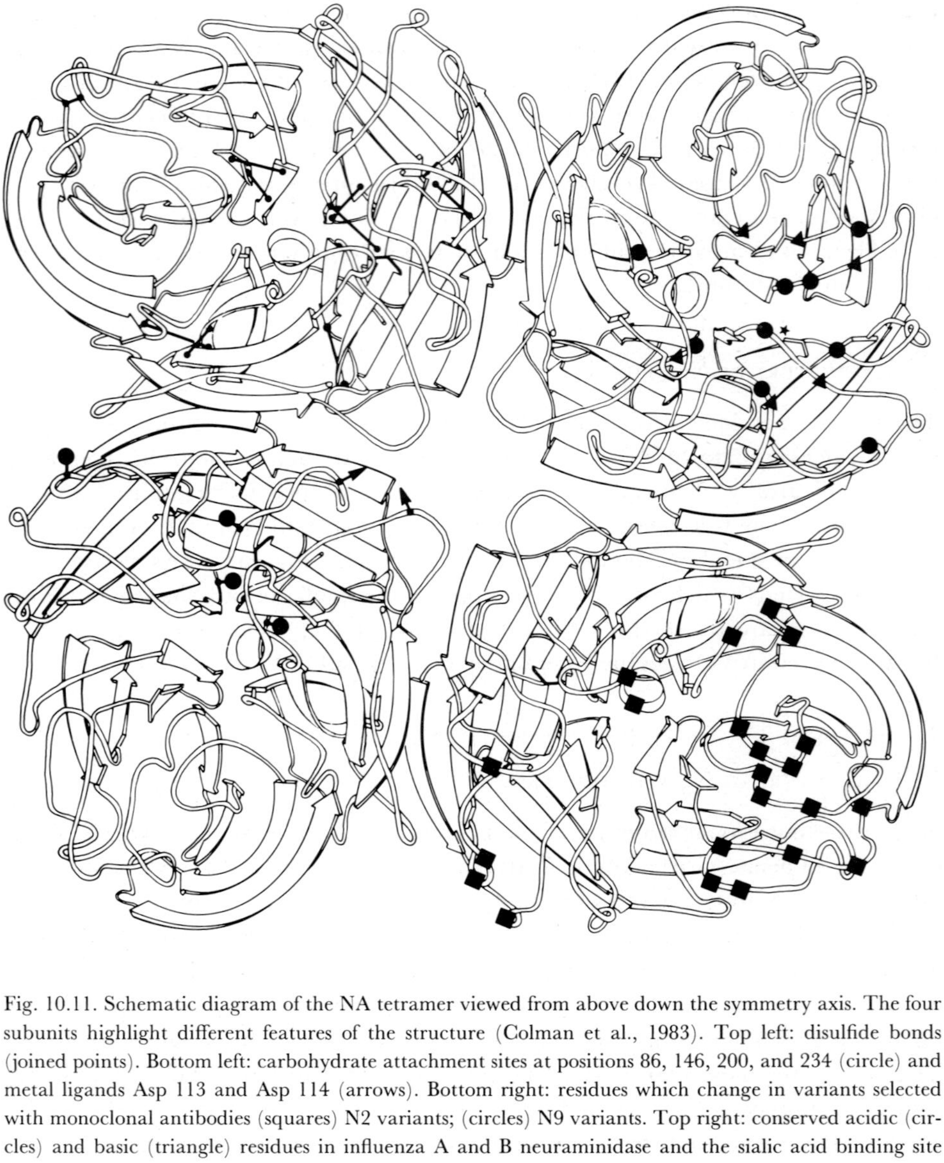

Fig. 10.11. Schematic diagram of the NA tetramer viewed from above down the symmetry axis. The four subunits highlight different features of the structure (Colman et al., 1983). Top left: disulfide bonds (joined points). Bottom left: carbohydrate attachment sites at positions 86, 146, 200, and 234 (circle) and metal ligands Asp 113 and Asp 114 (arrows). Bottom right: residues which change in variants selected with monoclonal antibodies (squares) N2 variants; (circles) N9 variants. Top right: conserved acidic (circles) and basic (triangle) residues in influenza A and B neuraminidase and the sialic acid binding site (star).

in infected cells. NB has not been found in virions. A protein analogous to NB has not been found in influenza A virus and this represents a major difference between the two virus types. No subtypes of influenza B virus exist and, in marked contrast to influenza A, influenza B has so far been found only in the human population.

Influenza B/Lee/40 neuraminidase has less than 25% sequence identity compared with either N2 or N9 neuraminidase in the head region. Conservation of several of the cysteine residues which form disulfide bonds in N2 and N9 neuraminidases and of several amino acid side-chains which line the active site pocket suggest that the polypeptides may be similarly folded, but this can only be confirmed by a full structure determination.

Crystals of neuraminidase heads from two different influenza B virus strains have been grown (Bossart et al., 1988). Neuraminidase crystals of influenza B/Hong Kong/8/73 were grown from solutions of potassium phosphate. The crystals are tetragonal prism, space group I422; the axes are a = 123 Å and c = 165 Å. Influenza B/Lee/40 neuraminidase crystals were grown from solutions of polyethylene glycol 4000. The crystals are tetragonal pyramids, space group $P4_12_12$ or its enantiomorph $P4_32_12$; the axes are a = 125 Å and c = 282 Å. On further analysis these crystals showed disorder, making the solution of the structure difficult if not impossible. However, when the B/Lee/40 NA is complexed with Fab fragments of a monoclonal antibody (see below), crystals are obtained which are well-ordered, and the route to the structure determination of influenza B NA may be through its complex with antibody.

11. Active center of neuraminidase

The influenza virus NA provides a unique opportunity to characterize an enzyme active center which is conserved amidst enormous variation in amino acid sequence. There is up to 75% sequence variation between NAs of influenza A and B, and hence those residues which are conserved in all known sequences are likely to be essential for enzyme function. In the structure determined at 2.9 Å resolution (Colman et al., 1983) several such conserved residues line the substrate binding pocket, and these have been used as targets for site-specific mutagenesis to delineate the mechanism of enzyme action (Lentz et al., 1987; Wei et al., 1987).

12. Antigenic properties of the neuraminidase

Antibodies to the NA do not neutralize virus infectivity directly, but if the cells in which virus is growing are bathed in antibody, the enzyme is inhibited, newly formed virus cannot leave the infected cell and the infection is effectively terminated. Escape mutants of N2 and N9 NA were obtained by growing the viruses in the presence of high concentrations of monoclonal antibodies to the NA (Webster et al., 1982, Webster et al., 1984). The presence of antibodies in the fluid surrounding the cells prevented these from yielding virus and these monoclonal antibodies could be used to select antigenic variants (escape mutants) of the NA. The NA genes of a number of the escape mutants were completely sequenced. Single nucleotide changes were found in each of the variants, which resulted in single-amino-acid sequence changes in the NA polypeptide (Lentz et al., 1984; Air et al., 1985b). Most of these sequence changes occurred on the top of the tetrameric NA head, on the rim around the active site crater, suggesting that neutralizing antigenic determinants (epitopes) were located in this region. Other epitopes almost certainly exist at the base of the tetramer, but escape mutants have not been isolated, presumably because antobodies binding in this region do not neutralize NA activity.

13. Structure of antigenic determinants (epitopes) on the NA

A great deal of information has been gathered concerning the structure of combining sites on antibodies and the interactions involved in the binding of small molecular weight haptens, but there is a considerable debate about the way in which antibodies recognize and bind to the surfaces of foreign proteins.

We define an 'epitope' as the region on the surface of a protein which binds a particular monoclonal antibody molecule. Regions on unfolded proteins which bind antibodies are not considered to be epitopes by this definition. The best evidence that an antibody is binding to native protein is if the biological activity of the protein is affected by antibody in a measurable way. Antibodies which affect biological activity presumably must be binding to the native protein.

It has been proposed that there are two kinds of epitopes on proteins, 'sequential' and 'conformational' (Sela, 1969). Of the five epitopes so far characterized by X-ray crystallography (Amit et al., 1986; Sheriff et al., 1987; Padlan et al., 1989; Colman et al., 1987b; Colman et al., 1989), all are 'conformational' (i.e., made up of several short amino acid sequences together in the three-dimensional structure but perhaps widely separated in the primary sequence).

We believe that sequential epitopes on proteins may be very rare and that most of the 'continuous' epitopes so far described reside, in fact, on unfolded, denatured protein molecules. In any case, it is doubtful that even segments of unfolded protein are truly continuous, with all amino acid side chains in contact with antibody.

In the past, many methods have been used in attempts to define antigenic determinants on protein molecules (reviewed by Benjamin et al., 1984). The methods have included the use of protein fragments to absorb antisera, the production of anti-peptide antisera and their reactions with intact proteins, proteolytic digestion of antigens in complexes with antibodies as a probe of protected peptide bonds within the epitope, and the protection of amino acids in the epitope from specific chemical derivatization. Other methods include the characterization of 'escape mutants' (variants which do not bind neutralizing monoclonal antibodies) and competitive binding studies of monoclonal antibodies for an antigen (Gerhard et al., 1981). Although data from some experiments have implicated particular amino acid residues as participants in particular epitopes, none has allowed a complete structural description of an epitope.

The first determination of the structure of an epitope on a protein molecule was the determination of the three-dimensional structure of a crystallized complex between lysozyme and the Fab fragment of a monoclonal antibody at 2.8 Å resolution (Amit et al., 1986). In this complex, 16 amino acids on the lysozyme and 17 on the antibody form a tightly packed interface from which water molecules are excluded. The interaction has been described as conforming to a 'lock-and-key' picture of antibody antigen interaction, in which, apart from some amino acid side chain movements, no structural changes occur in either the antibody or the antigen. Two more lysozyme-Fab complex structures, have since been determined at high resolution (Sheriff et al., 1987; Padlan et al., 1989). A synthetic peptide containing residues 40–60 of lysozyme (encompassing most of the HyHEL-5 epitope) neither bound to the HyHEL-5 antibody nor inhibited binding of antibody to the lysozyme (S. Smith-Gill, personal communication).

This result is not surprising since this peptide represented only part of the epitope and is missing four amino acids of the 14 in the contacting surface. It is established in several protein antigens that single amino acid changes in the contact surface of the antigen are sufficient to reduce binding to antibody to undetectable levels.

14. Crystalline neuraminidase-antibody complexes

We have grown crystals of monoclonal antibody Fab fragments complexed with influenza virus neuraminidase of the N9 subtype. A number of crystalline complexes were obtained. The first complex yielded small crystals only, which were unsuitable for X-ray diffraction analysis. This complex, between N9 NA and antibody 32/3 Fab (Fig. 10.12) could be grown as extremely thin, extensive two-dimensional platelet microcrystals (Fig. 10.13), but all attempts to grow three-dimensional crystals suitable for X-ray diffraction analysis have failed (Tulloch et al., 1986). However, recent advances in high-resolution electron microscopy suggest that considerable structural information may be gleaned from such crystals. Using new methods for electron imaging which cut down the effects of radiation-induced specimen movements (Bullough and Henderson 1987), and sophisticated methods of computer processing (Henderson et al., 1986), Bullough and Tulloch were able to extract phases from images to relatively high resolution (Bullough, 1988). Despite the large unit cell dimensions of these crystals (a = b = 159.5 Å) and the large solvent content (approximately 70% by volume), a projection structure was determined to 7 Å resolution (Fig. 14). The N9 NA-NC41 complex, which has been solved to atomic resolution (Colman et al., 1987b), forms crystals of the same space group and is believed to have similar packing (Fig. 10.15) (Tulloch et al., 1986). Comparison of the N9 NA-NC41 and NA-32/3 complexes has suggested that the 32/3 Fab may bind to the NA in the same general antigenic region as the NC41 Fab (Colman et al., 1987b). This is indeed consistent with the known serology (Webster et al., 1987), since 32/3

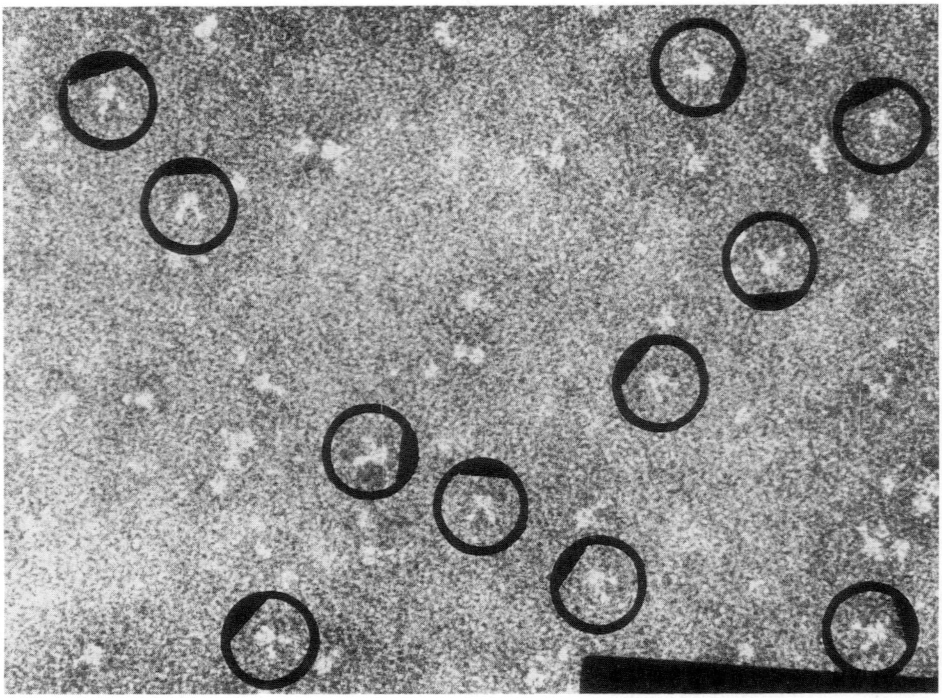

Fig. 10.12. Electron micrographs of isolated neuraminidase-antibody complexes. Edge projections of individual N9 NA-32/3 Fab complexes (examples circled), are interpreted here as single tetrameric NA heads viewed edge-on, each with 4 Fab molecules attached and subtending an angle of approx. 75° to the plane of the NA head. The upper 2 Fab arms eclipse the lower 2 in this projection. Specimens were negatively stained with neutral 2% potassium phosphotungstate.

can select NA escape mutants with amino acid substitutions at 329 and 370.

Crystals of five other complexes which diffract X-rays to beyond 3 Å have been obtained. These are NC41 Fab complexed with N9 NA from tern and whale viruses and with two variants of tern virus NA with sequence changes which reduce but do not abolish binding, and whale virus N9 NA with another antibody, NC10 Fab.

Fab fragments from four different monoclonal antibodies have also been complexed with influenza B virus neuraminidase (B/Lee/40) and the complexes have been crystallized. Three of the complex crystals are, so far, not suitable for X-ray diffraction studies, but the fourth (B/Lee/40 NA-B1 Fab) forms large crystals which diffract X-rays to 3 Å resolution (Laver et al., 1988), and structure analysis is underway.

14.1. Neuraminidase-NC41 FAB structure

The crystal structure of the N9-NC41 complex is reported to 2.9 Å resolution (Colman et al., 1987b).

Fig. 10.13. Electron micrograph of thin crystals of N9 NA complexed with Fab of antibody 32/3. Lattice image of a 2-dimensional microcrystal of N9 NA-32/3 Fab complex negatively stained with neutral 2% potassium phosphotungstate: (a) before and (b) after computer image processing. The unit cell, with a approx. 170 Å, is indicated.

C-α traces of antibody and antigen, shown in Fig. 10.16 on p. 212, indicate that the epitope is discontinuous, being comprised of five separate peptide segments involving about 17 amino acid residues. In addition it appears that complementarity determining region (CDR) H1 of the NC41 Fab may interact with carbohydrate covalently attached to Asn 200 of a neighboring subunit of the tetrameric antigen (Colman et al., 1989). Carbohydrates on influenza virus antigens are usually presumed to be excluded from the antigenic repertoire as 'self'. However, they can be antigenic in the context of a foreign protein structure, and their involvement in antibody binding would not be surprising. In the case of the influenza hemagglutinin, it has been demonstrated that carbohydrate can mask proteinaceous antigenic sites (Skehel et al., 1984).

A schematic diagram of the NA monomer is shown in Fig. 10.17. On this, the epitope recognized by NC41 antibody covers an area on the edge of the active site crater (Fig. 10.11) and involves several surface loops. The positions of amino acid sequence changes in escape mutants which affect binding of antibody NC41 are shown in Fig. 10.17. Changes at positions 367 (Ser to Asn), 369 (Ala to Asp), 370 (Ser to Leu), 372 (Ser to Phe), 400 (Asn to Lys) and 432 (Lys to Asn) abolish binding of antibody NC41 to neuraminidase whilst changes at positions 329 (Asn to Asp) and 368 (Ile to Arg) reduce binding of NC41 antibody (Webster et al., 1987; Air and Webster, unpublished results). All of these amino acids contribute to the surface area of the antigen that is buried when NC41 antibody binds. The peptide corresponding to the epicenter of the binding site (amino acids 366–373) has been syn-

Fig. 10.14. A 7 Å resolution projection structure of the N9 NA-32/3 Fab complex solved by electron diffraction. One unit cell (plane group p^4gm, a = b = 159.5 Å) is outlined in the center. The central complex protomer consists of an approximately square NA head of dimensions 90 Å × 90 Å. It is thought that an Fab arm is attached at each corner of the NA and is pointing up towards the viewer. All nearest neighbor complex protomers are thought to be facing the other way, with the Fab arms pointing into the plane of the paper. Fig. after Bullough (Bullough, 1988).

thesized. The peptide does not bind to antibody NC41, and does not compete with NA (Webster, Air and Laver, unpublished results).

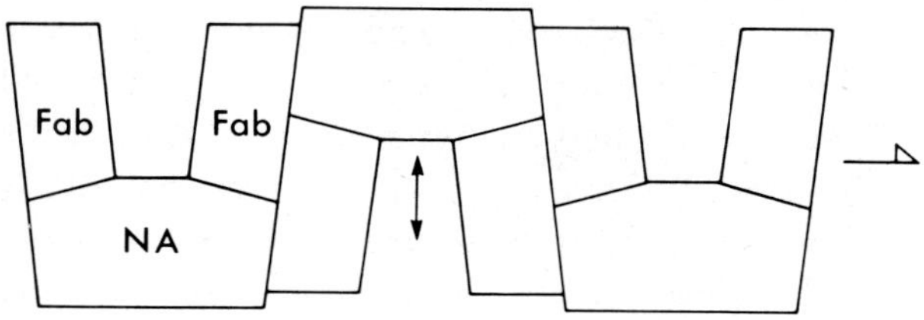

A

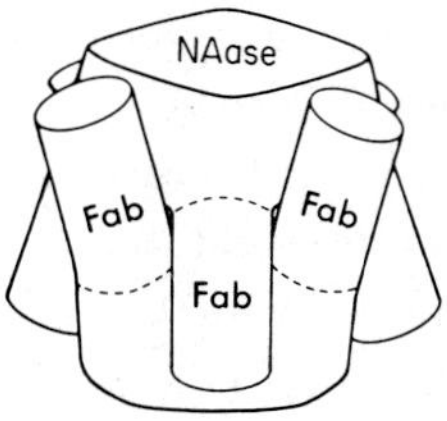

B

Fig. 10.15. Schematic views of the proposed arrangement of N9 NA-Fab complex protomers. (A) NA-32/3 Fab protomers in the plane of the 2-dimensional crystalline sheet. Lateral contact between nearest-neighbor protomers is mediated via a Fab fragment from the front side of one protomer and a Fab fragment from the rear side of its neighbor in this view. The relative position, along the vertical c-axis direction, of the central protomer with respect to its neighbors is not known (indicated by the vertical arrow), but the unit cell dimensions in the tetragonal plane demand that there be no relative position in which lateral contact between neighboring protomers is via NA heads alone. (B) The proposed packing of N9 NA-NC35 Fab complex in the c-axis direction of the crystals. The view is slightly inclined compared to that in (A).

15. Structure of escape mutants

How do single-amino-acid sequence changes in the loops abolish antibody binding? Remember, only one out of about 17 contact residues is altered in each mutant. One possible mechanism is that a single sequence change can drastically alter the conformation of the whole epitope, so that is is no longer recognized by antibody. The evidence suggests, however, that this is unlikely.

The structure of an escape mutant of influenza virus N2 NA (A/Tokyo/3/67), which was selected with monoclonal antibody S10/1 (Laver et al., 1982), has been determined (Varghese et al., 1988). Residue 368, which is lysine in the wild type, changed to glutamic acid in the mutant. This change abolished S10/1 antibody binding.

This experiment shows that the S10/1 epitope does indeed include amino acid residue 368, and that the change in the side chain has sufficed to effectively abolish binding of the antibody to the epitope. The single amino acid substitution of lysine to glutamic acid in the mutant introduces two free charged acid groups (Glu 368 and Asp 389) in the interface of the antibody and the antigen. The change in the epitope structure in this case is substantial, being two electronic charges plus attending shape changes. In some cases, however, escape mutants that have lost the capacity to bind antibody show more subtle chemical changes than are observed here, such as arginine for lysine (Laver et al., 1982; Air et al., 1985b).

A survey of single-amino-acid changes in escape mutants of influenza antigens (Caton et al., 1982; Wiley et al., 1981; Lentz et al., 1984; Air et al., 1985b; Webster et al., 1987) does not show any preference for size or character of the replacing or the replaced residue in the antigen, and indicates that even very subtle changes abolish the binding of an antibody for an antigen. The structure has been reported of only one other viral escape mutant, an influenza hemagglutinin (Knossow et al., 1984). In this case, the substitution of an aspartic acid for a glycine residue also caused only local structural alterations on the surface of the molecule.

Although the possibility of some escape mutants influencing antibody binding at a distance cannot be ruled out, the two influenza virus surface antigens studied so far both show that local structural changes suffice to effectively abolish antibody binding to an antigen.

16. Mechanism of inhibition of NA activity by antibody

Many monoclonal antibodies to the NA inhibit enzyme activity towards large substrates such as fetuin (50 kDa), presumably by blocking access of this large molecule to the active site. Enzyme activity towards small substrates such as *N*-acetyl neuraminyl lactose (0.5 kDa) is not inhibited at all by some of these monoclonal antibodies, even at the highest concentration which can be obtained (Fig. 10.18). Other antibodies, however, do inhibit enzyme activity towards the small substrate. NC41 antibody belongs to this category. Both the intact antibody and Fab fragments reduce the enzyme activity towards NANL by 75% (Fig. 10.19). What is the mechanism of this inhibition?

In the published structure of the N9 NA-NC41 Fab complex (Colman et al., 1987a) it appears that the active site is still accessible to small substrates, and that therefore inhibition of enzyme activity by NC41 Fab is not due to steric hindrance. The crystal structure of the complex of NC41 Fab and NA (Colman et al., 1987b) shows a very small shift in the position of an active site residue, Arg 371, which is known to be critical for enzyme activity since mutation to Lys resulted in loss of enzyme activity (Lentz et al., 1987). It is possible that this shift causes the lower enzyme activity. However, it is also possible that the enzyme active site may need to be mobile in order to carry out its catalytic function and that binding of NC41 antibody freezes the NA in an unfavorable configuration.

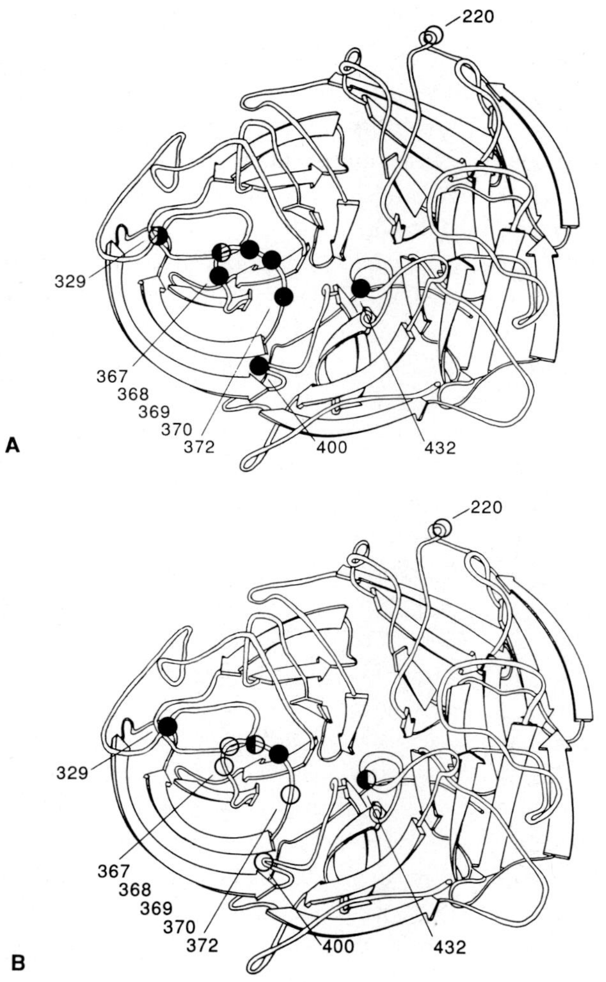

Fig. 10.17. Schematic diagram showing the chain fold in the neuraminidase monomer viewed down the 4-fold axis. This diagram has been revised from that previously published (Varghese et al., 1983, and shown in Fig. 10.3) in the region involving amino acid 329 (Tulip et al., 1989). Escape mutants, selected with monoclonal antibodies, which abolish (filled circles) or diminish (half-filled circles) binding by (A) antibody NC41 or (B) antibody NC10 are indicated, along with mutations which have no effect (open circles). The epitopes recognized by NC41 and NC10 antibodies involve the same loops (Tulip et al., 1989) but the recognition is very different (Laver et al., 1987; Webster et al., 1987). N2 numbering is used. The side chains of amino acids 368–370 point towards the viewer, while that of Arg 371 (an active site residue) points away and into the catalytic site located above and to the right of Cα371. As shown in (A), mutations at positions 367, 369, 370, 372, 400 and 432 abolish the binding of NC41 antibody to neuraminidase, whereas mutations at 368 and 329 reduce binding. A mutation at residue 220 (outside the NC41 epitope) has no effect on binding of NC41 to neuraminidase. Binding by antibody NC10 (B) is abolished by mutations at positions 329 and 370, reduced by mutations at 369 and 432, and unaffected by mutations at 367, 368, 372 as well as the change well outside the epitope at 220. Clearly, other sequence changes may also affect binding by NC41 and NC10, but these have not yet been identified.

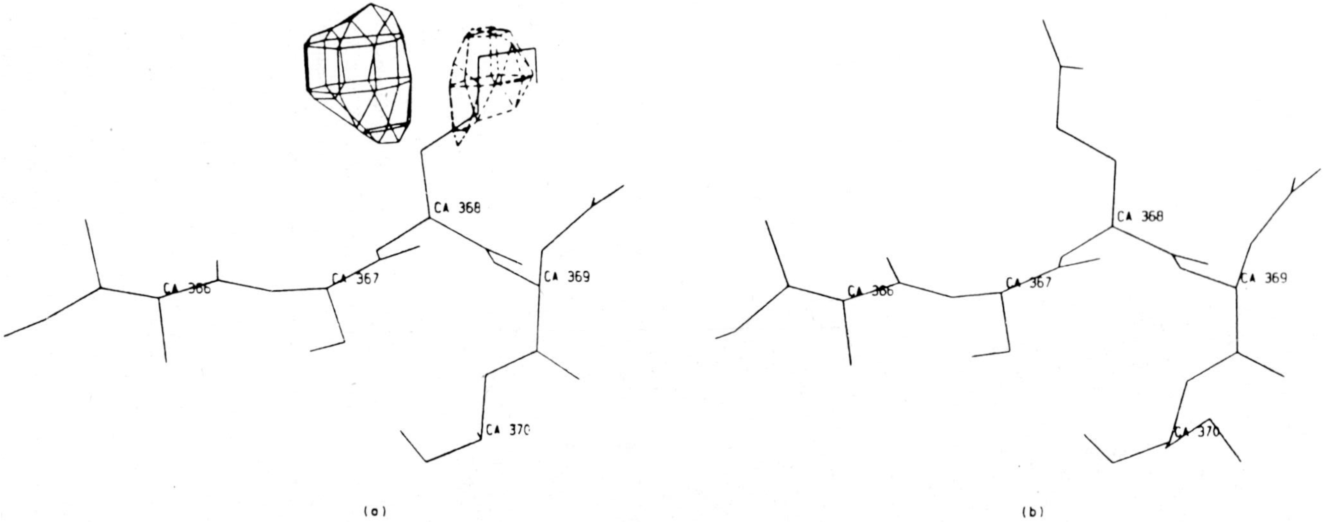

Fig. 10.18. An atomic model of amino acids in the neighborhood of residue 368 in neuraminidase of A/Tokyo/3/67. (a) Wild type NA with Lys at position 368 superimposed on an electron density difference map. (B) The refined atomic model of the same region in the mutant which has Glu at 368. (Varghese et al., 1988).

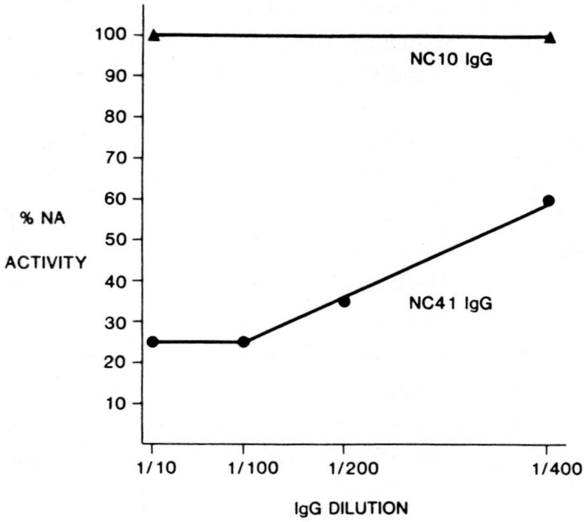

Fig. 10.19. NA inhibition curves for monoclonal antibodies NC41 and NC10 on N9 neuraminidase using *N*-acetyl neuraminyl lactose (0.5 kDa) as substrate. Both antibodies inhibited N9 NA activity almost totally (95% inhibition) when a large substrate, fetuin (50 kDa) was used.

17. Isosteric binding of NC41 antibody to a mutant neuraminidase

The NC41 Fab fragment crystallizes when complexed with the monoclonal variant of N9 known as Ox2, Asn 329 to Asp (Laver et al., 1987). Residue 329 is in the region of N9 NA where the structure is revised (Tulip et al., 1989), and it is at the edge of the epitope for NC41. These crystals are isomorphous with crystals of wild-type complex and diffraction data from them imply that the binding of NC41 to wild-type and variant neuraminidase is isosteric, although the binding affinity for the variant is less than for wild type (Laver et al., 1987; Colman et al., 1987a). When high-resolution data are made available, it may be possible to explain why the change from Asn to Asp reduces binding of antibody NC41.

18. Structure of a second N9 NA-FAB complex

The N9 neuraminidase used here was isolated from an influenza virus found in a whale in Maine, USA in 1984. It differs from the neuraminidase of the virus from terns (isolated in Australia in 1975 and used in the NC41 complex study) by 14 amino acids (Air et al., 1987), mostly on the underside of the globular neuraminidase head (Varghese et al., 1983). The NC10 antibody was raised against the tern virus neuraminidase (Webster et al., 1987), but binds an epitope where the tern and whale virus neuraminidases are identical in sequence. Crystals of tern virus N9 NA - NC10 Fab have also been grown but are disordered and unsuitable for structure analysis.

The structural data reported are not yet of high quality (Colman et al., 1989), but an α-carbon chain trace has been made for the V region of Fab NC10. No electron density is seen for the C-region of the Fab. The general shape of the tetrameric N9-NC10 Fab protomer is similar to that of the NC41 complex (and other complexes examined by electron microscopy, Tulloch et al., 1986). Packing of the protomers is as described for N9 NA-NC35 Fab crystals (Fig. 10.15, Tulloch et al., 1986), which share the same space group and similar cell dimensions with the N9-NC10 complex crystal. Two neuraminidase heads make close contact via the membrane-proximal surface. Two sets of Fab fragments interdigitate, interacting through the V-modules of the Fab. Crystal growth does not require the participation of the disordered C-module.

The epitope recognized by NC10 antibody is, to a large degree, common to that seen by the NC41 antibody (Webster et al., 1987). In the structure (Colman et al., 1989), contacts with surface loops around residues 330, 342, 369, 401 and 432 appear possible. These observations are consistent with studies showing lack of binding of NC10 to neuraminidase variants with sequence changes at positions 329, 369, 370 and 432 (Webster et al., 1987).

Although the two antibodies NC10 and NC41 are binding to the same region of the NA surface, the arrangements of the two antibodies on that surface are strikingly different, each being rotated about 90° from the position of the other (Fig. 10.20). Thus, whereas of all the NC41 CDR's, L1 is most remote from the antigen, L2 and H1 of NC10 are most distant from the epitope. Since the antibodies NC10 and NC41 have very different amino acid sequences, belonging to different families of heavy chains and different classes of light chains, this difference in binding configuration of the variable CDRs is not surprising. It will be interesting to compare the interactions of NA with each of the antibodies when high-resolution structural data are available.

19. Conclusions and future prospects

Neuraminidase molecules from two influenza virus A subtypes and from influenza B virus have been crystallized and the three-dimensional structures of the type A NAs have been determined to medium resolution, while that of type B is underway. Site-specific mutagenesis experiments have been used to partially define the role of some active site residues and future experiments using suitable vectors aim to obtain sufficient mutant NAs for crystallization and structure determinations. Refinement of these structures and knowledge of how the catalytic site functions will then be used in attempts to design inhibitors specific for the viral enzyme which can be used to control influenza in people or domestic animals. Influenza virus neuraminidase is also being used in experiments aiming to understand the structure of epitopes on protein molecules and the way in which an antibody is able to recognize one out of an almost infinite variety of epitopes in nature. Seven crystalline complexes, involving NA with Fab fragments are currently being analyzed. These are NC41 Fab complexed with N9 NA from tern and whale viruses and with two variants of tern virus NA with sequence changes which reduce but do not abolish binding. The other three complexes are N9 NA-NC10 Fab, N9 NA-32/3 Fab (thin crystals only) and type B influenza virus NA complexed with B1 Fab. Furthermore a number of escape mutants affecting the epitope recognized by NC41 antibody have been crystallized and their refined structures will tell us how single sequence changes are able to abolish antibody binding.

Although the experiments are far from complete, it appears that some areas of the surfaces of the HA and NA are more immunogenic than others, and it is possible that drift involves changes in immunogenicity as well as changes in specificity. It is not known if drift can continue indefinitely, or if it can reverse. We do not know how antigenic shift occurs (several mechanisms may operate), and although we may in due course understand, and hence predict, the course of antigenic drift, there are no indications as yet that we may foresee antigenic shifts.

We do not know why drift occurs in influenza viruses when other RNA viruses, such as measles, which yield variants with the same frequency as influenza when grown in the presence of monoclonal antibodies, do not vary antigenically in nature.

Drift in influenza A viruses follows a set path with changes accumulating over time and with older strains disappearing as new ones take over. Antigenic variants of influenza B virus appear to cocirculate, disappearing then and reappearing (Air et al., 1990). Influenza C virus is reported to hardly vary at all. Why are these groups of orthomyxoviruses so different in their antigenic properties? The three-dimensional structure of both surface antigens (hemagglutinin and neuraminidase) of influenza virus are known, and studies with monoclonal antibodies and with variants selected by these antibodies have identified some of the sites which change during antigenic drift.

The understanding of antigenic drift requires an understanding at the molecular level of the antigenic sites and the interaction of these sites with antibodies. At the present time, crystallographic analysis in conjunction with further antigenic and sequence analyses is the most likely way currently available by which these questions will be answered.

The structures of two epitopes of type A neuraminidase have been determined by X-ray crystallography and the structure of the corresponding epitope of escape mutants have been determined. Similar work with type B neuraminidase is currently underway.

Thus, studies on the influenza virus surface antigens are yielding a wealth of infor-

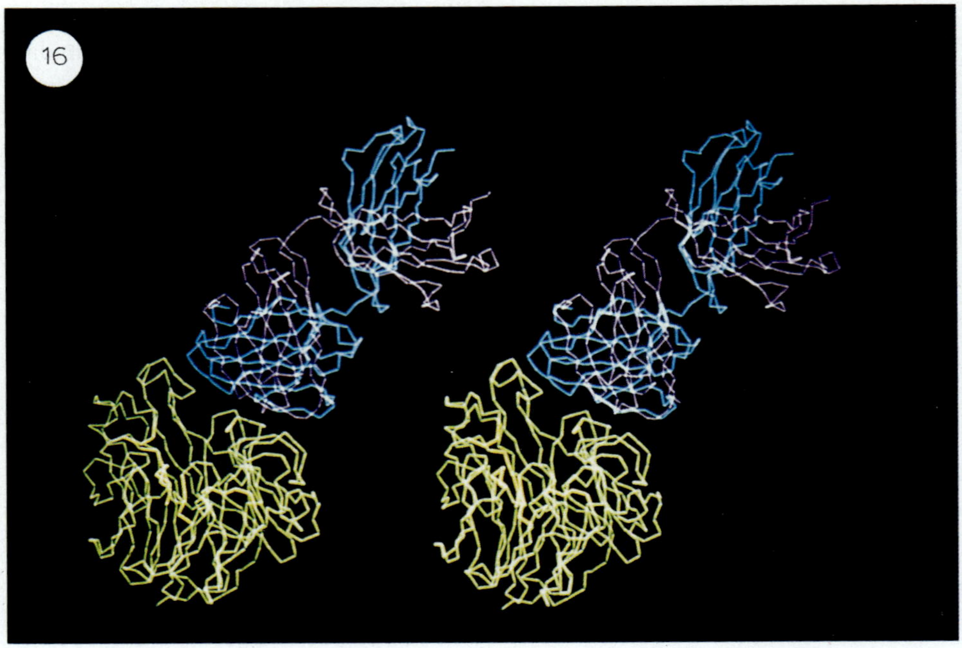

Fig. 10.16. Stereo image of one NA subunit complexed with one NC41 Fab. The NA is green, the antibody heavy chain purple and the light chain blue. The enzyme active center is facing the viewer, below and to the left of the antibody binding site.

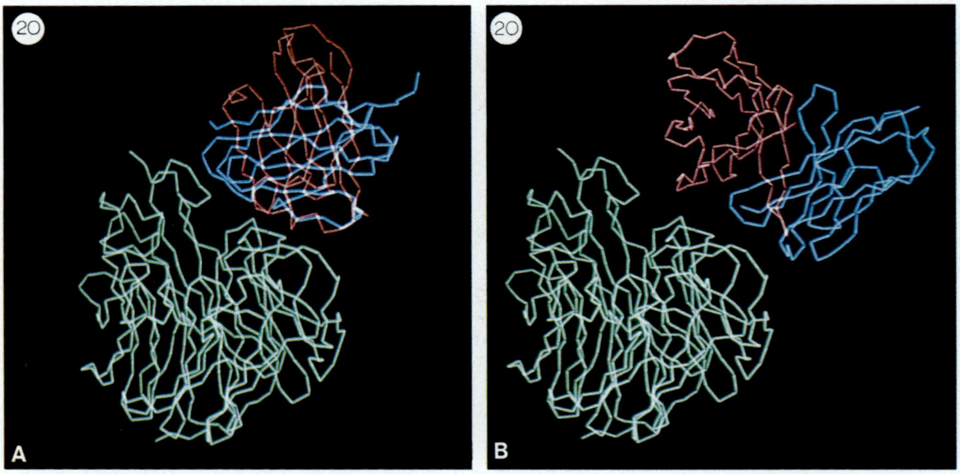

Fig. 10.20. Cα chain tracings of the complexes between (A) tern virus NA and antibody NC41 Fab and (B) whale virus NA and antibody NC10 Fab. Only the V-module is shown; in the case of the NC10 complex the C module is not visible in the electron density map (Colman et al., 1989). The two antibodies recognize much the same region on the NA, but are in very different orientations. Picture kindly provided by Bill Tulip.

mation about their structure, biological activity and antigenic properties, and ways in which they interact with receptors, inhibitors and antibodies. Moreover, the work is being done on components of an important human pathogen, and therefore some of the results obtained with these proteins, while yielding information of great interest, may also be directly applicable to alleviating a widespread and potentially devastating human disease.

20. Acknowledgements

The authors were supported in part by grant AI-18203 and AI-19084 from the National Institute of Allergy and Infectious Diseases, and by grant CA-13148 from the National Cancer Institute. The writing of this joint review was greatly assisted by international telephone facilities donated by the Australian Overseas Telecommunications Commission. The sections on the hemagglutinin were mostly taken from a review by Drs. Don Wiley and John Skehel (Wiley and Skehel, 1987) and we thank them and Annual Reviews Inc. for allowing us to use this material. Much of the material on the neuraminidase also appeared in a review (Air and Laver, 1989) and we thank Alan R. Liss for allowing us to reproduce this.

21. References

Air, G.M. (1981) *Proc. Natl. Acad. Sci.* USA 78, 7639–7643.

Air, G.M. and Compans, R.W. (1983) In *Genetics of Influenza Viruses* (P. Palese and D.W. Kingsbury, Eds.) Springer-Verlag, Vienna pp. 280–304.

Air, G.M. and Laver, W.G. (1989) *Proteins: Struct. Function Genet.* 6, 341–356.

Air, G.M., Ritchie, L.R., Laver, W.G. and Colman, P.M. (1985a) *Virology* 145, 117–122.

Air, G.M., Els, M.C., Brown, L.E., Laver, W.G. and Webster, R.G. (1985b) *Virology* 145, 237–248.

Air, G.M., Webster, R.G., Colman, P.M. and Laver, W.G. (1987) *Virology* 160, 346–354.

Air, G.M., Gibbs, A.J., Laver, W.G. and Webster, R.G. (1990) *Proc. Natl. Acad. Sci. USA* 87, 3884–3888.

Amit, A.G., Mariuzza, R.A., Phillips, S.E.V. and Poljak, R.J. (1986) *Science* 233, 747–753.

Baker, A.T., Varghese, J.N., Laver, W.G., Air, G.M. and Colman, P.M. (1987) *Proteins* 2, 111–117.

Benjamin, D.C., Berzofsky, J.A., East, I.J., Gurd, F.R.N., Hannum, C., Leach, S.J., Margoliash, E., Michael, J.G., Miller, A., Prager, E.M., Reichlen, M., Sercarz, E.E., Smith-Gill, S.J., Todd, P.E. and Wilson, A.C. (1984) *Annu. Rev. Immunol.* 2, 67–101.

Bergelson, L.D., Bukrinskaya, A.G., Prokazova, N.V., Shaposhnikova, G.I., Kocharov, S.L. and Fomina-Ageeva, E.V. (1982) *Eur. J. Biochem.* 128, 467–474.

Berton, M.T., Naeve, C.W. and Webster, R.G. (1984) *J. Virol.* 52, 919–927.

Blok, J., Air, G.M., Laver, W.G., Ward, C.W., Lilly, G.G., Woods, E.F., Roxburgh, C.M. and Inglis, A.S. (1982) *Virology* 119, 109–121.

Blok, J., and Air, G.M. (1982) *Virology* 121, 211–229.

Bossart, P.J., Babu, Y.S., Cook, W.J., Air, G.M. and Laver, W.G. (1988). *J. Biol. Chem.* 263, 6421–6423.

Both, G.W. and Sleigh, M.J. (1981) *J. Virol.* 39, 663–672.

Both, G.W., Sleigh, M.J., Cox, N.J. and Kendall, A.P. (1983) *J. Virol.* 48, 52–60.

Brand, C.M. and Skehel, J.J. (1972) *Nature* 238, 145–147.

Bukrinskaya, A.G., Kornilaeva, G.V., Vorkunova, N.K., Timofeeva, N.G., Shaposhnikova, G.I. and Bergelson, L.G. (1982) *Vopr. Virusol.* 1982, 661–666.

Bullough, P. (1988) *Ph.D. Thesis,* University of Cambridge.

Bullough, P. and Henderson, R. (1987) *Ultramicroscopy* 21, 223–230.

Buonagurio, D.A., Nakada, S., Desselberger, U., Krystal, M. and Palese, P. (1985) *Virology* 146, 221–232.

Caton, A.J., Brownlee, G.G., Yewdell, J.W. and Gerhard, W. (1982). *Cell* 31, 417–427.

Colman, P.M., Air, G.M., Webster, R.G., Varghese, J.N., Baker, A.T., Lentz, M.R., Tulloch, P.A. and Laver, W.G. (1987a) *Immunol. Today* 8, 323–326.

Colman, P.M., Laver, W.G., Varghese, J.N., Baker, A.T., Tulloch, P.A., Air, G.M. and Webster, R.G. (1987b) *Nature* 326, 358–363.

Colman, P.M., Tulip, W., Varghese, J.N., Tulloch, P.A., Baker, A.T., Laver, W.G., Air, G.M. and Webster, R.G. (1989). *Phil. Trans. Roy. Soc. Lond. Ser. B,* 323, 511–518.

Colman, P.M., Varghese, J.N. and Laver, W.G. (1983). *Nature* 303: 41–44.

Compans, R.W. and Choppin, P.W. (1975). In *Comprehensive Virology 4* (H. Fraenkel-Conrat and R.R. Wagner, Eds.) Plenum Press, New York. pp. 179–252.

Daniels, R.S., Douglas, A.R., Skehel, J.J. and Wiley D.C. (1983) *J. Gen. Virol.* 64, 1657–1662.

Daniels, R.S., Douglas, A.R., Skehel, J.J. and Wiley, D.C. (1985a) *Bull. WHO* 63, 273–277.

Daniels, R.S., Douglas, A.R., Skehel, J.J. Waterfield, M.D., Wilson, I.A. and Wiley, D.C. (1985b) In *The Origin of Pandemic Influenza Viruses.* (W.G. Laver, Ed) Elsevier, New York. pp. 1–7.

Daniels, R.S., Douglas, A.R., Skehel, J.J., Wiley, D.C., Naeve, C.W., Webster, R.G., Rogers, G.N. and Paulson, J.C. (1984) *Virology* 138, 174–177.

Daniels, R.S., Jeffries, S.A., Yates, P.A., Schild, G.C. Rogers, G.N., Paulson, J.C., Wharton, S.A., Douglas, A.R., Skehel, J.J. and Wiley, D.C. (1987) *EMBO J.* 6, 1459–65.

Daniels, R. Skehel, J.J. and Wiley, D.C. (1985c) *J. Gen. Virol.* 66, 457–464.

Els, M.C., Air, G.M., Murti, K.G., Webster, R.G., and Laver, W.G. (1985) *Virology* 142, 241–248.

Fang, R., Min Jou, W., Huylebrook, D., Devos, R. and Fiers, W. (1981) *Cell* 25, 315–323.

Fazekas de St. Groth, S. (1978). In *Topics in Infectious Diseases* Vol. 3 (W.G. Laver, H. Bachmayer and R. Weil, Eds.) Springer-Verlag, Vienna. pp. 25–48.

Fields, S., Winter, G. and Brownlee, G.G. (1981). *Nature* 290, 213–217.

Gerhard, W. and Webster, R.G. (1978) *J. Exp. Med.* 148, 383–392.

Gerhard, W., Yewdell, J., Frankel, M. and Webster, R.G. (1981) *Nature* 290, 713–717.

Gething, M. Bye, J., Skehel, J.J. and Waterfield, M.D. (1980) *Nature* 287, 301–306.

Gottschalk, A. (1957). *Biochim. Biophys. Acta* 23, 645–646.

Gottschalk, A. (1959). In *The Viruses* Vol. 3 (F.M. Burnet and W.V.M. Stanley, Eds.) Academic Press, New York. pp. 51–61.

Henderson, R., Baldwin, J.M., Downing, K.H., Lepault, J. and Zemlin, F. (1986) *Ultramicroscopy* 19, 147–178.

Herrler, G., Dürkop, I., Becht, H. and Klenk, H.D. (1988) *J. Gen. Virol.* 69, 839–846.

Herrler, G., Nagele, A., Meier-Ewart, H., Bhown, A.S. and Compans, R.W. (1981) *Virology* 113, 439–451.

Hewat, E.A., Cusack, S., Ruigrok, W.R.H. and Verwey, C. (1984) *J. Mol. Biol.* 175, 175–193.

Hinshaw, V.S., Air, G.M., Gibbs, A.J., Graves, L.,, Prescott, B. and Karunakaran, D. (1982) *J. Virol.,* 42, 865–872.

Hirst, G.K. (1941) *Science* 94, 22–23.

Hogle, J.M., Chow, M. and Filman, D.J. (1986) *Science* 229, 1358–1365.

Hovanec, D.L. and Air, G.M. (1984) *Virology* 139, 384–392.

James, M.N.G. and Sielecki, A.R. (1984) *Biochemistry* 24, 3701–3713.

Kawaoka, Y., Naeve, C. and Webster, R.G. (1984). *Virology* 139, 303–316.

Keil, W., Geyer, R., Dabrowski, J., Dabrowski, U., Niemann, H., Strim, S., and Klenk, H. (1985) *EMBO J.* 4, 2711–2720.

Knossow, M., Daniels, R.S., Douglas, A.R., Skehel, J.J., and Wiley, D.C. (1984). *Nature* 311, 678–680.

Knossow, M., Lewis, M., Rees, D., Wilson, I.A. and Wiley, D.C. (1986). *Acta Crystallogr.* B42, 627–632.

Krug, R.M. (1983) In *Genetics of Influenza Viruses* (P. Palese and D.W. Kingsbury Eds.) Springer-Verlag, Vienna. pp. 70–98.

Krystal, M., Elliott, R.M., Benz, E.W., Young, J.F. and Palese, P. (1982) *Proc. Natl. Acad. Sci.* USA 79, 4800–4804.

Krystal, M., Young, J.F., Palese, P., Wilson, I.A., Skehel, J.J. and Wiley, D.C. (1983) *Proc. Natl. Acad. Sci. USA* 80, 4527–4531; 81, 1261 (1984).

Lamb, R.A. (1983) In *Genetics of Influenza Viruses* (P. Palese and D.W. Kingsbury, Eds.) Springer-Verlag, Vienna. pp. 21–69.

Lamb, R.A. and Choppin, P.W. (1983) *Annu. Rev. Biochem.* 52, 467–506.

Laver, W.G., Air, G.M., Dopheide, T.A. and Ward, C.W. (1980) *Nature* 283, 454–457.

Laver, W.G., Air, G.M., Webster, R.G. and Markoff, L.J. (1982) *Virology* 122, 450–460.

Laver, W.G., Air, G.M., Webster, R.G., Gerhard, W., Ward, C.W. and Dopheide, T.A. (1979) *Virology* 98, 226–237.

Laver, W.G. and Webster, R.G. (1973) *Virology* 51, 383–391.

Laver, W.G., Luo, M., Bossart, P.J., Babu, Y.S., Smith, C., Accavitti, M.A., Tulloch, P.A. and Air, G.M. (1988) *Virology* 167, 621–624.

Laver, W.G., Webster, R.G. and Colman, P.M. (1987) *Virology* 156, 181–184.

Lentz, M.R., Air, G.M., Laver, W.G., and Webster, R.G. (1984) *Virology* 135, 257–265.

Lentz, M.R., Webster, R.G. and Air, G.M. (1987) *Biochemistry* 26, 5351–5358.

Lu, B.L., Webster, R.G., Brown, L.E. and Nerome, K. (1983) *Bull. WHO* 61, 681–687.

Moss, B.A., Underwood, P.A., Bender, V.J. and Whittaker, R.G. (1980) In *Structure and Variation in Influenza Virus* (W.G. Laver and G.M. Air, Eds.) Elsevier, Amsterdam. pp. 329–338.

Nakada, S., Creager, R.S., Krystal, M., Aaronson, R.P. and Palese, P. (1984) *J. Virol.* 50, 118–124.

Nakajima, K., Desselberger, U. and Palese, P. (1978). *Nature* 274, 334–339.

Natali, A., Oxford, J.S. and Schild, G.C. (1981) *J. Hyg.* 87, 185.

Newton, S.E., Air, G.M., Webster, R.G. and Laver, W.G. (1983). *Virology* 128, 495–501.

Padlan, E.A., Silverton, E.W., Sheriff, S., Cohen, G.H., Smith-Gill, S.J. and Davies D.R. (1989). *Proc. Natl. Acad. Sci. USA* 86, 5938–5942.

Palese, P., Tobita, K, Ueda, M. and Compans, R.W. (1974) *Virology* 61, 397–410.

Paulson, J.C. (1985). In *The Receptors* Vol. II. Academic Press, New York. pp. 131–219.

Pereira, M.S. (1982). In *Virus Persistence* (Soc. Gen. Microbiol. Symp. 33) (B.W.J. Mahy, A.C. Minson and G.K. Darby, Eds) Cambridge University Press, Cambridge. pp. 15–37.

Pfeifer, J.B. and Compans, R.W. (1984). *Virus Res.* 1, 281–296.

Raymond, F.L., Caton, A.J., Cox, N.J., Kendal, A.P. and Brownlee, G.G. (1983). *Nucleic Acids Res.* 11, 7191–7203.

Rogers, G.N., Daniels, R.S., Skehel, J.J., Wiley, D.C., Wang, X., Higa, H.H. and Paulson, J.C. (1985). *J. Biol. Chem.* 260, 7362–7367.

Rogers, G.N., Pritchett, T. and Paulson, J.C. (1983a). *Fed. Proc.* 42, 2181.

Rogers, G.N., Paulson, J.C., Daniels, R.S., Skehel, J.J., Wilson, I.A. and Wiley, D.C. (1983b) *Nature* 304, 76–78.

Rossmann, M.G., Arnold, E., Erickson, J.W., Frankenberger, E.A., Griffith, J.P., Hecht, H., Johnson, J.E., Kamer, G., Luo, M., Mosser, A.G., Rueckert, R.R., Sherry, B. and Vriend, G. (1986) *Nature* 317, 145–153.

Rota, P.A, Shaw, M.W. and Kendal, A.P. (1987). *Virology* 161, 265–275.

Scholtissek, C., Rohde, W., von Hoyningen, V. and Rott, R. (1978) *Virology* 87, 13–20.

Sela, M. (1969). *Science* 166, 1365.

Shaw, M.W., Lamb, R.A., Erickson, B.W., Breidis, D.J. and Choppin, P.W. (1982) *Proc. Natl. Acad. Sci. USA*, 72, 93–97.

Sheriff, S., Silverton, E.W., Padlan, E.A., Cohen, G.H., Smith-Gill, S.J., Finzel, B.C. and Davies, D.R. (1987). *Proc. Natl. Acad. Sci. USA* 84, 8075–8079.

Skehel, J.J., Daniels, R.S., Douglas, A.R. and Wiley, D.C. (1983) *Bull. WHO* 61, 671–676.

Skehel, J.J., Stevens, D.J., Daniels, R.S., Douglas, A.R., Knossow, M., Wilson, I.A. and Wiley, D.C. (1984) *Proc. Natl. Acad. Sci. USA* 81, 1779–1783.

Sleigh, M.J., Both, G.W. Underwood, P.A. and Bender, V.J. (1981) *J. Virol.* 37, 845–853.

Suzuki, Y., Matsunaga, M. and Matsumoto, M. (1985) *J. Biol. Chem.* 260, 1362–1365.

Tulip, W.R., Varghese, J.N., Webster, R.G., Air, G.M., Laver, W.G. and Colman, P.M. (1989) *Cold Spring Harbor Symposium on Immune Recognition.* 54. Cold Spring Harbor Laboratory, Cold Spring Harbor, NY, pp. 257–263.

Tulloch, P.A., Colman, P.M., Davis, P.C., Laver, W.G., Webster, R.G. and Air, G.M. (1986) *J. Mol. Biol.* 190, 215–225.

Underwood, A., Skehel, J.J. and Wiley, D.C. (1987). *J. Virol.* 61, 206–208.

Underwood, P.A. (1982). *J. Gen. Virol.* 62, 153–169.

Underwood, P.A. (1985). *Arch. Virol.* 84, 53–62.

Varghese, J.N., Laver, W.G. and Colman, P.M. (1983) *Nature* 303, 35–40.

Varghese, J.N., Webster, R.G., Laver, W.G. and Colman, P.M. (1988). *J. Mol. Biol.* 200, 201–203.

Verhoeyen, M., Fang, R., Min Jou, W., Devos, R., Huylebroeck, D., Saman, E. and Fiers, W. (1980). *Nature* 286, 771–776.

Verhoeyen, M., Van Rompuy, L., Min Jou, W., Huylebroeck, D. and Fiers, W. (1983). *Nucleic Acids Res.* 11, 4703–4712.

Wang, M.L., Skehel, J.J. and Wiley, D.C. (1986). *J. Virol.* 57, 124–128.

Ward, C.W. (1981) *Curr. Top. Microbiol. Immunol.* 94/95, 1–74.

Ward, C.W. and Dopheide, T.A. (1980) *Virology* 103, 37–53.

Ward, C.W. and Dopheide, T.A. (1981a) *ICN-UCLA Symp. Mol. Cell. Biol.* 21, 323–340.

Ward, C.W. and Dolpheide, T.A. (1981b). *Biochem. J.* 195, 337–340.

Webster, R.G., Air, G.M., Metzger, D.W., Colman, P.M., Varghese, J.N., Baker, A.T. and Laver, W.G. (1987). *J. Virol.* 61, 2910–2916.

Webster, R.G. and Berton, M.T. (1981). *J. Gen. Virol.* 54, 243–251.

Webster, R.G., Brown, L.E. and Laver, W.G. (1984) *Virology* 135, 30–42.

Webster, R.G., Hinshaw, V.S. and Laver, W.G. (1982) *Virology* 117, 93–104.

Webster, R.G., Reay, P.A. and Laver, W.G. (1988) *Virology* 164, 230–237.

Wei, X., Els, M.C., Webster, R.G. and Air, G.M. (1987) *Virology* 156, 253–258.

Weis, W., Brown, J.H., Cusack, S., Paulson, J.C., Skehel, J.J. and Wiley, D.C. (1988). *Nature* 333, 426–431.

WHO (1980) Memorandum. *Bull. WHO* 58, 585–591.

Wiley, D.C. (1985) In *Virology* (B.N. Fields, Ed) Raven Press, New York. pp.45–68.

Wiley, D.C. and Skehel, J.J. (1987). *Ann. Rev. Biochem.* 56, 365–394.

Wiley, D.C. and Skehel, J.J. and Waterfield, M. (1977) *Virology* 79, 446–448.

Wiley, D.C., Wilson, I.A. and Skehel, J.J (1981) *Nature* 289, 373–378.

Wilson, I.A., Skehel, J.J. and Wiley, D.C. (1981). *Nature* 289, 366–373.

Winter, G., Fields, S. and Brownlee, G.G. (1981). *Nature* 292, 72–75.

Wrigley, N.G. (1979). *Br. Med. Bull.* 35, 35–38.

Wrigley, N.G., Brown, E.B., Daniels, R.S., Douglas, A.R., Skehel, J.J. and Wiley, D.C. (1983). *Virology* 131, 308–314.

Yamashita, M., Krystal, M., Fitch, W.M. and Palese, P. (1988) *Virology* 163, 112–122.

Yewdell, J.W., Caton, A.J. and Gerhard, W. (1986). *J. Virol.* 57, 623–628.

Yewdell, J.W., Webster, R.G. and Gerhard, W.U. (1979). *Nature* 279, 246–248.

Eds. M.H.V. van Regenmortel and A.R. Neurath
Immunochemistry of viruses, II. The basis for serodiagnosis and vaccines.

Paramyxoviruses 11

RANJIT RAY[1] and RICHARD W. COMPANS[2]

[1]*Secretech Inc., Birmingham, AL 35205, and* [2]*University of Alabama at Birmingham, Department of Microbiology, Birmingham, AL 35294, U.S.A.*

1. Structure and replication of paramyxoviruses

The family paramyxoviridae is divided into three genera: paramyxoviruses, morbilliviruses and pneumoviruses (Matthews, 1982). The paramyxovirus genus includes Newcastle disease virus, parainfluenza types 1–4, simian virus 5 and mumps virus. Morbilliviruses includes measles, canine distemper and rinderpest viruses, whereas respiratory syncytial virus and pneumonia virus of mice are members of the pneumovirus genus.

The viruses are pleomorphic with a spike-covered, lipid-containing envelope enclosing a helical nucleocapsid. A negative sense viral genomic RNA molecule, nucleocapsid protein (NP) and two other proteins (P and L) associated with RNA-dependent RNA polymerase activity are components of the nucleocapsid. A nonglycosylated membrane or matrix protein (M) is also internal to the lipid bilayer. Two viral glycoproteins, the hemagglutinin-neuraminidase (HN) and fusion (F) proteins, form spike-like projections on the external surface of the viral envelope and are anchored through the lipid bilayer. These two envelope glycoproteins are known to be responsible for initiation of the infection process (Scheid et al., 1972; Homma and Ohuchi, 1973; Scheid and Choppin, 1974). Attachment of virions to neuraminic acid-containing receptors is mediated by the attachment protein HN (Scheid et al., 1972; Choppin and Compans; 1975). The F protein is synthesized as a precursor form Fo and subsequently cleaved by a host protease to yield a disulfide-linked F1–F2 complex which is essential for infectivity (Homma and Ohuchi, 1973; Scheid and Choppin, 1977). Most evidence with paramyxoviruses favors a mechanism of penetration and uncoating in which the F glycoprotein initiates fusion of the viral envelope with the host cell plasma membrane. Subsequently, the nucleocapsid is discharged into the cytoplasm to begin the replication cycle. The structure and replication of paramyxoviruses have been described in detail in other review articles (Choppin and Compans, 1975; Compans and Nakamura, 1978; Matsumoto, 1982; Orvell and Norrby, 1985; Kolakofsky and Roux, 1987; Kingsbury, 1989).

Whereas acute infections by some paramyxoviridae are cytopathic, other members of the family frequently cause noncytopathic infection in which infected cells may produce large amounts of virus with little or no morphological change. In such infections, there is no significant alteration of host cell biosynthesis, and the release of virus by budding provides a mechanism by which virions can be produced in large amounts with no damage to the cellular plasma membrane. Many paramyxoviruses have been reported to produce persistent, noncytopathic infection of cells in culture (Choppin and Compans, 1975). The specific properties of these cells vary considerably in different virus-cell systems and may or may not include the participation of defective interfering (DI) particles or temperature-sensitive virus mutants. Persistent infection by paramyxoviruses is also involved in chronic diseases of man. Persistent measles virus infection causes subacute sclerosing panencephalitis (SSPE) which may be due to failure of M protein synthesis in neural cells (Choppin et al., 1981), restriction in expression of the M protein by mutations causing defects in transcription, translation or protein stability (Cattaneo et al., 1988; 1989) and/or modulation of viral antigens by antibody (Oldstone and Fujinami, 1982). Persistent infections with paramyxoviruses have also been implicated in Paget's disease of bone (Mills et al., 1981; Goswami et al., 1984). In adult patients with chronic pulmonary disease, parainfluenza virus type 3 (PI3) was recovered on successive monthly occasions and it was suggested that this virus may persist in the abnormal respiratory tract (Gross et al., 1973). Persistent parainfluenza type 1 and type 3 virus shedding in healthy young adults was shown to occur from both symptomatic and asymptomatic subjects throughout the winter (Muchmore et al., 1981: Frank et al., 1981, Goswami et al., 1984). A recent review article (Oldstone, 1989) describes specific interactions of virus and lymphocytes in the context of various persistent infections.

Monkey kidney cells persistently infected with the paramyxovirus SV5 exhibit no significant cytopathic effect, and produce high yields of infectious virus for long periods of time (Choppin, 1964). There is no apparent involvement of defective interfering particles or production of interferon in this persistent infection. Cell cultures persistently infected with PI3 virus (Wechsler et al., 1987) were also shown to produce high levels of infectious virus. In both of these persistent paramyxovirus infections, infected cells do not show fusion with each other although they contained large amounts of the cleaved fusion protein. Cells persistently infected by mumps virus showed minimal cytopathology; however, widely dispersed foci of multinucleate cells were observed. Both defective interfering (DI) particles and infectious virus particles were found during the course of mumps persistent infection

(McCarthy et al., 1981; Andzhaparidze et al., 1982), and DI particles were implicated as its likely modulating factor. DI particles have been directly involved in establishment of persistent infection with Sendai virus (Kolakofsky and Roux, 1987).

Several aspects of the immunochemistry of paramyxoviruses have been reviewed previously(Orvell and Norrby, 1985) including the structural components of paramyxoviruses and their biological functions, antigenic characterization of the viral proteins using monoclonal antibodies, and intratypic antigenic relationships between viruses. The present chapter is focussed on more recent studies of the immune responses to paramyxoviruses as well as on current approaches being used for vaccine development. We have particularly emphasized studies with the human parainfluenza viruses and respiratory syncytial virus, two important causes of serious respiratory tract infections in infants (Chanock and McIntosh, 1985; McIntosh and Chanock, 1985) for which no effective vaccines are currently available.

2. Immune responses to paramyxoviruses

2.1. Antibody responses

2.1.1. Serum antibody response

The two paramyxovirus glycoproteins forming spikes on the external surface of the virion are responsible for adsorption to host cell receptors (Scheid et al., 1972) and for virus-induced membrane fusion leading to entry into cells (Homma and Ohuchi, 1973; Scheid and Choppin, 1974). Inhibition of virus attachment and/or penetration, or destruction of virus by lysis due to antibody plus complement, are possible mechanisms of virus neutralization by antibody. The importance of antibodies to these peplomer antigens for virus neutralization and induction of a protective immune response has been discussed in other reviews (Choppin and Scheid, 1980; Orvell and Norrby, 1985). A summary of recent studies on the protective immune response induced by individual paramyxovirus structural proteins is presented in Table 11.1.

Serological studies with children naturally infected with parainfluenza viruses have demonstrated that the antibody response to the HN glycoprotein progressively rises with the number of infections (Kasel et al., 1984). On the other hand, an antibody response to the F glycoprotein was detected only after repeated infections, which may be due to the lower immunogenicity of this protein. It was suggested that development of serum antibodies to both the glycoproteins of PI3 virus is correlated with protection from infection.

Human sera have been studied for cross-reactions between parainfluenza or mumps virus by complement fixation, hemagglutination-inhibition and enzyme-linked-immunoassay using purified viral envelope glycoprotein and nucleocapsid

TABLE 11.1

Protective immune responses induced by individual paramyxovirus structural proteins

Virus	Structural proteins	Types of immune response	References
Parainfluenza virus	HN	Neutralizing antibody	Kasel et al., 1984; Coelingh et al.; 1987; Patterson et al., 1987; Spriggs et al., 1987; Ray et al., 1985; 1988b.
		Cytotoxic T cells	Al-Ahdal et al. 1985; Randall and Young, 1988.
	F	Neutralizing antibody	Kasel et al.,1984; Patterson et al., 1987; Spriggs et al., 1987; Ray et al., 1988b; 1989
		Cytotoxic T cells	Al-Ahdal et al., 1985.
	NP	Cytotoxic T cells	Randall and Young, 1988.
	M	Cytotoxic T cells	Randall and Young, 1988.
Respiratory syncytial virus	G	Neutralizing antibody	Olmsted et al., 1986; Murphy et al., 1986a; Gimenez et al., 1987; Walsh et al., 1987; Wertz et al., 1987; Levine et al., 1988.
	F	Neutralizing antibody	Olmsted et al., 1986; Stott et al., 1986; Murphy et al., 1986a; Walsh et al., 1987.
		Cytotoxic T cells	Pamberton et al., 1987.
		T-helper cells	Openshaw et al., 1988
	NP	T-helper cells	Openshaw et al., 1988

preparations (Julkunen, 1984). All these methods showed marked cross-reactivity between PI1 and PI3 antibodies. Mumps virus infection was often shown to raise heterologous antibodies even against purified viral antigens. Heterologous antibody responses between PI1 and PI3 viruses and between PI1 and mumps viruses were also shown using rabbit antisera against viral envelope glycoproteins. The cross-reactive antibodies were found to be directed against both HN and F glycoproteins of the viruses. Antigenic relationships among paramyxoviruses were also recently examined by immunoprecipitation with specific antisera (Ito et al., 1987) and the results suggested that Sendai virus, PI1 and PI3 belong to one antigenic group, and PI2, PI4, mumps virus and SV5 belong to a different group. Although Newcastle disease virus belongs to a separate antigenic group, it also showed occasional cross-reactions with human paramyxoviruses. The M polypeptide was shown to exhibit antigenic individuality and was suggested to be the antigen best suited for serodiagnosis of infection by human paramyxoviruses. No cross-reaction was detected between measles virus and the paramyxoviruses.

Antibody-dependent cellular cytotoxicity (ADCC) mediated by human killer lymphocytes (K cells) appears to have an important role in rapidly killing virus-infected cellular targets. K-cell-mediated ADCC appeared at the same time following measles virus vaccination as did antiviral IgG (Sissons and Oldstone, 1980). Antibodies of the IgG isotype were found to be the major recognition factors for contact between the effector cells through the Fc receptors and the target cells.

Analyses of human serum antibody response to RSV infection have shown that infants recognize mainly the F and N proteins whereas adults also recognize the G protein (Ward et al., 1983; Gimenez et al., 1987). A different study with bovine RSV suggested that the presence or development of antibodies to the F and N proteins is not sufficient for protection or recovery from infection (Westenbrink et al., 1989). Most children infected in the presence of maternal antibodies responded to both G and F (Murphy et al., 1986) or mainly to the G and P proteins (Levine et al., 1988). The importance of antibodies in the respiratory tract for prevention of RSV infection was demonstrated by intranasal administration of neutralizing antibodies (Prince et al., 1987). Parenteral inoculation of purified human immunoglobulin in cotton rats or owl monkeys effected a reduction in the level of pulmonary virus at the height of RSV infection. The therapeutic effect of purified human immunoglobulin was found to be much greater when it was administered topically by the intranasal route than by parenteral inoculation. On the other hand, passive transfer of RSV antibodies in cotton rats was shown to suppress the antibody response to RSV G and F glycoproteins expressed by recombinant vaccinia viruses (Murphy et al., 1988). It was suggested that this phenomenon mimics the effects of maternally derived RSV antibodies on the immune response of young infants to infection with RSV. Pre-existing RSV antibodies in serum may therefore have an important role on the immune response to RSV glycoproteins during infection in infancy. It appears that the antibody response to both G and F in the respiratory tract is important for resistance to RSV infection.

The existence of two subgroups of RSV, designated A and B, has been reported based on reactivities with a panel of monoclonal antibodies (Mufson et al., 1985). At least four structural proteins showed significant differences in antigenic properties between the groups. Subgroup B was further divided into two distinct sets B1 and B2, based on antigenic differences in the G protein (Akerlind et al., 1988). Differences between the two envelope glycoproteins of RSV have also been shown from the reactivities of rabbit antisera (Walsh et al., 1987; Johnson et al., 1987). Antisera to fusion proteins of RSV A and B viruses reacted with nearly equivalent neutralization titers with the heterologous F protein. However, antiserum to the G protein neutralized the homologous strain of RSV to much higher titer than the heterologous virus strain. Antigenic relationships between human and bovine strains of RSV have also been studied using monoclonal antibodies (Orvell et al., 1987). The bovine strains shared many epitopes with the two human subgroups, but appeared to be antigenically distinct from the human viruses.

2.1.2. Secretory antibody response

Nasal secretory IgA antibodies seem to play a significant role in resistance to infection by parainfluenza viruses. The relative protective effects of antibodies in serum and in nasal secretions were studied in adult volunteers who were challenged intranasally with parainfluenza type 1 virus (Smith et al., 1966). Antibody in nasal secretions proved to be a better index of host resistance to infection than the level of serum antibody. It was suggested from another study in human volunteers that infection does not occur when high titers of neutralizing activity are detectable in the nasal secretions (Tremonti et al., 1968). However, upon primary infection, infants have been reported to produce nasal IgA antibodies with little or no virus-neutralizing activity (Yanagihara and McIntosh, 1980). In a recent study of immunization of hamsters, parainfluenza virus glycoproteins elicited significantly higher local glycoprotein-specific IgA response when administered intranasally as compared with systemic immunization. This was implicated in resistance to challenge infection with live virus (Ray et al., 1988). Similar observations were also noted with Sendai virus envelope glycoproteins (Guink et al., 1989).

2.2. Cell-mediated responses

Cell-mediated immune responses are an important defense mechanism of the host against viral infection (Yap and Ada, 1977) and are induced by various structural proteins of paramyxovirus (Table 11.1). Early studies suggested that the cell-mediated immune response was directed against the viral glycoproteins and syngeneic major histocompatibility complex antigens displayed on infected cell surfaces (Zinkernagel and Doherty, 1975; Doherty and Zinkernagel, 1976; Finberg et al., 1978; Koszinowski et al., 1977). However, it was subsequently observed that the internal viral proteins may also stimulate the cell-mediated immune response (Bennink et al., 1982; Townsend et al., 1984; Taylor and Askonas, 1986; Bangham et al., 1986; Randall et al., 1988) by being exposed as short peptides in association with MHC molecules as a target antigen presented at the infected cell surface (Townsend et al., 1986). The antigen processing pathways for presentation may vary among viruses depending upon the mode of virus entry and interaction with the MHC molecules (Long and Jacobson, 1989). The mechanisms of the cell-mediated immune response are discussed below.

2.2.1. Natural killer (NK) cells

One of the important cellular responses in recovery from virus infection is thought to be mediated by natural killer (NK) cells, which function before the appearance of cytotoxic T cells or antibody (Harfast et al., 1978; Santoli et al., 1978; Minato et al., 1979; Welsh and Hallenbeck, 1980; Mims and White, 1984). Virus-infected cell cultures were reported to be more susceptible to NK cell-mediated lysis due to activation of NK cells by released interferon (Santoli et al., 1978). The relatively resistant P815 mastocytoma cells were reported to become sensitive to NK cell-mediated cytolysis upon modification by Sendai HN and F glycoproteins (Abidi and Flanagan, 1984). Measles virus-infected cells were also shown to be efficiently lysed by unprimed lymphocytes, probably by NK cells, and the lysis occurred before the release of progeny virus (Casali and Oldstone, 1982). Lymphocytes from both seropositive and seronegative human donors lysed measles virus infected human fibroblasts within 4 h of incubation. This lysis was shown to be independent of HLA restriction type, specific antibodies or release of interferon. These studies indicate that NK cells may play a role in recovery from paramyxovirus infection.

2.2.2. Cytotoxic T lymphocytes

Evidence has been obtained that liposomes bearing HN and/or F glycoproteins of Sendai virus stimulate cytotoxic T lymphocytes (CTL) (Hale et al., 1980; McGee et al., 1980). It was shown that the glycoproteins and the major histocompatibility antigens must be present within the same membrane lipid bilayer for effective elicitation, recognition and lysis of target cells by anti-Sendai virus CTL (Hale et al., 1980). Efficient recognition and lysis of target cells by specific CTL or antibodies also required an active fusion protein. X-irradiated P815 or EL-4 cells were incubated with different amounts of MDBK cell-grown Sendai virus, treated or untreated with trypsin to cleave the F protein and tested for their ability to elicit anti-Sendai virus CTL. The results indicated that an active fusion protein quantitatively potentiated the ability of cells exposed to MDBK-grown Sendai virus to elicit anti-Sendai virus CTL. It was suggested that an active fusion protein was required for integration of the Sendai virus glycoproteins into the cell membrane by fusion of the viral envelope with the cell membrane (Hale et al., 1980).

Cells modified by insertion of HN and F together or separately were also used as target cells for CTL (Al-Ahdal et al., 1985). P815 mastocytoma cells modified with liposomes bearing HN and F showed susceptibility to cytolysis similar to that of actively infected cells. On the other hand, cells modified with liposomes bearing HN or F were also susceptible. The sum of reactivities of the anti-HN component and the anti-F component was close to that seen with HN- and F- bearing targets. That the cytolytic activity was mediated by cytotoxic T lymphocytes was confirmed by depletion with anti-Thy 1.2 antibody and complement.

It was suggested that protection of mice from challenge infection with the paramyxovirus SV5 involves the induction of humoral antibody and cytotoxic T cell responses (Randall and Young, 1988; Randall et al., 1988). Immunization with *Staphylococcus aureus* antibody-antigen complexes containing SV5 proteins induced cytotoxic T cells specific for SV5. Spleen cells from immunized mice were stimulated in vitro with SV5 and the ability of the spleen cells to lyse SV5-infected target cells was measured by ^{51}Cr release assays. *S. aureus*-antibody-M protein complexes were shown to be particularly effective in inducing cytotoxic T cells. Weaker but significant levels of cytotoxic T cell activity were also induced by similar HN and NP complexes.

Cytotoxic T cells are also generated following infection with RSV. T cells primed by subgroup A virus showed significant cross-reactivity with the virus from subgroup B (Bangham and Askonas, 1986). The internal nucleocapsid of RSV has been identified as a significant target antigen for cytotoxic T cells (Bangham et al., 1986). Mice primed intranasally with RSV were shown to exhibit RSV-specific helper T cell memory in the spleen (Openshaw et al., 1988). Helper T cell recognition was in part specific for the RSV strain used in priming and in part cross-reactive between RSV strains. The specificity of these T cell lines and clones was not reported. High levels of RSV-specific cytotoxic T cells were shown to be induced with a vaccinia virus recombinant expressing the fusion protein, although they showed little cross-reactivity between two different RSV strains (Pemberton et al., 1987). On the other hand, vaccinia virus recombinants expressing F or NP were shown to induce stronger virus-specific T helper cells than did the G glycoprotein (Openshaw et al., 1988).

Specifically induced cytotoxic T cells were also shown to be present in venous blood and cerebrospinal fluid of children with mumps meningitis (Kreth et al., 1982). The cytotoxic T cells were elicited by mumps virus infections and were not found during other systemic virus infections such as measles. However, the immune response to specific mumps virus proteins was not determined. The virus-specific cytotoxic T lymphocytes were suggested to be directly regulated by certain genes of the HLA B locus (Kress and Kreth, 1982).

Thus, the available evidence strongly suggests that cytotoxic T cells may play an important role in the protective immune response against paramyxovirus infection. Identification of specific viral proteins required for generation of cytotoxic T cells should lead to their inclusion in an effective subunit vaccine.

2.2.3. Delayed type hypersensitivity (DTH)

DTH is a T lymphocyte mediated inflammatory reaction characterized by infiltration of mononuclear cells. The delayed hypersensitivity effector cells (Td) respond to antigen by proliferating and releasing lymphokines, including interferons, attracting macrophages to the site, which may phagocytose the incoming virus or virus-antibody complexes (Mims and White, 1984). After processing, the viral antigens are presented on the surface of macrophages, dendritic cells or Langerhans cells. Two subpopulations of mouse Td lymphocytes were found to be stimulated both in vitro and in vivo by using infectious or ultraviolet light-inactivated Sendai virus (Ada et al., 1981). One T lymphocyte populations was K,D region-restricted and of the Lyt 1−, 2+, 3+ phenotype and could be selectively stimulated in vivo by low doses of infectious virus or in vitro by using spleen lymphocytes, tumor cells or LPS blasts as stimulator cells and responder cell population which contained no adherent cells. This subpopulation could be selectively elicited in mice by using very low doses of infectious challenge virus. The other population was I region-restricted, of the Lyt 1+, 2−, 3− phenotype and was selectively induced by fusion-negative (noninfectious) Sendai virus preparations presented by adherent cells, presumably macrophages. In contrast to the K,D region-restricted T cell population, I region-restricted Td cells require adherent cells for effective stimulation (Ertl, 1981). Exposure of fusion-positive (infectious) Sendai virus preparations presented by adherent cells to other IA antigen-expressing cells such as LPS blast cells did not result in the generation of I region-restricted T cells (Ertl, 1981). It was suggested (Ertl, 1981) that macrophages play an important role as stimulators for I region-restricted DTH responses in vivo. Intravenous injection of noninfectious fusion-negative Sendai virus induces antigen-specific suppressor T cells which inhibit the generation of K,D restricted T cells. This differential suppression of distinct T effector cells was suggested to be important in maintaining the balance between protective and/or immunopathological effects of the T cell-mediated immune reactions *in vivo*.

2.3. Virus persistence and the immune response

It is of interest to consider the possible mechanisms of virus persistence in vivo in the context of the immune response. In the case of persistent human infection by PI3 virus (Gross et al., 1973), high titers of serum antibody were observed, but there was no significant increase in local antibody in respiratory tract secretions as compared with patients who had recovered from an acute infection. Thus, persistent infection might be correlated with a defect in the mucosal immune response. Virus in the respiratory tract may be inaccessible to serum antibody, because PI3 is assembled at apical surfaces at epithelial cells (R.Ray and R.W. Compans, unpublished) and is therefore inaccessible to antibody present at basolateral (serosol) membrane domains.

Patients with SSPE have high levels of antibody in serum and cerebrospinal fluids, and apparently normal cellular immune responses (Sever, 1983). In this case, the failure of the immune system to clear the infection may be related to a defect in expression of viral envelope proteins on the surfaces of neural cells. Similarly, persistent infection by Sendai virus in the brains of mice appears to be associated with a defect in the expression of surface glycoproteins in neurons (Kristensson et al., 1983). The lack of expression of antigens on the cell surface may provide a mechanism for escape from immune surveillance by viruses in these persistently infected cells.

3. Approaches to vaccine development

Remarkable progress has been made in recent years in development of vaccines for respiratory syncytial and parainfluenza viruses. The approaches taken toward such development are summarized in Table 11.2.

TABLE 11.2

Current approaches for development of vaccines for respiratory syncytial and parainfluenza viruses

Virus	Approach	References
Human parainfluenza virus	Cold adapted-temperature sensitive PI3 virus vaccine	Crookshanks-Newman and Belshe, 1984; 1986
	Bovine PI3 virus vaccine	Coelingh et al. 1988
	Protease activation mutant of Sendai virus	Tashiro and Homma, 1983; 1985; Hsu et al., 1987; Murphy et al., 1988
	Glycoprotein subunit vaccine	Ray et al., 1985; 1988
	Subunit vaccine using recombinant protein	Coelingh et al., 1987; Ray et al., 1989
	Recombinant vaccinia virus	Spriggs et al., 1987; 1988.
Respiratory syncytial virus	Glycoprotein subunit vaccine	Walsh et al., 1987; Routledge et al., 1988
	Recombinant vaccinia virus	Trudel et al., 1989; Olmsted et al., 1986; Stott et al., 1986; Wertz et al., 1987; Johnson et al., 1987; Stott et al., 1987

3.1. Killed virus vaccine

Early attempts to develop an effective PI3 vaccine were unsuccessful. Intramuscular immunization with formalin-inactivated PI3 vaccines in infants of different age groups demonstrated an increase in virus-specific serum antibodies in vaccinees without significant protection against the disease (Fulginiti et al., 1967; Parrott et al., 1967).

In a murine model, protection of the upper respiratory tract from Sendai virus was shown to correlate with the presence of antiviral IgA antibodies in nasal washings, whereas protection of the lower respiratory tract correlated with antiviral IgG antibodies in bronchoalveolar lavage fluids (Nedrud et al., 1987). Intranasal immunization with inactivated virus alone conferred partial protection to the lower respiratory tract and no protection to the upper respiratory tract. However, oral immunization with inactivated virus plus cholera toxin as an adjuvant combined with intranasal administration of small amounts of inactivated virus was shown to prime for a protective immune response by eliciting IgA antibodies in the murine respiratory tract.

Enhancement of RSV disease following intramuscular administration to infants of formalin-inactivated RSV vaccine was noted in early studies (Chin et al., 1969; Fulginiti et al., 1969; Kapikian et al., 1969; Kim et al., 1969). The mechanism of enhancement of this disease was recently studied in cotton rats by intramuscular inoculation of formalin-inactivated RSV vaccine followed by subsequent challenge infection with live virus (Prince et al., 1986). Within 24 h after infection with RSV, preimmunized cotton rats developed pulmonary lesions that reached a maximum by day 4. Histologically, the lesions resembled an experimental pulmonary arthus reaction. It was suggested that the action of formalin on RSV may be responsible for this effect, because inoculation of live virus or virus heated in the absence of formalin did not induce enhanced immunopathology. In another study (Piedra et al., 1989), it was suggested that pathogenesis of pulmonary injury during natural RSV infection in the immunized host may be due primarily to prior sensitization to the virus. In the animal model, sensitization to non-viral tissue culture components also contributed to lung injury. It was suggested that the initial pulmonary injury may result from the interaction of pre-existing antibody to RSV and cell culture components; however, the late-appearing tissue damage might have resulted from cell-mediated inflamatory response rather than antibody-precipitated immune complexes.

3.2. Live, attenuated viruses

Previous studies have shown that intranasal administration of cold-adapted influenza virus vaccines is protective against illness during outbreaks of virus infection (Belshe and Van Voris, 1984; Belshe et al., 1984, Murphy and Webster, 1985). It was suggested that intranasally administered live attenuated virus vaccines provide a local immune response without lower respiratory tract involvement and may protect from subsequent natural infection. Cold-adapted and temperature-sensitive mutants of PI3 virus have been developed as candidate vaccines produced by serial passages at suboptimal temperature (Belshe and Hissom, 1982). Plaque purified viruses produced by serial passage at suboptimal temperature were tested for three phenotypic markers, temperature sensitivity, plaque size, and cold-adapted properties. Virus clones showing one or more of the above properties were evaluated for growth in weanling Syrian hamsters (Crookshanks-Newman and Belshe, 1984). All the clones exhibited some degree of attenuation in the lungs and nasal turbinates of the hamsters, and reduction of replication appeared directly related to the cold passage level of the virus. The cold-adapted virus mutants were used for intranasal vaccination to test their ability for induction of a protective immune response against wild-type PI3 virus infection (Crookshanks-Newman and Belshe, 1986). Results of virus recovery suggested an inverse relationship between the degree of protection induced by cold-adapted mutants and the number of passages of the virus in the cold. Assuming that there is not much antigenic variation within the same virus type, the cold-adapted mutants of human PI3 virus represent a promising approach for induction of a protective immune response against natural infection.

The possible use of bovine PI3 virus as a live virus vaccine is also under investigation (Coelingh et al., 1988). The levels of replication of bovine and human PI3 viruses and the cross-protective effect of the bovine virus were evaluated in non-human primates. These studies indicated that the replication of bovine PI3 is restricted in old monkeys, and that infection with the bovine strain induces significant resistance to subsequent challenge infection with human PI3 virus.

Penetration of paramyxoviruses into the host cell occurs by fusion of the viral membrane with the plasma membrane of the host cell and is mediated by the F protein (Choppin and Scheid, 1980). Activation of the F protein by proteolytic cleavage to disulfide linked F1 and F2 subunits is necessary for virus penetration and infectivity (Hsu et al., 1981; Hsu et al., 1983; Richardson and Choppin, 1983). A protease activation (pa) mutant of Sendai virus has been isolated in which the F protein was resistant to cleavage by trypsin or the protease present in mouse lung, but could be cleaved by chymotrypsin (Tashiro and Homma, 1983; 1985). The mutant virus showed a loss of virulence in mice, probably due to lack of proteolytic cleavage of the fusion glycoprotein and termination of infectivity after one step replication, and it induced resistance to wild-type virus infection. The use of pa mutants with known mutations around the cleavage site as vaccine candidates has also been studied (Hsu et al., 1987; Murphy et al., 1988). The pa mutants were attenuated in mice as indicated by an increase in their LD50 and a decrease in pulmonary histopathology of mutant-infected mice compared to mice infected with wild-type virus. The mutants were shown to be immunogenic and induced resistance to wild-type virus challenge.

3.3. Subunit vaccine

The available information previously suggested that human paramyxoviruses are monotypic (Orvell and Norrby, 1985). However, the neutralization domains of human PI3 virus F glycoprotein were recently found to vary among clinical isolates of the virus (Coelingh and Tierrney, 1989). A high degree of antigenic variation was observed in two of the four F neutralization domains of PI3 strains, resulting in co-circulation of antigenically different strains. However, it was suggested that variation in the F glycoprotein alone was unlikely to significantly influence the ability of a particular strain to reinfect seropositive children, since the conservation of the HN neutralization epitopes would render them susceptible to neutralization.

The HN protein of human PI3 virus appeared to be antigenically stable, as indicated by conservation of neutralizing epitopes (Coelingh et al., 1985) as well as amino acid sequence conservation (95%) in strains isolated over a 25-year interval

(Coelingh et al., 1988). Sharing of biochemical features and antigenic epitopes of the HN glycoprotein of human and bovine PI3 viruses was also reported (Coelingh et al., 1986; 1987a). However, comparison of nucleotide and predicted amino acid-sequences of the human and bovine PI3 HN, M, and F genes and gene products indicated a significant divergence of these two viruses, which was suggested to occur during separate evolution in the respective hosts (Coelingh et al., 1986; Sakai et al., 1987; Suzu et al., 1987). The majority of monoclonal antibodies against the HN glycoprotein of human PI3 virus failed to show any significant reactivity with bovine PI3 viruses in hemagglutination inhibition (HI) and virus neutralization tests (Ray and Compans, 1986) suggesting that these viruses show major antigenic differences. Other workers also suggested that the bovine virus strains and variants are antigenically quite different from human strains (Rydbeck et al., 1987). Extensive differences were detected in all four proteins, HN, F, NP and M, between strains of human and bovine origin. This further suggested that the differences between the two viruses exist in both glycosylated envelope proteins as well as in internal proteins.

The use of purified glycoproteins as vaccine is being investigated with human parainfluenza viruses as well as RSV. The hemagglutinin-neuraminidase (HN) and fusion (F) glycoproteins of human PI3 virus have been solubilized with octylglucoside, and reconstituted into lipid vesicles following removal of the detergent by dialysis. The efficacy of the glycoprotein preparation as a subunit vaccine was compared by intranasal or subcutaneous administration in hamsters (Ray et al., 1985; 1988a). Animals receiving intranasal immunizations with the glycoprotein preparation were found to be completely resistant to challenge infection. Only partial protection was observed, however, in animals immunized subcutaneously with the same dose of antigen. The glycoprotein-specific IgA response was significantly higher in intranasally immunized animals and was implicated in resistance to challenge infection. In subsequent studies, the HN and F glycoproteins were purified by immunoaffinity chromatography using specific monoclonal antibodies (Ray and Compans, 1987). The efficacy of immunization with the purified glycoprotein antigens either alone or in combination was tested by challenge infection with live PI3 virus (Ray et al., 1988b). Results of virus recovery from lungs and trachea demonstrated that although immunization with HN or F alone induced an antibody response to the respective glycoproteins, it did not provide a significant level of protection. However, immunization with a mixture of both the glycoproteins induced higher virus-neutralizing activity and afforded complete protection from challenge infection. It was suggested that both envelope glycoproteins of PI3 virus are essential components of a subunit vaccine (Ray et al., 1988c). It was also concluded that both glycoproteins are needed for development of subunit vaccines for measles virus (Varsanyi et al., 1987). Immune stimulating complexes of both H and F subunits were shown to confer protection against measles virus infection in subcutaneously immunized mice. However, passive transfer of anti-H mabs was shown to confer full protection against lethal infection in newborn mice, whereas anti-F monoclonal antibodies failed to protect the animals. On the other hand, no correlation between the degree of virus reduction in lungs and the level of neutralizing antibody was observed in mice following immunization with solid matrix-antibody-antigen complexes of simian virus 5 and subsequent intranasal infection with live virus. (Randall and Young, 1988; Randall et al., 1988). Vaccination of mice with the immune stimulating complexes containing either surface or internal SV5 proteins induced cytotoxic T cells; however, the greatest degree of protection was observed with the nucleoprotein or matrix protein.

The HN protein of PI3 virus has been expressed in insect cells using a baculovirus expression vector (Coelingh et al., 1987b). Cotton rats immunized intraperitoneally with recombinant HN showed a 100-fold reduction of virus replication in the upper respiratory tract compared to controls, and virus replication in the lungs was completely or almost completely suppressed (Coelingh et al., 1987b). The F glycoprotein of human PI3 virus has also been produced in insect cells using a baculovirus expression vector (Ray et al., 1989). Immunization of hamsters with the recombinant protein elicited antibody which neutralized virus infectivity and blocked fusion of virus-infected cells. The immunized animals showed a partial protective immune response to challenge infection with homologous virus.

Immunization of cotton rats with affinity purified attachment protein (G) or the fusion protein (F) of RSV was reported to confer complete pulmonary resistance and partial nasal resistance against challenge infection with live virus (Walsh et al., 1987). In a different study, mice immunized with either the purified F or G protein showed protection from subsequent virus challenge, whether they had developed neutralizing antibody or not (Routledge et al., 1988). Studies with immunostimulating complexes of RSV proteins suggested that neutralizing antibodies are more efficiently induced with the F protein of bovine than of human RSV (Trudel et al., 1989).

3.4. Recombinant virus vaccines

Live recombinant vaccinia virus vectors containing complete cDNA copies of different viral genes have been constructed for expression of the gene products. Vaccinia recombinants expressing surface glycoproteins of P13 virus, when inoculated intradermally into Patas monkeys, were shown to induce high titers of P13-specific serum neutralizing antibodies and induced significant resistance to challenge infection by P13 in both the upper and lower respiratory tract (Spriggs et al., 1988). Intradermal inoculation of cotton rats with vaccinia recombinants expressing HN or F was also shown to induce a vigorous immune response, and provided substantial protection against P13 virus infection (Spriggs et al., 1987). Animals receiving vaccinia-HN showed nearly complete protection of the lower respiratory tract and greatly reduced virus replication in the upper respiratory tract. On the other hand, animals receiving vaccinia-F also showed nearly complete protection of the lower respiratory tract, but less complete protection of the upper respiratory tract. The level of virus replication in cotton rats receiving both vaccinia-F and vaccinia-HN recombinants was shown to be similar to that seen in animals which received vaccinia-HN alone. However, a vaccinia recombinant expressing F of SV5 was shown to induce higher levels of neutralizing antibody in hamsters than did a vaccinia-HN recombinant (Patterson et al., 1987). Animals inoculated with vaccinia-HN were however found to be better protected from challenge infection with SV5 (Patterson et al., 1987).

Immunization of mice by the intranasal or intraperitoneal route with a vaccinia recombinant expressing the G protein of RSV resulted in complete protection of the lower respiratory tract from challenge infection with live RSV (Stott et al., 1986). Mice were also inoculated intraperitoneally with a vaccinia recombinant expressing the F protein and subsequently challenged with RSV. The lungs of the vaccinated mice were shown to be free of detectable RSV (Wertz et al., 1987). In another report, cotton rats inoculated intradermally with a vaccinia-F recombinant of RSV were reported to develop a six-fold higher titer of neutralizing antibodies than that induced by vaccinia-G (Olmsted et al., 1986). Immunization with vaccinia-F was shown to confer almost complete resistance to replication of RSV in the lower respiratory tract and a significant resistance in the upper respiratory tract. The degree of resistance conferred by vaccinia-F exceeded that induced by vaccinia G. In another study, the G protein of RSV was also shown to induce a less complete protective response to challenge infection by homologous virus and provided less complete cross-subgroup protection (Johnson et al., 1987). It appears that a variety of parameters including the route of vaccination, the subtype of challenge virus, and the species used may influence the antibody titers and extent of protection elicited by recombinant vaccinia viruses (Stott et al., 1987).

4. Concluding remarks

Protective immune responses generally depend upon cooperative interplay between cell types and different soluble mediators. T cells play a central role in coordinating the immune response. It appears that both humoral and cell-mediated immune (CMI) responses play an important role in protection from virus infection. Not much information is available on the CMI response following human parainfluenza virus infection. However, available evidence strongly suggests that virus-specific antibodies play a major protective role against parainfluenza virus infection in both the lower and upper respiratory tract. Nasal secretory IgA antibodies seem to play a more significant role in resistance. Relatively more information is available on both humoral and T cell-mediated immune responses during RSV infection and both of them appear to play a role in protection. The immune response of infants to RSV is complicated because of the presence of maternal antibody and the immaturity of the immune system. Similar complexity is not known to occur following parainfluenza virus infection.

There are no effective vaccines against human parainfluenza viruses or respiratory syncytial viruses. The preventive measure to be taken by vaccination should be considered at a very early age. The antigenic composition of human parainfluenza viruses appears to be stable, although some variations have recently been reported. Immunization of experimental animals suggested that both envelope glycoproteins of paramyxoviruses are important components of a subunit vaccine. The protective immune response against human parainfluenza virus appears to be type-specific, since immunization with envelope glycoproteins or primary infection with one virus type failed to induce protection against challenge infection with the other virus type.

Parainfluenza and RSV infections are limited to epithelial surfaces and do not spread systemically. There is opportunity for induction of local immune responses, including secretory IgA antibodies which have antiviral activity in the absence of complement. Secretory IgA antibodies may act as an initial immunological barrier to the entry of these viruses. The duration of a protective immunity is another important point to consider for an effective immunization schedule. It appears that stimulation of both systemic and local immune responses by immunization with the envelope glycoproteins or other suitable candidate vaccines may prevent the infection of both the upper and lower respiratory tract by parainfluenza virus. A subunit vaccine containing the two virus glycoprotein antigens is a rational approach for immunization of the vulnerable infant population with minimum risk potential. The two virus glycoproteins may be prepared by the conventional way or by recombinant DNA techniques. Live vector-based vaccines, or live attenuated parainfluenza vaccine, would certainly be most cost-effective; however, biological consequences, including initial pathology and chance of reversion to wild-type virus, require careful study. The situation with RSV is more complex because of the previous experience of potentiation of the disease resulting from vaccination. The oligosaccharide side chains on the envelope glycoproteins display considerable influence on the attachment and/or penetration functions of the infective virion (Lambert, 1988). Studies to understand the specific role of oligosaccharide-mediated virus-host cell interactions and the role of these oligosaccharides in immunity to RSV infections have also been suggested as areas for future investigation.

5. Acknowledgements

Research on parainfluenza viruses was supported by contracts NOI-AI62604 and N44-AI-82504 from the National Institute of Allergy and Infectious Diseases. We thank Dr. Laurent Roux for helpful comments and Ms. Joellynn Heaton and Ms. Betty Jeffrey for preparation of the manuscript.

6. References

Abidi, T.F. and Flanagan, T. D. (1984) *J. Virol.* 50, 380–386.
Ada, G.L., Leung, K.N., Ertl, H. (1981) *Immunol. Rev.* 58, 5–24.
Akerlind, B., Norrby, E., Orvell, C. and Mufson M.A. (1988) *J. Gen. Virol.* 69, 2145–2154.
Al-ahdal, M.N., Nakamura, I. and Flanagan, T.D. (1985) *J. Virol.* 54, 53–57.
Andzhaparidze, O.G., Boriskin, Y.S., Bogomolova, N.N. and Drynov, I.D. (1982) *J. Gen. Virol.* 63, 499–503.
Bangham, C.R.M. and Askonas, B.A. (1986) *J. Gen. Virol.* 67, 623–629.
Bangham, C.R.M., Openshaw, P.J.M., Ball, L.A., King, A.M.Q., Wertz, G.W. and Askonas, B.A. (1986) *J. Immunol.* 137, 3973–3977.
Belshe, R.B. and Hissom, F.K. (1982) *J. Med. Virol.* 10, 235–242.
Belshe, R.B. and Van Voris, L.P. (1984) *J. Infect. Dis.* 149, 735–740.
Belshe, R.B., Van-voris, L.P., Bartram, J. and Crookshanks, F.K. (1984) *J. Infect. Dis.* 150, 834–840.
Bennink, J.R., Yewdell, J.W. and Gerhard, W. (1982) *Nature* 296, 75–76.
Casali, P. and Oldstone, M.B.A. (1982) *Cell. Immunol.* 70, 330–334.
Cattaneo, R., Schmid, A., Eschle, D., Baczko, K., ter Meulen, V. and Billeter, M.A. (1988) *Cell* 55, 255–265.
Cattaneo, R., Schmid, A., Baczko, K., Liebert, U.G., ter Meulen, V., Udem, S.A. and Billeter, M.A. (1989) In *Cell Biology of Virus Entry* (R.W. Compans, A. Helenius and M.B.A. Oldstone, Eds.) Alan R. Liss, New York. pp. 425–434.
Chanock, R.M. and McIntosh, K. (1985) In *Virology* (B.N. Fields, D.M. Knipe, J.L. Melnick, R.M. Chanock, B. Roizman and R.E. Shope, Eds.) Raven Press, New York. pp. 1241–1253.
Chin, J., Magoffi, R.L., Shearer, L.A., Schieble, J.H. and Lennette, E.H. (1969) *Am. J. Epidemiol* 89, 449–463.
Choppin, P.W. (1964) *Virology* 23, 224–233.
Choppin, P. and Compans, R.W. (1975) in: *Comprehensive Virology* (H. Fraenkel-Conrat and R.R. Wagner, Eds.) Vol. 4, Plenum Press, New York, pp. 95–178.
Choppin, P.W. and Scheid, A. (1980) *Rev. Infect. Dis.* 2, 40–61.
Choppin, P.W., Richardson, C.D., Merz, D.C., Hall, W.W. and Scheid, A. (1981) *J. Infect. Dis.* 143, 352–363.
Coelingh, K.L.V., Winter, C.C. and Murphy, B.R. (1985) *Virology,* 143, 569–582.
Coelingh, K.J., Winter, C.C., Murphy, B.R., Rice, J.M., Kimball, P.C., Olmsted, R.A. and Collins, P.L. (1986) *J. Virol.* 60, 90–96.
Coelingh, K.L.V.W., Murphy, B.R., Rice, J.M., Kimball, P.C., Winter C.C. and Collins, P.L. (1987a) In *Biology of Negative Strand Viruses* (B. Mahy and D. Kolakofsky, Eds.) Elsevier, Amsterdam, pp. 391–396.
Coelingh, K.L.V., Murphy, B.R., Collins, P.L., LeBacq-Verheyden, A.M. and Battey, J.F. (1987b) *Virology* 160, 465–472.
Coelingh, K.L.V., Winter, C.C., Tierney, E.L., London, W.T. and Murphy B.R. (1988) *J. Infect. Dis.* 157, 655–662.
Coelingh, K.V. and Tierney, E.L. (1989) *J. Virol.* 63, 375–382.
Compans, R.W. and Nakamura, K. (1978) In *Virology and Rickettsiology* (G.P. Hsiung and R.H. Green, Eds.) Vol.I, Part 2. CRC Press, West Palm Beach, Fl pp. 361–385.
Crookshanks-Newman, F.K. and Belshe, R.B. (1984) *J. Med. Virol.* 13, 243–249.
Crookshanks-Newman, F.K. and Belshe, R.B. (1986) *J. Med. Virol.* 18, 131–137.
Doherty, P.C. and Zinkernagel, R.M. (1976) *Immunology* 31, 27–32.
Ertl, H.C.J. (1981) *Cell. Immunol.* 62, 38–49.
Finberg, R. Mescher, M. and Burakoff, S.J., (1978) *J. Exp. Med.* 148, 1620–1627.
Frank, A.L., Taber, I.H., Wells, C.R., Wells, J.M., Glezen, W.P. and Paredes, A. (1981). *Patterns of shedding of myxoviruses and paramyxoviruses in children. J. Infect. Dis.* 144, 433–441.

Fulginiti, V.A., Sieber, O.F., John, T.J., Askin, P. and Umlauf, H.J. (1967) *Pediatr. Res.,* 1, 50–58.

Fulginiti, V.A., Eller, J.J., Sieber, O.F., Joyner, J.W., Minamitani, M. and Meiklejohn, G. (1969) *Am. J. Epidemiol.* 89, 435–448.

Gimenez, H.B., Keir, H.M. and Cash, P. (1987) *J. Gen. Virol.* 68, 1267–1275.

Goswami, K.K.A., Cameron, K.R., Russell, W.C., Lange, L.S. and Mitchell, D.N. (1984) *J. Gen. Virol.* 65, 1881–1888.

Gross, P.A., Green, R.H. and Curnen, M.G.M. (1973) *Am. Rev. Respir. Dis.* 108, 894–898.

Guink, N.E., Kris, R.M., Goodman-Snitkoff, G., Small, P.A. and Mannino, R.J. (1989) *Vaccine* 7, 147–151.

Hale A.H., Lyles, D.S. and Fan, D.P. (1980) *J. Immunol.* 124, 724–731.

Harfast, B., Anderson, T. and Perimann, P. (1978) *J. Immunol.* 121, 755–761.

Homma, M. and Ohuchi, M. (1973) *J. Virol.* 12, 1457–1465.

Hsu, M.C., Scheid, A. and Choppin, P.W. (1981) *J. Biol. Chem.* 256, 3557–3563.

Hsu, M.C., Scheid, A. and Choppin, P.W. (1983) *Virology* 126, 361–369.

Hsu, M.C., Scheid, A. and Choppin, P.W. (1987) *Virology,* 156, 84–90.

Ito, Y., Tsurudome, M., Hishiyama, M. and Yamada, A. (1987) *J. Gen. Virol.* 68, 1289–1297.

Johnson, P.R., Olmsted, R.A., Prince, G.A., Murphy, B.R., Alling, D.W., Walsh, E.E. and Collins, P.L. (1987) *J. Virol.,* 61, 3163–3166.

Julkunen, I. (1984) *J. Med. Virol.* 14,177–187.

Kapikian, A.Z., Mitchell, R.H., Chanock, R.M. Shvedoff, R.A. and Stewart, C.E. (1969) *Am J. Epidemiol.* 89, 405–421.

Kasel, J.A., Frank, A.L., Keitel, W.A., Taber, L.H. and Glezen W.P. (1984) *J. Virol.* 52, 828–832.

Kim, H.W., Chanchola, J.G., Brandt, C.D., Pyles, G., Chanock, R.M., Jensen, K. and Parrott, R.H. (1969) *Am. J. Epidemiol.* 89, 422–434.

Kingsbury, D.W. (1989) *The Paramyxoviruses,* Plenum Press, in press.

Kolakofsky, D. and Roux, L. (1987) In *The Molecular Basis for Viral Replication,* pp. 277–297 (P.R. Bercoff, Ed.) Plenum Press, New York.

Koszinowski, U.H, Gething, M.J. and Waterfield, M.D. (1977) *Nature* 267, 160–163.

Kress, H.G. and Kreth, H.W. (1982) *J. Immunol.* 129, 844–849.

Kreth, H.W., Kress, L., Kress, H.G., Ott, H.F., and Eckert, G. (1982) *J. Immunol.* 128, 2411–2415.

Kristensson, K., Orvell, C., Lesstma, J. and Norrby, E. (1983) *J. Infect. Dis.* 147, 297–301.

Lambert, D.M. (1988) *Virology* 164, 458–466.

Levine, S., Dajani, A. and Klaiber-Franco, R. (1988) *J. Gen. Virol.* 69, 1229–1239.

Long, E.O. and Jacobson, S. (1989) *Immunol. Today* 10, 45–48.

Matsumoto, T. (1982) *Microbiol. Immunol.* 26, 285–320.

Matthews, R.E.F. (1982) *Intervirology* 17, 1–1181.

McCarthy, M., Wolinsky, J.S. and Lazzarini, R.A. (1981) *Virology* 114, 343–356.

McGee, M., Hale, A.H. and Panetti, M. (1980) *Eur. J. Immunol.* 10, 923–928.

McIntosh, K. and Chanock, R.M. (1985) *Virology* (B.N. Fields, D.M. Knipe, J.L. Melnick, R.M. Chanock, B. Roizman and R.E. Shope, Eds.) pp. 1285–1304. Raven Press, New York.

Minato, N., Bloom, B.R., Jones, C., Hollard, J. and Reid, L. (1979) *J. Exp. Med.* 149, 1117–1133.

Mills, B.G., Singer, F.R., Weiner, L.P. and Holst, P.A. (1981) *Proc. Natl. Acad. Sci.* 78, 1209–1213.

Mims, C.A. and White, D.O. (1984) *Viral Pathogenesis and Immunology*. Blackwell, Oxford. pp. 87–131.

Muchmore, H.G., Parkinson, A.J., Humphries, J.F., Scott, E.N., McIntosh, D.A., Scott, L.V., Cooney, M.K. and Miles, J.A.R. (1981) *Nature,* 289, 187–189.

Mufson, M. A., Orvell, C., Rafnar, B. and Norrby, E. (1985) *J. Gen. Virol.* 66, 2111–2124.

Murphy, B.R. and Webster, R.G. (1985) In: *Virology,* (B.N. Fields, D.M. Knipe, J.L. Melnick, R.M. Chanock, B. Roizman and R.E. Shope, Eds.) Raven Press, New York. pp. 1179–1239.

Murphy, B.R., Graham, B.S., Prince, G.A., Walsh, E.E., Chanock, R.M., Karzon, D.T. and Wright, P.F. (1986) *J. Clin. Microbiol.* 23, 1009–1014.

Murphy, B.R., Prince, G.A., Collins, P.L., Coelingh, K.V., Olmsted, R.A., Spriggs, M.K., Parrott, R.H., Kim, H.W., Brandt, C.D. and Chanock, R.M. (1988) *Virus Res.* 11, 1–15.

Nedrud, J.G., Liang, X., Hague, N. and Lamm. M.E. (1987) *J. Immunol.* 139, 3484–3492.

Oldstone, M.B.A. and Fuginami, R.S. (1982) In *Virus Persistence 33rd Symp. Soc. Gen. Microbiol.* (B.W.J. Mahy, Minson, A.C. and Derby, G.K., Eds.) Cambridge University Press, Cambridge.

Olmsted, R.A., Elango, N., Prince, G.A., Murphy, B.R., Johnson, P.R., Moss, B., Chanock, R.M. and Collins, P.L. (1986) *Proc. Natl. Acad. Sci. USA* 83, 7462–7466.

Oldstone, M.B.A. (1989) *Cell* 56, 517–520.

Openshaw, P.J.M., Pemberton, R.M., Ball, L.A., Wertz, G.W. and Askonas, B.A. (1988) *J. Gen. Virol.* 69, 305–312.

Orvell, C. and Norrby, E. (1985) In *Immunochemistry of Viruses. The Basis for Serodiagnosis and Vaccines* (M.V.H. Van Regenmortel and A.R. Neurath, Eds.) Elsevier, Amsterdam. pp. 241–264.

Orvell, C., Norrby, E. and Mufson, M.A. (1987) *J. Gen. Virol.* 68, 3125–3135.

Parrott, R.H., Kim, H.W., Arrobio, J.O., Canchola, J.G., Brandt, C.D., DeMeio, J.L., Jensen, K.E. and Chanock, R.M. (1967) In *First International Conference on Vaccines against Viral and Rickettsial Diseases of Man.* Plenum Press, New York, pp. 35–41.

Paterson, R.G., Lamb, R.A., Moss, B. and Murphy, B.R. (1987) *J. Virol,* 61, 1972–1977.

Pemberton, R.M., Cannon, M.J., Openshaw, P.J.M., Ball, L.A., Wertz, G.W. Askonas, B.A. (1987) *J. Gen. Virol.* 68, 2177–2182.

Piedra, P.A., Faden, H.S., Camussi, G., Wong, D.T. and Ogra, P.L. (1989) *Vaccine* 7, 34–38.

Prince, G.A., Jenson, A.B., Hemming, V.G., Murphy, B.R., Walsh, E.E., Horswood, R.L. and Chanock, R.M. (1986) *J. Virol.* 57, 721–728.

Prince, G.A., Hemming, V.G., Horswood, R.L., Baron, P.A. and Chanock, R.M. (1987) *J. Virol.* 61, 1851–1854.

Randall, R.E. and Young, D.F. (1988) *J. Gen. Virol.* 69, 2505–2516.

Randall, R.E., Young, D.F. and Southern, J.A. (1988) *J. Gen. Virol.* 69, 2517–2526.

Ray, R., Brown, V.E. and Compans, R.W. (1985) *J. Infect. Dis.* 152, 1219–1230.

Ray, R. and Compans, R.W. (1986) *Virology* 148, 232–236.

Ray, R. and Compans, R.W. (1987) *J. Gen. Virol.* 68, 409–418.

Ray, R. Glaze, B.J., Moldoveanu, Z. and Compans, R.W. (1988a) *J. Infect. Dis.* 157, 648–654.

Ray, R., Glaze, B.J. and Compans, R.W. (1988b) *J. Virol.* 62, 783–787.

Ray, R., Glaze, B.J., Burnett, T.L. and Compans, R.W. (1988c) In *Technological Advances in Vaccine Development* (L. Lasky, Ed.) Alan R. Liss, New York. pp. 1–10

Ray R., Galinski, M. and Compans, R.W. (1989) *Virus Res.* 12, 169–180.

Richardson, C.D. and Choppin, P.W. (1983) *Virology* 131, 518–532.

Routledge, E.G., Willcocks, M.M., Samson, A.C.R., Morgan, L., Scott, R., Anderson, J.J. and Toms, G.L. (1988) *J. Gen. Virol.* 69, 293–303.

Rydbeck, R., Love, A., Orvell. C. and Norrby, E. (1987) *J. Gen. Virol.* 68, 2153–2160.

Sakai et al. (1987) *Nucleic Acids Res.* 15, 2927–2944.

Santoli, D., Trinchieri, G. and Koprowski, H. (1978) *J. Immunol.* 121, 532–538.

Scheid, A., Caliguiri, L.A., Compans, R.W. and Choppin, P.W. (1972) *Virology* 50, 640–652.

Scheid, A. and Choppin, P.W. (1974) *Virology* 57, 475–490.

Scheid, A. and Choppin, P.W. (1977) *Virology,* 80, 54–66.

Sever, J.L. (1983) *Rev. Infect. Dis.* 5, 467–473.

Sissons, J.G.P. and Oldstone, M.B.A. (1980) *Adv. Immunol.* 29, 209–260.

Smith, C.B., Purcell, R.H., Bellanti, J.A. and Chanock, R.M. (1966). *New Engl. J. Med.* 275, 1145–1152.

Spriggs, M.K., Murphy. B.R., Prince, G.A., Olmsted, R.A. and Collins, P.L. (1987) *J. Virol.* 61, 3416–3423.

Spriggs, M.K., Collins, P.L., Tierney, E., London, W.T. and Murphy, B.R. (1988) *J. Virol.* 62, 1293–1296.

Stott, E.J., Ball, L.A., Young, K.K., Furze, J. and Wertz, G.W. (1986) *J. Virol.* 60, 607–613.

Stott, E.J., Taylor, G., Ball, L.A., Anderson, K., Young, K.K.-Y., King, A.M.Q. and Wertz, G.W. (1987) *J. Virol.* 61, 3855–3861.

Suzu, S., Sakai, Y., Shioda, T. and Shibuta, H. (1987) *Nucleic Acids Res.* 15, 2945–2958.

Tashiro, M. and Homma, M. (1983) *Infect. Immun.* 39, 879–888.

Tashiro, M. and Homma, M. (1985) *J. Virol.* 53, 228–234.

Taylor, P.M. and Askonas, B.A. (1986) *Immunology* 58, 417–420.

Townsend, A.R.M., Skehel, J.J., Taylor, P.M. and Palese, P. (1984) *Virology* 133, 456–459.

Townsend, A.R.M., Rothbard, J., Gotch, F.M., Bahadur, G., Wraith, D. and McMichael, A.J. (1986) *Cell* 44, 959–968.

Tremonti, L.P., Lin, J.L. and Jackson, G.G. (1968) *J. Immunol.* 101, 572–577.

Trudel, M., Nadon, F., Seguin, C., Simard, C. and Lussier, G. (1989) *Vaccine,* 7, 12–16.

Varsanyi, T.M., Morein, B., Love, A. and Norrby, E. (1987) *J. Virol.* 61, 3896–3901.

Walsh, E.E., Hall, C.B., Briselli, M., Brandriss, M.W. and Schlesinger, J.J. (1987) *J. Infect. Dis.* 155, 1198–1204.

Ward, K.A., Lambden, P.R., Ogilvie, M.M. and Watt, P.J. (1983) *J. Gen. Virol.* 64, 1867–1876.

Welsh, R.M., Jr. and Hallenbeck, L.A. (1980) *J. Immunol.* 124, 2491–2497.

Wechsler, S.L., Lambert, D., Galinski, M.S., Mink, M.A., Rochovansky, O. and Pons, M.W. (1987) *J. Gen. Virol.* 68, 1737–1748.

Wertz, G.W., Stott, E.J., Young, K.K.Y., Anderson, K. and Ball, L.A. (1987) *J. Virol.* 61, 293–301.

Westenbrink, F., Kimman, T.G. and Brinkhof, J.M.A. (1989) *J. Gen. Virol.* 70, 591–601.

Yanagihara, R. and McIntosh, K. (1980) *Infect. Immun.* 30, 23–28.

Yap, K.L. and Ada, G.L. (1977) *Immunology* 32, 151–159.

Yewdell, J.W., Franl, E. and Gerhard, W. (1981) *J. Immunol.* 126, 1814–1819.

Zinkernagel, R.M. and Doherty, P.C. (1975) *J. Exp. Med.* 141, 1427–1436.

Eds. M.H.V. van Regenmortel and A.R. Neurath
Immunochemistry of viruses, II. The basis for serodiagnosis and vaccines.

Arenaviridae and bunyaviridae — 12

K.E. WRIGHT[1], C. SCHMALJOHN[2], A. SCHMALJOHN[2] and M.J. BUCHMEIER[3]

[1]Department of Microbiology, University of Ottawa, Ottawa, Canada, [2]USAMRIID, Ft. Detrick, Frederick, MD, U.S.A. and [3]Research Institute of Scripps Clinic, Department of Neuropharmacology, 10666 North Torrey Pines Road, La Jolla, CA 92037, La Jolla, CA, U.S.A.

1. Arenaviridae

The arenaviridae are a group of 13 enveloped RNA animal viruses sharing common group-specific antigens as well as the unique morphological feature consisting of electron-dense 20-nm bodies within the virion. The presence of these dense bodies, which have subsequently been shown to consist of host cell derived ribosomes (Carter et al., 1973; Pederson, 1973; Pederson and Konigshofer, 1976; Farber and Rawls, 1975), suggested the group name from *arenosus* (L. Sandy) (Rowe et al., 1970a). The prototype virus of the family is lymphocytic choriomeningitis virus (LCMV). Serological analysis and geographic distribution of the viruses and their natural rodent hosts has led to a functional division of the arenaviridae into two groups. These are the Old World arenaviruses which include the African arenaviruses Lassa (LAS), Mopeia (MOP), Mobala (MOB) and LCMV, and the new world arenaviruses, Amapari (AMA), Tacaribe (TAC), Junin (JUN), Machupo (MAC), Tamiami (TAM), Parana (PAR), Pichinde (PIC), and Latino (LAT). All of these, with the possible exception of TAC, establish lifelong persistent infections in a single or limited species of rodent and are transmitted to man by direct contract with, or aerosolization of rodent excreta. No substantial evidence of an arthropod vector exists for the arenaviruses. Several members of the group are etiologic agents of severe human disease (Table 12.1).

2. Properties of Arenaviridae

2.1. Morphology, morphogenesis and virion composition

The arenaviridae are pleomorphic, enveloped viruses with a mean diameter of 110–

130 nm (Fig. 12.1). The envelope is formed from the host cell plasma membrane by budding, and contains sparse, 10-nm club-shaped projections composed of viral glycoproteins (Murphy et al., 1970; Gard et al., 1977; Vezza et al., 1977; Buchmeier et al., 1978).

TABLE 12.1

Arenaviruses

Virus	Natural host	Geographic distribution	Naturally occurring human disease
Old World			
Lymphocytic choriomeningitis (LCM)	*Mus musculis*	Americas, Europe	Undifferentiated febrile illness, aseptic meningitis rarely serious
Lassa	*Mastomys natalensis*	Africa	Lassa fever, mild to severe and fatal disease
Lassa-like viruses from Africa	*Mastomys, Praomys*	Mozambique, Zimbabwe, Cent Afr Rep	Unknown
New World			
Junin	*Calomys musculinus*	Argentina	Argentine hemorrhagic fever (AHF) rarely serious
Tacaribe	*Artibeus* bats	Trinidad, West Indies	None
Machupo	*Calomys musculinus*	Bolivia	Bolivian hemorrhagic fever (BHF)
Amapari	*Oryzomys gaeldi* *Neacomys guianae*	Brazil	None
Parana	*Oryzomys bucinnatus*	Paraguay	None
Tamiami	*Sigmodon hispidus*	USA (Florida)	Antibodies detected
Pichinde	*Oryzomys albigularis*	Colombia	None
Latino	*Calomys callosus*	Bolivia	Unknown
BeAn 293022 (proposed flexal virus)	*Oryzomys* spp.	Brazil	Unknown

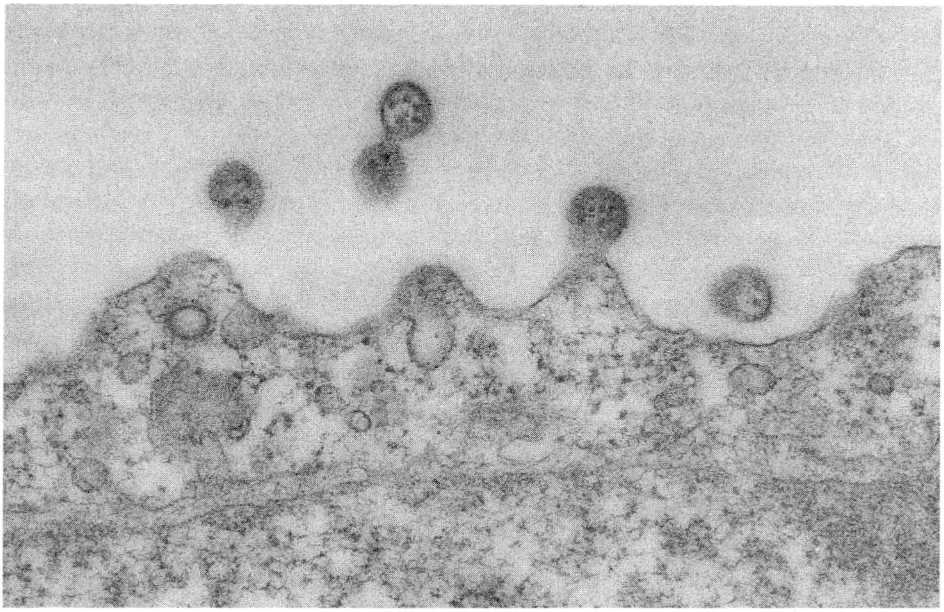

Fig. 12.1. LCM virus budding from the plasma membrane of infected BHK-21 cells. Typical 110-nm virions containing numerous electron-dense 20 nm particles are evident. (x 26 000).

Internal nucleocapsid structures have been difficult to visualize, but the presence of 20-nm electron-dense granules thought to be host cell ribosomes is characteristic (Murphy and Whitfield, 1975). Electron microscopic studies of PIC virions disrupted under controlled conditions have demonstrated a filamentous structure consisting of viral RNA and the nucleocapsid protein (N or NP) which resembled a string of beads 4 nm in diameter (Young and Howard, 1986).

Two strands of genomic RNA are present. These are a large (L) RNA (7.2 kb) and a smaller 3.4 kb RNA (S) (Vezza et al., 1977). The arenaviruses have been considered to be negative strand viruses (Carter et al., 1974), but the coding strategy of both the strands of RNA is ambisense, with both L and S containing positive and negative sense genes (Auperin et al., 1984a; Bishop and Auperin, 1987; Salvato et al., 1989). The L and S RNAs are unique except for a shared 3′ terminal sequence that is complementary to sequences at the 5′ termini (Auperin et al., 1984b; Romanowski et al., 1985; Southern and Bishop, 1987). The S RNA encodes the viral nucleocapsid protein in the negative sense at the 3′ end and the viral glycoprotein precursor, GP-C, in the positive sense at the 5′ end. Similarly, L encodes a 200-kDa polypeptide thought to be the virion RNA polymerase in the negative sense at the 3′ end and a newly described putative zinc binding protein, Z, in the positive sense at the 5′ end (Fig. 12.2).

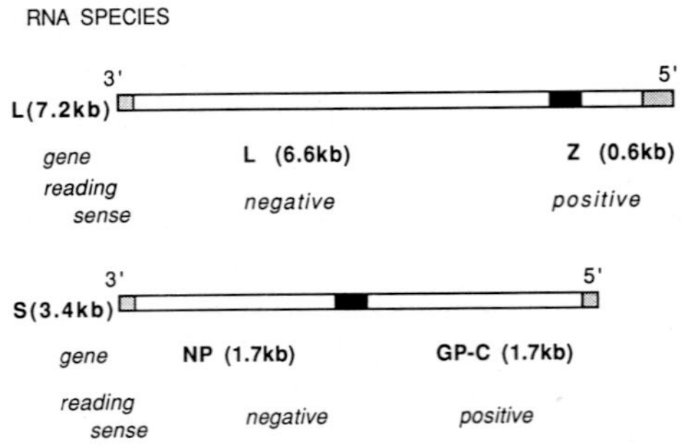

Fig. 12.2. Genetic organization of the LCMV, L and S RNAs. Four known genes, L, Z, NP and GP-C encode their respective proteins as described in the text. Black bars indicate non-coding intergenic hairpin regions and 3′ and 5′ non-coding regions of each RNA segment.

2.2. Serological relationships

Antigenic cross-reactions among the arenaviruses have been demonstrated in vitro using indirect immunofluorescence (IF), complement fixation (CF), ELISA and neutralization (Nt) assays. Antigens present on both Old and New World arenaviruses can be identified readily by immunofluorescence (Rowe et al., 1970b; Wulff et al., 1978; Buchmeier et al., 1981; Howard et al., 1985; Sanchez et al., 1989). Generally speaking, antisera directed against the New World viruses react with LCMV antigens, but anti-LCMV antiserum has a more restricted reactivity against New World viruses, staining AMA, minimally TAC, and not MAC virus (Rowe et al., 1970b). Monoclonal antibodies (mAb) raised against various arenaviruses have revealed epitopes on the viral proteins responsible for these cross reactivities. A monoclonal antibody which reacts with an epitope contained within amino acids 370–382 of the GP-C glycoprotein precursor of LCMV also reacts with LAS, MOP, MOB, PIC, TAC, JUN, AMA and PAR viruses (Buchmeier et al., 1980; Weber and Buchmeier, 1988). mAbs directed to either GP-1 or NP of LCMV do not cross-react with the New World viruses, but one mAb epitope detected on NP of PIC is shared by LAS and LCMV (Buchmeier et al., 1981). These data indicate that group-specific epitopes present on both new and Old World arenaviruses are located on internal proteins (NP) and surface glycoproteins.

Monoclonal antibodies also define subgroup and type-specific antigens. Subgroup-specific antigens appear to be restricted to NP, as mAb directed to NP but not GP of TAC and JUN react with viral antigen in cells infected with other New World viruses (Howard et al., 1985; Allison et al., 1984). Most of the anti-PIC monoclonals directed to NP are virus-specific, but three anti-NP mAb detect an epitope(s) on other New World viruses, namely TAM and PAR (Buchmeier et al., 1981). Within the Old World group, LCMV antiserum reacts with LAS and other African isolates, but with reduced titer compared to the reaction with homologous antigen. Antiserum to LAS reacts only weakly with LCMV (Peters et al., 1987). Similarily, a mAb specific for NP of LCMV reacts with LAS and MOP, both African arenaviruses, but does not react with any New World arenaviruses (Buchmeier et al., 1980, 1981). mAbs directed to LAS virus GP-2 and NP cross-react with other African viruses, demonstrating that group-specific epitopes are present on both proteins (Gonzalez et al., 1984). Reactivity of these mAb against New World viruses has not been reported. Group specific antigens are also detected by complement fixation (Wiebenga et al., 1964; McKenzie et al., 1965; Calisher et al., 1970, Casals et al., 1975), and this response is likely directed against epitopes on NP (Buchmeier et al., 1980).

Virus or strain-specific antigens probably exist on all three structural proteins. Neutralization epitopes on GP-1 are largely virus specific, but cross-reactivity has been reported between closely related viruses, such as TAC and JUN (Henderson and Downs, 1965; Weissenbacher et al., 1975, 1976; Howard et al., 1985).

All strains of LCMV, with the exception of certain clonal isolates of LCMV Armstrong, share a single neutralization epitope defined by both mAb and polyclonal antisera on GP-1 (Parekh and Buchmeier, 1986; Wright et al., 1989a). One isolate of LCMV Armstrong containing a mutation at position 173 of GP-C bears a second neutralization epitope that distinguishes it from other Armstrong isolates, and other strains of LCMV (Buchmeier and Parekh, 1987; Wright et al., 1989a). Non-neutralization, type-specific epitopes have also been described on the glycoproteins of LCMV (Buchmeier, 1984; Parekh and Buchmeier, 1986), TAC (Allison et al., 1984; Howard et al., 1985), and the African arenaviruses (Gonzalez et al., 1984).

Cross-reactions described by in vitro assays may have in vivo correlates. Guinea pigs and marmosets immunized with TAC virus developed low levels of neutralizing antibody to JUN that appeared late after immunization (Weissenbacher et al., 1975, 1976; Coto et al., 1980; Weissenbacher et al., 1982). When subsequently challenged with JUN, a secondary neutralizing anti-JUN response occurred, indicating that the animals were primed by cross-reacting antigen on TAC (Coto et al., 1980; Weissenbacher et al., 1982).

The significance of neutralizing antibody to LAS in vivo is unclear since infection of guinea pigs with MOP, MOB or LCMV does not induce cross-reactive neutralizing antibody to LAS, but these animals are protected against LAS challenge (Peters et al., 1987). Clearly the relative importance of B and T cell-mediated immune responses in LAS infection remains to be elucidated.

2.3. Natural hosts and diseases

Arenaviruses persist in nature by virtue of their ability to establish lifelong persistent infection in rodent hosts. These naturally infected hosts show few if any symptoms of infection, reproduce normally, and transmit infection to their offspring in utero or neonatally. Each virus is restricted in nature to specific species of rodents, although in the laboratory, other rodents can be infected with many of the arenaviruses (Peters et al., 1987). Horizontal transmission of virus also occurs but does not result in virus persistence. In both acute and persistent infections virus is excreted in the urine (Peters et al., 1987; McCormick, 1987).

2.3.1. Old World arenaviruses

Among the Old World arenaviruses, LAS and LCMV are known to cause human disease. Viruses antigenically related to LAS have been isolated in Central African Republic, Zimbabwe, Mozambique and South Africa (Wulff et al., 1977; Johnson et al., 1981; Gonzalez et al., 1983; McCormick, 1987). Some of these viruses persist in the same rodent host (*Mastomys natalensis*) as LAS, and are transmitted to humans as indicated by significant incidence of seropositivity, but no clinical disease has been associated with these infections (McCormick, 1987). LCMV infection produces a spectrum of illness ranging from subclinical, or mild influenza-like symptoms, to aseptic meningitis (Casals, 1984).

Lassa virus also produces a spectrum of illness in humans, but is largely restricted to West Africa, where in Sierra Leone infection rates have been estimated to range from 5% to 20% of the population. Of these, 10% to 25% will present with classical symptoms, 5% to 8% will be hospitalized, and 1% to 2% will die (McCormick, 1987).

The natural host of LAS is the rodent *Mastomys natalensis* (Monath et al., 1974). As with LCMV in the mouse, neonatal infection of the rodent results in lifelong persistence of the virus in liver, kidney, lymph nodes, lung and brain, with shedding in the urine, throat secretions and blood (Walker et al., 1975). Transmission to humans occurs from contact with excreta in the home, either by aerosol, ingestion or through cuts and abrasions in the skin (Monath, 1975; Casals, 1984). Person-to-person transmission is also possible and presents a substantial risk in the hospital setting (McCormick, 1987).

2.3.2. New World viruses

Two New World arenaviruses, JUN and MAC, cause disease in man. JUN virus is transmitted to humans from field rodents of the genus *Calomys,* causing Argentine hemorrhagic fever (AHF) (Weissenbacher et al., 1987). Persistently infected rodents excrete JUN in saliva and urine and human infection occurs through skin abrasions, conjunctiva or through respiratory mucosa from contaminated dust (Weissenbacher et al., 1987). About 30% of infections are subclinical. Of the remaining infections, the majority recover spontaneously, but total mortality can reach 16% in untreated patients (Peters, 1984; Casals, 1984). Like Lassa fever, all organ systems are affected, but hemorrhagic manifestations are more frequent. Treatment of acute phase patients with immune plasma has proven effective in reducing mortality to below 1%. Convalescence is lengthy but usually without sequelae, although a small number of patients experience transient late neurological symptoms (Peters, 1984). Pathological findings in fatal cases of AHV are not severe enough to indicate the cause of death. Most consistent is the observation of depletion of the bone marrow and lymphoid necrosis. Death appears to be the result of hypovolemic shock due to plasma leakage, suggesting a lesion in the vascular endothelium similar to that described in LAS (Casals, 1984, Peters, 1984).

Bolivian hemorrhagic fever (BHF) results from infection with Machupo virus. This infection is 'house-associated' and is contracted from persistently infected rodents of the species *Calomys callosus* in much the same way that Lassa fever virus is passaged to humans in Africa (Casals, 1984). Human to human transmission is rare. Clinically BHF is very similar to AHF, although inapparent infections with Machupo virus are rare. Mortality ranges from 5% to 30% (Casals, 1984). Machupo virus has largely been controlled through an aggressively applied program of rodent control (Johnson et al., 1966).

Other New World viruses persist in specific rodents in nature with the exception of TAC which was isolated from fructivorous bats of the genus *Artibeus* (Table 12.1). None of these viruses cause disease in humans, although inapparent infections with PIC virus have been documented in a laboratory setting (Buchmeier et al., 1974).

3. Potential targets of immune responses

3.1. Structure and morphology

3.1.1. Structural proteins

Quantitatively, the major structural protein of the arenaviruses is NP (60–68 kDa), which has been estimated to constitute 58% of total protein in virions (Vezza et al., 1977). In the virion, NP is complexed with genomic RNA in a ribonuclear protein (RNP) (Gard et al., 1977; Buchmeier et al., 1978; Young and Howard, 1983, 1986; Bruns et al., 1983a, 1986). It has been difficult to consistently observe a nucleocapsid structure in the arenaviruses by electron microscopy, however disruption of virus particles releases an RNP complex which has been examined (Pedersen and Konigshofer 1976; Buchmeier et al., 1978). The RNP of PIC is a linear array of globular nucleosomes each about 4–5 nm in diameter which are supercoiled into helical structures which form circles (Young and Howard, 1983, 1986). Several minor protein species have been observed in purified virions of various arenaviruses. These include a minor protein of 77 or 79 kDa associated with the RNPs of TAM and TAC (Gard et al., 1977; Saleh et al., 1979). A nonglycosylated protein of approximately 15 kDa is consistently observed in preparations of LCMV and may represent the gene product of the Z gene, a putative zinc finger protein found in LCMV (Salvato et al., 1989, Salvato and Shimomaye, 1989) and TAC (Iapalucci et al., 1988, 1989). Other nonglycosylated proteins have been described for LCMV (Bruns et al., 1983), JUN (Martinez Segovia and deMitri, 1977), LAS virus (Clegg and Lloyd,

1983) and PIC virus (Harnish et al., 1981). Some of these have been shown to be degradation products of NP while others are of unknown origin.

A 200-kDa L protein has been visualized for PIC (Harnish et al., 1981), LCMV (Bruns et al., 1983b; Buchmeier and Parekh, 1987), and other African arenaviruses (Gonzalez et al., 1984), and presumably is present in all the arenaviruses. This L protein is associated with the viral RNP complex, and has been shown by using antibodies generated to synthetic peptides to correspond to the product of the 6.6 kb L gene located on the large RNA (Singh et al., 1987; Salvato et al., 1989).

The arenaviruses contain a single glycoprotein gene (GP-C) which encodes a polypeptide of 75–76 kDa for LCMV (Buchmeier and Oldstone, 1979; Harnish et al., 1981; Saleh et al., 1979; Clegg and Lloyd, 1983, 1984; Lukasevich and Lemeshko, 1985). GP-C is cleaved post-translationally in LCMV, LAS, MOP, MOB and PIC viruses to yield two structural glycoproteins, GP-1 and GP-2 (G1 and G2) of approximately 44–55 and 35–41 kDa. In contrast, JUN, TAC and TAM viruses have only a single structural glycoprotein (Gard et al., 1977; Boersma et al., 1982). The basis for this difference is unclear, but the sequence of TAC virus GP-C (Franze-Fernandez et al., 1987) lacks the paired basic amino acid sequence, ARG-ARG or ARG-LYS, which provides the cleavage recognition sequence for LCMV, LAS and PIC viruses (Buchmeier et al., 1987).

3.2. Morphogenesis

Both the major glycoproteins, GP-1 and GP-2, of PIC (Vezza et al., 1977), LCMV (Buchmeier et al., 1978), and the single GP of TAC and TAM (Gard et al., 1977) are sensitive to proteolytic cleavage on the surface of virions. For viruses with two glycoproteins, they are present in the virion in equimolar amounts (Vezza et al., 1977), but their arrangement in the envelope is not known. It appears that for LCMV, GP-1 is more accessible on the surface of infected cells than GP-2, as the latter is difficult to detect either by surface iodination of infected cells or virions (Buchmeier and Oldstone, 1979; Bruns et al., 1983b) or by immunofluorescence using mAb specific for GP-2. Bruns and Lehmann-Grube (1983) have proposed that in the viral envelope, a single molecule of GP-2 is complexed with three molecules of GP-1, or that one GP-1 is linked to another larger glycoprotein (gp85). Such an arrangement seems unlikely since GP-1 and GP-2 are produced by GP-C cleavage and are represented in the virion in equimolar amounts. Furthermore, the origin of the gp85 glycoprotein described by these authors is unclear, and cannot be explained on the basis of the known genetic map of LCMV. Other studies have suggested that LCMV GP-1 and GP-2 are present in the virion in homo-oligomers with up to tetrameric complexes of GP-1 observable in SDS denatured, non-reduced virions (Wright et al., 1989a).

The GP-1 and GP-2 glycoproteins are produced by post-translational cleavage of the cell associated precursor GP-C. Cleavage is mediated by a cellular protease at a site defined by the sequence ARG-ARG at positions 262–263 of GP-C (Buchmeier et al., 1987). Full-length GP-Cs of LCMV (Buchmeier and Parekh, 1987) and TAC (Saleh et al., 1979) contain mannose-rich oligosaccharides, but the cleaved glycoproteins contain only complex oligosaccharides, indicating that they undergo further processing either prior to or just after cleavage. Recent studies with LCMV have shown that cleavage of GP-C occurs immediately after trimming of the oligosaccharide side chains, but before the molecule is transported to the plasma membrane. This data is consistent with cleavage occurring in the trans Golgi or immediate post-Golgi compartment (Wright and Buchmeier, unpublished data, 1989) (Wright et al., 1989b). A small fraction of GP-C becomes fully processed in the absence of cleavage, therefore transport is not dependent on cleavage. Others have reported full-length GP-C on the surface of radioiodinated LCMV infected cells (van der Zeijst et al., 1983a), and we observe on occasion full-length GP-C in purified virions (Parekh and Buchmeier, unpublished observations), hence cleavage may not be highly efficient. Experiments with PIC have shown that GP-C does not appear on the surface of cells infected with ts mutants in which cleavage fails to occur (Shivaprakash et al., 1988).

3.3. Protein functions

Although the arenaviruses are ambisense, an RNA-directed RNA polymerase is required for copying genomic RNA to functional messenger RNAs. Such an activity has been associated with RNP complexes, of PIC (Leung et al., 1979), and LCMV (Fuller-Pace and Southern, 1989). Although it has yet to be directly established, the L protein is likely to be RNA polymerase. L is associated with the RNP, its size is consistent with other RNA polymerases, and sequence comparisons of the predicted protein indicate segments conserved with other viral polymerases (Singh et al., 1987, Salvato et al., 1989).

A protein kinase activity has also been described associated with the RNP of LCMV (Howard and Buchmeier, 1983). This enzyme preferentially phosphorylates serine and threonine residues on NP in vitro. Attempts to find phosphorylated proteins in mature virions have been largely unsuccessful, but one report describes a soluble phosphorylated form of NP, termed p63E in LCMV (Bruns et al., 1986).

No hemagglutinating, neuraminidase or fusion activities have been associated with arenavirus glycoproteins. GP-1, the major surface glycoprotein, is responsible for binding to host cells. Monoclonal antibodies to two closely linked GP-1 epitopes, only one of which neutralized virus in vitro, inhibited virus binding to susceptible cells by up to 95% (Buchmeier, 1989, unpublished observations).

4. Mechanisms of immunity

4.1. Antibody-defined epitopes

4.1.1. Immune responses to virus infection

As for most viruses, the immune responses elicited depend on the route of infection, as well as on the age, genetic background and immune status of the host. Comprehensive studies have been reported for LCMV, and data exist for other arenaviruses as well. In the natural hosts, arenaviral infection occurs congenitally or at birth. Early attempts to find circulating antibody in mice chronically infected with LCMV were negative, and it was surmised that persistent infection was accompanied by immunologic tolerance to viral antigens (Buchmeier et al., 1980). However, it was subsequently demonstrated that hosts persistently infected with LCMV had both complexed (Oldstone and Dixon 1967, Buchmeier and Oldstone, 1979) and free antibody (Oldstone et al., 1980) to the virus. In LCMV carrier mice, anti-viral antibody was detected in the serum by indirect immunofluorescence and by immune precipitation (Buchmeier et al., 1980). Serum from *C. musculinus* persistently infected with JUN has high-titered levels of Nt activity that is ineffective in clearing virus (Weissenbacher et al., 1987). In contrast, *C. callosus* infected with MAC have no demonstrable Nt antibody (Johnson et al., 1973). These animals are not tolerant since antibody can be detected by indirect immunofluorescence (IF) (Johnson et al., 1973; Webb et al., 1973). LAS virus also may persist in the presence of antibody, but the majority of animals do not mount an antibody response (Walker et al., 1975).

Infection of adult animals with LCMV yields an acute disease which may be symptomatic or asymptomatic depending upon dose and route of infection and genetic background of the mouse. Infection by the peripheral route results in the development of antibody detectable by CF or ELISA by day 5 coincident with decline in virus titers (Buchmeier et al., 1980). Antibody titers remain high for several months (Buchmeier et al., 1980). Neutralizing antibody appears from 7 days to 3 weeks after infection, depending on the virus dose and the mouse strain (Kimmig and Lehmann-Grube, 1979). Both CF and Nt antibody persist for at least a year, and probably longer (Kimmig and Lehmann-Grube, 1979).

Other arenaviruses have not been studied in their natural hosts to the same extent as LCMV. Machupo induces antibody detectable by CF, Nt and IF in a proportion of infected adult *Calomys,* and the presence of Nt antibody was found to correlate with virus elimination (Webb et al., 1973; Johnson et al., 1973). Animals not producing Nt antibody became persistently infected (Johnson et al., 1973). Lassa virus also induced persistent infection in adult *Mastomys,* and CF but not Nt antibody was detected (Walker et al., 1975).

The arenaviruses differ in their ability to elicit, and their sensitivity to Nt antibody. LCMV induces Nt antibody in adult mice, but at low frequencies (Buchmeier and Parekh, 1987). Tacaribe, JUN and MAC all induce Nt ab in rodents and humans, but Nt antibody develops rather late in infection (Casals, 1984: Peters, 1984). Nt antibody to LAS virus appears so late in convalescence that it can have little effect in virus clearance (Casals, 1984) while PIC fails to induce a Nt response in any host (Howard, 1987).

4.1.2. Identification and expression of B cell epitopes

The availability of monoclonal antibodies to arenaviruses has made detailed mapping of B cell epitopes possible (Fig. 12.3) (Howard et al., 1985: Parekh and Buchmeier, 1986). Using competitive binding assays with a panel of mAbs directed against LCMV, it was demonstrated that Nt mAb recognized a single antigenic site on GP-1 (Parekh and Buchmeier, 1986). One isolate of LCMV Armstrong, ARM-4, expressed a second Nt site (Parekh and Buchmeier, 1986: Wright et al., 1989a). Competitive binding and neutralization kinetic studies of TAC indicated the existence of two neutralization sites on the glycoprotein of this virus (Howard et al., 1985).

In addition to the major neutralization site on GP-1, two topographically linked non-neutralization sites have been identified. Three closely linked epitopes have been described on GP-2 of LCMV. Two of these have been mapped to a single stretch of amino acid sequence, GP-C 370–382. These epitopes can be distinguished on the basis of critical internal contact residues and virus specificity (Weber and Buchmeier, 1988).

Extensive epitope mapping of NP has not been carried out, but virus cross-reactivity patterns observed with NP specific mAb suggest the existence of at least three

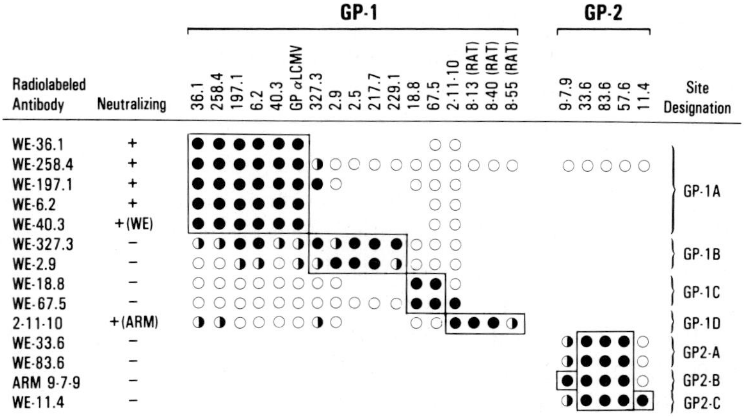

Fig. 12.3. Summary of mAb epitope mapping data for LCMV. Inhibition of binding of radiolabeled antibody by unlabeled competing antibody is indicated as follows, Filled circles, >80% inhibition by unlabeled antibody; half-filled circles, 40–80% inhibition; open circles <40% inhibition. Virus neutralizing antibodies are indicated. Reproduced from Parekh and Buchmeier, 1986, with permission.

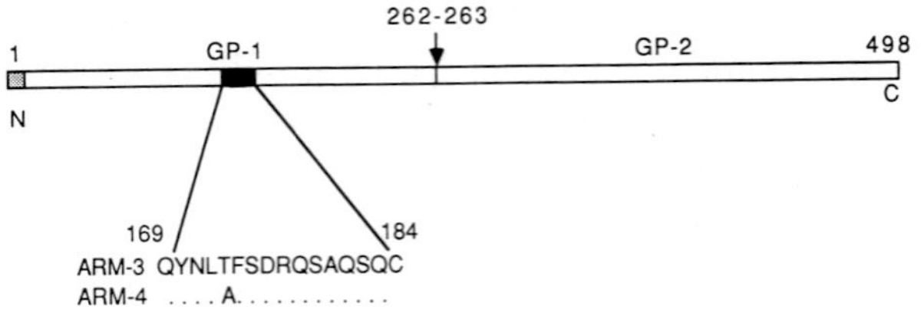

Fig. 12.4. Location of two antigenic sites on LCMV GP-C. A polymorphic neutralizing determinant has been shown to be expressed on mutant strain ARM-4 but is lost on ARM-3 correlating with an Ala → Thr substitution at position 173 of GP-C of these viruses (Wright, Salvato and Buchmeier, 1989).

antigenic sites on NP of PIC, one virus-specific, one group-specific and one that is shared with LCMV and LAS virus (Buchmeier et al., 1981). Complement-fixing antibody is directed to NP (Geschwender, 1976; Buchmeier et al., 1977), but it is unclear whether all sites on NP induce CF antibody.

Efforts to map Nt epitopes on GP-1 have been hindered by their conformational nature. Three of the four sites described on GP-1, including the neutralization site, are conformation-dependent, that is, destroyed by denaturation. Consequently epitope mapping using synthetic peptides has not been feasible. Despite this limitation, one epitope, GP-1D, recognized by mAb 2-11.10, has been partially localized. This epitope is disulfide-dependent, hence was destroyed by reducing agents. Using mutant strains of LCMV-Arm which differed in their binding of 2-11.10 we found that substitution of threonine for alanine or lysine at position 173 of GP-C abrogated binding activity (Fig. 12.4). Insertion of threonine completed a concensus *N*-linked glycosylation site ASN X THR at positions 171–173 and resulted in addition of an extra oligosaccharide side chain on GP-1 (Wright et al., 1989a). Thus, it is likely that antibody binding was blocked by the presence of this sugar chain.

4.1.3. Identification of protective epitopes

A role for antibody in prevention or clearance of arenavirus infections may be underestimated. Generally, it is thought that neutralizing antibody does not play a significant role in clearing primary virus infections, but under certain conditions such a role can be demonstrated in arenavirus infections. In JUN virus infections the treatment of choice for acutely ill patients has been passive transfer of immune plasma (Maitzegui et al., 1979). This procedure results in rapid reduction of virus burden and reduction in morbidity and mortality. Animal studies also suggest that passive administration of serum containing Nt antibody can protect animals against primary infection with JUN. Studies in animals also demonstrate a similar phenomenon with LAS virus (Fisher-Hoch and McCormick, 1987; Peters et al., 1987), but in human patients, transfer of Nt antibody has little therapeutic effect (McCormick et al., 1987).

Studies with LCMV have defined B cell epitopes that attenuate T-cell-mediated disease in adult LCMV infected mice (Wright and Buchmeier, 1989). Passively transferred mAb against both Nt and non-Nt epitopes on GP-1 protected mice from fatal immune-mediated disease after intracranial infection and did so by limiting growth of virus in the brain. mAbs directed against either GP-2 or NP were not protective. These results indicate that although T cells may be responsible for clearing virus during primary infection, pre-existence of antibody or rapid induction of appropriate B cell responses protect against secondary infections. These data indicate that the protective effect of antibody is not restricted to those antibodies that neutralize virus in vitro.

4.2. Cellular responses

4.2.1. Immune responses to virus infection

T-cell responses to viral products are a central feature of arenavirus pathobiology. In LCMV infection, for example, cytotoxic T-cells are clearly important in virus clearance, and when their generation is blocked, persistent infection ensues. On the other hand, these same T-cells, when focused in an inflammatory infiltrate in the brain, cause lethal choriomeningitis. Clearly it is desirable to harness the beneficial functions of CTL while minimizing tissue destruction.

T-cell responses to the arenaviruses have usually been most extensively studied in the mouse where a natural host virus relationship exists with LCMV. Adult mice infected with LCMV develop both delayed-type hypersensitivity (DTH), as demonstrated by footpad swelling, (Tosolini and Mims, 1971) and cytotoxic T cell responses (Zinkernagel and Doherty, 1974). Cytotoxic T cells are pivotal in mediating virus clearance after a peripheral infection of adult mice with LCMV, and absence of this response leads to lifelong virus persistence (Marker and Volkert, 1973; Byrne and Oldstone, 1984; Moskophidis et al., 1987). Induction of, and target cell recognition by, CTL map to the S RNA segment of LCMV, the segment encoding NP and GP-C (Riviere et al., 1986). Studies examining the cross-reactivity of CTLs elicited by various strains of LCMV indicate that all five strains of the virus share common CTL epitopes, but some also have unique determinants (Ahmed et al., 1984a). The pattern of cross-reactivity also varies with the MHC background of the mouse, indicating determinants are recognized selectively in the context of different MHC class I molecules (Ahmed et al., 1984). The characteristic pattern of cross-reactivity observed in H-2b and H-2d mice is also observed when examining cross-reactivity of cloned CTLs from each strain of mouse (Byrne et al., 1984). These data suggest that there are multiple CTL determinants, and that recognition of these varies with the genetics of the host.

These observations were confirmed using LCMV proteins expressed in vaccinia vectors. H-2^b mice recognize both NP and GP-C, but the majority of CTL clones from these mice were directed against the latter. On the other hand, H-2^d and H-2^q mice appear to generate CTL activity mainly to NP, with a small proportion of CTL clones recognizing GP-C (Oldstone and Whitton, 1989; Whitton et al., 1988a). Analysis using a series of vaccinia-GP recombinants truncated at the C-terminus has mapped more precisely the epitopes recognized by H-2^b mice (Whitton et al., 1988b). In C57B/6 (H-2^b) mice epitopes were recognized on NP, on GP-1 between residues 1 and 218, and on GP-2 between residues 272–293 (Whitton et al., 1988b). Using a series of synthetic peptides spanning this region as targets, the GP-2 epitope has been further mapped to amino acids 278–286. Responses to this determinant are restricted by D^b. The few available H-2^b CTL clones directed against NP recognize a determinant at the carboxy end of the molecule, between residues 301 and 558 (Oldstone and Whitton, 1989). In H-2^d mice, the major epitope lies on NP and has been mapped to residues 121–127 (Oldstone and Whitton, 1989). Thus H-2^d mice recognize structures different from those recognized by H-2^b mice. The NP epitope is specifically restricted by L^d.

Pichinde virus does not cause disease in mice, but does elicit a DTH reaction after footpad inoculation (Wright, 1989, unpublished data) and a strong CTL response after primary and secondary infection by other routes (Walker et al., 1984). The cytotoxic response is thought to be directed to GP-C, as infection of syngeneic cells with a vaccinia recombinant expressing PIC NP does not render the cells susceptible to lysis (Ozols et al., 1988). In certain strains of hamster, PIC induces DTH after inoculation in the footpad; in other strains of hamster such a response is missing and thought to be actively suppressed (Chan et al., 1983). Unlike the other viruses examined, JUN induces a good CTL response in mice, but does not elicit footpad swelling (Barrios et al., 1982). Cytotoxic cell responses to LAS virus, LCMV and MOP have been studied in guinea pigs (Jahrling and Peters, 1986; Peters et al., 1987). Such a response could be measured in splenocytes from guinea pigs inoculated 15 days previously with any of the 3 viruses.

4.2.2. Protective epitopes and epitopes inducing T cell-mediated disease

Primary infection with LAS virus, like LCMV, is thought to be cleared mainly by T-cell-mediated responses. Guinea pigs can be protected from challenge with LAS by transfer of immune splenocytes if the splenocytes are harvested early in infection (Peters et al., 1987), but not late. Recombinant vaccinia viruses expressing LAS GP-C or NP have been constructed and used in protection experiments. Immunity to both proteins protects guinea pigs from death although not from infection (Auperin et al., 1988; Auperin et al., 1987; Clegg and Lloyd, 1987). When non-human primates were vaccinated with the same constructs, only the virus expressing GP-C was protective. As observed with guinea pigs, monkeys became infected, but infection caused few if any symptoms (Auperin et al., 1988). Guinea pigs were protected by immunization with LAS GP-C but did not produce anti-LAS antibody (Auperin et al., 1987). These results suggest that protective T cell epitopes for guinea pigs lie on both GP-C and NP, whereas those for monkeys reside on GP-C.

Pichinde may also contain a protective epitope on NP. A vaccinia construct containing NP from PIC was able to modulate disease and delay death in MHA hamsters after challenge with PIC (Ozols et al., 1988). These results parallel earlier results showing that the transfer of IL-2 into susceptible hamsters prior to infection slowed death (Wright et al., 1987) and suggest that immunization with NP may activate helper T-cells that synthesize IL-2.

The observation that a vaccinia-LCMV-NP recombinant virus elicited a CTL response when inoculated into mice suggested that such a construct might confer protective immunity. Both H-2^d and H-2^b mice were protected from lethal intracerebral challenge with LCMV after vaccination with a construct containing full-length NP (Klavinskis et al, 1989a, 1989b). Immunization with a construct expressing truncated NP (residues 1–201) only protected H-2^d mice, demonstrating that the presence of a CTL epitope for H-2^d mice on the vaccine correlated with protection (Klavinskis et al., 1989a, b). No anti-viral antibody was detected in these mice.

T-cell epitopes of LCMV inducing disease appear to be identical to those which are protective. Cytotoxic clones specific for either GP (restricted by H-$2D^b$) or NP (restricted by H-$2L^d$) can induce central nervous system disease in immunosuppressed syngeneic mice infected 5 days earlier with LCMV (Klavinskis et al., 1989c). As few as $10^{2.3}$ cloned CTL induced disease when injected intracerebrally into infected mice. When cells were transferred to immunosuppressed mice at the time of infection, virus titers remained low in the brain, and the animals survived (Klavinskis et al., 1989c).

4.2.3. Modulation of T cell responses by virus

Infection of neonatal mice with LCMV results in lifelong persistent infection without demonstrable CTL response. T-cells in neonatal mice are susceptible to infection with the virus, whereas T-cells in adult mice are relatively resistant (Lehmann-Grube et al., 1983; Doyle and Oldstone, 1978). Helper T-cells are preferentially infected in young mice (Ahmed et al., 1987; Tishon et al., 1988; Oldstone et al., 1988). It has been suggested that immature T-cells required to provide help for generation of LCMV-specific CTL are infected and destroyed or functionally inactivated early in life (Lehmann-Grube et al., 1983; Ahmed et al., 1987; Oldstone et al., 1988). During the course of infection, variant viruses with the ability to induce persistent infection in immunocompetent adult mice are selected in lymphoid tissues. As in neonates, persistence is associated with failure to mount a CTL response (Ahmed et al., 1984b). Although induction of and recognition by virus-specific CTL maps to the S segment of the genome, and hence relies on expression of NP and/or GP-C (Riviere et al., 1986), failure to mount a CTL responses against variant viruses maps to the L genomic segment (Salvato et al., 1988). Thus differences between the variant and wild-type viruses may not be due to alterations of specific T cell epitopes, but rather to alterations in polymerase or other heretofore uncharacterized genes en-

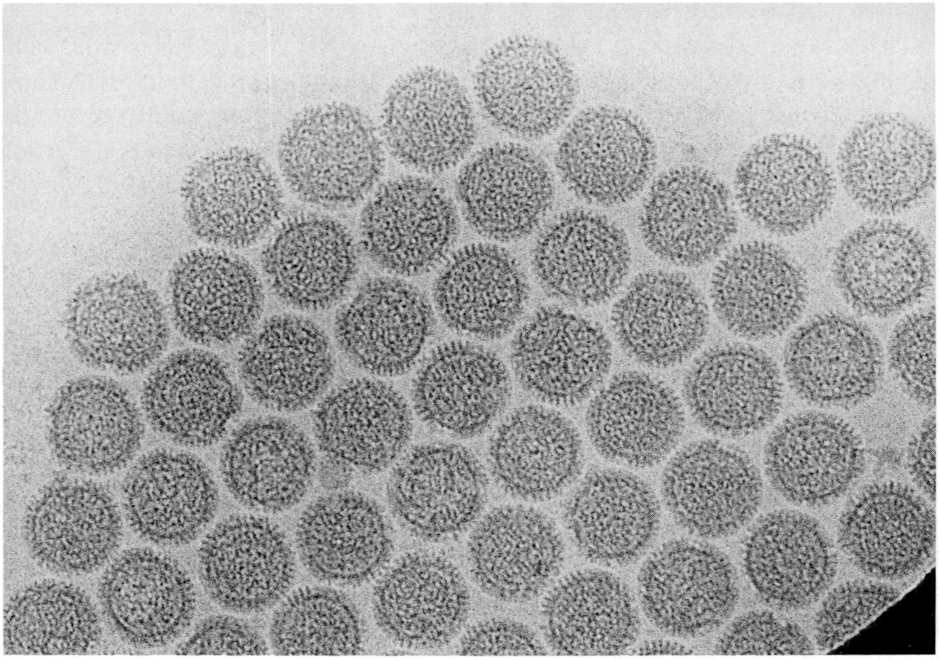

Fig. 12.5. Cryo-electron microscopy of Uukuniemi virus particles displaying surface glycoprotein projections. Particles are approximately 100 nm in diameter. Photograph courtesy M. Cyrklaff and C-H. von Bonsdorff.

coded on L. T-cells of the helper phenotype are preferentially infected in adult mice inoculated with an immunosuppressive variant of LCMV termed Clone 13, but not with wild-type, and total T cells numbers are decreased, so Clone 13 may express a different tropism than wild-type virus in its ability to infect mature T cells (Oldstone et al., 1988). Depletion of T-cells specific for LCMV is not lifelong. Carrier mice that have been cured of their infection by transfer of immune splenocytes are protected from a second challenge with the virus, and can generate LCMV-specific CTL (Jamieson and Ahmed, 1988).

Infection of macrophages with PIC is known to interfere with proliferative responses to macrophage growth factor (Friedlander et al., 1984), but it is not known whether macrophage antigen processing functions necessary for generation of immune responses are altered in any way. There have been suggestions that LCMV may depress some macrophage functions, such as phagocytosis (Gledhill et al., 1965) and lysozymal functions, but other studies indicate that infection with LCMV has no effect on macrophages (Mims and Wainwright, 1968; Oldstone et al., 1973, Schwartz et al., 1978).

4.2.4. Vaccines

At the present time, no safe vaccines exist for any of the pathogenic arenaviruses. Attempts have been made to develop live attenuated vaccines for MAC (Peters et al., 1987) and JUN (Guerrero et al., 1969). The latter was tested in human volunteers and appeared to be efficient in inducing neutralizing antibody (Ruggiero et al., 1974, 1981), but its use was discontinued because the virus had been passaged through heterodiploid cells and mouse brain.

A second generation live attenuated JUN vaccine designated candidate 1 is currently in field trials in Argentina. This strain, which was developed from the JUN strain XJ-44 by passage into FRhL-1 cells, is stably attenuated and tested without incident in over 100 human volunteers before initiating double blind placebo field trials in approximately 6000 individuals at risk in the endemic area of Argentina.

Recombinant approaches have been applied to efforts to vaccinate against Lassa. Vaccinia recombinants containing LAS proteins have been tested in guinea pigs and primates for efficacy against LAS challenge (Auperin et al., 1987; Clegg and Lloyd, 1987; Auperin et al., 1988). Only constructs containing GP-C were protective in monkeys, and although immunization reduced mortality, infection and viremia were not prevented (Auperin et al., 1988).

5. Bunyaviridae

Bunyaviridae is the largest known family of RNA animal viruses, with more than 250 serologically distinct members (Bishop et al., 1980; Karabatsos, 1978). The family has been divided into five genera based upon molecular and antigenic properties. Prototype viruses for the Bunyavirus, Hantavirus, Nairovirus, Phlebovirus and Uukuvirus genera, respectively are: Bunyamwera, Hantaan, Crimean-Congo hemorrhagic fever, sandfly fever Naples and Uukuniemi viruses (Bishop, 1985; Bishop et al., 1980; Schmaljohn and Dalrymple, 1983). Most viruses in the Bunyaviridae family are arthropod-borne (primarily by mosquitoes or ticks), with the exception of hantaviruses, which appear instead to be transmitted via aerosolized rodent excreta (Bishop et al., 1980; Lee et al., 1981) or by biting among rodents (Glass et al., 1988). Serious and fatal infections have been associated with certain bunyaviruses (e.g., LaCrosse encephalitis), hantaviruses (e.g., Korean hemorrhagic fever), nairoviruses (e.g., Crimean-Congo hemorrhagic fever), and phleboviruses (e.g., Rift Valley fever), but many are not known to infect humans. Although all viruses in the family share morphological and morphogenic properties, and are distinguished from other RNA viruses by a three-segmented, single-stranded genome, numerous differences in replication strategies and antigenic characteristics have been described for viruses in each genus.

6. Properties of Bunyaviridae

6.1. Morphology, morphogenesis and virion composition

Unique morphological features characteristic of viruses in each genus have been described (Martin et al., 1985), however, virions generally are 80–120 nm in diameter and appear as spherical particles. Penton-hexon clusters arranged in an icosahedral lattice were found on the surface of Uukuniemi virus (von Bonsdorff and Pettersson, 1975). The outer envelope of virion particles consists of a bilaminar membrane with integral surface projections of approximately 5–10 nm (Fig. 12.5). The internal composition of virions has been inferred from biochemical studies and is schematically illustrated in Fig. 12.6.

The three genome segments, designated as large (L), medium (M) and small (S), individually complex with nucleocapsid proteins to yield three separate ribonucleoprotein structures. Although at least one each of the L, M, and S ribonucleocapsids must be present in infectious virions, the ratio of those present varies (Bishop and Shope, 1979). Two virus-specified glycoproteins constitute the spike-like surface projections, one or both of which have transmembrane regions. No matrix (M) protein is found in virion particles, suggesting a direct interaction of the nucleocapsid and envelope proteins. Relative RNA and structural protein sizes of a representative member of each genus of the Bunyaviridae are illustrated in Fig. 12.7.

Morphogenesis occurs primarily in association with smooth membranes, particularly in the Golgi, and involves budding of morphologically complete virions into cytoplasmic vacuoles, transport of the vacuoles to the cell surface, and release of mature virions by exocytosis (Smith and Pifat, 1982).

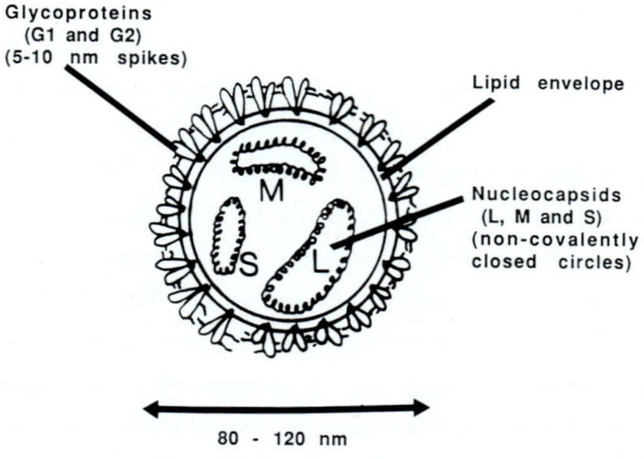

Fig. 12.6. Schematic representation of a Bunyaviridae virion. Particles contain three separate, non-covalently closed nucleocapsid structures consisting of the large (L), medium (M) or small (S) RNA segment complexed with nucleocapsid protein (N). A host-derived, bilaminar lipid envelope with integral virus-specified glycoprotein projections surrounds the nucleocapsids.

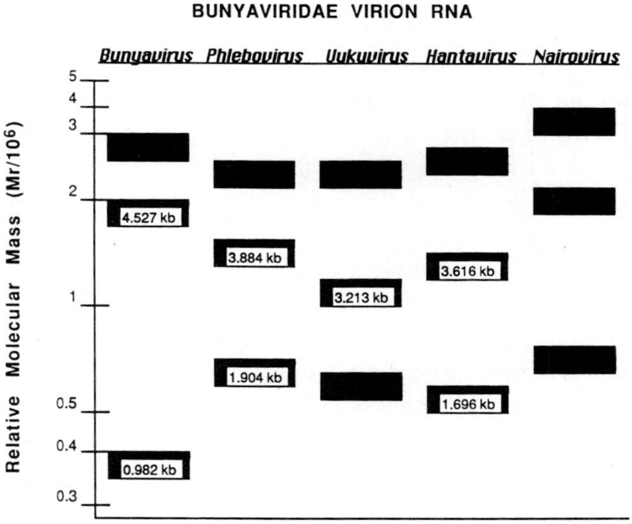

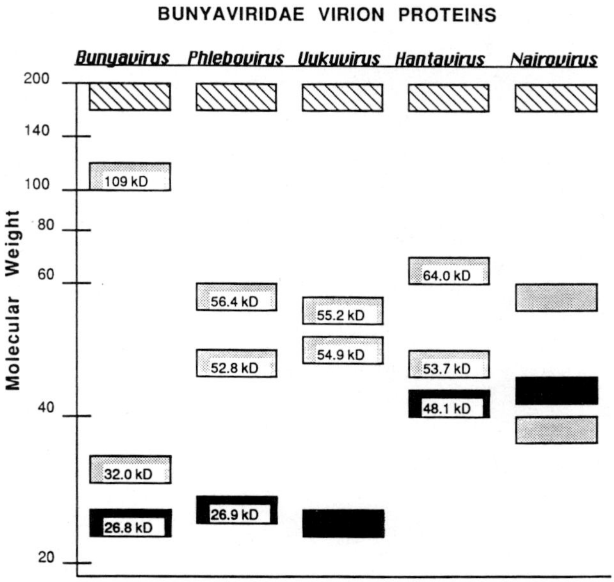

Fig. 12.7. Electrophoretic migration patterns of the three RNA genome segments and the structural proteins of a virus representative of each genus of the Bunyaviridae. Where sequence analysis is available, the number of nucleotides and the predicted molecular weights of proteins (non-glycosylated) are listed. Data were obtained from the following references: (1) Bunyavirus (Snowshoe hare virus) Akashi and Bishop, 1983; Fazakerley et al., 1988; (2) Phlebovirus (Rift valley fever virus) Collett et al., 1985; (Punta Toro virus) Emery and Bishop, 1987); (3) Uukuvirus (Uukuniemi virus) Ronnholm and Petterson, 1987; (4) Hantavirus (Hantaan) Schmaljohn et al., 1986a, 1987b); Nairovirus (Qalyub) Clerx and Bishop, 1981.

6.2. Coding assignments of viral genes

All viruses in the Bunyaviridae examined to date encode their nucleocapsid protein (N) in the S genome segment and their envelope proteins (G1 and G2) in the M genome segment. The coding strategies used to generate these proteins, however, differ dramatically among viruses in each genus (Fig. 12.8).

The simplest of the strategies described thus far is that of the hantaviruses for which known coding regions are strictly negative-sense. A single continuous open reading frame (ORF) located in the viral-complementary-sense S or M RNA, is used to encode the nucleocapsid protein (N) and the envelope proteins (G1 and G2), respectively (Schmaljohn et al., 1986a, 1987b). These structural proteins are apparently the only gene products of the hantavirus S and M segments.

Like hantaviruses, bunyaviruses encode their envelope glycoproteins in a continuous open reading frame in the viral complementary-sense RNA; however, at least

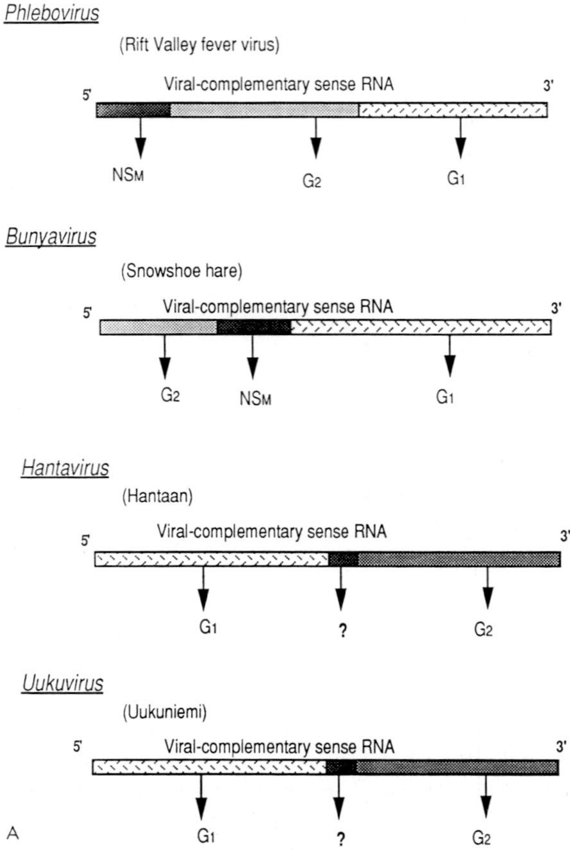

Fig. 12.8. Known coding assignments of the M and S genome segments of representative viruses in the Bunyaviridae. (A) The envelope glycoproteins (G1 and G2) of viruses in each genus examined to date

one non-structural protein (NS_M) is also encoded, and has been localized to the gene region between G1 and G2 coding sequences (Fazakerley et al., 1988). The bunyavirus S segment codes for N and a non-structural protein (NSS) using the same sequences but in two different overlapping reading frames of the viral-complementary sense RNA. (Akashi and Bishop, 1983; Akashi et al., 1984; Bishop et al., 1982; Cabradilla, et al., 1983, Gerbaud et al., 1987).

The continuous ORF of the viral-complementary-sense M segment of phleboviruses contains coding sequences for NS_M which precede those of G1 and G2. Coding sequences for the phlebovirus N protein are found in the viral-complementary sense RNA of the S segment and sequences for NS_s in a non-overlapping region of the viral-sense RNA. This strategy, which has thus far been reported for RNA viruses only in the S segments of phleboviruses, and viruses in the Arenaviridae family, has been termed 'ambisense' (Bishop, 1986). The S segment of uukuviruses also encodes N and NS_s, probably with an ambisense strategy, however, uukuvirus M segments,

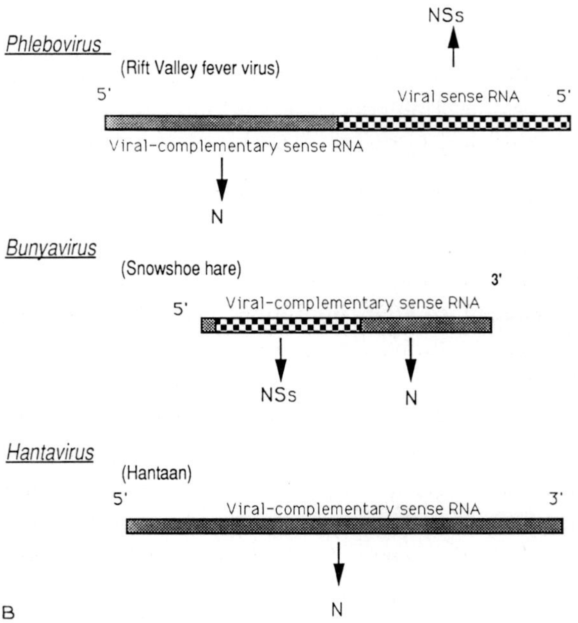

B

are encoded in a single continuous open reading frame in the viral-complementary sense RNA. Nonstructural protein coding regions (NS_M) have also been identified within the open reading frame for viruses in the phlebovirus and bunyavirus genera. The carboxy-terminal sequences of the G1 proteins of viruses in the hantavirus and Uukuvirus genera have not been defined, thus small amounts of NS_M coding information may exist between sequences encoding G1 and G2 as indicated by ? (B) Viruses in the phlebovirus, bunyavirus and hantavirus genera each encode their nucleocapsid protein (N) in the viral complementary senser RNA. Additionally, phleboviruses encode a nonstructural protein (NS_S) with non-overlapping sequences of the viral sense RNA. Bunyaviruses encode NS_S in an overlapping reading frame of the viral-complementary sense RNA. NS_S protein(s) have not been reported for hantaviruses.

differ from those of phleboviruses in that G1 and G2 are the only gene products (Ronnholm and Pettersson, 1987). The function(s) of neither NS_S nor NS_M proteins have yet been determined for any virus in the family.

Little information is available for the L genome segments of viruses in the Bunyaviridae. The L segments have long been presumed to encode the virion-associated viral transcriptase (L protein) which is required to copy the negative-sense RNAs to messenger-sense RNAs. Direct proof of this was obtained by preparing reassortant bunyaviruses with mixed infections of Tahyna and LaCrosse. The L genome segments of these two viruses could be differentiated by nucleic acid analysis and the L proteins by electrophoretic migration. By examination of the nucleic acids and proteins of reassortant viruses, it was clearly established that the L protein sorts with the L genome segment (Endres et al., 1989).

Coding strategy information has not yet been reported for any of the genome segments of viruses in the Nairovirus genus.

6.3. Antigenic variation

Serological relationships among viruses in different genera of the Bunyaviridae have not been reported, and with the exception of phleboviruses and uukuviruses which display a low degree of amino acid homology in their envelope proteins (Ronnholm and Pettersson, 1987), no inter-generic relationships can be discerned by comparison of available nucleic acid or predicted amino acid sequences. These data suggest different origins or an early evolutionary divergence of the members of this family.

The viruses within each genus have been subdivided into serogroups based upon cross-reactivities in tests such as hemagglutination inhibition, neutralization and complement fixation. The largest genus, the Bunyavirus genus, is the best characterized serologically, and includes 16 serogroups with more than 100 distinct viruses. Even within a particular serogroup viruses display antigenic differences detectable by monoclonal antibody (mAb) reactivities. An example of this diversity can be observed with viruses in the Bunyamwera serogroup, in that people infected with one virus in this group are not protected from infection with another virus in the same group (Gonzalez and Georges, 1988). At the genetic level, even individual isolates of the same virus are distinguishable. For example, numerous isolates of LaCrosse virus, obtained in the same or different geographic locales, either at the same or different times, all were found to be genetically distinct by oligonucleotide fingerprint analyses, suggesting that a high and frequent rate of mutation occurs (Beaty, et al., 1988).

To examine antigenic variation and to localize point mutations which may be important in immunity, Battles and Dalrymple (1988) sequenced gene regions of 22 isolates of RVFV which were known to encode epitopes involved in neutralization and protection (see below). The isolates had been collected over a 34 year time period from six African countries. They were able to identify specific base variations which resulted in amino acid coding differences and rendered isolates resistant to neutralization by mAbs defining the epitopes. Amino acid differences which correlated with loss or reduction of reactivity with the mAb in question involved changes of amino acid charge. In some cases, there was reduced mAb reactivity despite conversation of the peptide against which the mAb was known to react; presumably, this is explicable on the basis of amino acid changes in other juxtaposed regions of the viral glycoproteins. (reference crystallography of Ag-Ab interaction with influenza; Colman et al., 1987)

In addition to the resultant 'antigenic drift' such point mutations would incur, the segmented genomes of viruses in the family also allow for more dramatic 'antigenic shifts' attributable to segment reassortment. Reassortment among viruses in the California serogroup of bunyaviruses has been demonstrated in cell culture and in mosquitoes both in the laboratory and in nature. Reassortment was also obtained by interrupted feeding of mosquitoes on animals viremic with different viruses (Beaty et al., 1985). The ability to reassort, however, appears to be limited only to very closely related viruses. It has not been possible to obtain reassortment between viruses in different serogroups of the Bunyavirus genus, nor between viruses in different genera.

Host influences on viruses have also been suggested to play a role in the antigenic properties of bunyaviruses. Repeated passage of Bunyamwera and snowshoe hare viruses in cultured mosquito cells resulted in the loss of reactivity of certain monoclonal antibodies with the parent, mammalian cell culture passaged virus (James and Millican, 1986). Bunyaviruses, phlebovirus, uukuviruses and nairoviruses all replicate alternately in invertebrate and vertebrate hosts, and are generally noncytolytic in arthropod host cells, but cytolytic in vertebrate cells. Consequently, host factors may greatly influence the antigenic evolution of these viruses. In contrast, hantaviruses replicate only in vertebrate (usually rodent) hosts, often persistently, with little or no detectable cytopathology. Although numerous hantavirus strains can be differentiated by serological means, the extent and frequency of antigenic drift in hantaviruses has not been reported. Evidence exists that a particular strain of virus may be antigenically stable as suggested by studies in which antigenic differences between Hantaan virus (a rodent isolate) and two isolates from Korean hemorrhagic fever patients could not be detected with monoclonal or polyclonal antibodies. Furthermore, although the isolates came from different vertebrate hosts, and were obtained over an eight year time-frame, the predicted amino acid sequences of the envelope proteins of these viruses differed by less than 3% (Schmaljohn et al., 1988).

7. Antibody-mediated immunity

In general, neither structural nor nonstructural viral proteins are translocated to infected-cell surfaces in quantities sufficient to be demonstrated with antibodies. Thus, the principal targets of protective antibodies appear to be virion surface glycoproteins, with little or no contribution from cytolytic mechanisms requiring complement or accessory cells. Two exceptions were noted. The first was observed by immuno-electron microscopy in cells infected with a hantavirus isolated from rats (R22 virus). The infected cells displayed a thick layer of antigen, which was thought to be virus-specific, located on top of, but distinct from the plasma membrane. It is unclear if or how this material is involved in viral morphogenesis as mature virions could be observed budding intracellularly into Golgi vesicles as is usual for viruses in the Bunyaviridae (Hung et al., 1985). It was not determined if the antigen layer had any significance for immunity. The second potentially important exception was noted with the phlebovirus, Rift Valley fever virus (RVFV), which expressed envelope glycoproteins and could be seen budding not only in Golgi but at plasma membranes in primary hepatocytes; in other cell types RVFV exhibited the more typical absence of cell-surface proteins (Anderson and Smith, 1987).

7.1. Neutralization

Neutralization of virion particles, i.e., reaction of antibody and virions in a manner which significantly reduces the number of observable infectious particles, presumably requires accessibility of virion proteins to the neutralizing antibody. The envelope glycoproteins of viruses in the Bunyaviridae are at least partially exposed on the virion surface as indicated by reactivity of antibodies with both G1 and G2 in studies such as immunoelectron microscopy, radioimmune precipitation, or enzyme treatment of intact virions. In contrast, antibodies to N do not react with virions. No information is available concerning the precise location of the L protein, however, it is assumed not to be exposed on the virion surface.

Because of obvious structural constraints, not all portions of G1 and G2 can be accessible to antibodies that might mediate neutralization. The G1 protein of LAC was found to be more susceptible to digestion by proteolytic enzymes than was G2, suggesting that more of G2 may be sequestered within the viral envelope. After enzyme treatment, the modified LAC virions demonstrated reduced neutralization by polyclonal sera, indicating that G1 plays an important role in neutralization of virion particles (Kingsford et al., 1983).

The involvement of G1 and G2 epitopes in virion neutralization has been more accurately defined by use of mAbs, which have been described for representative viruses in each genus except the Uukuvirus genus. Not surprisingly, mAbs reactive with nucleocapsid proteins have been found devoid of neutralizing (NT), hemagglutination-inhibiting (HI), or protective activities, though they provide useful serological reagents by virtue of the fact that nucleocapsid proteins may carry either virus-specific or conserved, genus-specific epitopes. In contrast, some mAbs reactive with G1 and G2 will demonstrate NT and/or HI activities and may protect animals from infection or disease. Where examined, both G1 and G2 are antigenically complex, each possessing multiple epitopes that can be discriminated by mAb reaction patterns and mAb competition grouping. With phleboviruses (Keegan and Collett, 1986; Pifat et al. 1988,), hantaviruses (Dantas et al., 1986; Arikawa et al., 1989), and nairoviruses (J. Smith, personal communication), both G1 and G2 possess one or more epitopes defined by neutralizing or hemagglutination-inhibiting mAbs. With bunyaviruses, NT and HI activities have been attributed only to G1-specific mAbs, but G2-specific mAbs have not been obtained and the possibility that G2 bears immunologically relevant epitopes has not been formally excluded (Gonzalez-Scarano et al., 1982).

7.2. Humoral immunity

Humoral immunity is experimentally defined by the ability of transferred antibodies (as serum or monoclonal antibodies) to confer specific antiviral resistance to non-immune animals. It is not always synonymous with viral neutralization as it is commonly viewed, in that the protective mechanism(s) may be inadequately or even erroneously reflected by in vitro neutralization assays. This discrepancy was clearly illustrated by analyses of antibody-mediated protection with mAbs. For example, with the phlebovirus, Punta Toro virus (PTV), it was demonstrated that a mAb with very low neutralization activity (1:10) in vitro, nevertheless protected all animals from a lethal PTV challenge (Pifat et al., 1988). The ability of antibodies having poor neutralizing antibody in vitro to confer protection in vivo was previously noted with alphaviruses as well (Schmaljohn et al., 1982); however, the protective mechanism for viruses in the Bunyaviridae seems unlikely to correspond to the antibody-directed cytolysis observed with alphaviruses, but more likely involves virion opsonization (and clearance by cells of the reticuloendothelial system) or complement-mediated virolysis.

More confounding was the observation that certain mAbs or polyclonal monospecific antibodies exhibited good neutralizing activity in vitro but were ineffectual in confering protection in vivo. These were G2-specific in the case of PTV and specific for the homologous protein, G1, in RVFV (Smith et al., 1987; Pifat et al., 1988; Dalrymple et al., 1989). These observations could not be easily explained on the basis of immunoglobulin isotype, avidity, or other trivial possibilities. The same principle; i.e., that G2 of RVFV confers a greater level of antibody-mediated immunity than G1 despite equivalent or greater levels of neutralizing antibodies induced by the latter, was upheld by data involving mice immunized with vaccinia recombinants that expressed only either G2 or G1 (Dalrymple et al., 1989). These results demonstrate an inadequacy of conventional serology in predicting antibody activities in vivo and indicate the need for caution in extrapolation of in vitro principles to immunity.

7.3. Molecular definition of B cell epitopes

Competitive inhibition with panels of mAbs have provided rough estimates of the number and complexity of antigenic sites important in neutralization and protection of viruses in the Bunyaviridae. This methodology, however, does not permit localization of epitopes within the gene. Molecular techniques were employed with the phlebovirus RVFV which allowed determination of sequences encoding epitopes recognized by three neutralizing (and protective) and one non-neutralizing G2-specific mAbs (Keegan and Collett, 1986). Progressively smaller cDNAs, representing specific regions of the RVFV G2 gene, were expressed as β-galactosidase fusion proteins in *E. coli* and the expressed proteins were tested against individual mAbs. With this method it was possible to limit the three epitopes reactive with neutralizing mAbs to 11, 20, or 34 amino acids, and the epitope reactive with a non-neutralizing mAb to 14 amino acids within the G2 gene. The 20 and 34 amino acid epitopes overlapped, but the 20 amino acid epitope appeared to be structurally constrained (as indicated by the inability of the mAb to recognize denatured antigen) while the 34-amino-acid epitope was not (as indicated by mAb reactivity with both native and denatured protein). Synthetic peptides representing 9–14 amino acids within neutralizing sites were found to elicit antibody responses in animals, and in one case the resultant antibody displayed high titered neutralizing activity (Smith et al., 1987).

8. T cell responses

The roles that T cells play in immunity to viruses in the Bunyaviridae remain relatively poorly defined, perhaps because antibodies alone have proven sufficient to confer resistance in all cases examined. However, the prominent role of antibody-mediated resistance does not preclude a critical role for T cells in recovery from primary infection, in cross-reactive immunity, and possibly in immunopathology. Nor should the influence of T cells on quantitative and qualitative aspects of antibody responses be underestimated.

Through their work with hantaviruses, Asada et al.(1987, 1989) have provided the clearest evidence for the potential importance of T cells in resistance to a member of the Bunyaviridae. They showed that Hantaan-specific cytotoxic T lymphocytes (CTL) can be demonstrated if immune mouse lymphocytes are restimulated in vitro with Hantaan antigen. Similarly, they were able to demonstrate cross-reactive CTL by restimulating, with Hantaan, spleen cells from mice immunized with hantaviruses representing different serotypes from Hantaan (Prospect Hill and Puumala viruses). The bulk of the cytotoxic cells, assayed on Hantaan-infected peritoneal macrophages, bore a surface phenotype ($Thy1^+$, $L3T^-$, $Lyt2^+$) generally associated with class I-restricted CTL. Adoptive transfer of immune T cells, subsets of immune T cells, or serum to nonimmune mice were used to demonstrate that the abilities of mice to control Hantaan viremia were attributable not only to antibodies but also, in part, to T cells bearing either helper (Thy 1+, L3T4+, Lyt2−) or cytolytic (Thyl+, L3T4−, Lyt2+) phenotypes. In one instance of cross-protection among related but dissimilar viruses (protection against Hantaan by immunization with Puumala) their evidence suggested that T cells may be particularly relevant (Asada et al., 1989).

Except for one preliminary report (Balady, 1987), in which vaccinia virus recombinants expressing RVFV envelope glycoproteins were shown to elicit CTL, the proteins that elicit T cell responses to viruses in the Bunyaviridae have not been defined. In contrast with humoral immune mechanisms, it has been demonstrated in many viral systems that T lymphocytes are somewhat indifferent to the presence or absence of native viral proteins on cell surfaces, and might well react with unfolded or fragmented proteins representing any of the viral gene products. Consequently, any and perhaps all of the virus-encoded proteins of Bunyaviridae may eventually be found to elicit some degree of T cell responsiveness. The challenge will be to distinguish which of the responses are genuinely important in immunity. With recent advances that include not only gene sequencing and thereby the potential exploitation of synthetic peptides, but also the synthesis of individual proteins or their fragments in a variety of expression systems, a more thorough understanding of the cellular immunology of Bunyaviridae can be anticipated.

9. Approaches to new vaccines

Vaccine development for viruses in this family has been guided by the observation that traditional killed vaccines, where it has been possible to make and test them, have generally been effective in experimental systems, underscoring the dominant role of antibodies in protection. In addition to killed vaccines, subunit preparations, live-attenuated viruses, and more recently, recombinant-expressed proteins or live recombinant viruses have been investigated as potential vaccines for viruses in the Bunyaviridae. Studies of RVFV vaccine candidates best illustrate the continuing evolution of thought and experimentation in vaccines against members of the Bunyaviridae.

The first RVFV vaccine developed was a formalin-inactivated virion preparation. This vaccine elicits virus-neutralizing antibodies, is exceptionally effective in preventing disease and death in laboratory animals, and has been used with apparent efficacy in people at risk from natural or laboratory exposure (Randall et al., 1964; Eddy et al., 1981); however, it is expensive to produce, is not particularly immunogenic (three inoculations are recommended), and it carries the inherent risk of incomplete inactivation. A candidate live-attenuated RVFV vaccine was derived from its more virulent progenitor by chemical mutagenesis and subsequent selection of a virus that appears to be highly attenuated in its virulent properties, yet replicates sufficiently to evoke antibodies and solid immunity in experimental animals including domestic livestock, the main target of RVFV epidemics (Caplen et al., 1985; Morrill et al., 1987). Despite apparent multiple mutations resulting from successive passes in the presence of the chemical mutagen, (Caplen et al., 1985) such a vaccine carries the theoretical risk of reversion or reassortment and, even in the event that it proves a highly safe and successful vaccine, such a strategy for vaccine production may be difficult or impossible to reproduce with other viruses in the family.

Recombinant DNA based vaccine strategies are still under development, but offer exciting possibilities for disease prevention. Moreover, these strategies involve the significant advantage of avoiding propagation of large amounts of pathogenic viruses in the laboratory or vaccine production facility. As described above, *E. coli*-expressed, RVFV-β-galactosidase fusion proteins were used to definitively map three neutralization epitopes on G2 (Keegan and Collett, 1987). Collett et al. (1987), further tested the abilities of such proteins to serve as immunogens. Although results verified the protective potential of G2 immunization, the fusion proteins were otherwise disappointingly poor immunogens and protected only 70% of animals from a lethal challenge of RVFV even when high dosages of antigen were administered.

In contrast to the bacterial expression products, vaccinia recombinants which expressed both G1 and G2 were readily able to elicit antibodies and immunity in experimental animals and more importantly prevented abortion in pregnant sheep, a major consequence of RVFV infection of livestock (Collett et al., 1987; Dalrymple et al., 1989). The strain of vaccinia used to construct recombinants was found to play a role in successful induction of immunity in the sheep (but not in mice), suggesting that caution must be taken in projecting the success of a vaccine tested in laboratory animals to that obtained in naturally infected hosts. In agreement with results obtained by passive protection of animals with mAbs, when tested individually, recombinants expressing G2 were found to elicit a greater level of immunity than those expressing only G1 (Dalrymple et al. 1989). A potential advantage to the vaccinia expression system is the concomitant induction of CTL (Balady, 1987). However, it has the drawback of being intimately linked to the biology and immunology of the vaccinia virus vector, so that expression levels, vaccinia virus strain, and vaccinia-immune status may all be critical to vaccine efficacy.

Expression of the same RVFV genes by baculovirus recombinants (Autographa californica nuclear polyhedrosis virus), resulted in production of abundant amounts of G1 and G2 in lepidopteran insect cell cultures. The glycoproteins were biochemically and antigenically indistinguishable from authentic RVFV G1 and G2, and the expressed proteins were immunogenic in mice (Schmaljohn et al., 1989a). Recombinant-infected cells that contained both G1 and G2 elicited neutralizing antibodies in mice and conferred immunity to lethal challenge. G2 alone was apparently less immunogenic, but was also effective. Potential impediments with this vaccine strategy include the purification of viral antigens from insect cell contaminants, the possible influences of glycosylation differences between invertebrate and vertebrate cells, and the theoretical concern that some important immunological component would be absent in the response elicited by a subunit vaccine.

Because they contained only glycoprotein genes, (unlike killed or live-attenuated vaccines) both the vaccinia and baculovirus recombinant immunogens elicited no antibodies to the nucleocapsid protein and therefore provided a convenient serological marker to discriminate vaccinated from infected individuals.

Similar but less extensive approaches have been used experimentally to evaluate experimental vaccines against other viruses in the Bunyaviridae. Recombinant vaccinia viruses have been shown to express apparently authentic glycoproteins of another phlebovirus, PTV, however protective properties of the expressed proteins have not yet been reported (Matsuoka, et al., 1988). Hantaan glycoprotein and nucleocapsid genes have been expressed in both vaccinia and baculovirus recombinants (Schmaljohn et al. 1987a, 1989b); antigens produced in both systems were able to elicit antibodies in animals. Vaccinia recombinants containing both G1 and G2 were particularly effective at inducing neutralizing antibody responses, but it is otherwise difficult to assess Hantaan vaccine efficacy because of the current lack of a suitable animal model for human disease.

10. Conclusion

The arenaviruses, and particularly LCMV, have long provided investigators with biological models of exceptional interest and merit. Over the last decade the molecular biology of this group has advanced to the extent that meaningful questions addressing molecular pathogenesis can be approached. The next generation of studies will combine biological and molecular approaches to elucidate mechanisms of viral persistence, immunosuppression, inhibition of luxury function, and to provide a safe and effective vaccine for diseases like Lassa fever and Argentine hemorrhagic fever.

With Bunyaviridae, molecular biology is providing a particularly important set of tools for unifying traditional serology, cellular and molecular immunology, and vaccine development. With the Bunyaviridae tested to date, there appear to be no unusual barriers to the expression of viral antigens by either vaccinia or baculovirus recombinants. Rapid advances in immunochemistry, previously thwarted by poor growth characteristics or high biohazard potential of important members of the Bunyaviridae, can be also expected for this diverse family.

11. Acknowledgments

This work was supported by PHS grant AI 16102 and DOD contract C6234.

12. References

Ahmed, R., Byrne, J.A and Oldstone, M.B.A. (1984a) *J. Virol.* 51, 34–41.

Ahmed, R., Salmi, A., Butler, L.D., Chiller, J.M. and Oldstone, M.B.A. (1984b) *J. Exp. Med.* 60, 521–540.

Ahmed, R., King, C.-C. and Oldstone, M.B.A. (1987) *J. Virol.* 61, 1571–1576.

Akashi, H., and Bishop, D.H.L. (1983) *J. Virol.,* 45, 1155–1158.

Akashi, H., Gay, M., Ihara, T., and Bishop, D.H.L. (1984) *Virus Res.* 1, 51–63.

Allison, L.M., Salter, M., Buchmeier, M.J., Lewicki, H. and Howard, C.R. (1984) In *Segmented Negative Strand Viruses* (Compans, R.W. and Bishop, D.H.L., Eds.) Academic Press, New York. pp. 209–216.

Anderson, G.W., Jr. and Smith, J.F. (1987) *Virology,* 161, 91–100.

Arikawa, J., Schmaljohn, A.L., Dalrymple, J.M. and Schmaljohn, C.S. (1989) *J. Gen. Virol.* 70, 615–624.

Asada, H., Tamura, M., Kondo, K., Okuno, Y., Takahashi, Y., Dohi, Y., Nagai, T., Kurata, T. and Yamanishi, K. (1987) *J. Gen. Virol.* 68, 1961–1969.

Asada, H., Balachandra, K., Tamura, M., Kondo, K., and Yamanishi, K. (1989) *J. Gen. Virol.* 70, 819–825.

Auperin, D.D., Galinski, M. and Bishop, D.H.L. (1984a) *Virology* 134, 208–219.

Auperin, D.D., Romanowski, V., Galinski, M. and Bishop, D.H.L. (1984b) *J. Virol.* 52, 897–904.

Auperin, D.D., Esposito, J.J., Lange, J.V., Bauer, S.P., Knight, J., Sasso, D.S. and McCormick, J.B. (1987) *Virus Res.* 9, 233–248.

Auperin, D.D., Morrison, H.G., Esposito, J.J., Lange, J.V., Fisher-Hoch, S.P. and McCormick, J.B. (1988) *7th International Meeting on Negative Strand Viruses.*

Balady, M.A. (1987) *7th International Congress on Virology,* (Abstr. 64).

Barrios, H.A., Rondinon, S.N., Blejer, J.L., Giovaniello, O.A. and Nota, N.R. (1982) *Acta Virol.* 26, 156–164.

Battles, J.K. and Dalrymple, J.M. (1988) *Am. J. Trop. Med. Hyg.* 39, 617–631.

Beaty, B.J., Sundin, D.R. and Chandler, L.J. (1985) *Science* 230, 548–550.

Beaty, B.J., Trent, D.W. and Roehrig, J.T. (1988) In *The Arboviruses: Epidemiology and Ecology* (T. Monath, Ed.) Vol. I. CRC Press, Boca Raton FL. pp. 59–85.

Bishop, D.H.L. and Auperin, D.D. (1987) *Curr. Top. Microbiol. Immunol.* 133, 5–17.

Bishop, D.H.L. (1985) *Intervirology* 24, 79–93.

Bishop, D.H.L. (1986) *Adv. Virus Res.* 31, 1–51.

Bishop, D.H.L., and Shope, R.E. (1979) In *Comprehensive Virology* Vol. 14, (H. Frankel-Conrat and R.R. Wagner, Eds.) Plenum Press, New York. pp. 1–156.

Bishop, D.H.L., Calisher, C.H., Casals, J. Chumakov, M.P., Gaidamovich, S. Ya., Hannoun, C., Lvov, D.K., Marshall, I.D., Oker-Blom, N., Pettersson, R.F., Porterfield, J.S., Russell, P.K., Shope, R.E. and Westaway, E.G. (1980) *Intervirology* 14, 125–143.

Bishop, D.H.L., Gould, K.G., Akashi, H. and Clerx-van Haaster, C.M. (1982) *Nucleic Acids Res.* 10, 3703–3713.

Boersma, D.P., Saleh, F., Nakamura, K. and Compans, R.W. (1982) *Virology* 123, 452–456.

Bruns, M. and Lehmann-Grube, F. (1983) *J. Gen. Virol.* 64, 2157–2167.

Bruns, M., Cihak, J., Muller, G. and Lehmann-Grube, F. (1983a) *Virology* 130, 247–251.

Bruns, M., Peralta, L.M. and Lehmann-Grube, F. (1983b) *J. Gen. Virol.* 64, 599–611.

Bruns, M., Peralta, L.M. and Lehmann-Grube, F. (1986) *Virology* 151, 77–85.

Buchmeier, M.J. (1984) In *Segmented Negative Strand Viruses, Arenaviruses, Bunyaviruses and Orthomyxoviruses* (R.W. Compans and D.H.L. Bishop, Eds.) Academic Press, New York. pp. 219–227.

Buchmeier, M.J. and Oldstone, M.B.A. (1979) *Virology* 99, 111–120.

Buchmeier, M.J. and Parekh, B.S. (1987) *Curr. Top. Microbiol. Immunol.* 133, 41–57.

Buchmeier, M.J., Adams, E. and Rawls, W.E. (1974) *Infect. Immun.* 9, 821–823.

Buchmeier, M.J., Gee, S. and Rawls, W.E. (1977) *J. Virol.* 22, 175–186.

Buchmeier, M.J., Elder, J.H. and Oldstone, M.B.A. (1978) *Virology* 89, 133–145.
Buchmeier, M.J., Lewicki, H.A., Tomori, O. and Johnson, K.M. (1980) *Nature* 288, 486–487.
Buchmeier, M.J., Lewicki, H.A., Tomori, O. and Oldstone, M.B.A. (1981) *Virology* 113, 73–85.
Buchmeier, M.J., Southern, P.J., Parekh, B.S., Wooddell, K.M. and Oldstone, M.B.A. (1987) *J. Virol.* 61, 982–985.
Byrne, J.A. and Oldstone, M.B.A. (1984) *J. Virol.* 51, 682–686.
Byrne, J.A., Ahmed, R. and Oldstone, M.B.A. (1984) *J. Immunol.* 133, 433–439.
Cabradilla, C. D., Holloway, B.P., and Obijeski, J.F. (1983) *Virology,* 128, 463–468.
Calisher, C.H., Tzianabos, T., Lord, R.D. and Coleman, P.H. (1970) *Am. J. Trop. Med. Hyg.* 19, 520–526.
Caplen, H., Peters, C.J., and Bishop, D.H.L. (1985) *J. Gen. Virol.* 66, 2271–2277.
Carter, M.F., Murphy, F.A., Brunschweig, J.P., Noonan, C. and Rawls, W.E. (1973) *J. Virol.* 12, 33–37.
Carter, M.F., Biswal, N. and Rawls, W.E. (1974) *J. Virol.* 13, 577–583.
Casals, J. (1984) in *Viral Infections of Humans,* 2 ed. (A.F. Evans, Ed.), pp. 127–150. Plenum Press, New York.
Casals, J., Buckley, S.M. and Cedeno, R. (1975) *Bull. WHO* 52, 421–427.
Chan, M., Clark, D.A. and Rawls, W.E. Rawls (1983) *J. Immunol.* 130, 925–931.
Clegg, J.C.S. and Lloyd, G. (1983) *J. Gen. Virol.* 64, 1127–1136.
Clegg, J.C.S. and Lloyd, G. (1984) In *Segmented Negative Strand Viruses* (R.W. Compans and D.H.L. Bishop, Eds.) Academic Press, New York, pp. 341–347.
Clegg, J.C.S. and Lloyd, G. (1987) *Lancet* 8552, 186–188.
Clerx, J.P.M. and Bishop, D.H.L. (1981) *Virology* 108, 361–372.
Collett, M.S., Purchio, A.F., Keegan, K., Frazier, S., Hays, W.M., Anderson, D.K., Parker, M.D., Schmaljohn, C., Schmidt J. and Dalrymple, J.M. (1985) *Virology* 144,228–245.
Collett, M.S., Keegan, K., Hu, S.-L., Sridhar, P., Puchio, A.F., Ennis, W.H. and Dalrymple, J.M. (1987) In *The Biology of Negative Strand Viruses* (B. Mahy and D. Kolakofsky, Eds. Elsevier, Amsterdam. pp. 321–329.
Colman, P.M., Laver, W.G., Varghese, J.N., Baker, A.T., Tulloch, P.A., Air, G.M. and Webster, R.G. (1987) *Nature* 326, 358–362.
Coto, C.E., Damonte, E.B., Calello, M.A. and Weissenbacher, M.C. (1980) *J. Infect. Dis.* 141, 389–393.
Dalrymple, J.M., Hasty, S.E., Kakach, L.T., and Collett, M.S. (1989) In *Vaccines 89.* Cold Spring Harbor Laboratory, Cold Spring Harbor NY. pp. 000–000.
Dantas, J.R., Jr., Okuno, Y., Asada, H., Tamura, M, Takahashi, M., Tanishita, O Takahashi, Y., Kurata, T., and Yamanishi, K. (1986) *Virology* 151, 379–384.
Doyle, M.V. and Oldstone, M.B.A. (1978) *J. Immunol.* 121, 1262–1269.
Eddy, G.A., Peters, C.J., Meadors, G. and Cole, F.E. (1981) In *Contributions to Epidemiology and Biostatistics* (M.A. Klingberg, Ed.) Vol. 3. Karger, Basel. pp. 124–141.
Emery, V.C. and Bishop, D.H.L. (1987) *Virology* 156, 1–11.
Endres, M.J., Jacoby, D.R., Janssen, R.S., Gonzalez-Scarano, F., and Nathanson, N. (1989) *J. Gen. Virol.* 70, 223–228.
Farber, F.E. and Rawls, W.E. (1975) *J. Gen. Virol.* 38, 195–201.
Fazakerley, J.K., Gonzalez-Scarano, F., Strickler, J., Dietzschold, B., Karush, F. and Nathanson, N. (1988) *Virology* 167, 422–432.
Fisher-Hoch, S.P. and McCormick, J.B. (1987) *Curr. Top. Microbiol. Immunol.* 134, 231–239.
Franze-Fernandez, M.-T., Zetina, C., Iapalucci, S., Lucero, M.A., Bouissou, C., Lopez, R., Rey, O., Daheli, M., Cohen, G.N. and Zakin, M.M. (1987) *Virus Res.* 7, 309–324.
Fuller-Pace, F.V. and Southern, P.J. (1989) In *Genetics and Pathogenicity of Negative Strand Viruses* (B. Mahy, and D. Kolafofsky, Eds.) Elsevier, New York. In press.
Friedlander, A.M., Jahrling, P.B., Merrill, P. and Tobery, S. (1984) *Infect. Immun.* 43, 283–286.
Gard, G.P., Vezza, A.C., Bishop, D.H.L. and Compans, R.W. (1977) *Virology* 83, 84–95.

Gerbaud, S., Vialat, P., Pardigon, N., Wychowsku, C., Girard, M. and Bouloy, M. (1987) *Virus Res.* 8, 1–13.

Geschwender, H.H., Rutter, G. and Lehmann-Grube, F. (1976) *Med. Microbiol. Immunol.* 162, 119–131.

Glass, G.E., Childs, J.E., Korch, G.W. and LeDuc J.W. (1988) *Epidem. Infect.* 101, 459–472.

Gledhill, A.W., Bilbey, D.L.T. and Niven, J.S.F. (1965) *Br. J. Exp. Pathol.* 46, 433–442.

Gonzalez, J.P. and Georges, A.-J, (1988) In *The Arboviruses: Epidemiology and Ecology* (T. Monath, Ed.) Vol. 2. CRC Press, Boca Raton Fl. pp. 88–98.

Gonzalez, J.P., McCormick, J.B., Saluzzo, J.F., Herve, J.P., Georges, A.J. and Johnson, K.M. (1983) *Intervirology* 19, 105–112.

Gonzalez, J.P., Buchmeier, M.J., McCormick, J.B., Mitchell, S.W., Elliot, L.H. and Kiley, M.P. (1984) In *Segmented Negative Strand Viruses* (R.W. Compans and D.H.L. Bishop, Eds.) Academic Press, New York. pp. 201–208.

Gonzalez-Scarano, F., Shope, R.E., Calisher, C.H. and Nathanson, N. (1982) *Virology*, 132, 222–225.

Guerrero, L.B. de, Weissenbacher, M.C. and Parodi, A.S. (1969) *Medecine (Buenos Aires)* 29, 1–5.

Harnish, D.G., Leung, W-C. and Rawls, W.E. (1981) *J. Virol.* 38, 840–848.

Henderson, J.R. and Downs, W.G. (1965) *Am. J. Trop. Med. Hyg.* 14, 796–799.

Howard, C.R. (1987) *Curr. Top. Microbiol. Immunol.* 134, 117–130.

Howard, C.R. and Buchmeier, M.J. (1983) *Virology* 126, 538–547.

Howard, C.J., Lewicki, H., Allison, L., Salter, M. and Buchmeier, M.J. (1985) *J. Gen. Virol.* 66, 1383–1395.

Hung, T., Chou, Z., Zhao, T., Xia, S. and Hang, C. (1985) *Intervirology* 23, 97–108.

Iapalucci, S., Lopez, R., Rey. O., Franze-Fernandez, M.T., Lucero, M.A., Ochoa, A., Cohen, G.N. and Zakin, M.M. (1988) *7th International Meeting on Negative Strand Viruses.*

Iapalucci, S., Lopez, N., Rey, O., Zakin, M., Cohen, G. and Franze-Fernandez, M. (1989) *Virology* 173, 357–361.

Jahrling, P.B. and Peters, C.J. (1986) *Med. Microbiol. Immunol.* 175, 165–167.

Jamieson, B. and Ahmed, R. (1988) *Proc. Natl. Acad. Sci. USA* 85, 2265–2268.

James, W.S. and Millian, D. (1986) *J. Gen. Virol.* 67, 2803–2806.

Johnson, K.M., Kuns, M.L., Mackenzie, R.B., Webb, P.A. and Yunker, C.E. (1966) *Am. J. Trop. Med. Hyg.* 15, 103–106.

Johnson, K.M., Webb, P.A. and Justices, G. (1973) In *Lymphocytic Choriomeningitis and Other Arenaviruses* (F. Lehmann-Grube, Ed.) Springer-Verlag, Berlin. pp. 241–258.

Johnson, K.M., Taylor, P. Elliot, L.H. and Tomori, O. (1981) *Am. J. Trop. Med. Hyg.* 30, 1291–1294.

Karabatsos, N. (1978) *Am. J. Trop. Med. Hyg.* 27, 372–373.

Kimmig, W. and Lehmann-Grube, F. (1979) *J. Gen. Virol.* 45, 703–710.

Kingsford, L., Ishizawa, L.D. and Hill, D.W. (1983) *Virology* 129, 443–455.

Keegan, K. and Collett, M.S. (1986) *J. Virol.* 58, 263–270.

Klavinskis, L.S., Oldstone, M.B.A. and Whitton, J.L. (1989a) *Vaccines* 89, pp. 485–489, Cold Spring Harbor Laboratory.

Klavinskis, L.S. Whitton, J.L. and Oldstone, M.B.A. (1989b) *J. Virol.* 63, 4311–4316.

Klavinskis, L.S., Tishon, A. and Oldstone, M.B.A. (1989c) *J. Immunol.* 143, 2013–2016.

Lee, H.W., Lee, P.W., Baek, L.J., Song, C.K., and Seong, I.W. (1981) *Am. J. Trop. Med. Hyg.* 30, 1106–1112.

Lehmann-Grube, F., Tijerina, R., Zeller, W., Ahaturvedi, U.C. and Lohler, J. (1983) *J. Gen. Virol*, 64, 1157–1166.

Leung, W-C., Leung, M.F.K.L. and Rawls, W.E. (1979) *J. Virol.* 301, 98–107.

Lukashevich, I.S. and Lemeshko, N.N. (1985) *Arch. Virol.* 86, 85–99.

MacKenzie, R.B., Webb, P.A. and Johnson, K.M. (1965) *Am. J. Trop. Med. Hyg.* 14, 1079–1084.

Maitzegui, J.I., Fernandez, N.J. and de Damaliano A.J. (1979) *Lancet* ii, 1216–1217.

Marker, O. and Volkert, M. (1973) In *Lymphocytic Choriomeningitis Virus and Other Arenaviruses* (F. Lehmann-Grube, Ed.) Springer-Verlag, Berlin. pp. 207–216.

Martin, M.L., Lindsey-Regnery, H., Sasso, D.R., McCormick, J.B. and Palmer, E. (1985) *Arch. Virol.* 86, 17–28.

Martinez Segovia, Z.M. and deMitri, M.I. (1977) *J. Virol.* 21, 579–583.

Matsuoka, Y., Ihara, T., Bishop, D.H.L. and Compans, R.W. (1988) *Virology* 167, 251–260.

McCormick, J.B. (1987) *Curr. Top. Microbiol. Immunol.* 134, 69–78.

McCormick, J.B., King, I.J., Webb, P.A., Johnson, K.M., O'Sullivan, R., Smith, E.S. and Trippel, S. (1987) *J. Infect. Dis.* 155, 445–455.

Mims, C.A. and Wainwright, S. (1968) *J. Immunol.* 101, 717–724.

Monath, T.P. (1975) *Bull. WHO* 52, 577–592.

Monath T.P., Newhouse, V.F., Kemp, G.E., Setzer, H.W. and Cacciapouti, A. (1974) *Science* 185, 263–265.

Morrill, J.C., Jennings, G.B., Caplen, H., Turell, M.J., Johnson, A.J., and Peters, C.J. (1987) *Am. J. Vet. Res.* 48, 1042–1047.

Moskophidis, D., Cobbold, S.P., Waldmann, H. and Lehmann-Grube, F. (1987) *J. Virol.* 61, 1867–1874.

Murphy, F.A. and Whitfield, S.G. (1975) *Bull. WHO* 52, 409–419.

Murphy, F.A., Webb, P.A., Johnson, K.M., Whitfield, S.G. and Chappell, W.A. (1970) *J. Virol.* 6, 507–512.

Oldstone, M.B.A. and Dixon, F.A. (1967) *Science* 158, 1193–1194.

Oldstone, M.B.A. and Whitton, J.L. (1989) In *Genetics and Pathogenicity of Negative Strand Viruses* (B. Mahy and D. Kolakofsky, Eds.) Elsevier, New York. In press.

Oldstone, M.B.A., Tishon, A., Chiller, J., Weigle, W. and Dixon, F.J. (1973) *J. Immunol.* 110, 1268.

Oldstone, M.B.A., M.J. Buchmeier, M.V. Doyle and A. Tishon (1980) *J. Immunol.* 124, 831–838.

Oldstone, M.B.A., Salvato, M., Tishon, T. and Lewicki, H. (1988) *Virology* 164, 507–516.

Ozols, D.Y., Rosenthal, K.L, Rawls, W.E. and Harnish, D.G. (1988) *7th International Meeting on Negative Strand Viruses.*

Parekh, B.S. and Buchmeier, M.J. (1986) *Virology* 153, 168–178.

Pederson, I.R. (1973) *J. Virol.* 11, 416–421.

Pederson, I.R. and Konigshofer, E.P. (1976) *J. Virol.* 20, 14–21.

Peters, C.J. (1984) In *Textbook of Human Virology* (R.B. Belshe, Ed.) PSG Publishing, Littleton MA. pp. 513–545.

Peters, C.J., Jahrling, P.J., Liu, C.T., Kenyon, R.H., McKee, K.T., Jr. and Barrera Oro, J.G. (1987) *Curr. Top. Microbiol. Immunol.* 134, 5–68.

Pifat, D.Y., Osterling, M.C., and Smith, J.F. (1988) *Virology* 167, 442–450.

Ramos, B.A., Courtney, C.J. and Rawls, W.E. (1972) *J. Virol.* 10, 661–667.

Randall, R., Binn, L.N. and Harrison, V.R. (1964) *J. Immunol.* 93, 293–299.

Riviere, Y., Southern, P.J., Ahmed, R. and Oldstone, M.B.A. (1986) *J. Immunol.* 136, 304–307.

Romanowski, V., Matsuura, Y. and Bishop, D.H.L. (1985) *Virus Res.* 3, 101–114.

Ronnholm, R. and Pettersson, R.F., (1987) *Virology,* 160, 191–202.

Rowe, W.P., Murphy, F.A., Bergold, G.H., Casals, J., Hotchin, J., Johnson, K.M., Lehmann-Grube, F., Mims, C.A., Traub, E. and Webb, P.A. (1970a) *J. Virol.* 5, 651–652.

Rowe, W.P., Pugh, W.E., Webb, P.A. and Peters, C.J. (1970b) *J. Virol.* 5, 289–292.

Ruggiero, H.A., Cintora, F., Gonzalez-Cambaceres, C., Maglio, F., Milani, H., Magnoni, C., Astarloa, L. and Perez Izquierdo, F. (1974) *Prensa Med. Arg.* 61, 231–240.

Ruggiero, H.A., Magnoni, C., de Guerrero, L.B., Milano, H.A., Pedro Izquierdo, F., Milani, H.L. and Weber, E.L. (1981) *J. Med. Virol.* 7, 227–232.

Saleh, F., Gard, G.P. and Compans, R.W. (1979) *Virology* 93, 369–376.

Salvato, M., Shimomaye, E., Southern, P. and Oldstone, M.B.A. (1988) *Virology* 162, 517–522.

Salvato, M. and Shimomaye, E. (1989) *Virology* 173, 1–10.

Salvato, M., Shimomaye, E. and Oldstone, M.B.A. (1989) *Virology* 169, 377–384.

Sanchez, A., Pifat, D.Y., Kenyon, R.H., Peters, C.J., McCormick, J.B. and Kiley, M.P. (1989) *J. Gen. Virol.* 70, 1125–1132.

Schmaljohn, A.L., Johnson, E.D., Dalrymple, J.M. and Cole, G.A. (1982) *Nature* 297, 70–72.

Schmaljohn, C.S. and Dalrymple, J.M. (1983) *Virology* 131, 482–491.

Schmaljohn, C.S., Jennings, G., Hay, J. and Dalrymple, J.M. (1986a). *Virology,* 155, 633–643.

Schmaljohn, C.S., Hasty, S.E., Rasmussen, L., and Dalrymple, J.M. (1986b) *J. Gen. Virol.* 67,707–717.

Schmaljohn, C.S., Jennings, G.B., and Dalrymple, J.M. (1987a) In: *The Biology of Negative Strand Viruses* (B. Mahy and D. Kolakofsky) Elsevier, Amsterdam. pp. 116–121.

Schmaljohn, C.S., Schmaljohn, A.L., and Dalrymple, J.M. (1987b) *Virology,* 157,31–39.

Schmaljohn, C.S., Arikawa, J., Hasty, S.E., Rasmussen, L., Lee, H.W., Lee, P.W., and Dalrymple, J.M. (1988) *J. Gen. Virol.,* 69,1949–1955.

Schmaljohn, C.S., Parker, M.D., Ennis, W.H., Dalrymple, J.M., Collett, M.S., Suzich, J.A. and Schmaljohn, A.L. (1989a) *Virology,* 170,184–192.

Schmaljohn, C.S., Arikawa, J., Dalrymple, J.M., and Schmaljohn, A.L. (1989b) In *Genetics and Pathogenicity of Negative Strand Viruses* (B. Mahy, and D. Kolakofsky, Eds.) Elsevier, Amsterdam. pp.58–66.

Schwartz, R., Lohler, J. and Lehmann-Grube, F. (1978) *J. Gen. Virol.* 39, 565–570.

Shivaprakash, M., Harnish, D. and Rawls, W. (1988) *J. Virol.* 62, 4037–4043.

Singh, M.K., Fuller-Pace, F.V., Buchmeier, M.J. and Southern, P.J. (1987) *Virology* 161, 448–456.

Smith, J.F., and Pifat, D.Y. (1982) *Virology,* 121, 61–81.

Smith, J., Hodgson, L., Dombalagian, M. and Komoriya, A. (1987) *7th International Congress on Virology* (Abstr. 65).

Southern, P.J. and Bishop, D.H.L. (1987) *Curr. Top. Microbiol. Immunol.* 133, 19–39.

Tishon, A., Southern, P.J. and Oldstone, M.B.A. (1988) *J. Immunol.* 140–1280–1284.

Tosolini, F.A. and Mims, C.A. (1971) *J.Infect. Dis.* 123, 134–144.

Van der Zeist, B.A.M., Noyes, B.E., Mirault, M-E., Parker, B., Osterhous, A.D.M.E., Swyryd, E.A., Bleumink, N., Horzinek, M.C. and G.R. Stark (1983a) *J. Virol.* 48, 249–261.

Vezza, A.C., Gard, G.P., Compans, R.W. and Bishop, D.H.L. (1977) *J. Virol.* 23, 776–786.

von Bonsdorff, C.-H., and Pettersson, R. (1975) *J. Virol.* 16, 1296–1307.

Walker, C.M., Rawls, W.E. and Rosenthal, K.L. (1984) *J. Immunol.* 132, 469–474.

Walker, D.H., Wulff, H., Lange, J.V. and Murphy, F.A. (1975) *Bull. WHO* 52, 523–534.

Webb, P.A., Johnson, K.M., Peters, C.J. and Justines, G. (1973) In *Lymphocytic Choriomeningitis Virus and other Arenaviruses* (F. Lehmann-Grube, Ed.) Springer-Verlag, Berlin. pp. 313–321.

Weber, E.L. and Buchmeier, M.J. (1988) *Virology* 164, 30–38.

Weissenbacher M.C., Coto C.E. and Calello M.A. (1975/76) *Intervirology* 6, 42–49.

Weissenbacher, M.C., Coto, C.E., Calello, M.A., Rondinone, S.N., Damonte, E.B. and Frigerio, M.J. (1982) *Infect. Immun.* 35, 425–430.

Weissenbacher, M.C., Laguens, R.P. and Coto, C.E. (1987) *Curr. Top. Microbiol. Immunol.* 134, 79–116.

Whitton, J.L., Southern, P.J. and Oldstone, M.B.A. (1988a) *Virology* 162, 321–327.

Whitton, J.L., Gebhard, J.R., Lewicki, H., Tishon, A. and Oldstone, M.B.A. (1988b) *J. Virol.* 62, 687–695.

Wiebenga, N.H., Shelokov, A., Gibbs, C.J. and MacKenzie, R.B. (1964) *Am. J. Trop. Med. Hyg.* 13, 626–628.

Wulff, H., MacIntosh, B.M., Hammer, D.B. and Johnson, K.M. (1977) *Bull. WHO* 55, 441–444.

Wulff, H., Lange, J.V. and Webb, P.A. (1978) *Intervirology* 9, 344–350.

Wright, K.E. and Buchmeier, M.J. (1989) *J. Virol.*, submitted.

Wright, K.E., Rosenthal, K.L. and Rawls, W.E. (1987) *Arch. Virol.* 92, 197–209.

Wright, K.E., Salvato, M.S. and Buchmeier, M.J. (1989a) *Virology* 171, 417–426.

Wright, K.E., Spiro, R.C. and Buchmeier, M.J. (1989b) submitted for publication.

Young, P.R. and Howard, C.R. (1986) *Med. Microbiol. Immunol.* 175, 79–83.

Zinkernagel, R.M. and Doherty, P.C. (1974) Nature 248, 701–702.

Eds. M.H.V. van Regenmortel and A.R. Neurath
Immunochemistry of viruses, II. The basis for serodiagnosis and vaccines.

Togaviruses 13

J.T. ROEHRIG

Immunochemistry Branch, Division of Vector-Borne Viral Diseases, Center for Infectious Diseases, Centers for Disease Control, Public Health Service, U.S. Department of Health and Human Services, P.O. Box 2087, Fort Collins, CO 80522, U.S.A.

1. Alphaviruses

1.1. Alphavirus structure, replication, and classification

1.1.1. Structure and replication

The structure and replication of alphaviruses have been the subject of excellent reviews (Schlesinger and Schlesinger, 1986; Strauss and Strauss, 1986). In this review, the discussion of viral structure and replication is limited to alphavirus immunochemistry.

The alphavirus nucleocapsid contains a positive-sense, single-stranded RNA genome of approximately 12 000 nucleotides surrounded by multiple copies of a single capsid protein (C), 35 kDa (Fig. 13.1). The nucleocapsid is enclosed by a host cell-derived lipid envelope virally modified to contain trimers that include a dimer of glycoproteins E1, 50 to 55 kDa, and E2, 50–55 kDa (Fuller, 1987). In Semliki Forest (SF) virus, an additional viral protein E3 (14 kDa) is also associated with the viral envelope. The structural proteins are translated from a polycistronic message encoded by the 3′ one-third of the genome. The nucleocapsid is synthesized, then it functions as an autoprotease and cleaves itself from the nascent polypeptide chain. The precursor (PE2) of the E2 and E3, and the E1 are then synthesized. The PE2/E1 cleavage appears to occur co-translationally, but the cleavage of PE2 to E2 and E3 occurs late. The PE2 cleavage is essential for alphavirus assembly.

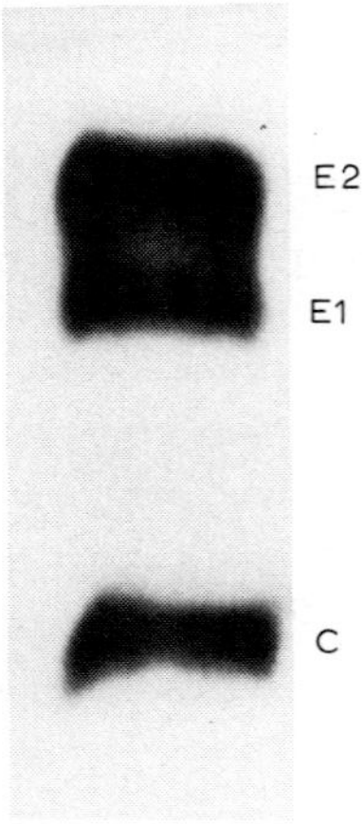

Fig. 13.1. Polyacrylamide gel profile of structural proteins of VEE virus strain TC-83.

1.1.2. Classification

Alphaviruses can be classified into six serocomplexes using neutralization (N) assays with polyclonal antisera (Table 13.1). The antigenic classification scheme has been validated by nucleic acid sequencing of six alphaviruses (Garoff et al., 1980; Rice and Strauss, 1981; Dalgarno et al., 1983; Kinney et al., 1986; Chang and Trent, 1987; Hahn et al., 1988).

Comparative analyses of the amino acid sequences of the alphavirus structural proteins identify common characteristics. All extramembranal cysteine residues are conserved among all alphaviruses on both the E1 and E2 glycoproteins (Fig. 13.2). All alphaviruses have a highly conserved, hydrophobic region on the E1 glycoprotein (amino acids 80–102) that is high in glycine content. The overall glycoprotein spike conformation is similar for all alphaviruses, and antigenic diversity is probably a function of sequence microheterogeneity and not a function of dramatic glycoprotein conformation shifts. Recent expression of Venezuelan equine encephalomyelitis

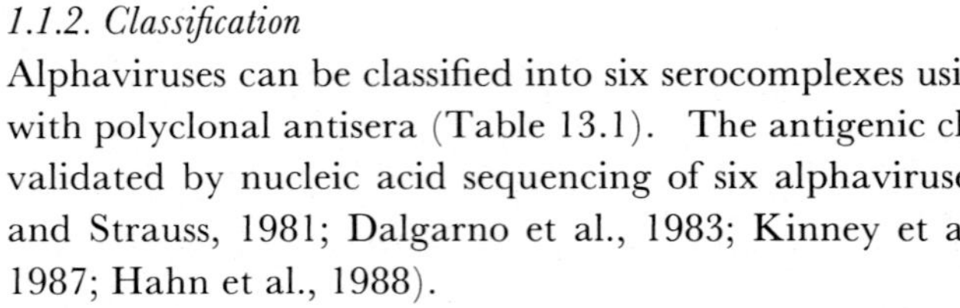

Fig. 13.2. Conserved extramembranal cysteine residues on the alphavirus glycoproteins using VEE virus strain TC-83 as an example. The scale numbers amino acid (10^2). Cysteine residues are keyed. Shaded areas correspond to transmembranal domains. The putative fusion (F) and neutralization (N) domains are highlighted.

TABLE 13.1

Serologic classification of alphaviruses[a]

Complex	Species
Western equine	Western equine encephalitis Sindbis Highlands J Aura Fort Morgan
Eastern equine	Eastern equine encephalitis
Venezuelan equine	Venezuelan equine encephalitis
Ndumu	Ndumu
Middleberg	Middleberg
Semliki Forest	Semliki Forest Chikungunya Getah Mayaro

[a] Adapted from Calisher et al., 1980.

(VEE) virus genes in vaccinia demonstrated that the structural proteins alone elicit a fully protective antiviral response (Kinney et al., 1988a, 1988b).

1.2. Antigenic composition of the E1 glycoprotein

1.2.1. E1 epitope definition

Information about the immunochemistry of the E1 glycoprotein primarily comes from studies with SF, VEE, Sindbis (SIN), and western equine encephalitis (WEE) viruses. E1 glycoprotein isolated from nonionic detergent-lysed SIN or WEE virions and purified by isoelectric focusing to preserve conformational integrity, elicits hemagglutination inhibition (HI) but not N antibody (Dalrymple et al., 1976; Trent and Grant, 1980).

Many laboratories have used monoclonal antibodies (mAbs) to examine the antigenic structure of the E1 glycoprotein (for review, see Roehrig, 1986). These studies have associated with E1 the biologic activities of N, HI, prophylactic protection from virus challenge, curing of disease, hemolysis, and enhancement of virus infectivity (Chanas et al., 1982; Clegg et al., 1983; Stanley et al., 1986). mAb analyses have identified 3 to 6 epitopes on the E1 glycoprotein, and their general biologic characteristics are listed in Table 13.2. The E1 epitopes can be spatially arranged into 1–6 domains by competitive binding assay (CBA, Fig. 13.3). Unfortunately,

TABLE 13.2

Biological characteristics of epitopes defined on the E1 glycoprotein of various alphaviruses

Virus(strain)	Epitope	HI	N	Protection
SIN(AR339)[a]	a	–	–	+
	b	–	–	+
	c	+	+	n.d.[f]
	d	+	–	n.d.
	e	+	–	+
SIN(AR339)[b]	a	–	–	+
	b	–	+	+
	c	+	+	+
	e	+	–	+
	f	+	+	+
	g	+	+	+
SF(MRS MP 192/7)[c]	a	–	+	+
	b	–	+	+
	c	+	–	+
	d	+	–	–
	e	–	–	+
	f	–	+	+
VEE(TC-83)[d]	a	–	–	–
	b	+	+	+
	c	–	–	+
	d	+	–	+
WEE(McMillan)[e]	a1	+	–	+
	a2	+	–	–
	b1	+	–	+
	b2	–	–	n.d.
	c	–	–	–
	d	+	–	+
	e1	+	–	–
	e2	–	–	–

[a] Schmaljohn et al., 1982; 1983.
[b] Mendoza et al., 1988.
[c] Boere et al., 1983; 1984.
[d] Roehrig et al., 1982b; Mathews and Roehrig, 1982.
[e] Hunt and Roehrig, 1985
[f] n.d. = not determined

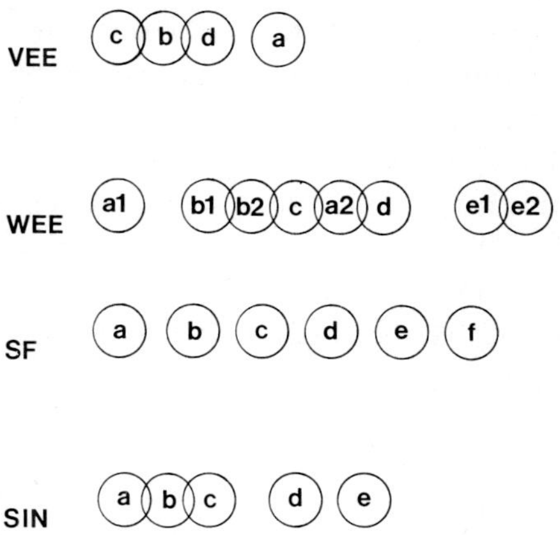

Fig. 13.3. Competition binding maps of alphavirus E1 glycoproteins. Overlapping circles signify competition between mAbs defining the letter epitope. Data are from the following papers: VEE virus, Roehrig et al., 1982b; WEE virus, Hunt and Roehrig, 1985; SF virus, Boere et al., 1984; SIN virus, Schmaljohn et al., 1983.

CBA analysis is imprecise because results depend on how many mAbs are isolated, the relative avidity of each mAb, and the CBA procedures used. E1 can elicit both neutralizing and non-neutralizing mAbs that can prophylactically protect animals from virus challenge (Schmaljohn et al., 1982; Mathews and Roehrig 1982; Boere et al., 1983; Stanley et al., 1986; Mendoza et al., 1988). When careful quantitation was performed, less neutralizing mAb than non-neutralizing mAb was required for prophylactic protection. Non-neutralizing anti-E1 mAbs appear to be more efficient than neutralizing ones at curing animals of SIN virus encephalitis. Presumably, non-neutralizing mAbs are more efficient because they recognize E1 epitopes on the cell surface and can therefore mediate destruction of virus-infected cells (Mendoza et al., 1988).

1.2.2. E1-mediated cross-protection among alphaviruses

Analysis of amino acid sequences suggest that E1 is the most conserved envelope protein (Strauss and Strauss, 1986). These results are consistent with the observation that purified E1 elicits alphavirus cross-reactive polyclonal and monoclonal antibody (Trent et al., 1979; Roehrig et al., 1982a; Hunt and Roehrig, 1985). mAbs elicited by alphavirus group common epitopes do not neutralize virus infectivity, but if given in sufficient quantity, they do protect animals from challenge with homologous or heterologous alphaviruses (Mathews and Roehrig, 1982; Boere et al., 1983). The significance of the E1 group common epitopes has been further demonstrated using SIN and SF viruses. Mice hyperimmunized with SIN virus were resistant to SF virus challenge in the absence of virus-neutralizing antibody. This protection appeared to be linked to cross-reactive cytotoxic T cells (Peck et al., 1979). After extended hyperimmunization, cross-reactive, antibody-dependent, complement-mediated, cellular cytoxicity (ADCMCC) of virus-infected cells also could be demonstrated (Gates et al., 1982; Smith et al., 1985; Wust et al., 1987).

While cross-protection between SF and SIN virus can be mediated by passive transfer of extended hyperimmune antisera, this is not so for all alphaviruses. The same antisera will passively protect animals from WEE virus challenge (a close relative of SIN virus), but it fails to passively protect animals from challenge with eastern equine encephalitis (EEE) or VEE viruses (J.H. Mathews, unpublished observation, 1988). The in vivo significance of laboratory-defined alphavirus cross-protection requires further study.

1.2.3. Contribution of conformation to E1 epitopes

The E1 determinants that encode alphavirus cross-protection by extended hyperimmune antisera are present on only the virus-infected cell surface. Apparently, when the virus buds these determinants are lost, and they are undetectable in the virion. Attempts have been made to determine the maturational requirements necessary for ADCMCC of virus-infected cells using SIN virus temperature-sensitive (ts) mutants. Cells infected with mutant ts20 (which fails to bud at restricted temperature and has a block in the terminal cleavage of PE2 to E2 and E3) were 50% less susceptible to E1-mediated ADCMCC than wild type (wt) virus-infected cells (Wolcott et al., 1982, 1985). This mutation alters the conformation of the glycoprotein spike and maps to an amino acid sequence change (His to Leu) at position 291 in the E2, which indicates that a change in the E2 glycoprotein that blocks PE2 cleavage, effectively alters E1 immunogenicity (Strauss et al., 1987).

Additional contributions of conformation to E1 antigenicity have been identified using mAbs. Anti-SIN and anti-WEE E1 mAbs identified some epitopes that were expressed on the surface of the virus-infected cell but were hidden or cryptic on the infectious virion (Schamljohn et al., 1982, 1983; Wolcott et al., 1984; Hunt and Roehrig, 1985). These epitopes failed to elicit virus-neutralizing antibody. Virus-neutralizing anti-E1 mAbs reacted with epitopes expressed on the surface of the infectious virion. El epitopes demonstrated a hierarchy of conformational stability. Some epitopes were stable after virion lysis with nonionic detergent but were denatured by incubation in sodium dodecyl sulfate (SDS) or β-mercaptoethanol (Roehrig et al., 1982a; Hunt and Roehrig, 1985). Two of the four epitopes identified on the E1 of VEE virus were stable to reduction and alkylation, which indicates that they are probably linear epitopes (J.T.R., unpublished observations, 1988).

1.2.4. E1 and membrane fusion

The fusion activity of alphaviruses has been studied extensively (White and Helenius, 1980; Väänänen and Kääriäinen, 1980, 1981; Marsh et al., 1983; Helenius, 1984; Kielan and Helenius, 1985, 1986; Kielan et al., 1986; Schmid et al., 1989). Three lines of evidence implicate E1 as the viral glycoprotein active in membrane fusion: the E1 has a highly conserved, hydrophobic amino acid sequence (80–102) rich in glycine residues that is consistent with fusion domains identified for other envelope viruses (Fig. 13.4); isolated E1 protein elicits HI antibody and can mediate pH-dependent fusion of erythrocytes; and cells that express only the E2 of SIN virus do not fuse when exposed to low pH. Similarly, only the hemagglutination reactive VEE virus E1^{d} epitope changes reactivity when incubated at low pH (J.T.R., unpublished observations, 1989). Recent data with recombinant SIN viruses isolated from an engineered infectious clone corroborate these observations and conclusively associate E1 with membrane fusion. The fusion characteristics of one strain of SIN virus were transferred to another only by exchange of E1 glycoprotein sequences (Boggs et al., 1989). Furthermore, E1 residues 72 and 313 appeared to be involved in defining E1 fusion characteristics.

1.3. Antigenic composition of the E2 glycoprotein

1.3.1. E2 epitope definition

Early work with SIN, WEE, and VEE viruses identified purified E2 as the glycoprotein most efficient in eliciting virus neutralizing antibody (Dalrymple et al., 1976; Trent et al., 1979; Trent and Grant, 1980; Kinney et al., 1983). Probably due to the interest in virus N, subsequent work with mAbs has led to a good understanding of the biologic and functional activity of the E2. Data derived from four alphaviruses (Ross river [RR], SIN, SF, and VEE) describe a similar picture of the E2 antigenic structure (Table 13.3). Eight epitopes were described on the E2 of VEE virus. These epitopes elicited a variety of antibody reactivities, but four (c,f,g,h) were most efficient at eliciting virus-neutralizing antibody and were spatially mapped into the same domain by CBA (Roehrig and Mathews, 1985). mAbs specific for this critical N domain were most efficient in protecting animals from virus challenge. Similar results were derived for both SF and SIN virus when careful antibody quantitations were performed (Boere et al., 1983; Mendoza et al., 1988). Protection by these virus antibodies was not dependent on complement, but when antibody $F(ab')_2$ fragments were prepared, they lost their ability to protect animals but retained their N capacity (Mathews et al., 1985; Boere et al., 1985).

```
INFLUENZA A       1 GLFGAIAGPIQGGWQ      15
INFLUENZA B       1 .F..IA-.FL..-..      13
SENDAI F          1  F...VI.I.AL.PAT     15
FLAVIVIRUS E     98  DR.WGN.CGLF.KGSIDT 115
ALPHAVIRUS E1    80 VFT.VYPFMWG.AYCFCDT  98
```

Fig. 13.4. Putative fusion regions of some enveloped viruses. Data are derived from the following papers: influenza A HA_2, influenza B HA_2, and Sendai F proteins, Gething et al., 1978; flavivirus E protein, Roehrig et al., 1989; and alphavirus E1 protein, Kinney et al., 1986.

TABLE 13.3

Biological characteristics of epitopes defined on the E2 glycoprotein of various alphaviruses

Virus(Strain)	Epitope	HI	N	Protection
RR (T48)[a]	a	n.d.[e]	+	n.d.
	b	n.d.	+	n.d.
	c	n.d.	+	n.d.
SIN (AR339)[b]	ab	+	+	+
	c	+	+/−	+/−
	d	+/−	−	−
SF (MRS MP 192/7)[c]	a	−	−	+
	b	+	−	−
	c	−	+	+
	d	−	+	+
	e	+	−	−
VEE(TC-83)[d]	a	−	−	−
	b	−	−	−
	c	+	+	+
	d	−	−	n.d.
	e	−	−	n.d.
	f	−	+	−
	g	−	+	−
	h	−	+	−

[a] Vrati et al., 1988.
[b] Mendoza et al., 1988.
[c] Boere et al., 1983; 1984.
[d] Roehrig et al., 1982b; Mathews and Roehrig, 1982; Roehrig and Mathews, 1985.
[e] n.d. = not determined.

1.3.2. Monoclonal antibody neutralization escape variants

Indirect placement of E2 epitopes has been determined by selection and sequencing of N escape variants (Stec et al., 1986; Strauss et al., 1987; Vrati et al., 1988; Johnson et al., 1990). VEE virus N escape variants clustered between amino acids 180 and 207, thereby corroborating the close spatial arrangement of the epitopes first identified by CBA (Table 13.4). A similar region was defined on SIN virus (Table 13.4). The N domain identified on RR virus appears to be closer to the carboxy-terminus (216–251). For all alphaviruses, this domain is highly variable in sequence and displays the antigenic characteristics of hydrophilicity, surface accessibility, and molecular flexibility (Fig. 5).

A second N domain was identified by mAb analysis of SIN virus mutants selected for rapid penetration into susceptible cells. The epitope, identified as E2[c] (not to be

TABLE 13.4

Variation selected in mAb N escape variants

Virus(strain)	E2 Epitope	Position number	Amino acid change
VEE(TC-83)[a]	c	182	Ser → Arg
	f	183	Gly → Val
	g	199	Glu → Lys
	h	207	Ile → Phe
SIN(AR339)[b]		181	n.d.[d]
		184	n.d.
	a	190	Lys → Met or Asn
		214	n.d.
	b	216	Lys → Asn or Glu
RR(T48)[c]	a	216	Thr → Ile
	b1	232	His → Arg
		234	Lys → Ile, Gln, Asn, or Glu
	b2	246	Asp → Asn or Val
		247	n.d.
		248	Thr → Pro
		251	Arg → Ser

[a] Johnson et al., 1990.
[b] Strauss et al., 1987.
[c] Vrati et al., 1988.
[d] n.d. = not determined.

confused with VEE virus E2[c] which is probably equivalent to SIN E2[a] or E2[b]), was defined by an amino acid sequence change at E2 residue 114 (Olmsted et al., 1984, 1986; Davis et al., 1986). mAbs that neutralized this mutant bound to but failed to neutralize the infectivity of wt SIN virus unless goat anti-mouse antibody was added. Therefore, this site appears not to be involved in N of wt SIN virus.

It is interesting, however, that viruses that demonstrate one sequence change at amino acid 114 lose neurovirulence. In contrast, studies with an artificially mutagenized SIN virus infectious clone correlated attenuation with multiple amino acid substitutions (Lustig et al., 1988; Polo et al., 1988). Multiple changes also were associated with attenuation of RR virus (Faragher et al., 1988). This varied characterization of SIN virus attenuation could be attributable to the low actual neurovirulence of wt SIN virus. The neurovirulent SIN virus strain is neuroadapted from wt SIN virus, which is not neurovirulent itself. The most accurate definition of alphavirus neurovirulent will probably come from studies with truly neurovirulent viruses. There are only five amino acid non-conservative changes between the virulent wt

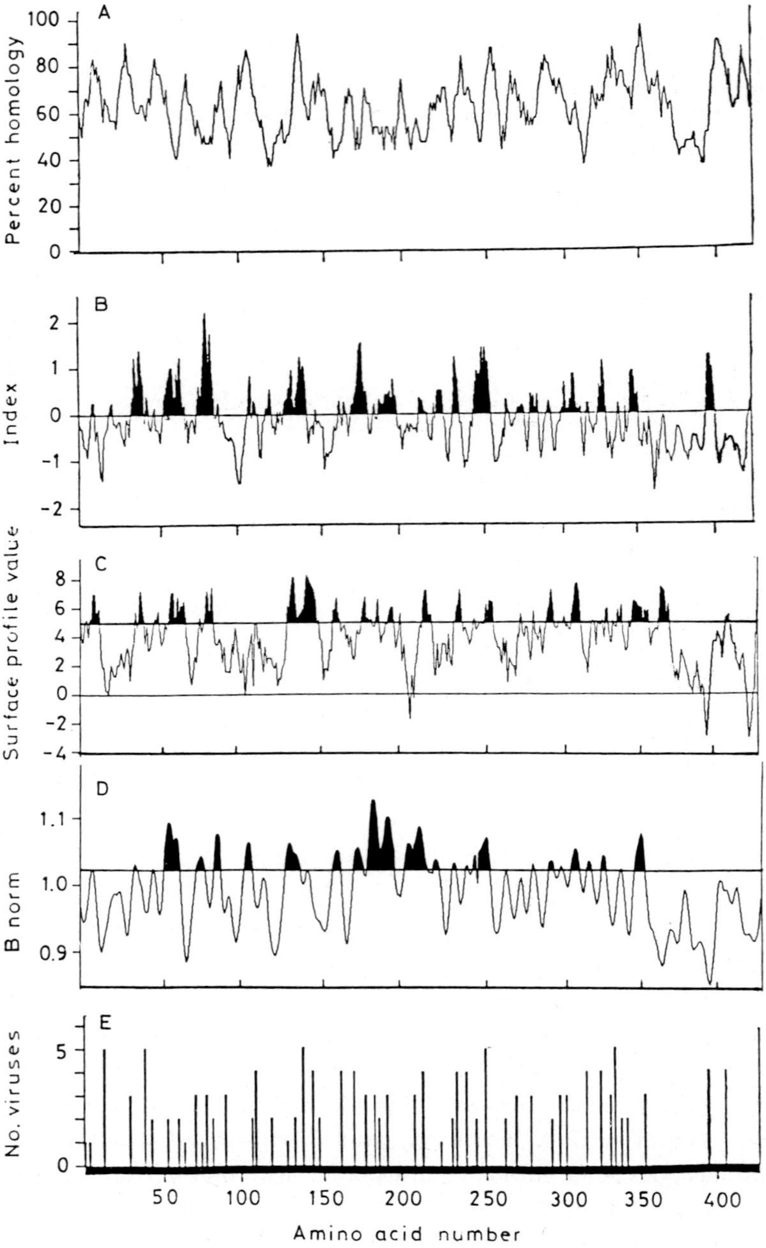

Fig. 13.5. Analysis of the consensus sequence of alphavirus E2 glycoprotein. Panel A: Homology profile using the amino acids sequence of SIN (Rice and Strauss, 1981), SF (Garoff et al., 1980), EEE (Chang and Trent, 1987), RR (Dalgarno et al. 1983), and VEE (Kinney et al., 1986) viruses. Variation from a derived consensus sequence is averaged over a 6 amino acid window. Panel B: hydrophilicity analysis using the values of Parker et al., 1986. Shaded areas are hydrophilic. Panel C: surface accessibility analy-

VEE virus and its attenuated vaccine derivative (Kinney et al., 1986; Johnson et al., 1986; Kinney et al., 1989). All of these changes occur in the E2. It is probably not coincidental that one of these changes occurs at residue 120 (similar to the 114 change in SIN) and that the attenuated vaccine demonstrates rapid penetration kinetics. These results are consistent with the observations that wt virulent VEE virus also can be attenuated by selecting for rapidly penetrating variants (Johnston and Smith, 1983)

1.3.3. Synthetic peptides and protein conformation

Attempts to mimic the E2 antigenicity with synthetic peptides have met with mixed success. Investigations with polyclonal antisera implied that recognition of some B-cell epitopes required proper E2 conformation, however all E2 mAb reactivity was retained in immunoblots with SDS denatured and reduced virus (Kaluza et al., 1980; Kaluza and Pauli, 1980; Roehrig et al., 1985). Immunoblotting of anti-VEE $E2^{c}$ mAb localized to a 12-kDa fragment prepared by enzymatic digestion with trypsin and chymotrypsin. The discrepancy in these data was probably due to reformation of disulfide linkages during the blotting process; most E2 epitopes were altered when reduced virus was alkylated before gel separation and blotting (J.T.R., unpublished observations, 1988). Peptides that elicit antiviral antibody, regardless of sequence, do not elicit neutralizing antibody (A. Hunt, written communication, 1988). A peptide from the amino-terminus of the VEE virus E2 does, however, protect mice from virus challenge while not eliciting neutralizing antibody. Antipeptide antibodies also define significant spatial overlap in the E2. Antipeptide antibody, elicited by the amino-terminal 25 amino acids, blocks virus binding of mAbs specific for the critical N domain (amino acids 180–207). This spatial overlap is consistent with that demonstrated for RR virus (Vrati et al., 1986): the deletion of amino acids 55–61 prohibited the binding of mAbs that define the critical N domain (amino acids 216–251).

1.3.4. Anti-idiotypic antibodies

Problems with models of antigenic regions also have been found with anti-idiotypic polyclonal and monoclonal antibodies to virus-neutralizing mAbs specific for VEE virus. Anti-idiotypic antibodies that specifically bound idiotype, and successfully interfered with idiotype binding to virus, did not elicit authentic anti-anti-idiotypic antibody (Ab3) with neutralizing or protective capabilities (Roehrig et al., 1985). Similar results have been reported for SF virus; however no information is currently available on the ability of these anti-idiotypes to induce authentic Ab3 (Oosterlaken

---

←

sis of Parker et al., 1986. Shaded areas are predicted to be accessible on the surface of the glycoprotein. Panel D: molecular flexibility analysis of Karplus and Schultz, 1985. Shaded areas are predicted to have molecular flexibility. Panel E: shared β-turns between 5 alphaviruses, predicted by the Rose (1978) modification of the Chou and Fasman analysis (1978).

et al., 1988a, 1988b). These results indicate that mimicking the E2 N domain in an artificial antigen is not a straightforward process.

1.4. Mechanisms of alphavirus neutralization

mAbs specific for VEE virus proteins have been used to define in vitro mechanisms of virus N (Roehrig et al., 1988). The results of this study corroborate previous results with WEE and VEE viruses (Cremer et al., 1964, 1966, 1975; Hahon, 1969, 1970). The ability of mAbs to neutralize virus infectivity correlated well with the ability to block viral attachment to susceptible cells. The N kinetics of these mAbs was quite rapid; 99% of virus was neutralized within one minute at high and low antibody doses. mAbs specific for the critical N domain on the E2 were also able to neutralize infectivity post-attachment, provided that the virion did not penetrate the cell membrane. One mAb stabilized the binding of virus to Vero cells but still neutralized virus infectivity.

1.5. Alphavirus T cell determinants

1.5.1. Helper T cell determinants

Evidence of T helper (T_h) cell activity in alphavirus immunity was first demonstrated by Burns et al. (1975). In nude mice, SIN virus elicited only a transient IgM response. Normal littermates mounted a normal antiviral IgG and anamnestic response. Subsequent studies determined that T cells were necessary to induce antigen-specific cellular proliferation (Rabinowitz and Alder, 1973). The murine T cell proliferative response to immunization with SIN, RR, and Getah viruses appeared to be virus-specific (Aaskov et al., 1983). A similar but more detailed analysis with the Western Hemisphere alphaviruses WEE, EEE, and VEE demonstrated a virus-specific T_h -cell response measured by antiserum depletion of T cell subsets, or by IL-2 production (Mathews et al., 1989). Specificity closely parallelled antibody cross-reactivity when viruses within the VEE serocomplex were studied. The T_h -cell response peaked at 14–21 days post-inoculation and was antigen dose-dependent.

Clear definition of T_h -cell determinants is based on the observation that the epitopes appear to be linear and to reside in amino acid sequences that are amphipathic α-helices (Rothbard, 1986; Berzofsky, 1988). Possible location of T_h-cell determinants on the E1 and E2 of VEE virus is shown in Fig. 13.6. A T_h -cell determinant reactive in BALB/c mice has recently been identified in the first 25 amino acids of VEE virus E2 (J.H. Mathews, written observations, 1989).

1.5.2. Cytotoxic T cell determinants

Unlike T_h -cell determinants, the determinants reactive with the cytotoxic T-cell (T_c) subset appear to be cross-reactive (Mullbacher and Blanden, 1978; Mullbacher

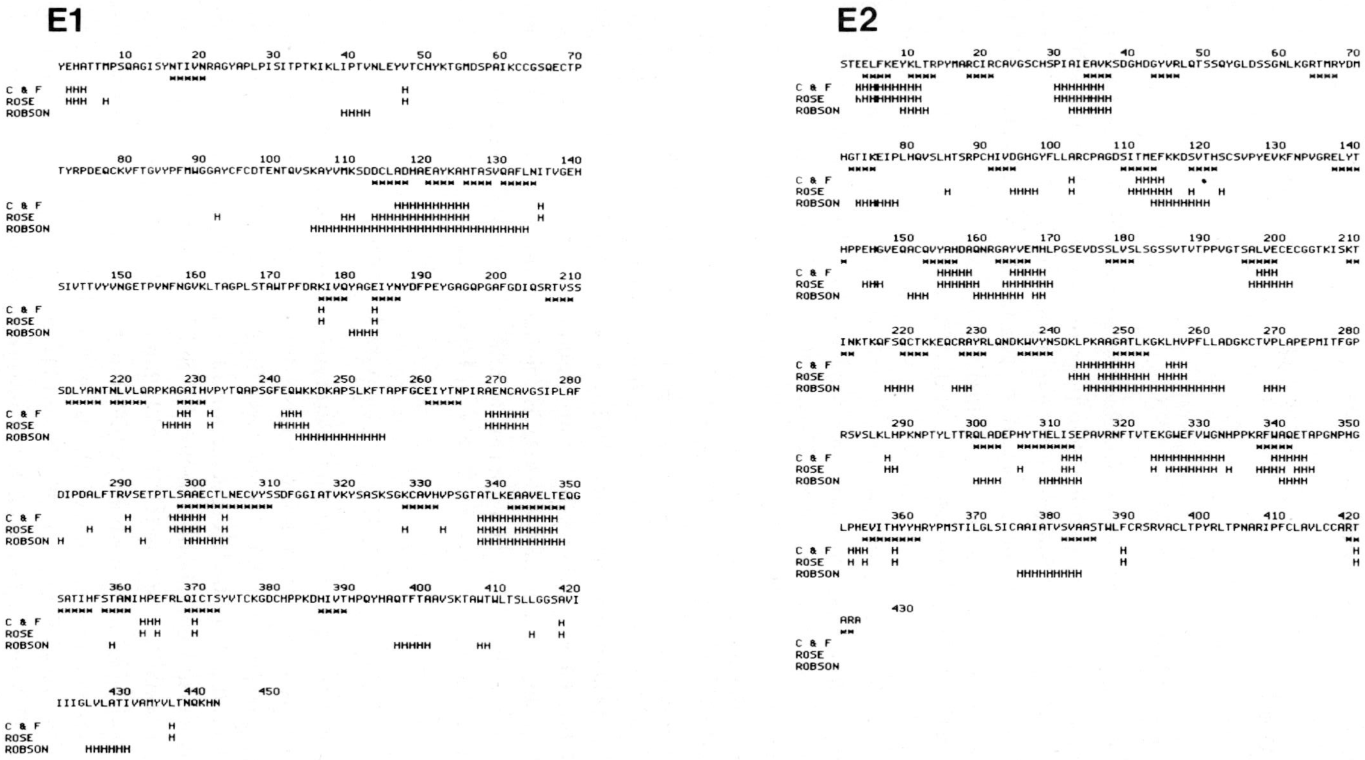

Fig. 13.6. Predicted T-cell determinants of the VEE virus glycoproteins. The sequence was scanned for amphipathic domains (*) using the Rothbard motif (1978). Secondary structure analysis used three techniques: C@F (Chou and Fasman, 1978), ROSE (Rose, 1978), and ROBSON (Garnier et al., 1978) has been used to predict α-helical structure (H). Potential T-cell domains are those areas where amphipathicity and α-helicity overlap.

et al., 1979). These results have been derived from studies with SF, Bebaru (a member of the SF serocomplex), and SIN viruses.

2. Rubiviruses

2.1. Rubella virus structure and replication

Rubella virus (RV) is the only member of the *Rubivirus* genus in the family Togaviridae. Recent investigations clearly define RV replication and structure and solidify its inclusion in the Togaviridae. Similar to that of alphaviruses, the rubella virion is composed of three structural proteins: capsid (C, 30 kDa, E2, 45–48 kDa, and E1, 63 kDa). Multiple copies of the C protein enclose a single stranded, positive-sense RNA approximately 12 000 nucleotides in length (Oker-Blom et al., 1984). The capsid is enclosed in a lipid envelope modified by the insertion of the two glycoproteins, E1 and E2, which occur as heterodimers. During viral replication, as with alphaviruses, the structural proteins are translated from a 24S subgenomic RNA transcribed from the 3′-end of the virion RNA. The structural proteins are transcribed in the following order: C, E2, E1 (Oker-Blom, 1984). Two species of E2 (E2a and E2b) can be found in the virion, but a third smaller species of E2 can be identified in infected cells. It appears that E2a and E2b are glycosylation variants with identical primary amino acid sequences (Oker-Blom et al., 1983; Kalkkinen et al., 1984; Petterson et al., 1985). Nucleotide sequencing of the RV 24S RNA identified the gene sequence of the structural proteins (Frey et al., 1986; Nakhasi et al., 1986; Vidgren et al., 1987; Clarke et al., 1987). One significant difference between RV and alphaviruses is the absence of the 6-kDa E1 signal sequence in RV. The carboxy-terminal 6 kDa of the E2 protein appears to fulfill the signal sequence function of the E1 (Hobman et al., 1988). RV also lacks an E2 precursor protein.

2.2. Antigenicity of rubella virus envelope glycoproteins

Historically, RV was thought to be a single serotype because infection or vaccination with RV conferred life long immunity; however, antigenic variation has been identified in the N response to RV (Gould and Butler, 1980). People immunized with RV vaccine produce antibodies specific for all structural proteins (Cusi et al., 1988). mAbs have been prepared for both the E1 and E2 glycoproteins. At least six epitopes have been identified on the RV E1 (Waxham and Wolinsky, 1983, 1985a, 1985b); none appear to elicit N antibody, while four elicit HI antibody. Three spatially distinct domains were identified on the E1 of the HPV-77 strain of RV (Dorsett et al., 1985; Green and Dorsett, 1986). N and HI activity of E1 were dissociable for this RV strain. One mAb specific for E2 demonstrated virus N (Green and Dorsett, 1986). The contribution of each virus glycoprotein to protection or recovery from rubella has not been defined. Similarly, identification of RV T cell determinants has not been reported.

3. Conclusions

Alphaviruses are peculiar agents of human and veterinary disease. Because mosquitoes are required for transmission in natural epidemics, efforts have focused on vector control for disease intervention and prevention. These efforts have had limited success due to the resilience of the mosquito. The progress described here shows that highly effective, genetically stable, reproducible alphavirus vaccines can be prepared. An understanding of the protective immune response and of the factors that define alphavirus virulence, should help to focus rapid development of recombinant vaccine candidates. Efforts will be guided by current studies with artificial subunit alphavirus immunogens and alphavirus T-cell determinants. A special effort should be mounted against RV. The similarity in structure and replication of RV and of alphaviruses should allow for a rapid understanding of RV pathogenesis and immunity. Both the molecular determinants of RV attenuated vaccine teratogenesis and the true contribution and function of E2 and RV replication and immunity must be defined. Continued effort and progress will make this a truly exciting time in togavirus immunochemistry.

4. Acknowledgements

I want to thank my colleagues for promptly supplying me with recent research results for this manuscript. I also want to acknowledge members of the Division of Vector-Borne Viral Diseases, Centers for Disease Control, particularly those of the Immunochemistry Branch, for their significant contributions to the understanding of arbovirus immunochemistry.

5. References

Aaskov, J.G., Dalglish, D.A., Davies, C.E.A., Tucker, M. and Donaldson, M.D. (1983) *Aust. J. Exp. Biol. Med.* 61, 529–540.

Berzofsky, J.A. (1988) *Vaccine* 6, 89–93.

Boere, W.A.M., Benaissa-Trouw, B.J., Harmsen, M., Kraaijeveld, C.A. and Snippe, H. (1983) *J. Gen. Virol.* 64, 1405–1408.

Boere, W.A.M., Harmsen, M., Vinje, J., Benaissa-Trouw, B.J., Kraaijeveld, C.A. and Snippe, H. (1984) *J. Virol.* 52, 575–582.

Boere, W.A.M., Benaissa-Trouw, Harmsen, T., Erich, T., Kraaijeveld, C.A. and Snippe, H. (1985(*J. Virol.* 54, 546–551.

Boggs, W.M., Hahn, C.S., Strauss, E.G., Strauss, J.H. and Griffin, D.E. (1990) *Virology* 169, 485–488.

Calisher, C.H., Shope, R.E., Brandt, W.E., Casals, J., Karabatsos, N.K., Murphy, F.M., Tesh, R.B. and Wiebe, M.E. (1980) *Intervirology*14, 229–232.

Chanas, A.C., Gould, E.A., Clegg, J.S.C. and Varma, M.G.R. (1982) *J. Gen. Virol.* 58, 37–46.

Chang, G.J. and Trent, D.W. (1987) *J. Gen. Virol.* 68, 2129–2142.

Chou, P.Y. and Fasman, G.D. (1978) *Annu. Rev. Biochem.* 47, 251–276.

Clarke, D.M., Loo, T.W., Hui, I., Chong, P. and Gillam, S. (1987) *Nucleic Acids Res.* 15, 3041–3057.

Clegg, J.C.S., Chanas, A.C. and Gould, E.A. (1983) *J. Gen. Virol.* 64, 1121–1126.

Cremer, N.E., Riggs, J.L., Fujimoto, F.Y., Hagens, S.J., Ota, M.I. and Lennette, E.H. (1964) *J. Immunol.* 93, 283–292.

Cremer, N.E., Lennette, E.H., Hagens, S.J. and Fujimoto, F.Y. (1966) *J. Immunol.* 96, 284–288.

Cremer, N.E., Riggs, J.L. and Lennette, E.H. (1975) *Immunochemistry* 12, 597–601.

Cusi, M.G., Rossolini, G.M., Cellesi, C. and Valensin, P.E. (1988) *Arch. Virol.* 101, 25–33.

Dalgarno, L., Rice, C.M. and Strauss, J.H. (1983) *Virology* 129, 170–187.

Dalrymple, J.M., Schlesinger, S. and Russell, P.K. (1976) *Virology* 69, 93–103.

Davis, N.L., Fuller, F.J., Dougherty, W.G., Olmsted, R.A. and Johnston, R.E. (1986) *Proc. Natl. Acad. Sci.* USA 83, 6771–6775.

Dorsett, P.H., Miller, D.C., Green, K.Y. and Byrd, F.I. (1985) *Rev. Infect. Dis.* 7(sl), S150–S156.

Faragher, S.G., Meek, A.D.J., Rice, C.M. and Dalgarno, L. (1988) *Virology* 163, 509–526.

Frey, T.K., Marr, L.D., Hemphill, M.L. and Dominguez, G. (1986) *Virology* 154, 228–232.

Fuller, S.D. (1987) *Cell* 48, 923–934.

Garnier, J., Osguthorpe, D.J. and Robson, B. (1978) *J. Mol. Biol.* 120, 97–120.

Garoff, H., Frischauf, A.-M., Simons, K., Lehrach, H. and Delius, H. (1980) *Nature* 288, 236–241.

Gates, D., Brown, A. and Wust, C.J. (1982) *Infect. Immun.* 35, 248–255.

Gething, M.J., White, J.M. and Waterfield, M.D. (1978) Proc. *Natl. Acad. Sci. USA* 75, 2737–2740.

Gould, J.J. and Butler, M. (1980) *J. Gen. Virol.* 49, 423–426.

Green, K.Y. and Dorsett, P.H. (1986) *J. Virol.* 57, 893–898.

Hahn, C.S., Lustig, S., Strauss, E.G. and Strauss, J.H. (1988) *Proc. Natl. Acad. Sci. USA* 85, 5997–6001.

Hahon, N. (1969) *J. Gen. Virol.* 4, 77–88.

Hahon, N. (1970) *J. Gen. Virol.* 6, 361–372.

Helenius, A. (1984) *Biol. Cell.* 51, 181–186.

Hobman, T.C., Shukin, R. and Gillam, S. (1988) *J. Virol.* 4259–4264.

Hunt, A.R. and Roehrig, J.T. (1985) *Virology* 142, 334–346.

Johnson, B.J.B., Kinney, R.M., Kost, C.L. and Trent, D.W. (1986) *J. Gen. Virol.* 67, 1951–1960.

Johnson, B.J.B., Brubaker, J.R., Roehrig, J.T. and Trent, D.W. (1989) *Virology*, in press.

Johnston, R.E. and Smith, J.F. (1988) *Virology* 162, 437–443.

Kaluza, G. and Pauli, G. (1980) *Virology* 102, 300–309.

Kaluza, G., Rott, R. and Schwarz, R.T. (1980) *Virology* 102, 286–299.

Karplus, P.A. and Schultz, G.E. (1985) *Naturwissenschaften* 72, 212–213.

Kalkkinen, N., Oker-Blom, C., and Pettersson, R.F. (1984) *J. Gen. Virol.* 65, 1549–1557.

Kielan, M.C. and Helenius, A. (1985) J. *Cell Biol.* 101, 2284–2291.

Kielan, M.C. and Helenius, A. (1986) In *The Togaviridae and Flaviviridae* (S. Schlesinger, and M.J. Schlesinger, Eds.) Plenum Press, New York. pp. 91–119.

Kielan, M.C., Marsh, M. and Helenius, A. (1986) *EMBO J.* 5, 3103–3109.

Kinney, R.M., Trent, D.W. and France, J.K. (1983) *J. Gen. Virol.* 64, 135–147.

Kinney, R.M., Johnson, B.J.B., Brown. V.L. and Trent, D.W. (1986) *Virology* 152, 400–413.

Kinney, R.M., Esposito, J.J., Johnson, B.J.B., Roehrig, J.T., Mathews, J.H., Barrett, A.D.T. and Trent, D.W. (1988a) *J. Gen. Virol.* 69, 3005–3013.

Kinney, R.M., Esposito, J.J., Mathews, J.H., Johnson, B.J.B., Roehrig, J.T., Barrett, A.D.T. and Trent, D.W. (1988b) *J. Virol.* 62, 4697–4702.

Kinney, R.M., Johnson, B.J.B., Welch, J.B., Tsuchiya, K.R. and Trent, D.W. (1989) *Virology* 170, 19–30.

Lustig, S., Jackson, A.C., Hahn, C.S., Griffin, D.E., Strauss, E.G. and Strauss, J.H. (1988) *J. Virol.* 62, 2329–2336.

Marsh, M., Bolzau, E. and Helenius, A. (1983) *Cell* 32, 931–940.

Mathews, J.H. and Roehrig, J.T. (1982) *J. Immunol.* 129, 2763–2767.

Mathews, J.H. and Roehrig, J.T. (1989) *J. Gen. Virol.* 70, 2877–2886.

Mathews, J.H., Roehrig, J.T. and Trent, D.W. (1985) *J. Virol.* 5, 594–600.

Mendoza, Q.P., Stanley, J. and Griffin, D.E. (1988) *J. Gen. Virol.* 70, 3015–3022.

Mullbacher, A. and Blanden, R.V. (1978) *Immunogenetics* 7, 551–561.

Mullbacher, A., Marshall, I.D. and Blanden, R.V. (1979) *Scan. J. Immunol.* 10, 291–296.

Nakhasi, H.L., Meyer, B.C. and Liu, T-Y. (1986) *J. Biol. Chem.* 261, 16616–16621.

Oker-Blom, C., Kalkkinen, N., Kaariainen, L. and Pettersson, R.F. (1983) *J. Virol.* 46, 964–973.

Oker-Blom, C. (1984) *J. Virol.* 51, 354–358.

Oker-Blom, C., Ulmanen, I., Kaariainen, L. and Pettersson, R.F. (1984) *J. Virol.* 49, 403–408.

Olmsted, R.A., Baric, R.S., Sawyer., B.A. and Johnston, R.E. (1984) *Science* 225, 424–426.

Olmsted, R.A., Meyer, W.J. and Johnston, R.E. (1986) *Virology* 148, 245–254.

Oosterlaken, T.A.M., Harmsen, M., Tangerman, C., Schielen, P., Kraaijeveld, C.A. and Snippe, H. (1988a) *J. Immunol. Methods* 115, 255–261.

Oosterlaken, T.A.M., Kraaijeveld, C.A., Snijders, A., Harmsen, M., Benaissa-Trouw and Snippe, H. (1988b) *Arch. Virol.* 100, 115–120.

Parker, J.M.R., Guo, D. and Hodges, R.S. (1986) *Biochemistry* 25, 5425–5432.

Peck, R., Wust, C.J. and Brown, A. (1979) *Infect. Immun.* 25, 320–327.

Pettersson, R.F., Oker-Blom, C., Kalkkinen, N., Kallio, A., Ulmanen, I., Kaariainen, L., Partanen, P. and Vaheri, A. (1985) *Rev. Infect. Dis.* 7(sl), S140–S149.

Polo, J.M., Davis, N.L., Rice, C.M., Huang, H.V. and Johnston, R.E. (1988) *J. Virol.* 62, 2124–2133.

Rabinowitz, S.G. and Alder, W.H. (1973) *J. Immunol.* 110, 1345–1353.

Rice, C.M. and Strauss, J.H. (1981) *Proc. Natl. Acad. Sci. USA* 78, 2062–2066.

Roehrig, J.T., Gorski, D. and Schlesinger, M.J. (1982a) *J. Gen. Virol.* 59, 421–425.

Roehrig, J.T., Day, J.W. and Kinney, R.M. (1982b) *Virology* 118, 269–278.

Roehrig, J.T. and Mathews, J.H. (1985) *Virology* 142, 347–365.

Roehrig, J.T., Hunt, A.R. and Mathews, J.H. (1985) In *High Technology Route to Virus Vaccines* (G.R. Dreesman, J.G. Bronson, and R.C. Kennedy, Eds.) *American Society for Microbiology*, Washington DC. pp. 142–153.

Roehrig, J.T. (1986) In *The Togaviridae and Flaviviridae* (S. Schlesinger, and M.J. Schlesinger, Eds.) Plenum Press, New York. pp. 251–278.

Roehrig, J.T., Hunt, A.R., Kinney, R.M. and Mathews, J.H. (1988) *Virology* 171, 49–60.

Roehrig, J.T., Hunt, A.R., Johnson, A.J. and Hawkes, R.A. (1989) *Virology*, in press.

Rose, G.D. (1978) *Nature* 272, 586–590.

Rothbard, J.B. (1986) *Ann. Inst. Pasteur,* 137E, 519–525.

Schlesinger, M.J. and Schlesinger, S. (1986) In *The Togaviridae and Flaviviridae* (S. Schlesinger and M.J. Schlesinger, Eds.) Plenum Press, New York pp. 121–148.

Schmaljohn, A.L., Johnson, E.D., Dalrymple, J.M. and Cole, G.A. (1982) *Nature* 297, 70–72.

Schmaljohn, A.L., Kokubun, K.M. and Cole, G.A. (1983) *Virology* 130, 144–154.

Schmid, S., Fuchs, R., Kielan, M., Helenius, A. and Mellman, I. (1989) *J. Cell Biol.* 108, 1291–1300.

Smith, C., Wolcott, J.A., Wust, C.J. and Brown, A. (1985) *J. Virol.* 53, 198–204.

Stanley, J., Cooper, S.J. and Griffin, D.E. (1986) *J. Virol.* 58, 107–115.

Stec, D.S., Waddell, A., Schmaljohn, C.S., Cole, G.A. and Schmaljohn, A.L. (1986) *J. Virol.* 57, 715–720.

Strauss, E.G. and Strauss, J.H. (1986) In *The Togaviridae and Flaviviridae* (S. Schlesinger and M.J. Schlesinger, Eds.) Plenum Press, New York. pp. 35–80.

Strauss, E.G., Schmaljohn, A.L., Griffin, D.E. and Strauss, J.H. (1987) *UCLA Symp Mol Cell Biol New Ser.* 54, 356–378.

Trent, D.W., Clewley, J.P., France, J.A. and Bishop, D.H.L. (1979) *J. Gen. Virol.* 43, 365–381.

Trent, D.W. and Grant, J.A. (1980) *J. Gen. Virol.* 47, 261–282.

Väänänen, P. and Kääriäinen, L. (1980) *J. Gen. Virol.* 46, 467–475.

Väänänen, P., Gahmberg, C.G. and Kääriäinen, L. (1981) *Virology* 110, 366–374.

Vidgren, G., Takkinen, K., Kalkkinen, N., Kääriäinen, L. and Pettersson, R.F. (1987) *J. Gen. Virol.* 68, 2347–2357.

Vrati, S., Faragher, S.G., Weir, R.C. and Dalgarno, L. (1986) *Virology* 151, 222–232.

Vrati, S., Fernon, C.A., Dalgarno, L. and Weir, R.C. (1988). *Virology* 162, 346–353.

Waxham, M.N. and Wolinsky, J.S. (1983) *Virology* 126, 194–203.

Waxham, N.M. and Wolinsky, J.S. (1985a) *Virology* 143, 153–165.

Waxham, M.N. and Wolinsky, J.S. (1985b) *Rev. Infect. Dis.* 7(s1), S133–S139.

White, J. and Helenius, A. (1980) *Proc. Natl. Acad. Sci. USA* 77, 3273–3277.

Wolcott, J.A., Gates, D.W., Wust, C.J. and Brown, A. (1982) *Infect. Immun.* 36, 704–709.

Wolcott, J.A., Wust, C.J. and Brown, A. (1984) *J. Virol.* 49, 379–385.

Wolcott, J.A., Wust, C.J. and Brown, A. (1985) *J. Gen. Virol.* 66, 1167–1170.

Wust, C.J., Crombie, R. and Brown, A. (1987) *Proc. Soc. Exp. Biol. Med.* 184, 56–63.

Eds. M.H.V. van Regenmortel and A.R. Neurath
Immunochemistry of viruses, II. The basis for serodiagnosis and vaccines.

Flaviviruses

14

F.X. HEINZ[1] and J.T. ROEHRIG[2]

[1] Institute of Virology, University of Vienna, Vienna, Austria, and [2] Immunochemistry Branch, Division of Vector-Borne Viral Diseases, Centers for Disease Control, Ft. Collins, CO, U.S.A.

Flaviviruses are spherical, lipid enveloped viruses that contain a positive strand RNA genome. Due to morphological similarities, flaviviruses were previously classified as a genus of the family Togaviridae. More recent data on virion structure, genome structure, replication and translation strategy, however, have revealed the uniqueness of these viruses and led to the establishment of the new family Flaviviridae (Westaway et al., 1985) with yellow fever virus being the prototype species.

At least 68 serologically related flaviviruses have been recognized to date (Calisher et al., 1989) of which about 70% are known to be transmitted by arthropods (mosquitoes or ticks) to their vertebrate hosts. Many flaviviruses are pathogenic for man and animals; several of them cause diseases that are prevalent and severe enough to represent a significant public health problem in endemic or epidemic areas (for review see Monath, 1986).

Established and effective vaccines are available against yellow fever (live attenuated), Japanese encephalitis and tick-borne encephalitis (inactivated and purified virus derived from infected mouse brain and infected primary chick embryo cells, respectively) (reviewed by Stephenson, 1988).

1. Genome organization and virion structure

The nucleotide sequence of the flavivirus RNA genome, first established for yellow fever virus by Rice et al. (1985), reveals a single long open reading frame of about 10 200 nucleotides coding for all structural and nonstructural proteins. The coding region starts with the sequence for the core (C), prM (precursor of the membrane protein M), and the envelope (E) structural proteins followed by a series of non-structural proteins ordered as follows: NS1, NS2A, NS2B, NS3, ns4a, NS4B, and NS5. The amino and carboxy-termini of all the structural proteins (Nowak et al., 1989) and the amino-termini of all nonstructural proteins except ns4a (Rice et al., 1986a; Speight et al., 1988; Chambers et al., 1989) have so far been identified for individual flaviviruses by protein sequencing.

The individual proteins are thought to be formed by co- or post-translational cleavage of a large precursor protein. As deduced from the specific amino acid sequences found at the cleavage sites, it is proposed that cellular proteases (including signalase) and at least one virus-specified protease are involved in the processing of the precursor protein (for review see Rice et al., 1986b).

Electron micrographs of negatively stained flaviviruses reveal spherical particles with a diameter of about 50 nm which carry a fringe of fine surface projections (Fig. 14.1A). In some instances, 7-nm ring-shaped elements with a 2- or 3-nm central hole were revealed at the virion surface (Smith et al., 1970; Wengler et al., 1987). These could be isolated by proteolytic cleavage and detergent treatment (Wengler et al., 1987). Completely processed, mature virions contain only the three structural

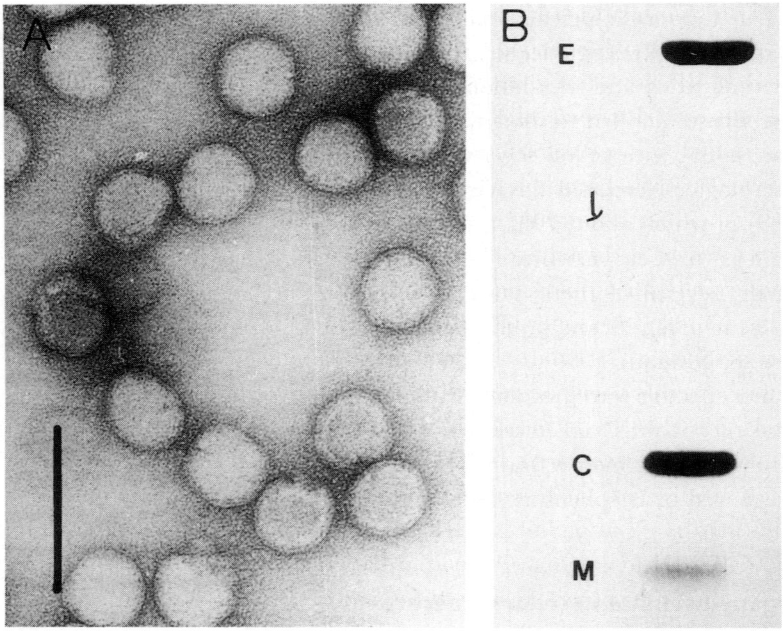

Fig. 14.1. (A) Electron micrograph of TBE virions purified from the supernatant of infected chick ambryo cells. Bar represents 100 nm. (B) SDS-PAGE of purified TBE virions revealing the three structural proteins E, C, and M. Reproduced from Heinz (1986) with permission

proteins E, C, and M with approximate molecular weights of 50 000–60 000, 15 000, and 8 000, respectively (Fig. 14.1B). The genome RNA together with protein C forms an isometric nucleocapsid whereas both E and M are associated with the lipid envelope by their carboxy-terminal ends (Fig. 14.2). The E protein of most flaviviruses is glycosylated as revealed by the incorporation of labeled sugars (Shapiro et al., 1971; Westaway, 1975), the reactivity with Concanavalin A (Stohlman et al., 1976; Clegg, 1982; Winkler et al., 1987a) and the presence of sequences that are potential *N*-glycosylation sites. At least one strain of West Nile virus (Wengler et al., 1985) and Kunjin virus (Coia et al., 1988) however, contain an unglycosylated E protein that is devoid of potential *N*-glycosylation sites. The M protein is generated by proteolytic cleavage (presumably by a Golgi protease) of its glycosylated precursor protein, prM. This precursor protein is a major constituent of immature virus particles found in infected cells (I forms) which can be released by the incubation with Tris (T forms) (Fig. 14.2A). This cleavage, which seems to be associated with virus release, does not necessarily occur at all individual protein molecules present in the virus membrane since M together with prM were found in Japanese encephalitis virus (Kimura-Kuroda and Yasui, 1983), dengue 2 virus (Henchal et al., 1985) and West Nile virus (Nowak et al., 1989) purified from the supernatant of infected cells (Fig. 14.2B).

As deduced from genomic sequence analyses (reviewed by Rice et al., 1986) as well as amino-terminal and carboxy-terminal protein sequence analyses (Nowak et al., 1989), the biosynthesis of the structural proteins (and the first nonstructural protein NS1) involves internal signal sequences that lead to the transfer of the nascent polypeptide into the lumen of the endoplasmic reticulum. Hydrophobic anchor sequences near the carboxy-terminal ends followed by charged amino acids stop the membrane transfer. The amino-termini of prM, E, and NS1 are generated by signalase. The carboxy-termini of both prM (M) and E still contain the internal signal sequence for the subsequent protein so that both are believed to contain a double membrane anchor with the carboxy-terminus located outside the membrane (Fig.

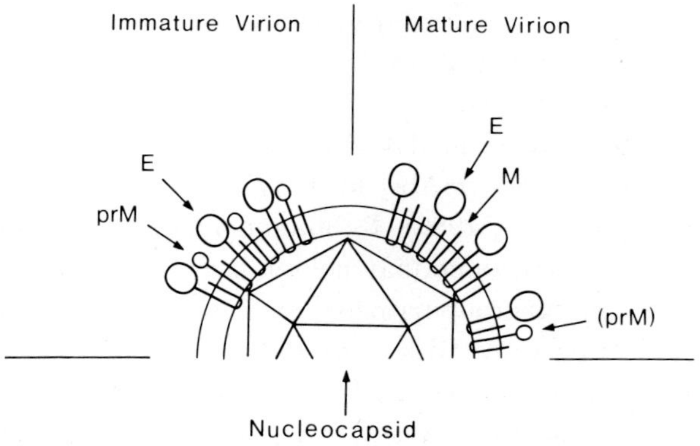

Fig. 14.2. Schematic model of a flavivirus. (A) immature (intracellular) form; (B) mature virions.

14.2). The virion associated C protein has lost its carboxy-terminal membrane anchor presumably by the action of a virus-encoded protease (Nowak et al., 1989). The proteins E and M seem to be closely associated in the viral membrane since both are found in complexes isolated after solubilization with Triton X-100 (Heinz and Kunz, 1980). Cross-linking experiments of protease-treated, octylglucoside solubilized E protein from West Nile virus revealed the presence of trimeric complexes suggesting that this protein forms a similar structure in the viral membrane (Wengler et al., 1987), although the preferential formation of oligomeric structures was not observed upon cross-linking of whole-tick-borne encephalitis virus (Heinz and Kunz, 1980).

2. Relationships among flaviviruses

Since most studies on immunological cross-reactivities using polyclonal immune sera were performed by hemagglutination inhibition (HI) and neutralization (NT) assays, the established antigenic relationships generally relate to the E protein which is the target for both HI and NT antibodies. As originally revealed by HI tests (Casals, 1957) all flaviviruses share common antigenic determinants. By the more discriminating cross-neutralization assay, however, flaviviruses can be antigenically subdivided into several serocomplexes or subgroups, each containing more closely related viruses. An extensive cross-neutralization study involving 42 flaviviruses (seven subgroups) was performed by DeMadrid and Porterfield (1974) and was more recently confirmed and extended to 66 flaviviruses (eight subgroups) by Calisher et al. (1989). The most important subgroups—with respect to disease potential are (i) the Japanese encephalitis subgroup: Japanese encephalitis (JE), Murray Valley encephalitis (MVE), West Nile (WN), St. Louis encephalitis (SLE), Kunjin (KUN), Usutu, Kokobera, Stratford, Alfuy, (ii) the Dengue subgroup: dengue (DEN) types 1, 2, 3, 4) and (iii) the tick-borne encephalitis (TBE) subgroup: tick-borne encephalitis Western (TBE-W.) and Far Eastern (TBE-F.E.) subtype, Louping Ill, Omsk hemorrhagic fever, Kyasanur Forest disease, Langat, Negishi. Interestingly, the prototype YF virus is not sufficiently cross-reactive by NT to warrant inclusion in any of the established serocomplexes.

The complexity of the antigenic structure of the flavivirus E protein was emphasized by application of monoclonal antibodies (mAbs) to study antigenic relationships among flaviviruses (reviewed by Heinz, 1986). By immunofluorescence and other binding assays, reactivity patterns were found that had not been previously recognized and did not correspond to the subgrouping based on cross-neutralization using polyclonal sera. Even neutralization assays with mAbs did not necessarily correlate with the established reactivity pattern. Several of these studies revealed the interesting phenomenon that the functional activity of a mAb may be different for different virus types or even strains, although the epitope—as revealed by antibody binding—is identical on both viruses. Paradoxically certain antibodies may be functionally active (e.g., in neutralization) only with a heterologous virus but not with the homologous virus which was used to raise the antibody (reviewed in Heinz, 1986; Cammack and Gould, 1986; Kimura-Kuroda and Yasui, 1986; Hawkes et al., 1988).

The last few years have led to a burst of sequence information on viral proteins that was deduced from the cDNAs of a number of different flaviviruses. This allows for the first time the correlation of antigenic relationships with sequence homologies of individual proteins (Table 1). For those flaviviruses sequenced so far the amino acid sequence homology data by and large reflect the serocomplexes established by cross-NT using polyclonal sera. It is striking that among the structural proteins the E protein reveals the highest degree of conservation, whereas the internal C protein is the most variable one. Based on the amino acid homology data of the C proteins, an evolutionary tree can be constructed (Fig. 14.3) (Mandl C., 1988) that represents a valuable supplement to the classification of flaviviruses and is concordant with the established system of serocomplexes, types and subtypes within the family Flaviviridae. There is a 20–30% homology between members of different subgroups, the isolated nature of YF virus is confirmed, and closer relationships among certain members of the same subgroup are revealed (60–70% homology). WN and KUN virus may be classified as two different subtypes of the same virus, since they share 87% of the amino acids, compared to 86% homology between the Western and Far Eastern subtype of TBE virus.

TABLE 14.1

Amino acid sequence homologies of flaviviruses

Subgroups compared[a]	Percent homology[b]			
	Protein C	prM minus M[c]	Protein M	Protein E
SG1-SG1	86	97	88	96
SG1-SG3	19–25	29–33	25–31	41–44
SG1-SG7	20–28	26–31	20–29	41
SG1-SGY	32–33	33	22–25	45
SG3-SG3[d]	57–72 (87)	61–77 (91)	61–83 (96)	72–83 (93)
SG3-SG7	30–40	34–44	29–41	46–53
SG3-SGY	25–29	33–39	35–40	44–46
SG7-SG7	66–68	69–77	59–69	62–69
SG7-SGY	25–29	31–34	37–39	41–46

[a] SG1: TBE-W., TBE-F.E.; SG3: JE, MVE, WN, SLE, KUN; SG7: DEN-1, DEN-2, DEN-4; SGY: YF

[b] Calculated using the Beckman Microgenie software package (Version 4.0)

[c] Part of protein prM that is not contained in protein M

[d] Numbers in brackets represent the homology data of the viruses WN vs. KUN

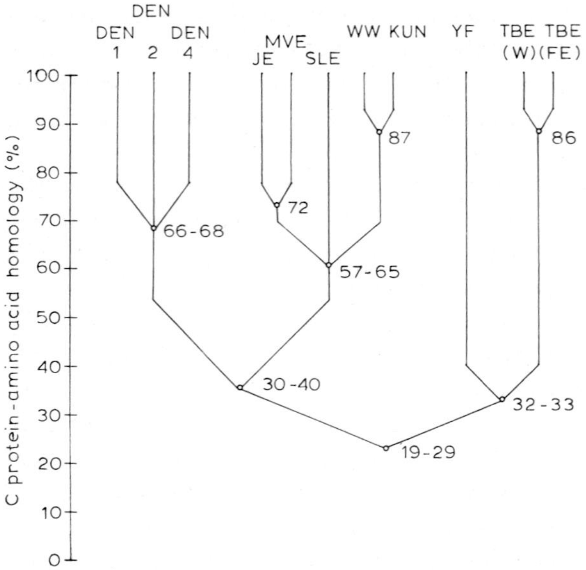

Fig. 14.3. Evolutionary tree of flaviviruses based on amino acid sequence homologies of the C protein.

Amino acid variation is not uniformly distributed, conserved and variable sequences tending to alternate along the protein sequences. However, the hydrophobicity profiles for the whole coding region of all flaviviruses sequenced so far, are strikingly similar suggesting a structural homology even in the absence of amino acid sequence homology.

3. Role of virus proteins in immune protection

Consistently with its role as the major virion surface component, the E protein plays a dominant role in the generation of neutralizing antibodies and the induction of a protective immunity. This was revealed by animal immunization experiments using isolated subviral components of defined protein composition (Kitano et al., 1974; Della Porta and Westaway, 1977; Heinz et al., 1981; Takegami et al., 1982) as well as neutralization and passive protection experiments with E protein-specific mAbs (reviewed by Roehrig, 1986, and Heinz, 1986). Studies with Japanese encephalitis virus indicated that low titer neutralizing antibodies were also induced by the M protein, which had been isolated by SDS-PAGE and allowed to renature by the removal of SDS using ion exchange chromatography (Takegami et al., 1982). Part of the M protein is therefore probably exposed on the surface of virions which is consistent with its accessibility for the action of proteases (Heinz and Kunz, 1979; Wengler et al., 1987).

Paradoxically, one of the nonstructural proteins (NS1) may also play an important role in immune protection from flavivirus infections. Antibodies to this protein, which was shown to correspond to the originally described soluble complement fixing (SCF) antigen (Smith and Wright, 1985), were found in human dengue patients (Falker et al., 1973), TBE patients (Guirakhoo, F., personal communication) and animals experimentally infected with TBE virus (Hambleton et al., 1983). This is very likely a characteristic of the immune response to flaviviruses in general. With the availability of NS1-specific mAbs it was possible to demonstrate that some of these, depending on their epitope specificity, were capable of conferring immunity against infection in experimental animals without having neutralizing activity (Schlesinger et al., 1985; Gould et al., 1986). This phenomenon has been correlated with complement-mediated lysis of infected cells that express NS1 at the plasma membrane (Schlesinger et al., 1985).

The induction of a protective immunity by NS1 was also directly shown by immunization with immunoaffinity-purified preparations of NS1 derived from YF (Schlesinger et al., 1986) and DEN-2 (Schlesinger et al., 1987) virus-infected cells which conferred resistance to virus challenge in experimental animals. These findings have important implications for current and future developments of new flavivirus vaccines, especially against DEN viruses (Brandt, 1988). By in vitro studies with polyclonal and monoclonal antibodies, it has been shown for several flaviviruses and also for alpha-viruses, poxviruses, bunyaviruses, rhabdoviruses, coronaviruses, herpesviruses, and reoviruses (reviewed by Porterfield, 1986) that non-neutralizing and neutralizing anti-virion antibodies at subneutralizing concentrations may enhance the infection of Fc-receptor or complement receptor-bearing cells. This phenomenon has therefore been termed 'antibody-dependent enhancement of infection' (ADE). Since ADE has been supposed to play a role in the pathogenesis of dengue hemorrhagic fever and dengue shock syndrome (Halstead, 1988) there are concerns about the use of the E protein as a component of dengue vaccines, e.g.,by recombinant DNA technologies. NS1 which is not a component of the virion and cannot induce infection-enhancing antibodies therefore represents a major target of current vaccine development programs (Brandt, 1988).

4. The antigenic structure of the E protein

The E protein is of central interest to several areas of flavivirus research owing to its many functional properties. These include receptor binding, hemagglutination of erythrocytes at acid pH, the capability of inducing a protective immune response and the putative involvement in an intraendosomal acid-catalyzed fusion step necessary for infection. Therefore the elucidation of its structural organization and the identification of functionally active domains represent an important key to the understanding of flavivirus biology. A major step forward was achieved by the application of mAbs for the analysis of the antigenic structure by the identification of individual epitopes and their functional and structural properties as well as their topological relationships (reviewed by Roehrig 1986, and Heinz, 1986). More recently additional, extensive epitope maps were published by Cammack and Gould (1986) for YF virus, Kimura-Kuroda et al. (1986) and Cecilia et al. (1988) for JE virus, Hawkes et al. (1988) for MVE virus, and by Guirakhoo et al. (1989) for TBE virus. These studies provided a large amount of information which in most cases, however, has not been correlated to structural and sequence data. Based on the absolute conservation of all 12 cysteines in addition to practically superimposable hydrophilicity plots one is justified in assuming a common structural organization shared by all flavivirus E proteins. Since the epitope map of TBE virus has been further developed into a structural E protein model (Mandl et al., 1989), this virus will be used as a basis in an attempt to provide a more general picture of the antigenic structure of flavivirus E proteins.

Fig. 14.4 shows an epitope map of the TBE virus E protein revealing serological specificities, functional activities, and topological relationships of 19 different epitopes. As deduced from mutual antibody blocking studies most of these epitopes cluster to form three major antigenic domains, termed A, B, and C, whereas three mAbs define isolated epitopes (i1, i2, and i3). The three antigenic domains are not homogeneous with respect to serological specificity and functional activity. There is a clear dissociation of functional activities at the epitope level (i.e., hemagglutination inhibition, neutralization and passive protection) and all different combinations (e.g.,HI^+, NT^+; HI^+, NT^-; HI^-, NT^+) were described in mAb studies with different flaviviruses (reviewed in Heinz, 1986; Kimura-Kuroda and Yasui, 1986; Cecilia et al., 1988). In several instances mAbs were identified that were even protective

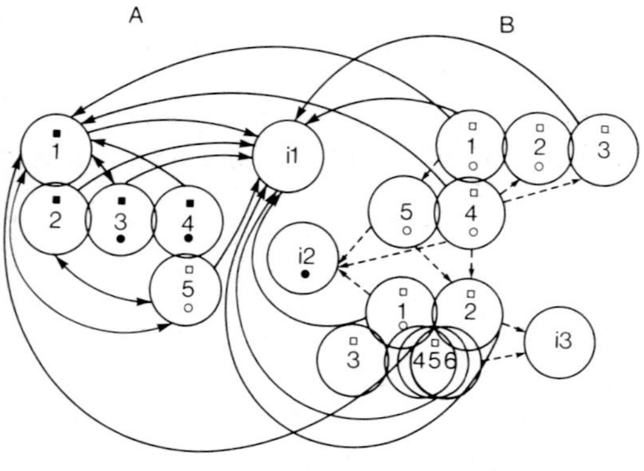

Fig. 14.4. Epitope model of the TBE virus glycoprotein E showing three antigenic domains (A, B, C) and three isolated epitopes (i1, i2, i3). Overlapping circles indicate mutual blocking, dotted arrows one-way blocking, and solid arrows enhancement of binding as determined in competitive binding assays. Functional activities: (■) strong HI activity; (□) weak HI activity; (●) strong NT activity; (○) weak HI activity. Reproduced from Guirakhoo et al. (1989) with permission

in vivo but non-neutralizing in vitro (Gould et al., 1986: Brandriss et al., 1986). Domain A-specific neutralizing and protective mabs seem to have higher specific activities than those directed to other sites.

Competitive binding studies with TBE virus (Heinz et al., 1984; Guirakhoo et al., 1989) and other flaviviruses (Schlesinger et al., 1984; Henchal et al., 1985; Kimura-Kuroda and Yasui, 1986; Cecilia et al., 1988) revealed that certain mAbs may not only block but can even enhance the binding of mAbs to distant epitopes indicated by the arrows in Fig. 14.4. This phenomenon is based on an enhancement of antibody avidity (Heinz et al., 1984) and is also of relevance in functional terms, since cooperative interactions of monoclonal antibodies were also shown to operate in neutralization and/or passive protection experiments for WN virus (Peiris et al., 1982), TBE virus (Heinz et al., 1986), and JE virus (Kimura-Kuroda and Yasui, 1986).

The epitope model shown in Fig. 14.4 was integrated into a structural E protein model (Fig. 14.5) (Mandl et al., 1989). This was designed by compiling data on the structural characterization of each epitope (e.g., sensitivity or resistance to denaturation, acid pH, dependency on intact disulfide bridges), on the comparative sequence analysis of mAb-escape mutants, on the amino-terminal sequence analysis of immune-reactive fragments, on the comparative sequence analysis of other flaviviruses and the assignment of disulfide bridges determined for WN virus by Nowak and Wengler (1987). This combined approach yielded information on structural features of protein domains and allowed conclusions to be drawn on the folding of the polypeptide chain that modify and extend a previously proposed protein organization into regions R1, R2, R3, L1, and L2 (Nowak and Wengler, 1987).

The different antigenic domains revealed by epitope mapping were shown to have different structural properties (Guirakhoo et al., 1989) and therefore are likely to correspond to different structural entities of the protein. These can be described as follows (Fig. 14.5): at least part of domain A is apparently composed of quite distant parts of the polypeptide chain as deduced from the location of amino acid exchanges in mab-resistant mutants selected by the mutually competing neutralizing mAbs A3, A4, and A5. In addition, these epitopes share their sensitivity to acid pH and SDS. The resistance to SDS but sensitivity to reduction and carboxymethylation, competitive blocking with mAb A3 and the analysis of immunoreactive fragments provide evidence that the disulfide bridge-stabilized part of domain A is involved in the flavivirus group-reactive epitopes A1 and A2. The amino acid exchanges found in two mAb-resistant mutants (A3a and A3b) selected independently by mAb A3 are separated by three amino acids and are located near a potential glycosylation site found in the sequence of dengue viruses. This position matches precisely the position of a neutralization epitope of YF virus (also determined by sequencing of mAb-resistant mutants) (Lobigs at al., 1987) and is near an antigenic site between residues 76 and 93 in DEN-1 virus (identified by expression in *Escherichia coli*) (P. Mason, personal communication; Brandt, 1988).

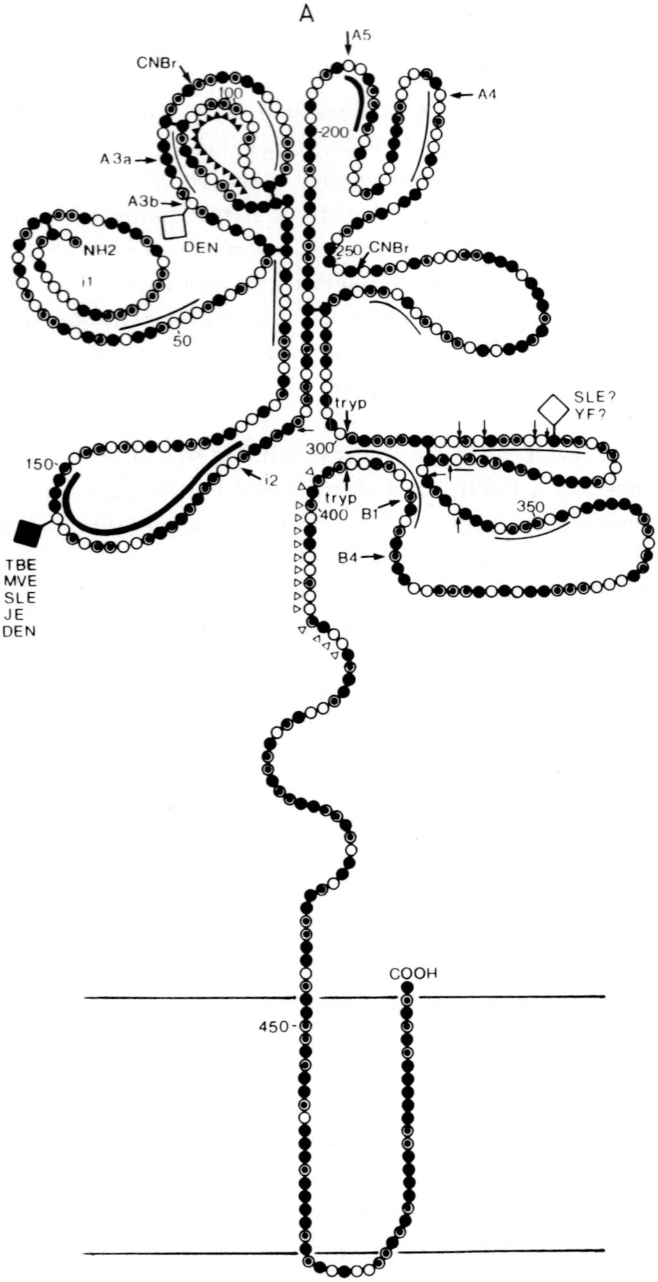

Fig. 14.5. Model of the TBE virus protein E. Open circles represent hydrophilic amino acid residues (Arg, Lys, Asn, Asp, Gln, Glu, His), dotted circles show intermediate amino acid residues (Pro, Tyr, Ser, Trp, Thr, Gly), and solid circles show hydrophobic amino acid residues (Ile, Val, Leu, Phe, Cys, Met, Ala). Position numbers are shown every 50 amino acids. Cysteine residues forming disulfide bridges are connected by solid lines. Arrows depict cleavage sites that liberate immunereactive fragments by the use of

A stretch of amino acids between residues 98 and 111 within the disulfide-bridge stabilized part of domain A represents the most highly conserved sequence of several conserved regions in the E protein. These may be involved in important functional activities such as membrane fusion or receptor binding.

The first 50 amino acids do not seem to be part of any of the three domains but carry the isolated epitope i1, as revealed by bacterial expression studies (Tuma et al., in preparation).

All epitopes of domain B are located on a tryptic fragment (residues 301–395) which contains a single disulfide bridge near its amino-terminus which is essential for antigenic reactivity. The amino acid exchanges in the mAb-resistant mutants B1 and B4, however, are located near the carboxy-terminus of this fragment. As revealed by denaturation/renaturation studies (Winkler et al., 1987b) this part of the E protein has a strong tendency for renaturation and probably represents an independently folding protein domain. The significance of this structure for flavivirus E proteins in general can be assumed from identical conclusions derived from a completely different approach applied to JE virus, namely the antigenic analysis of defined fragments expressed as fusion proteins in *E. coli* (Mason et al., submitted).

Domain C seems to be confined to a single hypervariable loop (see Fig. 5) that defines type- and subtype-specific antigenic determinants and carries the single carbohydrate side chain of the TBE virus E protein and potential *N*-glycosylation sites of several other flaviviruses. For TBE virus the carbohydrate does not seem to be essential for antibody binding but probably plays a role in conferring denaturation resistance to some of these epitopes (Guirakhoo et al., 1989).

Consistent with their functions as antigenic sites in the immune response to the native E protein, domains B and C were shown to be surface-exposed in WN virus by their accessibility to proteolytic cleavage (Wengler et al., 1987). As can be seen from the model the proposed antigenic domains also contain variable and hypervariable sequences that are likely to be part of structures exposed on the surface of the native protein.

←

trypsin (tryp) and cyanogen bromide (CNBr). Small arrows indicate potential cleavage sites within these fragments that are not utilized. Two solid lines stand for the lipid membrane that is spanned by two transmembrane regions of protein E. The polypeptide chain is folded to indicate the antigenic domains A, B, and C, which are designated by large capital letters. Arrows together with the names of neutralizing mAbs depict the locations of the mutations identified in the respective antigenic variants of TBE virus by sequence analysis. A line of solid triangles indicates the almost perfectly conserved sequence within domain A. A line of open triangles marks the region of a potential T-cell determinant. A solid diamond represents the carbohydrate side chain of TBE virus. The MVE, SLE, and JE viruses and DEN viruses have potential *N*-glycosylation sites at the homologous position. YF and SLE viruses have such a site within domain B, DEN viruses an additional site within domain A. The homologous positions of TBE virus are shown by open diamonds. The thin solid lines indicate variable and the thick solid lines hypervariable sequences as deduced from a sequence comparison of 15 flaviviral E proteins. Reproduced from Mandl et al. (1989) with permission

The synthetic peptide approach was also used to study the antigenic structure of the E protein of MVE (Roehrig et al., 1989a, b) and DEN-4 virus (Markoff et al., 1988). Based on computer predictions of potential antigenicity MVE peptides were synthesized, several of which reacted with hyperimmune ascitic fluids. These were derived from the amino-terminal part (one of them induced neutralizing antibodies), from areas corresponding to domain A, and domain B. For DEN-4 virus a series of overlapping peptides covering the whole E protein sequence was synthesized and reacted with immune sera and human convalescent sera. Different reactivities were observed for individual sera but there was an accumulation of reactivities with the amino-terminal peptides and peptides corresponding to domains A and B. Additional reactivities were also found with sequences between domains A and B indicating the usefulness of applying different approaches for the elucidation of antigenically active sequences.

As deduced from studies using the whole E protein, most of the antigenic sites recognized by antibodies are dependent on the native conformation. This is consistent with the general view of the antigenic structure of determinants recognized by B cells and is also substantiated by the finding that a significant neutralizing immune response with different E protein preparations of WN virus could only be achieved when all six disulfide bridges were present (Wengler and Wengler, 1989).

Although studies on the T-lymphocyte response to flaviviruses and viral proteins have been initiated (Kurane et al., 1989; Rothman et al., 1989), at present no data are available on the identification of functional antigenic sites recognized by T-cells. Predictions based on the potential to form amphipathic α-helices, however, indicated a possible T-cell antigenic site between residues 398 and 413 for TBE virus that is structurally conserved for other flavivirus E proteins too (Mandl et al., 1989). Using a different computer program, T-cell epitopes in MVE virus were predicted within the amino-terminal 50 amino acids, in sequences located between antigenic domains C and A, domain A, and domain B (Roehrig et al., 1989a, b).

5. The antigenic structure of NS1

The NS1 protein was first described as 'soluble complement fixing' (SCF) antigen in dengue-virus-infected tissue culture supernatants and in serum of infected mice (Brandt et al., 1970a, b) and was also shown to be expressed on the surface of flavivirus-infected cells (Cardiff and Lund, 1976). NS1 is a glycoprotein and sequence analyses of several flaviviruses revealed potential *N*-glycosylation sites that seem to be conserved to a certain extent within serological subgroups. Like the E protein, NS1 also contains 12 conserved cysteine residues that have not yet been assigned to disulfide bridges. The functional role of NS1 is not known but it was hypothesized to be involved in virus maturation (Rice et al., 1986). More recent data on the structural analysis revealed that NS1 can be present as both a membrane-associated and a soluble form (P. Young, personal communication; Lee et al., 1989) which may be derived by different proteolytic cleavages at their carboxy-terminal ends (Mason et al., 1987; Speight et al., 1988; Chambers et al., 1989). Final resolution of this question depends on C-terminal sequence analyses of different NS1 preparations. By competitive mAb binding studies, Henchal et al. (1987) identified a large serotype-specific domain of five epitopes on dengue 2 NS1 that overlapped with a more cross-reactive epitope. Some of the serotype-specific mAbs were partially protective in passive mouse immunization experiments without apparent correlation between CF activity and protective capacity. As in studies with the E protein, synergistic interactions between mAbs were observed and solid protection (such as that observed with neutralizing polyclonal sera) was achieved only with a combination of two mAbs that each fixed complement to a high titer. Consistent with previous findings (Schlesinger et al., 1985) CF activity may therefore be involved in the mechanism of protection by anti-NS1 antibodies.

So far, data on the structural location of epitopes in NS1 are not available. However, most of them seem to be conformation-dependent, since they lose their reactivity upon denaturation and reduction (P. Young, personal communication) and most of the mAbs did not react in immunoblots with bacterially expressed NS1 fragments (Putnak et al., 1988). In addition it was recently shown for several flaviviruses that NS1 exists as a stable dimer (Winkler et al., 1988) in both its soluble and its membrane associated form (P. Young, personal communication), and sequences in the amino-terminal part of NS1 may be responsible for oligomerization (Putnak, personal communication). It was shown with DEN-2 virus that only the dimeric form but not the monomeric form despite a high-titered antibody response can induce a protective immune response (P. Young, personal communication). Dimerization therefore apparently causes some structural rearrangement in NS1 — as also revealed by dimer-specific mAbs — that is essential for a protective activity.

6. New approaches to the development of flavivirus vaccines.

Successful conventional flavivirus vaccines are in widespread use for the prevention of YF, JE, and TBE virus (reviewed by Stephenson, 1988). A promising attenuated DEN-2 vaccine is undergoing clinical trials (Bhamarapravati et al., 1987) but trials with candidate vaccines of other dengue types have been less than satisfactory (Innis et al., 1988). Due to the great medical impact of dengue in developing countries, WHO is sponsoring a specific vaccine development program, which focuses on new technologies such as identification of protective epitopes, expression of proteins by recombinant DNA methods, synthetic peptides as immunogens and molecular definition of virulence with a view to obtain defined attenuated strains (Brandt, 1988). A major effort is also devoted to the use of protein NS1 as an immunogen to circumvent the possible hazard of inducing ADE by immunization with the E protein.

With regard to the preparation of immunogens by recombinant DNA technology bacterial systems seem to be less appropriate, as an E protein expressed in such a system did not provide protection in experimental animals, despite a good antibody response (Mason et al., submitted). Some protective effect was observed with a bacterially expressed fragment of NS1 in the YF virus system (Cane and Gould, 1988). There is evidence that disulfide bridges of the E protein are not properly formed in bacteria (Tuma et al., in preparation), which, however, seem to be essential in order to obtain a good neutralizing immune response (Wengler and Wengler, 1989). Eukaryotic expression systems are much more promising and protective activity in experimental animals was demonstrated with E and NS1 proteins expressed by recombinant Baculovirus vectors (Zhang et al., 1988; Lai et al., 1989) and by immunization with vaccinia virus recombinants expressing E and NS-1 (Bray et al., 1989).

Immunization experiments with synthetic peptides from the MVE-E protein revealed protective activity with peptides containing amino acids 35–50, 122–141, and 230–251 (Roehrig et al., 1989a, b) and it was suggested that strong peptide antigenicity correlated closely with the predicted presence of a helper T-cell epitope. Peptide sequences of the DEN-2 E-protein were also engineered into Hepatitis B core particles that were capable of inducing an antiviral immune response (D. Trent, personal communication). In view of the problem of inducing ADE it is of interest to note that Cecilia et al. (1988) identified an epitope in the JE virus E protein that was defined by a strongly neutralizing and protective mAb without ADE activity. The location of this potentially interesting epitope has not been determined so far.

Major efforts will be necessary to assess the structural prerequisites of immunogenicity, to use expression systems that yield native proteins, and to elucidate the ADE problem at the molecular level. For the synthetic peptide approach the combination of specific helper T-cell sites and functional B-cell sites will be essential.

7. Conclusions

Due to the benefit of modern molecular biological and immunological techniques, major progress towards a more complete understanding of the antigenic and functional properties of protective flavivirus proteins has been achieved. These studies are more advanced in the case of the envelope protein E. For this protein, data obtained by epitope mapping with monoclonal antibodies, comparative sequence analyses of mAb-resistant mutants, structural characterization of epitopes, expression and antigenic analysis of E-protein fragments have been combined with the assignment of disulfide bridges and processed into a structural model. The further definition of protective epitopes recognized by antibodies and the identification of functional T-helper cell epitopes should be an important focus in the future, especially with a view to developing peptide vaccines. It is desirable that epitopes involved in

ADE are resolved at the molecular level and hopefully these are distinct from those involved in protection.

The finding that one of the nonstructural proteins, NS1, can induce a protective immune response opens new avenues for developing novel vaccines, especially for dengue, and has therefore boostered research activities on this protein. Structural studies however, are less advanced as compared to the E protein, because NS1 is not as easily purified, disulfide bridge assignments have not yet been performed and epitope mutants cannot be generated, as it does not induce neutralizing antibodies.

A great effort is currently being devoted to the expression of both E and NS1 proteins by recombinant DNA technology. Since, as expected, most antigenic determinants recognized by antibodies are conformation dependent, the analysis of the structure of the expressed proteins as compared to the native proteins will be a major concern.

Ultimately it is to be hoped that the conformation of both proteins will be resolved by X-ray crystallography and that sites involved in the immune response and sites governing their functional activities can be defined in the three-dimensional protein structure.

8. Acknowledgements

We thank Christian Kunz, Christian Mandl and Farshad Guirakhoo for valuable discussions in the course of preparing this article and Noel Barrett and Tim Skern for critically reading the manuscript. We are also grateful to Susanne Pfauser for her expertise and patience in typing the manuscript.

9. References

Bhamarapravati, N., Yoksan, S., Chayaniyayothin, T., Angsubphakorn, S. and Bunyaretvej, A. (1987) *Bull. WHO* 65, 189–195.

Brandriss, M.W., Schlesinger, J.J., Walsh, E.E. and Briselli, M. (1986) *J. Gen. Virol.* 67, 229–234.

Brandt, W.E., Cardiff, R.D. and Russell, P.K. (1970a) *J. Virol.* 6, 500–506.

Brandt, W.E., Chiewsilp, D., Harris, D.L. and Russell, P.K. (1970b) *J. Immunol.* 105, 1565–1568.

Brandt, W.E. (1988) *J. Infect. Dis.* 157, 1105–1111.

Bray, M., Falgout, B., Zhao, B. and Lai, C.-J. (1989) In *Vaccines 89: Modern Approaches to New Vaccines Including Prevention of AIDS*. Cold Spring Harbor Laboratory, pp. 357–362.

Calisher, C.H., Karabatsos, N., Dalrymple, J.M., Shope, R.E., Porterfield, J.S., Westaway, E.G. and Brandt, W. (1989) *J. Gen. Virol.* 70, 37–43.

Cammack, N. and Gould, E.A. (1986) *Virology* 150, 333–341.

Cane, P.A. and Gould, E.A. (1988) *J. Gen. Virol.* 69, 1241–1246.

Cardiff, R.D. and Lund, J.K. (1976) *Infect. Immun.* 3, 1699–1709.

Casals, J. (1957) *Trans. N Y Acad. Sci.* 19, 219–235.

Cecilia, D., Gadkari, D.A., Kedornath, N. and Ghosh, S.N. (1988) *J. Gen. Virol.* 69, 2741–2747.

Chambers, T.J., McCourt, D.W. and Rice, C.M. (1989) *Virology* 169, 100–109.

Clegg, J.C.S. (1982) *Anal. Biochem.* 127, 389–394.

Coia, G., Parker, M.D., Speight, G., Byrne, M.E. and Westaway, E.G. (1988) *J. Gen. Virol.* 69, 1–21.

Della-Porta, A.J. and Westaway, E.G. (1977) *Infect. Immun.* 15, 874–882.

DeMadrid, A.T. and Porterfield, J.S. (1974) *J. Gen. Virol.* 23, 91–96.

Falker, W.A., Diwan, A.R. and Halstead, S.B. (1973) *J. Immunol.* 111, 1804–1809.

Gould, E.H., Buckley, A., Barrett, A.D.T. and Cammack, N. (1986) *J. Gen. Virol.* 67, 591–595.

Guirakhoo, F., Heinz, F.X. and Kunz, C. (1989) *Virology* 169, 90–99.

Halstead, S.B. (1988) *Science* 239, 476–481.

Hambleton, P., Stephenson, J.R., Baskerville, A. and Wiblin, C.N. (1983) *Infect. Immun.* 40, 995–1003.

Hawkes, R.A., Roehrig, J.T., Hunt, A.R. and Moore, G.A. (1988) *J. Gen. Virol.* 69, 1105–1109.

Heinz, F.X. and Kunz, C. (1979) *Arch. Virol.* 60, 207–216.

Heinz, F.X. and Kunz, C. (1980) *J. Gen. Virol.* 46, 301–309.

Heinz, F.X., Tuma, W. and Kunz, C. (1981) *Infect. Immun.* 33, 250–257.

Heinz, F.X., Mandl, C.W., Berger, R., Tuma, W. and Kunz, C. (1984) *Virology* 133, 25–34.

Heinz, F.X. (1986) In *Advances in Virus Research* (K. Maramorosch, F.A. Murphy and A.J. Shatkin, Eds.) Vol. 31. Academic Press, New York. pp. 103–168.

Heinz, F.X., Mandl, C.W., Winkler, G., Tuma, W. and Kunz, C. (1986) *Vaccines 86: New Approaches to Immunization,* Cold Spring Harbor Laboratory, Cold Spring Harbor, NY, pp. 387–392.

Henchal, E.A., McCown, J.M., Burke, D.S., Segumi, M.C. and Brandt, W.E. (1985) *Am. J. Trop. Med. Hyg.* 34, 162–169.

Henchal, E.A., Henchal, L.S. and Thaisomboonsuk, B.K. (1987) *J. Gen. Virol.* 68, 845–851.

Innis, B.L., Eckels, V.H., Kraiselburd, E., Dubois, D.R., Meadors, G.F., Gubler, D.J., Burke, D.S. and Bancroft, W.H. (1988) *J. Infect. Dis.* 158, 876–880.

Kimura-Kuroda, J. and Yasui, K. (1986) *J. Gen. Virol.* 67, 2663–2672.

Kitano, T., Suzuki, K. and Yamaguchi, T. (1974) *J. Virol.* 14, 631–639.

Kurane, I., Innis, B.L., Nisalak, A., Hoke, C., Nimmanitya, S., Meager, A. and Ennis, F. (1989) *J. Clin. Invest.* 83, in press.

Kyte, J. and Doolittle, R.F. (1982) *J. Mol. Biol.* 157, 105–132.

Lai, C.J., Zhang, Y.M., Falgout, B., Bray, M., Chanock, R. and Eckels, K.H. (1989) In *Vaccines 89: Modern Approaches to New Vaccines Including Prevention of AIDS.* Cold Spring Harbor Laboratory, Cold Spring Harbor, NY. pp. 351–356.

Lee, J.M., Crooks, A.J. and Stephenson, J.R. (1989) *J. Gen. Virol.* 70, 335–343.

Lobigs, M., Dalgarno, L., Schlesinger, J.J. and Weir, R.C. (1987) *Virology* 161, 474–478.

Mandl, C.W. (1988) *Ph.D. Thesis, University of Vienna.*

Mandl, C.W., Guirakhoo, F., Holzmann, H,. Heinz, F.X. and Kunz, C. (1989) *J. Virol.* 63, 564–571.

Markoff, L.J., Bray, M., Lai, C.-J., Chanock, R.M., Eckels, K., Summers, P., Gentry, M.K., Houghton, R.A. and Lerner, R.A. (1988) *Vaccines 88: New Chemical and Genetic Approaches to Vaccination.* Cold Spring Harbor Laboratory, Cold Spring Harbor, NY, pp. 161–165.

Mason P.W., McAda, P.C., Dalrymple, J.M., Fournier, M.J., and Mason, T.L. (1987) *Virology* 158, 361–372.

Monath, T.P. (1986) in *The Togaviridae and Flaviviridae* (S. Schlesinger, and M.J. Schlesinger, Eds.), pp. 375–453.

Nowak, T., and Wengler, G. (1987) *Virology* 156, 127–137.

Nowak, T., Färber, P.M., Wengler, G. and Wengler, R. (1989) *Virology* (in press).

Peiris, J.S.M., Porterfield, J.S. and Roehrig, J.T. (1982) *J. Gen. Virol.* 58, 283–289.

Porterfield, J.S. (1986) In *Advances in Virus Research* (K. Maramorosch, F.A. Murphy, and A.J. Shatkin, Eds.) Vol. 31. Academic Press, New York. pp. 335–355.

Putnak, J.R., Charles, P.C., Padmanabhan, R., Ine, K., Hoke, C. and Burke, D.S. (1988) *Virology* 163, 93–103.

Rice, C.M., Lenches, E.M., Eddy, S.R., Shin, S.J., Sheets, R.L. and Strauss, J.H. (1985) *Science* 229, 726–733.

Rice, C.M., Aebersold, R., Teplow, D.B., Pata, J., Bell, J.R., Vorndam, A.V., Trent, D.W., Brandriss, M.W., Schlesinger, J.J. and Strauss, J.H. (1986a) *Virology* 151, 1–9.

Rice, C.M., Strauss, E.G. and Strauss, J.H. (1986b) In *The Togaviridae and Flaviviridae* (Schlesinger, S., and Schlesinger, M.J., eds.), Plenum Press, New York and London. pp. 279–326.

Roehrig, J. (1986) In *The Togaviridae and Flaviviridae* (Schlesinger, S., and Schlesinger, M.J., eds.) Plenum Press, New York and London. pp. 251–278.

Roehrig, J.T., Hunt, A.R., Johnson, A.J. and Mathews, J.H. (1989a) In *Vaccines 89: Modern Approaches to New Vaccines Including Prevention of AIDS*. Cold Spring Harbor Laboratory, Cold Spring Harbor, NY, pp. 347–350.

Roehrig, J.T., Hunt, A.R., Johnson, A.J., and Hawkes, R.A. (1989b) *Virology* 171, 49–60.

Rothman, A.L., Kurane, I., Zhang, Y.-M., Lai, C.-J., and Ennis, F.A. (1989) submitted.

Schlesinger, J.J., Walsh, E.E. and Brandriss, M.W. (1984) *J. Gen. Virol.* 65, 1637–1644.

Schlesinger, J.J., Brandriss, M.W. and Walsh, E.E. (1985) *J. Immunol.* 135, 2805–2809.

Schlesinger, J.J., Brandriss, M.W. Cropp, C.B. and Monath, T.P. (1986) *J. Virol.* 60, 1153–1155.

Schlesinger, J.J., Brandriss, M.W. and Walsh, E.E. (1987) *J. Gen. Virol.* 68, 853–857.

Shapiro, D., Brandt, W.E., Cardiff, R.D. and Russell, P.K. (1971) *Virology* 44, 108–124.

Smith, G.W. and Wright, P.J. (1985) *J. Gen. Virol.* 66, 559–571.

Smith, T.J., Brandt, W.E., Swanson, J.L., McCown, J.M. and Buescher, E.L. (1970) *J. Virol.* 5, 524–532.

Speight, G., Coia, G., Parker, M.D. and Westaway, E.G. (1988) *J. Gen. Virol.* 69, 23–34.

Stephenson, J.R. (1988) *Vaccine* 6, 471–480.

Stohlman, S.A., Eylar, O.R. and Wisseman, C.L.jr. (1976) *J. Virol.* 18, 132–140.

Takegami, T., Miyamoto, H., Nakamura, H. and Yasui, K. (1982) *Acta Virol.* 26, 312–320.

Wengler, G., Castle, E., Leidner, U., Nowak, T. and Wengler, G. (1985) *Virology* 147, 264–274.

Wengler, G., Wengler, G., Nowak, T. and Wahn, K. (1987) *Virology* 160, 210–219.

Wengler, G. and Wengler, G. (1989) *J. Gen. Virol.*, in press.

Westaway, E.G. (1975) *J. Gen. Virol.* 27, 283–292.

Westaway, E.G., Brinton, M.A., Gaidamovich, S.Y., Horzinek, M.C., Igarashi, A., Kääriäinen, L., Lvov, D.K., Porterfield, J.S., Russell, P.K. and Trent, D.W. (1985) *Intervirology* 24, 183–192.

Winkler, G., Heinz, F.X. and Kunz, C. (1987a) *Virology* 159, 237–243.

Winkler, G., Heinz, F.X. and Kunz, C. (1987b) *J. Gen. Virol.* 68, 2239–2244.

Winkler, G., Randolph, V.B., Cleaves, G.R., Ryan, T.E. and Stollar, V. (1988) *Virology* 162, 187–196.

Zhang, Y.M., Hayes, E.P., McCarty, T.C., Dubois, D.R., Summers, P.L., and Eckels, K.H., Chanock, R.M. and Ching-Juh Lai (1988) *J. Virol.* 62, 3027–3031.

Eds. M.H.V. van Regenmortel and A.R. Neurath
Immunochemistry of viruses, II. The basis for serodiagnosis and vaccines.

Lentiviruses 15

S.D. PUTNEY[1] and R.C. MONTELARO[2]

[1]Repligen Corporation, One Kendall Square, Building 700, Cambridge, MA 02139, U.S.A. and [2]Department of Biochemistry, Louisiana State University, Baton Rouge, LA 70806, U.S.A.

1. Introduction

The lentivirus subfamily of retroviruses currently includes visna virus of sheep, the caprine arthritis-encephalitis virus (CAEV) of goats, equine infectious anemia virus (EIAV) of horses, and the immunodeficiency viruses of man (HIV), nonhuman primates (SIV), cats (FIV) and cows (BIV). This taxonomic group is based upon similarities in virion morphology, protein serologic properties, genomic organization and nucleotide sequences, host cell specificities, and patterns of pathogenesis. The first lentivirus discovered, EIAV, was also the first mammalian virus reported with a specific disease association (in 1904). (Vallee and Carre, 1904) With the isolation and characterization of visna virus and CAEV during the 1970s, the lentiviruses became established as an independent retrovirus subfamily distinguished from the numerous oncoviruses primarily by the former's lack of association with tumor development. Despite the intriguing pathogenesis patterns displayed by these animal lentiviruses, until recently they remained of interest primarily to veterinarians. This status changed dramatically in 1983 with the identification of HIV as the causative agent of AIDS. The search for AIDS animal models rapidly led to the isolation of SIV (Daniel et al., 1985; Murphy-Corb et al., 1986), BIV (Gonda et al., 1988), and FIV (Pedersen et al., 1987). It is probable that more animal lentiviruses will be identified in the future.

Lentiviruses constitute a major challenge to contemporary infectious disease research both in human and veterinary medicine. Lentiviruses utilize an array of strategies to maintain a persistent infection and to evade host defense systems. Many of these persistence mechanisms are correlated with specific, and often unique structural and immunologic properties of the lentivirus proteins. This review summarizes the biochemical and immunologic features of lentivirus proteins that are associated with important aspects of virus replication and persistence. The viruses primarily described are HIV, visna, and EIAV which have been the most thoroughly studied to date; other lentiviruses are discussed where appropriate to demonstrate general or unique immunochemical features.

2. Lentivirus antigens: structural and biochemical properties

A general model of a lentivirus virion is presented in Fig. 15.1. Overall the virus has a diameter of approximately 100 nm with a lipid envelope enclosing an oblong core containing two copies of the positive single-stranded RNA genome. The viral genome contains three major genes (5′-*gag-pol-env*-3′) encoding the structural proteins of the virus. A summary of the nomenclature used for the lentivirus structural proteins is presented in Table 15.1 using the standardized retrovirus nomenclature system proposed by Leis et al. (Leis et al., 1988). In general, lentiviruses contain structural components that are analogous to those found in oncoviruses.

The *gag* gene encodes the 3–4 internal proteins of the virus including the matrix (MA) protein immediately beneath the lipid bilayer, the major capsid shell protein (CA), and the basic ribonucleoprotein (NC) associated with the RNA genome. Each virus particle contains several thousand copies of each of the *gag* proteins. (Parekh et al., 1980) EIAV contains an additional *gag* core protein (p9), while SIV and HIV-2 virions contain a novel core protein, p14, that is encoded by an alternate reading frame called *sid* or *vpx* (Peterlin and Luciw, 1988; Henderson et al., 1988).

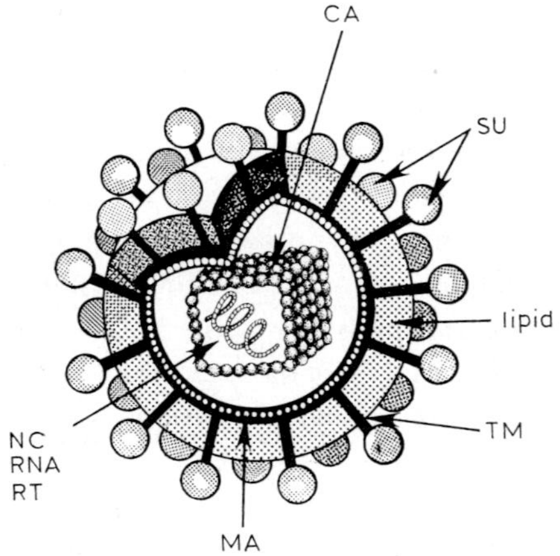

Fig. 15.1. General model for a lentivirus. The protein nomenclature is described in Table 15.1.

TABLE 15.1

Lentivirus structural proteins

Virus	SU	TM	MA	CA	NC	RT	PR	Other
HIV-1	gp120	gp41	p17	p24	p7	p66	p12	—
SIV	gp120	gp32	p16	p28	p8	?	?	p14(vpx)
EIAV	gp90	gp45	p15	p26	p11	p60	?	p9(gag)
CAEV	gp135		p19	p28	p16	?	?	?
Visna	gp135		p16	p27	p14	?	?	?

[a] Adapted from Leis et al. (1988) SU, surface protein; TM, transmembrane protein; MA, matrix protein; CA, capsid protein; NC, nucleocapsid protein; RT, reverse transcriptase; PR, protease.

The location and function of these latter proteins are unknown, although they are present at levels of several thousand copies per virus particle.

The *pol* gene encodes the reverse transcriptase (RT) enzyme of which 5–10 copies are associated with the viral RNA in each virion. The *env* gene encodes the surface proteins of the virus including the highly glycosylated surface glycoprotein (SU) and the sparsely glycosylated hydrophobic transmembrane protein (TM). There are several hundred copies of each env protein per virus particle, perhaps arranged in a symmetrical pattern in the lipid bilayer as trimers of the SU and TM glycoprotein (Ozel et al., 1988).

In addition to the virion structural proteins, lentiviruses also encode several nonstructural proteins that are not observed with oncoviruses. These nonstructural proteins have been characterized most thoroughly in HIV where in certain segments of the genome all three possible reading frames are utilized to produce different proteins. The HIV nonstructural proteins include tat, rev, and nef which appear to mediate either positive or negative regulatory functions during virus replication, but that are not incorporated into the virus particle (Gallo et al., 1988; Peterlin and Luciw, 1988).

Numerous studies indicate that both the structural and nonstructural lentivirus proteins are immunogenic during persistent infection (see below). Thus, all lentivirus-encoded proteins need to be considered in evaluating the immunologic properties of these viruses. The following sections provide a brief summary of the salient structural and functional features of lentivirus proteins.

2.1. Envelope proteins

The SU and TM glycoproteins form the surface 'knob and spike' projections located on the surface of the viral envelope. These proteins are the primary targets for host immune responses and they mediate a number of important biological functions including binding to receptors and membrane fusion. The SU and TM proteins are synthesized initially by translation of a spliced subgenomic mRNA to produce a polyprotein with SU as the amino-terminal component and TM as the carboxy-terminal component. In infected cells, the envelope polyprotein remains membrane associated, is rapidly glycosylated, and then processed by proteolytic cleavage into mature SU and TM proteins upon insertion into the plasma membrane. Lentivirus envelope polyproteins typically contain a common cleavage site immediately following a characteristic stretch of basic residues at the end of the SU protein. Only visna virus differs from this pattern in that the visna *env* polyprotein is assembled without being cleaved. The SU and TM proteins remain associated by noncovalent interactions, eventually forming the trimer complex found in mature virus particles. The SU protein is loosely bound to the cell membrane or viral envelope and is readily shed from the surface of the cell or virus. In contrast, the TM protein is bound tightly, presumably traversing the lipid bilayer one or more times.

The SU protein of lentiviruses mediates the early stages of virus infection includ-

TABLE 15.2

Biochemical properties of lentiviral SU proteins[a]

Property	HIV	SIV	EIAV	Visna
Total residues	516	527	434	656
Signal sequence length	30	22	5	100
N-linked glycosylation sites [b]	5.1	4.0	2.8	4.3
Serine/threonine residues[b]	16.9	17.8	12.9	12.6
Basic residues[c]	12.1	13.3	11.5	15.5
Acidic residues[c]	8.2	10.7	9.7	9.9
Proline residues[c]	4.5	4.6	4.4	4.5
Cysteine residues[c]	3.7	4.2	4.4	4.1
α-Helix[d]	47%	42%	48%	48%
β-Sheet[d]	28%	28%	26%	26%
Turns[d]	17%	24%	20%	19%
Random[d]	9%	7%	7%	7%

[a] Sequences obtained from Bionet data base.

[b] Summary of potential glycosylation sites expressed per 100 amino acids.

[c] Values are expressed per 100 residues. Basic residues, lysine, arginine, and histidine; acidic residues, aspartic and glutamic.

[d] Predicted secondary structure potentials were determined by the Chou-Fasman algorithm from Bionet and expressed as the percent of the total.

ing recognition and binding to cell surface receptors. The respective SU proteins of different lentiviruses appear to share only limited amino acid sequence identity, perhaps reflecting the differences in host cell specificity. Despite this lack of obvious sequence identity, however, the lentivirus SU proteins do have remarkable similarities in their biochemical properties. A summary of selected biochemical parameters for several lentivirus SU proteins is presented in Table 15.2. It is interesting to note that the various SU proteins display similar patterns of amino acid content, potential *N*-linked glycosylation sites, and predicted secondary structures. These structural similarities suggest that functional and immunologic studies performed with one lentivirus may be applicable to other lentivirus systems.

HIV gp120 is the most thoroughly characterized SU protein. The HIV gp 120 contains 516 amino acids constituting a peptide backbone of about 60 kDa. However, during *env* polyprotein processing, the gp 120 is extensively glycosylated at a variety of the more than 20 potential *N*-linked glycosylation sites found in the amino acid sequence; the exact pattern of *N*-linked glycosylation sites may vary among different HIV isolates. Extensive glycosylation is a characteristic of all lentivirus SU proteins. The functional role of glycosylation in HIV gp120 is uncertain. The oligosaccharides of CAEV gp135 have been shown to confer protease resistance to the virus particle, to protect the virus from neutralizing antibody, and to influence viral infectivity and virulence *in vivo* (Huso et al., 1988; O. Narayan, personal communication). Deglycosylation of HIV gp120 can enhance the reactivity of the protein with antibody from certain AIDS patients, while the deglycosylation of EIAV gp90 fails to alter significantly the reactivity of this protein with serum from infected horses (Montelaro et al., 1988). Most recently, it has been demonstrated that HIV gp120 glycosylation affects the level of viral infectivity and cytopathicity, but does not prevent virus replication or host cell infectibility (Montefiori et al., 1988). Thus, the role of gp120 oligosaccharides in HIV immunologic and biological properties remains an important issue to be examined.

Comparisons of the nucleotide sequences of the *env* gene of various HIV isolates indicates specific areas of conserved and variable sequences. Overall, the nucleotide sequence among the *env* genes of HIV isolates from different patients may vary by 15–20% and the deduced amino acid sequence by 10–12% (Hahn et al., 1986). Many of the immunogenic segments of HIV gp120 have been localized to variable sections of the glycoprotein. Thus, an important question in HIV research is a direct correlation of specific amino acid sequence variations with the observed variations in the biologic and antigenic properties of different virus isolates. Although the lentivirus SU proteins may differ extensively, it is clear that significant modifications of antigenic properties can be mediated by alterations in only a few amino acids out of a total of several hundred. For example, sequential isolates of EIAV recovered from a single experimentally infected pony differ in only 3–4 amino acids of at total of 434 residues (Payne et al., 1987). Yet these isolates differ markedly in neutralization patterns with various serum samples taken during the course of infection in the pony and in binding assays with a panel of EIAV gp90-specific monoclonal antibodies. A similar pattern of variation is observed in sequential HIV isolates taken from an individual AIDS patient over a period of time (Hahn et al., 1986).

In light of the highly variable nature of HIV gp120, the elucidation of biologically important sites located in conserved segments of the glycoprotein is critical to designing potential antiviral protocols. In the case of HIV gp120, certain epitopes have been identified in conserved domains, in addition to the CD4 receptor binding domain (discussed below). A striking observation is the conservation among HIV-1 isolates of the locations of the cysteine residues in gp120 (Hahn et al., 1986). There is a similar conservation of cysteine residues among the gp90 sequences of various EIAV isolates and the gp135 sequences of different visna virus strains (Payne et al., 1987; Braun et al., 1987). This marked conservation of cysteine residues suggests their critical role in SU protein structure and function.

The TM protein of lentiviruses mediates the membrane fusion function observed with this virus, and can be the target for neutralizing antibodies and cell-mediated immune responses. In contrast to the SU protein of lentiviruses, the TM protein is only sparsely glycosylated, hydrophobic, and relatively conserved in amino acid sequences among isolates of a particular lentivirus. Recently a general model has been proposed for the amino-terminal segment of TM protein of HIV and for other retroviruses (Gallaher et al., 1989). The scheme of the HIV gp41 molecule presented in Fig. 15.2 demonstrates the conserved structural features proposed for the amino-terminal 200 amino acids of TM proteins. Overall the TM protein is projected to be a fibrous structure with a significant degree of sidedness; all of the immunodominant sites are located at the apex and on one side of the molecule and the potential *N*-linked glycosylation sites on the other side. Several other common features are also evident in this model. (1) The laterally extended amino terminus contains the fusion peptide, including the hydrophobic core LFLGFL with potential as a membrane insertional hairpin similar to the hydrophobic core of signal sequences. This is followed by an internally looping structure rich in serine and threonine (5/11) that is postulated to interact with gp120, and by (2) an ascending amphipathic helix that is a predicted T-cell epitope. This extended helix leads to the apex of the molecule containing an immunodominant epitope that is perhaps stabilized by the characteristic conserved vicinal cysteines. The apex is followed by an extended α-helix containing the potential glycosylation sites. The descending helix terminates in a bend also rich in serine and threonine (4/7 residues) and believed to interact with gp120. (3) It is a highly charged region which can form an α-helix stabilized by multiple ion pairs. After a bending tryptophan-rich region external to the lipid bilayer, there appears to be a hydrophobic membrane spanning α-helix that is terminated by a sharp turn just as the protein emerges from the lipid bilayer into the cell cytosol. The remaining sequence of HIV gp41 leading to the carboxyl terminus contains hydrophobic regions which can form amphipathic helices that may facilitate association of individual envelope proteins to form the trimer structure found in mature virions (Venable et al., 1989).

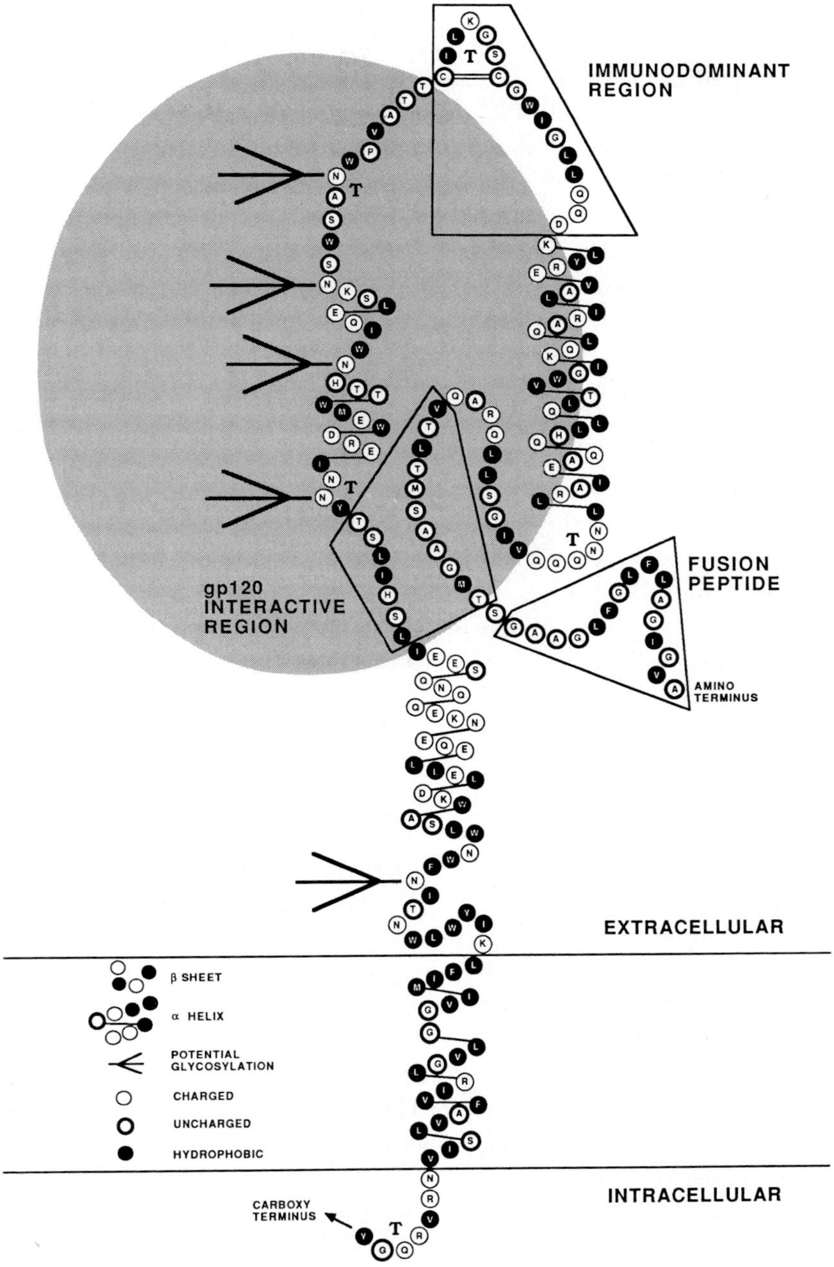

Fig. 15.2. Model for the structure of the amino-terminal portion of HIV gp41 adapted from Gallaher et al. (Gallaher et al., 1989).

2.2. Core proteins

The lentivirus core proteins are translated from the full-length genomic RNA to produce a core polyprotein with the protein order of MA-CA-NC. The precursor polyprotein is transported to the site of virus assembly at the plasma membrane where it is processed by the viral protease (PR) during budding into the native structural proteins found in virus particles. Although most of the core polyprotein is completely cleaved, small amounts of the uncleaved precursor or intermediate cleavage products can be found in purified virus and on the surface of infected cells. The significance of surface gag polyprotein is uncertain at this time, however, it may serve as a CMI target in certain oncovirus systems. The core proteins of lentiviruses constitute the predominant protein components of the virus particle constituting up to 90% of the virus protein.

The 15–17-kDa MA protein, located immediately beneath the envelope lipid bilayer is myristylated at its amino-terminus, and may be phosphorylated. The virus particle contains enough MA protein to form a continuous monolayer, which is believed to be noncovalently associated with the envelope TM protein. The MA protein is mostly sequestered inside the viral envelope. However, there is some evidence that HIV p17 may be exposed at the virion surface and accessible to neutralizing antibodies (Naylor et al., 1987; Argos, 1989).

The CA protein is the most abundant protein in lentiviruses and it is the major component of the core shell. Lentivirus CA proteins have a mass of 24–28 kDa and may be modified by phosphorylation, as in the case HIV p24. The CA proteins of various lentiviruses display significant sequence homology reflected by serological cross-reactivities. Recently an interesting model has been proposed for the structure of lentivirus CA proteins based on a putative homology with the picornavirus VP2 coat protein (Argos et al., 1989). According to this model, the CA is protein is postulated to fold into an eight-stranded β barrel structure with a characteristic 'exposed' PUFF region. By analogy to the known environments of picornavirus VP2 proteins, probable antigenic sites are predicted, mostly involving the protein sequences connecting the β strands. Although certain predictions made from this model correlate with CA antigenic segments predicted or demonstrated by others, the assignment of antigenicity to specific structural motifs remains speculative at this time. It is interesting to note, however, that the C-terminal 125 amino acids of HIV p24 and EIAV p26 are the primary target for CA-specific antibodies in serum from AIDS patients or EIAV-infected horses, respectively (Ghrayeb et al., 1986; R.C. Montelaro, unpublished observations).

The third major *gag* protein common to all lentiviruses is the highly basic NC protein that is bound to the viral RNA, presumably to form a helical ribonucleoprotein complex. The NC proteins of lentiviruses vary in molecular mass from 7–11 kDa. They are all rich in lysine and arginine and contain a cys-his box characteristic for various nucleic acid binding proteins and perhaps essential for reverse transcription and assembly (Meric and Goff, 1989).

In addition to the *gag* proteins described above, lentiviruses also contain small amounts of RT associated with the ribonucleoprotein complex and of PR in an undetermined location in the viral core. In the case of SIV and HIV-2, the virus particle contains several thousand copies (an amount equivalent to *gag* proteins) of a 14-kDa nucleic acid binding protein that is encoded by the *sid* or *vpx* gene (Henderson et al., 1988). The exact location and role of p14 is not known.

A decade ago in evaluating virus immunogenocity little attention was given to the internal proteins of lipid enveloped viruses, including retroviruses. During the past several years, however, the elucidation of immune responses to the matrix and nucleocapsid proteins of VSV, influenza, and rabies viruses as being critical to the establishment of protective immunity (Puddington et al., 1986; Dietzschold et al., 1987b; Townsned et al., 1986) has generated a new perspective for the potential immunogenic role of lentivirus core proteins. The conserved nature of lentivirus core proteins, in contrast to the variable envelope proteins, makes the former an important potential component for inducing broadly protective immune responses. This possibility is being actively pursued in several laboratories.

2.3. Nonstructural proteins

During replication, lentiviruses produce several regulatory proteins from multispliced subgenomic mRNAs. The regulatory proteins are found both in the cytoplasm and nucleus of infected cells, but not in purified virus preparations. HIV-1 regulatory proteins include the tat, vif, nef, rev, and U proteins (Gallo et al., 1988; Peterlin and Luciw, 1988) In general, these proteins are relatively small (6–20 kDa) scarce, and frequently bind nucleic acid. All of the nonstructural proteins appear to be immunogenic in HIV infected individuals. The role of the corresponding antibodies in affecting virus replication, if any, is unknown. However, the fact that certain regulatory proteins may be produced in infected cells in the absence of virus production or of the expression of *env* and *gag* genes creates an important potential for serving as immune targets during latent stages of infection. In this regard, it is interesting to note that immunization of monkeys with the yellow fever virus nonstructural protein NS1 resulted in protection from fatal hepatitis after virus challenge (Schlesinger et al., 1986).

3. Lentivirus infections

3.1. Characteristics

Lentiviruses typically establish persistent infections and induce chronic diseases displaying a variety of clinical symptoms and progressions of pathogenesis (Haase, 1986). Although the name lentivirus indicates a 'slow' virus infection, the kinetics of virus replication and disease development varies greatly. For example, EIAV induces clinical symptoms in horses within several weeks post-infection. Equine infectious anemia is characterized by recurrent high levels of viremia associated with periodic disease cycles involving severe erythrolytic anemia, immune-complex glomerulonephritis, edema, diarrhea, and lethargy. However, the disease is seldom fatal and horses routinely establish immunologic control of virus replication by one year post-infection, remaining inapparently infected for the remainder of their life. Visna virus and CAEV typically induce disease within several months post-infection and establish relatively low levels of viremia. Infected animals remain infected for life and develop progressive cases of encephalopathy. Although immunosuppression was not originally considered a hallmark of these early animal lentiviruses, more recent studies reveal a variety of immune disorders in infected animals.

HIV and SIV induce very similar diseases characterized primarily by severe immune suppression, opportunistic infections, and a general progressive wasting leading ultimately to death. In contrast to the animal lentivirus infections described above after the initial infection cycle, HIV and SIV infections appear to involve a relatively long latent period prior to the development of disease. In the case of SIV, this latent period varies from 6–12 months depending on the particular isolate, while HIV infections may remain latent for 5–7 years post-infection. Viremia levels of HIV and SIV are typically 100-fold lower than observed with viruses such as EIAV. Although immune suppression is the hallmark of HIV and SIV infections, encephalopathy can also be an important clinical manifestation. In fact, certain isolates of SIV induce only an encephalitis in experimental infections of rhesus macaques, indicating a strain specificity for pathogenesis (M. Murphey-Corb, personal communication). The kinetics and characteristics of FIV and BIV infections are currently being studied.

On the surface it may appear that the lentiviruses display little uniformity in their patterns of replication or pathogenesis. There are, however, significant common features in lentivirus infections. During natural infections, lentiviruses predominantly establish cytopathic infections in cells of the immune system. All lentivirus infections involve cells of the monocyte-macrophage lineage as targets, and certain lentivirus infections (HIV and SIV) also involve the specific CD4 positive subset of lymphocytes as host cells. Thus, many of the varied manifestations of lentivirus infections are most likely caused by differences in virus replication and their effect on specific host targets.

3.2. Persistence strategies

A schematic representation of a lentivirus replication cycle is summarized in Fig. 15.3. The life cycle of a lentivirus provides an array of mechanisms for eluding host immunosurveillence and maintaining a persistent infection. After envelope protein-mediated binding to the target cell and entry to the cell cytoplasm, the viral RNA

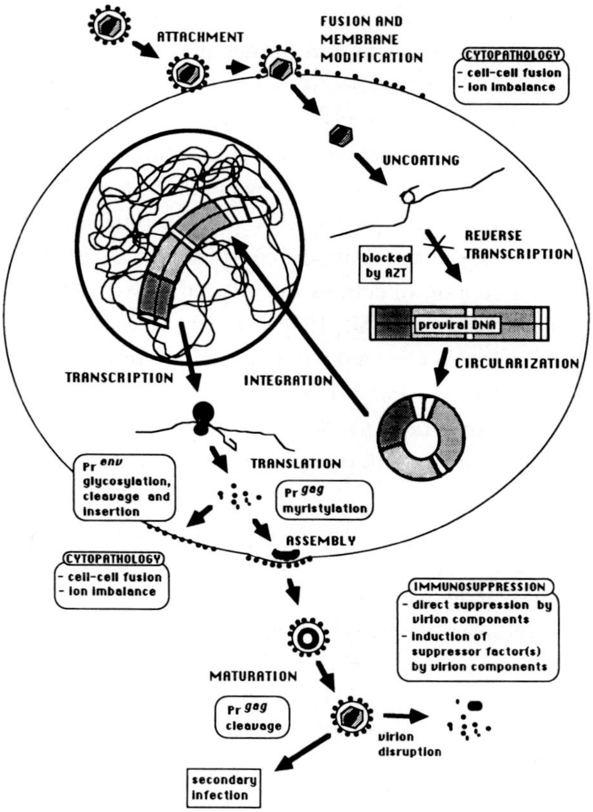

Fig. 15.3. Schematic of a lentivirus replication cycle.

genome is reverse transcribed into proviral DNA which integrates into the host cell genome and can be eliminated only by death of the cell. The integrated provirus may be expressed immediately or remain latent for years. Thus, infection of replicating bone marrow cells or immune progenitor cells can result in a population of inapparently infected mature immune cells in which virus replication can be induced. The dependence of virus gene expression on the state of cell differentiation has been clearly demonstrated with CAEV which is latent in monocytes but activated during differentiation to macrophages (Narayan et al., 1983).

This genetic strategy for survival is complemented by a variety of escape mechanisms involving lentivirus protein structure and function. Once virus production is induced in infected cells, the spread of virus to new target cells can be mediated by free virus or by cell-to-cell transfer facilitated by cell membrane fusion. The latter mechanism of transmission is capable of sequestering virus from detection by the host immune system and can occur in the presence of virus-specific immune responses (Gupta et al., 1989).

The production of free virus from infected cells is also shielded by several other escape mechanisms. For example, lentivirus antigens can induce high levels of antibodies; however, the level of neutralizing antibodies is relatively low during all stages of infection. Goats infected with CAEV produce antibodies to all viral proteins, but these antibodies are ineffective in neutralizing the virus (Narayan et al., 1984). In fact, neutralizing antibodies are difficult to elicit, even when strong adjuvants are used together with preparations of disrupted CAEV or with purified viral proteins. In the case of EIAV, circulating virus-antibody complexes isolated from the blood of naturally infected animals remain infectious in cultured cells (McGuire et al., 1972). Infections by visna virus (Narayan et al., 1978) and HIV (Robert-Guroff et al., 1985; Weiss et al., 1985) are also characterized by relatively low levels of neutralizing antibodies. As discussed in section III, although the titers of neutralizing antibodies elicited by lentivirus infections are low, these antibodies and cytotoxic lymphocytes are important mechanisms that can contain lentivirus infections.

Lentiviruses also undergo relatively rapid genomic variation during replication which alters the structure and antigenic properties of the virus and allows it to circumvent established immune responses (Clements et al., 1988). This results in antigenic variants that can elude immunosurveillance and propagate even in the presence of fully competent host immune systems. The propensity for antigenic variation during persistent infections has been documented for visna virus, EIAV, HIV, SIV, and CAEV (Narayan et al., 1978; Narayan et al., 1977; Montelaro et al., 1984; Payne et al., 1987; Hahn et al., 1986; McGuire et al., 1988). Lentivirus isolates recovered from different infected hosts may differ in overall nucleotide and deduced amino acid sequences by as much as 15–20%. In addition, these viral isolates can differ significantly in their biologic and antigenic properties. The observed variations are located predominantly in the *env* gene, while minor variations are observed in the *gag* and *pol* sequences. However, only minor variation in the envelope glycoprotein can alter antigenic properties. Antigenic variants of lentiviruses recovered from a single infected animal over a period of months or years reveal variations of only 1–2%, representing changes of only 4–5 amino acids of a total of several hundred residues in the SU proteins of the virus. In fact, distinct neutralization variants may differ by a single amino acid in the envelope glycoprotein (Reitz et al., 1988). To date, no two independent isolates of a particular lentivirus appear to be identical; isolates may be distinguished by nucleotide sequence or by biologic and antigenic properties. Lentivirus isolated sequentially from a single infected host contains a large mixture of genetically, and presumably antigenically, distinct variants that evolve in parallel and coexist during persistent infection.

Lentivirus infections can also result in various levels of immunosuppression in the host (Haase, 1986) and this can be another mechanism contributing to virus persistence. This immunosuppression can be caused by the depletion of essential immune cell populations that serve as targets for virus replication, or the suppression can be mediated by the lentivirus envelope protein. For example, there are several immunosuppressive activities that are mediated by HIV gp120 binding to CD4 lymphocytes. Native gp120 activates resting CD4 lymphocytes as measured by an increase in intracellular levels of inositol triphosphate and calcium and elevates interleukin-2 receptor levels (Kornfeld et al., 1988). Antigen-specific CD4 lymphocyte responses of cells from normal donors are inhibited by gp120 (Diamond et al., 1988; Chirmule et al., 1988) Native and recombinant gp120 suppresses PHA-induced lymphocyte blastogenesis and has other effects on functions of immune cells from uninfected donors (Mann et al., 1987; Shalaby et al., 1987). In addition to these effects, gp120, either bound or processed and presented on an APC, could become a target for destruction of uninfected CD4-positive cells by antibodies and effector lymphocytes of HIV-infected individuals (Kulaga et al., 1988; Lyerly et al., 1987).

This battery of escape mechanisms constitutes a formidable challenge to natural protective systems of the host and to the development of effective vaccines against lentivirus infections. However, there is reasonable hope that elucidation of the functional and immunologic properties of lentivirus proteins can provide the foundation for designing effective vaccines. The feasibility of accomplishing the immunologic management of lentiviruses, despite their array of escape mechanisms, is indicated by the fact that horses infected with EIAV routinely establish control of virus replication and disease by progressive changes in immune responses during the first year post-infection. It is also likely that protective immune responses are elicited by other lentiviruses, such as HIV, and that these are responsible for the extended asymptomatic period characteristic for HIV infection.

4. Immune response to lentivirus infection

4.1. Humoral immunity and neutralizing antibodies

Infection by lentiviruses results in both humoral (antibody-mediated) and cellular immune responses against viral antigens. Antibodies that react with core, envelope, reverse transcriptase and other viral enzymatic and regulatory gene products are often found in infected hosts. For example, part of the evidence leading to the discovery of HIV as the cause of AIDS was the presence of virus-specific antibodies in serum of AIDS patients. Western blotting revealed antibodies to the p24 core antigen (Kalyanaraman et al., 1984; Schupbach et al., 1984), its precursor p55 (Schupbach et al., 1984), the transmembrane envelope protein (gp41) (Sarngadharan et al., 1984) and the polymerase gene product (Schupbach et al., 1984). Approximately 80% of infected, asymptomatic people have antibodies to the p24 core antigen (Ghrayeb et al., 1986; Kalyanaraman et al., 1984) and 80% have antibodies to the p66 (unprocessed) and p51 (processed) polymerase gene products (Veronese et al., 1986). Subsequent analysis of the viral genome and of regions critical for infectivity revealed the presence of regulatory and enzymatic gene products. Serological studies have shown that sera of infected people contain antibodies to these proteins. Infected people, to varying degrees, have antibodies to the tat (Goh et al., 1986; Aldovini et al., 1986; Barone et al., 1986), protease (Kramer et al., 1986), vif (Lee et al., 1986; Change et al., 1985b; Kan et al., 1984), nef (Allan et al., 1985; Franchini et al., 1986b; Franchini et al., 1986a), rev (Goh et al., 1987) and U proteins (Matsuda et al., 1988) of HIV-1. The HIV-2 and SIV specific gene product X also elicits an immune response in infected hosts (Henderson et al., 1988; Franchini et al., 1988; Kappes et al., 1988).

Because immune responses elicited by the envelope protein are generally the most prevalent and most relevant for vaccine development, the envelope has been the subject of most studies with regard to mapping antibody binding sites. The immunodominant epitopes on the HIV envelope have been extensively studied. Virtually 100% of infected individuals develop antibodies to a domain in the amino-terminal half of gp41 of HIV-1 (Change et al., 1985a; Wang et al., 1986; Ghrayeb et al., 1989), of HIV-2 (Norrby et al., 1987) and of SIV (Johnson et al., 1988), respectively. Another immunodominant epitope is at the carboxy-terminus of gp120 (Palker et al., 1987).

Epitopes on the envelope protein of EIAV have been identified. Using monoclonal antibodies (Hussain et al., 1987; Hussain et al., 1988) and recombinant proteins containing fragments of the envelope (Payne et al., in press), three epitopes on the external transmembrane glycoprotein, gp45 were identified. Competitive binding experiments using these monoclonal antibodies and immune sera from infected animals show that these epitopes are recognized during EIAV infection. Immunodominant epitopes, recognized by sera from all tested animals, are located at the carboxy-terminal end of gp90 and the amino-terminal half of gp45 (Payne et al., in press). These immunodominant regions are in the same relative positions in the HIV-1 envelope.

An important line of research is the delineation of epitopes on the virus envelope to which an antibody will bind and interfere with the infection process (i.e., neutralizing antibody). The most important form of protective immunity against viral particles is neutralizing antibody. Neutralization could theoretically be accomplished by antibodies that block binding of the virus to its receptor or that interfere with steps subsequent to attachment but prior to integration, such as virus penetration or uncoating. Several mechanisms of virus neutralization have been elucidated. Examples are: (1) antibody can cross-link virus particles which leads to alteration of protein conformation and neutralization (Icenogle et al., 1983; Emini et al., 1983); (2) a monoclonal antibody to the retrovirus alpha-1 protein blocks cell attachment (Lee et al., 1981); (3) mouse mammary tumor virus is neutralized by steric hindrance by antibody bound to determinants adjacent to those that bind to cell surface receptors (Massey and Schochetman, 1981); (4) attachment to cell surface occurs without penetration when influenza virus is neutralized by antibody (Possee et al., 1982; Taylor and Dimmock, 1985a; Taylor and Dimmock, 1985b); and (5) West Nile and rabies viruses are neutralized by antibodies that inhibit the intraendosmal acid-catalyzed fusion step leading to virus uncoating (Dietzschold et al., 1987a; Gollins and Porterfield, 1986).

Visna virus is neutralized by two different mechanisms depending on the cell type (Kennedy-Stoskopf and Narayan, 1986). The site of neutralization of fibroblast infection is at the cell membrane where antibodies prevent cell attachment, whereas for macrophages the neutralization occurs after binding of virus. Virions are phagocytized and uncoated but RNA is not transcribed. Virus binds to macrophages much faster than the antibodies bind to virus and this may explain why infection persists in the presence of neutralizing antibody.

Although HIV neutralizing antibodies are found in a majority of infected people (Robert-Guroff et al., 1985; Weiss et al., 1985), the mechanism(s) of neutralization has not been elucidated. A possible mechanism of antibody-mediated neutralization of HIV or SIV is prevention of attachment to the virus receptor, CD4. Antibodies that block binding of virus or of purified gp120 to CD4 are found in a majority of HIV infected people (Skinner et al., 1988a; Schnittman et al., 1988). However, there is no correlation between the CD4 binding inhibition titer and neutralizing antibody titer in such sera (Schnittman et al., 1988). Monoclonal antibodies that block binding of purified gp120 to CD4 have been isolated but are unable to neutralize virus infectivity (Lasky et al., 1987; Dowbenko et al., 1988). These data suggest that CD4 blocking antibodies do not neutralize infectivity. The reason for this may be the large binding constant for the reaction between gp120 and CD4 (which is greater than 10^9) (Skinner et al., 1988a; Lasky et al. 1987).

The principal mode of neutralization of HIV-1, at least by antibody elicited by purified envelope subunits, is not prevention of attachment to CD4. Using immunization with recombinant envelope proteins or with peptide subunits, the principal neutralization determinant has been mapped to a domain on the carboxy-terminal half of gp120 (Fig. 15.4) (Putney et al., 1986; Rusche et al., 1988; Goudsmit et al., 1988a; Palker et al., 1988; Kenealy et al., 1989). Peptides from this domain elicit antibodies that both neutralize cell-free virus and prevent cell fusion. The action of such antibodies is blocked by a peptide from the sequence of this determinant (Rusche et al., 1988). This determinant does not appear to act in the binding of gp120 to CD4, since antibodies bound to this determinant do not affect gp120 binding to CD4-bearing cells (Skinner et al., 1988a; Linsley et al., 1988) and insertion mutations in the domain do not prevent HIV-1 binding to CD4 (Kowalski et al., 1987).

This domain is from a region of the envelope protein which is highly variable between independent isolates of HIV-1. Fig. 15.5 shows the amino acid sequence of this region from several independent HIV-1 isolates (Starcich et al., 1986). Whereas the sequence of gp120 differs as much as 20–25% between IIIB (an original HIV-1 isolate (Ratner et al., 1985)) and other isolates, the neutralizing determinant differs

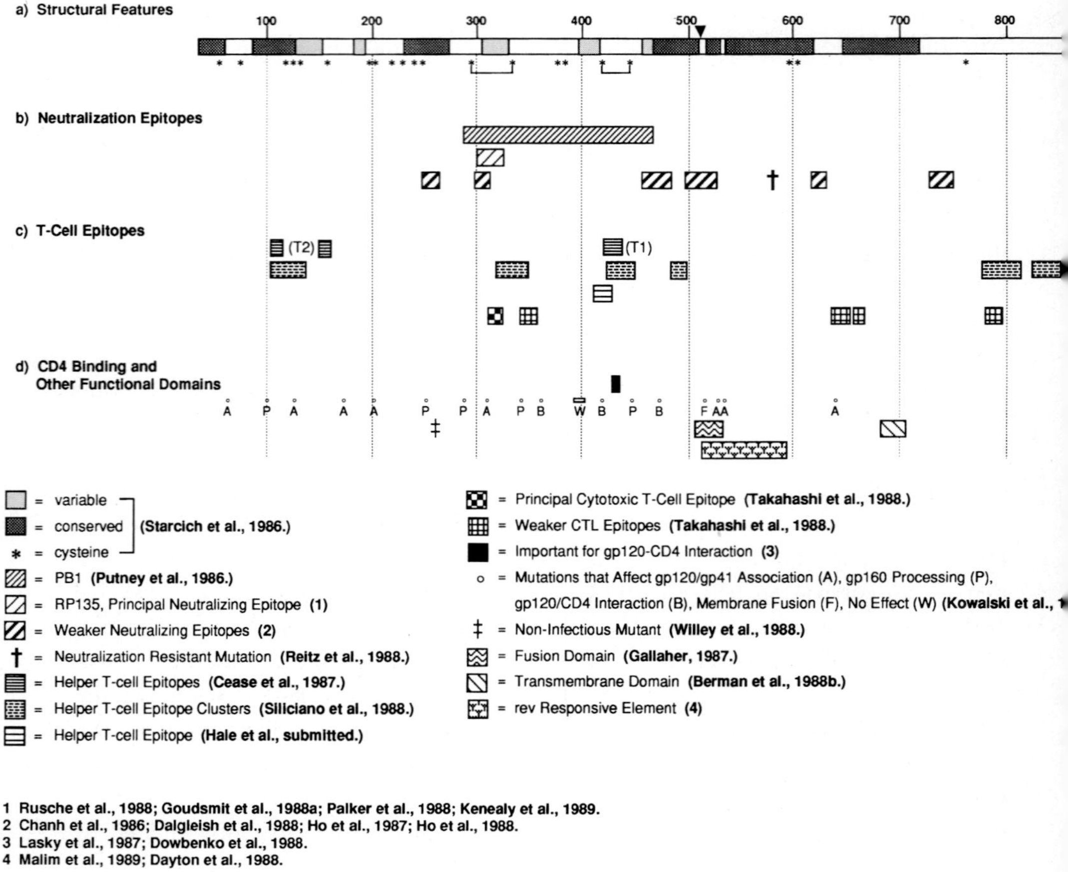

Fig. 15.4. Map of HIV-1 envelope showing the external envelope, gp120, and the transmembrane envelope, gp41. The cysteines, conserved in gp120 in all HIV-1 isolates, are shown by asterisks (*). Cysteines joined by disulfied bonds (T. Gregory, personal communication) are so shown. Shown are location of b) naturalizing epitopes, c) T-cell epitopes, and d) regions involved in CD4 binding and other functions. Shown are the locations of mutations affecting CD4 binding (B), gp120-gp41 association (A), gp160 processing (P) and membrane fusion (F). Also shown in (d) is the location of a mutation that renders virus non-infectious but that does not interfere with envelope synthesis, or processiong the stability of gp41 or the binding of gp120 to CD4 (Willey et al., 1988).

as much as 40–50%. There are, however, conserved elements within this region such as the flanking cysteines that are present in all HIV-1 isolates and that are joined to each other by a disulfide cross bridge (T. Gregory, personal communication). Therefore, this determinant is contained in a loop in native gp120. Several monoclonal antibodies that bind to this determinant have been characterized (Linsley et al., 1988; Matsushita et al., 1988; Skinner et al., 1988b; Kinney-Thomas et al., 1988; Fung et al., 1987). These antibodies neutralize cell free virus and prevent fusion of

		RP135		
III_B (BH10)	C T R P N N N T R K S I R I Q R	G P G	R A F V T I G K I G N M R Q A H C	III_B (BH10)
RF	C - - - - - - - - - S - T K	- - -	- V I Y A T - Q I - - D I - K - - C	RF
MN	C - - - - Y - K - - - - H I	- - -	- - - Y - T K N I - - T I - - - - C	MN
SC	C - - - - - - - T R S - H I	- - -	- - - Y A T - D I - - D I - - - - C	SC
WMJ-2	C - - - Y - - V - R S L S I	- - -	- - - R - R E I - - I I - - - - C	WMJ-2
LAV-MAL	C - - - G - - - - R G - H F	- - -	Q - L Y - T - I V - D I - R - Y C	LAV-MAL
SF-2	C - - - - - - - - - S - Y I	- - -	- - - H - T - R I - - D I - K - - C	SF-2
NY5	C - - - - - - - K - G - A I	- - -	- T L Y A R E - I - - D I - - - - C	NY5
Z3	C - - - G SDKKI - Q S - R I	- - -	K V - Y A K - G I T - - - - C	Z3
WMJ1	C - - - - - - V - R R H - H I	- - -	- - - Y - G E I R - - I - - - - C	WMJ1
WMJ3	C - - - - D I A - R - - H I	- - -	- - - Y - G K I - - - I - - - - C	WMJ3
Z6	C - - - Y K - - - Q S T P I	- L -	Q - L Y -TRGRT- I - - - - - C	Z6
LAVELI	C A - - Y Q - - - Q - T P I	- L -	Q S L Y - TRSRS I - - - - - C	LAVELI
CDC451	C - - - - - H - - - - V T L	- - -	- V W Y - T - E I L - - I - - - - C	CDC451
CDC42	C - - - - - - - - - - V T L	- - -	- V W Y - T - E I L - - I - - - - C	CDC42
BAL	C - - - - - - - - - S - H I	- - -	- - - Y - T - E I - - D I - - - - C	BAL

Fig. 15.5. Amino acid sequences of the principal neutralizing determinant (RP135) from several HIV-1 isolates. The flanking cysteines are conserved in all isolates and are joined by a disulfide cross-bridge (T. Gregory, personal communication) and this region is a loop in the native envelope. The central gly-pro-gly is conserved in the majority of isolates. The relatively conserved amino acids are boxed.

infected cells with uninfected CD4 cells. Because of the hypervariability of this domain, antibodies elicited by proteins or peptides comprised of the amino acid sequence from one isolate do not neutralize HIV-1 isolates with different sequences of the RP determinant (Rusche et al., 1988; Palker et al., 1988).

The function of the neutralizing determinant is unknown. The RP determinant domain may act as an 'umbrella' to cover more conserved neutralizing epitopes and prevent them from eliciting group-common neutralizing antibodies able to neutralize HIV-1 strains of different sequences. Although the sequence of this determinant is variable it is believed to serve a function in infectivity. For example, we have constructed an infectious proviral clone in which the RP determinant is deleted. (R. Grimaila, E. Emini, T. Matthews, S. Putney, unpublished). Transfection with this vector gives packaged virus particles that are unable to infect CD4 target cells. In addition, a change of the Gly-Pro-Gly sequence to Gly-Ala-Gly at the tip of the loop in an infectious molecular clone of III_B greatly reduces viral infectivity (D. Looney and F. Wong-Staal, personal communication). The fact that the Gly-Pro-Gly, the sequences of the base of the loop, and the overall length of the loop are generally conserved indicates that there is some constraint on the extent of mutation that can be tolerated.

Immunization with purified envelope subunits such as gp120, gp160 or with synthetic peptides containing the RP determinant induces virus strain type-specific neutralizing antibodies (Putney et al., 1986; Rusche et al., 1988; Palker et al., 1988;

Looney et al., 1988; Matthews et al., 1986; Weiss et al., 1986; Nara et al., 1988) that can be blocked with peptides from the RP determinant (Rusche et al., 1988). This indicates that this neutralization determinant is the only one contained in these subunit immunogens or that more conserved neutralizing epitopes are buried in gp160/gp120 and inaccessible to the immune system. In contrast, immune sera from chimpanzees or humans infected with HIV neutralize a number of different isolates. Antisera from chimpanzees (Goudsmit et al., 1988b) or from a laboratory worker infected with the III_B virus isolate (Weiss et al., 1988) taken within a year after seroconversion neutralize only the III_B isolate. This neutralization can be competitively blocked with a peptide containing the neutralizing determinant (T. Matthews and D. Bolognesi, personal communication). The peptide can no longer effectively compete with virus-neutralizing antibodies from sera taken later than one year after infection. Antisera taken eighteen months after seroconversion neutralize several other isolates. After this transition from type-specific to group-common neutralization, these antisera behave as do many immune sera from seropositive individuals which neutralize multiple isolates. These data suggest that the nature of the neutralizing antibody response changes during the course of infection.

In contrast to the inability of an envelope subunit to elicit cross-neutralizing antibodies, group-common neutralizing antibodies have been elicited by immunization of a human volunteer with recombinant vaccinia virus expressing gp160 followed by boosting with formalin-fixed autologous cells infected with this recombinant vaccinia virus (Zagury et al., 1988). However, Group-common neutralization is observed only after this boost. It is possible that the vaccinia virus in combination with the boost with infected cells more readily present conformationally dependent neutralizing epitopes to the immune system. Therefore, it may be possible to design a subunit immunogen to do the same.

There are two possible explanations for the broadening of the neutralizing antibody response to recognize diverse isolates. One is that conserved segments of the envelope or other viral proteins contain neutralization epitopes. Since envelope subunits are unable to elicit measurable titers of group-common neutralizing antibodies, it is possible that conserved epitopes are formed by a particular conformation the envelope adopts in the virus or in virus infected cells. Some conserved segments of the envelope have been reported to elicit low titers of neutralizing antibodies (Fung et al., 1987; Chanh et al., 1986; Dalgleish et al., 1988; Ho et al., 1988; Ho et al., 1987) and it is possible that one or more of these regions contribute to conformational epitopes. Antibodies to a determinant on the p17 core antigen have also been reported to neutralize in a group-common fashion (Naylor et al., 1987; Papsidero et al., 1989). The second possibility is that the group-common neutralizing antibody response observed is due to a population of antibodies directed to the type-specific neutralizing domain of different virus species. These species could arise by immunological selection for mutations in the hypervariable neutralizing domain to create a population of similar but distinct virus species unaffected by the pre-existing immunity. This would result in a series of viruses, each of which gives rise to neutralizing antibodies of limited specificity. The sum of the action of this population of type-specific antibodies would be group-common neutralization.

Variation in the sequence of gp120, and particularly in the RP determinant, is found between sequential isolates taken from a single infected person (Hahn et al., 1986; Fisher et al., 1988; Saag et al., 1988). The amino acid sequence variability of gp120 of these isolates is on the order of 6% and the differences are located primarily in the hypervariable domains. This variation does not occur during in vitro passage of the virus, suggesting that immune selection plays a role in the variability (Saag et al., 1988).

The phenomenon of type-specific and group-common neutralizing antibodies is well documented in the EIAV and visna virus systems (Narayan et al., 1978; Rwambo et al., in press). Infection of ponies with an EIAV inoculum induces a clinical episode within 15–20 days and antibodies to the p26 core antigen and the gp90 outer envelope antigen are detected soon thereafter (Rwambo et al., in press). These antibodies fail to neutralize the infecting strain and therefore probably do not play a role in the cessation of initial viremia. After the second clinical episode, type-specific neutralizing antibodies are detected and these antibodies neutralize the parental virus but not virus isolates following subsequent clinical outbreaks.

Analysis of sequential isolates taken from ponies experimentally infected with a standard EIAV strain show that virus isolates are neutralized only by antisera taken during a subsequent disease outbreak. Virus isolates are not neutralized by sera taken before the disease cycle. In general, however, the range of viruses neutralized by immune sera increases with each disease episode suggesting a 'maturation' of neutralizing antibody response. Sequential virus isolates show sequence variation in the envelope (particularly in the outer envelope) and such variation appears to be a mechanism by which the virus escapes neutralization. Using both disrupted and whole viruses as immunogens, a panel of monoclonal antibodies to the envelope of EIAV was generated (Hussain et al., 1987; Hussain et al., 1988). Several of these neutralized virus infectivity and these neutralizing antibodies reacted with variable segments of gp90. These antibodies reacted with the homologous isolate but not with heterologous isolates.

Similar types of studies (Narayan et al., 1978) show that sheep infected with plaque-purified visna virus develop neutralizing antibodies to the parental virus and viruses isolated during the first year of infection. Viruses isolated after the first year, however, are not neutralized by the early sera. Sera taken three to four years post-infection neutralize all virus isolates but with different titers, indicating a broadening of the neutralizing antibody response. A panel of monoclonal antibodies has been generated to the outer envelope of visna virus (gp135) and these identified five partially overlapping epitopes (Stanley et al., 1987). These epitopes are of two types: overt, or exposed on the native virus and covert, not recognized on the virus but recognized on purified gp135. Although these antibodies do not neutralize, antibodies recognizing overt epitopes recognize the parental virus but not sequential isolates suggesting drift in the sequence of these epitopes.

These data closely resemble the development of the neutralizing antibody response in HIV infected people in that the early antibodies are type-specific and later antibodies are group-common in their neutralization patterns. The variants of EIAV and visna virus isolated later in infection in the above experiments show patterns of variable and conserved regions in the envelope amino acid sequence. The broadening of the response may be due to two possibilities. (1) Exposure of the infected animal to a multitude of different virus isolates generated by immune selection. This results in a spectrum of type-specific neutralizing antibodies; or (2) generation of an immune response to an (initially) immunologically silent conserved neutralization epitope that is the target of group-common neutralizing antibody. Evidence for the second explanation is that sheep hyperimmunized with plaque-purified detergent disrupted visna virus develop group-common neutralizing antibodies (Narayan et al., 1987). It is likely that the detergent alters the conformation of gp135 to expose neutralization determinants inaccessible in infectious virus particles.

One way to explore neutralization epitopes is to select viruses resistant to neutralizing antibodies and to determine the mutations responsible for the resistant phenotype. In one study using HIV-1, the effect of mutations on the susceptibility to neutralizing antibody of a molecular clone of the III_B strain was examined and a mutant virus resistant to neutralization by human antisera was selected (Robert-Guroff et al., 1986). The envelope protein was sequenced and the only change, which was confirmed to confer the neutralization resistant phenotype, was an Ala to Thr replacement in the amino-terminal half of gp41 (Fig. 15.4) (Reitz et al., 1988). Synthetic peptides or envelope fragments containing this region of gp41 are unable to elicit or bind neutralizing antibodies and the mutant virus is neutralized by antibodies to the RP loop. It is likely that the neutralization resistance is due to a conformational change in gp120, induced by the mutation, that affects distant neutralization epitopes. It is also possible that this region contributes to part of a neutralization determinant formed by one or more gp120 domains, distant in the linear sequence, that are proximal in the three dimensional structure. Neutralization resistant mutants of visna virus have also been selected by in vitro culturing with neutralizing antibody (Dubois-Dalcq et al., 1979; Narayan et al., 1981).

4.2. Cellular immunity

Cellular immunity is a broad term describing specific and non-specific killer, cytotoxic, and helper T-cells and other cells that act directly or indirectly to eliminate viral pathogens. T-cell-mediated anti-viral responses are considered the most important form of cellular immunity because they are antigen-specific and are activated by subsequent viral challenge. In contrast to antibodies which recognize a large number of sites on a particular antigen, T-cell recognition is limited to a small number of immunodominant sites. T-cells recognize foreign antigens only when specific determinants of the antigens are presented on the surface of another cell, termed an antigen presenting cell (APC). Recognition only occurs in association with proteins encoded by the major histocompatibility complex locus (MHC). Helper T-cells recognize antigen in the context of class II MHC molecules and are generally $CD4^+$, whereas CTL recognize antigen on APCs in the presence of class I MHC molecules and can be either $CD8^+$ or $CD4^+$. When activated, helper T-cells secrete IL-2, IL-3, gamma interferon, and other lymphokines that lead to B-cell, T-cell and macrophage activation, whereas cytotoxic T-cells lyse cells bearing the foreign antigen in the context of the class I MHC molecule. Helper T-cells are found in HIV infected humans and chimpanzees and are elicited by envelope subunits and by other viral proteins (Zagury et al., 1988; Eichberg et al., 1987; Zarling et al., 1986; Zarling et al., 1987; Walker, C.M. et al., 1988).

Experiments involving adoptive transfer of CTL to unimmunized animals followed by virus challenge suggest that CTL specific for surface or internal viral antigens may serve as a protective host defense. For example, adoptive transfer to mice of either uncloned or cloned CTL specific for influenza A virus results in decreased virus titers and mortality when animals are subsequently challenged with infectious virus (Lin and Askonas, 1981; Yap et al., 1978). The effect is seen only when the recipients are of the same MHC I locus as the donor. Similarly, CTL from mice infected with lymphocytic chonomeningitis virus, when transferred to syngeneic (but not allogeneic) recipients, cause complete clearance of virus from animals previously infected (Byrne and Oldstone, 1984). CTL have also been shown to play a protective role in murine leukemia virus infection (Chesebro and Wehrly, 1976; Plata et al., 1987b).

Lentivirus epitopes that elicit cellular immunity were most extensively studied with HIV-1. CTL specific for the HIV-envelope, core, and reverse transcriptase have been identified in HIV infected people (Plata et al., 1987a; Walker et al., 1987; Walker, B.D. et al., 1988; Sethi et al., 1988). Envelope and core-specific CTL were elicited by HIV infection of gibbon apes (Lusso et al., 1988). Envelope-specific CTL were elicited by immunization with recombinant vaccinia virus expressing gp160 (Zarling et al., 1987). Because HIV is a chronic infection, the immune system is continually exposed to viral antigens. It is probable that CTL are present, at varying levels, throughout the infection as indicated by results of Hoffenbach et al.(1989) who found CTL and CTL precursor cells with high frequency in HIV infected individuals. It is possible that CTL reduce the level of virus infected cells and thereby lower the virus burden. An inverse correlation has been noted (Walker et al., 1987; Hoffenbach et al., 1989) between the severity of disease symptoms and the level of envelope-specific CTL.

The properties of a peptide or epitope essential for recognition by T-cells are not fully understood and algorithms have been developed to predict the location of T-cell determinants (Rothbard, 1986; Berzofsky et al., 1987). One study used an algorithm to predict amphipathicity of gp160 to locate two immunodominant helper T-cell epitopes (Cease et al., 1987). The corresponding peptides, denoted T1 and T2, stimulate proliferation and lymphokine release from lymphocytes of mice immunized with native or recombinant gp120 (Fig. 15.4). Conversely, the T2 peptide induces T cell immunity to native gp120. T2 and T1 were recognized by T cells from a number of humans immunized with vaccinia virus expressing gp160 and boosted with recombinant envelope (Berzofsky et al., 1988). This indicates that T1 and T2 function as T-cell epitopes in humans. Both T1 and T2 are in relatively conserved regions of gp120 and T2 overlaps a segment of gp120 required for interaction with CD4 (see below). In addition to T1 and T2, six additional regions of gp160, mapped using overlapping peptides, have been found to contain clusters of T helper cell epitopes (Fig. 15.4) (Hale et al., submitted for publication). Four of the six regions are conserved and all six are recognized by T cells of mice of either three of four MHC types.

In other experiments, $CD4^+$ $CD8^-$ T-cell clones responding to gp120 were isolated from HIV seronegative human donors (Siliciano et al., 1988). The T-cell epitope on the linear gp120 sequence recognized by one of these clones was mapped using synthetic peptides and fragments from an acetic acid digest gp120. The sequence recognized by this clone overlaps with, but is not identical to, T2 (Fig. 15.4). Although this region has a relatively conserved polypeptide sequence among different HIV-1 isolates, a single amino acid substitution abolishes recognition by this clone.

The same gp120 specific $CD4^+$ $CD8^-$ clones are cytolytic for both autologous CD4 peripheral blood lymphocytes and an autologous $CD4^+$ lymphocyte clone preincubated with gp120. The uptake of gp120 by $CD4^+$ clones and hence their lysis is prevented by soluble CD4 and by monoclonal antibodies that block binding of gp120 to CD4. This and similar studies (Lanzavecchia et al., 1988; Lyerly et al., 1987) provide evidence for a mechanism of $CD4^+$ cell depletion in HIV infection in which gp120 shed from the virus is picked up by uninfected $CD4^+$ cells and presented as antigen in association with class II MHC. These cells would then become targets for class II restricted killing by $CD4^+$ cytotoxic cells specific for gp120.

A target within gp120 for MHC class I restricted CTL has been mapped (Takahashi et al., 1988) and this epitope is within the principal neutralizing determinant discussed earlier (Fig. 15.4). $H\text{-}2^d$ mice immunized with vaccinia virus expressing gp160 generate $CD8^+$ $CD4^-$ lymphocytes that recognize an epitope within the neutralizing determinant. These CTL responses were detected in mice of only one of four MHC class I haplotypes tested. This suggests that the CTL repertoire elicited by the envelope is limited. Because of the variability of the sequence in this epitope, this CTL response is virus isolate-specific (Takahashi et al., 1989 a,b).

Although the epitopes recognized by human class I or II restricted CTL have not yet been determined, envelope-specific CTL clones isolated from seropositive donors are both virus type-specific or group-common in their target selection (Hoffenbach et al., 1989; Koenig et al., 1988). This suggests that these epitopes are from both variable and conserved regions of the envelope. The response of T-cells to the envelope, therefore, appears to be broader than that of neutralizing antibody which recognizes principally one epitope.

4.3. Receptor binding

The receptor for HIV-1, HIV-2 and SIV is the CD4 molecule. Virus receptors have not yet been identified for other lentiviruses. Tropism for cells of the lymphoid and macrophage/monocyte lineages is the result of the specific interaction of the HIV-1, HIV-2 or SIV outer envelope with CD4. CD4-specific monoclonal antibodies block HIV infection and prevent binding of purified gp120 to CD4 (Dalgleish et al., 1984; Klatzmann et al., 1984; McDougal et al., 1986). In addition, gp120-CD4 complexes can be immunoprecipitated from virus infected cells (McDougal et al., 1986). Human cell lines, such as HeLa, become susceptible to HIV infection when transfected with CD4 expression vectors (Maddon et al., 1986). In contrast, murine cell lines transfected with the human CD4 gene express CD4 and bind virus, but virus replication does not ensue. Since HIV replicates in human but not murine cells when they are transfected with the HIV-1 provirus genome, it is possible that a host factor (s), other than CD4, is necessary for a step after virus binding for replication to occur, HIV infection of CD4-bearing cells is blocked by soluble CD4 but neither soluble CD4 or anti-CD4 antibodies inhibit infection of glial or muscle cells (Clapham et al., 1989). This suggests that virus entry into $CD4^-$ cell types is not CD4-mediated and suggests the existence of another receptor (s).

The regions of HIV-1 gp120 that interact with CD4 have been determined. Monoclonal antibodies to gp120 that block binding to CD4 were generated (Lasky et al., 1987; Dowbenko et al., 1988). These antibodies bind to gp120 with an association constant of about 10^8 M^{-1}, more than an order of magnitude lower than the gp120-CD4 association constant. Therefore, such antibodies are unable to block HIV infection of CD4 bearing cells. A peptide from gp120 that binds to one of these antibodies was isolated and sequenced. Although this peptide cannot bind to CD4, several mutations introduced into this domain (Fig. 15.4) resulted in gp120 unable to bind to CD4. The gp120 binding regions of all three blocking antibodies were mapped using proteolytic fragments of gp120 and gp120 fragments expressed using the λgt11 expression system (Dowbenko et al., 1988).

In separate experiments in vitro mutagenesis was used to insert linkers and delete sequences randomly throughout the envelope protein (Kowalski et al., 1987). The mutants were introduced into a vector that expressed gp160 after transfection into $CD4^+$ cells. Mutations in three locations at the carboxy-terminal half of gp120 interfere with binding to CD4 (Fig. 15.4). These positions are separated by approximately 50 amino acids, and the CD4 binding determinant identified by mapping blocking monoclonal antibodies falls midway between the second and third of these positions.

The regions of gp120 associated with CD4 binding are relatively conserved in amino acid sequence between HIV-1, HIV-2, and SIV isolates. This is understandable considering that each of these molecules binds CD4. The regions of gp120 defined by these mutations have an overall hydrophobic composition suggesting that they are not accessible to antibodies. This is consistent with the fact that no monoclonal antibodies that neutralize viral infectivity by preventing binding of gp120 to CD4 have been identified.

4.4. Fusion peptide and mechanism of membrane fusion

After binding to the receptor, entry of enveloped viruses can be envisioned as occurring by either direct fusion of the viral envelope with the cell plasma membrane (fusion from without) or by internalization of the virus-receptor complex by receptor-mediated endocytosis (fusion from within). Visna virus infects cells in culture by both pathways (Harter and Choppin, 1967). At low multiplicities of infection fusion from within is observed and at multiplicities of greater than 3, fusion from without is the principal route. Fusion from without can be blocked by antisera elicited by the purified outer envelope glycoprotein, gp135 (Crane et al., 1988) and the kinetics of neutralization by neutralizing antibodies and of fusion inhibition are different. This indicates that the determinant on gp135 responsible for neutralization and for fusion are different. Antisera that neutralized a parental virus but not an antigenic variant derived from in vivo passage of the parental virus prevents fusion of both viruses. This indicates that, whereas neutralization epitopes on visna virus are variable, the fusion epitopes are conserved.

In contrast to visna virus HIV appears to induce fusion only from without. This was demonstrated by two studies showing that endocytosis of CD4 is not necessary for HIV infection. (1) Cells expressing CD4 with mutations in the CD4 cytoplasmic domain that impair the ability of CD4 to undergo endocytosis are infected as efficiently as those expressing wild-type CD4 (Bedinger et al., 1988; Maddon et al., 1988) and (2) neutralization of the low pH environment of the endosomal compartment does not interfere with viral entry (McClure et al., 1988; Stein et al., 1987). This means that the low pH dependent conformational changes that facilitate direct virus fusion with endosomal membranes that is necessary for the entry of some enveloped viruses is not necessary for entry of HIV. Current thinking holds, therefore, that direct fusion of the virus envelope or the membrane of a virus-infected cell with the membrane of an uninfected CD4-bearing cell occurs. After fusion, the contents of the virus or the infected cell are released into the cytoplasm of the uninfected cell.

The amino-terminus of HIV-1 gp41 plays a role in membrane fusion and has homology with the amino-terminus of the F_1 glycoprotein of paramyxoviruses (Gallaher, 1987). The F_1 protein is generated by cleavage of a precursor, Fc, by a trypsin like enzyme to expose the amino-terminus of F_1 which is hydrophobic and contains the sequence phe-x-gly. This amino-terminus of F_1 is responsible for both virus envelope-cell fusion and infected cell-cell fusion. The amino-terminus of HIV gp41, which is generated after cleavage of gp160, is also hydrophobic and contains phe-leu-gly. Mutations that affect the ability of the envelope to induce syncytia formation map to this region (Kowalski et al., 1987). Mutations in the amino-terminus of the SIV transmembrane protein also abolish the ability of the envelope to induce fusion (Bosch et al., 1989).

A model for the role of gp41 in initiating membrane fusion, reflecting this data, has been proposed (Kowalski et al., 1987) (Fig. 15.6). Binding of gp120 to CD4 leads to a conformational change in gp120 that allows the amino-terminus of gp41

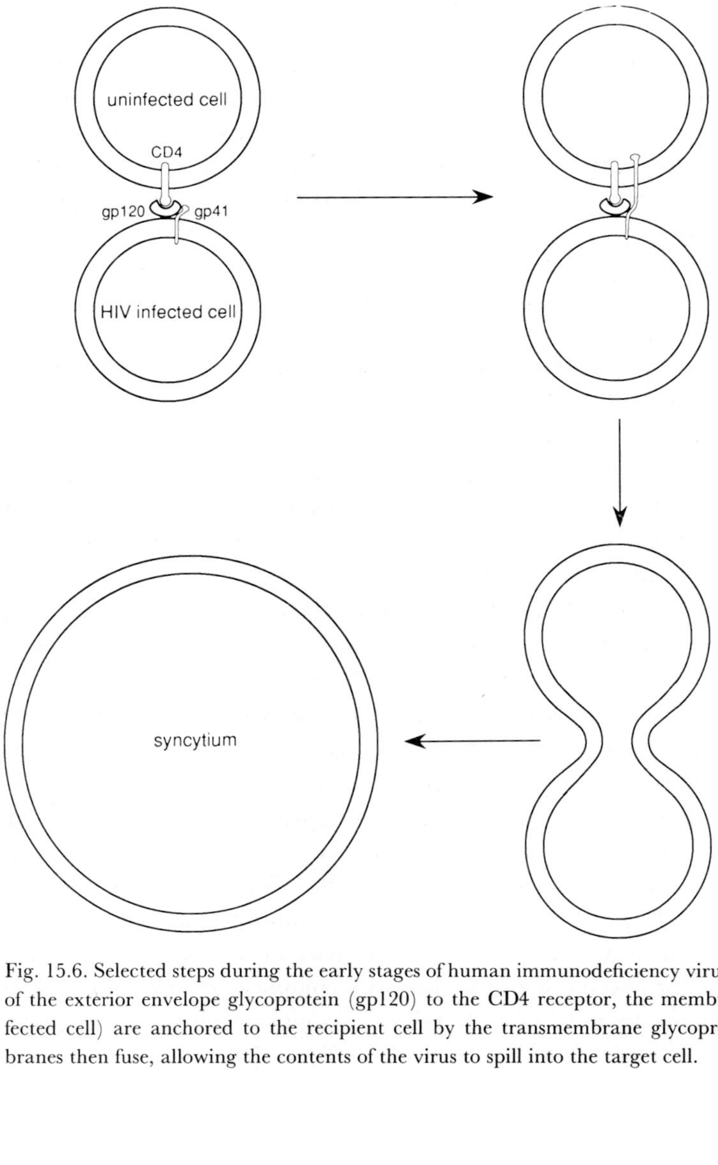

Fig. 15.6. Selected steps during the early stages of human immunodeficiency virus infection. After binding of the exterior envelope glycoprotein (gp120) to the CD4 receptor, the membranes of the virus (or infected cell) are anchored to the recipient cell by the transmembrane glycoprotein (gp41). The membranes then fuse, allowing the contents of the virus to spill into the target cell.

to penetrate the membrane of the CD4 bearing cell. Whether there is a protein that acts as a receptor for gp41 has not been determined. Once gp41 spans the gap between the membranes, direct membrane fusion occurs and the contents of the virus particle or infected cell are released into the uninfected cell. Similar models for the role of the envelope glycoproteins in fusion have been proposed for paramyxoviruses (White et al., 1983).

This model suggests that antibodies that bind the amino-terminus of gp41 may neutralize viral infectivity. Antibodies elicited by a peptide consisting of the C-terminus of gp120 and the N-terminus of gp41 have been reported to have a weak neutralizing effect (Chanh et al., 1986). It is possible that this segment of gp41 is buried in the envelope structure and therefore inaccessible to antibodies. The amino-terminus of gp41 appears to directly associate with the amino-terminal half of gp120 because mutations in either of these two regions disrupt this association (Kowalski et al., 1987). gp120 produced by these mutations is in the supernatant rather than associated with the cell. Unlike visna virus where antibodies that prevent fusion do not neutralize virus, antibodies bound to the HIV neutralizing RP determinant also inhibit fusion. It is possible that antibodies bound to this determinant either interfere with a gp120 conformational change that results in membrane insertion of gp41 or directly block gp41 insertion.

5. Lentivirus vaccines: problems and potential

Oncovirus (bovine and feline leukemia) and lentivirus (EIAV, visna) infections have been of practical concern to veterinarians for the past twenty years, and most early vaccine attempts were in these animal systems employing classical approaches. The relatively recent discoveries of human retroviruses associated with leukemia and AIDS have generated a new urgency for the development of effective control procedures and a reevaluation of previous animal retrovirus trials.

In the case of oncoviruses there have been extensive efforts to develop effective vaccines in the murine leukemia virus (MuLV) model and for natural infections by feline leukemia virus (FeLV). Antigen preparations for oncovirus vaccines have consisted primarily of inactivated whole virus, purified native or recombinant proteins, and synthetic peptides. Despite the major efforts expended, only limited success has been realized, and no practical vaccine has given 100% protection against challenge with live virus. For example, mice can be successfully immunized against Friend MuLV induced-disease by vaccination with either whole killed virus, purified glycoprotein (either individual components or a gp70-p15E multimer), or cloned *env* polyprotein in vaccinia (Hunsman et al., 1975; Hunsman et al., 1981; Earl et al., 1986). Unfortunately, the vaccination procedures are complicated and rigorous and can also be hindered by the induction of endogenous murine retroviruses (Hunsman, 1985). Caution in evaluating these early MuLV vaccine studies is required, since the effectiveness of a vaccine trial was typically based on the prevention of MuLV clinical symptoms, rather than infection. Thus, it is not clear whether any of the 'successful' vaccines completely blocked MuLV infection.

In contrast to MuLV, initial FeLV vaccines prepared from inactivated whole virus or purified glycoprotein (gp70) proved relatively ineffective in preventing FeLV infection (Schaller et al., 1977; Salerno et al., 1987). Vaccine trials using recombinant vaccinia virus expressing FeLV gp70 resulted in low antibody production and no protection of vaccinated cats against virus challenge (Gilbert et al., 1987). Limited success has been reported using a FeLV vaccine prepared from the membrane fraction of FeLV-infected cells in culture (Lewis et al., 1981). This vaccine is commercially available (Leukocell trademark) and has gained widespread use during the past 5 years. Most recently, it has been reported that cats can be protected against FeLV-induced disease by vaccination with viral gp70 incorporated into immunostimulating complexes (ISCOMS) (Osterhaus et al., 1985; Osterhaus et al., 1989). Although the protection of FeLV ISCOM immunized cats could be correlated with high levels of neutralizing antibody in most vaccine recipients, some animals appeared to be protected in the absence of neutralizing antibody, suggesting alternate mechanisms for protection. In addition, the FeLV ISCOM preparation elicited a clear booster response in seropositive animals under field conditions. This latter result supports the concept that vaccine preparations may be used to boost specific immune responses in clinically healthy seropositive individuals, with the intention of producing additional protection against the development of clinical symptoms. Such an approach has been proposed with HIV (Salk, 1987).

A clear example of a successful retrovirus vaccine is the inactivated whole virus vaccine used to prevent infection by the type-D simian AIDS retrovirus (Marx et al., 1986). All monkeys immunized with formalin-inactivated type-D virus in threonyl muramyl dipeptide (MDP) adjuvant were protected against virus challenge immediately after completion of the immunization regimen. The immunized animals remained virus free and displayed high levels of anamnestic responses when given a booster dose one year later, indicating the establishment of a strong immunologic memory. The boosted animals were still protected against virus challenge (M. Gardner, personal communication).

Interestingly the first retrovirus vaccine attempt was not with an oncovirus, but with the lentivirus EIAV. Kono and his colleagues working with EIA first reported the occurrence of retrovirus antigenic variation in 1973 and postulated that this variation was the basis for the unique nature of infectious anemia (Kono et al., 1973). As part of his EIAV studies, Kono evaluated attenuated EIAV vaccines for their ability to protect horses from infection by homologous and heterologous strains of virus. The results of these studies (Kono et al., 1970) indicated that the attenuated virus vaccine provided protection against the induction of disease by homologous, but not heterologous strains of EIAV. Based on these pioneering vaccine trials, veterinarians abandoned any hope for an EIAV vaccine and adopted a control program

TABLE 15.3

Vaccine and immunoglobulin efficacy experiments

	Immunogen				Antibodies		
	gp120[a] native	gp120[b] recombinant	gp160[c] vaccinia	gp41[d] peptide	IgG-infected[e] human	IgG-IIIB infected[f] chimpanzee	0.5β[f] monoclonal
Immunogen strain	IIIB	IIIB	LAV-1	IIIB	WA	IIIB	IIIB
Binding antibody	αgp120	αgp120	αgp120, gp41	αgp41	αenv, core	αenv, core RT	αgp120
Neutralizing antibody	+	+	−	+/−	+	+	+
ADCC	+	ND	ND	ND	+	ND	ND
T-helper response	+	+	+	ND	NA	NA	NA
CTL response	ND	ND	+	ND	NA	NA	NA
Challenge strain	IIIB	IIIB	LAV-1	NY-5	IIIB	IIIB	IIIB

ND, not determined; NA, not applicable

[a] Arthur et al., personal communication, 1987.

[b] Berman et al., 1988.

[c] Hu et al., 1987.

[d] Kennedy et al., 1987.

[e] Prince et al., 1988.

[f] Emini et al., in preparation.

based on the identification of seropositive animals using the Coggins immunodiffusion assay (Issel and Coggins, 1979), the first retrovirus diagnostic.

However, recent studies indicate that an inactivated EIAV vaccine in muramyl dipeptide (MDP) adjuvant can provide protection against challenge by homologous and heterologous strains of virus (Issel et al., 1989). In these studies, a group of Shetland ponies were immunized at three-week intervals with 5–9 doses of inactivated EIAV in MDP adjuvant and then challenged over the next year sequentially with the avirulent vaccine strain of EIAV, then a virulent variant derived from the vaccine strain, and finally a highly virulent field isolate containing a mixture of serotypes. The vaccinated animals were protected from clinical disease in all cases and appeared to be virus free, as determined by the ability to recover virus in cell culture and to detect EIAV DNA in PCR analyses of leukocytes taken post-challenge. These results provide some degree of hope for the development of effective lentivirus vaccines including vaccines against AIDS.

Chimpanzees are the only animals that can be reliably infected with HIV. Therefore, they are an important model for HIV vaccine experiments. Although HIV infects chimpanzees, it does not cause disease even after extended periods of time. Thus this model can only evaluate the capacity of a vaccine to prevent infection; it cannot be used to assess any vaccine effect on the development of clinical symptoms. Six immunization/challenge experiments designed to test the ability of immunogens to protect chimpanzees from HIV infection have been performed (Table 15.3). Four of these were active immunizations using, in two cases gp120 (Arthur, L., Nara, P., Fischinger, P. Personal communication, 1987), in one case a synthetic peptide of gp41 (Kennedy et al., 1987) and in one case vaccinia virus expressing gp160 (Hu et al., 1987). Two additional experiments were done by administering HIV-specific antibodies (Prince et al., 1988; Emini et al., in preparation).

One experiment used a recombinant vaccinia virus expressing gp160 of the LAV-1 strain of HIV-1 (Hu et al., 1987). After the initial immunization and one boost, low gp120 specific antibody titers and no neutralizing antibodies were detected. The immunized animals did have helper and cytotoxic T-lymphocytes specific for LAV-1 infected cells. Two of the active immunization experiments used as immunogen gp120 purified from either HTLV-III_B (III_B) infected cells (Arthur et al., 1987) (native gp120) or from a CHO cell line producing recombinant gp120 (Berman et al., 1988). Both proteins were formulated with aluminum hydroxide which is the only adjuvant approved for human use. Five immunizations with native gp120 or four doses of recombinant gp120 elicited antibodies that bound gp120 as assayed by both radioimmunoprecipitation and Western blotting and neutralized the III_B virus. The neutralization titers of the antibodies were approximately 10 × lower than those elicited using Freund's adjuvant with the same immunogen. In one early experiment (Kennedy et al., 1987), two chimpanzees were immunized with four doses of a synthetic peptide consisting of amino acids 735–752 of the III_B sequence. Complete Freund's followed by incomplete Freund's adjuvant was used and the animals developed gp41-binding antibody but a very low titer of neutralizing antibody to the NY-5 strain of HIV-1.

For the passive immunization experiment (Prince et al., 1988), plasmas from seropositive donors were screened for neutralizing antibodies to HTLV-III_B, and samples from approximately twenty people that had relatively high levels of neutralizing antibodies were pooled. After treatment to inactivate infectious virions, the IgG fraction was isolated and had a neutralization titer five to ten-fold higher that that in the unfractionated plasma. Subsequent experiments (see below) showed that there was no antibody present to the principal neutralization determinant of the III_B isolates. One day after receiving the immunoglobulin, the animals were challenged with 100 $TCID_{50}$ of HTLV-III_B.

When animals immunized with any of these preparations were challenged, they seroconverted as rapidly as the unimmunized controls and there were no statistically significant differences from controls in the ability to isolate virus. No analysis has yet been done to determine whether there are any differences, particularly in the ability to be neutralized by the immune antisera, between the virus recovered from the infected animals and the virus used for the challenge. One explanation for the disappointing results, therefore, is that the animals were protected from infection by a subset of the viruses in the challenge inoculum and were infected by variants within the population that were resistant to neutralization. This is of particular concern because the virus challenge inoculum was isolated from a seropositive individual and such viruses consist of a population of similar viruses differing in the sequence of gp120. This is particularly true for the III_B virus isolate from which four independent molecular clones have been sequenced which differ in amino acid sequence of the envelope by several percent (Ratner et al., 1985).

Another possible reason for the infection is that the binding and neutralizing antibody titers at the time of challenge were too low so that only a fraction of the challenge virus was neutralized. This is of concern particularly in the active immunization experiments where the neutralization titers were relatively low. It will be important in subsequent experiments to use an adjuvant or modified immunogens able to elicit higher titers of virus-neutralizing antibodies.

The results of these experiments leave doubt as to whether neutralizing antibody, which can prevent infection in vitro, can prevent infection in vivo. An experiment was therefore performed (Emini et al., 1990) to assess the ability of neutralizing antibodies, when complexed with infectious virion, to prevent infection following inoculation of the virion-antibody complexes (Table 15.3). The experiment consisted of incubating HIV-1 III_B virus with antibody for 90 minutes, inoculating uninfected chimpanzees with the antibody-virus complex and monitoring for subsequent seroconversion to viral antigens. Immunoglobulins used in this experiment include an IgG fraction isolated from serum an uninfected chimpanzee (-control) a mouse monoclonal antibody (0.5β) which neutralizes the III_B isolate and binds to the RP135 determinant (Matsushita et al., 1988) and an IgG fraction isolated from polyclonal antisera from a chimpanzee which had been infected with the III_B isolate for approximately 1 year.

The chimpanzee that was inoculated with the IgG fraction taken from the uninfected chimpanzee seroconverted in six weeks which is consistent with other animals that had been challenged with the III_B virus. The animal inoculated with the virus preincubated with the 0.5β monoclonal antibody did not seroconvert until 12 weeks and showed a significantly lower antibody titer than the control animal. The chimpanzee receiving the IgG fraction taken from the III_B infected chimpanzee has shown no signs of seroconversion even after 60 weeks. All of the neutralizing antibody in the IgG fraction which was protective can be absorbed with the RP135 peptide.

These results show that the RP135-directed monoclonal antibody delays infection and strongly suggests that the RP135-directed neutralizing antibodies in the polyclonal antisera protected against infection. Although encouraging, these experiments do not prove rigorously that RP135-directed antibodies can protect against infection and this must await an immunization/challenge experiment with an RP135 immunogen. The experiment does, however, show that antibody is sufficient to prevent infection and also shows that, at least in the presence of high levels of neutralizing antibodies, antibody-dependent enhancement of infection does not occur.

In addition to the chimpanzee, the SIV/rhesus monkey system is a model for HIV vaccine development. In contrast to the HIV/chimpanzee system, SIV induces an immunodeficiency disease in rhesus macaques that is fundamentally the same as observed in human AIDS. Current SIV vaccine trials include inactivated virus preparations, purified envelope proteins in MDP or ISCOMS, and envelope proteins expressed in vaccinia virus. Although the initial attempts with these vaccines were unsuccessful in eliciting protective immune responses, more recent studies suggest that protection can be accomplished using inactivated SIV (M. Murphy-Corb and R. Desrosiers, personal communications). Although the immune targets eliciting the protective response have not yet been determined, success of these experiments gives hope that effective human lentivirus vaccines can be developed.

6. Acknowledgements

Ronald C. Montelaro is supported by Public Health Service grants AI6560, AI2580, CA49296, and AI 28243 from the National Institutes of Health. We thank Stephanie Palms and Stephanie Robert for preparation of the manuscript.

7. References

Aldovini, A., Debouck, C., Feinberg, M.B., Rosenberg, M., Arya, S.K. and Wong-Staal, F. (1986) *Proc. Natl. Acad. Sci. USA* 83, 6672–6676.

Allan, J.S., Coligan, J.E., Lee, T., McLane, M.F., Kanki, P.J., Groopman, J.E. and Essex, M. (1985) *Science* 230, 810–813.

Argos, P. (1989) *EMBO J.* 8, 779.

Arthur, L.O., Pyle, S.W., Nara, P.L., BJW, Jr., Bonda, M.A., Kelliher, J.C., Gilden, R.V., Robey, W.G., Bolognesi, D.P., Gallo, R.C., Fischinger, P.J. (1987) *Proc. Natl. Acad. Sci. USA* 84, 8583–8587.

Barone, A.D., Silva, J.J., Ho, D.D., Gallo, R.C., Wong-Staal, F. and Chang, N.T. (1986) *J. Immunol.* 137, 669–673.

Bedinger, P., Moriarty, A., von Borstel, R.C., II, Donovan, N.J., Steimer, K.S. and Littman, D.R. (1988) *Nature* 334, 162–165.

Berman, P.W., Groopman, J.E., Gregory, T., Clapham, P.R., Weiss, R.A., Ferriani, R., Riddle, L., Shimasaski, C., Lucas, C., Lasky, L.A., Eichberg, J.W. (1988a) *Proc. Natl. Acad. Sci. USA* 85, 5200–5204.

Berman, P.W., Nunes, W.M. and Haffar, O.K. (1988b) *J. Virol.* 62, 3135–3142.

Berzofsky, J.A., Bensussan, A., Cease, K.B., Bourge, J.F., Cheynier, R., Lurhuma, Z., Salaun, J.J., Gallo, R., Shearer, G.M. and Zagury, D. (1988) *Nature* 334, 706–708.

Berzofsky, J.A., Cease, K.B., Cornette, J.L., Souge, J.L., Margalit, H., Berkower, I.J., Good, M.F., Miller, L.H. and Delisi, C. (1987) *Immunol. Rev.* 98, 9–52.

Bosch, M.L., Earl, P.L., Fargnoli, K., Picciafuoco, S., Giombini, F., Wong-Staal, F. and Franchini, G. (1989) *Science* 244, 694–697.

Braun, M., Clements, J. and Gonda, M. (1987) *J. Virol.* 61, 4046–4054.

Byrne, J.A. and Oldstone, M.B.A. (1984) *J. Virol.* 51, 682–686.

Cease, K.B., Margalit, H., Cornette, J.L., Putney, S.D., Robey, W.G., Ouyang, C., Streicher, H.Z., Fischinger, P.J., Gallo, R.C., DeLisi, C. and Berzofsky, J.A. (1987) *Proc. Natl. Acad. Sci. USA* 84, 4249–4253.

Chang, N.T., Chanda, P.K., Barone, A.D., McKinney, S., Rhodes, D.P., Tam, S.H., Shearman, C.W., Huang, J. and Chang, T.W. (1985a) *Science* 228, 93–96.

Chang, N.T., Huang, J., Ghrayeb, J., McKinney, S., Chanda, P.K., Chang, T.W., Putney, S.D., Sarngadharan, M.G., Wong-Staal, F. and Gallo, R.C. (1985b) *Nature* 315, 151–154.

Chanh, T.C., Dressman, G.R., Kanda, P., Linette, G.P., Sparrow, J.T., Ho, D.D. and Kennedy, R.C. (1986) *Eur. Mol. Biol. J.* 5, 3065–3071.

Chesebro, B. and Wehrly, K. (1976) *J. Exp. Med.* 143, 85–99.

Chirmule, N., Kalyanaraman, V., Oyaizu, N. and Pahwa, S. (1988) *J. AIDS* 1, 425–430.

Clapham, P.R., Weber, J.N., Whitby, D., McIntosh, K., Dalgleish, A.G., Maddon, P.J., Deen, K.C., Sweet, R.W. and Weiss, R.A. (1989) *Nature* 337, 368–370.

Clements, J.E., Gdovin, S.L., Montelaro, R. and Narayan, O. (1988) *Annu. Rev. Immunol.* 6, 139–159.

Crane, S.E., Clements, J.E. and Narayan, O. (1988) *J. Virol.* 62, 2680–2685.

Dalgleish, A.G., Beverley, P.C.L., Clapham, P.R., Crawford, D.H., Greaves, M.F. and Weiss, R. (1984) *Nature* 312, 763–768.

Dalgleish, A.G., Chanh, T.C., Kennedy, R.C., Kanda, P., Clapham, P.R. and Weiss, R. (1988) *Virology* 165, 209–215.

Daniel, M.D., Letvin, N.L., King, N.W., Kannagi, M., Sehgal, P.K., Hunt, R.D., Kanki, P.J., Essex, M. and Desrosiers, R.C. (1985) *Science* 228, 1201–1204.

Dayton, A.I., Terwilliger, E.F., Potz, J., Kowalski, M., Sodroski, J.G., Haseltine, W.A. (1988) *J. AIDS* 1, 441–452.

Diamond, D.C., Sleckman, B.P., Gregory, T., Lasky, L.A., Greenstein, J.L., Burakoff, S.J. (1988) *J. Immunol.* 141, 3715–3717.

Dietzschold, B., Tollis, M., Lafon, M., Wunner, W.H. and Koprowski, H. (1987a) *Virology* 161, 29–36.

Dietzschold, B., Wand, H., Rupprecht, C.E., Celis, E., Tollis, M., Ertl, H., Hebes-Katz, E. and Koprowski, H. (1987b) *Proc. Natl. Acad. Sci. USA* 84, 9165–9169.

Dowbenko, D., Nakamura, G., Fennie, C., Shimasaki, C., Riddle, L., Harris, R., Gregory, T. and Lasky, L. (1988) *J. Virol.* 62, 4703–4711.

Dubois-Dalcq, M., Narayan, O. and Griffin, D.E. (1979) *Virology* 92, 353–366.

Earl, P.L., Moss, B., Morrison, R.P., Wehrly, K., Nishio, J. and Chesebro, B. (1986) *Science* 234, 728–731.

Eichberg, J.W., Zarling, J.M., Alter, H.J., Levy, J.A., Berman, P.W., Gregory, T., Lasky, L.A., McClure, J., Cobb, K.E., Moran, P.A., Hu, S.L., Kennedy, R.C., Chanh, T.C. and Dreesman, G.R. (1987) *J. Virol.* 61, 3804–3808.

Emini, E.A., Ostapchuk, P. and Wimmer, E. (1983) *J. Virol.* 48, 547–550.

Emini, E.A., Nara, P.L., Schleif, W.A., Lewis, J.A., Davide, J.P., Lee, D.R., Kessler, J., Conley, S., Matsushita, S., Putney, S.D., Gerety, R.J. and Eichberg, J.W. (1990) *J. Virol.*, in press.

Fisher, A.G., Ensoli, B., Looney, D., Rose, A., Gallo, R.C., Saag, M.S., Shaw, G.M., Hahn, B.H. and Wong-Staal, F. (1988) *Nature* 334, 444–447.

Franchini, G., Robert-Guroff, M., Ghrayeb, J., Chang, N.T. and Wong-Staal, F. (1986a) *Virology* 155, 593–599.

Franchini, G., Robert-Guroff, M., Wong-Staal, F., Ghrayeb, J., Kato, I., Chang, T.W. and Chang, N.T. (1986b) *Proc. Natl. Acad. Sci. USA* 83, 5282–5285.

Franchini, G., Rusche, J.R., O'Keefe, T.J. and Wong-Staal, F. (1988) *AIDS Res. Hum. Retroviruses* 4, 243–250.

Fung, M.S.C., Sun, C., Sun, N.-C., Chang, N.T. and Chang, T.W. (1987) *Bio/Technology* 5, 940.

Gallaher, W.R. (1987) *Cell* 50, 327–328.

Gallaher, W.R., Ball, J.M., Garry, R.F., Griffin, M. and Montelaro, R.C. (1989) *AIDS Res. Human Retroviruses* 5, 431.

Gallo, R., Wong-Staal, F., Montagnier, L., Haseltine, W.A. and Yoshida, M. (1988) *Nature* 333, 504–505.

Ghrayeb, J., Kato, I., McKinney, S., Huang, J.J., Chanda, P.K., Ho, D.D., Sarangadharan, M.G., Chang, T.S. and Chang, N.T. (1986) *DNA* 5, 93–99.

Ghrayeb, W.T., Kaplan, P.M., Mueller, W.T. and Putney, S.D. (1989) In *HIV Detection by Genetic Engineering Methods* (P.A. Luciw and K.S. Steimer, Eds.) Marcel Dekker, New York. pp. 41–57.

Gilbert, J., Pedersen, N. and Nunberg, J. (1987) *Virus Res.* 7, 49–67.

Goh, W.C., Rosen, C., Sodroski, J., Ho, D.D. and Haseltine, W.A. (1986) *J. Virol.* 1986, 181–184.

Goh, W.C., Sodroski, J.G., Rosen, C.A. and Haseltine, W.A. (1987) *J. Virol.* 61, 633–637.

Gollins, S.W. and Porterfield, J.S. (1986) *Nature* 321, 244–246.

Gonda, M.A., Braun, M.J., Carter, S.G., Kost, T.A., Bess, J.W. and Arthur, L.O. and Van der Maaten, M. (1988) *Nature* 330, 388–391.

Goudsmit, J., Debouck, C., Meloen, R.H., Smit, L., Bakker, M., Asher, D.M., Wolff, A.V., Gibbs, C.J., Jr. and Gajdusek, D.C. (1988a) *Proc. Natl. Acad. Sci. USA* 85, 4478–4482.

Goudsmit, J., Thiriart, C., Smit, L., Bruck, C. and Gibbs, C.J. (1988b) *Vaccine* 6, 229–232.

Gupta, P., Balachandran, R., Ho, M., Enrico, A. and Rinaldo, C. (1989) *J. Virol.* 63, 2361.

Haase, A. (1986) *Nature* 322, 130–136.

Hahn, B.H., Shaw, G.M., Taylor, M.E., Redfield, R.R., Markham, P.D., Salahuddin, S.Z., Wong-Staal, F., Gallo, R.C., Parks, E.S. and Parks, W.P. (1986) *Science* 232, 1548–1553.

Hale, P.M., Cease, K.B., Hougton, R.A., Ouyang, C., Putney, S.D., Javaherian, K., Margalit, H., Cornette, J.L., Spouge, J.L., DeLisi, C. and Berzofsky, J.A. (1989) *Int. Immunol.* 1, 409–415.

Harter, D.H. and Choppin, P.W. (1967) *Virology* 31, 279.

Henderson, L.E., Sowder, R.C., Copland, T.D., Benveniste, R.E. and Oroszlan, S. (1988) *Science* 241, 199–202.

Ho, D.D., Kaplan, J.C., Rackauskas, I.E. and Gurney, M.E. (1988) *Science* 239, 1021–1023.

Ho, D.D., Sarngadharan, M.G., Hirsch, M.S., Schooley, R.T., Rota, T.R., Kennedy, R.C., Chahn, T.C. and *V.L.*, S. (1987) *J. Virol.* 61, 2024–2028.

Hoffenbach, A., Langlade-Demoyen, P., Dadaglio, G., Vilmer, E., Michel, F., Mayaud, C., Autran, B. and Plata, F. (1989) *J. Immunol.* 142, 452–462.

Hu, S.L., Fultz, P.N., McClure, H.M., Eichberg, J.W., Thomas, E.K., Zarling, J., Singhal, M.C., Kosowski, S.G., Swenson, R.B., Anderson, D.C. and Todaro, G. (1987) *Nature* 328, 721–723.

Hunsmann, G. (1985) *Cancer Res* 45, 4691–4693.

Hunsmann, G., Moenning, V. and Schafer, W. (1975) *Virology* 66, 327.

Hunsmann, G., Schneider, J. and Schulz, A. (1981) *Virology* 113, 602.

Huso, D., Narayan, O., and Hart, G. (1988) *J. Virol.* 62, 1974.

Hussain, K.A., Issel, C.J., Schnorr, K.L. and Montelaro, R.C. (1987) *J. Virol.* 61, 2956–2961.

Hussain, K.A., Issel, C.J., Schnorr, K.L., West, M.D., Rwambo, P. and Montelary, R.C. (1988) *Arch. Virol.* 98, 213–224.

Icenogle, J., Schiwen, J., Duke, G., Gilbert, S., Rueckert, R. and Anderegg, J. (1983) *Virology* 127, 412–425.

Issel, C. and Coggins, L. (1979) *J. Am. Vet. Med. Assoc.* 174, 727–733.

Issel, C.J., McManus, J., Hagus, S., Adams, W.V. and Montelaro, R.C. (1989), in preparation.

Johnson, P.R., Parks, D.E., Norrby, E., Lerner, R.A., Purcell, R.H. and Chanock, R.M. (1988) *AIDS Res. Human Retroviruses* 4, 159.

Kalyanaraman, V.S., Cabradilla, C.D., Getchell, J.P., Narayanan, R., Braff, E.H., Chermann, J., Barré-Sinoussi, F., Montagnier, L., Spira, T.J., Kaplan, J., Fishbein, D., Jaffe, H.W., Curran, J.W. and Francis, D.P. (1984) *Science* 225, 321–323.

Kan, N.C., Franchini, G., Wong-Staal, F., DuBois, G.C., Robey, W.G., Lautenberger, J.A. and Papas, T.S. (1986) *Science* 231, 1553–1555.

Kappes, J.C., Morrow, C.D., Lee, S., Jameson, B.A., Kent, S.B.H., Hood, L.E., Shaw, G.M. and Hahn, B.H. (1988) *J. Virol.* 62, 3501–3505.

Kenealy, W.R., Matthews, T.J., Ganfield, M., Langlois, A.J., Waselefsky, D.M. and Petteway, S.R. (1989) *AIDS Res. Human Retroviruses* 5, 173–182.

Kennedy, R.C., Kanda, R., Dressman, G.R., Eichberg, J.W., Chanh, T.C. (1987) In *Vaccines '87* (Chanock, Lerner, Brown and Ginsberg, eds.) Cold Spring Harbor Laboratory, Cold Spring Harbor NY. pp. 250–255

Kennedy-Stoskopf, S. and Narayan, O. (1986) *J. Virol.* 59, 37–44.

Kinney-Thomas, E., Weber, J.N., McClure, J., Clapham, P.R., Singhal, M.C., Shriver, M.K. and Weiss, R. (1988) *AIDS* 2, 25–29.

Klatzmann, D., Champagne, E., Chamaret, S., Gruest, J., Guetard, D., Hercend, T., Gluckman, J.C. and Montagnier, L. (1984) *Nature* 312, 767–768.

Koenig, S., Earl, P., Powell, D., Pantaleo, G., Merli, S., Moss, B. and Fauci, A.S. (1988) *Proc. Natl. Acad. Sci. USA* 85, 8638–8642.

Kono, Y., Kobayasi, K. and Fukunaga, Y. (1970) *Natl. Inst. Anim. Health O (Tokyo)* 10, 113–122.

Kono, Y., Kobayasi, K. and Fukunga, Y. (1973) *Arch. Virol.* 41, 1–10.

Kornfeld, H., Cruikshank, W.H., Pyle, S.W., Berman, J.S., Center, D.M. (1988) *Nature* 335, 445–448.

Kowalski, M., Potz, J., Basiripour, L., Dorfman, T., Goh, W.C., Terwilliger, E., Dayton, A., Rosen, C., Haseltine, W. and Sodroski, B. (1987) *Science* 237, 1351–1355.

Kramer, R.A., Schaber, M.D., Skalka, A.M., Ganguly, K., Wong-Staal, F. and Reddy, E.P. (1986) *Science* 231, 1580–1584.

Kulaga, H., Folks, T.M., D.H, Rutledge, R. and Kindt, T.J. (1988) *IV International Conference on AIDS* Book 1, 179.

Lanzavecchia, A., Roosnek, E., Gregory, T., Berman, P. and Abrignani, S. (1988) *Nature,* 530–532.

Lasky, L.A., Nakamura, G., Smith, D.H., Fennie, C., Shimasaki, C., Patzer, E., Berman, P., Gregory, T. and Capon, D.J. (1987) *Cell* 50, 975–985.

Lee, P.W.K., Hayes, E.C. and Joklik, W.K. (1981) *Virology* 108, 134–146.

Lee, T., Coligan, J.E., Allan, J.S., McLane, M.F., Groopman, J.E., and Essex, M. (1986) *Science* 231, 1546–1549.

Leis, J., Baltimore, D., Bishop, J.M., Coffin, J., Fleissner, E., Goff, S.P., Oroszlan, S., Robinson, H., Skalka, A.M., Temin, H. and Vogt, V. (1988) *J. Virol.* 62, 1808–1809.

Lewis, M.G., Mathes, L.E. and Olsen, R.G. (1981) *Infect. Immun.* 34, 888–894.

Lin, Y.L. and Askonas, B.A. (1981) *J. Exp. Med.* 154, 225–234.

Linsley, P.S., Ledbetter, J.A., Kinney-Thomas, E. and Hu, S.L. (1988) *J. Virol.* 62, 3695–3702.

Looney, D.J., Fisher, A.G., Putney, S.D., Rusche, J.R., Redfield, R.R., Burke, D.S., Gallo, R.C. and Wong-Staal, F. (1988) *Science* 241, 357–359.

Lusso, P., Markham, P.D., Ranki, A., Earl, P., Moss, B., Dorner, F., Gallo, R.C. and Krohn, K. (1988) *J. Immunol.* 141, 2467–2473.

Lyerly, H.K., Matthews, T.J., Langlois, A.J., Bolognesi, D.P. and Weinhold, K.J. (1987) *Proc. Natl. Acad. Sci. USA* 84, 4601–4605.

Maddon, P.J., Dalgleish, A.G., McDougal, J.S., Clapham, R.P., Weiss, R.A. and Axel, R. (1986) *Cell* 42, 333–348.

Maddon, P.J., McDougal, J.S., Clapham, P.R., Dalgleish, A.G., Jamal, S., Weiss, R.A. and Axel, R. (1988) *Cell* 54, 865–874.

Malim, M.H., Hauber, J., Le S., Maizel, J.V. and Cullen, B.R. (1989) *Nature* 338, 254–256.

Mann, D.L., Lasane, F., Popovic, M., Arthur, L.O., Robey, W.G., Blattner, W.A. and Newman, M.J. (1987) *J. Immunol.* 138, 2640.

Marx, P.A., Pedersen, N.C., Lerche, N.W., Osborn, K.G., Lowenstine, L.J., Lackner, A.A., Maul, D.H., Kwang, H.S., Kluge, J.D., Zaiss, C.P., Sharpe, V., Spinner, A.P., Allison, A.C. and Gardner, M.B. (1986) *J. Virol.* 60, 431–435.

Massey, R.J. and Schochetman, G. (1981) *Virology* 115, 20–32.

Matsuda, Z., Chou, M.J., Matsuda, M., Huang, J.H., Chen, Y.M., Redfield, R., Mayer, K., Essex, M. and Lee, T.H. (1988) *Proc. Natl. Acad. Sci. USA* 85, 6968–6972.

Matsushita, S., Robert-Guroff, M., Rusche, J., Koito, A., Hattori, T., Hoshino, H., Javaherian, K., Takatsuki, K. and Putney, S. (1988) *J. Virol.* 62, 2107–2114.

Matthews, T.J., Langlois, A.J., Robey, W.G., Chang, N.T., Gallo, R.C., Fischinger, P.J. and Bolognesi, D.P. (1986) *Proc. Natl. Acad. Sci. USA* 83, 9709–9713.

McClure, M.O., Marsh, M. and Weiss, R.A. (1988) *Eur. Mol. Biol. J.* 7, 513–518.

McDougal, J.S., Kennedy, M.S., Sligh, J.M., Cort, S.P., Mawle, A. and Nicholson, J.K.A. (1986) *Science* 231, 382–385.

McGuire, T., Crawford, T. and Henson, J. (1972) *Immunol. Commun.* 1, 545–551.

McGuire, T.C., Norton, L.K., O'Rourke, K.I. and Cheevers, W.P. (1988) *J. Virol.* 62, 3488–3492.

Meric, C. and Goff, S.P. (1989) *J. Virol.* 63, 1558–1568.

Montefiori, D.C., Robinson, W.E. and Mitchell, W.M. (1988) *Proc. Natl. Acad. Sci. USA* 85, 9248–9252.

Montelaro, R.C., Parekh, B., Orrego, A. and Issel, C.J. (1984) *J. Biol. Chem.* 259, 10539–10544.

Montelaro, R.C., Robey, W.G., West, M.D., Issel, C.J. and Fischinger, P.J. (1988) *J. Gen. Virol.* 69, 1711–1717.

Murphey-Corb, M., Martin, L.N., Rangan, S.R.S., Baskin, G.B., Gormus, B.J., Wolf, R.H., Andes, W.A., West, M. and Montelaro, R.C. (1986) *Nature* 321, 435–437.

Nara, P.L., Robey, W.G., Pyle, S.W., Hatch, W.C., Dunlop, N.M., Bess, J.W., Kelliher, J.C., Arthur, L.O. and Fischinger, P.J. (1988) *J. Virol.* 62, 2622–2628.

Narayan, O., Clements, J.E., Griffin, D.E. and Wolinsky, J.S. (1981) *Infect. Immun.* 32, 1045–1050.

Narayan, O., Clements, J.E., Kennedy-Stoskopf, S., Sheffer, D. and Royal, W. (1987) In *Contributions to Microbiology and Immunology* (J.M. Cruse and R.E. Lewis, Eds.) Vol. 8. Karger, Basel. pp. 60–76.

Narayan, O., Griffin, D. and Clements, J.F. (1977) *Science* 197, 376–378.

Narayan, O., Griffin, D.E. and Clements, J.E. (1978) *J. Gen. Virol.* 41, 343–352.

Narayan, O., Kennedy-Stoskopf, S., Sheffer, D., Griffin, D.E. and Clements, J.E. (1983) *Infect. Immun.* 41, 33–67.

Narayan, O., Sheffer, D., Griffin, D., Clements, J. and Hess, J. (1984) *J. Virol.* 49, 349.

Naylor, P.H., Naylor, C.W., Badamchian, M., Wada, S., Goldstein, A.L., Wang, S., Sun, D.K., Thornton, A.H. and Sarin, P.S. (1987) *Proc. Natl. Acad. Sci. USA* 84, 2951–2955.

Norrby, E., Biberfeld, G., Chiodi, F., von Gegerfeldt, A., Naucler, A., Parks, E. and Lerner, R. (1987) *Nature* 329, 248–250.

Osterhaus, A., Weijer, K., Uytdehaag, F., Jarret, O., Sundquist, B. and Morein, B. (1985) *J. Immunol.* 135, 591–598.

Osterhaus, A., Weijer, K., Uytdehaag, F., Knell, P., Jarrett, O., Akerblom, L. and Morein, B. (1989) *Vaccine* 7, 137–141.

Ozel, M., Pauli, G. and Gelderblom, H. (1988) *Arch. Virol.* 100, 255–266.

Palker, T.J., Clark, M.E., Langlois, A.J., Matthews, T.J., Weinhold, K.J., Randall, R.R., Bolognesi, D.P. and Haynes, B.F. (1988) *Proc. Natl. Acad. Sci. USA* 85, 1932–1936.

Palker, T.J., Matthews, T.J., Clark, M.E., Cianciolo, G.J., Randall, R.R., Langlois, A.J., White, G.C., Safai, B., Synderman, R., Bolognesi, D.P. and Haynes, B.F. (1987) *Proc. Natl. Acad. Sci. USA* 84, 2479–2483.

Papsidero, L.D., Sheu, M. and Ruscetti, F.W. (1989) *J. Virol.* 63, 267–272.

Parekh, B., Issel, C. and Montelaro, R. (1980) *Virology* 107, 520–525.

Payne, S.L., Fang, F.D., Lui, C.P., Dhruva, B., Rwambo, P., Issel, C.J. and Montelaro, R.C. (1987) *Virology* 161, 321–331.

Payne, S.L., Rushlow, K., Issel, C.J. and Montelaro, R.C. (1989) *J. Virol.* 172, 609–615

Pedersen, N.C., Ho, E., Brown, M.L. and Yamamoto, J.K. (1987) *Science* 235, 790–793.

Péterlin, B.M. and Luciw, P.A. (1988) *AIDS* 2, 529–540.

Plata, F., Autran, B., Martins, L.P., Wain-Hobson, S., Raphael, M., Mayaud, C., Denis, M., Guillon, J.M. and Debre, P. (1987a) *Nature* 328, 348–351.

Plata, F., Langlade-Demoyen, P., Abastado, J.P., Berbar, T. and Kourilsky, P. (1987b) *Cell* 48, 231.

Possee, R.D., Schild, G.C. and Dimmock, N.J. (1982) *J. Gen. Virol.* 58, 373–386.

Prince, A.M., Horowitz, B., Baker, L., Shulman, R.W., Ralph, H., Valinsky, J., Cundell, A., Brotman, B., Boehle, W., Rey, R., Piet, M., Reesink, H., Lelie, N., Tersmette, M., Miedema, F., Barbosa, L., Nemo, G., Nastala, C.l., Allan, J.S., Lee, D.R. and Eichberg, J.W. (1988) *Proc. Natl. Acad. Sci. USA* 85, 6944–6948.

Puddington, L., Bevan, M.J., Rose, J.K. and LaFrançois, L. (1986) *J. Virol.* 60, 708–717.

Putney, S.D., Matthews, T.J., Robey, W.G., Lynn, D.L., Robert-Guroff, M., Mueller, W.T., Langlois, A.J., Ghrayeb, J., Petteway, S.R., Jr., Wienhold, K.J., Fischinger, P.J., Wong-Staal, F., Gallo, R.C. and Bolognesi, D.P. (1986) *Science* 234, 1392–1395.

Ratner, L., Haseltine, W., Patarca, R., Livak, K.J., Starcich, B., Josephs, S.F., Doran, E.R., Rafalski, A.N., Whitehorn, E.A., Baumeister, K., Ivanoff, L., Petteway, S.R., Pearson, M.L., Lautenberger, J.A., Papas, T.S., Ghrayeb, J., Chang, N.T., Gallo, R.C. and Wong-Staal, F. (1985) *Nature* 313, 277–284.

Reitz, M.S., Jr., Wilson, C., Naugle, C., Gallo, R.C. and Robert-Guroff, M. (1988) *Cell* 54, 57–63.

Robert-Guroff, M., Brown, M. and Gallo, R.C. (1985) *Nature* 316, 72–74.

Robert-Guroff, M., Reitz, M.S.J., Robey, W.G. and Gallo, R.C. (1986) *J. Immunol.* 137, 3306–3309.

Rothbard, J. (1986) *Ann. Inst. Pasteur* 137E, 518–526.

Rusche, J.R., Javaherian, K., McDanal, C., Petro, J., Lynn, D.L., Grimaila, R., Langlois, A., Gallo, R.C., Arthur, L.O., Fischinger, P.J., Bolognesi, D.P., Putney, S.D. and Matthews, T.J. (1988) *Proc. Natl. Acad. Sci. USA* 85, 3198–3202.

Rwambo, P.M., Issel, C.J., Adams, W.V., Hussain, K.A., Miller, M. and Montelaro, R.C. (1990) *Arch. Virol.,* in press.

Saag, M.S., Hahn, B.H., Gibbons, J., Li, Y., Parks, E.S., Parks, W.P. and Shaw, G.M. (1988) *Nature* 334, 440–444.

Salerno, R., Lehman, E., Larson, V. and Hellerman, R. (1978) *J. Natl. Cancer Inst.* 61, 1487–1497.

Salk, J. (1987) *Nature* 327, 473–476.

Sarngadharan, M.G., Popovic, M., Bruch, L., Schupbach, J. and Gallo, R.C. (1984) *Science* 224, 506–508.

Schaller, J., Hoover, E. and Olsen, R. (1977) *J. Natl. Cancer Inst.* 59, 1441–1450.

Schlesinger, J.J., Brandriss, M.U., Croppe, C.B. and Monath, T.P. (1986) *J. Virol.* 60, 1153–1155.

Schnittman, S.M., Lane, H.C., Roth, J., Burrows, A., Folks, T.M., Kehrl, J.H., Koenig, S., Berman, P. and Fauci, A.S. (1988) *J. Immunol.* 141, 4181–4186.

Schupbach, J., Popovic, M., Gilden, R.V., Gonda, M.A., Sarngadharan, M.G. and Gallo, R.C. (1984) *Science* 224, 503–505.

Sethi, K.K., Naher, H. and Stroehmann, I. (1988) *Nature* 335, 178–181.

Shalaby, M.R., Krowka, J.F., Gregory, T.J., Hirabayashi, S.E., McCabe, S.M., Kaufman, D.S., Stites, D.P. and Ammann, A.J. (1987) *Cell* 110, 140.

Siliciano, R.F., Lawton, T., Knall, C., Karr, R.W., Berman, P., Gregory, T. and Reinherz, E.L. (1988) *Cell* 54, 561–575.

Skinner, M.A., Langlois, A.J., McDanal, C.B., McDougal, J.S., Bolognesi, D.P. and Matthews, T.J. (1988a) *J. Virol.* 62, 4195–4200.

Skinner, M.A., Ting, R., Langlois, A.K., Weinhold, K.J., Lyerly, H.K., Javaherian, K. and Matthews, T.J. (1988b) *AIDS Res. Human Retroviruses* 4, 187–197.

Stanley, J., Bhaduri, L.M., Narayan, O. and Clements, J.E. (1987) *J. Virol.* 61, 1019–1028.

Starcich, D.H., Hahn, G.H., Shaw, G.M., McNeely, P.D., Modrow, S., Wolf, H., Parks, E.S., Parks, W.p., Josephs, S.F., Gallow, R.C., Wong-Staal, F. (1986) *Cell* 45, 637–648.

Stein, B.S., Gowda, S.D., Lifson, J.D., Penhallow, R.C., Bensch, K.G. and Engleman, E.G. (1987) *Cell* 49, 659–668.

Takahashi, H., Cohen, J., Hosmalin, A., Cease, K.B., Houghton, R., Cornette, J.L., DeLisi, C., Moss, B., Germain, R.N. and Berzofsky, J.A. (1988) *Proc. Natl. Acad. Sci. USA* 85, 3105–3109.

Takashashi, H., Houghten, R., Margulies, D.H., Putney, S.D., Moss, B., Germain, R. and Berzofsky, J.A. (1989) *J. Exp. Med.* 170, 2023–2035.

Taylor, H.P. and Dimmock, N.J. (1985a) *J. Exp. Med.* 161, 198–209.

Taylor, H.P. and Dimmock, N.J. (1985b) *J. Gen. Virol.* 66, 903–908.

Townsned, A.R.M., Rothland, J., Gotch, F.M., Bahadur, G., Warith, D. and McMichael, A.J. (1986) *Cell* 44, 959–968.

Vallee, H. and Carre, H. (1904) *C.R. Acad. Sci.* 139, 331–333.

Venable, R.M., Pastor, R.W., Brooks, B.R. and Carson, F.W. (1989) *AIDS Res. Human Retroviruses* 5, 7–22.

Veronese, F.D.M., Copeland, T.D., Devico, A.L., Rahman, R., Oroszlan, S., Gallo, R.C. and Sarngadharan, M.G. (1986) *Science* 231, 1289–1291.

Walker, B.D., Chakrabarti, S., Moss, B., Paradis, T.J., Flynn, T., Durno, A.G., Blumberg, R.S., Kaplan, J.C., Hirsch, M.S. and Schooley, R.T. (1987) *Nature* 328, 345–348.

Walker, B.D., Flexner, C., Paradis, T.J., Fuller, T.C., Hirsch, M.S., Schooley, R.T. and Moss, B. (1988) *Science* 240, 64–66.

Walker, C.M., Steimer, K.S., Rosenthal, K.L. and Levy, J.A. (1988) *J. Clin. Invest.* 82, 2172–2175.

Wang, J.J.G., Steel, S., Wisniewolski, R. and Wang, C.Y. (1986) *Proc. Natl. Acad. Sci. USA* 83, 6159–6163.

Weiss, R.A., Clapham, P.R., Cheingsong-Popov, R., Dalgleish, A.G., Carne, C.A., Weller, I.V.D. and Tedder, R.S. (1985) *Nature* 316, 69–71.

Weiss, R.A., Clapham, P.R., Weber, J.N., Dalgleish, A.G., Lasky, L.S. and Berman, P.W. (1986) *Nature* 324, 572–575.

Weiss, S.H., Goedert, J.J., Gartner, S., Popovid, M., Waters, D., Markham, P., diMarzo, G., Gail, M.H., Barkley, W.E., Gibbons, J., Gill, F.A., Leuther, M., Shae, G.M., Gallo, R.C. and Blattner, W.A. (1988) *Science* 239, 68–71.

White, J., Kielian, M. and Helenius, A. (1983) *Q. Rev. Biophys.* 16, 151–195.

Willey, R.L., Smith, D.H., Lasky, L.A., Theodore, T.S., Earl, P.L., Moss, B., Capon, D.J. and Martin, M.A. (1988) *J. Virol.* 62, 139–147.

Yap, K.L., Ada, G.L. and McKenzie, I.F.C. (1978) *Nature* 273, 238.

Zagury, D., Bernard, J., Cheynier, R., Desportes, I., Leonard, R., Fouchard, M., Reveil, B., Ittele, D.,

Lurhuma, Z., Mbayo, K., Wane, J., Salaun, J.J., Goussars, B., Dechazal, L., Burney, A., Nara, P. and Gallo, R.C. (1988) *Nature* 332, 728–731.
Zarling, J.M., Eichberg, J.W., Moran, P.A., Mcclure, J., Sridhar, P. and Hu, S.L. (1987) *J. Immunol.* 139, 988–990.
Zarling, J.M., Morton, W., Moran, P.A., McClure, J., Kosowski, S.G. and Hu, S.L. (1986) *Nature* 323, 344–346.

Eds. M.H.V. van Regenmortel and A.R. Neurath
Immunochemistry of viruses, II. The basis for serodiagnosis and vaccines.

Rhabdoviruses 16

E. CELIS

Cellcor Therapies, 200 Wells Avenue, Newton, MA 02159, U.S.A.

In general, rhabdoviruses are enveloped, bullet-shaped, have a negative-stranded RNA genome (complementary to mRNA), and replicate in the cytoplasm of the infected cells (Kucera and Myrvik, 1985). With respect to their structure, rhabdoviruses have two main components, an envelope and a nucleocapsid composed of a ribonucleoprotein (RNP) complex. The two antigens found on the viral envelope are the glycoprotein (G), which forms the characteristic spikes of these viruses, and the non-glycosylated matrix protein (M). The nucleocapsid on the other hand, contains the viral nucleic acid associated to a nucloprotein (N), an RNA transcriptase (L), and a phosphoprotein referred to as NS (because it was originally thought to be non-structural). Although a detailed review of the antigenic structure of rhabdoviruses was recently published in the first volume of these series (Wunner et al., 1985b), it is important to mention here that the G antigen is responsible for the specific binding of the virus to its putative cellular receptor during infection, and that antibodies (Abs) to G in general will inhibit infection and thus are referred to as virus-neutralizing Abs (VNA). The family Rhabdoviridiae, which is composed of approximately 80 different viruses, is divided into two genera, Vesiculovirus and Lyssavirus. Vesicular stomatitis virus represents the prototype of the Vesiculovirus genus and rabies virus has been the most studied constituent of the Lyssavirus genus. Some of the current notions pertaining to the immune response to vesicular stomatitis and rabies viruses will be reviewed in this chapter, in particular the recognition of viral antigens by T lymphocytes and the participation of these cells in protection against infection and disease. Special emphasis will be given to the role that internal viral antigens play in the induction of protective immune responses.

1. Vesicular stomatitis virus

Although cattle, horses and swine are the species most affected by vesicular stomatitis virus (VSV), the immune response to this vesiculovirus has been mostly studied in laboratory mice. Several serotypes of VSV have been described, two of which, VSV-Indiana (VSV-Ind) and VSV-New Jersey (VSV-NJ), have been extensively studied. Protective immunity to VSV infection may be conferred by VNA, or in some cases by cytotoxic T lymphocytes (Zinkernagel and Rosenthal, 1981).

1.1. Humoral immune responses to VSV

Early studies determined that the G of VSV contain those epitopes responsible for the induction of VNA, which will protect mice against infection (Kelley et al., 1972; Dietzschold et al., 1974). It was reported recently that enhanced secondary VNA responses were observed when interferon-γ (IFN-γ) was included in the primary immunization schedules of mice or cattle injected with either purified G or whole inactivated VSV (Anderson et al., 1988). In addition, the same studies demonstrated that cattle receiving IFN-γ during primary immunization were more resistant to a live virus challenge than those animals vaccinated without IFN-γ. The role of IFN-γ in enhancing Ab responses could be due at least in part to the capacity of this T cell-derived lymphokine to increase the expression of major histocompatibility complex (MHC) class II surface molecules on antigen-presenting cells (Basham and Merijan, 1983; Skoskiewicz et al., 1985), thus increasing the stimulation and function of antigen-specific helper T cells.

Using a large number of G-specific monoclonal Abs (mAbs) in competition experiments, several neutralizing and non-neutralizing epitopes have been described for VSV (Volk et al. 1982; LeFrançois and Lyles, 1983). Furthermore, some of the amino acids of G which participate in the formation of these epitopes were identified using escape mutants selected with some of the mAbs (Vandepol et al., 1986; Luo et al., 1988). In many instances (as with rabies virus, see below), single amino acid differences were found between the mAb-selected escape mutant and the wild-type VSV isolate. The substituted amino acid residue may be responsible for either altering a linear antibody epitope, or on the other hand, for modifying a conformational epitope of the viral G molecule. These two possibilities may be distinguished by performing antibody binding experiments using synthetic peptides from the region of G that includes the amino acid substitution found in the escape mutant.

1.2. Cellular immune responses to VSV

In addition to VNA, cytotoxic T lymphocytes (CTL) appear to also have an important role in protective immune responses to VSV. As described in more detail in Chapter 4 of this book, CTL and T helper (Th) cells recognize antigen on cell surfaces of accessory cells which act as either target cells for CTL or as antigen-presenting cells (APC) for Th lymphocytes. The work done by many laboratories over the past 15 years has helped significantly in our understanding of the mechanism of antigen recognition by T lymphocytes (Zinkernagel and Doherty, 1974; Benacerraf, 1978; Rosenthal and Shevach, 1973; Kappler and Marrack, 1976; Unanue, 1984). The current view is that cell surface molecules encoded by the MHC function as receptors for peptides derived from processed or degraded antigen molecules, and that the T cell receptor for antigen interacts with this MHC-peptide complex (Thomas et al., 1981; Lamb et al., 1982; Bastin et al., 1987; Babbitt et al., 1985; Buus et al., 1987). Most CTL express the CD8 surface marker and recognize peptides associated with MHC class I molecules such as H-2K, H-2D and H-2L in mice, or HLA-A, -B or -C in humans. Th cells on the other hand, express the CD4 surface marker and interact with peptides that become complexed to MHC class II molecules (also called Ia molecules) which are encoded by the I-A and I-E genes in the mouse and by the HLA-DR, -DP and -DQ genes in humans.

The recognition of VSV antigens by T cells is probably one of the most crucial events in the development of protective immunity. Using a temperature-sensitive mutant of VSV that produces a slow lethal infection in the central nervous system of nude mice (but not in normal immunocompetent mice), Doll and Johnson (1988) reported that adoptive transfer of T cells abrogated disease (but not infection) in the recipient nude mice. As with most antigens, the production of Abs to VSV requires the participation of Th cells (Leist et al., 1987), which induce the antigen-stimulated B lymphocytes to become antibody-producing plasma cells. On the other hand, CTL play a major role in protective immune responses by killing virus-infected cells before the mature virions are assembled, thus preventing the dissemination of infectious particles. Some studies performed several years ago, suggested that the majority of VSV-specific CTL recognized antigenic determinants present in the viral G (Zinkernagel et al., 1978; Hale et al., 1978). Recent reports however, indicate that the frequency of CTL directed against internal VSV antigens, such as N, is more predominant than originally thought and probably these CTL are the main component of the cytolytic cellular immune response to VSV (Yewdell et al., 1986; Puddington et al., 1986). Similar observations were made in the influenza virus system (Bastin et al., 1987; Yewdell et al., 1985). The recognition of internal viral antigens by T cells is of relevance for the design of vaccines that attempt to achieve broad cross-protective immunity since fewer amino acid variations occur in these viral components as compared with the viral G (Wunner et al., 1985b). As shown below, most cross-reactive rabies-specific Th cells recognize epitopes of the N molecule.

2. Rabies and rabies-related viruses

Rabies virus (RV) has the capacity to infect many warm-blooded animals, including humans. The infection, which generally occurs from a bite of an infected animal, is of grave consequence resulting in death in the majority of cases (for a general review see Baer, 1985). Control and prevention of RV infection can be achieved by either pre- or post-exposure vaccination. In humans most vaccinations against RV are performed in individuals that have already been in contact with an infected animal (post-exposure treatment). As originally shown by Pasteur (1885), infection can be prevented in individuals exposed to a rabid animal, by prompt administration of a vaccine consisting of inactivated (or attenuated) virus. On the other hand, prevention in domestic animals, or in humans at high risk of infection (e.g., veterinarians, state health officials and laboratory workers), is usually achieved by vaccination prior to contact with the virus (pre-exposure treatment). There are many types of vaccines against rabies ranging from those prepared using virus grown in animal nerve tissues (e.g., Semple vaccine) to the preferable, but more expensive tissue culture vaccines (e.g., Human Diploid Cell Vaccine or HDCV) (for a recent review see Celis et al., 1989b). Rabies vaccination induces both virus-specific VNA and strong CTL responses (Baer, 1985; Wiktor et al., 1977; Turner, 1985), however as discussed below, the individual role of these immune effector mechanisms in conferring protective immunity remains unclear.

2.1. Humoral immune responses to rabies and rabies-related viruses

As with VSV, the G molecule of RV is solely responsible for the induction of VNA (Cox et al., 1977; Crick and Brown, 1969). VNA seem to protect against viral infection by either blocking the binding of virus to its (not yet identified) cell surface receptor or by inhibiting the endosomal acid-catalyzed membrane fusion step that is required for virus uncoating (Dietzschold et al., 1987). Abs to the internal antigens of RV, although non-neutralizing, also participate in the induction of protective immune responses (see sections 2.2 and 2.3 of this chapter). Studies using mAbs specific for the G or RV, and escape mutants to neutralization by these Abs have indicated the existence of at least three major neutralizing regions (I, II and III) in the G molecule of laboratory strains of RV (for a detailed review see Wunner et al., 1985b). Some escape mutants selected with mAbs specific for the antigenic site II of RV show significant decrease of pathogenicity in mice (Prehaud et al., 1988). Such mutants, if stable, could be used as live oral vaccines for both domestic and wild animals.

Until recently, all the RV-specific mAbs with virus-neutralizing activity appeared to recognize conformational determinants in the viral G. In other words, these monoclonal VNA did not react with denatured G preparations of RV (Dietzschold et al., 1982; Wunner et al., 1985a). There is now evidence however, that linear (non-conformational) epitopes for rabies-specific VNA exist in the G molecule. Two recently isolated RV-specific mAbs were shown both to have high virus-neutralizing activity and to bind by Western blot analysis to reduced and alkylated G, and to relatively small fragments of this viral envelope antigen (Bunschoten et al., 1989; Wunner et al., 1988a). These results are indicative that it may be possible to design a synthetic and inexpensive vaccine with the capacity to induce VNA against RV.

2.2. Cellular immune responses to RV

At the present time it is unknown whether CTL or Th are more important for resistance to RV infection. Although it is known that both cytotoxic and helper T-cell responses are induced simultaneously after vaccination, in the next two sections of this chapter these two types of cellular immune responses will be reviewed separately only for the purpose of clarity.

2.2.1. Cytotoxic T cell responses

In addition to VNA, virus-specific T lymphocytes have an important role in protective immunity against RV. Early studies showed that vaccination of mice with β-propiolactone-inactivated RV induced the production of virus-specific CTL (Wiktor et al., 1977). Furthermore, a vaccinia-rabies G (V-RG) recombinant virus and purified whole or fragmented rabies G were all shown to stimulate RV-specific CTL (Macfarlan et al., 1986; Wiktor et al., 1984). These findings indicated that some of the antigenic epitopes recognized by these CTL were present in the viral G molecule. It is interesting to note that broadly cross-reactive RV-specific CTL clones, which were stimulated by whole virus, did not kill V-RG recombinant virus-infected target cells, suggesting that these T cells recognized antigenic determinants of the viral internal antigens.

Recent observations indicate that CTL alone are not sufficient to confer protective immunity against RV. The first line of evidence was derived from experiments where mice that were vaccinated with a vaccinia recombinant virus containing a mutated rabies G gene, corresponding to a single amino acid substitution (V-RG_{LEU}), were not protected against a lethal viral challenge with RV (Celis et al., 1989b). Animals vaccinated with the non-mutated V-RG recombinant virus produced significant levels of VNA and resisted a lethal challenge with live RV (Celis et al., 1989b; Wiktor et al., 1984). On the other hand, the V-RG_{LEU} recombinant virus mutant failed to induce VNA but stimulated a comparable CTL response as the non-mutated V-RG recombinant virus, indicating that G-specific CTL alone were not protective under these circumstances (Celis et al., 1989b). A summary of these results is presented in Table 16.1. The second series of observations suggesting that CTL alone cannot protect against a lethal RV infection were done by in vivo T cell depletion experiments. Mice which were depleted of either their CTL or Th cells by administration of high amounts of mAbs specific for these subsets of T cells (Dialynas et al., 1983; Sarmiento et al., 1980) were vaccinated with inactivated RV and challenged 7 days later with live virus (Celis et al., 1989b). As shown in Table 16.2 the vaccinated mice which were depleted of their CTL survived the lethal chal-

TABLE 16.1

Immunogenicity and protective activity of vaccinia-rabies glycoprotein recombinant viruses in mice[a]

Vaccine[b]	VNA, GMT (range) IU/ml[c]		CTL[d] activity	Survivors after lethal challenge[e]
	Day 90	Day 96		
V-RG	27 (13.5–40.5)	97.2 (81.0–243.0)	48	9/10
V-RG_{LEU}	≤0.1	3.6 (3.0–6.0)	52	0/10
Vaccinia	≤0.1	3.5 (1.0–6.0)	2	0/10

[a] Derived from data recently published (Celis et al., 1989b).

[b] Groups of 10 female BALB/c mice were immunized intravenously with 10^6 pfu of vaccine and rested for 3 months.

[c] On day 90, mice were bled and serum VNA titers determined. Results are expressed as the geometric mean titers (GMT) with their range in international units (I.U.)/ml. On the same day, all mice were boosted with 10 μg/ml purified and β-propiolactone-inactivated ERA virus. Serum VNA titers were determined 6 days after the booster.

[d] Rabies-specific T cell-mediated cytotoxic responses were measured in a parallel experiment with a 5-h ^{51}Cr release assay using RV (ERA)-infected P815 (H-2^d) mastocytoma cells as target cells and purified T cells from the spleens of the virus immune mice as effector cells. Effector T cells were incubated for 7 days with syngeneic RV-infected spleen cells before the cytotoxic assays were performed. Values represent the % specific ^{51}Cr released by the target cells at an effector/target ratio of 20. Similar results were observed with V-RG- and V-RG_{leu}-infected P815 cells (data not shown). Vaccinia-infected P815 cells were not killed by these T cells (data nor shown).

[e] In a parallel experiment, mice were lethally challenged in the hind legs with $10^{7.5}$ LD_{50} of CVS-24 rabies virus on day 6 instead of receiving the ERA vaccine.

lenge with RV. Control experiments demonstrated that the anti-CTL mAb-treated mice were unable to produce either rabies-specific or allogeneic (anti-H-2^b) CTL (data not shown). On the other hand, the Th cell-depleted vaccinated animals, which in turn produced a normal rabies-specific CTL response (data not shown), succumbed to the same lethal challenge (Table 16.2). It is clear from these experiments that induction of CTL alone may not be enough to protect an individual against a lethal RV infection. However, one cannot exclude the possibility that CTL in combination with other immune effector mechanisms could participate in protective immune responses.

While most CTL are restricted by MHC class I molecules (Zinkernagel and Doherty, 1974), there is also evidence that some virus-specific CTL recognize antigen on target cells in the context of MHC class II (Ia) molecules (Yasukawa and Zarling, 1984; Kaplan et al., 1984). Human T cell lines and clones specific for RV antigens which express the surface CD4 differentiation marker (characteristic for helper/inducer T lymphocytes) have been shown to exert significant Ia-restricted cytolytic activity (Celis et al., 1988a; Celis et al., 1988b). Furthermore, mouse T/T hybridomas that recognized rabies antigens presented by Ia molecules also exhibited cyto-

TABLE 16.2

Effect of T cell depletion on the immune response and survival to rabies virus challenge

MAb treatment[a]	Vaccine[b]	VNA, GMT (range), IU/ml[c]		Survivors after lethal challenge[d]
		Day 9	Day 26	
Experiment 1				
Anti-Th	+	0.22 (0.1–0.7)	≤0.1	3/10
None	+	8.4 (6.0–18.0)	4.6 (2.0–9.0)	10/10
None	–	0.22 (0.2–0.6)	0.65 (0.3–1.0)	3/10
Experiment 2				
Anti-Th	+	ND[e]	ND	0/5
Anti-CTL	+	ND	ND	5/5
None	+	ND	ND	5/5
None	–	ND	ND	0/5

[a] Groups of 10 (Expt. 1) or 5 (Expt. 2) female BALB/cJ mice were inoculated i.p. with 0.5 ml of ascites containing mAbs GK1.5 (anti-Th or anti-L3T4; Dialynas et al., 1983) or 2.43 (anti-CTL or anti-Lyt 2.2; Sarmiento et al., 1980) on days 0, 7 and 14. Control mice recieved 0.5 ml PBS.

[b] On day 2, mice were vaccinated with 10 μg/ml purified and β-propiolactone-inactivated ERA virus (+) or with PBS (–).

[c] VNA titers were determined on day 26 on mice that survived the lethal challenge.

[d] Mice were challenged on day 9, in the hind legs with $10^{7.5}$ mouse intracerebral LD_{50} of rabies CVS-24 virus.

[e] Not determined.

lytic activity against RV antigen-pulsed target cells (Celis et al., 1989a) (see section 2.2.2). It remains unclear whether $CD4^+$ CTL play a role in rabies immune responses by killing infected cells that express Ia molecules, or if the function of these T cells is to down-regulate immune responses by eliminating APC and rabies-specific B lymphocytes (Braakman et al., 1987).

2.2.2. Helper T cell responses: fine specificity and MHC restriction

To study in more detail the role of cellular immunity, in particular of Th cells, in resistance to rabies infection, in vitro model systems have been developed to analyze the reactivity of sensitized lymphocytes to rabies viral antigens. The in vitro production of VNA by immune human B lymphocytes was shown to require both the presence of T cells and low concentrations of rabies antigen (Uytdehaag et al., 1983). Several studies have shown that in vitro proliferative T cell responses to RV antigens do not always correlate with titers of VNA in sera of human vaccine recipients (Nicholson et al., 1979; Celis et al., 1988b). Results obtained with immune mouse T cells demonstrated proliferative responses of these cells to intact and fragmented purified rabies G, and to synthetic peptides containing sequences of the G molecule (Macfarlan et al., 1984).

A systematic approach to the study of the recognition of rabies antigens by T lymphocytes was undertaken with the isolation of antigen-specific T cell lines and T cell clones which were maintained in tissue culture for several months (Celis et al., 1986). Rabies specific $CD4^+$ T cell lines and clones were obtained from human rabies vaccine recipients by repetitive stimulation of their immune peripheral blood mononuclear cells (PBMC) with inactivated RV in the presence of interleukin-2 (IL-2) and autologous irradiated-APC. The rabies-reactive human T cell lines were shown to proliferate specifically to purified G from RV and to RNP, and to produce IFN-γ as a result of antigen stimulation (Celis et al., 1986). Furthermore, recognition of antigen by the rabies-specific T cell line was restricted by the products of the HLA-DR and DQ genes (Celis et al., 1986). An additional and probably important observation was that rabies-specific Abs enhanced both the antigen-mediated proliferative response and the production of IFN-γ by the rabies-specific T cells (Celis et al., 1985b). This potentiating effect by antigen-specific Abs was observed earlier with hepatitis B virus-reactive T cells and was shown to depend on the Fc portion of the IgG molecules (Celis et al., 1984). The most likely explanation for the enhancement of T cell responses by Abs of the IgG class is that Ab-complexed antigen is captured by Fc receptor-bearing APC more efficiently than antigen alone, thus increasing antigen presentation to T cells (Celis et al., 1985a). These observations appear to be relevant to prevention of rabies virus infection in humans. It is well known that post-exposure treatment of severely bitten individuals must include both vaccine and immune antiserum (Baer, 1985). Thus, it is likely that anti-rabies Abs besides having VNA activity could enhance T cell effector functions that decrease viral infection, such as the production of IFN-γ (Wiktor et al., 1972; Wiktor et al., 1976). Furthermore, as discussed below (section 2.3), Abs specific for internal viral antigens such as N could participate in protective immune responses via the enhancement of T cell reactivity to antigen.

A more detailed analysis of the recognition of RV antigens by $CD4^+$ human T cells was recently reported using T cell clones, purified preparations of several strains of viruses, synthetic peptides containing RV antigenic sequences and mouse L cells transfected with human MHC genes as APC (Celis et al., 1988a; Celis et al., 1988b). A summary of these findings is presented in Table 16.3. A large number of T cell clones (a total of 21), which were isolated from a recipient of RV vaccine strain PM (Imovax, Merieux Institute), were studied for their ability to recognize various laboratory RV strains (PM, CVS, ERA), and the RV-related strains Duvenhage 6 (DUV6; isolated from a bat in Poland), and Mokola (Celis et al., 1988b). A low number of the T cell clones (2/21) were shown to react only with the immunizing RV strain (PM). A larger number of T cell clones reacted with both PM and CVS viruses (4/21), or recognized all three RV laboratory strains (8/21). Some of the T cell clones were highly cross-reactive since in addition to reacting with the three laboratory RV strains, the clones also recognized the rabies-related virus DUV6 (2/21) or in other cases responded to both DUV6 and Mokola viruses (5/21).

Furthermore, from the results presented in Table 16.3, it is also apparent that while most T cell clones with low degrees of cross-reactivity recognized antigenic determinants of the G molecule, most of the broadly cross-reactive T cell clones were specific for the N antigen of RV. These results are in agreement with the finding that G is the least conserved molecule of the major RV antigens (Wunner et al., 1988b).

Using a series of overlapping synthetic peptides of amino acid sequences of G and N molecules of ERA strain of RV, and mouse L cells transfected with and expressing human class II MHC products, the fine specificity and genetic restriction of some of these T cell clones were deduced (Celis et al., 1988a). As shown in Table 16.3, most of the T cell clones were restricted by either the DR7 or the DR4 MHC molecules. One N-specific T cell clone reacted with antigen presented by the DQw8

TABLE 16.3

General characteristics of rabies-specific human T_h clones[a]

T cell clone[b]	T-cell proliferative responses to		
	Virus strains[c]	Viral antigen[d]	Restricting MHC
B1	P	G	?
C16	P	G	DR7
B6	P, C	RNP	DR4
C2, B7, C18	P, C	G	?
B8	P, C, E	N (13D)	DR7
C6, 2C5	P, C, E	G (G11.3)	DR7
C11	P, C, E	NS (NS.1)	DPw4
C12, B2, C13	P, C, E	G	?
A1	P, C, E, D	N (21D)	DR4
C10	P, C, E, D	G (G11.3)	DR7
C9	P, C, E, D, M	N (13D)	DR7
B3	P, C, E, D, M	N (21D)	DR4
B4	P, C, E, D, M	?	?
B5	P, C, E, D, M	N (2D)	DQw8
2C4	P, C, E, D, M	N (33D)	DR4

[a] Summary of recent results (Celis et al., 1988a; Celis et al., 1988b; Celis, unpublished results), in which purified viruses, viral antigens and synthetic peptides were tested for their capacity to induce T-cell proliferative responses in the presence of antigen-presenting cells (APC).

[b] T_h-cell clones from a human rabies vaccine recipient [DR 4/7 (Dw 15/7); DPw4; DQ w2/w8] were selected from an IL-2-dependent T cell line by limiting dilution.

[c] These viruses at a concentration of 5 μg/ml were tested: P, PM; C, CVS; E, ERA; D, Duvenhage-6; M, Mokola.

[d] Purified rabies RNP, NS and G, (2 μg/ml), and VR-G-infected APC were used to define the major specificities of the T cell clones; names in parenthesis represent the synthetic peptides that defined the reactivity of the T cells (see Table 16.4 for amino acid sequences and residue positions).

molecule and another T cell clone with the rabies NS molecule in the context of DPw4 (Celis et al., 1988a). Three DR7-restricted, G-specific T cell clones (C6, 2C5 and C10, Table 16.3) proliferated to a peptide which occupies residue positions 285–299 of the G molecule (Table 16.4). In addition, five rabies N synthetic peptides were also shown to contain epitopes that induced proliferative responses in DR7, DR4 and DQw8-restricted T cell clones (Table 16.4). Furthermore, recent experiments indicated the existence of more than a dozen antigenic determinants recognized by CD4$^+$ human T cells throughout the N molecule of RV (Celis et al., 1989b). The antigenic determinant of rabies NS molecule recognized by the DPw4-restricted T cell clone C11 was found to reside with in amino acid positions 101–120 of the RV phosphoprotein (Table 16.4).

Analysis of the fine specificity and MHC restriction of RV-reactive T cells has also

TABLE 16.4

Peptides that define the specificity of rabies-reactive T cell clones[a]

Peptide name	MHC restriction	T cell[b] clones	Amino acid[c] sequence
(A) Rabies G peptides			
G11.3	DR7	C6, C10, 2C5	DALESIMTTKSVSFR 285 299
G11.1	I-E^d	2B50	TTKSVSFRRLSHRKLVPGFGKAYTIFNKTLM 292 323
G10	I-A^d	2C19	EECLDALESIMTTK 281 294
(B) Rabies N peptides			
2D	DQw8,I-A^d	B5,1C2,2B15	NNQVVSLKPEIIVDQ 11 25
13D	DR7	B8,C9	ALTGGMELTRDPTVP 121 135
21D	DR4	A1,B3	PNFRFLAGTYDMFFS 201 215
33D	DR4	2C4	YVSVSSNHQARPNSF 424 438
(C) Rabies NS peptide			
NS.1	DPw4	C11	GVQIVRQMRSGERILTIWSQ 101 120

[a] Summary of results showing the epitopes recognized by human T_h clones and by mouse T/T hybridomas (see Table 16.3 and Celis et al., 1989 for more details).

[b] Human CD4+ Th clones and mouse T/T hybridomas (underlined) that showed reactivity towards synthetic peptides derived from RV amino acid sequences.

[c] Amino acid sequences deduced from the nucleotide sequence of RV ERA strain (Wunner et al., 1989). Numbers under the sequences represent the amino acid positions of the first and last residue of each peptide.

been performed with immortalized mouse T/T hybridomas (E. Celis et al., 1989a). BALB/c mice (H-2^d) were immunized with inactivated RV (ERA strain) and their splenic T cells were fused to the HAT sensitive T-cell lymphoma BW5147.G.1.4 with polyethylene glycol. The resultant hybridomas were screened for their ability to produce IL-2 as a result of stimulation with rabies viral antigens in the presence of APC. Rabies-reactive T/T hybridomas were further studied for their fine specificity and MHC restricting elements using synthetic peptides and L cells transfected and expressing the I-A^d and I-E^d genes as a source of APC. The summary of these studies shown in Table 16.4 indicates that three antigenic determinants recognized by mouse T cells were identified together with their restricting class II MHC molecules (Celis et al., 1989a). It is of interest that both of the RV murine T cell epitopes present in the G molecule, one (G11.1) of which is restricted by I-A^d and the other (G10) by I-E^d, are localized near (but are not identical) to the determinants that stimulate the human DR7-restricted T cell clones (G11.3; Table 16.4). Similarly, other mouse T/T hybridomas were shown to recognize an antigenic determinant restricted by the I-A^d molecule present in a synthetic peptide (2D) derived from the sequence of residues 11–25 of the rabies N molecule, which was also recognized by a human T cell clone (B5) shown to be restricted by DQw8 (Tables 3 and 4). The overall results suggest that in rabies viral antigens there are regions whose secondary structures allow interactions with MHC molecules regardless of the species. It is important to note that most of the RV-derived T-cell immunogenic peptides contain amino acid sequences that conform with recently proposed models designed to predict T cell epitopes (Rothbard, 1986; Margalit et al., 1987).

Studies such as these, which are aimed towards the identification of T cell epitopes, in particular those which are highly cross-reactive between rabies and rabies-related viruses, and their corresponding MHC restricting molecules have important consequences for the design of new synthetic vaccines. Linking linear B cell epitopes that induce rabies-specific VNA, such as the ones recently identified (Bunschoten et al., 1989; Wunner et al., 1988a) (see section 2.1) with well characterized T cell epitopes could provide effective synthetic peptide vaccines as recently shown in other systems (Borras-Cuesta et al., 1987; Milich et al., 1988).

2.3. Role of internal RV antigens in protective immunity

For many years immune resistance to RV infection was considered to rely mostly on the presence of high serum levels of virus-specific VNA (Baer, 1985). Recent observations however, point to the importance of the RV nucleocapsid antigens, specifically the N molecule, for induction of protective immunity. These experiments demonstrated that immunization with purified rabies RNP protected mice and raccoons against a peripheral lethal challenge with RV (Dietzschold et al., 1987). Injection with RNP induced significant amounts of anti-RNP Abs but no VNA, indicating that other immune effector mechanisms besides VNA play a crucial role in protection against RV. The observation that the major component of rabies RNP was the viral N molecule, suggested that the immune response to this internal viral antigen may induce protection against RV infection in the absence of VNA. Reports by other groups using recombinant DNA-derived rabies N as immunogens have confirmed these observations. Some of these studies reported that immunization of mice with a vaccinia-rabies N recombinant virus protected the animals against a lethal challenge with RV (Bellini et al., 1988). Recently it was also shown that C3H mice immunized with a synthetic hybrid peptide-containing epitopes for both N-specific Th and B lymphocytes produced high titers of anti-RNP Abs and induced protection to a peripheral lethal RV challenge (Dietzschold et al., 1989).

The exact mechanism by which the immune response to rabies N protects against RV infection is currently unknown, but several possibilities exist: RNP immunization induces N-specific Th cells and CTL that may inhibit, or delay infection of nerve cells until the viral challenge, acting as a booster immunization, induces VNA. Moreover, production of IFN-γ by anti-rabies immune T lymphocytes (Celis et al., 1986) could be an important factor in protection, since it has been shown that the infectivity of rabies virus is significantly reduced by this lymphokine (Wiktor et al., 1972). Abs to the rabies N antigen may also participate in the protective response by enhancing T cell reactivity to rabies antigens and by producing IFN-γ (Celis et al., 1985b). In support of some of these hypotheses are the recent observations that RV RNP-immune mice produced significantly more IFN-γ than control (non-immune) mice after a lethal viral challenge (Dietzschold et al., 1989). Furthermore, administration of anti-IFN (α, β and γ) Abs to RNP-immune mice decreased significantly the number of survivors after a challenge with live RV (Dietzschold et al., 1989).

3. Conclusions (implications for vaccine development)

It is no longer questioned that virus-specific T cells play an important role in protective immune responses against a rhabdovirus infection. With this fact in mind, the design of a new generation of safe and efficient vaccines, which utilize modern approaches such as recombinant DNA technology and synthetic peptides must take into consideration the regions of the viral antigens that elicit broadly cross-reactive cellular immune responses. The delineation of T cell epitopes for both Th and CTL together with the identification of their MHC restricting elements, are important prerequisites that should be fulfilled in order to select a particular subunit vaccine that will be intended to protect either humans or animals against life-threatening infections by pathogenic viruses such as RV.

4. References

Anderson, K.P., Fennie, E.H. and Yilma, T. (1988) *J. Immunol.* 140, 3599–3604.

Babbitt, B.P., Allen, P.H., Matsueda, G., Haber, E. and Unanue, E.R. (1985) *Nature* 317, 359–361.

Basham, T.Y. and Merijan, T.C. (1983) *J. Immunol.* 130, 1492–1494.

Bastin, J., Rothbard, J., Davey, J., Jones, I. and Townsend, A. (1987) *J. Exp. Med.* 165, 1508–1523.

Baer, G. (1985) *In Virology* (B.N. Fields, Ed.) Raven Press, New York. pp 1133–115.

Bellini, W.J., Sumner, J.W., Fekadu, M., Shaddock, J.H. and Esposito, J.J. (1988) In *Proceedings of The 22nd Joint Viral Diseases Panel Meeting of the Japan-U.S. Cooperative Medical Science Program*, Tokyo, Japan. pp. 51–52.

Benacerraf, B. (1978) *J. Immunol.* 120, 1809–1812.

Borras-Cuesta, F., Petit-Camurdan, A. and Fedon, Y. (1987) *Eur. J. Immunol.* 17, 1213–1215.

Braakman, E., Rotteveel, F.T.M., van Bleek, G., van Seventer, G.A. and Lucas C.J. (1987) *Immunol. Today* 8, 265–267.

Bunschoten, H., Gore, M., Claassen, I.J.T.M., Uytdehaag, F.G.C.M., Dietzschold, B., Wunner, W.H. and Osterhaus, A.D.M.E. (1989) *J. Gen. Virol.* 70, 291–298.

Buus, S., Sette, A., Colon, S.M., Miles, C. and Grey, H.M. (1987) *Science* 235, 1353–1358.

Celis, E., Karr, R.W., Dietzschold, B., Wunner, W.H. and Koprowski, H. (1988a) *J. Immunol.* 141, 2721–2728.

Celis, E., Kato, I., Miller, R.W. and Chang, T.W. (1985a) *Hepatology* 5, 744–751.

Celis, E., Miller, R.W., Wiktor, T.J., Dietzschold, B. and Koprowski, H. (1986) *J. Immunol.* 136, 692–697.

Celis, E., Ou, D., Dietzschold, B., Otvos L., Jr. and Koprowski, H. (1989a) *Hybridoma* 8, 263–275.

Celis, E., Ou, D., Dietzschold, B. and Koprowski, H. (1988b) *J. Virol.* 62, 3128–3134.

Celis, E., Rupprecht, C.E. and S.A. Plotkin (1989b) In *The Molecular Approach to New and Improved Vaccines* (M.M. Levine and G.C. Woodrow, Eds.) In Press.

Celis, E., Wiktor, T.J., Dietzschold, B. and Koprowski, H. (1985b) *J. Virol.* 56, 426–433.

Celis, E., Zurawski, V.R., Jr. and Chang, T.W. (1984) *Proc. Natl. Acad. Sci. USA* 81, 6846–6850.

Cox, J.H., Dietzschold, B. and Schneider, L.G. (1977) *Infect. Immun.* 16, 754–759.

Crick, J. and Brown, F. (1969) *Nature* 222, 92.

Dialynas, D.P., Quan, Z.S., Wall, K., Pierres, A., Quintans, J., Loken, M.R., Pierres, M. and Fitch, F.W. (1983) *J. Immunol.* 131, 2445–2451.

Dietzschold, B., Schneider, L.G. and Cox, J.H. (1974) *J. Virol.* 14, 1–7.

Dietzschold, B., Gore, M., Ertl, H., Celis, E., Otvos Jr. L. and Koprowski, H. (1989) In *Genetics and Pathogenicity of Negative Strand-Viruses* (B.W.J. Mahy and D. Kolakofsky, Eds.) Elsevier, Amsterdam, pp. 295–305.

Dietzschold, B., Tollis, M., Lafon, M., Wunner, W.H. and Koprowski, H. (1987) *Virol.* 161, 29–36.

Dietzschold, B., Wang, H., Rupprecht, C.E., Celis, E., Tollis, M., Ertl, H., Heber-Katz, E. and Koprowski, H. (1987) *Proc. Natl. Acad. Sci. USA* 84, 9165–9169.

Dietzschold, B., Wiktor, T.J., Macfarlan, R. and Varrichio, A. (1982) *J. Virol.* 44, 595–602.

Doll, S.C. and Johnson, T.C. (1988) *J. Gen. Virol.* 69, 1969–1977.

Hale, A.H., Witte, O.N., Baltimore, D. and Eisen, H.N. (1978) *Proc. Natl. Acad. Sci. USA* 75, 970–974.

Kaplan, D.R., Griffith, R., Braciale, V.L. and Braciale T.J. (1984) *J. Immunol.* 88, 193–206.

Kappler, J.W. and Marrack, P.C. (1976) *Nature* 262, 797.

Kelley, J.M., Emerson, S.U. and Wagner, R.R. (1972) *J. Virol.* 10, 1231–1235.

Kucera, L.S. and Myrvik, Q.N. (1985) In *Fundamentals of Medical Virology*. Lea and Febiger, Philadelphia. pp. 136–140.

Lamb. J.R., Eckles, D.D., Lake, P., Woody, J.N. and Green N. (1982) *Nature* 300, 66–69.

LeFrançois, L. and Lyles, D.S. (1983) *J. Immunol.* 130, 394–398.

Leist, T.P., Cobbold, S.P., Waldmann, H., Aguet, M. and Zinkernagel, R.M. (1987) *J. Immunol.* 138, 2278–2281.

Luo, L., Li, Y., Snyder, R.M. and Wagner, R.R. (1988) *Virology* 163, 341–348.

Macfarlan, R.I., Dietzschold, B. and Koprowski, H. (1986) *Mol. Immunol.* 23, 733–741.

Macfarlan, R.I., Dietzschold, B., Wiktor, T.J., Kiel, M., Houghten, R., Lerner, R.A., Sutcliffe, J.G. and Koprowski, H. (1984) *J. Immunol.* 133, 2748–2752.

Margalit, H., Spouge, J.L., Cornette, J.L., Cease, K.B., Delisi, C. and Berzofsky, J.A. (1987). *J. Immunol.* 138, 2213–2229.

Milich, D.R., Hughes, J.L., McLachlan, A., Thorton, G.B. and Moriarty, A. (1988). *Proc. Natl. Acad. Sci. USA* 85, 1610–1614.

Pasteur, L. (1885) *C.R. Acad. Sci. (Paris)* 101, 765–772.

Prehaud, C., Coulon, P., Lafay, F., Thiers, C. and Flamand, A. (1988) *J. Virol.* 62, 1–7.

Puddington, L., Bevan, M.J., Rose, J.K. and LeFrançois, L. (1986) *J. Virol.* 60, 708–717.

Rosenthal, A.S. and Shevach, E.M. (1973) *J. Exp. Med.* 138, 1194–1212.

Rothbard, J.B. (1986) *Ann. Inst. Pasteur/Virol.* 137E, 518–526.

Sarmiento, M., Glasebrook, A.L. and Fitch, F.W. (1980) *J. Immunol.* 125, 2665–2672.

Skoskiewicz, M.J., Colvin, R.B., Schneeberger, E.E. and Russell, P.S. (1985) *J. Exp. Med.* 162, 1645–1664.

Thomas, J.W., Danho, W., Bullesbach, E., Fohles, J. and Rosenthal, A.S. (1981) *J. Immunol.* 126, 1095–1100.

Turner, G.C. (1985) *Ann. Inst. Pasteur/Virol.* 136E, 453–460.

Unanue, E.R. (1984) *Ann. Rev. Immunol.* 2, 395–428.

Uytdehaag, F.G.C.M., Osterhaus, A.D.M.E., Loggen, H.G., Bakker, R.H.J., van Asten, J.A.A.M., Kreeftenberg, J.G., van der Martel, P. and van Steenis, B. (1983) *J. Immunol.* 131, 1234–1239.

Vandepol, S.B., LeFrançois, L. and Holland, J.J. (1986) *Virology* 148, 312–325.

Volk, W.A., Snyder, R.M., Benjamin, D.C. and Wagner, R.R. (1982) *J. Virol.* 42, 220–227.

Wiktor, T.J., Doherty, P.C. and Koprowski, H. (1977) *Proc. Natl. Acad. Sci. USA* 74, 334–338.

Wiktor, T.J., Koprowski, H., Mitchell, J.R., and Merigan, T.C. (1976) *J. Infect. Dis.* 133, A260.

Wiktor, T.J., Macfarlan, R.I., Reagan, K.J., Dietzschold, B., Curtis, P.J., Wunner, W.H., Kieny, M.P., Lathe, R., Lecocq, J.P., Mackett, M., Moss, B. and Koprowski, H. (1984) *Proc. Natl. Acad. Sci. USA* 81, 7194–7198.

Wiktor, T.J., Postic, B., Ho, M. and Koprowski, H. (1972) *J. Infect. Dis.* 126, 408–418.

Wunner, W.H., Dietzschold, B., Macfarlan, R.I., Smith, C.L., Golub, E. and Wiktor, T.J. (1985a) *Ann. Inst. Pasteur/Virol.* 136E, 353–362.

Wunner, W.H., Dietzschold, B. and Wiktor, T.J. (1985b) *In Immunochemistry of Viruses. The Basis for Serodiagnosis and Vaccines.* (M.H.V. Van Regenmortel and A.R. Neurath, Eds.) Elsevier, Amsterdam. pp. 367–388.

Wunner, W.H., Gore, M. and Dietzschold, B. (1988a) *Proceedings of The 22nd Joint Viral Diseases Panel Meeting, of the Japan-U.S. Cooperative Medical Science Program,* Tokyo, Japan. p. 50.

Wunner, W.H., Larson, J.K., Dietzschold, B. and Smith C.L. (1988b) *Rev. Infect. Dis.* 10S, 771–784.

Yasukawa, M. and Zarling, J.M. (1984) *J. Immunol.* 133, 422–427.

Yewdell, J.W., Bennink, J.R., Mackett, M., LeFrançois L., Lyles, D.S. and Moss B. (1986) *J. Exp. Med.* 163, 1529–1538.

Yewdell, J.W., Bennink, J.R., Smith, G.L. and Moss B. (1985) *Proc. Natl. Acad. Sci. USA* 82, 1785–1789.

Zinkernagel, R.M., Althage, A. and Holland J. (1978) *J. Immunol.* 121, 744–748.

Zinkernagel, R.M. and Doherty, P.C. (1974) *Nature* 248, 701–702.

Zinkernagel, R.M. and Rosenthal, K.L. (1981) *Immunol. Rev.* 58, 131–155.

Eds. M.H.V. van Regenmortel and A.R. Neurath
Immunochemistry of viruses, II. The basis for serodiagnosis and vaccines.

Coronaviruses

17

W. SPAAN[1], D. CAVANAGH[3] and M.C. HORZINEK[2]

[1]Department of Virology, Institute of Medical Microbiology, Faculty of Medicine, Leiden University, P.O. Box 320, 2300 AH Leiden, The Netherlands; [2]Institute of Virology, Veterinary Faculty, State University of Utrecht, Yalelaan 1, 3508 TD Utrecht, The Netherlands, and [3]Division of Molecular Biology, AFRC Institute for Animal Health, Houghton Laboratory, Houghton, Huntingdon, Cambs., PE17 2DA, U.K.

1. Classification of coronaviruses

The family Coronaviridae, of which infectious bronchitis virus of chicken (IBV) has been designated the prototype (Matthews, 1982), was recognized as a separate virus group more than 20 years ago (Tyrrell et al., 1968). At that time, the distinction was based solely on morphology: in negatively stained preparations the virions appear surrounded by a fringe of petal-shaped, widely spaced surface structures, the peplomers, which radiate from the membrane like the spines of a crown-of-thorns (latin corona = crown, wreath). In the eighties, coronavirology gained momentum, and international symposia started to be held at three-year intervals. Numerous reviews have appeared, the most recent one by the present authors (Spaan et al., 1988). Today, coronaviruses are known: (1) to be constructed from essentially 3 polypeptides, residing in the peplomer, the membrane and the nucleocapsid, respectively; (2) to posses a single-stranded RNA genome of positive polarity and a size of ⩾30 kilobases; the sequence of genes is 5′-replicase-peplomer-membrane-nucleocapsid protein-3′ for all coronaviruses, with variations to this theme afforded by interspersed sequences unique to individual viruses; (3) to use a 'nested set' of 3′-coterminal subgenomic mRNAs with a common leader sequence; of each mRNA only the unique 5′-positioned sequences are translated; (4) to generate this set through a leader-primed independent initiation of transcription at specific sites on a negative-stranded template; (5) to mature by budding from intracellular membranes; (6) to recombine at high frequency, giving rise to variants with different biological, pathogenic and pathogenetic properties; (7) to occur in four antigenic clusters with the prototypes MHV (I), TGEV (II), IBV (III) and TCV (IV; see Table 17.1), as determined by inclusive serologic assays; (8) to cause acute, but also persistent infections, sometimes with neurologic conditions or immunopathology.

New viruses are described time and again, the reason for naming them 'corona-like' being mainly morphological resemblance. Since coronaviruses are notoriously fastidious agents to isolate in cell culture, biochemical evidence for a conclusive taxonomic assignment is difficult to obtain. However, virus from intestinal contents has been used for purification of particles to a degree that made immunoblotting analysis and antigenic comparison possible (Zhou et al., 1988).

2. Diseases caused by coronaviruses

The organ systems attacked by coronaviruses are listed in Table 1; the reader is referred to Wege et al. (1982) for an exhaustive treatment of the subject.

It is obvious from this table that coronaviruses are enteric pathogens in most species. Since they are notoriously labile at low pH, passage through the stomach is the major obstacle in pathogenesis; this explains why — in the absence of lactogenic immunity — coronaviral diarrhea is common in neonates where gastric acidity is only developing. Cytopathology caused in the intestinal epithelium is a plausible explanation for the disease picture in these cases. While it is true that the alimentary tract is the primary site of virus replication in many infections, the genetic flexibility of coronaviruses facilitates changes in tissue tropism, with subsequent alterations to pathogenesis and epidemiology. A recent example is the emergence of a respiratory variant of TGE virus in Western Europe (Pensaert et al., 1986). This virus is indistinguishable from strains of classical TGEV by conventional serology, but does not replicate in the gut upon oronasal inoculation. Natural infection and seroconversion is not accompanied by virus excretion with the feces. The speed and pattern of spread in the pig population rather suggests aerogenic transmission (for a review see Pensaert, 1989).

Another example of a primary enteric infection followed by virus dissemination to other organs is FIP. Doublets of enteric and peritonitis-causing feline coronaviruses exist (e.g. FECV-UCD/FIPV-UCD1; FECV-79-1683/FIPV-79-1146) which are indistinguishable antigenically but very different in their disease-producing potential (for a review see Pedersen, 1987). FIPV grows in macrophages (Pedersen, 1976) but appears to enter the organism after a phase of replication in the gut (Hayashi et al., 1982). There is field evidence that FIPV strains may arise from FECV in closed colonies. In a recent experimental comparison of feline coronaviruses it was shown that virulent strains infect more macrophages, produce higher titers in them, and spread more readily to other macrophages; in addition, their infectivity was enhanced by antibody from sensitized cats (Stoddart 1988, personal communication). Immune enhancement is central to FIP pathogenesis and is due to anti-peplomer responses, as has been proven in vivo with the aid of a vaccinia recombinant (Spaan et al., 1989).

Mouse hepatitis virus is certainly the coronavirus whose pathogenic potential was most extensively studied. The first isolate, MHV-JHM, was obtained from a para-

TABLE 17.1

Natural disease conditions caused by coronaviruses

Condition	Host	Symptoms	Virus	Antigenic cluster	Abbreviation
Enteric	Man	diarrhea necrotizing enterocolitis	human enteric coronavirus	I	HECV
	Dog	diarrhea	canine coronavirus	II	CCV
	Cat	diarrhea	feline enteric coronavirus	II	FECV
	Cattle	diarrhea	bovine coronavirus	I	BCV
	Pig	diarrhea	transmissible gastro-enteritis virus	II	TGEV
	Pig	diarrhea	porcine epidemic diarrhea virus	II	PEDV
	Mouse	diarrhea hepatitis encephalitis	mouse hepatitis virus	I	MHV
	Turkey	diarrhea	turkey enteric coronavirus	IV	TCV
Respiratory	Man	common cold	human coronavirus (OC43) (229E)	I II	HCV
	Rat	pneumonia rhinitis	rat coronavirus	I	RCV
	Chicken	bronchitis	infectious bronchitis virus	III	IBV
Neurologic	Pig	encephalomyelitis	hemagglutinating encephalitis virus	I	HEV
Immune, generalized	Cat	granulomas polyserositis	feline infectious peritonitis virus	II	FIPV

lysed mouse (Cheever et al., 1949), and its neuropathogenic properties — especially chronic demyelination — have received much attention. However, MHV is also an important enteropathogenic agent which is epitomized by the name 'lethal intestinal virus of infantile mice' (LIVIM; Kraft, 1962). Again macrophages play a key role in the pathogenesis; they have been found to harbor virus when encountered in the lamina propria of the lower intestine (Hierholzer et al., 1979) and to represent one of the determinants of hepatotropism.

3. Virion structure

Coronaviruses possess three major structural proteins: a nucleocapsid protein (N), a small integral membrane glycoprotein (M, E1) and a large spike glycoprotein (S, E2); for the sake of uniformity we will use the letters N, M and S in this review. While all coronaviruses contain these proteins, a subset (the MHV strains JHM and S, HEV, HCV-OC43 and BCV) is now recognized to possess an additional glycopolypeptide (gp65) which is unrelated to S or M.

4. Genome organization and expression

The genomic RNA of coronaviruses is the largest known among RNA viruses, approximately 27–30 kb. The genome is organized into six or seven regions, each containing one or more ORFs which are separated by junction sequences that contain the signal(s) for the transcription of multiple subgenomic mRNAs.

In recent years much sequence data have been obtained, amongst others the complete sequence of IBV genomic RNA, and of approximately 12 kb and 8.3 kb (extending from the 3′ end) of the MHV and TGEV genomic RNAs, respectively (Fig. 17.1; for references see Spaan et al., 1988).

From the sequence data several ORFs could be deduced. The 5′ two thirds of the IBV genome (region F) encode non-structural protein(s), probably the replicase/transcriptase. The ORFs encoding the structural viral proteins have been identified; from their location a consensus gene order 5′-S-M-N-3′ can be inferred. The number and location of the other ORFs are different between IBV, MHV and TGEV (Fig.1).

Coronaviruses have elaborated more than one strategy to produce their proteins. Multiple subgenomic mRNAs which form a 3′ nested set and contain a common leader sequence are synthesized to position internal genes at the unique 5′ end of a mRNA. Though most of these mRNAs are functionally monocistronic, several do contain more than one ORF within the unique region. In these mRNAs initiation of translation at downstream ORFs may occur, e.g. at the AUG codon of the second ORF of MHV mRNA5 and of the third ORF of IBV mRNA D. Finally, the expres-

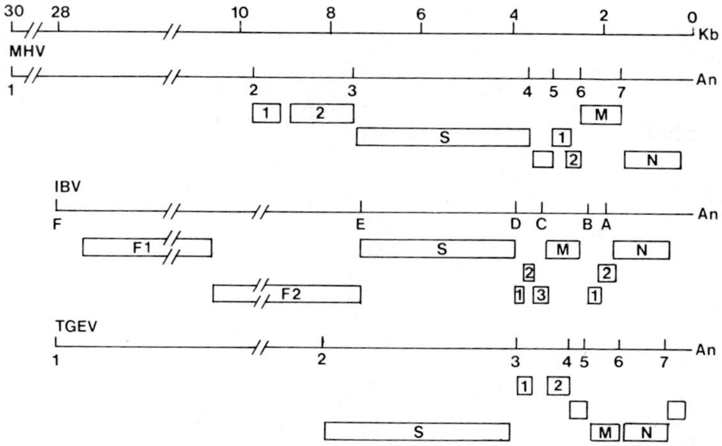

Fig. 17.1. The organization and expression of the MHV, IBV and TGEV genomes. The genomes are presented from 5′ (left) to 3′ (right) by horizontal lines; short vertical lines show the location of the conserved junction sequences that contain the signal(s) for the transcription of multiple subgenomic mRNAs. The numbers (1–7) and letters (A–F) below the junction regions indicate the nomenclature of the corresponding mRNAs for MHV or TGEV and IBV, respectively. The mRNAs (not indicated) form a 3′ coterminal nested set and their length can be deduced from the nucleotide kb map at the top of the figure. The symbol An indicates the poly (A) tract at the 3′ end of the genome. Open reading frames are shown as open rectangles, and the position on the genome of the ORFs encoding the replicase/transcriptase, the peplomer, the small integral membrane and nucleocapsid protein are indicated by F1 and F2, S, M and N, respectively. The rectangles which include numbers (1–3) indicate that the unique regions (enclosed by the junction sequences) of the corresponding mRNAs contain more than one ORF (for references see Spaan et al., 1988).

sion of the 5′-two thirds of the IBV and MHV genomes involves ribosomal frameshifting and post-translational cleavage of a polyprotein, respectively (reviewed by Spaan et al., 1988).

5. Primary structure and function of structural proteins

5.1. The nucleocapsid protein

The number of amino acids in the basic *N protein* ranges from 382–455 and has been determined by cloning and sequencing for MHV strains A59 (Armstrong et al., 1983) and JHM (Skinner and Siddell, 1983), IBV strains Beaudette and M41 (Boursnell et al., 1985), for TGEV (Kapke and Brian, 1986) and BCV (Lapps et al., 1987) (Table 17.2).

In addition to its role in encapsidating genomic RNA and facilitating its incorporation into virions (by the formation of ribonucleoprotein, RNP) the N protein has

TABLE 17.2

Properties of virion proteins

	IBV	MHV	FIPV	TGEV	BCV
Nucleocapsid protein (N)					
M_r (× 1000)	45	50	–[a]	43	49
No. of amino acids	409	455	–	382	448
Matrix glycoprotein (M)					
M_r (× 1000)	25	26	–	30	26
No. of amino acids:					
total, mature protein	224	227	–	245	229
hydrophilic N-terminus	21	24	–	29	23
membrane-embedded domain	76	81	–	88	81
luminal C-terminus	127	122	–	128	125
No. potential glycosylation sites	1,2	4	–	1	6
Spike protein (S)					
M_r (× 1000)	128	137	159	158	–
No. of amino acids[b]					
total, with signal	1162	1235[c] 1324[d]	1452	1447	–
in S1, including signal sequence	537	628[c] 717[d]	spike not cleaved		–
in S2,	625	607[c] 606[d]			
in signal sequence	18	–	–	16	–
No. potential glycosylation sites	28	21	35	32	–
gp65/130					
Present in virion	–	JHM+ A59−	–	–	+

[a] Information not available
[b] Approximate, because of strain variation
[c] MHV-JHM
[d] MHV-A59

been implicated in the process of RNA replication. Addition of antiserum raised against N (but not against S or M) to an in vitro replication system inhibited the synthesis of genome-sized RNA by 90% (Compton et al., 1987) and co-immunoprecipitated MHV-specific RNAs found in infected cells (Baric et al., 1988). Using a Northwestern blotting assay a specific interaction between the N protein and the area of nucleotide 56–65 of the leader sequence of MHV was recently demonstrated (Stohlman et al., 1988).

5.2. The small envelope glycoprotein

Nucleotide sequencing of the M gene of MHV-A59 and MHV-JHM (Armstrong et al., 1984; Pfleiderer et al., 1986), IBV-Beaudette and IBV-6/82 (Boursnell et al., 1984; Binns et al., 1986a), BCV (Lapps et al., 1987) and TGEV (Laude et al., 1987) has revealed that the number of amino acids in the M protein ranges from 229–245. The M proteins vary also in the extent of glycosylation, the type of linkage of the glycans (*O*-linked or *N*-linked) and the degree to which *N*-linked high-mannose (simple) glycans have been converted to complex glycans (Table 17.2). Computer predictions of the secondary structure have led to a model in which approximately 10% of the N-terminal part of the molecule is exposed at the outer surface of the virus membrane (see references above and Rottier et al., 1986). This view is supported by experimental evidence (Rottier et al., 1984; Cavanagh et al., 1986a). The next 80 or so residues, approximately one third of the molecule, form three hydrophobic α-helices, which span the membrane three times. The C-terminal half of the protein has neither strong hydrophobic nor hydrophilic properties and is located in the interior of the virus particle.

In the presence of tunicamycin MHV and IBV formed virions which lacked S but contained M and RNP (Holmes et al., 1981; Rottier et al., 1981; Stern and Sefton, 1982a) and virion budding occurred at the site of M accumulation (Tooze et al., 1984). These data indicate that M is necessary for virus maturation and that it determines the site at which virus particles are assembled. The intracellular accumulation of the M protein is a property of the protein itself (Rottier and Rose, 1987; Machamer and Rose, 1987). A mutant M protein of IBV possessing only the first transmembrane domain accumulated intracellularly while another one with only the third domain was transported to the plasma membrane (Machamer and Rose, 1987). The domain of M which interacts with the RNP is unknown; it has been shown that M itself has an affinity for RNA (Sturman et al., 1980).

5.3. The peplomer protein (S)

The *S protein* gene has been sequenced for several strains of IBV (Binns et al., 1985; Niesters et al., 1986; Binns et al., 1986b), MHV-JHM (Schmidt et al., 1987) and MHV-A59 (Luytjes et al., 1987), TGEV (Rasschaert and Laude, 1987; Jacobs et al., 1987) and FIPV (de Groot et al., 1987a). These S glycoproteins possess an overall hydrophobic hydropathicity profile, an N-terminal signal sequence, a C-terminal hydrophilic sequence preceded by a membrane-spanning domain, and a large number of potential *N*/-linked glycosylation sites (21–35), most of which appear to be glycosylated. In contrast, there are large differences in the number of amino acids of S between different coronaviruses and different isolates of a given virus, ranging from about 1160 to 1452 (Table 17.2).

In some coronaviruses S is cleaved to form two glycopeptides, S1 and S2. The order of S1 and S2 within the S0 precursor glycoprotein of IBV and MHV is N-terminus : S1 : S2 : C-terminus (in MHV 90B and 90A are equivalent to S1 and S2, respectively; Binns et al., 1985; Cavanagh et al., 1986b; Luytjes et al., 1987). For IBV and MHV the cleavage site is adjacent to several basic amino acids. The S0 protein of TGEV and FIPV lacks such pairs of basic residues.

The extent to which S is cleaved into S1 and S2 depends on the virus in question and the cells infected. Most authors have reported the S protein of IBV (Stern and Sefton, 1982b; Cavanagh 1983a,b; Cavanagh and Davis, 1987) and BCV (Hogue et al., 1984; Deregt et al., 1987) to be in a cleaved form while none of S of TGEV, FIPV and CCV (Garwes and Reynolds, 1981; Horzinek et al., 1982) and only a little of HCV (Hogue and Brian, 1986) occurs cleaved. Cleavage of MHV S protein varies from 0% to 100%, depending on the virus strain and cell type (Sturman et al., 1985; Sugiyama et al., 1986).

Sedimentation studies have indicated that the peplomer of IBV is an oligomer comprising two or three molecules of S1 + S2 (Cavanagh, 1983b). Interpeptide disulfide bonds are not involved in maintaining quaternary structure, permitting S1 to be removed from virions by urea treatment with S2 left in place. This led to the proposal that the outer, bulbous part of S might be formed largely by S1, with S2 being anchored by its C-terminus in the virus envelope (Cavanagh, 1983b). Computer-aided analysis of the S protein primary structure of IBV, MHV and FIPV (De Groot et al., 1987b) has identified two heptad repeats in the C-terminal domain which indicate an intra-chain coiled coil structure. These results have been confirmed for TGEV (Rasschaert and Laude, 1987). The major repeat suggests a helix occupying more than half the length of the peplomer. In the oligomer the major helices are probably involved in an inter-chain coiled coil, reminiscent of structures in the hemagglutinin glycoprotein trimer of influenza virus.

It has been assumed that attachment of virions to cells is mediated by the S protein or by both S and gp65 (in MHV-JHM, MHV-S, HEV, BCV and HCV-OC43, which possess the latter). Removal from IBV of S1 (but not of S2) by urea abolished both infectivity and hemagglutinating activity (HA; Cavanagh and Davis, 1986). The inference that HA of IBV is mediated by S1 is supported by the finding that monoclonal antibodies (mAbs) to S1 inhibit HA (Mockett et al., 1984).

The S protein induces membrane fusion. This has been demonstrated in three ways: first, antiserum to S but not to M (Sturman et al., 1985) and mAbs specific for S (Collins et al., 1982; Wege et al., 1984)) inhibited cell fusion after infection with MHV (fusion from within); second, MHV-A59 grown in 17 Cl 1 line of spontaneously transformed BALB/c 3T3 cells did not cause fusion of L cells when added at high multiplicity (fusion from without) unless S had previously been cleaved by trypsin (Sturman et al., 1985); third, vaccinia virus recombinants containing either the MHV S gene or the FIPV S gene were able to induce cell fusion (Vennema et al., submitted). Although S is responsible for the induction of membrane fusion, its cleavage is not always a necessary condition. In FIPV and TGEV, a cleaved S protein has not been observed, yet replication results in formation of syncytia (De Groot et al., 1987c). Similarly, HCV-OC43 and a small plaque variant of MHV-A59 (Sawicki, 1987) possess uncleaved S but nevertheless undergo multiple cycles of replication. The requirement of S cleavage for membrane fusion may therefore depend not only on the virus strain and host cell membrane properties (Frana et al., 1985) but also on the type of fusion in question — from within, from without or virus-endosome fusion.

5.4. The GP65 surface protein

Early studies showed that HCV-229E, HCV-OC43, HEV, MHV-JHM, and BCV possess a glycopolypeptide in addition to S and M (see Siddell et al., 1983; Makino et al., 1983). More recent studies with BCV (King et al., 1985; Deregt et al., 1987), HCV-OC43 (Hogue and Brian, 1986) and diarrhea virus of infant mice (Sugiyama et al., 1986) have confirmed that these viruses contain a glycopolypeptide of about 65 kDa (gp65) which, in the absence of mercaptoethanol, runs as a dimer of 130–140 kDa (gp130–140) in polyacrylamide gel electrophoresis. Siddell (1982) has shown by tryptic peptide fingerprinting that gp65 is structurally unrelated to S.

The nucleotide sequence of the gp65 gene of BCV predicts a 47-kDa polypeptide of 424 amino acids (Parker et al., 1989). A potential signal and transmembrane anchor sequence are located at the N-terminal and C-terminal end of the protein, respectively. In vitro translation of the cloned gp65 gene in the presence of microsomes gave rise to a 62-kDa product which comigrated with gp65 protein from virus-infected cells (Parker et al., 1989).

In those coronaviruses which possess gp65 it is this protein, probably as a dimer, which is associated with HA (see Siddell et al., 1983; King et al., 1985). However, not all strains of MHV which posses the gp65 cause HA (Talbot, 1989). Hemagglutination studies indicated that HCV-OC43 and BCV recognize O-acetylated sialic acid or a similar derivative as the red blood cell receptor-like influenza C viruses (Vlasak et al., 1988a). It will be important to examine if viral binding in vivo also involves modified sialic acids. BCV also exhibits an acetyl esterase receptor destroying activity similar to the enzymatic activity found in influenza C viruses (Vlasak et al., 1988b). This activity is associated with the gp65 spike protein and inactivation of the esterase activity dramatically reduced viral infectivity (Vlasak et al., 1988b).

5.5. Relationships among coronaviruses

A range of serological approaches have been used to assess the relationships among coronaviruses: neutralization, hemagglutination-inhibition, complement fixation and gel diffusion tests (Bradburne, 1970; Horzinek et al., 1982; Gerna et al., 1981): fluorescent antibody (Pedersen et al., 1978; Pensaert et al., 1981; Zhou et al., 1988); indirect immunoperoxidase antibody (Gerna et al., 1981); immunoblotting (Horzinek et al., 1982; Hogue et al., 1984; Zhou et al., 1988); two-dimensional immunoelectrophoresis (Schmidt and Kenny, 1981); enzyme-linked immunosorbent assay (MacNaughton et al., 1981; Horzinek et al., 1982; Zhou et al., 1988); radioimmunoprecipitation (Horzinek et al., 1982; Zhou et al., 1988). These have resulted in the placing of coronaviruses into four antigenic groups (Table 17.1). (Relationships among coronaviruses of the same animal species are discussed in section 9.1).

As discussed in section 5.4 several coronaviruses contain a 65-kDa glycopolypeptide. This group corresponds to antigenic group I. Further evidence for the close relationship of viruses in this group has emerged from nucleotide sequencing. Thus the N protein of BCV shows greater homology with MHV (70%; 72% base homology than with that of TGEV (group II) (29%; 37% base homology) and IBV (group III)(29%; base homology 43%) (Kapke and Brian, 1986; Lapps et al., 1987). The N proteins of BCV and HCV-OC43 have 97.5% homology (Kamahora et al., 1989), in keeping with their extremely similar genomic RNA fingerprints (Lapps and Brian, 1985). Similarly, MHV M protein shares 86% homology with BCV but only 38% and 35% with TGEV and IBV, respectively (Laude et al., 1987; Lapps et al., 1987).

With respect to the S protein most amino acid sequence conservation is in the C-terminal half, with overall homologies of 35, 30 and 29% for IBV-FIPV, IBV-MHV and MHV-FIPV, respectively (De Groot et al., 1987b); the N-terminal domains exhibit very little sequence conservation.

6. Antigenic properties of the virion proteins

The coronavirus virion proteins are very interesting antigenically. Thus not only do S and gp65 induce virus neutralizing antibody but also some anti-M antibodies neutralize virus in the presence of complement (Collins et al., 1982; Woods et al., 1987). Moreover, passively administered monoclonal antibodies to each of the S, M and N proteins can protect to some degree against otherwise lethal challenges with MHV.

Studies with several coronaviruses have indicated that S, despite being the minor virion protein in terms of the number of molecules per virion (Cavanagh, 1981), is more antigenic than the N and M proteins: MHV (Hasony and MacNaughton, 1981; Hasony and MacNaughton, 1983); HCV-229E and HCV-OC43 (Schmidt and Kenny, 1981; MacNaughton et al., 1981); IBV (Cavanagh et al., 1984). However, this was not always the case. Horzinek et al. (1982) reported that the antibody response to M of TGEV was greater than that to S and Mockett (1985) showed that infection by IBV induced similar titers of antibody to S and M. The method used to assess the antibody response may influence the results. Nude mice, raised in a specific pathogen-free colony and which were free of detectable anti-S MHV antibody, produced peak titers of anti-virus antibody, as assessed by an ELISA, at 14–28 days after infection with MHV (Talbot et al., 1985). Kinetics of production of antibodies with different properties in response to infection by IBV are intriguing. Thus hemagglutination-inhibiting (HI) antibodies were first detected at 9 days and peaked at about 12 days, whereas virus neutralizing (VN) antibodies first appeared at about 15 days and were not maximal until after 35 days (Mockett and Darbyshire, 1981). The HI antibodies induced by primary infection by IBV are much more strain specific than those resulting from a second infection (Brown and Dracewill, 1988).

That the S glycoprotein induces VN antibody and HI antibody in the case of IBV, has been shown directly by immunization with spike protein from TGEV (Garwes et al., 1979) and IBV (Cavanagh et al., 1984), with recombinant FIPV S protein (De Groot et al., 1989), with a vaccinia virus recombinant containing the IBV S (Tomley et al., 1987) and the FIPV S gene (Spaan et al., 1989) and also with urea-released S1 of IBV (Cavanagh et al., 1986a). In another approach, neutralizing mAbs (nmAbs) were analyzed by radioimmunoprecipitation or ELISA using purified proteins of MHV (Collins et al., 1982; Wege et al., 1984; Talbot et al., 1984; Talbot and Buchmeier, 1985), IBV (Mockett et al., 1984; Niesters et al, 1987), TGEV (Laude et al., 1986; Jimenez et al., 1986), BCV (Deregt and Babiuk, 1987), and FIPV (Fiscus and Teramoto, 1987a,b).

The 65-kDa glycopolypeptide, present on some coronaviruses, also induces VN antibody (Deregt and Babiuk, 1987). The M protein of TGEV not only induces antibody which neutralizes in the presence of complement, it also induces the release of interferon by peripheral blood lymphocytes (Charley and Laude, 1988). Two anti-M mAbs were able to inhibit this induction.

Mice which had been inoculated intraperitoneally with purified S, but not N or M, from MHV-3 were protected from lethal challenge (Hasony and MacNaughton, 1981). However, Nakanaga et al. (1986) have reported that an anti-N mAb, which was not neutralizing *in vitro,* some anti-M mAbs (Fleming et al., 1987), as well as some anti-S mAbs directed against one (Wege et al., 1984) and two (Talbot et al., 1984) antigenic sites of MHV-JHM were able to passively protect mice against lethal challenge. Variants of MHV-JHM selected by nmAbs had reduced neurovirulence (Buchmeier et al., 1984; Fleming et al., 1986; Dalziel et al., 1986). The low level of protection inducible by inactivated IBV was abolished when S1 had been removed by urea treatment (Cavanagh et al., 1986a). Gough et al., (1983a,b) released a subunit from TGEV by sonication which had a density of 1.02 g/ml in sucrose and a size of approximately 25 kDa, as estimated by gel filtration. This material induced nAbs in pigs, protected them against challenge with virulent TGEV and induced antibody in gilts which, when transferred to suckling offspring, resulted in reduced morbidity and mortality after challenge. Further characterization of the 25-kDa entity has not been published.

Vaccination with synthetic peptides homologous to residue 993–1002 (Talbot et al., 1988) and 848–856 (M. Koolen and W. Spaan, unpublished) of S protein of MHV did protect mice against a lethal coronavirus-induced encephalitis.

7. Antibody-dependent enhancement of infection

Feline infectious peritonitis (FIP) is a progressive debilitating condition affecting domestic and wild Felidae, which is characterized by a fibrinous polyserositis (peritonitis, pleuritis), disseminated pyogranulomas and vasculitis. For more detailed information about clinico-pathologic findings and FIP diagnosis the reader is referred to reviews by Pedersen (1983a,b) and Barlough and Stoddart (1986).

Although extensively studied, the pathogenesis of FIP is not well understood. However, the infection of cells of the monocyte-macrophage lineage (Pedersen, 1976; Ward, 1970) appears to be the central pathogenetic phenomenon. Whereas FECV remains confined to the digestive tract and usually does not spread further than the intestinal epithelium and regional lymph nodes (Pedersen et al., 1981; 1984), FIPV rapidly disseminates to other organs via blood-borne monocytes (Weiss and Scott, 1981a).

Apparently the humoral immune response is not protective. In fact, the presence of antibody directed against FCV proteins often leads to an accelerated, fulminating course of the disease, i.e. clinical signs and lesions develop earlier and the mean survival time is significantly reduced ('early death') (Pedersen and Boyle, 1980; Weiss et al. 1980; Weiss and Scott, 1981b). These findings may be explained by antibody-dependent enhancement (ADE) of virus infection, a phenomenon previously described for representatives of the flaviviridae, alphaviridae, bunyaviridae, poxviridae, rhabdoviridae, herpesviridae and reoviridae (for a review see Halstead, 1988). Binding of antibody results in the formation of infectious virus-antibody complexes, which attach to the surface of the macrophage (via Fc-receptors) with higher efficiency than virus without antibody (for a review see Porterfield, 1986). Experimental evidence has recently been obtained that this mechanism is also involved in FIP. The number of feline peritoneal macrophages that could be infected with FIPV UCD1 in vitro was 12-fold increased when the virus was pre-incubated with serum from an FIP-sensitized cat. This enhancement was also found using the HPLC-purified IgG-fraction of the serum. It could be prevented by blocking the macrophage Fc receptor with aggregated IgG or by blocking the Fc portion of the enhancing IgG with protein A (Cheryl Stoddart, personal communication).

It is important to note that antibody-mediated early death is not always observed. Early death did not occur in kittens which had been vaccinated with either FECV UCD, FECV 79-1683 or FIPV UCD2 prior to a lethal challenge with FIPV 79-1146 (Fiscus et al. 1987; Pedersen et al., 1984; Pedersen and Floyd, 1985); of two kittens vaccinated with FIPV UCD4 and challenged with FIPV 79-1146, only one developed enhanced FIP (Pedersen and Floyd, 1985). Development of a vaccine will not be an easy task. It will probably require a better understanding of the complex immune-mediated phenomena and identification of the viral proteins involved. Molecular cloning of the FIPV genome and the isolation of viral genes provides powerful tools to tackle these problems.

Since the coronavirus spike protein S mediates attachment of the virus to the host cell, is involved in membrane fusion and elicits neutralizing antibodies (see above), it is considered to be a good candidate for the development of a coronavirus vaccine. However, in the case of FIPV it is important to know wether or not S is involved in ADE.

In order to study a possible role of the spike protein in protection and pathogenesis the gene encoding the FIPV spike protein was inserted into the vaccinia virus genome. Low levels of FIPV neutralizing antibodies were obtained following immunization with the recombinant vaccinia virus. A challenge infection with FIPV of the immunized kittens resulted in early death in all animals (Spaan et al., 1989).

8. Identification of antigenic domains

mAbs have been used in competition assays to determine the number of antigenic sites or domains on the S, gp65 and M proteins. Greater resolution of the antibody binding sites of the S protein has been achieved by comparison of the amino acid sequences of several IBV S proteins, sequence analysis of escape mutants of IBV, analysis of recombinants of MHV and the reactivity of bacterial fusion proteins and synthetic peptides to antibodies of MHV and IBV.

Inoculation of chickens with IBV lacking the S1 (N-terminal) subunit of S, and with S1 alone, had shown that some VN antibodies were induced by the S1 subunit (Cavanagh et al., 1986c). Using competition ELISA Koch et al. (1989) deduced two antigenic sites on the S1 (N-terminal) subunit of the S protein of IBV.

Comparison of the amino acid sequence of S1 of IBV strains M41, Beaudette and H120, all belonging to the Massachusetts serotype revealed the presence of two hypervariable regions (HVR) located at amino acid positions 38–51 and 99–115 from the N-terminus of S1 (excluding the signal sequence; Niesters et al., 1986; Kusters et al., 1989). A total of 32 amino acid differences were detected in a comparative analysis of nine different Mass isolates, some of which have been used as a vaccine strain (Cavanagh et al., 1988). Most of the substitutions occurred in both HVR1 and 2. Direct evidence that HVR 1 contains residues that are associated with epitopes that induce neutralizing antibodies was obtained by sequence analysis of escape mutants. These mutants were selected by growing wild type or chemically mutagenized virus in the presence of a neutralizing monoclonal antibody. A single nucleotide change resulting in the substitution of a proline residue located in HVR 1 by a histidine residue was found in all mutants (Cavanagh et al., 1988).

Some weakly neutralizing mAbs reacted with S2 (Koch et al., 1986). These mAbs were much less dependent on epitope configuration than those binding to S1. By using beta-galactosidase fusion proteins and synthetic peptides an immunodominant region located at the N-terminus of S2 of IBV has recently been identified (Lenstra et al., 1989; Kusters et al., submitted). This region covers several overlapping epitopes which bind to weak neutralizing monoclonal antibodies and are conserved in several serotypes (Kusters et al., submitted).

Using mAbs in competition ELISAs, Talbot et al. (1984) detected four antigenic sites (A–D) on the S protein of MHV-JHM. Subsequently site D was differentiated into two sites, designated D and E (Talbot and Buchmeier 1985). Sites A, B and C bound VN mAbs whereas sites D and E bound only non-neutralizing antibody. In addition, mAbs to sites A and B could passively protect mice from challenge (Talbot et al., 1984). Wege et al. (1984) distinguished six epitope groups on MHV-JHM, designated A–F. Neutralizing antibodies were associated with sites A and B, and site A antibodies could also inhibit membrane fusion and could passively protect. Using fusion proteins generated by expression of fragments of the MHV-A59 S gene and synthetic peptides, Luytjes et al. (1989) have identified the epitope reactive with a site A mAb. This mAb reacted with an epitope comprising 11 amino acid residues situated 1/5 from N-terminus of S2 (generated by cleavage of the precursor protein). Using recombinants, Makino et al. (1987) determined that the epitopes for two neutralizing mAbs (Fleming et al., 1987) were in the carboxy-terminal 1/3 of S. Talbot et al. (1988) synthesized a decapeptide homologous to residues situated 4/5 from the N-terminus of SO, 3/5 from the N-terminus of S2. Immunization of mice with this peptide induced a measure of protection against an otherwise lethal challenge. Taken together, these results show that MHV has sequences associated with fusion and with VN and protective antibody which are located in the stalk part of S.

Two (A and B) antigenic sites, each associated with VN mAbs, have been detected, by competition assay, on the S protein of BCV by Deregt and Babiuk (1987). Most recently Deregt et al. (1989) used trypsin to generate a 37-kDa fragment of S which contained sites A and B.

Four antigenic sites have been identified on the S protein of TGEV (Delmas et al., 1986; Jimenez et al., 1986; Correa et al., 1988). Most of the VN mAbs were associated with site A or A/B. Interestingly, mutants resistant to site A mAbs occurred at very low frequency and site A was highly conserved among TGEV isolates (Jimenez et al., 1986; Laude et al., 1986), suggesting that this site has an essential function.

Experiments with protein fragments and prokaryotic expression products showed that sites B, A/C and D of Correa et al. (1988) were contained within residues 325–378, 1–280 and 379–529, respectively (Correa et al., 1989). A synthetic peptide comprising the first 18 residues of the mature S reacted with an antiserum raised against a 28-kDa fragment of S which contained sites A and C.

Competition experiments have indicated that gp65 protein of BCV has at least three antigenic sites; the antibody that they induce has VN activity (Deregt and Babiuk, 1987). Two non-overlapping antigenic sites have been detected by mAbs on the M protein of MHV (Talbot et al., 1984) and TGEV (Rasschaert et al. 1987). In both viruses one site comprised two or more overlapping epitopes. In the case of TGEV two of these epitopes were associated with mAbs which block interferon production and neutralized virus in the presence of complement.

9. Evolution of coronaviruses

9.1. Antigenic variation

Antigenic variation of IBV has been studied for decades because variants have often been of economic importance in the poultry industry. More recently mAbs have been used to differentiate a recently emerged porcine coronavirus from classical TGEV, again of economic importance, while mAbs are also proving useful to study the basis of pathogenicity in MHV and FIPV.

Following the first isolation of IBV in the USA in the 1930s and its designation as the Massachusetts serotype, the next three decades in the USA saw the isolation of different serotypes, as defined by VN and HI tests. These caused disease despite the use of Massachusetts serotype vaccines, hence the intense interest shown in them (Hopkins, 1974);. European isolates were considered to be primarily of the Massachusetts serotype until the late 1970s, since when many new serotypes have been isolated and defined by VN test (Davelaar et al., 1983; Cook, 1984; Cook and Huggins, 1986). An understanding of how these serotypes evolve has recently begun to emerge form sequencing studies. On the one hand some serotypes differ by up to 46% of the amino acids in S1, the S subunit which determines serotype (Niesters et al., 1986; Binns et al., 1986b; Niesters, 1987; Kusters et al., 1989), indicating that some serotypes diverged a long time ago. In contrast, some serotypes differ by less than 2% of their S1 amino acids (Cavanagh et al., in preparation) which suggest recent divergence. This finding also suggests that VN antibodies are induced by very few antigenic sites. Overall, sequencing has shown that the amino-terminal 25% or so of the S sequence is the most variable, a finding mirrored by a comparison of the S proteins of TGEV and FIPV (Jacobs et al., 1987). These sequences are known to contain major antigenic sites, including those which induce VN antibody (Cavanagh et al., 1988; Correa et al., 1989).

Monoclonal antibodies have shown that TGEV does not show extensive antigenic variation (Laude et al., 1986; Jimenez et al., 1986) and have recently been used to differentiate TGEV from a closely related virus, porcine respiratory coronavirus (PRCV). This disease emerged in Europe in 1986. As its name implies PRCV grows principally in the respiratory tract; unlike TGEV, it produces mild or no clinical signs and is a major problem in the pig industry because antigenically the two viruses are very similar and cannot be distinguished by conventional serological techniques. This affects the granting of export licenses. Recently some mAbs raised in response to the S protein of TGEV have been found to differentiate between PRCV and TGEV (Garwes et al., 1988; Callebaut et al., 1988). Interestingly, these mAbs were non-neutralizing and were against antigenic sites B, C and D and not to site A, which is highly conserved among TGEV isolates and apparently essential for S function (Delmas et al., 1986; Jimenez et al., 1986). Whether the difference between PRCV and TGEV detected by the mAbs are responsible for the different tissue tropisms of these two viruses is not known but is under study, as are other areas of the genome. Attempts to differentiate virulent and avirulent strains of TGEV have been reported (Welch and Saif, 1988).

The S protein of FIPV isolates has been found to vary antigenically more than the M and N proteins (Fiscus and Teramoto, 1987a). Virulent isolates could be differentiated from an avirulent isolate using anti-S mAbs and biological differences between these two antigenic groups have been studied in vitro (Fiscus and Teramoto, 1987b).

Polyclonal sera have been used to establish antigenic relationships between murine coronavirus (Childs et al., 1983; Talbot et al., 1985). Spike protein antigenic site A was conserved among 6 strains of MHV analyzed but mAbs to the 4 other antigenic sites differentiated strains, of various pathogenicities, to some extent (Talbot and Buchmeier, 1985). Fleming et al. (1983), using mAbs, found that M was strongly conserved antigenically, S little conserved, and N conserved to an intermediate extent. Interestingly, mAbs to N proved most useful in delineating antigenic relationships among MHV strains, the antigenic groups correlating with pathotypes. In addition, it is clear that the S protein plays a major role in the type of disease produced by MHV strains. mAbs to S antigenic sites A and B of MHV-JHM protected mice against lethal challenge and converted the infection from fatal encephalomyelitis to demyelination (Buchmeier et al., 1984). VN mAbs to site B selected variants in vitro which had similarly changed pathogenicity (Dalziel et al., 1986; Wege et al., 1988). One mAb, for which the target antigenic site was in the C-terminal 1/3 of S (Makino et al, 1987) also selected a variant with reduced neurovirulence (Fleming et al., 1986). Following infection with MHV-JHM, variants have been recovered from the central nervous system. Some of these variants had insertions (Taguchi et al., 1985) or deletions (Morris et al., 1989) in the S protein. Finally, S of MHV-A59 has a similar sequence to that of MHV-JHM except that A59 has a stretch of 89 amino acids which are not present in the S1 of JHM (Luytjes et al., 1987). Thus, as in studies of the N (Boursnell et al., 1985), M (Cavanagh and Davis, 1988) and S proteins of IBV, the genome of MHV clearly has a great propensity for change with consequences for its evolution. One contributing factor for this change is recombination.

9.2. RNA recombination

The RNA of coronaviruses can undergo recombination (Lai et al., 1985). A high frequency of recombination was detected in cells infected with a ts-mutant of MHV strain A59 and wild-type JHM virus at the non-permissive temperature (Makino et al., 1986). Multiple recombination sites have been detected at the 5′ end of MHV genomic RNA (Keck et al., 1987). Although sequence data of the cross-over sites are not available, the close relationship between the parental genomic RNAs suggests homologous recombination.

Results from studies of poliovirus recombination strongly suggest a copy-choice (or polymerase jumping) mechanism (Kirkegaard and Baltimore, 1986). The ability of the coronaviral polymerase to switch templates during the discontinuous mRNA transcription supports the copy-choice concept. The pool of leader RNA containing incomplete transcripts that have been detected in MHV infected cells (Baric et al., 1987) can be the result of a discontinuous and non-processive replication mechanism. These RNA intermediates may play an important role in homologous recombination but also in the generation of defective interfering RNAs (Keck et al., 1987; Makino et al., 1984 and 1985).

Recombination is an important feature of coronavirus evolution. Not only has recombination between MHV strains been shown to occur in vitro (see above) but also in mouse brain (Keck et al., 1988). Comparison of IBV M protein sequences (Cavanagh and Davis, 1988) with the S1 sequences of the same strains (Niesters 1987) suggests that some IBV field strains are recombinants. The high sequence divergence at the N terminus between the spike proteins of TGEV and FIPV could be the result of a RNA recombination event (Jacobs et al., 1987). Sequence analysis of mRNA2 of MHV-A59 has revealed a high similarity between the predicted amino acid sequence of a second open reading frame located at its 5′ end and the HA1 subunit of the influenza C spike protein. This could be the result of a nonhomologous recombination event (Luytjes et al., 1988). Also, uptake of genetic information via non-homologous recombination may account for FIPV ORF-2 located at the 3′ end of the genome which is absent in TGEV (De Groot et al., 1988).

These features and the differences among coronaviruses with respect to the number and order of the genes (De Groot et al., 1987c), the length of the intergenic region between the ORFs encoding D3 and M of IBV (Cavanagh and Davis, 1988) and the presence of a translatable gp65 ORF (Parker et al., 1989) all testify to the inconstant nature of coronavirus genomes. This is likely to be a major area of future research.

10. References

Armstrong, J., Smeekens, S. and Rottier, P. (1983) *Nucleic Acids Res.* 11, 883–891.

Armstrong, J., Niemann, H., Smeekens, S., Rottier, P. and Warren, G. (1984) *Nature* 308, 751–752.

Baric, R.S., Shieh, C.K., Stohlman, S.A. and Lai, M.M.C. (1987) *Virology* 156, 342–354.

Baric, R.S., Nelson, G.W., Fleming, J.O., Deans, R.J., Keck, J.G., Casteel, N. and Stohlman, S.A. (1988) *J. Virol.* 62, 4280–4287.

Barlough, J.E. and Stoddart, C.A. (1986) In *Contemporary Issues in Small Animal Practice 3,* (F.W. Scott, Ed.) Churchill Livingstone, New York. pp. 93–108.

Binns, M.M., Boursnell, M.E., Cavanagh, D., Pappin, D.J. and Brown, T.D. (1985) *J. Gen. Virol.* 66, 719–726.

Binns, M.M., Boursnell, M.E., Tomley, F.M. and Brown, T.D. (1986a) *Nucleic Acids Res.* 14, 5558–5558.

Binns, M.M., Boursnell, M.E., Tomley, F.M. and Brown, D.K. (1986b) *J. Gen. Virol.* 67, 2825–2831.

Boursnell, M.E., Brown, T.D. and Binns, M.M. (1984) *Virus Res.* 1, 303–313.

Boursnell, M.E., Binns, M.M., Foulds, I.J. and Brown, T.D. (1985) *J. Gen. Virol.* 66, 573–580.

Bradburne, A.F. (1970) *Arch. Ges. Virusforsch.* 31, 352–364.

Brown T.D. and Dracewill, B. (1988) *Avian Pathol.* in press.

Buchmeier, M.J., Lewicki, H.A., Talbot, P.J. and Knobler, R.L. (1984) *Virology* 132, 261–270.

Callebaut, P., Correa, I., Pensaert, M., Jimenez, G. and Enjuanes, L. (1988) *J. Gen. Virol* 69, 1725–1730.

Cavanagh, D. (1981) *J. Gen. Virol.* 53, 93.

Cavanagh, D. (1983a) *J. Gen. Virol.* 64, 1787–1791.

Cavanagh, D. (1983a) *J. Gen. Virol.* 64, 2577–2583.

Cavanagh, D. and Davis, P.J. (1986) *J. Gen. Virol.* 67, 1443–1448.

Cavanagh, D. and Davis, P.J. (1987) *Avian Pathol.* 16, 1–13.

Cavanagh, D. and Davis, P.J. (1988) *J. Gen. Virol.* 69, 621–629.

Cavanagh, D., Darbyshire, J.H., Davis, P., and Peters, R.W. (1984) *Avian Pathol.* 13, 573–583.

Cavanagh, D., Davis, P.J., Darbyshire, J.H. and Peters, R.W. (1986a) *J. Gen. Virol.* 67, 1435–1442.

Cavanagh, D., Davis, P.J. and Pappin, D.J. (1986b) *Virus Res.* 4, 145–156.

Cavanagh, D., Davis, P.J., Pappin, D.J., Binns, M.M., Boursnell, M.E. and Brown, T.D. (1986c) *Virus Res.* 4, 133–143.

Cavanagh, D., Davis, P.J. and Mockett, A.P.A. (1988) *Virus Res.* 11, 141–150.

Charley, B. and Laude, H. (1988) *J. Virol.* 62, 8–11.

Cheever, F.S., Daniels, J.B., Pappenheiner, A.M. and Bailey, O.T. (1949) *J. Exp. Med.* 90, 181–194.

Childs, J.C., Stohlman, S.A., Kingsford, L. and Russell, R. (1983) *Arch. Virol.* 78, 81–88.

Collins, A.R., Knobler, R.L., Powell, H. and Buchmeier, M.J. (1982) *Virology* 119, 358–371.

Compton, S.R., Rogers, D.B., Holmes, K.V., Fertch, D., Remenick, J. and McGowan, J.J. (1987) *J. Virol.* 61, 1814–1820.

Cook, J.K.A. (1984) *Avian Pathol.* 13, 733–741.

Cook, J.K.A. and Huggins, M.B. (1986) *Avian Pathol.* 15, 129–138.

Correa I., Jimenez, G., Sune, C., Bullido, M.J. and Enjuanes L. (1988) *Virus Res.* 10, 77–94.

Correa I., Bullido, M.J., Sune, C., Gebauer, F., Baay, M.F.D., Zwaagstra, K.A., Posthumus, W.P.A., Lenstra, J.A. and Enjuanes L. (1989) *J. Gen. Virol.,* in press.

Dalziel, R.G., Lampert, P.W., Talbot, P.J. and Buchmeier, M.J. (1986) *J. Virol.* 59, 463–471.

Davelaar, F.G., Kouwenhoven, B. and Burger, A.G. (1983) *Clin. Vet.* 106, 7–11.

De Groot, R.J., Maduro, J., Lenstra, J.A., Horzinek, M.C., van der Zeijst, B.A.M. and Spaan, W.J. (1987a) *J. Gen. Virol.* 68, 2639–2646.

De Groot, R.J., Luytjes, W., Horzinek, M.C., van der Zeijst, B.A.M., Spaan, W.J. and Lenstra, J.A. (1987b) *J. Mol. Biol.* 196, 963–966.

De Groot, R.J., ter Haar, R.J., Horzinek, M.C. and van der Zeijst, B.A.M. (1987c) *J. Gen. Virol.* 68, 995–1002.

De Groot, R.J., Andeweg, A.C., Horzinek, M.C. and Spaan, W.J.M. (1988) *Virology* 167, 370–376.

De Groot, R.J., van Leen, R.W., Dalderup, M.J.M., Vennema, H., Horzinek, M.C. and Spaan, W.J.M. (1989) *Virology* 171, 493–502.

Delmas, B., Guelfi, J. and Laude, H. (1986) *J. Gen. Virol.* 67, 1405–1418.

Deregt, D. and Babiuk, L.A. (1987) *Virology* 161, 410–420.

Deregt, D., Sabara, M. and Babiuk, L.A. (1987) *J. Gen. Virol.* 68, 2863–2877.

Deregt, D., Parker, M.D., Cox, G.C. and Babiuk, L.A. (1989) *J. Gen. Virol.* (in press).

Fiscus, S.A. and Teramoto, Y.A. (1987a) *J. Virol.* 61, 2607–2613.

Fiscus, S.A. and Teramoto, Y.A. (1987b) *J. Virol.* 61, 2655–2657.

Fiscus, S.A., Rivoire, B.L. and Teramoto, Y.A. (1987) *Adv. Exp. Med. Biol.* 218, 559–568.

Fleming, J.O., Stohlman, S.A., Harmon, R.C., Lai, M.M., Frelinger, J.A. and Weiner, L.P. (1983) *Virology* 131, 296–307.

Fleming, J.O., Trousdale, M.D., El-Zaatari, F.A., Stohlman, S.A. and Weiner, L.P. (1986) *J. Virol.* 58, 869–875.

Fleming, J.O., Trousdale, M.D., Bradbury, J., Stohlman, S.A. and Weiner, L.P. (1987) *Microb. Pathol.* 3, 9–20.

Frana, M.F., Behnke, J.N., Sturman, L.S. and Holmes, K.V. (1985) *J. Virol.* 56, 912–920.

Garwes, D.J. and Reynolds, D.J. (1981) *J. Gen. Virol.* 52, 153–157.

Garwes, D.J., Lucas, M.H., Higgins, D.A., Pike, B.V. and Cartwright, S.F. (1979) *Vet. Microbiol.* 3, 179–190.

Garwes, D.J., Stewart, F. and Cartwright, S.F. (1988) *Vet. Rec.* 122, 86–87.

Gerna, G., Cereda, P.M., Grazia Revello, M., Cattaneo, E., Battaglia, M. and Torsellini Gerna, M. (1981) *J. Gen. Virol.* 54, 91–102.

Gough, P.M., Ellis, C.H., Frank, C.J. and Johnson, C.J. (1983a) *Antiviral Res.* 3, 211–221.

Gough, P.M., Frank, C.J., Moore, D.G., Sagona, M.A. and Johnson, C.J. (1983b) *Vaccine* 1, 37–41.

Halstead, S.B. (1988) *Science* 239, 476–481.

Hasony, H.J. and MacNaughton, M.R. (1981) *Arch. Virol.* 69, 33–41.

Hasony, H.J. and MacNaughton, M.R. (1983) *FEMS Lett.* 16, 107–110.

Hayashi, T., Watanabe, Y. Nakayama, H. and Fujiwara, K. (1982)

Hierholzer, J.C., Broderson, J.R. and Murphy, F.A. (1979) *Infect. Immun.* 24, 508–522.

Hogue, B.G. and Brian, D.A. (1986) *Virus Res.* 5, 131–144.

Hogue, B.G., King, B. and Brian, D.A. (1984) *J. Virol.* 51, 384–388.

Holmes, K.V., Doller, E.W. and Sturman, L.S. (1981) *Virology* 115, 334–344.

Hopkins, S.R. (1974) *Avian Dis.* 18, 231–239.

Horzinek, M.C., Lutz, H. and Pedersen, N.C. (1982) *Infect. Immun.* 37, 1148–1155.

Jacobs, L., De Groot, R., van der Zeijst, B.A.M., Horzinek, M.C. and Spaan, W. (1987) *Virus Res.* 8, 363–371.

Jimenez, G., Correa, I., Melgosa, M.P., Bullido, M.J. and Enjuanes, L. (1986) *J. Virol.* 60, 131–139.

Kamahora, T., Soe, L.H. and Lai, M.M.C. (1989) *Virus Res.* 12, 1–9.

Kapke, P.A. and Brian, D.A. (1986) *Virology* 151, 41–49.

Keck, J.G., Stohlman, S.A., Soe, L.H., Makino, S. and Lai, M.M. (1987) *Virology* 156, 331–341.

Keck, J.G., Matsushima, G.K., Makino, S., Fleming, J.O., Vannier, D.M., Stohlman, S.A. and Lai, M.M. (1988) *J. Virol.* 62, 1810–1813.

King, B., Potts, B.J. and Brian, D.A. (1985) *Virus Res.* 2, 53–59.

Kirkegaard, K. and Baltimore, D. (1986) *Cell* 47, 433–443.

Koch, G., Hartog, L., Kant, A., van Roozelaar, D. and de Boer, G.F. (1986) *Isr. J. Vet. Med.* 42, 89–97.

Koch, G., Kusters, J.G., van der Zeijst, B.A.M., Postumus, W.P.A. and Meloen, R.H. (1989) *Proc. 1st Int. Symp. Infect. Bronchitis.* Rauischolzhausen (F. R. G.), p. 145.

Kraft, L.M. (1962) *Science* 137, 282–283.

Kusters, J.G., Niesters, H.G.M., Lenstra, J.A., Horzinek, M.C. and van der Zeijst, B.A.M. (1989) *Virology* 169, 217–221.

Lai, M.M., Baric, R.S., Makino, S., Keck, J.G., Egbert, J., Leibowitz, J.L. and Stohlman, S.A. (1985) *J. Virol.* 56, 449–456.

Lapps, W. and Brian, D.A. (1985) *Arch. Virol.* 86, 101–108.

Lapps, W., Hogue, B.G. and Brian, D.A. (1987) *Virology* 157, 47–57.

Laude, H., Chapsal, J.M., Guelfi, J., Labiau, S. and Grosclaude, J. (1986) *J. Gen. Virol.* 67, 119–130.

Laude, H., Rasschaert, D. and Huet, J.C. (1987) *J. Gen. Virol.* 68, 1687–1693.

Lenstra, J.A., Kusters, J.G., Koch, G. and van der Zeijst, B.A.M. (1989) *Mol. Immunol.* 26, 7–15.

Luytjes, W., Sturman, L.S., Bredenbeek, P.J., Charité, J., van der Zeijst, B.A.M., Horzinek, M.C. and Spaan, W.J. (1987) *Virology* 161, 479–487.

Luytjes, W., Bredenbeek, P.J., Noten, A.F.H., Horzinek, M.C. and Spaan, W.J.M. (1988) *Virology* 166, 415–422.

Luytjes, W.L., Geerts, D., Posthumus, W., Meloen, R. and Spaan, W. (1989) *J. Virol.* 63, 1408–1412.

Machamer, C.E. and Rose, J.K. (1987) *J. Cell Biol.* 105, 1205–1214.

MacNaughton, M.R., Madge, M.H. and Reed, S.E. (1981) *Infect. Immun.* 33, 734–737.
Makino, S., Taguchi, F., Hayami, M. and Fujiwara, K. (1983) *Microbiol. Immunol.* 27, 445–454.
Makino, S., Taguchi, F., and Fujiwara, K. (1984) *Virology,* 133, 9–17.
Makino, S., Fujioka, N. and Fujiwara, K. (1985) *J. Virol.* 54, 329–336.
Makino, S., Keck, J.G., Stohlman, S.A. and Lai, M.M. (1986) *J. Virol.* 57, 729–737.
Makino, S., Fleming, J.O., Keck, J.G., Stohlman, S.A. and Lai, M.M. (1987) *Proc. Natl. Acad. Sci. USA* 84, 6567–6571.
Matthews, R.E.F. (1982) *Classification and Nomenclature of Viruses.* Karger, Basel.
Mockett, A.P. (1985) *J. Virol. Methods.* 12, 271–278.
Mockett, A.P.A. and Darbyshire, J.H. (1981) *Avian Pathol.* 10, 1–10.
Mockett, A.P., Cavanagh, D. and Brown, T.D. (1984) *J. Gen. Virol.* 65, 2281–2286.
Morris, V.L., Tieszer, C., Mackinnon, J. and Percy, D. (1989) *Virology* 169, 127–136.
Nakanaga, K., Yamanouchi, K. and Fujiwara, K. (1986) *J. Virol.* 59, 168–171.
Niesters, H.G., Lenstra, J.A., Spaan, W.J., Zijderveld, A.J., Bleumink-Pluym, N.M., Hong, F., van Scharrenburg, G.J., Horzinek, M.C. and van der Zeijst, B.A.M. (1986) *Virus Res.* 5, 253–263.
Niesters, H.G., Bleumink-Pluym, N.M., Osterhaus, A.D., Horzinek, M.C. and van der Zeijst, B.A.M. (1987) *Virology* 161, 511–519.
Niesters, H.G. (1987) *Molecular epidemiology of infectious bronchitis virus.* PhD Thesis, University of Utrecht.
Parker, M.D., Cox, G.J., Deregt, D., Fitzpatrick, D.R. and Babiuk, L.A. (1989) *J. Gen. Virol.* 70, 155–164.
Pedersen, N.C. (1976) *Am. J. Vet. Res.* 37, 567–572.
Pedersen, N.C. (1983a) *Feline Pract.* 13, 5–14.
Pedersen, N.C. (1983b) *Feline Pract.* 13, 13–19.
Pedersen, N.C. (1987) In *Virus Infections of Carnivores* (M.J. Appel and M.C. Horzinek, Eds.) Elsevier, Amsterdam. pp. 267–286.
Pedersen, N.C. and Boyle, J.F. (1980) *Am. J. Vet. Res.* 41, 868–876.
Pedersen, N.C. and Floyd, K. (1985) *Compend. Contin. Educ. Pract. Vet.* 7, 1001–1011.
Pedersen, N.C., Ward, J. and Mengeling, W.L. (1978) *Arch. Virol.* 58, 45–53.
Pedersen, N.C., Boyle, J.F., Floyd, K., Fudge, A. and Barker, J. (1981) *Am. J. Vet. Res.* 42, 368–377.
Pedersen, N.C., Evermann, J.F., McKeirnan, A.J. and Ott, R.L. (1984) *Am. J. Vet. Res.* 45, 2580–2585.
Pensaert, M.B. (1989) In *Virus Infections of Porcines* (M.B. Pensaert and M.C. Horzineks Eds.) Elsevier, Amsterdam. pp. 154–165.
Pensaert, M.B., Debouck, P. and Reynolds, D.J. (1981) *Arch. Virol.* 68, 45–52.
Pensaert, M., Callebaut, P. and Vergote, J. (1986) *Vet. Q.* 8, 257–261.
Pfleiderer, M., Skinner, M.A. and Siddell, S.G. (1986) *Nucleic Acids Res.* 14, 6338–6338.
Porterfield, J.S. (1986) *Adv. Virus Res.* 31, 335–355.
Rasschaert, D. and Laude, H. (1987) *J. Gen. Virol.* 68, 1883–1890.
Rasschaert, D., Guelfi, J. and Laude, H. (1987) *Biochimie* 69, 591–600.
Rottier, P.J. and Rose, J.K. (1987) *J. Virol.* 61, 2042–2045.
Rottier, P.J.M., Horzinek, M.C. and van der Zeijst B.A.M. (1981) *J. Virol.* 40, 350–357.
Rottier, P., Brandenburg, D., Armstrong, J., van der Zeijst B.A.M. and Warren, G. (1984) *Proc. Natl. Acad. Sci. USA* 81, 1421–1425.
Rottier, P.J., Welling, G.W., Welling-Wester, S., Niesters, H.G., Lenstra, J.A. and van der Zeijst, B.A.M. (1986) *Biochemistry* 25, 1335–1339.
Sawicki, S.G. (1987) *Adv. Exp. Med. Biol.* 218, 169–174.
Schmidt, I., Skinner, M. and Siddell, S. (1987) *J. Gen. Virol.* 68, 47–56.
Schmidt, O.W. and Kenny, G.E. (1981) *Infect. Immun.* 32, 1000–1006.
Siddell, S.G. (1982) *J. Gen. Virol.* 62, 259–269.
Siddell, S., Wege, H. and ter Meulen, V. (1983) *J. Gen. Virol.* 64, 761–776.
Skinner, M.A. and Siddell, S.G. (1983) *Nucleic Acids Res.* 11, 5045–5054.

Spaan, W., Cavanagh, D. and Horzinek, M.C. (1988) *J. Gen. Virol.* 69, 2939–2952.
Spaan, W.J.M., Vennema, H., De Groot, R. et al. (1989) In *Vaccines '89: Modern Approaches to New Vaccines Including Prevention of AIDS* (R.A. Lerner, H. Ginsberg, R.M. Chanock, and F. Brown, Eds.) Cold Spring Harbor Laboratory, Cold Spring Harbor NY. pp. 491–496.
Stern, D.F. and Sefton, B.M. (1982a) *J. Virol.* 44, 794–803.
Stern, D.F. and Sefton, B.M. (1982b) *J. Virol.* 44, 804–812.
Stohlman, S.A., Baric, R.S., Nelson, G.N., Soe, L.H., Welter, L.M. and Deans, R.J. (1988) *J. Virol.* 62, 4288–4295.
Sturman, L.S., Holmes, K.V. and Behnke, J. (1980) *J. Virol.* 33, 449–462.
Sturman, L.S., Ricard, C.S. and Holmes, K.V. (1985) *J. Virol.* 56, 904–911.
Sugiyama, K., Ishikawa, R. and Fukuhara, N. (1986) *Arch. Virol.* 89, 245–254.
Taguchi, F., Siddell, S.G., Wege, H. and ter Meulen, V. (1985) *J. Virol.* 54, 429–435.
Talbot, P.J. (1989) *Intervirology* 30, 117–120.
Talbot, P.J. and Buchmeier, M.J. (1985) *Virus Res.* 2, 317–328.
Talbot, P.J., Salmi, A.A., Knobler, R.L. and Buchmeier, M.J. (1984) *Virology* 132, 250–260.
Talbot, P.J., Salmi, A.A., Knobler, R.L. and Buchmeier, M.J. (1985) *J. Immunol.* 134, 1217–1224.
Talbot, P.J., Dionne, G. and Lacroix, M. (1988) *J. Virol.* 62, 3032–3036.
Tomley, F.M., Mockett, A.P., Boursnell, M.E., Binns, M.M., Cook, J.K., Brown, T.D. and Smith, G.L. (1987) *J. Gen. Virol.* 68, 2291–2298.
Tooze, J., Tooze, S. and Warren, G. (1984) *Eur. J. Cell Biol.* 33, 281–293.
Tyrrell, D.A.J., Almeida, J.D., Berry, D.M., Cunningham, C.H., Hamre, D., Hofstad, M.S., Malluci, L. and McIntosh, K. (1968) *Nature* 220–650.
Vlasak, R., Luytjes, W., Leider, J., Spaan, W. and Palese, P. (1988a) *J. Virol.* 62, 4686–4690.
Vlasak, R., Luytjes, W., Spaan, W. and Palese, P. (1988b) *Proc. Natl. Acad. Sci. USA* 85, 4526–4529.
Ward, J.M., (1970) *Virology* 41, 191–194.
Wege, H., Siddell, S. and ter Meulen, V. (1982) *Curr. Top. Microbiol. Immunol.* 99, 165–200.
Wege, H., Doerries, R. and Wege, H. (1984) *J. Gen. Virol.* 65, 1931–1942.
Wege, H., Winter, J. and Meyermann, R. (1988) *J. Gen. Virol* 69, 87–98.
Weiss, R.C., Dodds, W.J. and Scott, F.W. (1980) *Am. J. Vet. Res.* 41, 663–671.
Weiss, R.C. and Scott, F.W. (1981a) *Am. J. Vet. Res.* 42, 382–390.
Weiss, R.C. and Scott, F.W. (1981b) *Comp. Immun. Microbiol. Infect. Dis.* 4, 175–189.
Welch, S. and Saif, L.J. (1988) *Arch. Virol.* 101, 221–235.
Woods, R.D., Wesley, R.D. and Kapke, P.A. (1987) *Adv. Exp. Med. Biol.* 218, 493–500.
Zhou, Y., Ederveen, J., Egberink, H., Pensaert, M. and Horzinek, M.C. (1988) *Arch. Virol.* 102, 63–71.

Eds. M.H.V. van Regenmortel and A.R. Neurath
Immunochemistry of viruses, II. The basis for serodiagnosis and vaccines.

Reoviruses 18

J.A. COHEN[1], W.V. WILLIAMS[2], D.B. WEINER[2,3] and M.I. GREENE[3]

Departments of [1]Neurology, [2]Medicine and [3]Pathology and Laboratory Medicine, University of Pennsylvania School of Medicine, Hospital of the University of Pennsylvania, 3400 Spruce St., Philadelphia, PA, U.S.A.

1. Introduction

The mammalian reoviruses have been utilized to study molecular aspects of virus-host interaction. Several features of the virus have facilitated these studies. First, with one exception, the gene segments are monocistronic. The segmented nature of the genome and the propensity of the virus to reassort upon mixed infection have allowed mapping of phenotypic characteristics to particular gene segments and to the polypeptides encoded by them. Second, to a great extent, the pathogenesis of reovirus infection and the nature of the virus-host interaction are determined by a single viral gene segment, S1, and its product, the σ1 outer capsid protein. Third, the σ1 protein is the principal target of the host immune response.

Relative to other reovirus proteins, the functional and structural features of the σ1 protein, particularly of serotype 3, have been most thoroughly studied. The functional domains of the σ1 protein and the associated structural features have been elucidated by studies employing complementary approaches including molecular cloning of the S1 gene segment, expression of the cloned protein, in vitro mutagenesis of the cloned gene to create protein variants, the development of neutralizing antibody-resistant mutants, the isolation of anti-idiotypic antibodies which mimic the protein, and the use of synthetic peptide analogues. This review will focus primarily on molecular aspects of the type 3 σ1 protein and its interaction with somatic and immune receptors.

2. Reovirus structure and infectious cycle

The mammalian reoviruses, which are members of the reoviridae, are ubiquitous enteric viruses and common human pathogens. The virion consists of an icosahedral outer capsid surrounding the inner capsid and viral genome. The genome consists of 10 double stranded RNA segments: three large segments (designated L1–3), three medium segments (designated M1–3), and four small segments (designated S1–4). Three serotypes have been distinguished on the basis of reactivity with neutralizing and anti-hemagglutination antibodies: strain 1 (prototype Lang), strain 2 (prototype Jones), and strain 3 (prototype Dearing). Despite overall structural similarity, the three serotypes demonstrate different routes of entry and dissemination, exhibit different patterns of tissue and cellular tropism, and mediate different diseases.

Reoviruses gain entry into the target cell by receptor-mediated endocytosis. After binding, virus-receptor complexes are taken up into phagocytic vesicles which rapidly fuse with lysosomes (Silverstein and Dales, 1968). Uncoating occurs within the lysosome. Subviral particles directly penetrate the lysosomal membrane, enter the cytoplasm, and initiate the replicative phase of the infectious cycle (Borsa et al., 1979). Lysosomotropic agents such as NH_4Cl inhibit reovirus growth and replication (Maratos-Flier, 1986). Reoviruses are typically cytopathogenic. After synthesis of viral components, the virions are assembled in cytoplasmic inclusions, known as viral factories, and are released by lysis of the target cell.

The reoviruses are non-enveloped viruses, and virion budding from intracellular or plasma membranes has not been observed. However, reovirus-infected cells can serve as the targets of serotype-specific cytotoxic T-cells (T_C) (Finberg et al., 1979), suggesting that viral components are expressed on the surface of these cells. Automated cytofluorometry using the neutralizing antibody, 9B.G5, demonstrates that the $\sigma 1$ protein is expressed on infected cells (Ertl et al., 1982; Kauffman et al., 1983b). Also, anti-$\sigma 1$ antibodies block T_C-mediated lysis (Finberg et al., 1981).

Reoviruses typically produce acute lytic infections of target cells but under certain conditions can produce persistent productive infections both in culture (Ahmed and Graham, 1977; Bell and Ross 1966; Matsuzaki et al., 1986) and in vivo (Spandidos and Graham, 1976). BALB/c 3T3 fibroblasts persistently infected with reovirus appear normal at the light microscopic level but continue to release infectious virus (Verdin et al., 1986). Ultrastructural analysis demonstrates viral factories, free virus in the cytoplasm, and virus in association with lysosomes. Persistently infected cells may express viral determinants including the $\sigma 1$ protein on their surface (Matsuzaki et al., 1986; Sharpe and Fields 1985).

3. The $\sigma 1$ outer capsid protein

3.1. Functional aspects

The $\sigma 1$ protein encoded by the S1 gene segment is located at the 12 vertices of the icosahedral virion outer capsid and is associated with the tips of the core spikes formed by the λ2 polypeptide. By serving as the cell attachment protein (Lee et al., 1981b; Weiner and Fields, 1977), the σ1 protein defines the pathways of viral entry and dissemination in the host, the cellular tropism, and the virulence of the virus (Kauffman et al., 1983b; Spriggs et al., 1983a; Weiner et al., 1977; Weiner et al., 1980a). It also functions as the viral hemagglutinin (Weiner et al., 1978). Through its interaction with receptor structures on the target cell surface (reovirus type 3 receptor, Reo3R), the σ1 protein mediates inhibition of target cell DNA synthesis and cell growth (Sharpe and Fields, 1981) and alteration of cellular differentiation (Cohen et al., unpublished data). Although the intact reovirus is a complex antigen, the σ1 protein is the principal serotype-specific target antigen of both humoral (Weiner and Fields, 1977; Weiner et al., 1978) and cellular (Finberg et al., 1979) immune responses. Within the structure of the σ1 protein, the domains which interact with somatic and immune receptors are restricted (Bassel-Duby et al., 1986; Bruck et al., 1986; Nagata et al., 1987; Williams et al., 1988; Williams et al., 1989a; Williams et al., 1989b). These features have permitted detailed analysis of the molecular interactions between reovirus and the host organism.

Several lines of evidence demonstrate that the σ1 protein mediates viral attachment to target cells. The antigen recognized by neutralizing antibodies maps to the S1 gene segment (Weiner and Fields, 1977). Serotype-specific neutralizing anti-reovirus antibodies specifically immunoprecipitate the σ1 protein (Burstin et al., 1982; Spriggs et al., 1983b) and prevent infection by blocking viral binding (Lee et al., 1981b). The σ1 protein produced by expression of the cloned S1 gene in *E. coli* binds to L cell targets and competes with intact reovirus (Masri et al., 1986; Pelletier et al., 1987). Thus, reoviruses appear to attach to target cells via interaction of the σ1 outer capsid protein with cell surface structures.

The σ1 protein also mediates reovirus binding to erythrocytes and functions as the viral hemagglutinin. Type 1 virus preferentially agglutinates human erythrocytes, while type 3 virus agglutinates bovine erythrocytes. This phenotypic characteristic maps to the S1 gene segment (Weiner et al., 1978). Hemagglutination-inhibiting antibodies specifically immunoprecipitate the σ1 protein (Burstin et al., 1982; Spriggs et al., 1983b). Purified σ1 protein is capable of agglutinating erythrocytes (Masri et al., 1986; Yeung et al., 1987). This activity is blocked by antibodies which block the hemagglutinating activity of intact virus. Thus, the σ1 protein mediates reovirus binding to receptor sites on both target cells and erythrocytes.

3.2. Antigenic features

Although the reovirus particle is a complex antigen, the anti-reovirus immune response is directed against a small number of epitopes on a small number of proteins. As for other viruses, the capsid proteins are the main target of immune response. Although a humoral immune response to the σ3 and λ2 proteins can be detected, the primary target of both the humoral (Weiner and Fields 1977; Weiner et al., 1978) and cell-mediated (Finberg et al., 1979) immune responses to reovirus is the σ1 outer capsid protein. The limited heterogeneity of the anti-reovirus immune response has allowed precise molecular dissection of the specificity of this response.

The σ1 protein is the dominant antigen in the humoral response to reovirus. Distinct antigenic epitopes, which appear to correspond to functional domains of the σ1 protein, have been delineated by examining the specificity pattern of serotype 3 σ1-specific monoclonal antibodies. By determining the ability of a panel antibodies to neutralize virus and to inhibit virus-induced hemagglutination (Brustin et al., 1982) and by determining the ability of antibodies to compete with one another for binding to virus (Spriggs et al., 1983b), four groups of antibodies can be distinguished. Group 1 antibodies have neutralizing activity but not hemagglutination-inhibition activity. Group II antibodies inhibit hemagglutination but do not neutralize virus. Group III antibodies have both activities, and group IV have neither. Group I and II antibodies both compete with group III but not with one another. These results suggest that the neutralizing and hemagglutination domains are functionally and topographically distinct. Group III antibodies probably bind to an intermediate location and are able to interact with both regions. Group IV antibodies appear to interact with a distinct domain. Lee and coworkers have reported similar findings with a distinct panel of monoclonal antibodies (Lee et al., 1981a).

The neutralizing antibody, 9B.G5, is of particular interest. This antibody abrogates reovirus type 3 infectivity without affecting reovirus type 1 (Burstin et al., 1982; Spriggs et al., 1983b). The neutralizing activity of the 9B.G5 antibody results from its ability to bind to the type 3 σ1 protein at or near the cell attachment domain and inhibit viral adsorption to target cells. Several findings suggest that the epitope recognized by the 9B.G5 antibody is, in part, conformational and may involve amino acid residues in three distinct regions of the carboxy-terminus of the σ1 protein. First, although 9B.G5 specifically immunoprecipitates the type 3 σ1 protein, it does not recognize the protein on Western blot (Pelletier et al., 1987). Second, mutant σ1 proteins with deletions spanning either amino acids 222–364 or 364–455 no longer bind 9B.G5 antibody (Nagata et al., 1987). Third, the amino acid sequence 317–332 of the type 3 σ1 protein exhibits substantial sequence similarity to a combined determinant formed by the heavy and light chain antigen-binding domains of the anti-reovirus receptor antibody, 87.92.6. Synthetic peptides corresponding to this region of shared sequence inhibit the binding of 9B.G5 to reovirus type 3 (Williams et al., 1988). Finally, monoclonal antibody resistant mutants grown in the presence of 9B.G5 display amino acid substitutions in two other distinct regions of the σ1 protein, amino acids 340 and 420 (Bassel-Duby et al., 1986). Although these variants do not bind neutralizing antibody, they maintain the ability to bind to target cells. Thus, the epitope recognized by 9B.G5 includes amino acids in the 317–332 segment plus residues 340 and 420. Which residues are involved in antibody-σ1 contact and which are necessary for the maintenance of the necessary σ1 conformation for antibody binding has yet to be determined.

3.3. Molecular cloning of the S1 gene

The serotype 3 S1 gene segment, which encodes the σ1 protein, has been molecularly cloned and characterized by several laboratories (Bassel-Duby et al., 1985; Cashdollar et al., 1985; Nagata et al., 1984). It encompasses 1416 nucleotides and contains two overlapping open reading frames. Both initiation sites appear to be functional. The longer open reading frame encompasses 1365 bases, from nucleotide 13–1377. The predicted protein product contains 455 amino acids and is 49 kDa. This value corresponds closely to the M_r of the σ1 protein determined by SDS-PAGE (Smith et al., 1969).

The first 18 amino acids of σ1 appear to comprise a signal peptide which presumably is cleaved from the processed mature protein. However, direct sequence information is unavailable because the amino-terminus is blocked. This portion of the protein contains a heptapeptide repeat pattern between amino acids 28–158, in which the first and fourth amino acids are hydrophobic. The remainder of the amino-terminal residues are predominantly hydrophilic. Prolines at residues 3 and 176 flank the repeat region. This sequence is typical of that of an α helical coiled coil.

The carboxy-terminal portion of the molecule is relatively hydrophilic with clusters of hydrophobic residues. It contains 12 prolines and multiple amino acids with aromatic side chains. The predicted tertiary structure includes a mixture of short α helical stretches, β sheets, random coils, and multiple turns. Overall, these characteristics predict a globular structure. Biochemical studies of the σ1 protein isolated from reovirus-infected L cells and of cloned σ1 expressed in *E. coli* suggest that in the virion the polypeptide exists as a multimer (Bassel-Duby et al., 1987).

3.4. Expression of the cloned σ1 protein and in vitro mutagenesis

Subcloning of the cloned type 3 S1 gene into expression vectors has facilitated functional studies of the σ1 protein. The protein produced in *E. coli* retains its antigenicity and is recognized by anti-σ1 antibodies (Masri et al., 1986; Pelletier et al., 1987). Furthermore, it is functionally active and binds to L cell targets, competes with intact virus for target cell binding, and binds to erythrocytes.

Mutant σ1 proteins with substantial deletions have been utilized to analyze the functional domains of the protein (Fig. 18.1) (Nagata et al., 1987). Conveniently placed restriction sites were used to divide the S1 gene into four approximately equal segments to generate in-frame deletion mutants. The resultant mutant σ1 proteins were expressed in *E. coli* and assayed for ability to bind to L cell targets, recognition by anti-σ1 antibodies, and glycophorin binding. While mutants with deletions involving the amino-terminal half of the protein exhibited preserved cell binding, mutants missing either of the two domains in the C-terminal half no longer bound to L cells. These findings confirm that target cell binding involves amino acid

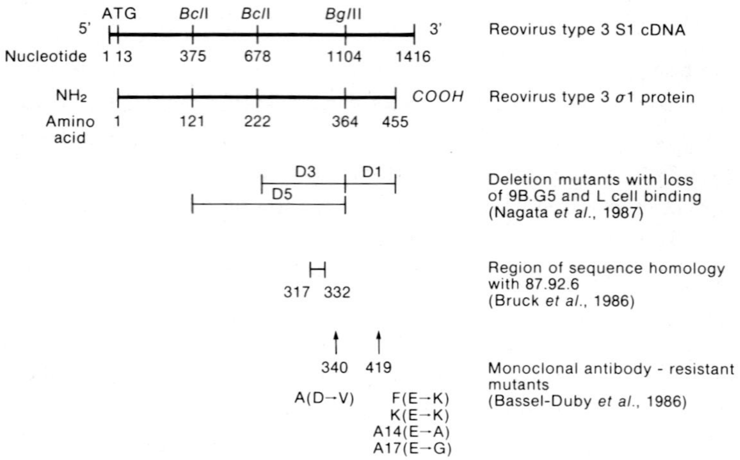

Fig. 18.1. Reovirus type 3 σ1 protein domains involved in binding to immune and somatic receptors.

residues residing in the carboxy half of the protein. They also suggest that the cell binding domain is formed by residues in both domains 3 and 4.

3.5. Monoclonal antibody resistant mutants

Studies of neutralizing antibody-resistant viral mutants have also provided information regarding the functional domains of the σ1 protein. Three variants (A, F, and K), isolated by growing reovirus type 3 in the presence of the neutralizing antibody, 9B.G5, exhibit greatly reduced antibody binding, altered cellular tropism, and reduced virulence (Spriggs and Fields, 1982; Spriggs et al., 1983a). Additional mutants (variants A14 and A17) have been developed by growing reovirus type 3 in the presence of the A2 antibody (Bassel-Duby et al., 1986). Variants F and K are nearly totally resistant to neutralization by 9B.G5. At very high antibody concentrations minimal neutralization of variant K has been observed. Binding studies demonstrate greatly reduced but detectable binding of this antibody to the virus by solid phase radio-immunoassay (RIA) (Fig. 18.2). Variant A demonstrates only partial resistance to neutralizing antibody. All of these viruses bind to L cells.

As outlined in Fig. 18.1, sequencing of the S1 gene segments of these viruses demonstrate point mutations leading to single amino acid substitutions at two distinct locations (Bassel-Duby et al., 1986). Variants F, K, A14, and A17 exhibit substitutions at residue 420 in domain 4. Interestingly, variant A is altered at position 340 in domain 3. The variant K S1 gene has been subcloned and utilized to express the mutated σ1 protein. This protein is not immunoprecipitated by the 9B.G5 antibody used to generate the variant virus. However, like intact variant K virus, it does bind to target cells.

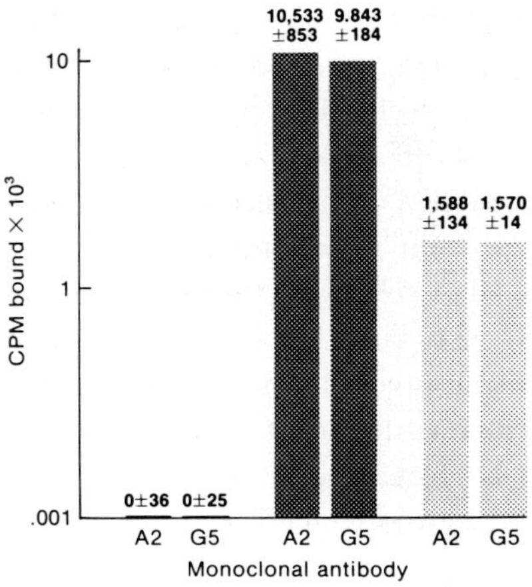

Fig. 18.2. Mutant reovirus type 3 strains demonstrate reduced antibody binding. The binding of radiolabeled G5 and A2 anti-α1 antibodies was determined by solid phase RIA employing microtiter plates coated with intact reovirus type 1 (solid bars), reovirus type 3 (darkly stippled bars), or variant K (lightly stippled bars). No binding to reovirus type 1 was observed. Binding to variant K was greatly reduced relative to wild type virus but detectable.

These findings further confirm that the residues which interact with neutralizing antibodies and with the reovirus binding site on the target cell surface are in the carboxy-terminal portion of the σ1 protein. Although these mutants exhibit loss of sensitivity to neutralization by 9B.G5, they bind to L cells. Thus, the neutralization and receptor binding domains of the σ1 protein, though related, do not coincide exactly.

4. Antibodies and peptides which bind to the reovirus type 3 receptor

4.1. Development of anti-idiotypic anti-receptor antibodies

The finding that 9B.G5 neutralizes type 3 reovirus by interacting with or near the amino acid residues which contact the cell surface receptor suggested a potential strategy by which antibody probes for the reovirus type 3 receptor could be developed. Anti-idiotypic antibodies were isolated which react with 9B.G5 in the hope of generating anti-Reo3R antibodies. Several lines of evidence demonstrate that both the xenogeneic rabbit antiserum anti-ID3 (Nepom et al., 1982a) and the mouse monoclonal antibody 87.92.6 (Noseworthy et al., 1983) generated in this way are immunologic, structural, and functional mimics of the σ1 protein and recognize a reovirus type 3 binding site important in the infectious process. Both forms of anti-Reo3R antibody have tissue and cellular binding patterns similar to that of intact virus (Dichter et al., 1986; Nepom et al., 1982a, Noseworthy et al., 1983). The binding parameters of anti-Reo3R antibody and intact reovirus type 3 are comparable.

There are approximately 79 000 binding sites for 87.92.6 on mouse R1.1 thymoma cells with a $K_d = 9.6 \times 10^{-10}$ M (Gaulton et al., 1985), in good agreement with earlier studies of the binding of intact reovirus to these cells (Epstein et al., 1984a). The binding of 87.92.6 to R1.1 and rat B104 neuroblastoma cells is blocked by reovirus type 3 (Gaulton et al., 1985; Kauffman et al., 1983a). Anti-receptor antibodies block viral binding to neurons in culture and prevent infection (Dichter et al., 1986). Analysis of membrane extracts of R1.1 and B104 cells by SDS-PAGE and immunoblotting demonstrates that reovirus and 87.92.6 bind to proteins of similar electrophoretic mobility (Co et al., 1985a). Immunoprecipitation studies of both cells lines using reovirus type 3 or anti-Reo3R antibodies demonstrate a single glycoprotein with M_r = 65 kDa and a heterogeneous isoelectric point of 5.8–6.0. Thus, these antibodies appear to recognize cell surface structures which function as receptors for reovirus type 3.

The 87.92.6 antibody also mimics structural features of the viral σ1 protein. The nucleic acid and deduced amino acid sequences of the type 3 σ1 protein (Bassel-Duby et al., 1985; Cashdollar et al., 1985; Nagata et al., 1984) and of the 87.92.6 anti-Reo3R antibody (Bruck et al., 1986) have been determined. Substantial sequence similarity exists between a combined determinant comprised of the second complementarity determining regions (CDR II) of the heavy and light chain variable regions (V_H and V_L) and amino acids 317–332 of the type 3 σ1 protein (Fig. 18.3). Therefore, we postulated that these regions of sequence similarity represent the cell binding domains of the type 31 protein and its internal image antibody, 87.92.6. Recent studies confirm that this is the case. Briefly, synthetic peptides corresponding to the shared amino acid sequences compete with both intact reovirus type 3 and 87.92.6 antibody for binding to the 9B.G5 neutralizing antibody and to the cell surface Reo3R (Williams et al., 1989a). These peptides also reproduce Reo3R

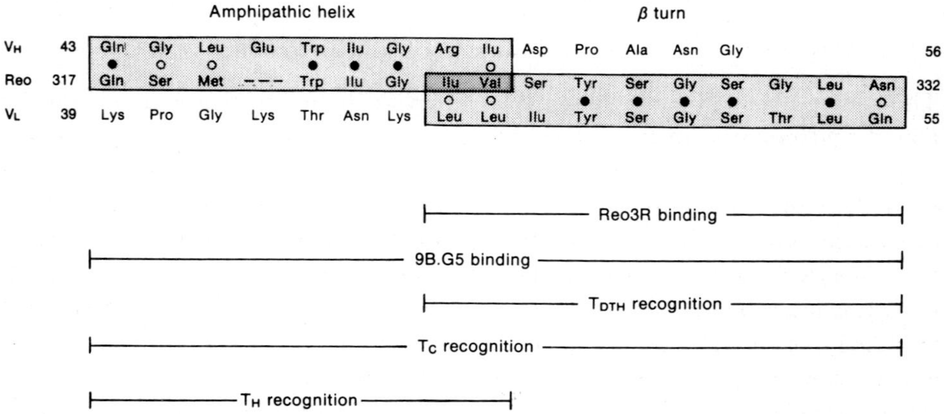

Fig. 18.3. Comparison of the sequences of reovirus type 3 σ1 and 87.92.6 antibody and predicted functional domains.

down-modulation and effects on cell growth induced by reovirus type 3 binding (Williams et al., 1989a, Cohen et al., unpublished data).

4.2. Reovirus receptor-binding peptides

Amino acids 317–332 of the type 3 $\sigma 1$ protein exhibits striking sequence similarity to a combined determinant formed by the second complementarity determining regions of the 87.92.6 heavy and light chain variable regions (Bruck et al., 1986). Amino acids 43–51 of the V_H CDR II are similar to amino acids 317–324 of the $\sigma 1$ protein. Amino acids 46–55 of the V_L CDR II are similar to amino acids 323–332 of the $\sigma 1$ protein. Use of synthetic peptides corresponding to these regions has permitted the mapping of functional characteristics to specific amino acid residues within the peptide regions.

Of particular interest is the sequence, tyr-ser-gly-ser, present in the putative binding regions of both the $\sigma 1$ and V_L proteins. Application of the Chou and Fasman algorithm (Chou and Fasman, 1974) to the $\sigma 1$ and 87.92.6 sequences predicts that this region folds into a reverse turn in both proteins. This observation focused attention on this region as a potentially important sequence in the interaction of the $\sigma 1$ protein and its immunologic mimic, 87.92.6, with anti-$\sigma 1$ antibodies and with the Reo3R. The region of sequence shared by the $\sigma 1$ protein and with 87.92.6 V_H CDR II is predicted to form an amphipathic helix. This structure is a common feature of determinants recognized by helper T-cells (Benjamin et al., 1984). To test these hypotheses, five synthetic peptides were constructed: V_L peptide corresponds to amino acids 39–55 of 87.92.6 V_L CDR II. V_H peptide corresponds to amino acids 43–56 of 87.92.6 V_H CDR II. The V_H-V_L peptide was constructed by adding amino-terminal cysteine residues to the V_H and V_L peptides (V_LSH and V_HSH) allowing them to be covalently coupled. V_L dimer is a divalent form of V_L. Reo peptide corresponds to amino acids 317–332 of the $\sigma 1$ protein, but this peptide is extremely hydrophobic restricting its use to solid phase assays.

5. Studies of antibody-$\sigma 1$ interactions

In competitive solid phase RIA employing microtiter wells coated with the 9B.G5 antibody, V_L peptide effectively and specifically inhibits the binding of radiolabeled reovirus type 3 (Williams et al., 1988). V_H-V_L peptide also is effective, but no more so on a molar basis than V_L. V_H peptide does not affect the interaction of reovirus type 3 with 9B.G5. Direct interaction of peptides with 9B.G5 has also been demonstrated. In solid phase RIA employing radiolabeled 9B.G5 antibody and microtiter plates coated with peptide, specific binding of antibody to Reo, V_L, and V_H-V_L peptides but not V_H peptide can be demonstrated (William et al., 1988). In competitive binding studies, the V_H-V_L peptide is a more potent inhibitor than is V_L peptide in preventing 9B.G5 binding to Reo, V_L, and V_H-V_L peptides. Interestingly, V_H peptide apparently can interact with 9B.G5 in the liquid phase, as it also can specifically inhibit binding of antibody to the other peptides. These studies suggest that interaction of the $\sigma 1$ protein with the 9B.G5 antibody is mediated principally by amino acids 323–332 of $\sigma 1$, corresponding to the V_L domain of 87.92.6 but may also involve residues in the V_H domain.

Infection leads to the production of serotype-specific anti-reovirus antibodies (Weiner and Fields, 1977). Neutralizing anti-reovirus type 3 antibodies can also be induced by immunization with the internal image anti-idiotypic antibody, 87.92.6 (Gaulton et al., 1986). Antibodies reactive with the V_L, V_H-V_L, and Reo peptides can be demonstrated after immunization with intact reovirus (Williams et al., 1989b). Conversely, immunization with V_L peptide coupled to chicken serum albumin or with Reo peptide alone leads to the production of antibodies which bind to and neutralize reovirus type 3. These antibodies also recognize the V_L, V_H-V_L, and Reo peptides in solid phase RIA. Immunization with V_H-V_L peptide, not coupled to carrier protein, leads to the production of antibodies which bind to V_H-V_L and V_L peptides but not to V_H. Immunization with V_H or V_L peptides alone does not lead to the production of anti-peptide antibodies. Taken together, these results suggest that the V_L peptide domain represents the dominant B-cell epitope in vivo. The V_H peptide domain may be required to maintain the V_L domain in a particular conformation. Alternatively, it may serve an additional role in the interaction with the immune system i.e.,interaction with antigen-specific helper T-cells cells (T_H) which provide B-cell help.

6. Interaction of 87.92.6 with 9B.G5 and Reo3R

The interactions of 87.92.6 with 9B.G5 and Reo3R have also been studied (Williams et al., 1988). In competitive RIA employing microtiter plates coated with 87.92.6, both V_H and V_L peptides inhibit 9B.G5 binding. However, V_H-V_L peptide is more effective. This finding indicates that both the V_H CDR II and the V_L CDR II are likely to be involved in this idiotype anti-idiotype interaction. However, when the binding of 87.92.6 to R1.1 cell surface Reo3R is studied, a different pattern is observed. While V_H-V_L peptide is more potent inhibitor of 87.92.6 binding than V_L peptide, V_H peptide does not alter 87.92.6-Reo3R interaction. Thus, both the V_L and V_H domains are involved in the 87.92.6-9B.G5 interaction. In contrast, although the V_H domain may function to maintain the necessary conformation of V_L, the latter appears to be more important in the 87.92.6-Reo3R interaction.

7. Cellular immune response to reoviruses

The physiologic role of cell-mediated immunity in clearance of virus remains unclear. The cytotoxic T-cell (T_C) response to reovirus is MHC class I-restricted and is generally serotype-specific. However, there is some serotype cross-reactivity. After immunization with reovirus under appropriate conditions, T-cell proliferative and DTH responses can also be demonstrated (Green and Weiner, 1980; Weiner et al., 1980c). Under other conditions antigen-specific suppressor cells (T_S) are generated. These responses are largely MHC class II-restricted and mediated by Thy1(+) lymphocytes. UV-irradiated mice are deficient in Ia-bearing adherent spleen cells and have depressed anti-reovirus type 1 T_C, DTH, and T-cell proliferative responses (Letvin et al., 1981). Nevertheless, they clear virus normally after p.o. or i.p. inoculation. Thus, the protective role of cellular immunity to reovirus remains to be confirmed.

7.1. Cytotoxic response

Several lines of evidence suggest that a major serotype-specific target antigen of T_C is the σ1 protein. The serotype-specificity is linked to the S1 gene segment (Finberg et al., 1979). Also, T_C-target cell interaction is blocked in a serotype-specific manner by anti-σ1 antibodies (Finberg et al., 1981). In some studies monoclonal antibodies which recognize the neutralizing domain (A2 and G5) block T_C lysis of target cells (Finberg et al., 1982). Antibodies directed against the hemagglutinin domain (F4, F7) or other structural domains (B11, H3) do not block lysis. Other studies indicate that additional epitopes of reovirus type 3 can serve as targets for T_C.

The 87.92.6 antibody represents an internal image of the type 3 σ1 domain recognized by the neutralizing antibody, 9B.G5. Interestingly, reovirus type 3-specific T_C can also lyse 87.92.6 hybridoma cells (Ertl et al., 1982; Sharpe et al., 1985). This lysis is antigen-specific and MHC class I-restricted. Between 27–39% of T_C clones which lyse reovirus type 3-infected P815 targets also lyse 87.92.6 hybridoma cells. Thus, the epitope shared by the σ1 protein and the 87.92.6 antibody is a dominant epitope in the T_C response. Lysis can be blocked by preincubation of targets with 9B.G5 (Ertl et al., 1982) or of effectors with 87.92.6 (Sharpe et al., 1985). Thus, serotype-specific T_C lysis is blocked by anti-H2 antibodies, anti-reovirus antibodies, and anti-idiotypic anti-Reo3R antibodies. These results suggest that T_C generated after virus infection could lyse cells expressing anti-idiotypic structures.

Reovirus type 3-specific T_C which lyse 87.92.6 hybridoma cells stain uniformly by automated cytofluorometry with 87.92.6 (Sharpe et al., 1985). Neither reovirus-specific T_C which do not lyse the hybridoma cells nor Sendai virus-specific T_C stain. Thus, a proportion of type 3-specific T_C express an idiotope shared with the 9B.G5 antibody that is recognized by 87.92.6.

7.2. DTH and helper T cell responses

The T_H and T_{DTH} responses to reovirus are generally serotype-specific and directed against antigenic determinants of the σ1 protein, but there is a significant level of cross-reactivity (Sharpe et al., 1984; Weiner et al., 1980c). Anti-idiotypic antibody is an effective mimic of intact virus in the generation of the T_{DTH} response. Mice immunized in the footpads with 0.1 μg of purified 87.92.6 antibody in saline in the absence of adjuvant exhibit an anti-reovirus DTH response (Sharpe et al., 1984). This response is relatively serotype-specific, since a small response to type 1 virus is elicited. 87.92.6 is able to prime for a response to the recombinant virus 1HA3 (expressing the type 3 σ1 on a type 1 background) but less so for a response to 3HA1 (expressing the type 1 σ1 on a type 3 background) but less so far a response to 3HA1 (expressing the type 1 σ1 on a type 3 background), strongly suggesting that the response is primarily directed against the σ1 on a type 3 background), strongly suggesting that the response is primarily directed against the σ1 protein.

Reovirus type 3-specific T-cell hybridomas became persistently infected after exposure to live type 3 virus (Matsuzaki et al., 1986). When cultured in the presence of uninfected syngeneic accessory cells, the persistently infected T-cell hybridomas secrete interleukin-2. Presumably, virus and/or viral components are released by these cells into the culture medium, are taken up and processed by accessory cells, and are presented to the hybridoma cells. This auto-stimulatory phenomenon is both reovirus-specific and MHC class II-restricted. These findings have important implications for autoimmunity in association with viral infection.

7.3. Suppressor T cell response

As for other antigens (Eichmann, 1974), quantitative and qualitative parameters of the immune response to reovirus are determined by the dose and the route of immunization. Inoculation of adult mice with live or UV-inactivated reovirus type 1 s.c. and i.p. and live virus i.v. leads to a prominent DTH response on subsequent challenge. In contrast, immunization with UV-inactivated virus i.v. leads to serotype-specific tolerance due to the generation of Thy1(+) T_S cells (Greene and Weiner, 1980). Serotype-specific tolerance can also be generated by p.o. administration of UV-inactivated type 1 virus (Rubin et al., 1981). Tolerance can be adoptively transferred by Thy1(+) mesenteric lymph node cells and splenic lymphocytes. The serotype-specificity maps to the S1 gene segment, indicating that the principal target antigen is the σ1 protein. However, only viruses with the type 1 M2 gene segment and its product, the μ1C outer capsid protein, can generate tolerance after p.o. inoculation. This is most likely due to the resistance to digestion this protein confers. This observation illustrates how viral proteins other than the target antigen can affect the immune response to virus.

Interaction of type 3 virus with spleen cells in vitro also leads to the generation of suppressor cells (Fontana and Weiner, 1980). Live and UV-inactivated reovirus type 3 but not serotype 1 inhibit the in vitro proliferative response of mouse spleen cells to concanavalin A. Studies using recombinant viruses confirm that this property maps to the type 3 S1 gene segment. The inhibition of the proliferative response is due to the generation of Thy1(+) T_S cells. The generation but not the function of such cells can be inhibited in this system by the addition of anti-reovirus antibodies. As discussed above, T-cells involved in the proliferative response also may express the somatic Reo3R, suggesting the possibility that the growth inhibition may be due in part to direct interaction of ligands with this receptor.

7.4. σ1 DOMAINS INVOLVED IN CELLULAR IMMUNITY

Immunization of mice with irradiated 87.92.6 hybridoma cells leads to the production of T_C which are capable of lysing reovirus type 3-infected target cells (Sharpe et al., 1984). Reovirus type 3-specific T_C are also capable of lysing targets passively coated with V_H-V_L oligopeptide sequences but not targets coated with either the V_H or V_L peptides alone (Williams et al., 1989b). Thus, reovirus-specific T_C appear to recognize a combined epitope corresponding to the V_H plus V_L domains in the context of MHC class I determinants.

Immunization with V_L peptide or Reo peptide but not V_H peptide primes mice for a DTH response to intact reovirus type 3 but not type 1 (Williams et al., 1989b). The target antigen is the type 3 σ1 protein, as demonstrated by the finding that these mice respond to the 1HA3 recombinant virus but not to 3HA1. Similarly, immunization with 1HA3 but not 3HA1 virus leads to a DTH response to V_L peptide. Thus, the V_L domain appears to be the dominant epitope involved in the DTH response to reovirus type 3.

In contrast, the V_H domain sequence appears to define the predominant helper T-cell determinant. As described above, the humoral response to the Reo peptide appears to be directed principally against the V_L domain. The V_H domain appears to be necessary for the generation of T-cell help for antibody production. After immunization of mice with V_H or V_L peptides alone, lymph node cells exhibit a selective in vitro proliferative response to the immunizing peptide. However, after immunization with Reo peptide, lymph node cells proliferate in response to Reo peptide and V_H but not V_L. After immunization with 87.92.6 antibody, lymph node cells exhibit proliferative and interleukin-2 responses to intact 87.92.6 antibody and V_H peptide but not to V_L peptide. Similarly, after systemic infection with reovirus type 3, spleen cells exhibit a proliferative response to 87.92.6 antibody and V_H peptide but no response to V_L peptide. Thus, the V_H domain appears to play an important role in the T_H recognition of the type 3 σ1 protein and of its immunologic mimic, 87.92.6.

8. Interaction of reoviruses with target cells

8.1. The somatic cell receptor

Several lines of evidence suggest that the Reo3R recognized by anti-receptor antibodies and the β_2 adrenergic receptor are structurally related but probably distinct proteins. Anti-Reo3R antibodies specifically immunoprecipitate β_2AR affinity purified from calf lung (Co and al., 1985b). Beta adrenergic ligands bind to immunoprecipitated Reo3R in a specific manner (Co et al., 1985b; Liu et al., 1988). The proteins have a similar apparent M_r of 65 000, pI = 5.87–6.0, and tryptic digest maps. Expression of the two receptors is closely linked. Mouse R1.1 thymoma cells, hamster DDT_1 smooth muscle cells, and rat L6 myoblasts express substantial numbers of Reo3R and β_2AR (Greene et al., unpublished data). The tissue distribution of Reo3R, as defined by immunohistochemical staining of tissue sections, corresponds to that of β_2AR, as defined by pharmacologic binding assays and northern blot analysis (Cohen et al., unpublished data). In cultures of embryonic rat hypothalamic neurons, the cell populations expressing immunoreactive Reo3R and β_2AR overlap (Ventimiglia et al., 1987). Thus, the Reo3R and β_2AR are closely related structurally and antigenically and have similar distribution patterns.

The exact relationship between the Reo3R and β_2AR remains unclear. Reo3R ligands and catecholamines do not compete for binding to either the Reo3R or the β_2AR (Co et al., 1985b; Liu et al., 1988; Sawutz et al., 1987). The binding of reovirus type 3 and anti-Reo3R antibodies does not lead to cyclic AMP accumulation (Choi and Lee, 1988; Gaulton and Greene, 1989). The binding of Reo3R ligands also does not affect isoproterenolol-induced cyclic AMP accumulation (Sawutz et al., 1987). Beta adrenergic agonists neither reproduce nor alter Reo3R-mediated inhibition of cell growth (Gaulton and Greene, 1989). Thus, if the β_2AR functions as a serotype-specific reovirus binding site, the virus and catecholamines appear to utilize distinct binding domains. In addition to the β_2AR, other cell surface proteins may be recognized by anti-Reo3R antibodies and may function as serotype-specific reovirus binding sites. For example, rat B104 neuroblastoma cells and mouse L fibroblasts express substantial numbers of Reo3R, as indicated by binding studies utilizing anti-Reo3R antibodies and intact reovirus type 3, but express very few β_2AR (Co et al., 1985a; Epstein et al., 1981; Gaulton and Greene, 1989; Sawutz et al., 1987). We hypothesize that these proteins are also members of the rhodopsin-like family of proteins with multiple hydrophobic transmembrane segments. Studies to identify these proteins are in progress.

Although distinct, the proteins serving as somatic Reo3R, the reovirus type 3-specific B cell receptor (typified by the 9B.G5 antibody), and MHC class I and II-restricted reovirus type 3-specific T-cell antigen receptors must share certain structural features. All bind peptides derived from the type 3 $\sigma 1$ protein and its internal image mimic, the 87.92.6 antibody. Studies described above demonstrate that the immunologic and functional mimicry of the σ1 protein by 87.92.6 results from structural mimicry at the level of primary amino acid sequence. It will be of great interest to compare the predicted amino acid sequences of the proteins which bind the σ1 protein.

8.2. Factors determining viral tropism

The mammalian reoviruses have proven particularly useful in the study of the mechanisms determining viral tropism. The reovirus serotypes display distinct target cell specificities which reflect the high affinity, serotype-specific interaction of the viral σ1 protein with target cell surface receptor structures (Epstein et al., 1984a; Lee et al., 1981b; Nepom et al., 1982b). Serotype-specific differences in the σ1 protein and the differential expression of serotype-specific receptor structures on potential target cells determine, in large part, the cellular tropism pattern of the virus. Within the neonatal rodent central nervous system, type 1 virus infects ependymal cells lining the ventricular system leading to a benign hydrocephalus (Kilham and Margolis, 1969; Margolis and Kilham, 1969). In contrast, reovirus type 3 infects neurons and glia leading to an acute panencephalitis (Dichter and Weiner, 1984; Margolis et al., 1971; Raine and Fields, 1973). The pattern of cellular tropism within the central nervous system correlates with the serotype of the S1 gene segment (Weiner et al., 1980a). The use of recombinant viruses has allowed precise genetic mapping of the tropism pattern. Recombinants which contain the type 3 S1 segment on a type 1 background, designated 1HA3, display the tropism of type 3. The reciprocal recombinant displays type 1 tropism. Thus, the serotype-specific cellular tropism maps to the S1 gene segment which encodes the σ1 outer capsid protein.

Binding studies have indicated that serotypes 1 and 3 utilize distinct receptors on most cells. Within the central nervous system, ependymal cells express receptors for reovirus type 1. Neurons and glial express receptors for serotype 3 (Dichter and Weiner, 1984; Nepom et al., 1982a; Noseworthy et al., 1983; Ventimiglia et al., 1987). Mouse L fibroblasts bear distinct receptors for reovirus type 1 and 3 (Epstein et al., 1984a). Intestinal epithelial cells bear receptors for reovirus type 1 but not 3 (Weiner et al., 1988).

Anti-receptor antibodies recognize the reovirus binding site expressed on both neurons in primary cultures and splenic lymphocytes (Dichter et al., 1986; Epstein et al., 1984b; Nepom et al., 1982a; Noseworthy et al., 1983; Ventimiglia et al., 1987), demonstrating that these proteins are antigenically related. Also, biochemical studies of the Reo3R expressed by the murine thymoma line, R1.1, and the rat neuroblastoma cell line, B104, demonstrate that the cell surface proteins recognized by these antibodies have similar molecular weights (Co et al., 1985a). Thus, lymphocytes and neuronal cells express biochemically similar proteins serving as serotype-specific reovirus binding sites.

8.3. Structural aspects of reovirus receptor-ligand interactions

Analysis of the interaction of the σ1 protein with the Reo3R of R1.1 cells demonstrates that the V_L peptide is the most efficient inhibitor of reovirus type 3 binding to the Reo3R (Williams et al., 1988). V_H-V_L peptide also is effective, but no more so on a molar basis than V_L. V_H peptide does not compete with intact virus for Reo3R.

As described above, Variant K was derived from reovirus type 3 grown in the presence of the 9B.G5 neutralizing antibody (Spriggs and Fields 1982). Substitution of G to A at nucleotide 1267 of the S1 gene produces a Glu to Lys change at residue 419 of the σ1 protein (Bassel-Duby et al., 1986). Although variant K exhibits reduced 9B.G5 antibody binding, it competes with type 3 virus for L cell receptor sites (Williams et al., 1988). Like type 3 virus, the binding of variant K to L cells is inhibited by V_L peptide. Thus, it appears that although residue 419 of the type 3 σ1 protein is involved in the binding to 9B.G5, residues 317–332 are more important in the interaction of the σ1 with Reo3R.

8.4. Receptor down-modulation

The synthetic peptides also have been utilized to analyze the structural requirements to produce functional consequences of ligand-Reo3R interaction. The binding of reovirus type 3 to both B- and T-lymphocytes leads to capping and internalization of the Reo3R (Epstein et al., 1981). A divalent form of V_L peptide capable of Reo3R cross-linking can down-modulate Reo3R from the target cell surface (Williams et al., 1989a). Monomeric V_L peptide binds but does not cross-link Reo3R molecules and does not induce down-modulation from the cell surface.

8.5. Proliferation inhibition

Infection of mouse L fibroblasts with reovirus type 3 leads to rapid and profound inhibition of host cell DNA synthesis with relative sparing of RNA and protein synthesis (Gomatos and Tamm, 1963; Sharpe and Fields, 1981). A decrease in the DNA synthetic rate is observed within 8–10 h after infection, prior to any visible cytopathic effects. By 24 h after infection, the DNA synthetic rate is 15–35% of mock infected control cells. Several studies support the hypothesis that interaction of the type 3 σ1 protein with the Reo3R produces the observed metabolic effect independent of subsequent lytic infection. The binding of type 1 virus, which infects L cells equally well, has no effect on host cell DNA synthesis. The ability to inhibit target cell growth and DNA synthesis has been genetically linked to the type 3 S1 gene segment through the use of recombinant viruses (Sharpe and Fields, 1981). UV-inactivated reovirus type 3, which is able to bind but has greatly reduced infectivity, retains the growth inhibiting effect (Gaulton and Greene, 1989; Sharpe and Fields, 1981). The binding of reovirus type 3 to B104 and R1.1 cells, which express the

Reo3R but do not support productive reovirus infection, produces significant growth inhibition. Finally, anti-Reo3R antibodies and synthetic peptides corresponding to the binding regions of the σ1 protein and 87.92.6 anti-Reo3R antibody are potent inhibitors of the growth of a variety of Reo3R-bearing cells (Gaulton and Greene, 1989; Nepom et al., 1982b; Sharpe and Fields, 1981; Williams et al., 1989a; Cohen et al., unpublished data).

Cross-linking of receptor molecules appears to be necessary as F(ab) fragments of anti-Reo3R antibodies bind to the cell surface but do not inhibit growth (Gaulton and Greene, 1989). When monovalent F(ab) fragments are made functionally polyvalent by the addition of secondary anti-mouse Ig antibody, the growth inhibitory effect is restored. V_L peptide is similarly effective but only in a form expected to produce Reo3R cross-linking (V_L dimer) (Williams et al., 1989a).

The intracellular second messenger mediating this metabolic effect is unknown. Despite similarities between the Reo3R and the β_2AR, changes in intracellular cyclic AMP concentration do not appear to be involved. The cyclic GMP pathway appears to be more relevant (Gaulton and Greene, 1989). Of great potential interest is the finding that BALB/c 3T3 cells persistently infected with reovirus type 3 grow normally in serum but exhibit a decreased proliferative response to epidermal growth factor and decreased surface expression of epidermal growth factor receptor (Verdin et al., 1986). It is presently unknown which viral product mediates this effect. Also, it is unknown whether reduced expression of epidermal growth factor receptor or other cell surface receptors is similarly involved in the growth inhibition following interaction of reovirus type 3 with Reo3R during acute infection.

Using a panel of peptide analogues, we have begun to determine more precisely which amino acid residues and which side chains are critical for interaction of ligand with the Reo3R to produce the growth inhibitory effect. V_L peptide variants missing the hydroxyl groups on the serine and threonine residues at positions 12 and 15 respectively have reduced ability to inhibit cell growth. Removal of the hydroxyl groups from the tyrosine and serine groups at positions 11 and 14 has little effect. Interestingly, substitution of alanine for glycine at position 13 augments the growth inhibitory action. Thus, in addition to receptor cross-linking, key amino acid side chains at certain positions appear to be critical for the induction of Reo3R-mediated anti-proliferative effects.

8.6. Reovirus receptor-mediated stimulation of glial differentiation

The nervous system is an important target tissue for reovirus infection. Several studies have demonstrated that in primary cultures of CNS cells, mature neurons, oligodendrocytes, and both type 1 and type 2 astrocytes express immunoreactive Reo3R but ependymal cells do not (Dichter and Weiner, 1984; Nepom et al., 1982; Noseworthy et al., 1983; Ventimiglia et al., 1987). Ependymal cells express receptors for reovirus type 1 (Nepom et al., 1982; Noseworthy et al., 1983). O-2A glial progenitors do not express receptors for either reovirus type 3 (Ventimiglia et al., 1987) or type 1 (Cohen et al., unpublished data). Reo3R appears in a developmentally regulated manner, at an early stage of oligodendrocyte differentiation prior to the appearance of the myelin-specific markers, galactocerebroside (GalC) and myelin basic protein (MBP) (Cohen et al., unpublished data).

Because of the developmentally regulated expression pattern of the Reo3R during oligodendrocyte differentiation and because of the biochemical and antigenic similarity of the Reo3R to the β_2AR, we hypothesized that perturbation of the Reo3R would alter oligodendrocyte differentiation. Recent studies confirm this prediction. The addition of anti-Reo3R antibodies to developing glial cultures isolated from neonatal rat optic nerve stimulates the early appearance of both GalC and MBP (Fig. 18.4A). Divalent Reo3R-binding peptides are similarly effective (Fig. 18.4B). Monomeric V_L peptide, which binds but does not cross-link Reo3R, is ineffective. These studies confirm that the same σ1 and 87.92.6 domains involved in Reo3R binding and down-modulation and in proliferation inhibition are also likely involved in Reo3R-mediated alteration of cellular differentiation. These studies also

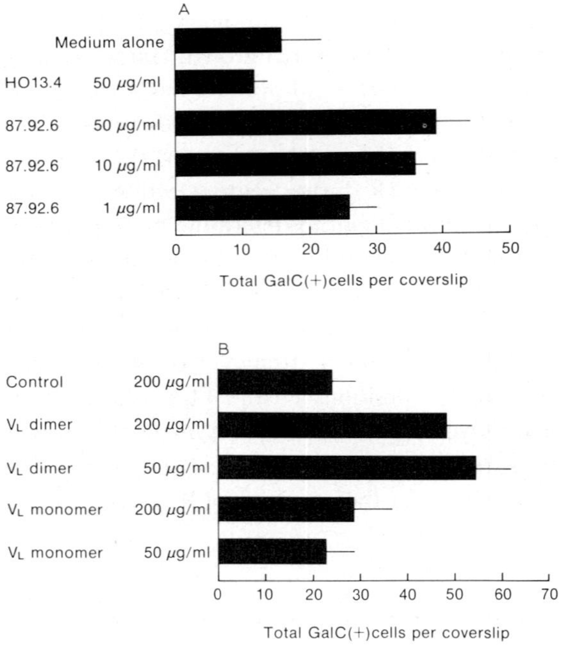

Fig. 18.4. Stimulation of oligodendrocyte differentiation by antibodies and peptides which bind the reovirus type 3 receptor. Single cell suspensions were prepared from neonatal rat optic nerves and cultured in defined medium plus anti-Reo3R antibody or Reo3R-binding peptide. After 1 day of culture the number of cells expressing GalC was determined by immunofluorescence microscopy. Panel A: 1–50 μg/ml 87.92.6 antibody stimulates a 3–4 fold increase in the number of GalC(+) cells relative to medium alone or isotype-matched control antibody. Panel B: V_L peptide in a dimeric form is similarly effective. Monomeric V_L peptide and a control peptide produce no change.

further confirm the requirement for Reo3R cross-linking for receptor-mediated alteration of target cell macromolecule synthesis.

9. Summary and conclusions

Studies mapping the functional domains of the putative binding regions of the type 3 σ1 protein and its immunologic mimic, 87.92.6, confirm the importance of residues 317–332 of the type 3 σ1 protein in the interaction with immune receptors and with the somatic Reo3R (Fig.s 1 and 3). The 87.92.6 antibody can be considered 'ligand-like', and the 9B.G5 antibody can be considered 'receptor-like'. However, there appear to be important differences between the 87.92.6 antibody and the σ1 protein and between the 9B.G5 antibody and the Reo3R. Interaction of the σ1 protein with the Reo3R appears to be mediated primarily by the V_L domain (σ1 amino acids 323–332). In contrast, the interaction of 9B.G5 with the σ1 protein also appears to involve amino acid residues 340 and 419 and may involve the V_H domain (residues 317–324). The interactions of 87.92.6 with both 9B.G5 and Reo3R appear to involve both the V_H and V_L CDR II domains of 87.92.6.

Amino acids 323–332 of the σ1 protein, the residues exhibiting the closest similarity to amino acids 46–55 of the V_L peptide, appear to comprise the domain recognized by cells mediating the DTH response and the domain most important in σ1-Reo3R interaction. Amino acid residues 317–324, the domain corresponding to amino acids 43–51 the V_H peptide, appear to form the predominant epitope recognized by T_H cells. T_C appear to recognize the combined V_H plus V_L epitope. Alternatively, the V_H segment may serve to stabilize the V_L segment in the conformation necessary for interaction with the 9B.G5 antibody and the T_C antigen receptor.

Through the use of several complementary experimental approaches, the structural aspects of the interaction of the reovirus type 3 σ1 protein with both somatic and immune receptor structures have been profitably explored. Future studies utilizing additional peptide analogues and additional σ1 mutants produced by in vitro mutagenesis of the S1 gene will allow finer dissection of the structural constraints on ligand—receptor binding in this system.

10. Acknowledgements

This work was supported by a grant to M.I.G. from the Lucille Markey Charitable Trust. J.A.C. is the recipient of a Clinician Investigator Development Award from NINDS. W.V.W. is the recipient of an NIH post-doctoral fellowship and awards from the McCabe and Measey Foundations.

11. References

Ahmed, R. and Graham, A.F. (1977) *J. Virol.* 23, 250–262.

Bassel-Duby, R., Jayasuriya, A., Chatterjee, D., Sonenberg, N., Maizel, J.V. Jr. and Fields, B.N. (1985) *Nature* 315, 421–423.

Bassel-Duby, R., Spriggs, D.R., Tyler, K.L. and Fields, B.N. (1986) *J. Virol.* 60, 64–67.

Bassel-Duby, R., Nibert, M.L., Homcy, CJ., Fields, B.N. and Sawutz, D.G. (1987) *J. Virol.* 61, 1834–1841.

Bell, T.M. and Ross, M.G.R. (1966) *Nature* 212, 412–414.

Benjamin, D.C., Berzovsky, J.A., East, I.J., Gurd, F.R.N., Hannum, C., Leach, S.J., Margoliash, E., Michael J.G., Miller, A., Prager, E.M., Reichlin, M., Sercarz, E.E., Smith-Gill, S.J., Todd, P.E. and Wilson, A.C. (1984) *Annu. Rev. Immunol.* 2, 67–101.

Borsa, J., Monash, B.D., Sargent, M.D., Copps, T.P., Lievaart, P.A. and Szekely, J.G. (1979) *J. Gen. Virol.* 45, 161–170.

Bruck, C., Co, M.S., Slaoui, M., Gaulton, G.N., Smith, T., Fields, B.N., Mullins, J.I. and Greene, M.I. (1986) *Proc. Natl. Acad. Sci.* USA 83, 6578–6582.

Burstin, S.J., Spriggs, D.R. and Fields, B.N. (1982) *Virology* 117, 146–155.

Cashdollar, L.W., Chmelo, R.A., Wiener, J.R. and Joklik, W.K. (1985) *Proc. Natl. Acad. Sci. USA* 82, 24–28.

Chou, P.Y. and Fasman, G.D. (1974) *Biochemistry* 13, 222–245.

Co, M.S., Gaulton, G.N., Fields, B.N. and Greene, M.I. (1985a) *Proc. Natl. Acad. Sci. USA* 82, 1491–1498.

Co, M.S., Gaulton, G.N., Tominaga, A., Homcy, C.J., Fields, B.N. and Greene, M.I. (1985b) *Proc. Natl. Acad. Sci. USA* 82, 5315–5318.

Dichter, M.A. and Weiner, H.L. (1984) *Ann. Neurol.* 16, 603–610.

Dichter, M.A., Weiner, H.L., Fields, B.N., Mitchell, G., Noseworthy, J., Gaulton, G. and Greene, M.I. (1986) *Ann. Neurol.* 19, 555–558.

Eichmann, K. (1974) *Eur. J. Immunol.* 4, 296–302.

Epstein, R.L., Powers, M.L. and Weiner, H.L. (1981) *J. Immunol.* 127, 1800–1803.

Epstein, R.L., Powers, M.L., Bogart, R.B. and Weiner, H.L. (1984a) *Virology* 133, 46–55.

Epstein, R.L., Finberg, R., Powers, M.L. and Weiner, H.L. (1984b) *J. Immunol.* 133, 1614–1617.

Ertl, H.C.J., Greene, M.I., Noseworthy, J.H., Fields, B.N., Nepom, J.T., Spriggs, D.R. and Finberg, R.W. (1982) *Proc. Natl. Acad. Sci. USA* 79, 7479–7483.

Finberg, R., Weiner, H.L., Fields, B.N., Benacerraf, B. and Burakoff, S.J. (1979) *Proc. Natl. Acad. Sci. USA* 76, 442–446.

Finberg, R., Weiner, H.L., Burakoff, S.J. and Fields, B.N. (1981) *Infect. Immun.* 31, 646–649.

Finberg, R., Spriggs, D.R. and Fields, B.N. (1982) *J. Immunol.* 129, 2235–2238.

Fontana, A. and Weiner, H.L. (1980) *J. Immunol.* 125, 2660–2664.

Gaulton, G., Co, M.S. and Greene, M.I. (1985) *J. Cell. Biochem.* 28, 69–78.

Gaulton, G.N., Sharpe, A.H., Chang, D.W., Fields, B.N. and Greene, M.I. (1986) *J. Immunol.* 137, 2930–2936.

Gaulton, G.N. and Greene, M.I. (1989) *J. Exp. Med.* 169, 197–211.

Gomatos, P.J. and Tamm, I. (1963) *Biochim. Biophys Acta* 72, 651–653.

Greene, M.I. and Weiner, H.L. (1980) *J. Immunol.* 125,283–287.

Kauffman, R.S., Noseworthy, J.H., Nepom, J.T., Finberg, R., Fields, B.N. and Greene, M.I. (1983a) *J. Immunol.* 131, 2539–2541.

Kauffman, R.S., Wolf, J.L., Finberg, R., Trier, J.S. and Fields, B.N. (1983b) *Virology* 124, 403–410.

Kilham, L. and Margolis, G. (1969) *Lab. Invest.* 21, 183–188.

Lee, P.W.K., Hayes, E.C. and Joklik, W.K. (1981a) *Virology* 108, 134–146. Lee, P.W.K., Hayes, E.C. and Joklik, W.K. (1981b) *Virology* 108, 156–163.

Letvin, N.L., Kauffman, R.S. and Finberg, R. (1981) *J. Immunol.* 127, 2334–2339.

Liu, J., Co, M.S. and Greene, M.I. (1988) *Immunol. Res.* 7, 232–238.

Maratos-Flier, E., Goodman, M.J., Murray, A.H. and Kahn, C.R. (1986) *J. Clin. Invest.* 78, 1003–1007.

Margolis, G. and Kilham, L. (1969) *Lab. Invest.* 21, 189–198.

Margolis, G., Kilham, L. and Gonatas, N.K. (1971) *Lab. Invest.* 24, 91–100.

Masri, S.A., Nagata, L., Mah, D.C.W. and Lee, P.W.K. (1986) *Virology* 149, 83–90.

Matsuzaki, N., Hinshaw, V.S., Fields B.N. and Greene, M.I. (1986) *J. Virol.* 60, 259–266.

Nagata, L., Masri, S.A., Mah, D.C.W. and Lee, P.W.K. (1984) *Nucleic Acids Res.* 12, 8699–8710.

Nagata, L., Masri, S.A., Pon, R.T. and Lee, P.W.K. (1987) *Virology* 160, 162–168.

Nepom, J.T., Weiner, H.L., Dichter, M.A., Tardieu, M., Spriggs, D.R., Gramm, C.F., Powers, M.L., Fields, B.N. and Greene, M.I. (1982a) *J. Exp. Med.* 155, 155–167.

Nepom, J.T., Tardieu, M., Epstein, R.L., Noseworthy, J.H., Weiner, H.L., Gentsch, J., Fields, B.N. and Greene, M.I. (1982b) *Surv. Immunol. Res.* 1, 255–261.

Noseworthy, J.H., Fields, B.N., Dichter, M.A., Sobotka, C., Pizer, E., Perry, L.L., Nepom, J.T. and Greene, M.I. (1983) *J. Immunol.* 131, 2533–2538.

Pelletier, J., Nicholson, R., Bassel-Duby, R., Fields, B.N. and Sonenberg, N. (1987) *J. Gen. Virol.* 68, 135–145.

Raine, C.S. and Fields, B.N. (1973) *J. Neuropathol Exp. Neurol.* 32, 19–33.

Rubin, D., Weiner, H.L., Fields, B.N. and Greene, M.I. (1981) *J. Immunol.* 127, 1697–1701.

Sawutz, D.G., Bassel-Duby, R. and Homcy, C.J. (1987) *Life Sci.* 40, 399–406.

Sharpe, A.H. and Fields, B.N. (1981) *J. Virol.* 38, 389–392.

Sharpe, A.H., Gaulton, G.N., McDade, K.K., Fields, B.N. and Greene, M.I. (1984) *J. Exp. Med.* 160, 1195–1205.

Sharpe, A.H. and Fields, B.N. (1985) *N. Engl. J Med.* 312, 486–497.

Sharpe, A.H., Gaulton, G.N., Ertl, H.C.J., Finberg, R.W., McDade, K.K., Fields, B.N. and Greene, M.I. (1985) *J. Immunol.* 134, 2702–2706.

Silverstein, S.C. and Dales, S. (1968) *J. Cell. Biol.* 36, 197–230.

Smith, R.E., Zweerink, H.J. and Joklik, W.K. (1969) *Virology* 39, 791–810.

Spandidos, D.A. and Graham, A.F. (1976) *J. Virol.* 20, 234–247.

Spriggs, D.R. and Fields, B.N. (1982) *Nature* 297, 68–70.

Spriggs, D.R., Bronson, R.T. and Fields, B.N. (1983a) *Science* 220, 505–507.

Spriggs, D.R., Kaye, K. and Fields, B.N. (1983b) *Virology* 127, 220–224.

Ventimiglia, R, Greene, M.I. and Geller, H.M. (1987) *Proc. Natl. Acad. Sci. USA* 84, 5073–5077.

Verdin, E.M., Maratos-Flier, E., Carpentier, J-L. and Kahn, C.R. (1986) *J. Cell. Physiol.* 128, 457–465.

Weiner, D.B., Girard, K., Williams, W., McPhillips, T. and Rubin, D.H. (1988) *Microb. Pathol.* 5, 29–40.

Weiner, H.L., Drayna, D., Averill, D.R. and Fields, B.N. (1977) *Proc. Natl. Acad. Sci. USA* 74, 5744–5748.

Weiner, H.L. and Fields, B.N. (1977) *J. Exp. Med.* 146, 1305–1310.

Weiner, H.L., Ramig, R.F., Mustoe, T.A. and Fields, B.N. (1978) *Virology* 86, 581–584.

Weiner, H.L., Powers, M.L. and Fields, B.N. (1980a) *J. Infect. Dis.* 141, 609–616.

Weiner, H.L., Ault, K.A. and Fields, B.N. (1980b) *J. Immunol.* 124, 2143–2148.

Weiner, H.L., Greene, M.I. and Fields, B.N. (1980c) *J. Immunol.* 125, 278–282.

Williams, W.V., Guy, H.R., Rubin, D.H., Robey, F., Myers, J.N., Kieber-Emmons, T., Weiner, D.B. and Greene, M.I. (1988) *Proc. Natl. Acad. Sci. USA* 85, 6488–8492.

Williams, W.V., Moss, D.A., Fieber-Emmons, T., Cohen, J.A., Myers, J.N., Weiner, D.B. and Greene, M.I. (1989a) *Proc. Natl. Acad. Sci. USA,* 86, 5537–5541.

Williams, W.V., London, S.D., Weiner, D.B., Wadsworth, S., Berzovsky, J.A., Robey, F., Rubin, D.H. and Greene, M.I. (1989b) *J. Immunol.,* in press.

Yeung, M.C., Gill, M.J., Alibhai, S.S., Shahrabadi, M.S. and Lee, P.W.K. (1987) *Virology* 156, 377–385.

Eds. M.H.V. van Regenmortel and A.R. Neurath
Immunochemistry of viruses, II. The basis for serodiagnosis and vaccines.

Hepadnaviruses 19

A.R. NEURATH[1] and Y. THANAVALA[2]

[1]The Lindsley F. Kimball Research Institute of The New York Blood Center, New York, NY 10021 and [2]Roswell Park Memorial Institute, Buffalo, NY 14263, U.S.A.

1. Introduction

The family of Hepadnaviridae (Gust et al., 1986) comprises related viruses infecting distinct species: human hepatitis B virus (HBV); woodchuck hepatitis B virus (WHBV or WHV); ground squirrel hepatitis B virus (GSHBV or GSHV) infecting Beechey and tree squirrels (Marion et al., 1980; Feitelson et al., 1986); duck hepatitis B virus (DHBV) infecting domestic Pekin (Mason et al., 1980) and wild ducks (Cova et al., 1986); and heron hepatitis B virus (HHBV) (Sprengel et al., 1988).

The genomes isolated from hepadnaviruses are partially double stranded circular DNAs. The minus strand of the DNA (complementary to mRNA) is full-length (approximately 3200 nucleotides long) and contains a nick. The plus strand varies in length from about 1700–2800 nucleotides. The two strands form an open circle which is maintained by base pairing involving about 200 nucleotides between the overlapping 5′-ends. A protein functioning as a primer is covalently linked to the 5′-end of the minus strand DNA. Four distinct (overlapping and read in three possible frames) open reading frames (ORFs) have been identified on the minus DNA strands of HBV, WHBV and GSHBV. The ORFs are designated as P, X, preC/C and preS/S coding, respectively, for proteins involved in virus replication, and for the virus nucleocapsid and envelope (env) proteins. The genomes of DHBV and HHBV have a similar genetic map except that ORF X and preC/C are fused so that there are three instead of four ORFs. The mechanism of hepadnavirus DNA replication differs from that of other DNA viruses and involves an RNA intermediate and its reverse transcription occurring inside immature virus nucleocapsid progeny core particles. Cloned hepadnavirus DNAs initiate infection when injected intrahepatically into the homologous species (Sprengel et al., 1984; Seeger et al., 1984b, 1987; Neurath and Kent, 1988). Transfection of human hepatoma cell lines with cloned human or animal hepadnavirus DNAs resulted in virus replication and production of fully infectious virus (Sureau et al., 1986, 1988; Yaginuma et al., 1987; Tsurimoto et al., 1987; Acs et al., 1987; Sells et al., 1987; Galle et al., 1988; Hirsch et al., 1988). This indicates that neither virus proteins nor specific features of viral genomes (nicks and/or gaps) are essential for the initiation of infection.

Although the major target organ for hepadnaviruses is the liver, and the viruses cause acute and chronic hepatitis and cirrhosis, and are associated with primary hepatocellular carcinoma, cells other than hepatocytes may also become infected with hepadnaviruses. These extrahepatic sites include the pancreas, kidney, bile duct, spleen, thymus, bone marrow, endothelial cells and peripheral blood lymphocytes (PBL) (Romet-Lemonne et al., 1983; Elfassi et al., 1984; Korba et al., 1986, 1987, 1989; Lieberman et al., 1987; Jilbert et al., 1987; Neurath and Kent, 1988; Ogston et al., 1989; Roisman et al., 1989). Lymphoid cells of the bone marrow were the first cells in which WHBV DNA was detectable following experimental infection, indicating the involvement of the host lymphoid system in hepadnavirus infections (Korba et al., 1989).

In contrast with the reproducible experimental transmission of hepadnavirus infection or with the transfection of hepatocytes by hepadnavirus DNAs leading to infection, the success in in vitro infection of hepatocytes or of other cells by hepadnaviruses was more limited. The infection of primary duck hepatocyte cultures (established 0–17 days post-hatching) appeared to be most reproducible and resulted in the infection of about 10% of the cells (Tuttleman et al., 1986; Uchida et al., 1988). The infection of adult human hepatocytes cultured in the presence of dimethyl sulfoxide by HBV was reported recently (Gripon et al., 1988). Studies on natural infection of ducks with DHBV suggested that host functions appearing during embryonic development may be required for virus replication and assembly (Urban et al., 1985).

The major structural components of hepadnavirus particles are the env proteins, consisting of three distinct but related species designated according to their size as small (S), middle (M) and large (L) proteins and a single protein (hepatitis B core antigen; HBcAg) representing the subunit of hepadnavirus nucleocapsids (Fig. 19.1). In addition, the nucleocapsids contain: (a) the reverse transcriptase/DNA polymerase, essential for virus replication and for repair of single stranded regions in the DNA; (b) a protein covalently linked to the 5′-end of the DNA minus strand, essential for virus replication; and (c) a protein kinase. The replication of hepadnaviruses in vivo and in vitro is associated with the production of (a) subviral particles (spherical particles approx. 20 nm in diameter and tubular forms approx. 20 nm wide) consisting of hepadnavirus env proteins and designated as hepatitis B surface antigens (HBsAg, WHBsAg, GSHBsAg, DHBsAg and HHBsAg, respectively, for the distinct hepadnaviruses) and (b) soluble protein(s) related to HBcAg and designated as hepatitis B e antigen (HBeAg). The production of these proteins is in excess over the synthesis of virions and appears to be associated with the pathogenesis and chronicity of hepadnavirus infections. In addition to proteins which become assembled into virions or are related to virus proteins, other nonstructural proteins play a role in hepadnavirus replication. These include X-protein, the product of ORF X (except DHBV and HHBV) and proteases, involved in post- translational processing of hepadnavirus primary gene products, the nature of which has not been firmly established.

Results of immunochemical studies on HBV structural components and on M- and L- proteins of hepadnavirus envelopes have been summarized in earlier reviews (Neurath and Kent, 1985, 1988). This chapter will focus on widening the scope of review to hepadnaviruses in general and on more recent immunochemical studies concerning hepadnavirus gene products.

2. Products of ORF preC/C: nucleocapsid core protein and e antigen (HBeAg)

The nucleocapsid core protein and HBeAg were originally considered to be the translational product of the same gene (ORF C), HBeAg being produced from hep-

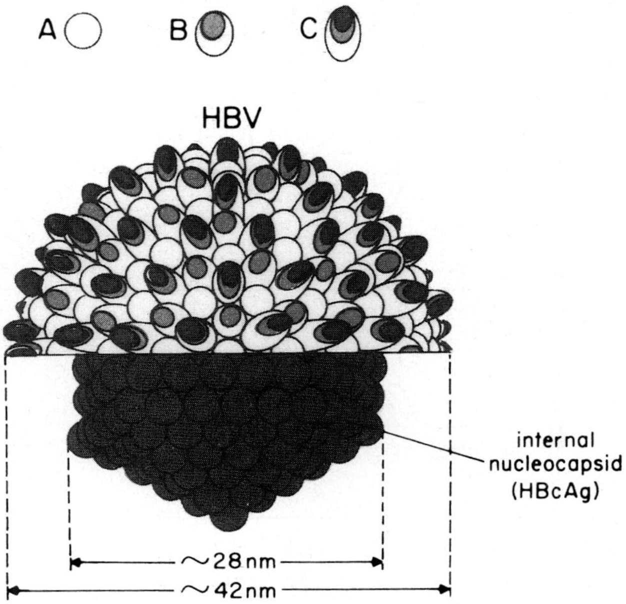

Fig. 19.1. Schematic model of hepadnavirus particles. Individual subunits containing (A) S-protein only; (B) M-protein and (C) L-protein are shown on top of the figure. The virus particles contain an internal nucleocapsid shown in the bottom split-open section. S-protein corresponds to white areas, preS2 sequences to gray areas, and preS1 sequences to black areas.

adnavirus nucleocapsids by detergent ± protease treatment (Neurath and Kent, 1985). Although HBeAg can indeed be generated from viral nucleocapsids artificially by these means, it is not generated in this way in vivo or in various in vitro tissue culture systems. These two related hepadnavirus proteins are translated from distinct mRNAs.

There are two in-phase ATG start codons on ORF preC/C of all hepadnaviruses (Fig. 19.2). Approximately 3.5 kb mRNA transcripts having common 3′ termini and initiated at distinct 5′-ends around the first ATG codon direct the translation of ORF preC/C products (Fig. 19.2). Transcripts initiated between the two ATG codons give rise to a translation product initiated at the second ATG and corresponding to the nucleocapsid core protein of 21–22 kDa. The sequence of nucleocapsid core proteins of HBV, WHV (GSHV) and DHBV (HHBV) contains 185 (183; depending on the antigenic subtype of HBV), 188 and 305 amino acids, respectively (Seeger et al., 1984a; Neurath and Kent, 1985; Sprengel et al., 1985; 1988; Cohen et al., 1988). The C-terminal region of the nucleocapsid core protein is in all cases an arginine-rich sequence, probably involved in interaction with hepadnavirus genomes and susceptible to proteolytic cleavage leading to generation of HBeAg (see below). About 66% of the amino acid sequence of the HBV nucleocapsid protein

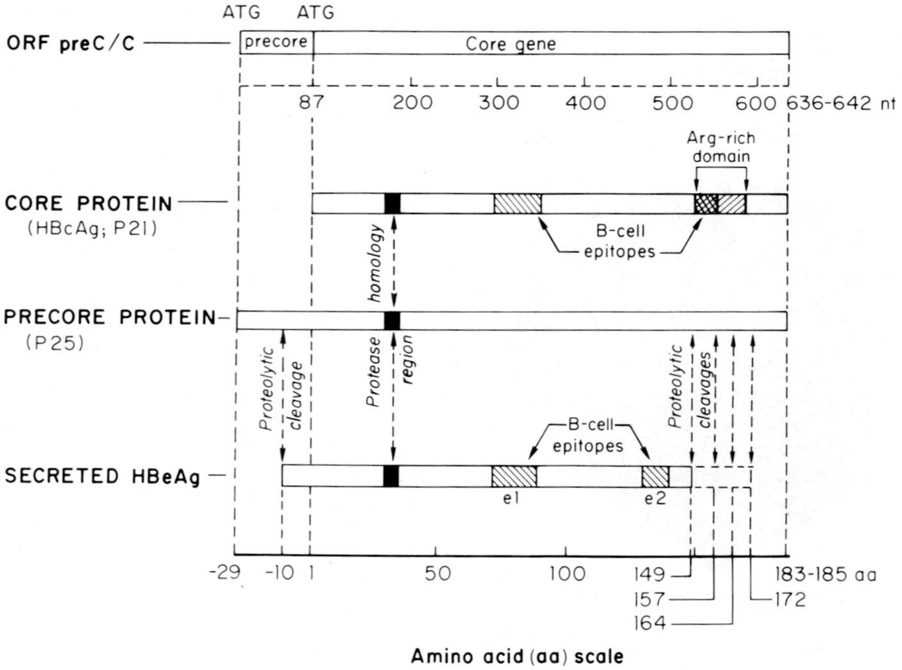

Fig. 19.2. Products coded for by the HBV DNA open reading frame (ORF) C. The length of the ORF and of the translation products differs for distinct antigenic subtypes of HBV. Derived from data of Bruss and Gerlich, 1988; Nassal, 1988; Salfeld et al., 1989; Colucci et al., 1988; and Machida et al., 1989.

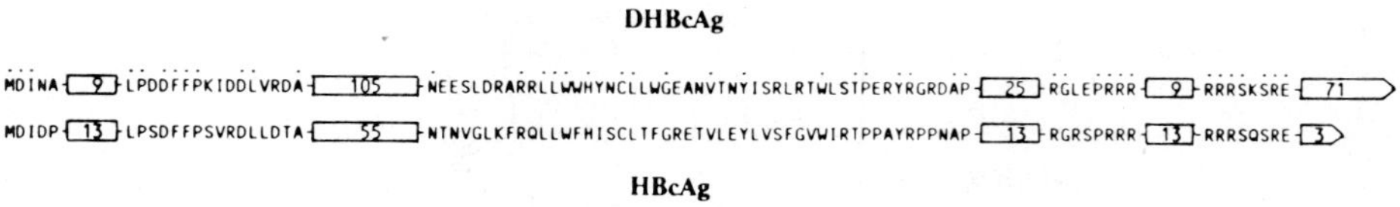

Fig. 19.3. Comparison of HBcAg and DHBcAg sequences. Identical amino acids are indicated by an asterisk. Non-homologous regions are shown as boxes containing the number of amino acids in this region. Reprinted with permission from Sprengel et al., 1985.

is homologous with the corresponding sequences of WHV and GSHV. There is much less homology between the nucleocapsid proteins of HBV (HBcAg) and DHBV or HHBV (DHBcAg, HHBcAg) (Fig. 19.3). Sequences with no homology to HBcAg are interspersed between sequences of DHBcAg (HHBcAg) having partial homology with HBcAg (Fig. 19.4). The middle portion of DHBcAg and HHBcAg, having the least homology with HBcAg (Fig. 19.4), share sequence homology with X gene products of HBV (WHV, GSHV). Thus, X-like gene product sequences are present in the core gene products of DHBV and HHBV which lack ORF X (Feitelson and Miller, 1988). Unlike the sequences of nucleocapsid proteins of mammalian hepadnaviruses, DHBcAg and HHBcAg sequences have one or two potential *N*-glycosylation sites (Mandart et al., 1984; Sprengel et al., 1985, 1988; Schlicht et al., 1987b). Hepadnavirus nucleocapsid proteins are not secreted and accumulate intracellularly in an assembled form (procapsids) (Ponzetto et al., 1984; Roossinck and Siddiqui, 1987; Ohori and Matsuda, 1989) The proteins are phosphorylated at serine residues within the C-terminal one-third of the protein sequence by a phosphokinase probably of cellular origin (Roossinck and Siddiqui, 1987; Pugh et al., 1989; Schlicht et al., 1989a). Phosphorylation of the nucleocapsid protein followed by dephosphorylation is probably essential for hepadnavirus maturation and infectivity.

The translational products of approx. 3.5 kb hepadnavirus mRNAs initiating upstream from the first ATG in ORF preC/C are designated as pre-core proteins (Fig.

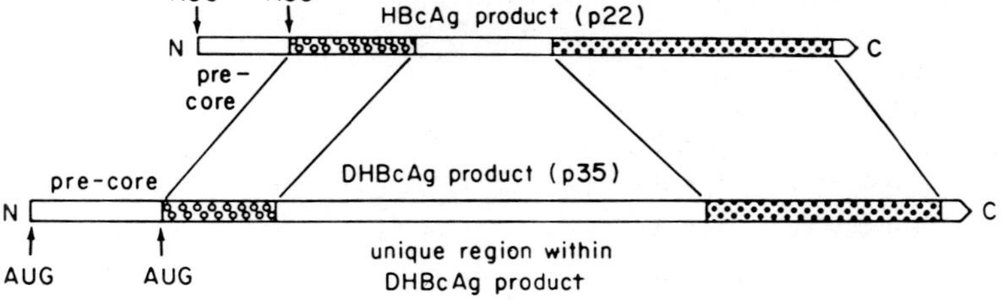

Fig. 19.4. Schematic representation of sequence relationship between translational products of C-genes of HBV (top) and DHBV (bottom). Homologous regions are stippled, open boxes indicate non-homologous regions. Reprinted with permission from Feitelson and Miller, 1988.

			HBeAg →
HBV adw2 adw	M Q L F H L C L I I S C T	C P T V Q A	S K L C L G W L W G
HBV ayw adr	M Q L F H L C L I I S C S	C P T V Q A	S K L C L G W L W G
HBV ayr	M Q L F P L C L I I S C S	C P T V Q A	S K L C L G W L W G
WHV	M Y L F H L C L V F A C V P	C P T V Q A	S K L C L G W L W G
GSHV	M Y L F H L C L V F A C V P	C P T V Q A	S K L C L G W L W D

Fig. 19.5. Comparison of the precore amino acid sequences of different hepadnaviruses. Derived from DNA sequencing data of Valenzuela et al., 1980; Ono et al., 1983; Galibert et al., 1979; Bichko et al., 1985; Kobayashi and Koike, 1984; Fujiyama et al., 1983; Okamoto et al., 1986; Cohen et al., 1988; and Seeger et al., 1984a. Arrow indicates proteolytic cleavage generating the N-terminus of HBeAg.

19.2) and have a molecular mass of approx. 25 000 and approx. 33 000 for mammalian and avian hepadnaviruses, respectively. The nucleocapsid core proteins are not translated from this mRNA since initiation of translation is more efficient at the first than at the second ATG. The 25-kDa precore protein and an intermediate product of 22 kDa, generated by proteolytic cleavage, are not secreted and accumulate intracellularly. The N-terminal 19(20) residue sequence of the 29(30)-amino-acid-long precore sequence of mammalian hepadnaviruses represents a signal peptide targeting the precore protein into the secretory pathway. The signal peptide sequence is cleaved off by a peptidase generating a secreted protein (HBeAg) having 10 additional amino acids at the N-termini as compared with hepadnavirus nucleocapsid proteins (Fig. 19.2,5). Additional proteolytic cleavages occur near the C-terminus (at residue 149 of HBcAg and at related sites within WHBcAg, GSHBcAg and DHBcAg and at additional sites) (Fig. 19.2, 6). resulting in secretion of HBeAg (Takahashi et al., 1983; Ou et al., 1986; Uy et al., 1986; Roossinck and Siddiqui, 1987; Weimer et al., 1987; Bruss and Gerlich, 1988; Garcia et al., 1988; Nassal, 1988; Standring et al., 1988; Jean-Jean et al., 1989). The precore sequence of avian hepadnaviruses is 43 amino acids long and has about 20% homology with the precore sequence of mammalian hepadnaviruses (Ou et al., 1986). The precore sequence is not essential for hepadnavirus assembly and replication as demonstrated by mutational analysis of hepadnavirus gene functions (Roossinck et al., 1986;

		149 ↓	
All HBV subtypes	— T T V V		R R R —
WHV	— H T V I		R R R —
GSHV	— H T V I		R R R —
DHBV	— T T V V	Y G	R R R —

Fig. 19.6. Proteolytic cleavage site within the precore protein sequence corresponding to the C-terminus of HBeAg. Analogous sites within the WHV, GSHV and DHBV precore proteins are also shown (Takahashi et al., 1983). Additional cleavage sites were also identified (Bruss and Gerlich, 1988; Nassal, 1988).

Chang et al., 1987; Junker et al., 1987; Schlicht et al., 1987b). Deletion of the precore region prevents the generation and secretion of HBeAg which is unnecessary not only for in vitro virus replication but also for in vivo infection. Therefore, it is assumed that production of HBeAg plays a role in virus pathogenesis and persistence.

The protease inhibitor pepstatin caused the intracellular accumulation of the 25-kDa precore protein and the 22-kDa intermediate and prevented secretion of HBeAg (Jean-Jean et al., 1989). This suggested the involvement of an aspartyl pro-

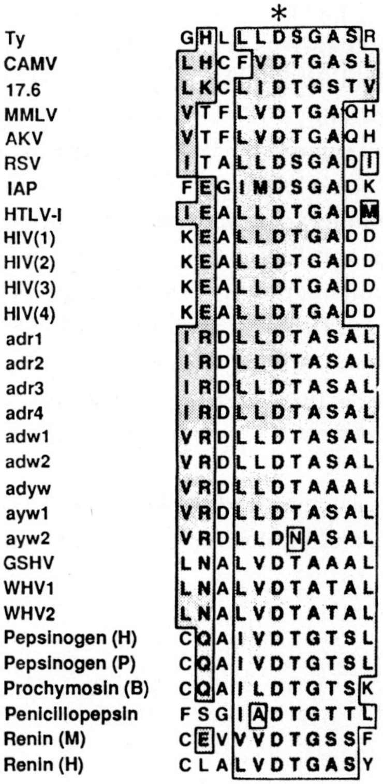

Fig. 19.7. Proposed active sites of hepadnavirus proteases located within the sequences of HBcAg proteins and their comparison with sequences of other cellular and viral aspartylproteases. Shown, from top to bottom, are: yeast transposable element (Ty), cauliflower mosaic virus (CAMV) of plants, *Drosophila* transposable element 17.6, Moloney murine leukemia virus (MMLV), AK virus (AKV), Rous sarcoma virus (RSV), intracisternal A particle (IAP) of mice, human T-cell leukemia virus type I (HTLV-I), four sequences of human immunodeficiency virus (HIV), human hepatitis B virus of surface antigen subtype adr (four sequences), adw (two sequences, adyw, and ayw (two sequences), ground squirrel hepatitis virus (GSHV), two sequences of woodchuck hepatitis virus (WHV), human (H) and porcine (P) pepsinogen, bovine (B) prochymosin, penicillopepsin, and murine (M) and human (H) renin. The asterisk denotes the invariant aspartic acid (D) residue found at the active site of carboxyl proteases. Identical and similar amino acids are enclosed in boxes. The sequence corresponding to HBV subtype ayr (Okamoto et al., 1986) is identical to the HBV adr sequences. Reprinted with permission from Miller, 1987.

tease in the generation of HBeAg. The presence in the sequence of hepadnavirus HBcAg of a segment homologous with aspartyl proteases (Fig. 19.7; Miller, 1987) suggested that a product derived from ORF preC/C may be involved in proteolytic processing resulting in generation of HBeAg. However, results of site-directed mutagenesis disagree with this proposal (Nassal et al., 1989).

Differences between the lengths of amino acid sequences corresponding to hepadnavirus nucleocapsid core proteins, precore proteins and HBeAg (Fig. 19.2) profoundly affect the folding of these proteins and their assembly into higher order structures, even in the absence of viral genomes (pregenomes). The folding of HBcAg was predicted based on primary sequence alignment of hepadnavirus nucleocapsid proteins with the VP3 capsid protein of Mengovirus. The sequence homology suggested an 8-stranded antiparallel β barrel fold for HBcAg (Fig. 19.8; Argos and Fuller, 1988). This protein appears to represent one of 180 capsomers forming a 28-nm core particle capsid according to icosahedral symmetry (P = 3). The individual capsomers have a diameter of approx. 4 nm (Onodera et al., 1982). Recombinant hepadnavirus nucleocapsid core protein assembles spontaneously into similar structures in the absence of viral nucleic acid (Cohen and Richmond, 1982; Miyanohara et al., 1986). Precore protein assembles into similar structures which, however, are labile and dissociate when exposed to 1 M NaCl (Miyanohara et al., 1986). On the other hand, HBeAg (secreted from infected cells and defined as described in Fig. 19.2) does not assemble into similar high-ordered structures and is found in serum of infected individuals predominantly in the form of dimers (Neurath et al., 1979) or associated with IgG (Neurath and Kent, 1985).

Sequence differences between nucleocapsid core proteins and HBeAg and self-assembly of the former protein are the basis for understanding the immunological pro-

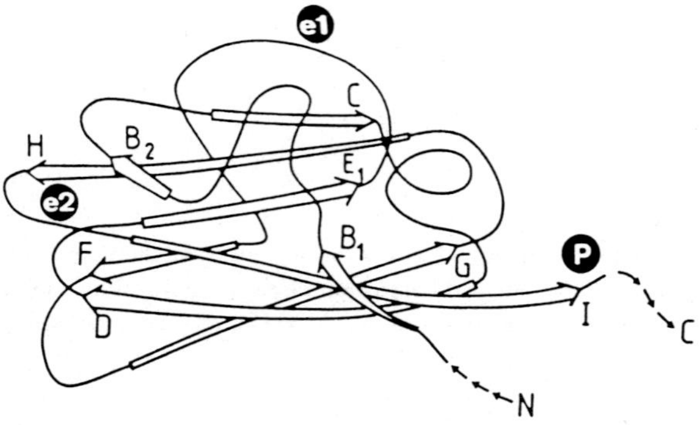

Fig. 19.8. The predicted folding of HBcAg protein. The positions of β-strands are indicated by letter B to I starting from the N-terminus. The location of B-cell epitopes e1 and e2 are indicated. Sites of serine phosphorylation are indicated by P. The nucleic acid-containing interior of the nucleocapsid lies below the figure plane; the external surface of the nucleocapsid corresponds to the upper portion of the structure. The 5-fold axis in a T = 3 icosahedron would be at the left of the figure. N and C indicate the N- and C-termini of the protein sequence. Reprinted with permission from Argos and Fuller, 1988.

perties of these two distinct entities. The nucleocapsid core protein undergoing self-assembly is designated as hepatitis core antigen (HBcAg). Monoclonal antibodies (mAb) raised against HBcAg cross-reacted with each other but did not recognize HBeAg. Anti-HBc positive sera from humans infected with HBV inhibited the reaction of the mAbs with HBcAg. These results indicated the presence of a single immunodominant conformational epitope on HBcAg (Waters et al., 1986a; Ferns and Tedder, 1986). mAbs generated by immunization with sodium dodecylsulfate (SDS) treated HBcAg, leading to dissociation of the nucleocapsid, did not recognize HBcAg and reacted with either of two distinct antibody binding sites on HBeAg. The entire panel of anti-HBe mAbs could be divided into two groups, each of which was cross-reactive and recognized, apparently, a single epitope. These two classes of mAbs together inhibited the reaction between HBeAg and human polyclonal anti-HBe by about 65%, indicating the presence of additional anti-HBe specificities in the human antibodies (Ferns and Tedder, 1984). Anti-HBc mAbs or polyclonal anti-HBc failed to inhibit the reaction between anti-HBe mAbs and HBeAg indicating the presence of immunologically non-cross-reactive epitopes on these two distinct but related products of ORF preC/C. On the other hand, immunization with either HBcAg or HBeAg resulted not only in the production of homologous antibodies but also of cross-reactive antibodies recognizing HBeAg and HBcAg, respectively. The level of these cross-reactive antibodies in polyclonal antisera was much lower than that of antibodies recognizing the immunizing antigen (Neurath and Strick, 1979; Milich et al., 1988b). This indicates that HBcAg and HBeAg share common non-immunodominant epitopes.

To define B cell epitopes on HBcAg and HBeAg, segments of ORF preC/C were expressed in *E. coli* and probed with a panel of polyclonal antisera and mAbs. Results of these studies revealed a single conformational determinant on HBcAg encompassing residues (73–85) and an overlapping linear HBeAg (e1) determinant (Salfeld, 1985; Salfeld et al., 1989) (Table 19.1; Fig. 19.2). A second HBeAg B cell epitope (e2) was assigned to residues (130–140) but required the intramolecular participation of the extended sequence between amino acids 10 and 140 for its expression, i.e., it appeared to be conformational (Table 19.1; Fig. 19.2, 8). In complementary studies, peptides derived from the sequence of HBcAg were tested for reactivity with anti-HBc. Peptides (73–85) and (107–118) were reported to inhibit the reaction between HBcAg and anti-HBc. The latter peptide also elicited anti-HBc antibodies (Colucci et al., 1988) (Table 19.1). The described major epitopes of HBcAg and HBeAg are within regions of ORF perC/C which display variability between distinct hepadnaviruses (Schödel et al., 1989). Nevertheless, the nucleocapsids of HBV and WHV cross-reacted immunologically (Werner et al., 1979; Millman et al., 1982; Stannard et al., 1983). Additional B cell epitopes encompassing HBcAg residues (150–159) from the arginine-rich C-terminal region of HBcAg were identified (Machida et al., 1989; Table 19.1). This B cell epitope appears to be inaccessible to antibodies as reported by Machida et al., 1989 in contradiction with results

TABLE 19.1

Contiguous B- and T-cell epitopes identified on the nucleocapsid (HBcAg) and e(HBeAg) antigens of HBV[a]

Epitopes		
B-cell[b]	T-cell[c]	H-2 haplotypes of murine T-cells recognizing HBcAg peptides[d]
	1–20	**q**
	28–52	**s**
c(73–85),c(74–89),e1(76–89)	70–94	**d**
	85–100	**d**
c(107–118)	**100–120**	**d**,**q**,f
	120–140	**s**,b
	120–131	**s**
e2(130–140)	**129–140**	b
	none	*p*
	none	**k**
c(150–159)		

[a] Derived from: Salfeld, 1985; Salfeld *et al.*, 1989; Colucci *et al.*, 1988; Milich *et al.*, 1987b,c; 1988; Machida *et al.*, 1989.

[b] HBcAg and HBeAg epitopes are indicated by c, e1 and e2, respectively. The numbering of amino acid residues starts at the first M in the sequence of HBcAg. The sequence is described in Neurath and Kent, 1985.

[c] Immunodominant epitopes are printed in bold type.

[d] H-2 haplotypes of high, intermediate and low mouse strain responders to HBcAg are indicated in bold, regular and italics, respectively.

reported by others (Schlicht et al., 1987b). This region of HBcAg also undergoes phosphorylation/dephosphorylation (Fig. 19.8).

Using a series of synthetic peptides, it was possible to locate T cell recognition sites within the sequence of HBcAg (Milich et al., 1987b). HBcAg-specific T cells from a variety of murine strains recognized multiple but distinct epitopes (Table 19.1). Immunization of mice with the appropriate peptides elicited T cells recognizing native HBcAg. It is of interest to note that some of these T cell epitopes overlapped B cell epitopes reported by others (Table 19.1).

Although HBcAg and HBeAg are equally immunogenic (and cross-reactive) at the T cell level, HBcAg elicits much higher levels of antibodies in comparison with HBeAg (Milich et al., 1988b). This can be explained by the observation that HBcAg is both a T cell independent and T cell-dependent antigen since it elicits antibody responses in athymic mice (Milich and McLachlan, 1986) unlike HBeAg. Consequently, HBcAg-nonresponder strains of mice were not identified. These properties make HBcAg a suitable carrier for epitopes of the HBV env protein or of other unrelated viruses (Milich et al., 1987c; Clarke et al., 1987; F. Brown personal communication).

Based on the principle of intermolecular/intrastructural help (Neurath et al., 1987a), HBcAg-specific T helper (T_h) cells provide help to B cells to produce not only anti-HBc but also antibodies to the envelope of HBV when HBcAg-primed animals are challenged with HBV. HBcAg can be replaced by appropriate HBcAg derived synthetic peptides to accomplish the same effect (Milich et al., 1987c). These findings led to the design of synthetic immunogens consisting of T cell epitopes derived from HBcAg and B cell epitopes derived from the preS2 region of the HBV env protein. The hybrid immunogens elicited antibodies recognizing M-protein of the HBV envelope. Immunization with the hybrid immunogen primed T cells in a preS2 nonresponder strain of mice and provided T_h cell functions for anti-envelope antibody production in vivo (Milich et al., 1988a).

HBcAg is a potent immunogen not only for immunized experimental animals but also in humans and animals infected with hepadnaviruses. Consequently, anti-HBc is one of the markers for hepadnavirus infection (Neurath and Kent, 1985). Therefore, it is not surprising to find evidence for anti-HBc immunity at the T cell level in infected humans. Anti-HBc-specific T lymphocytes were detected in peripheral blood in patients with chronic active hepatitis type B (Vento et al., 1985; Ferrari et al., 1986). These cells belonged to the $CD4^+$ (helper/inducer) population. HBcAg-specific T lymphocytes were also detected in hepatic lymphomononuclear cell infiltrates in patients with chronic active hepatitis B. Several HBcAg-specific T cel lines were established from these infiltrates which did not contain T lymphocytes specific for HBV env proteins. These results (Ferrari et al.,. 1987a,b; 1988) indicated that HBcAg is an important sensitizing antigen within the infected liver, in agreement with earlier findings that epitopes coded for by ORF preC/C are exposed on the surface of HBV-infected hepatocytes (Huang and Neurath, 1979; Vento et al., 1985). The HBcAg-specific T lymphocytes belonged to both the $CD4^+$ and $CD8^+$ subsets. The $CD4^+$ cells provided HBcAg-specific help to autologous B cells in producing anti-HBc antibodies *in vitro*. The $CD8^+$ subset had suppressor activity inhibiting both the proliferation of autologous HBcAg-specific $CD4^+$ cells and $CD4^+$ cell augmented anti-HBc production by autologous B cells.

The cytotoxic effect of peripheral blood mononuclear cells from patients with acute HBV infection on infected autologous liver cells was also studied (Mondelli et al., 1987). HBcAg-specific cytotoxicity was ascribed to both T and non-T cells. This indicates that HBcAg is a target for cytotoxic T cells on infected hepatocytes. Such cytotoxic cells appear to play an important role in the pathogenesis of HBV infections. Since HBcAg and HBeAg T cell epitopes are cross-reactive, HBeAg present in serum of infected individuals is expected to have the capacity to modulate the cytotoxic effect of anti-HBc-specific T cells and to contribute to persistence of hepadnavirus infections. This may explain the importance of HBeAg secretion in the natural history of hepadnavirus infections, although HBeAg plays no role in virus assembly and infectivity.

The observation that immunization with HBcAg can prime an immune response to HBV env proteins and that HBcAg-specific cell-mediated cytotoxicity can contribute to eliminating HBV-infected hepatocytes suggested the possibility that preimmunization with HBcAg may modulate or prevent subsequent infection with HBV. This expectation was borne out by immunization of chimpanzees with HBcAg (HBeAg) resulting in partial protection against infection (Murray et al., 1987). The incorporation of HBcAg into a vaccine might be economically feasible since it is possible to obtain high levels of recombinant HBcAg in *Saccharomyces cerevisiae* (Miyanohara et al., 1986; Kniskern et al., 1986).

3. Products of ORF preS/S

As is the case with ORF preC/C, there are several protein products coded for by ORF preS/S, all of which become part of hepadnavirus envelopes. Proteins coded for by each of the hepadnavirus ORFs preS/S are related, share a common C-terminal portion, but differ in their N-termini. Synthesis of these proteins is regulated at both the transcriptional and translational level. The proteins are translated from mRNAs which can initiate either outside ORF preS/S or within this ORF. The abundance of the respective mRNAs differs and is determined by distinct efficiencies of the respective promoters. In addition, a single mRNA species may give rise to more than one protein due to the presence of more than one AUG initiation codon and the distinct efficiencies of translation initiated at these codons (Neurath and Kent, 1988; Price et al., 1988). The smallest protein product of ORF preS/S, designated as S-protein, corresponds to the 3′ region of the ORF and is initiated at the last ATG initiation codon. A larger protein initiated at the next upstream ATG is designated as middle (M) protein, and the largest protein initiating at the first ATG in ORF preS/S is designated as long (L) protein. The relationship of these three proteins to each other is evident from Fig. 19.9. The region spanning the amino acid sequence between the first and second initiator methionine (M) is designated as preS1. The region spanning the amino acid sequence between the second and third initiator M is designated as preS2. The envelope of hepadnaviruses is composed of L-, M-, and S-proteins (Fig. 19.1). These proteins are also present in subviral particles, S-protein being the most abundantly represented and L-protein being the least abundantly represented.

In analogy with HBV, there are three ATG initiation codons in ORF preS/S of all other mammalian hepadnaviruses. These viruses have similarly organized virus envelopes differing from each other in the size of the distinct env proteins (Table 19.2). The spacing and number of ATG initiation codons in ORF preS/S of avian hepadnaviruses is different, resulting in protein species distinct in size from those of mammalian hepadnaviruses (Table 19.2). The protein species present in hepadnavirus envelopes can be discerned by polyacrylamide gel electrophoresis either by stain-

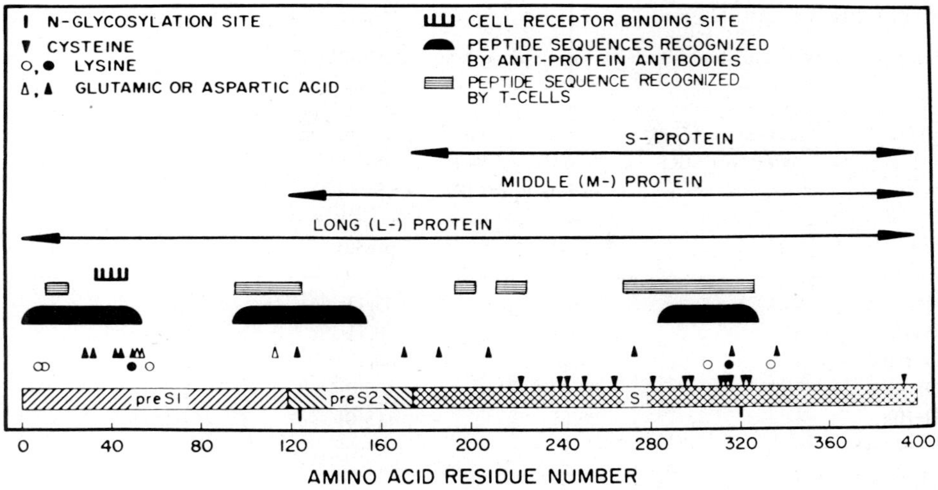

Fig. 19.9. Schematic representation of HBV env proteins and their relatedness. Closed and open symbols correspond to conserved and variable residues, respectively, in sequences of distinct HBV env protein subtypes. Amino acid residues are numbered starting with the first methionine in the preS1 sequence. Residue 175 corresponds to the first methionine in the S-protein sequence (Fig. 19.11).

ing for total protein or by immunoblotting using antisera specific for the preS1, preS2 and S regions of the viral envelope. This is exemplified for HBV in Fig. 19.10. Each of the env proteins exists in forms differing in extent of glycosylation. Therefore, each of the protein species is represented by at least two bands. The number of potential *N*-glycosylation sites on each of the proteins is indicated in Table 19.2.

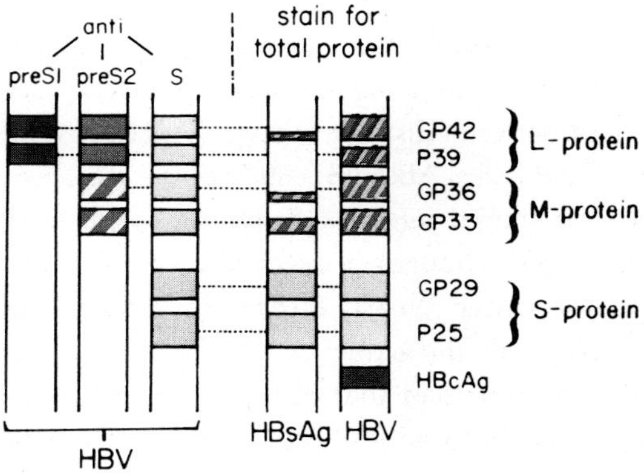

Fig. 19.10. Schematic representation of results describing the separation of HBV env proteins by polyacrylamide gel electrophoresis (PAGE) under denaturing conditions. The proteins of HBV or of subviral HBsAg particles were detected either by a silver stain (two right-hand lanes) or transferred to nitrocellulose membranes and subsequently detected by labeled antibodies specific for preS1 (black), preS2 (dark gray) or S (light grey) sequences, respectively (only results with HBV are shown (three left-hand lanes)).

TABLE 19.2

Size and glycosylation of hepadnavirus envelope proteins coded for by ORF preS/S

Virus	ATG start codons per ORF	Number of amino acid residues and [number of *N*-glycosylation sites] in regions of ORF preS/S		
		preS1	preS2	S
HBV	3	119(108)[a][1][c]	55[1]	226[1]
WHBV	3	149(144)[a][1][d]	60[1]	222[1]
GSHBV	4	146[0]	60[1]	222[1]
DHBV	7(6)[a]	88(52)[b][1][d]	109[2(1)[a]]	167[1]
HHBV	4	50[0]	113[0]	169[1]

[a] Differences due to subtype-dependent variations
[b] Differences due to uncertainty in establishing whether translation starts at 1st or 2nd ATG
[c] *N*-Glycosylation site not utilized
[d] Not present in all subtypes

Before hepadnavirus genomes were sequenced, S-protein was considered to be the only env protein of hepadnaviruses. Subviral particles consisting exclusively of S-protein are sufficient for eliciting protection against hepadnavirus infection. For these historical reasons, the immunochemical properties of S-protein and of the preS regions of M- and L-protein, respectively, will be dealt with separately.

3.1. Hepadnavirus S-proteins

S-proteins of hepadnaviruses infecting distinct species share considerable homology in primary amino acid sequence (Fig. 19.11). The homology between WHBsAg and GSHBsAg is approx. 90%, i.e., higher than between distinct subtypes of HBV. The homology between DHBsAg and HHBsAg is 83.5%. About 67–69% of the sequence of WHBsAg and GSHBsAg, respectively, is shared with the sequences of distinct HBsAg subtypes. About 13–15% of amino acid changes between the sequences of WHBsAg (GSHBsAg) and HBsAg are conservative. Nearly 50% of residues in the sequence of DHBsAg (HHBsAg) are shared with the sequence of HBsAg. The sequence of avian S-proteins is considerably shorter than that of mammalian hepadnavirus S-proteins due to a long deletion between residues (100–157), representing the most hydrophilic region within S-proteins of mammalian hepadnaviruses (Neurath and Kent, 1985) and containing the immunodominant B and T cell epitopes as determined by studies on the HBV S-protein (Fig. 19.9). Noticeable is the conservation of cysteine (C) residues; 12/14 (11/14) of C residues are shared between S-proteins of HBsAg and WHBsAg (GSHBsAg). C residues at positions 76 and 221 ap-

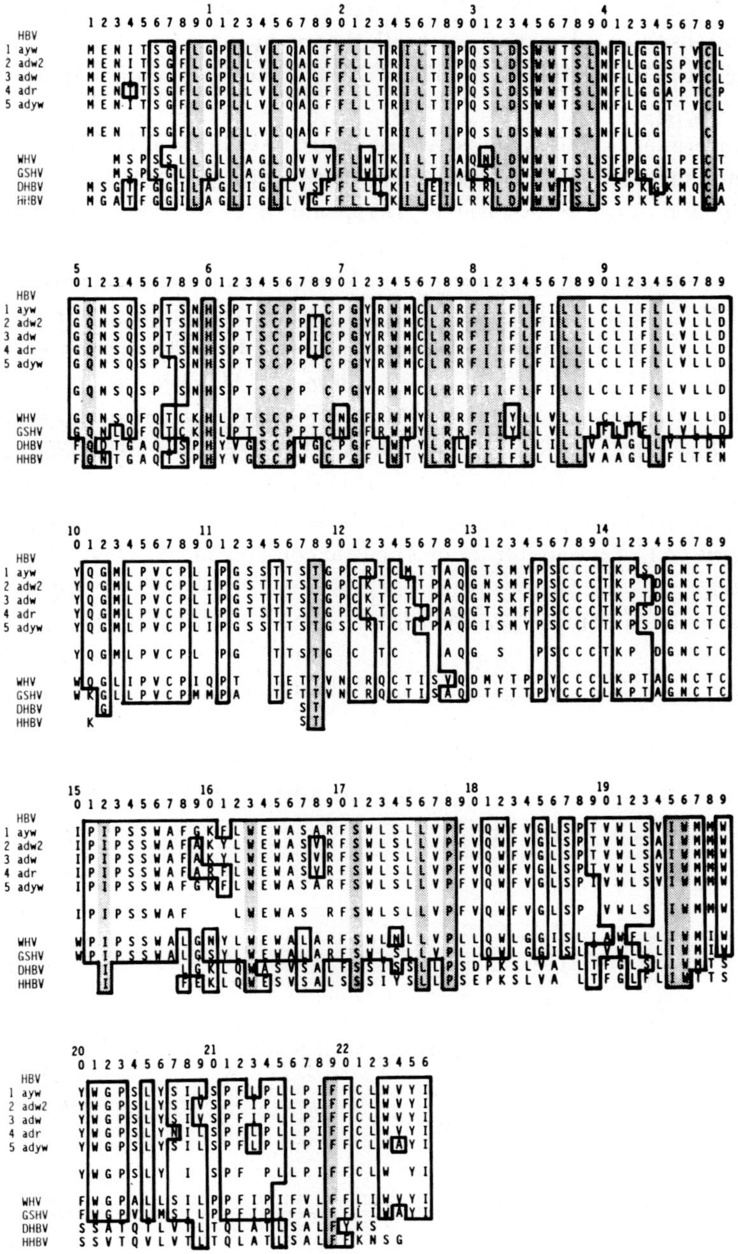

Fig. 19.11. Comparison of S-protein amino acid sequences of selected subtypes of HBV (Neurath and Kent, 1985), and of WHV (Galibert et al., 1982), GSHV (Seeger et al., 1984a), DHBV (Mandart et al., 1984), and HHBV (Sprengel et al., 1988). Consensus sequence for HBV is in line 6. Boxed areas indicate sequences shared between any of the HBV subtypes and WHV and/or GSHV (DHBV, HHBV). Shaded areas indicate sequences common for all hepadnaviruses.

pear in HBsAg only. However, there is a S→C change at position 58 in the sequence of WHBsAg and GSHBsAg in comparison with other hepadnavirus S-proteins. The first three C residues at positions 48, 65 and 69 are shared between mammalian and avian hepadnaviruses.

The homology in primary amino acid sequence between HBsAg, WHBsAg and GSHBsAg is reflected in serological cross-reactivity between these S-proteins as determined by polyclonal antisera (Werner et al., 1979; Marion et al., 1980; Feitelson et al., 1981; Millman et al., 1982; Stannard et al., 1983) and by monoclonal antibodies (Cote et al., 1982; Cote and Gerin, 1983). A relationship between HBsAg, WHBsAg and GSHBsAg was also discerned by comparative tryptic peptide mapping (Gerlich et al., 1980; Feitelson et al., 1981). It is noteworthy that sequences S (139–147), (representing dominant B and T cell epitopes of HBsAg (see below)), corresponding to distinct mammalian hepadnaviruses are related, except amino acid replacements at residues 140 (T→L) and 144 (D→A). This sequence also contains the single *N*-glycosylation site within mammalian hepadnavirus S-proteins (residue 146; Fig. 19.11). The *N*-linked oligosaccharides of HBsAg consist of two major sialylated components (A1, A2) and a trace of a neutral component (N) (Fig. 19.12) (Hanaoka et al., 1986).

Related epitopes between HBsAg and WHBsAg explain the elicitation of anti-

<u>N</u>

Gal β1→4GlcNAc β1→2Man α1 ↘
6
Man β1→4GlcNAc β1→4GlcNAc
3
Gal β1→4GlcNAc β1→2Man α1 ↗

<u>A 1</u>

Gal β1→4GlcNAc β1→2Man α1 ↘
6
Man β1→4GlcNAc β1→4GlcNAc
3
NeuAc α2→6Gal β1→4GlcNAc β1→2Man α1 ↗

<u>A 2</u>

NeuAc α2→6Gal β1→4GlcNAc β1→2Man α1 ↘
6
Man β1→4GlcNAc β1→4GlcNAc
3
NeuAc α2→6Gal β1→4GlcNAc β1→2Man α1 ↗

Fig. 19.12. Proposed structures of *N*-linked saccharide chains of HBV S-protein.

HBs in chimpanzees immunized with WHsAg and the partial protection of these chimpanzees upon challenge with HBV (Cote et al., 1986). On the other hand, woodchucks immunized with HBsAg developed anti-HBs which did not cross-react with WHsAg. These animals developed acute infections upon challenge with WHV.

The host range of hepadnaviruses is determined by their env proteins. Despite similarities in S-protein, WHV and GSHV are infectious only for the homologous host. To determine the contribution of distinct regions of the virus env proteins to the host range of these viruses, the infectivity for ground squirrels of viruses containing recombinants of GSHV and WHV DNAs was assayed. GSHV recombinants having 76% and 29%, respectively, of their S-protein sequence replaced by the WHV S-protein sequence proved to be infectious. These results suggested that the major host range determinants are not encoded by the S gene and are thus probably located within the preS region of hepadnavirus env proteins (Seeger et al., 1987).

Unlike secretion of HBeAg which requires a cleavable signal peptide sequence, hepadnavirus S-proteins (HBsAg) are secreted without removal of a signal sequence or any proteolytic cleavage. S-Protein is initially synthesized as an integral membrane protein. The transmembrane form of S-protein is stepwise converted through intermediate stages into an assembled, secreted lipoprotein complex consisting of approximately 100 S-protein molecules (containing both glycosylated and nonglycosylated forms of S-protein), and having a diameter of approx. 22 nm. This process occurs without involvement of any other viral protein (Simon et al., 1988) and is accomplished due to two signal sequences within S-protein. The first signal encompasses residues (7–22) and confers a secretory phenotype to downstream domains of S-protein. The second signal, located between residues 84 and 136, probably centered around residues (80(84)–98), is involved in the translocation of downstream sequences and confers a transmembrane orientation to S-protein (Eble et al., 1987). Both signal sequences are hydrophobic and relatively conserved within mammalian hepadnaviruses and, to a lesser extent, within all hepadnaviruses.

Dominant B cell epitopes on HBsAg depend on the maintenance of disulfide bonds. Complete reduction and alkylation of HBsAg leads to an antigen morphologically indistinguishable from native HBsAg which, however, only partly cross-reacts immunologically with native HBsAg and acquires B cell epitopes which are not present on the native particle (Neurath, 1989; Neurath and Strick, 1989). All cysteine residues in HBsAg are involved in disulfide bonds. Two to four disulfide bonds are intermolecular (probably involving residues 121, 124 and 221, and possibly residue 48), while the other 10 cysteine residues are presumed to be involved in intramolecular disulfide bonds (Guerrero et al., 1988). The formation of intermolecular disulfide bonds is essential for the immunogenicity of HBsAg (Wampler et al., 1985) and prevents the dissociation of polymeric HBsAg into S-protein monomers after treatment with SDS. The formation of the approx. 22nm HBsAg particles is independent of whether S-protein is or is not glycosylated. No evidence exists for the preferential dimerization of glycosylated and nonglycosylated S-protein subunits. Both the N-terminus and the C-terminus of S-protein is exposed on HBsAg particles, as is the central hydrophilic region (Guerrero et al., 1988). Site-directed mutagenesis of cysteine 121 to serine resulted in HBsAg with altered immunological properties, as determined by reactivity with mAb to group-specific *a*-determinants. The altered HBsAg was not recognized by mAb recognizing subtype-specific *y*-determinants (in which arginine (R) 122 plays an essential role; Fig. 19.11, and see below) (Antoni and Peterson, 1988). It is of interest to note that WHsAg, GSHBsAg, DHBsAg and HHBsAg are structurally similar to HBsAg, yet the arrangement of disulfide bonds in these antigens cannot be identical to that of HBsAg since the two non-human mammalian antigens lack cysteine 221 and the avian antigens lack cysteines 121, 124 and 221, presumed to be involved in intermolecular disulfide bonds. Further studies will be required to establish the location of cysteine residues involved in intermolecular disulfide bonds within non-human hepadnavirus S-protein complexes.

3.2. HBV SUBTYPES

The approx. 22nm HBsAg particles are the most abundant HBV antigen present in sera of infected individuals and S-protein is the most abundant constituent of these particles. Serological reactions in which human or animal antibodies to HBsAg (anti-HBs) and HBsAg were used revealed four major antigenic subtypes which were further subdivided (as indicated in parentheses) to yield the following subtypes: *adw (adw2, adw4); adr (adrq*$^-$*, adrq*$^+$*); ayw (aywl, ayw2, ayw3, ayw4)* and *ayr* (Neurath and Kent, 1985). In this scheme, the letter *a* denotes antigenic specificities common to all HBV subtypes; the letters *d/y* and *w/r*, respectively, denote pairs of mutually exclusive specificities characteristic for the antigenic subtypes; numbers 1–4 further subdivide the *w* specificity, and the presence or absence of *q*-determinants furthers subdivides the *adr* subtypes. These subtypes are the result of antigenic differences between S-proteins and are directly related to amino acid replacements within the S-protein sequence. The *a* specificity can be attributed to more than a single epitope (Neurath and Kent, 1985). Discontinuous epitopes with *a* specificity elicit virus-neutralizing antibodies (Neurath and Kent, 1985; Iwarson et al., 1985; Waters et al., 1986b).

Subtype specificities *d/y* and *w/r* can be attributed to amino acid replacements K/R at positions 122 and 160, respectively, in the S-protein sequence (Fig. 19.11, Table 19.3; Okamoto et al., 1987a,b). Thus, single point mutations at appropriate positions in gene S result in alterations of HBV subtype specificity. Rare mutants lacking *w/r* subtype specificities in wild-type HBV or generated by site-directed mutagenesis have been found. These mutants have neither K nor R at position 160; instead, there is asparagine (N) at this position. A wild-type HBV mutant lacking the *d/y* specificity has also been found. Unexpectedly, this mutant had a K residue at position 122, but glutamic acid (E) and alanine (A), respectively, at residues 144 and 145 (compare with sequence in Fig. 19.11). Site-directed mutagenesis of the virus genome resulting in amino acid changes E → D and A → G at position 144 and 145 imparted *d* specificity to this virus isolate. This indicates that amino acid changes distal from residue 122 can affect *d*/*y* subtype specificity (Tachibana et al., 1989; Okamoto et al., 1989). The relationship between amino acid alterations within the S-protein sequence and HBV subtypes (excluding rare isolates) is summarized in Table 19.3. Extensive sequencing of HBV genomes corresponding to identical antigenic subtypes revealed, in some cases, sequence divergence ⩾8%, i.e., similar to differences between distinct subtypes. Therefore, antigenic subtyping does not reflect true genotypic variations in coding sequences for S-protein (Okamoto et al., 1988). The sequence variability between distinct HBV genomes is considerable in comparison with genomes of other DNA viruses. The estimated rate of nucleotide substitutions in HBV DNA is 10^4-fold higher than that in DNAs of other DNA viruses but about 10^2-fold lower in comparison with the envelope gene of the human immunodeficiency virus-1.

TABLE 19.3

Variations in the amino acid sequences coded for by the S gene that correspond to allelic determinants and subtypes.

	Subtype												Variation	
No.†	*ayr*	*adr*						*ayw*			*adw*		Allelic	Subtypic
3	S	N	N	N	N	S	S	N	N	N	N	N		
4	T	T	T	T	T	T	T	I	I	I	I	I	*w*/*r*	
44	G	G	G	G	E	E	G	G	G	G	G	G		
45	A	A	A	A	A	A	A	T	T	T	S	S		
46	P	P	P	P	P	P	P	T	T	T	P	P		*yw*
47	T	T	T	T	T	T	T	V	V	V	V	V	*w*/*r*	
49	P	P	L	P	P	P	P	L	L	L	L	L		
57	T	T	T	T	T	T	T	T	T	I	T	T		
68	T	I	I	I	I	I	I	T	T	T	I	I	*d*/*y*	
110	L	L	L	L	L	L	L	I	I	I	I	I	*w*/*r*	
112	G	G	R	G	G	G	G	G	G	G	G	G		
113	T	T	T	T	T	T	T	S	S	S	S	S	*w*/*r*	
114	S	S	S	S	S	S	S	S	S	S	T	T		*dw*
120	P	P	P	P	P	P	P	P	P	S	P	P		
122	R	K	K	K	K	K	K	R	R	R	K	K	*d*/*y*	
125	T	T	T	T	T	T	T	M	T	T	T	T		
126	I	I	I	I	I	I	I	T	T	T	T	T	*w*/*r*	
127	P	P	P	P	P	P	P	T	P	P	P	P		
131	T	T	T	T	T	T	T	T	T	I	N	N		
133	M	M	M	M	M	M	M	M	M	M	M	K		
134	F	F	F	F	F	F	F	Y	Y	Y	F	F		*yw*
143	S	S	S	S	S	S	S	S	S	S	T	T		*dw*
159	A	A	A	A	A	A	A	G	G	G	A	A		*yw*
160	R	R	R	R	R	R	R	K	K	K	K	K	*w*/*r*	
161	F	F	F	F	F	F	F	F	F	F	Y	Y		*dw*
165	W	W	W	G	W	W	W	W	W	W	W	W		
168	V	V	V	V	V	V	V	A	A	A	V	V		*yw*
184	V	V	V	V	V	A	A	V	V	V	V	V		
189	T	T	T	T	T	T	T	T	T	I	T	T		
194	A	V	V	V	V	V	V	V	V	V	A	A		
207	N	N	N	N	N	N	N	S	S	S	S	S	*w*/*r*	
209	L	L	L	L	L	L	L	L	L	L	V	V		*dw*
213	L	L	L	L	L	L	L	L	L	L	I	I		*dw*
224	V	V	V	V	V	V	V	V	V	A	V	V		

First vertical column indicates amino acid positions from the N-terminus of S-protein. Reprinted with permission from Okamoto *et al.*, 1986.

3.3. B cell epitopes on S-protein

The dimer of S-protein containing the appropriate intra- and inter-subunit disulfide bonds is probably the minimum structure carrying all dominant discontinuous epitopes on S-protein (Neurath and Kent, 1985). In attempts to define epitopes on S-protein, antisera to synthetic peptides from the S-protein sequence were assayed for reactivity with HBsAg particles. Antibodies to most of these peptides, except to those derived from the most hydrophobic segments of the S-protein sequence (residues (82–95) and (186–211), reacted with HBsAg (Neurath and Kent, 1985). However, only peptides derived from a more restricted region of the S-protein sequence (109–147) were shown to elicit virus-neutralizing antibodies (Fig. 19.9). In particular, immunization with peptide (110–137) (cyclized by a disulfide bond between cysteines 122/124 and 137) was reported to protect chimpanzees against challenge with infectious HBV. Similarly, peptide (109–137) from the WHV S-protein elicited in woodchucks antibodies protective against WHV challenge (Gerin et al., 1985; 1986). A cyclic peptide S(139–147) bound antibodies to intact S-protein with an affinity similar to that determined for the reaction between HBsAg particles and human anti-HBs. The same peptide, when polymerized with glutaraldehyde, elicited antibodies reacting with native HBsAg with affinity values similar to those found for antibodies from individuals immune to hepatitis B (Neurath, 1989). The sequence (139–147) appears to be a segment of a dominant epitope, since the peptide was able to adsorb a major portion of anti-S-protein antibodies from polyclonal anti-HBs antisera. The peptide also encompasses a dominant T cell epitope (see below).

To establish the role of disulfide bonds between particular cysteine residues within the sequence S(121–151) in eliciting anti-peptide antibodies cross-reactive with native S-protein, a series of peptides with this sequence having five out of the seven cysteines replaced by serines was synthesized, cyclized and used to prepare antisera. The highest levels of antibodies recognizing HBsAg were obtained with peptides expected to have disulfide bonds between residues (121–138), (124–138) and (139–147) (Neurath, 1989; Neurath and Strick, 1989).

3.4. T cell epitopes on S-protein

To rationally design second-generation recombinant and/or synthetic HBV vaccines, an understanding of the regulation of the immune response at both the B and T cell level is essential. Thus far we have considered responses to S-protein at the B cell level. This section reviews the identification and characterization of T cell recognition sites within S-protein using appropriate examples from both the mouse and human system.

Peripheral blood lymphocytes (PBL) from donors who had been vaccinated with three doses of the plasma-derived vaccine from Merck, Sharp and Dohme (MSD), were induced to proliferate by HBsAg mostly only after in vitro restimulation with HBsAg (Celis et al., 1984). The antigen-reactive lymphocytes from one donor with the highest proliferation index in response to HBsAg were cloned by limiting dilution. Several clones were obtained which grew well and proliferated selectively when exposed to HBsAg of both *ad* and *ay* subtypes. Phenotypically, all the clones were $CD3^+$, $CD4^+$, $CD8^-$ indicating that they may belong to the helper/inducer class. In fact, subsequent experiments confirmed that two of the clones also provided help in the production of anti-HBs antibodies by an autologous cell population enriched for HBsAg-binding cells.

Overall, the degree of in vitro lymphocyte proliferation varied greatly among different individuals and did not correlate with titers of anti-HBs antibodies. Similar observations were made by one of the authors (Y. Thanavala et al., unpublished data) using both PBL and isolated $CD4^+$ cells from vaccinated donors and from individuals who became anti-HBs-positive following recovery from HBV infection.

Celis et al. (1984) reported that without in vitro restimulation with HBsAg the response of fresh PBL obtained from a panel of vaccinated donors was generally very low. They suggested that this might have resulted from a low frequency of HBsAg-reactive T cells in the circulation of these individuals. In contrast, in our laboratory (Pride et al., 1988; Thanavala et al., 1989) we have always observed good HBsAg-specific proliferation of T cells from all vaccinated donors (after the third dose of vaccine). Celis et al. (unpublished results) found in some instances that higher T cell proliferative responses were observed when lymphocytes were collected from the donors later (2–6 months after the third vaccine dose), possibly indicating that the frequency of HBsAg-reactive cells in the circulation changes during the course of immunization and that these changes may vary among individuals. The absence of a T cell proliferative response in vitro by PBL from some individuals might have been due to the presence of suppressor cells or of factors that inhibited the growth of HBsAg-reactive cells. These potential problems can be circumvented by performing all in vitro proliferation experiments using $CD4^+$ T lymphocytes rather than unfractionated cells.

Using a number of synthetic peptides, Celis et al. (1988) examined the antigenic sites on HBsAg recognized by human cloned T cell lines and found a major T cell determinant near the N-terminal end of S-protein. Of ten overlapping peptides tested only one peptide S(4–33) gave strong in vitro proliferative responses when tested against three cloned T cell lines and three short-term T cell lines (stimulated previously only once with HBsAg in vitro). The cloned T cells were from a single donor albeit independently prepared at distantly separated time points. A T cell clone obtained from another vaccinated donor responded to HBsAg but failed to respond to S(4–33) but did proliferate modestly with peptide S(202–222). The response to this peptide was only 14% of the total response to HBsAg itself.

Studies were also undertaken with smaller overlapping peptides from the S(4–33) region. The results suggested the existence of only one T cell determinant in this region. Peptides S(4–33), S(14–33), S(19–33) and S(19–28) all gave equivalent proliferative responses, at approximately the same concentrations of peptides used. With peptide S(24–33), 100–500 times more peptide was required to efficiently stimulate the T cell clone. Peptide S(4–25) failed to trigger any proliferation at all. By comparing results obtained with all peptides, it is evident that residues S(26–28) are critical for this particular T cell epitope.

Two algorithms have been proposed to identify the amino acid sequence of antigens likely to be involved in T cell recognition:

(a) DeLisi, Berzofsky and colleagues (DeLisi and Berzofsky, 1985; Margalit et al., 1987) have observed that a large number of T lymphocytes appear to react with helical amphipathic structures contained within the sequences of protein antigens.

(b) The studies of Rothbard, McMichael, Townsend and colleagues (Townsend et al., 1986; Bastin et al., 1987) point to the existence of a 4–5-amino acid sequence motif in a large number of T cell epitopes. The motif is composed of a glycine or a charged amino acid residue in the first position, followed by two or three hydrophobic residues, which are followed in turn by either a glycine, a polar or a charged residue in the last position.

The peptide S(4–33) has a sequence that agrees with both predictions. By using a computer program (AMPHI), DeLisi and Berzofsky determined that residues S(23–39) have the highest amphipathic score within S-protein. Consistent with the Rothbard model is the fact that all of the peptides of the S(4–33) region that stimulate T cells contain the proposed four amino acid sequence motif (Fig. 19.11). The pertinent residues are RILT. Peptide S(4–25), which failed to stimulate any HBsAg-reactive T cells, lies mostly outside of the helical amphipathic region S(23–39) and lacks two of the four residues of the motif sequence. It should be noted that amino acid residues bordering the RILT sequence also play an important role in this epitope.

Studies in our own laboratory (Y. Thanavala) have shown that using peripheral blood lymphocytes of vaccine recipients (immunized with HeptaVax B or RecombiVax HB from MSD), good in vitro T cell proliferative responses to both HBsAg and a synthetic peptide corresponding to residues S(139–147) are always observed (see section 3.5). The peptide driven responses are however only elicited after the third vaccine dose. However, early during immunization HBsAg-specific responses at the T cell level are already apparent, the first detectable response always coinciding with the appearance of anti-HBs antibodies (Thanavala et al., unpublished results).

Milich et al. (1985) reported proliferation of lymphocytes from two human vaccinees by native HBsAg and by peptides S(110–137), S(38–52) and S(48–81). A number of other peptides tested (from both the N- and C-termini of S-protein) failed to elicit any in vitro stimulatory response.

The response of human T cells appears to be strickly regulated by the molecular form of HBsAg. Celis et al. (1988) reported that formic acid denatured HBsAg and peptide S(4–33) were both three orders of magnitude more efficient in eliciting proliferative T cell responses than native HBsAg. Thus there is a clear difference between T cell and B cell recognition of S-protein. It has been well established that both group specific and subtype-specific antibodies to S-protein react marginally or not at all with denatured forms of HBsAg (section 3.3). Yet T cell recognition is more efficient in the same antigen system using denatured or 'processed' (peptide) antigen forms. Similar evidence is also found using mouse T cells.

Murine humoral responses to HBsAg are regulated by H-2 linked Ir genes (Milich and Chisari, 1982). HBsAg high-responder and non-responder haplotypes were identified. Using representative high and non-responder mouse strains, studies were undertaken to assess specific in vitro T lymphocyte proliferative responses to HBsAg S-protein. Lymph-node cells obtained from immunized high-responder ($H2^q$) C3H.Q strain mice demonstrated HBsAg-specific in vitro T cell proliferative responses to both the *a* and *d* determinants (Milich et al., 1983). The in vitro response to the group specific *a* determinant was of lower magnitude and occurred later that the *d*-specific response (this paralleled the serum antibody response seen in vivo in this strain). When identical studies were performed using lymph node cells from the HBsAg non-responder SJL mouse, the earliest significant HBsAg-specific lymphocyte proliferation was only observed at later time intervals (day 20 following immunization). On further examination it was found that the proliferating cells were primarily B cells. However, the B cell proliferation was T cell-dependent, since anti-Thy 1.2 treatment also abrogated all the B cell-mediated proliferation. Footpad immunization (but not intraperitoneal injection) of this mouse strain with HBsAg does result in a delayed anti-HBs response which is subtype-specific. Thus, non-specific T cell help generated in the draining lymph node after footpad immunization with HBsAg in complete Freund's adjuvant (CFA) appears sufficient to elicit both B cell proliferation and anti-HBs production (only subtype-specific) in this non-responder mouse strain. Therefore, by analogy it might be possible to circumvent anti-S non-responsiveness in humans by strategies that enhance non-specific T helper function.

To examine in detail the role of H-2 linked Ir genes in regulating the response to S-protein at the T cell level, HBsAg-specific T cell proliferative responses of the H-2 congenic B10.T(6R) responder strain, the B10.S(7R) non-responder strain and (responder × non-responder) F_1 mice were analyzed. T cells from B10.T(6R) mice demonstrated significant *d*-specific proliferation and also some *a*-specific proliferation. Much higher concentrations of antigen were needed to stimulate the *a*-specific response. HBsAg-specific T cell proliferation was not detectable in the B10.S(7R) mice (Milich et al., 1984). The in vitro proliferation studies were strengthened by measuring in vitro IL-2 production of HBsAg-primed lymph node cells. The subtype-specific IL-2 response was superior to the group-specific response and this paralleled the T cell proliferation response pattern (Milich et al., 1985b). Thus for example, HBsAg primed SJL/J lymph node T cells produced minimal IL-2 after challenge with HBsAg which is consistent with the lack of a T cell proliferation response in vitro of this strain and its classification as a non-responder following i.p. immunization with HBsAg.

Overall, the HBsAg-specific T cell proliferative response of H-2 congenic and intra H-2 recombinant strains generally paralleled in vitro anti-HBs production in terms of the Ir genes involved and the hierarchy of response status among H-2 haplotypes. The correlation was however not absolute. Several strains capable of producing group-specific anti-S antibodies did not demonstrate a group-specific T cell proliferative response to HBsAg. This is exemplified by the lack of *a*-specific T-cell proliferative responses in the B10.A(4R) and B10.S(8R) strains which do produce low levels of anti *a*-specific antibodies after primary in vivo immunization. Furthermore B10.BR and B10.A strains also fail to produce *a*-specific T cell responses in vitro but produce anti-*a*-specific antibodies upon secondary immunization *in vivo*. These observations suggest that subtype-specific proliferating T cells can provide help to group-specific (and subtype-specific) B cell clones to produce group-specific antibody *in vivo*. The alternative possibility is that non-proliferating group-specific T helper cells also exist. The particular assay system employed by Milich et al. does not allow for their discrimination. Such a dichotomy between T helper and T proliferating cells has been described for other antigen systems (Peterson et al., 1983: Shevach and Rosenthal, 1973.

Lymphocytes from the F1 mice showed a proliferative response that was equivalent to that obtained using responder parental T cells. Furthermore only antigen presenting cells (APC) of the parental responder or the F1 itself functioned efficiently. No T cell response was observed if parental non-responder strain APC were used (Milich et al., 1985b). These results point to a defect in S nonresponder mice at the level of T cell-APC interation.

The studies of Milich et al., (1985a,b) suggest that there are clear differences between T and B cell recognition sites on S-protein and that epitopes on S-protein recognized by T and B cells do not necessarily overlap. Most antibody binding sites are located on the C-terminal half of S-protein (Fig. 19.9), while T cell sites were found on both the C- and N- terminal halves of the sequence. Similarily as with human T cells, murine T cell activation was induced by denatured S-protein. Reduced and alkylated HBsAg, subunit polypeptide P25, its tryptic fragments (both N- and C-terminal fragments) and a number of synthetic peptides from S-protein were all capable of eliciting in vitro T cell proliferation in murine lymph node cells. Using a series of H-2 congenic strains, multiple T cell sites were found and the particular site recognized depended on the H-2 genotype of the mouse strain. Thus for example in HBsAg immunized C3H.Q mice, peptides S(38–52) and S(95–109) both elicited in vitro T cell proliferation. Both these peptides are located on the N-terminal tryptic fragment of the P25 polypeptide (the N-terminal tryptic fragment was designated P25-1 and corresponded to amino acid residues 1–122 and the C-terminal tryptic fragment was designated P25-2 and corresponded to residues 123–226). In the congenic C3H mouse, peptides S(140–154) and S(110–137) were effective. In summarizing the results of T cell responses obtained in a number of H-2 congenic strains, three types of H-2 genotype-dependent T cell responses were observed: (1) P25-1 specific responses predominate; (2) P25-2 specific responses predominate; (3) P25-2 specific responses were exclusively obtained.

The first type of response was exemplified in strains classified as high to intermediate responders which show group and subgroup-specific T cell proliferation and primary anti-S antibody responses. The second type of response was seen in intermediate to low responder strains which demonstrate subtype-specific T cell responses and primary anti-HBs production in vivo (the primary group-specific response tends to be variable). The third type of response occurred in mice which show subtype-specific T cell proliferation and secondary subtype-specific antibody responses. These animals do not mount primary or secondary group-specific antibody production.

Comparison of B and T cell epitopes defined by studies with synthetic peptides derived from the sequence of ORF preC/C (Table 19.1) and of S-protein indicate that dominant B and T cell epitopes overlap. Similarly, B and T cell epitopes derived from the preS region of HBV env protein (see below) are located at adjacent positions with a probable partial overlap. This is in agreement with a recent finding suggesting that sequences forming part of (conformation-dependent) antibody recognition sites and linear sequences recognized by T cells correspond to overlapping segments of virus env proteins and that antigen-specific B memory cells may be instrumental in selection of the repertoire of T cells (Barnett et al., 1989; Graham et al., 1989). These findings cast doubt on the usefulness of synthetic peptides containing B and T cell epitopes from non-adjacent regions of virus env proteins or from distinct structural components of the same virus (envelope-nucleocapsid) (Milich, 1987; Milich et al., 1987c; Milich et al., 1988a) if peptides mimicking adjacent B and T cell epitopes of a protein can be synthesized.

3.5. Mimicking of S-protein B and T cell epitopes by anti-idiotypic antibodies

The hypervariable regions of immunoglobulins functioning as antigen combining sites are themselves capable of provoking an antibody response. These antigenic determinants on immunoglobulins are designated 'idiotypes', and antibodies directed against them 'anti-idiotypes' (anti-Id). In circumstances where there is a close complementarity of shape between antigen and idiotype, and between idiotype and anti-Id, it can be predicted that anti-Id would represent an 'internal image' of the original antigen.

'Internal image' anti-Ids are expected to mimic specific B-cell epitopes on antigens. This has been also well documented for B cell epitopes on S-protein. Immunization of rabbits with immunochemically purified antibodies to S-protein led to the production of anti-Ids recognizing a common idiotype on human and animal anti-HBs (Kennedy and Dreesman, 1983; Kennedy et al., 1984a). The idiotype-anti-Id reaction was inhibited by a cyclic synthetic peptide S(122–137). Reduction and alkylation abolished the inhibitory activity of this peptide, indicating that the anti-Id mimicked a conformation-dependent group *a* epitope on S-protein (Kennedy et al., 1983a). Preimmunization of experimental animals with the anti-Id enhanced the immune response to S-protein and to the synthetic peptide S(122–137) indicating that anti-Id can prime an immune response to epitopes on S-protein (Kennedy et al., 1983b; 1984b).

Others (Thanavala et al., 1986a) used for the production of monoclonal anti-Ids a syngeneic monoclonal idiotype which was raised in BALB/c mice to HBsAg and recognized an *a* determinant on S-protein. Four of the monoclonal anti-Ids produced were further characterized. They were able to inhibit the binding of HBsAg to the idiotype, suggesting that they recognize idiotypic determinants very close to, if not identical with the paratopic regions of the combining site.

If the anti-Id is behaving as an internal image, then, like antigen it should bind polyclonal anti-HBs sera raised in different species. Two of the mouse anti-Ids (4D4 and 2F10) reacted with all polyclonal anti-HBs sera from distinct species tested. Similar results were reported for rabbit anti-Id (Kennedy et al., 1984a). Selective adsorption of antibodies to the *a* but not the *d* or *y* determinants from the polyclonal antisera resulted in removal of antibodies reacting with the anti-Ids showing that the internal image reactivity was restricted to a part of HBsAg as would be predicted since the anti-Id should mirror only one of the epitopes on the original antigen (Thanavala et al., 1986a).

This epitope encompasses residues S (139–147) since the internal image anti-Ids inhibited the binding of both monoclonal and polyclonal anti-HBs antisera to the cyclical and linear forms of this peptide (Thanavala et al., 1986b). The binding of pooled human anti-HBs IgG and of a monoclonal anti-HBs to the cyclical form of another peptide S (124–137) was not inhibited by the anti-Ids, thus providing evidence for the specificity of the anti-Ids. The binding of the linear S (139–147) peptide by the monoclonal and polyclonal anti-HBs was more readily inhibited by the anti-Id than was the binding of the cyclical peptide. This is consistent with a lower affinity of anti-HBs for the linear form. These results suggest that the cyclized peptide has a reasonable similarity to the native determinant on S-protein.

The murine response to *a* determinants of S-protein is H-2 restricted. It was of interest to determine if this haplotype restriction also applied to the anti-HBs response induced by the monoclonal anti-Id. Therefore, mice of two responder (BALB/c, SWR/J) and one non-responder (B10.M) strain were immunized with either HBsAg or with the monoclonal anti-Id 2F10 coupled to keyhole lympet hemocyanin (KLH) and absorbed on alum. The non-responder B10.M strain injected with HBsAg did not make an anti-HBs response but the SWR/J and BALB/c mice did make a strong and sustained antibody response, confirming earlier findings by others. However, the B10.M mice injected with the monoclonal anti-Id did make anti-HBs. The response was specific for the *a* determinant and the anti-HBs antibodies carried the idiotype to which the immunizing anti-Id had been raised.

Although it has been shown that B and T cells of the same antigen specificity can bear serologically similar idiotypic determinants, such T cells do not transcribe IgV_H gene segments related to those found in B cells (Yanagi et al., 1984). Nonetheless, antibodies directed to idiotypic determinants on specific antigen-binding antibodies can influence specific T cell reactivity. Support for this concept comes from a series of well documented experiments (Sy et al., 1980; Eichman and Rajewsky, 1975; Thomas et al., 1983). Thus, despite the absence of Ig on T cells, anti-Id appears to bind to some activation-related structures on certain T cell subsets. Consequently, it was of interest to determine whether internal image anti-Id was capable of stimulating T cells. Mice were immunized with the purified monoclonal anti-Id in the hind footpads and the draining lymph nodes were collected. Nylon wool purified T cells were obtained and stimulated in vitro for varying periods of time with either 2F10, and $IgG_1\kappa$ control antibody, HBsAg or medium alone. The results shown in Table 19.5 show that the purified T cells were stimulated to proliferate not only by the anti-Id used for immunization but also by the antigen which it mimics, i.e.,HBsAg. In reciprocal experiments, mice were immunized with HBsAg. Results shown in Table 19.4 indicate that the internal image anti-Id efficiently stimulated in a specific manner T cells from animals immunized with HBsAg.

The synthetic peptide S (139–147) was also used in these experiments. T cells from mice primed in vivo with anti-Id or HBsAg were stimulated in vitro by this peptide (Table 19.4,5). These results indicate that the anti-Id mimicking an *a* determinant of S-protein can indeed prime appropriate T cells specific for this determinant. The sharing of idiotypes of related specificity between T and B cells would permit the use of the same anti-Id to activate both B and T cells for an anamnestic recognition of similar epitopes on HBV.

PBL or $CD4^+$ cells from individuals who had been vaccinated against HBV or who had been infected with HBV and recovered are stimulated in in vitro culture by both HBsAg and the monoclonal anti-Id 2F10 but not by control monoclonal antibodies.

3.6. The preS region of hepadnavirus env proteins

The following properties distinguish the preS sequences of hepadnaviruses from S-protein: high hydrophilicity; high content of charged amino acid residues; and absence of cysteines (Neurath and Kent, 1988). These properties lead one to predict that preS domains of hepadnavirus env proteins should be exposed on the surface of virus particles. Absence of discontinuous epitopes generated by disulfide bonds suggested the presence of continuous epitopes within the preS domains of hepadnavirus env proteins, a prediction born out by the successful mimicking of preS epitopes with synthetic peptide analogues.

PreS sequences of hepadnavirus env proteins are much less conserved than sequences of S-proteins. Only approx. 72% of amino acid residues within the preS sequences of distinct HBV subtypes are conserved. About 2/3 of the amino acid variations are non-conservative (Neurath and Kent, 1988). Comparisons between preS sequences of WHV and GSHV indicate that about 70% of the residues are conserved. Unlike with S-proteins of mammalian hepadnaviruses (Fig. 19.11), there is

TABLE 19.4

Results of *in vitro* T-cell proliferation assays with lymph node cells from mice immunized with HBsAg

Stimulus in vitro	μg/well	cpm[a]
2F10 anti-Id	5	22082
	10	36848
	20	43842
	50	66066
2F10 Heavy Chain	20	43812
2F10 Light Chain	20	4089
Control Antibody	5	989
	10	1293
	20	1238
	50	1248
HBsAG	0.1	49628
	0.5	41533
Synthetic Peptide S(139–147)	0.05	21799
	0.1	27107
Control Peptide	0.05	408
	0.1	340
Medium		761

[a] Numbers correspond to cpm of [^{3}H]thymidine incorporated into cellular DNA (mean from four separate experiments) within 120 h.

very little homology between preS sequences of human and animal hepadnaviruses. The highest homology (61.5%) between these distinct hepadnaviruses occurs within a 26-residue stretch in the preS1 sequence (corresponding to residues (92–117) of HBV). Despite the lack of homology in primary amino acid sequence between preS regions of distinct hepadnaviruses, predictive algorithms indicate that the secondary structure of these domains is conserved, suggesting functional similarities between these regions of the env protein (Schaeffer and Sninsky, 1984; Neurath and Kent, 1988).

The reactivity of HBV (HBsAg) with antisera to peptides from the preS sequence established that most of the preS region is exposed on the surface of these particles (Neurath and Kent, 1988), except possibly the most hydrophobic segment approximately between residues 55–85 in the preS1 region. The expression of preS sequences on env proteins of WHV, GSHV, DHBV and HHBV was also established (Persing et al., 1986a,b; Schlicht et al., 1987a; Neurath and Kent, 1988; Yokosuka et al., 1988; Cheung et al., 1989; Wen et al., 1990). The preS regions of distinct hepadnaviruses are immunogenic and elicit antibody responses in the respective host species in the course of natural infection. These responses usually precede responses to S-protein and are considered to be important in initiating the clearance of the respective viruses essential for recovery from disease. In the case of DHBV, S-protein appeared to be a significantly weaker immunogen as compared with epitopes on the preS region (Schlicht et al., 1987a; Cheung et al., 1989). This agrees with the deletion in the DHBV S-protein sequence of residues (102–157) (Fig. 19.11) carrying immunodominant S-epitopes on mammalian hepadnaviruses. On the other hand, immunization of a heterologous species (BALB/c mice and guinea pigs) with HBsAg containing S-, M- and L-protein indicated that preS1, preS2 and S regions of the HBV env protein are about equally immunogenic, the level of antibodies generated being proportional to the molar proportions of these envelope domains in the immunogens used (Heermann et al., 1987).

TABLE 19.5

Results of *in vitro* T-cell proliferation assays with lymph node cells from mice immunized with anti-Id

In vitro stimulus	μg/well	cpm[a]
2F10 anti-Id	5	20547
	10	41505
	20	58990
	50	76053
Control Ab	5	558
	10	589
	20	120
	50	412
HBsAg	0.1	52817
	0.5	41541
Synthetic Peptide S(139–147)	0.1	29015
Medium		752

[a] Numbers correspond to cpm of [^{3}H]thymidine incorporated into cellular DNA (means from triplicate experiments) within 120 h.

The preS1 and preS2 sequences of all hepadnavirus env proteins (except HHBV) contain at least one glycosylation site (Table 19.2). However, only residue 123 in the preS2 sequence is glycosylated in HBV particles generated during natural infection. Saccharide chains linked to this residue of the preS2 sequence are of the high mannose type (Stibbe and Gerlich, 1982). The preS2 sequence in M-protein is always glycosylated. On the other hand, the preS2 sequence within L-protein is never glycosylated, suggesting that the presence of the preS1 sequence interferes with the glycosylation of the preS2 sequence. This observation may be important for understanding the immunological cross-reactivity at the T cell level between HBV env proteins and peptides from the preS2 sequence (see below).

Unlike S- and M- protein, L-protein is myristoylated at the penultimate G residue in the preS1 sequence (Persing et al., 1987; Lanford et al., 1989). All other hepadnaviruses share with the preS1 sequence of HBV the N-terminal MG sequence essential for myristoylation. The signal for myristoylation appears in the DHBV env protein at the second M residue in the preS1 sequence, this residue probably representing the N-terminus (Neurath and Kent, 1988). This suggests that preS1 sequences of all hepadnaviruses are myristoylated. The myristoylation of the preS1 sequence may play a role in inhibition of secretion of particles containing S-, M- and L-protein (Persing et al., 1986a; Neurath and Kent, 1988). However, the preS1 sequence itself, without myristoylation, seems to affect the morphogenesis of HBsAg particles, since the incorporation of preS1 sequences, or portions thereof, into HBsAg increases the probability of formation of tubular forms as compared with spherical approx. 22-nm particles (Hemmerling et al., 1989; Kuroda et al., 1989b). The content of L-protein in tubular forms is usually higher than that of the spherical particles (Heermann et al., 1987).

The preS1 sequence contains the receptor binding site for hepatocytes located within residues preS(21–47) (Neurath et al., 1986). Antibodies to this segment were expected to be protective based on the principle of the attachment blockade pathway of virus-neutralization. There is a limited degree of amino acid variability within the preS(21–47) sequence since 4/27 residues are not conserved among HBV subtypes (Table 19.6). These amino acid changes do not affect the binding of HBV to hepatocytes. However, the inhibition of interaction between HBV and hepatocytes by antibodies depends on their subtype specificities (Neurath and Strick, 1989). This suggests that the virus-neutralizing activity of antibodies recognizing the receptor binding site on L-protein may depend on a match between virus and antibody subtypes.

The preS2 sequence of the HBV env protein carries a low affinity binding site for human serum albumin (HSA) or for chemically modified albumin (Neurath et al., 1988). It was suggested that HSA may mediate the attachment of HBV to hepatocytes since these cells also contain a receptor for HSA (polymerized HSA) and that immunization with HBV proteins containing the hepatocyte binding site or with appropriate synthetic peptide analogues would elicit an autoimmune response to liver

TABLE 19.6

Amino acid variability within the preS (15–47) and preS (120–153) regions of HBV env proteins assignable to serologic subtypes of HBV

HBV subtype	Residue number											
	PreS1						PreS2					
	18	19	25	35	38	39	121	126	127	130	132	141
adw	V	P	L	G	S	T	Q	A	L	A	Q	L
adw[2]	V	P	F	G	S	N	Q	A	F	T	Q	L
adyw	T	S	F	R	T	N	H	T	F	A	Q	F
adr	V	P	F	G	S	N,H	Q	T	F	A	L	F
ayr	V	P	F	G	S	N	Q	T	F	A	L	F
ayw	T	S	F	R	T	A	Q	T	F	T	Q	F

Reprinted with permission from Neurath et al., 1987b.

cell receptors. However, none of these suggestions could be confirmed by experimental results: (i) Albumin binding sites are not detectable on WHV, indicating that albumins are not the ligands mediating the attachment of hepadnaviruses to host cells (Pohl et al., 1986); (ii) Immunization of experimental animals with immunogens containing preS2 sequences or fragments thereof does not lead to the generation of anti-HSA antibodies or of antibodies recognizing epitopes on hepatocytes (Neurath et al., 1988c); (iii) There is no evidence for the existence of a defined albumin receptor on the surface of hepatocytes (Reed and Burrington, 1989).

3.7. B cell epitopes in the preS region of hepadnavirus env proteins

Two complementary approaches were used to define B cell epitopes in the preS region of the HBV env protein: (a) the reactivity of polyclonal and monoclonal antibodies to L- or M-protein (i.e., HBV or HBsAg) with synthetic peptides derived from the preS1 and preS2 sequence was followed; and (b) the reactivity of antisera to peptides from the preS1 and preS2 sequence with L- and M-protein was determined.

The goal of these experiments was to: (i) identify segments within the preS sequence which are obligatory for antibody binding; (ii) determine the influence of flanking sequences on the binding of antibodies; and (iii) determine the effect of subtype-specific amino acid replacements on recognition of preS epitopes by antibodies. The success of this approach depended on the contiguous nature of B cell epitopes on the preS sequence of the HBV env protein. This requirement was met with polyclonal and most monoclonal antibodies studied. However, one of the preS1-specific mAbs studied did not recognize any of the peptides from the preS1 sequence, and apparently recognized discontinuous epitopes (Heermann et al., 1987; own unpublished data). One of the preS2-specific mAbs recognized preS2 sequences in M-protein but not in L-protein. Deglycosylation abolished the reactivity of this mAb with M-protein, suggesting that the antibody recognized exclusively the glycosylated preS2 sequence (Heermann et al., 1988). This mAb also failed to recognize any of the peptides from the preS2 sequence (own unpublished data). Other mAbs and all polyclonal antibodies studied proved to be suitable for mapping contiguous epitopes on the preS sequence using synthetic peptides.

Peptides from the entire preS sequence of the HBV env protein were synthesized and their reactivity with rabbit polyclonal anti-HBV was studied. This antiserum recognized peptides (1–21) (the sequence (1–11) is absent in HBV subtypes *adw* and *ayw* (Neurath and Kent, 1988)), (12–32), (32–53), (94–117) from the preS1 sequence and (120–145) from the preS2 sequence. Peptides (53–73), (74–93) and (153–171) were not recognized (Neurath and Kent, 1988; own unpublished data). This indicates that dominant B cell epitopes are localized in the N-terminal and C-terminal regions of the preS1 sequence and in the N-terminal region of the preS2 sequence (Fig. 19.9). Antisera to the same peptides were also prepared and tested for reactivity with HBV env proteins. All sera except anti-preS(74–93) reacted with HBV, suggesting that nearly the entire preS sequence is exposed on the HBV envelope and that antibodies with specificities missing in the repertoire of antibodies generated by immunization with the protein can be prepared by immunization with synthetic peptides.

Antibodies present in sera from humans exposed to HBV react preferentially with peptides preS(12–32) preS(12–47) and preS(120–145), preS(120–153) (Neurath and Kent, 1988; Neurath et al., 1988a).

Antisera to synthetic peptides preS(1–13) and preS(130–143) from the preS1 sequence of the GSHV L-protein (Table 19.2) specifically recognized the L-protein of this virus (Seeger et al., 1987). An antiserum to the peptide preS(1–13) (numbering starting at the second ATG codon of the preS sequence) of the DHBV env protein reacted specifically with L-protein of this hepadnavirus (Wen et al., 1989). These results suggest that preS domains of hepadnaviruses other than HBV can also be successfully mimicked by synthetic peptides.

To delineate with more precision the dominant preS2 epitope within the preS(120–153) sequence, overlapping shorter peptides and peptides with single amino acid replacements were synthesized and their reactivity with polyclonal and monoclonal antibodies was evaluated. Among three overlapping peptides, preS(120–134), preS(128–139) and preS(134–145), the middle peptide was optimally recognized by anti-HBV. However, the latter peptide had decreased antigenicity as compared with the peptide preS(120–145). A series of peptides corresponding to the sequence preS(128–153) *adw2* with each individual residue replaced with alanine (A) were synthesized. These peptides were tested for reactivity with polyclonal and monoclonal anti-preS2-specific antibodies. Replacement of any residues within the sequence (132–139) by A completely eliminated the reactivity with a monoclonal antibody (mAb F376) and diminished the reactivity with polyclonal antisera. These results indicated that the sequence preS(132–139) is absolutely essential for recognition by mAb F376 (Neurath et al., 1988a). The shortest sequences recognized by other mAbs corresponded to preS(133–139) and preS(137–143) (Milich et al., 1986a). Although preS(132–139) was the minimum-length sequence recognized by mAb F376, the optimum inhibition of the reaction of ^{125}I-labeled mAb F376 with recombinant HBsAg containing M-protein occurred only with longer peptides preS(128–153) or preS(128–145). Thus, it is important to distinguish between a minimum-length partial epitope and a 'full-length' epitope recognized by antibodies (own unpublished results obtained in collaboration with P. Briand and M.H.V. Van Regenmortel). This distinction is essential for the development of synthetic vaccines. To determine the N-terminal boundary of synthetic peptides optimally recognized by anti-HBV, a series of peptides with increasing lengths having a constant preS(145) C-terminus and an N-terminus corresponding the residues (120–138) were synthesized. Maximum binding of polyclonal anti-HBV antibodies was observed with peptide preS(124–145) and with additional peptides elongated at the N-terminus. These results suggested that only the four N-terminal amino acids (residues preS(120–123)) do not seem to play a role in binding of anti-HBV and might be dispensable in the design of a synthetic preS2 B cell epitope. However, these four N-terminal amino acids constitute part of a T cell epitope (see below), indicating that the peptide preS(120–145) is required to mimic adequately the entire preS2 sequence. However, addition of residues preS(146–153) improves the recognition of synthetic analogues by mAb F376 (Neurath et al., 1987b), suggesting that the peptide preS(120–153) is the immunogen of choice for incorporation into a polyvalent synthetic hepatitis B vaccine. The contribution of N-terminal residues of the preS2 sequence to B cell epitopes is also supported by the results of epitope mapping for mAb F124. This mAb recognizes preS(120–153) peptides with a threonine (T) and not alanine (A) at position 126 (Table 19.6).

Within the preS1 sequence, adjacent sequences preS(1–21), preS(12–32) and preS(32–53) as well as the sequence preS(94–117) were recognized by polyclonal rabbit anti-HBV. Within the repertoire of rabbit and human anti-HBV, antibodies recognizing the peptide preS(12–32) appeared to be the most abundant (Neurath and Kent, 1988). Unlike antibodies to synthetic peptides preS(1–21) and preS(12–32), only antibodies to a peptide elongated towards the C-terminus, preS(21–47), blocked the interaction between HBV and hepatocytes, indicating that residues (32–47) are important in eliciting antibodies recognizing the HBV receptor binding site. The peptide preS(21–47) was poorly immunogenic unlike peptides (1–21) and (12–32), suggesting the essential role of residues (12–20) for immunogenicity of peptides derived from the N-terminal region of the preS1 sequence. This is in agreement with more detailed studies on the localization of T cell epitopes on the preS1 sequence (see below). Attempts to define short segments of the preS1 region recognized by murine polyclonal antibodies led to the definition of five distinct antibody binding sites corresponding to the residues (16–27), (32–53), (41–53), (94–105) and (106–117) (Milich et al., 1987a). A mAb F35.25 against purified HBV subtype *ad* was specific for the sequence preS(32–53) (Petit et al., 1989). This mAb reacted preferentially with L-protein of the *ad* subtype, indicating that amino acid replacements at residues 35, 38, 39, 48 and 51 may contribute to subtype specificity. Because of the importance of residues within the (12–20) sequence for immunogenicity of preS1 peptides and the involvement of residues (21–47) in the HBV receptor binding activity and the virus-neutralizing activity of anti-(21–47) (Neurath et al., 1989a), the peptide preS(12–47) is considered as a candidate component for a synthetic hepatitis B immunogen. Experiments in chimpanzees have demonstrated that this peptide elicits protection against HBV (Neurath et al., 1989a; Thornton et al., 1989). Recent data (obtained in collaboration with Briand and Van Regenmortel) suggest the potential of a shorter peptide, preS(13–41), for active immunization (methionine at residue 12 is absent in the preS1 sequence of HBV subtypes *adw* and *ayw* as the result of post-translational myristoylation of glycine at residue 13).

Synthetic peptides considered for vaccination must elicit antibodies which recognize to a similar extent the peptide itself and the native protein which they are expected to mimic. The synthetic peptides should not elicit significant subpopulations of antibodies recognizing the peptide but not the protein from which the peptide sequence is derived. Peptides preS(120–145) and preS(120–153) elicited predominantly antibodies which recognized equally well the preS sequence within the peptide and M-protein (Neurath et al., 1987c; Neurath et al., 1988a). The affinity constants (K) for the reaction between anti-peptide antibodies and HBsAg were of similar magnitude as K for the reaction between anti-peptide antibodies and the synthetic peptide, as well as K for the reaction between HBsAg and preS-specific antibodies elicited by HBsAg. Similarly, antisera to peptides from the preS1 region recognized the preS1 sequence within L-protein with high fidelity (Neurath et al., 1987c).

There is a considerable amino acid variability between B cell epitopes on the preS1 and preS2 regions from distinct serological subtypes of HBV (Table 19.6), raising the question whether preS sequences are sufficiently related immunologically to have the potential of inducing cross-protective immunity. The effect of variability is, for example, reflected in the observation that mAb F376 reacted preferentially with the sequence preS(128–153) corresponding to the HBV subtype *ayw* as compared with subtype *adw*2. This difference can be ascribed solely to an amino acid substitution at position 141 (F → L; Table 19.6). On the other hand, antisera to synthetic peptides carrying the immunodominant epitopes of the preS1 and preS2 sequence, respectively, and corresponding to two HBV subtypes, *adw*2 and *ayw*, each recognize preS1 and preS2 epitopes on all serological subtypes of the HBV env protein. Thus, the sequence variability within the preS1 and preS2 region does not represent a serious obstacle for the development of synthetic peptide or genetically engineered hepatitis B preS immunogens for vaccination. This conclusion is confirmed by protection of chimpanzees immunized with synthetic peptides preS(120–145) and preS(120–174) against challenge with a heterologous subtype of HBV (Thornton et al., 1987, 1989; Emini et al., 1989). Immunization of chimpanzees with synthetic peptides preS(12–47), preS(94–117) and preS(120–145), respectively, derived from the sequence of an *ay* subtype elicited antibodies to L- and M-proteins, respectively, of both subtypes *d* and *y* (Thornton et al., 1989).

Treatment with formaldehyde has been used to inactivate residual HBV or other infectious agents possibly present in hepatitis B immunogens to assure the safety of the final product. Such treatment decreases the antigenic activity of preS2 epitopes to 1/3 and drastically reduces the recognition of preS1 sequences by antibodies (Ohnuma et al., 1986). For this reason, the conventional formaldehyde inactivation procedure does not seem to be appropriate for immunogens containing M- and L-protein and heat inactivation has been proposed instead.

3.8. T cell epitopes in the preS region of hepadnavirus env proteins

The observations that preS (12–32) and preS (120–145) were highly immunogenic without conjugation to carrier molecules, suggested that they contained epitopes which stimulated T cells to provide appropriate help to antigen-specific B cells. Therefore, studies were undertaken to identify the T cell recognition sites within the preS1 and preS2 regions.

When SJL/J mice (high responders to preS (12–32)) were immunized with HBsAg containing S-, M- and L-protein (HBsAg/P39) their lymph node T cells responded to in vitro stimulation with a number of synthetic peptides at relatively high concentrations. Thus, peptides preS(1–21), preS(12–32) and preS(94–117) elicited dose-dependent T cell proliferative responses. These results indicate that the overlapping sequence between peptides preS (1–21) and preS (12–32), i.e., residues (12–32) encompass a T cell site and that peptide preS(94–117) contains another site. The reciprocal experiments of priming in vivo with peptides preS(12–32) and preS(94–117) and challenging in vitro with HBsAg/P39 confirmed the localization of T cell epitopes within these two sequences. The responses elicited were preS1-specific as no stimulation was elicited in vitro with HBsAg particles containing only S-protein (HBsAg/P25) or S + M protein (HBsAg/P33) (Milich et al., 1986b).

Additional experiments were aimed at examining further the fine specificity of T cell recognition of peptides preS(12–32) and preS(94–117). A T cell recognition site was mapped to the preS(12–21) sequence and residues 14, 18 and/or 19 were shown to be most critical. Within the peptides preS(94–117) two non-overlapping sites were located corresponding to residues preS(94–15) and preS(106–117) (Milich et al., 1987).

In addition to stimulating T cells *in vitro,* these peptides from the preS1 region also provided T cell help for HBsAg-specific antibody production. Thus, priming mice with the peptide preS(12–32) and subsequent challenge with HBsAg/P39 resulted in antibody production to determinants within the preS1 and preS2 region and in an early subtype specific anti-S response. This was followed by later group-specific anti-S responses. Therefore, peptide preS(12–32) was able to prime for antibody production to preS1 and preS2 and S region determinants. The special relevance of these results is that SJL/J mice are non-responders to the S region of HBsAg but priming with preS(12–32) was able to circumvent this non-responsiveness. Priming with peptide preS(94–117) induced preS1 and preS2 region-specific antibodies upon challenge with HBsAg/P39 but no S region responses were elicited. Priming with peptide preS(94–117) resulted in the production of different preS1-specific antibodies than did priming with peptide preS(12–32). This suggests that T cell fine specificity is directing B cell specificity by virtue of preferential T cell helper function for distinct B cell clones recognizing determinants on the same polypeptide. What somewhat limits the value of peptide preS(12–32) as a key preS1 T cell determinant is the fact that this sequence is recognized only by animals of the H-2^s and H-2^f haplotypes whereas the peptide preS(94–117) was recognized by mice of a large number of haplotypes.

The definition of preS epitopes recognized by human T cells is essential for improving hepatitis B vaccines. Toward this goal, Jin et al. (1988) characterized three human cloned T cell lines derived from the peripheral blood lymphocytes of hepatitis B vaccine recipients. Clone LL1 recognized HBsAg in association with HLA-A11 (i.e.,MHC class I restricted) and clones SR45 and SR54 recognized HBsAg in association with HLA-DR7 only (i.e.,MHC class II). Clone LL1 was stimulated by L protein of HBsAg (preS1 + S2 + S) but not by S-protein. Clones SR54 and SR45 responded to epitopes on S-protein.

Further delineation of the exact region in preS1 recognized by clone LL1 was achieved using a number of overlapping synthetic peptides corresponding to the preS1 region. Peptides preS(1–28), preS(12–32) and preS(21–47) gave maximal stimulation comparable to that with yeast derived L-protein indicating that the entire determinant recognized by the human T cell clone is contained in these peptides. The three peptides overlap at residues preS(21–28). It was concluded that the epitope recognized by clone LL1 is located in this sequence (PLGFFPDH). Presumably the peptide preS(21–28) contains distinct sites for MHC association (with HLA-A11 in this instance) and for binding by the T cell receptor of clone LL1. This site is close to but distinct from the preS1 site (12–21) identified by Milich et al., (1986b) as the preS1 epitope recognized by mouse T cells in association with MHC class II antigens. As discussed earlier, T cells specific for preS(12–21) could provide T helper function to B cells specific for S-protein. The clone LL1 recognizes the site corresponding to residues preS(21–28) in association with class I antigens and its ability to provide helper function is as yet not known. The peptide preS(21–28) fits well with the G-hydrophobic-hydrophobic-P charged amino acid motif proposed by Rothbard and Taylor (1988).

Another issue addressed in the experiments of Jin et al., (1988) is whether immune T cells can respond to infected cells that express viral antigens endogenously. The *endosomal* pathway has frequently been implicated in the processing of *exogenous* native proteins for MHC Class II associated antigen presentation. This pathway is readily inhibited by chloroquine or leupeptin. The *non-endosomal* pathway is primarily thought to process *endogenously* expressed viral proteins. This pathway is insensitive to pretreatment with chloroquine. The two pathways were traditionally regarded as mutually exclusive. However, results using the preS1-specific T cell clone LL1 (Class I restricted) suggest that the non-endosomal pathway is readily accessible to both exogeneous and endogenous antigens. Similarly, using clone SR54 (Class II restricted), it was observed that the endosomal pathway is also accessible to both forms of antigen. Thus, two distinct pathways exist for processing antigen that will eventually stimulate T cells like clone LL1 and SR54. Each T cell sees the products of one pathway or the other (but not both). The key difference from previous work is the finding that exogenous antigen is not limited to the endosomal pathway but can also be processed via the non-endosomal pathway. Inhibition of the endosomal pathway has no effect on non-endosomal processing of exogenous antigen. Thus, the two pathways are separate and operate in parallel with each other. The recognition of endogenous antigen implies that a non-infectious viral component vaccine can elicit active T cell immunity against infection with live virus.

Additional studies established the location of T cell epitopes within the preS2 sequence. When mice of the H-2^q haplotype were primed in vivo with the synthetic peptide preS(120–145), isolated lymph node cells proliferated in vitro when stimulated with the homologous peptide but not with M-protein (=HBsAg/P33). A shorter peptide preS(120–132) elicited weaker responses than preS(120–145). The peptide preS(133–145) also elicited T cell proliferation when used at very high concentrations. The in vitro T cell response elicited by the truncated or full-length peptides was strongly subtype dependent. T-cells from mice that were primed in vivo with HBsAg/P33 failed to respond in in vitro proliferation assays with synthetic preS(120–145), suggesting that the T cell epitopes on the synthetic peptide and on the preS2 region of M-protein are distinct (Milich et al., 1986c).

However other findings (Neurath et al., 1985) are not consistent with a total lack of cross-reactivity between T cell epitopes on the synthetic peptide preS(120–145) and on preS2 sequences of HBV envelope proteins. SJL/J mice injected with three doses of preS(120–145) and then immunized with HBsAg containing preS2 sequences produced enhanced antibody responses as compared with control mice that had not been preimmunized with the synthetic peptide. In addition, very recent evidence from Thanavala et al. in collaboration with A. R. Neurath confirms the relevance of the preS(120–145) sequence for T cell recognition of HBsAg. Thus, lymphocytes obtained from chimpanzees that had been immunized with a recombinant vaccine containing S proteins and the major epitopes of the preS region (Hemmerling et al., 1989), proliferated strongly in vitro when stimulated with peptides preS(120–145), preS(12–47) and S(139–147). Similar results were obtained using $CD4^+$ lymphocytes (from human HBV carriers) stimulated in vitro with preS(120–145) and preS(12–47), respectively.

To determine the usefulness for immunization of hybrid peptides containing 'minimum-length' B- and T-cell epitopes from non-adjacent regions, peptides preS(12–25)-preS(132–139) and preS(120–131)-preS(26–32) were synthesized and tested for antigenicity and immunogenicity. Results of assays with antisera to peptides preS(15–47) and preS(120–153) confirmed that B cell epitopes were present in the preS(132–139) and preS(26–32) segments of the hybrid peptides. However, anti-HBV recognized the two hybrid peptides by about one order of magnitude less efficiently than the respective longer peptides preS(15–47) and preS(120–153). This indicated the importance of sequences flanking the preS(132–139) and preS(26–32) segments for recognition by anti-HBV. Each of the hybrid peptides elicited $\leqslant 1/1000$ of the level of antibodies recognizing L- and M-protein, respectively, in comparison with peptides preS(15–47) and preS(120–153) (Neurath et al., 1988b). These results suggest that T-B cell epitope shuffling should be avoided and that peptides of reasonable length ($\leqslant 40$ amino acids) containing adjacent (overlapping) B and T cell epitopes should be given preference for the design of synthetic immunogens.

4. Gene products of ORF P: the genome linked protein and reverse transcriptase

4.1. The genome linked protein

Genomes of all hepadnaviruses have a protein covalently linked to the 5′-end of the DNA minus strand. This protein is essential for initiation of minus strand DNA synthesis. Results of experiments with antisera to fusion proteins or to synthetic peptides containing sequences coded for by ORF P of DHBV indicated that the translational product of ORF P becomes initially covalently attached to the DNA minus strand and is subsequently proteolytically processed. After processing, the N-terminal quarter of the ORF P gene product remains covalently linked to the DNA minus strand by a phosphodiester linkage between a tyrosine residue of the protein and DNA (Bosch et al., 1988; Bartenschlager and Schaller, 1988). Similar studies with hepadnaviruses other than DHBV have not yet been reported. The N-terminal quarter of the deduced amino acid sequence of P gene products is relatively constant between different hepadnaviruses (Fig. 19.13.). Therefore, the attachment of the genome linked protein to hepadnavirus DNAs probably occurs by similar mechanisms. There are no tyrosine residues near the N-terminus of the DHBV P gene product, indicating that an internal tyrosine must be involved in linkage to DNA. Tyrosine residues within a probable consensus sequence are likely to be involved in the protein-DNA linkage (Fig. 19.14.).

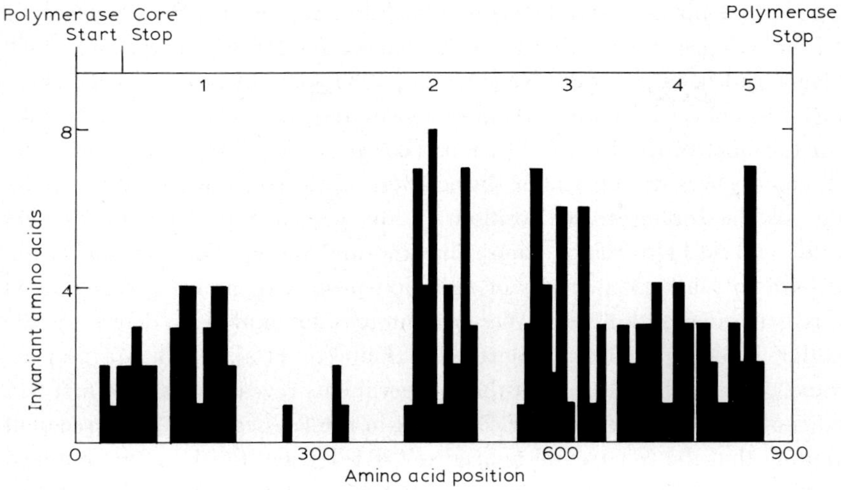

Fig. 19.13. Conserved regions of the hepadnavirus ORF-P translational products. The predicted amino acid sequences of 14 hepadnaviruses (nine from HBV, two from WHBV, one from GSHV and two from DHBV) were aligned and are presented as a histogram where the number of identical amino acids found at the same position within a block of 10 amino acids is plotted as a function of amino acid position along ORF-P. Reprinted with permission from Miller, 1988.

4.2. Reverse transcriptase/DNA polymerase/ribonuclease H

Retroviral reverse transcriptase (RT) is expressed as a nucleocapsid-polymerase fusion protein by ribosomal frameshifting during translation of the overlapping regions on viral mRNA. This is not the case for hepadnavirus RTs which are generated by initiation of translation at an ATG codon on the genome-length mRNA serving as the template for reverse transcription. The first ATG codon in ORF P is essential for generation of RT activity (Chang et al., 1989; Schlicht et al., 1989b).

	61
HBV	F T G L Y S
	64
WHV	I T G L Y S
	64
GSHV	I T G L Y S
	92
DHBV	L S G L Y Q
	92
HHBV	L T G L Y Q

Fig. 19.14 Possible consensus sequence containing the tyrosine residue involved in covalent linkage of the genome-linked protein to the 5′-end of hepadnavirus minus strand DNA. Deduced from DNA sequencing data (Neurath and Kent, 1985; Sprengel et al., 1984; 1988; Cohen et al., 1988; Seeger et al., 1984a).

The proteins functioning in hepadnavirus replication are generated by proteolytic cleavage of the polyprotein resulting from translation of ORF P (Bosch et al., 1988; Bartenschlager and Schaller, 1988; Nassal et al., 1989). Unlike retrovirus RT, hepadnavirus RT has never been detected in free form and was found exclusively associated with the nucleocapsid core of virions (Bosch et al., 1988; Radziwill et al., 1988). RT activity was detected after dissociation of nucleocapsid cores and polyacrylamide gel electrophoresis at positions in the gel corresponding to 70 kDa (major band) and 90 kDa (minor band) (Bavand and Laub, 1988; Bavand et al., 1989). Antisera to synthetic peptides or to fusion proteins containing segments of ORF P were generated and used for Western immunoblot analysis to determine the region of ORF P coding for hepadnavirus RT (Feitelson et al., 1988; Mack et al., 1988; Bavand et al., 1989). Results of these experiments revealed that the first 314 amino acid residues coded for by ORF P are not in the RT sequence, in agreement with the finding that the N-terminal quarter of ORF P codes for the genome linked protein and that highly variable sequence segments in ORF P (Fig. 13.) represent a non-essential spacer containing proteolytic cleavage sites (Bartenschlager and Schaller, 1988). On the other hand, antisera to segments (502–644), (781–795) and (822–838) reacted with RT, indicating that it is derived from the C-terminal portion of HBV ORF P. This region contains the most highly conserved sequences within ORF P of distinct hepadnaviruses (Fig. 19.13.). Region 2 (Fig. 19.13), the most highly conserved region in hepadnavirus ORF P, shares considerable homology with the pol sequence of retroviruses (Toh et al., 1983; Miller, 1988). RNAase H is probably coded for by the extreme C-terminal portion of ORF P (Toh et al., 1983, Bartenschlager and Schaller, 1988).

5. X-protein

ORF X of HBV, WHV, and GSHV has the capacity to code for a protein of about 16 kDa. ORF X is not present in the DHBV and HHBV genomes. In vitro mutagenesis of HBV ORF X demonstrated that expression of X-protein is not obligatory for production of HBV in cell culture (Yaginuma et al., 1987). Comparisons of the predicted amino acid sequences of ORF X corresponding to HBV subtypes indicated 80% amino acid homology (Lo et al., 1988). The overall homology between predicted X-protein sequences of HBV, WHV and GSHV is only approx. 34%.

Biosynthesis of X protein during in vivo hepadnavirus infections was most convincingly demonstrated by the detection of anti-X antibodies in human, woodchuck and ground squirrel sera using as antigens recombinant or in vitro translated X-protein or synthetic peptides corresponding to residues (100–115) and (144–154) of HBV X-protein (Moriarty et al., 1985; Kay et al., 1985; Elfassi et al., 1986; Meyers et al., 1986; Persing et al., 1986b; Pfaff et al., 1987; Hess et al., 1988; Koike et al., 1988; Liang et al., 1988; Theilmann et al., 1988)

Antibodies to X-protein were most frequently detected in sera of patients with primary hepatocellular carcinoma, liver cirrhosis or chronic active hepatitis. However, it has been reported recently (Hess et al., 1988) that anti-X was detected at high frequency in sera of patients with previously resolved HBV infection as well as in patients with acute hepatitis B, suggesting that anti-X antibodies are induced early and frequently during HBV infection as a result of X-protein expression *in vivo*. These authors also suggested that immune responses to X-protein may contribute either to antiviral immunity or to pathogenesis of liver disease.

The detection of a 28-kDa protein in tissues infected with HBV by antisera to peptides derived from the HBV X-protein sequence was reported (Moriarty et al., 1985). However, others (Pfaff et al., 1987) failed to detect X gene products in HBV-infected tissues. Kaneko and Miller (1988) reported the detection of 0.65 kb X region-specific transcripts in the nucleus of liver cells infected with WHV. These transcripts represented less than 1% of total virus-specific RNA. A significant portion of the transcripts was not fully polyadenylated. In addition, a small amount of X-specific 0.9 kb fully polyadenylated RNA was found in these tissues. On the other hand, cells transfected with recombinant HBV DNA synthesized X-protein detectable by immunoflourescence in the periphery of nuclei (Pugh et al., 1986; Siddiqui et al., 1987; Chisaka et al., 1987). X-protein functions as a trans-activating factor stimulating homologous and heterologous viral promoters/enhancers (Spandau and Lee, 1988; Siddiqui et al., 1989; Twu and Robinson, 1989).

6. Prophylactic immunogens derived from hepadnavirus env proteins

6.1. Recombinant vaccines

The production of recombinant hepatitis B vaccines is described in Chapters 8 and 9. Therefore, this section will focus on the design of hepatitis B immunogens expressing a multitude of protective epitopes from different regions of the env protein. Hepadnavirus particles consisting of S-, M- and L-protein express on their surface a mosaic of protective epitopes (Fig. 19.1). It is expected that the most efficient vaccines will have to either mimic this epitope mosaic or improve on the mosaic expression of epitopes leading to an optimal and genetically least restricted protective immune response. To accomplish this goal, the following problems have to be considered: (a) efficient biosynthesis, assembly and secretion of constituent env protein epitopes into particles resembling serum-derived HBsAg (approx. 22-nm spheres and tubular forms); (b) resistance of synthesized env proteins to intracellular and extracellular proteolytic degradation; (c) prevention of excessive glycosylation potentially resulting in blocking of some epitopes leading to decreased immunogenicity; (d) surface exposure of all protective epitopes, i.e., prevention of masking of epitopes from one region of the env protein with sequences from another region of the protein; (e) suitability for mass vaccination (oral immunization?).

TABLE 19.7

Selected recombinant HBV envelope proteins applicable for vaccination against hepatitis B

Expression system	Promoter	HBV env protein sequence expressed	Physical Form	Reported Yields	References
Mammalian cells (CHO)	Foreign	S	~22 nm sph[d]	~1μg/10^6 cells/day	1
Mammalian cells (CHO)	Foreign	S+M (2:1)	~22 nm sph	~1μg/10^6 cells/day	2
Mammalian (mouse) cells	Foreign	S+M+L (8:2:1)	~22 nm sph	~1μg/10^6 cells/day	3,4
Mammalian (mouse) cells	HBV	M or preS(134-162)-S	~22 nm sph	Not reported	5
Mammalian (mouse) cells	HBV	preS(9-36)-S, preS(9-67)-S	~22 nm sph and tubular	Not reported	5
Baculovirus-insect cells	Foreign	S+M(1:2.5)	~22 nm sph	~0.7μg/ml/5 days	6
Baculovirus-insect cells	Foreign	S,M	~22 nm sph	~9μg/ml/4 days	7
Baculovirus-insect cells	Foreign	both S-protein and HBcAg	~22 nm sph	S: ~0.4 μg/ml/5 days HBcAg: ~5μg/ml/5 days	8
Baculovirus-insect cells	Foreign	S	~22 nm sph	90/μg/ml/2 days[a]	9
Baculovirus-insect cells	Foreign	L	~22 nm sph	6/μg/ml/2 days[a]	9
Baculovirus-insect larvae	Foreign	S[b]	~22 nm sph	150/μg/larva	10
Adenovirus	Foreign	S or M[c]	~22 nm sph	0.3–2.8/μg/ml/5 days	11–13
Adenovirus	Foreign	L	Not secreted/particles not formed		13
Vaccinia virus	Foreign	M	~22 nm sph	not reported	14
Vaccinia virus	Foreign	L	Not secreted/particles not formed		15

The consideration of these problems will be based on results obtained with recombinant proteins listed in Table 19.7. (a) Data presented in Table 19.7 and results summarized recently (Neurath and Kent, 1988) show that recombinant L-protein is minimally secreted and assembles into approx. 22-nm spherical particles only when co-expressed with M- and/or S-proteins which become the predominant constituents of the assembled particles. Therefore, the abundant expression of preS1 epi-

TABLE 19.7 (continued)

Expression system	Promoter	HBV env protein sequence expressed	Physical Form	Reported Yields	References
Yeast (*Saccharomyces cerevisiae*)	Foreign	S	~22 nm sph	~50μg/ml	16
Yeast (*Saccharomyces cerevisiae*)	Foreign	M	~22 nm sph	preS2 sequence proteolytically degraded	17–19
Yeast (protease-deficient strain)	Foreign	M	~22 nm sph	Not reported	20
Yeast (*Saccharomyces cerevisiae*)	Foreign	L	Few, if any, particles formed		21,22
(Ty1-based vectors integrating at chromosomal loci)	Foreign	Coassembled S+M+L	~22 nm sph	Not reported	23
Methylotrophic yeast (*Pichia pastoris*)	Foreign	S	~22 nm sph	400μg/ml	24
Salmonella	Foreign	S (122–137)-flagellin preS2 (120–145)-flagellin	Fusion proteins	Not reported	25
Salmonella	Foreign	*E. coli* heat-labile enterotoxin-HBV	Fusion proteins	Not reported	26,27
E. coli	Foreign	preS2(132–145)-periplasmic maltose binding protein Mal 1 E	Fusion protein	Not reported	28
E. coli	Foreign	preS(132–145)-Lam B outer membrane protein	Fusion protein	Not reported	29

[a] Intracellular, not secreted
[b] preS sequences not detected
[c] The secreted HBsAg particles consisted of M- and S-protein
[d] sph = spherical particles

1, Patzer et al. (1986). 2, Michel et al. (1984). 3, Samanta and Youn (1989). 4, Youn and Samanta et al. (1989). 5, Hemmerling et al. (1989). 6, Price et al. (1988). 7, Kang et al. (1987). 8, Takehara et al. (1988). 9, Lanford et al. (1989). 10, Price et al. (1989). 11, Ballay et al. (1985). 12, Morin et al. (1987). 13, Molnar-Kimber et al. (1988). 14, Cheng and Moss (1987). 15, Cheng et al. (1986). 16, Bitter et al. (1988). 17, Valenzuela et al. (1985a). 18, Kuroda et al. (1989b). 19, Imamura et al. (1987). 20, Langley et al. (1988). 21, Dehoux et al. (1986). 22, Kniskern et al. (1988). 23, Jacobs et al. (1989). 24, Cregg et al. (1987). 25, Wu et al. (1989). 26, Schödel and Will (1989). 27, Schödel et al. (1989). 28, Leclerc et al. (1989). 29, Charbit et al. (1987).

topes in a particulate form cannot be accomplished without redesigning the recombinants. The production, secretion and assembly of L-protein is controlled at several levels. When HBV env protein mRNAs are initiated from endogenous HBV promoters, three distinct mRNA species coding for S-, M- and L-protein are generated. The latter mRNA is the least abundant (1–2%) (Neurath and Kent, 1988). Transcription directed by the preS1 promoter is negatively regulated by downstream DNA sequences containing the preS2 promoter region. Deletion mutagenesis studies localized this region to DNA sequences coding for the preS(97–122) segment of the HBV env protein (Bulla and Siddiqui, 1989). Omission of this sequence, or portions thereof, would be expected to increase the level of transcription of a messenger coding for a truncated protein lacking the C-terminal portion of the preS1 sequence. Immunodominant T and B cell epitopes on the preS1 sequence are located on its N-terminal region (Neurath and Kent, 1988; Neurath et al., 1988a). Thus, high level expression of a mRNA for a truncated L-protein with most dominant T and B cell epitopes may be accomplished.

The N-terminus of hepadnavirus preS1 sequences contains a signal for myristoylation (see above). The myristoylation of L-protein, occurring in mammalian cells and yeast, may interfere with efficient secretion and assembly of this protein. This problem may potentially be eliminated by additional truncation of the HBV *env* gene removing the MG myristoylation signal. This has indeed been accomplished by generating recombinants expressing preS1 epitopes from the N-terminal portion of the preS1 sequence but missing the first eight N-terminal residues (ref. 5 in Table 19.7).

The production of both M- and S-protein from M-protein mRNA can be ascribed to alternative initiation of translation at either the first or second AUG on this mRNA (refs. 2–4, 6, 11 in Table 19.7).

Cumulatively, these results indicate that it would be desirable to exert more control over the level and coassembly of S-, M- and L-proteins produced by recombinant DNA techniques.

(b) Particles consisting of M-protein (M + S protein) are susceptible to proteolytic degradation resulting in loss of preS2 epitopes. This effect was most striking for M-particles produced in yeast (refs. 17–19 in Table 19.7) This problem was alleviated by expression of M-protein in protease deficient strains of yeast (ref. 20 in Table 19.7). Attempts were also made to minimize this problem by preparing mutants in which arginine (R) residues at positions preS 135 and preS 137 were replaced by glutamine (Q). Such mutations resulted in production of more stable particles at a ten-fold higher yield in comparison with particles containing the unaltered sequence (Kuroda et al., 1989). However, these attempts to improve yields ignored the essential role of these R residues in antigenicity and immunogenicity of M-protein (Neurath et al., 1988a,b; Langley et al., 1988).

(c) The glycosylation of HBV env proteins produced in yeast is different from that observed for serum-derived HBsAg or for HBsAg produced in mammalian cells. S-protein produced in yeast is always nonglycosylated unlike that produced in mammalian cells which can be either nonglycosylated or *N*-glycosylated. M- and L-protein produced in yeast can be either glycosylated or nonglycosylated, unlike mammalian cell-derived M-protein which is always glycosylated (one or two *N*-glycosylation sites) and L-protein which is nonglycosylated or glycosylated in the S-protein sequence (Neurath and Kent, 1988). *N*-linked chains on yeast M- and L-proteins are of the high mannose type. In addition, unlike HBsAg synthesized by mammalian cells, yeast produced particles also contain *O*-linked saccharide chains. There is a single *N*-linked high mannose type core oligosaccharide chain attached to both M- and L-protein at residue preS 123. The *O*(serine/threonine)-linked chains are also present, although less abundantly, on molecules that do not carry *N*-glycans (Langley et al, 1988; Rutgers et al., 1988; Rutgers et al., 1990). HBsAg derived from insect cells infected with recombinant baculoviruses is also modified with *N*-linked high mannose polysaccharides (Lanford et al., 1989). Consideration has to be given to the possibility that differences in glycosylation may affect the immune response to HBV env proteins.

(d) Studies on anti-S-protein antibody binding to HBsAg consisting of S-protein only and of S-protein and preS sequences, respectively, indicated that preS sequences can mask S epitopes (ref. 19 in Table 19.7; Heermann et al., 1987; Rutgers et al., 1988). Similarly, preS2 domains within assembled L-protein are masked by preS1 sequences (Heerman et al., 1987). Attachment of peptides derived from the preS1 sequence by a hydrophobic tail to particles consisting of S-protein and M-protein also resulted in decreased accessibility of preS2 epitopes to antibodies (Neurath et al., 1989b). These findings suggest that recombinant HBsAg particles consisting exclusively of S-, M- and L-protein, respectively, probably do not represent the optimal immunogen for vaccination due to limited surface exposure of all epitopes. Such limitations may be overcome by one of the following designs.

(i) Construction of recombinant particles consisting of S-protein and of dominant preS1 and preS2 epitopes, deleting portions of preS sequencs which are not essential for eliciting virus-neutralizing antibodies. Recombinants expected to have such properties have been prepared (ref. 5 in Table 19.7) and are being evaluated in clinical trials. (ii) Coassembly of S-, M- and L-proteins generating particles with the surface mosaic of epitopes resembling HBV (Fig. 19.1). This can be accomplished with Ty*1*-based vectors which integrate at several chromosomal loci by homologous recombination with resident transposons in *Saccharomyces cerevisiae.* Using such vectors, several copies of expression cassettes encoding S-, M- and L-protein were introduced into Saccharomyces cerevisiae. This approach opens the possibility to produce particles containing all HBV env protein epitopes in predetermined ratios (ref. 23 in Table 19.7). (iii) Immunization with a mixture of synthetic peptides mimicking immunodominant epitopes on the HBV env protein represents an alternative approach to elicit multispecific virus-neutralizing and protective antibodies.

(e) Eradication of hepatitis B would be simplified by using oral vaccines. Fusion proteins consisting of HBV env epitopes and either flagellin or E. coli heat labile enterotoxin (ref. 25–27 in Table 19.7) may possibly result in such vaccines. The lack of immune responses to HBV elicited by such fusion proteins (ref. 26, 27 in Table 19.7) may possibly be circumvented by more appropriate selection of preS sequences inserted into the fusion protein adhering to results of preS epitope mapping (see above).

The world-wide eradication of hepatitis B will depend on the price of hepatitis B vaccines. For this reason, the yield of immunogens prepared by distinct recombinant DNA approaches is of utmost importance. The methylotrophic yeast *Pichia pastoris* expression system (ref. 24 in Table 19.7) may provide an economically feasible system for the manufacture of large quantities of protective immunogens against hepatitis B.

6.2. Recombinant HBsAg as carrier for foreign epitopes

The propensity of S- and M-proteins to assemble into particles containing about 100 protein subunits suggested that these proteins may serve as a vehicle for the presentation of other antigens. Since S-protein carries the domains responsible for self assembly, hybrid proteins were prepared by inserting sequences of foreign antigens into the preS region of the HBV env protein. The corresponding hybrid genes were engineered and expressed in Saccharomyces cerevisiae. Taking this approach, HBsAg particles containing a fragment of herpes simplex virus 1 gD glycoprotein were prepared. The hybrid particles elicited anti-HBs and anti-HSV-1 gD antibodies (Valenzuela et al., 1985b). Similarly, a fragment of the gene coding for the human immunodeficiency virus 1 (HIV-1) env glycoprotein gp120 was inserted into the preS region of the HBV env protein gene and expressed in Chinese hamster ovary (CHO) cells. The recombinant protein assembled into particles which elicited anti-HBs and anti-HIV-1, including virus-neutralizing antibodies to the latter virus (Michel et al., 1988). However, the level of anti-HIV-1 antibodies elicited by the hybrid particles was by two to three orders of magnitude less than the level of antibodies elicited with synthetic peptides derived from the HIV-1 env protein sequence (Neurath et al., 1990).

Attempts were also made to insert foreign epitopes into the sequence of S-protein without affecting its assembly into particles. This was accomplished by inserting the foreign sequences into hydrophilic domains within the S-protein sequence, i.e., between residues 113 and 114 or 50 and 51, respectively (Delpeyroux et al., 1987). The insertion of a neutralization epitope from the VP1 poliovirus capsid protein between residues 113 and 114 of S-protein, led to the production of HBsAg particles eliciting virus-neutralizing antibodies to poliovirus (Delpeyroux et al., 1986). However, this insertion resulted in a 10^4-fold drop in immunogenicity of S-protein, suggesting that insertion of amino acid sequences between residues 113 and 114 of S-protein resulted in severe distortion of S-protein structure (see Fig. 19.11). This problem was overcome by co-transfection of mammalian cells with distinct plasmids carrying both the modified and unmodified S genes. The resulting hybrid particles containing modified and unmodified S-protein elicited high levels of anti-HBs and of poliovirus neutralizing antibodies (Delpeyroux et al., 1988).

Recombinant HBsAg particles consisting of S- and M-protein were also used as carriers for peptides from the preS1 sequence of the HBV env protein and from the gp120 sequence of HIV-1 (Neurath et al., 1989b). Long chain fatty acids or mercaptans were covalently linked to the synthetic peptides. Peptides with the attached hydrophobic tails adsorbed to the HBsAg particles resulting in hybrid immunogens expressing a multitude of epitopes.

6.3. Synthetic peptide immunogens

The dominant B cell epitopes on S-protein are discontinuous (see above), and synthetic peptides from the S-protein sequence elicit anti-S-protein antibodies at levels which are lower by at least two orders of magnitude than the levels of homologous anti-peptide antibodies (Neurath., 1989). Therefore, it appears to be premature to consider linear peptides derived from the S-protein sequence as candidate components of hepatitis B vaccines. Further progress in delineating discontinuous S-protein epitopes will be required to develop synthetic immunogens. On the other hand, selected peptides from the preS sequence are promising as potential prophylactic immunogens for the following reasons.

(i) The peptides mimic with high fidelity B and T cell epitopes on the native protein (see above). (ii) Humans develop anti-preS1 and anti-preS2 immune responses in the course of acute hepatitis B (Neurath and Kent, 1988; Alberti et al., 1988; Brahm et al., 1988; Coursaget et al., 1988; Ise et al., 1988; Heijtink et al., 1989), after vaccination with either serum-derived HBsAg (Neurath and Kent, 1988; Coursaget et al., 1988) or with recombinant particles containing S- and M-protein (Adamowicz et al., 1988; Hemmerling et al., 1989). The corresponding immune responses at the B and T cell levels were detected using synthetic peptides from the preS1 and preS2 region. (iii) Antibodies to these peptides are virus-neutralizing (Neurath and Kent, 1988; Neurath, 1989; Neurath et al., 1989a). (iv) Chimpanzees immunized with the peptides are protected against HBV challenge (Neurath, 1989; Neurath et al., 1989a; Thornton et al., 1987, 1989; Emini et al., 1989).

6.4. Antitidiotypic antibodies as immunogens

Internal image anti-Ids can act as surrogate immunogens with potential for vaccination (Roitt et al., 1981; Nisonoff and Lamoyi, 1981; Sacks and Sher, 1983; Ertl et al., 1984; Sharpe et al., 1984). There are certain niches where anti-Ids, either alone or in conjunction with nominal antigen, or its subunits, may facilitate the design of a new generation of vaccines. Their major potential may be for individuals who: (i) are low or nonresponders to nominal antigen (as in the case of HBV, as discussed earlier); (ii) have diseases which cause a profound dysfunction of the immune response (as in leishmaniasis, leprosy and possibly AIDS).

For vaccination (which may represent only 'priming') to be successful, it is not sufficient to prime the immune system with a single protective B-cell epitope. Challenge of the host with the living organism will, in most cases, fail to produce a secondary protective response because of the lack of an anamnestic recognition of a helper T cell epitope. The sharing of idiotypes between B and T cells of related antigen specificity would permit the use of anti-Ids to activate both B and T cells for an anamnestic recognition of similar epitopes on the infectious organism (see section 3.5).

Protection of chimpanzees against HBV infection accomplished by immunization with anti-Id, mimicking an epitope encompassing residues S(122–137), provides direct evidence for the potential of anti-Ids as components of hepatitis B vaccines (Kennedy et al., 1986a).

7. HBV-related variant viruses

The detection of HBV DNA by nucleic acid hybridization but not of HBsAg in serum and liver biopsies of patients with hepatitis has been reported by several groups (Brèchot et al., 1985; Wands et al., 1986; Lai et al., 1989). Infectious agents present in such sera appeared to differ from the known HBV subtypes since preimmunization with HBsAg did not prevent the transmission of hepatitis (Wands et al., 1986; Coursaget et al., 1987). These results suggested the existence of genetic variants of HBV which acquired env protein epitopes distinct from those present on known HBV subtypes. Other results (see above) indicated the rare emergence of HBV variants lacking the *d*/*y* and/or *w*/*r* subtype-specific determinants. However, these variants still shared common group-specific *a*-determinants known to elicit protective antibodies. Thus, it seems possible that HBV variants with altered *a*-determinants may also arise. Supportive direct evidence for the emergence of such variants comes from recent studies of McMahon et al. (1989). HBV variants became detectable in the serum of a chronic HBV carrier with hepatocellular carcinoma. The patient received a liver transplant followed by therapeutic doses of a human anti-HBs mAb. After several months, HBsAg failing to react with the mAb was detected in the serum of the treated patient. Fragments of HBV DNA, isolated from the serum, were replicated by the polymerase chain reaction and sequenced. The sequences indicated amino acid changes at positions which are conserved between distinct HBV subtypes (Fig. 19.11). There was an amino acid change at position 145 from glycine (G) to arginine (R). The G residue is conserved not only between distinct HBV subtypes but between all mammalian hepadnaviruses. Therefore, this single amino acid replacement may cause the disappearance of one of the group-specific *a*-determinants. Similar amino acid variations may occur in nature and may explain the emergence of HBV variants replicating in vivo in the presence of anti-*a* antibodies. The emergence of HBV variants should be considered in the development of strategies for world-wide immunization against hepatitis B.

8. References

Acs, G., Sells, M.A., Purcell, R.H., Price, P., Engle, R., Shapiro, M. and Popper, H. (1987) *Proc. Natl. Acad. Sci. USA* 84, 4641–4644.

Adamowicz, P., Tron, F., Vinas, R., Mevelec, M.N., Diaz, I., Couroucé, A.M. Mazert, M.C., Lagarde, D. and Girard, M. (1988) In *Viral Hepatitis and Liver Disease* (A.J. Zuckerman, Ed.) Alan R. Liss, New York. pp. 1087–1090.

Alberti, A., Pontisso, P., Tagariello, G., Cavalletto, D., Chemello, L. and Belussi, F. (1988) *Lancet* i, 1421–1424.

Antoni, B.A. and Peterson, D.L. (1988) in *Viral Hepatitis and Liver Disease* (A.J. Zuckerman, Ed.) Alan R. Liss, New York. pp. 313–317.

Argos, P. and Fuller, S.D. (1988) *EMBO J.* 7, 819–824.

Ballay, A., Levrero, M., Buendia, M.-A., Tiollais, P. and Perricaudet, M. (1985) *EMBO J.* 4, 3861–3865.

Barnett, B.C., Graham, C.M., Burt, D.S., Skehel, J.J. and Thomas, D.B. (1989) *Eur. J. Immunol.* 19, 515–521.

Bartenschlager, R. and Schaller H. (1988) *EMBO J.* 7, 4185–4192.

Bastin, J., Rothbard, J., Davey, J., Jones, I. and Townsend, A. (1987) *J. Exp. Med.* 165, 1508–1523.

Bavand, M.R. and Laub, O. (1988) *J. Virol.* 62, 626–628.

Bavand, M., Feitelson, M. and Laub, O. (1989) *J. Virol.* 63, 1019–1021.

Bichko, V., Pushko, P., Dreilina, D., Pumpen, P. and Gren, E. (1985) *FEBS Lett.* 185, 208–212.

Bitter, G.A., Egan, K.M., Burnette, W.N., Samal, B., Fieschko, J.C., Peterson, D.L., Downing, M.R., Wypych, J. and Langley, K.E. (1988) *J. Med. Virol.* 25, 123–140.

Bosch, V., Bartenschlager, R., Radziwill, G. and Schaller, H. (1988) *Virology* 166, 475–485.

Brahm, J., Vento, S., Rondanelli, E.G., Ranieri, S., Fagan, E.A., Williams, R. and Eddleston, A.L.W.F. (1988) *J. Med. Virol.* 24, 205–209.

Brèchot, C., Degos, F., Lugassy, C., Thiers, V., Zafrani, S., Franco, D., Bismuth, H., Trépo, C., Benhamou, J.-P., Wands, J., Isselbacher, K., Tiollais, P. and Berthelot, P. (1985) *New Engl. J. Med.* 312, 270–276.

Bruss, V. and Gerlich, W.H. (1988) *Virology* 163, 268–275.

Bulla, G.A. and Siddiqui, A. (1989) *Virology* 170, 251–260.

Celis, E., Kung, P.C. and Chang, T.W. (1984) *J. Immunol.* 132, 1511–1516.

Celis, E., Ou, D. and Otvos, L. Jr. (1988) *J. Immunol.* 140, 1808–1815.

Chang, C., Enders, G., Sprengel, R., Peters, N., Varmus, H.E. and Ganem, D. (1987) *J. Virol.* 61, 3322–3325.

Chang, L.-J., Pryciak, P., Ganem, D. and Varmus, H.E. (1989) *Nature* 337, 364–368.

Charbit, A., Sobczak, E., Michel, M.-L., Molla, A., Tiollais, P. and Hofnung, M. (1987) *J. Immunol.* 139, 1658–1664.

Cheng, K.-C. and Moss, B. (1987) *J. Virol.* 61, 1286–1290.

Cheng, K.-C., Smith, G.L. and Moss, B. (1986) *J. Virol.* 60, 337–344.

Cheung, R.C., Robinson, W.S., Marion, P.L. and Greenberg, H.B. (1989) *J. Virol.* 63, 2445–2451.

Chisaka, O., Araki, K., Ochiya, T., Tsurimoto, T., Hiranyawasitte-Attatippaholkun, W., Yanaihara, N. and Matsubara, K. (1987) *Gene* 60, 183–189.

Clarke, B.E., Newton, S.E., Carroll, A.R., Francis, M.J., Appleyard, G., Syred, A.D., Highfield, P.E., Rowlands, D.J. and Brown, F. (1987) *Nature* 330, 381–384.

Cohen, B.J. and Richmond, J.E. (1982) *Nature* 296, 677–678.

Cohen, J.I., Miller, R.H., Rosenblum, B., Denniston, K., Gerin, J.L. and Purcell, R.H. (1988) *Virology* 162, 12–20.

Colucci, G., Beazer, Y., Cantaluppi, C. and Tackney, C. (1988) *J. Immunol.* 141, 4376–4380.

Cote, P.J. Jr., and Gerin, J.L. (1983) *J. Virol.* 47, 15–23.

Cote, P.J. Jr., Dapolito, G.M., Shih, J.W.-K. and Gerin, J.L. (1982) *J. Virol.* 42, 135–142.

Cote, P.J. Jr., Shapiro, M., Engle, R.E., Popper, H., Purcell, R.H. and Gerin, J.L. (1986) *J. Virol.* 60, 895–901.

Coursaget, P., Bourdil, C., Adamowicz, P., Chotard, J., Diop Mar, I., Yvonnet, B., Mevelec., M.N., Barrès, J.-L., N'Doye, R. and Chiron, J.P. (1987) *Lancet* ii, 1354–1358.

Coursaget, P., Adamowicz, P., Bourdil, C., Yvonnet, B., Buisson, Y., Barrès, J.-L., Saliou, P., Chiron, J.-P. and Diop Mar, I. (1988) *Vaccine* 6, 357.

Cova, L., Lambert, V., Chavallier, A., Hantz, O., Fourel, I., Jacquet, C., Pichoud, C., Boulay, J., Chomel, B., Vitvitski, L. and Trépo, C. (1986) *J. Gen. Virol.* 67, 537–547.

Cregg, J.M., Tschopp, J.F., Stillman, C., Siegel, R., Akong, M., Craig, W.S., Buckholz, R.G., Madden, K.R., Kellaris, P.A., Davis, G.R., Smiley B.L., Cruze, J., Torregrossa, R., Velicelebi, G. and Thill, G.P. (1987) *Biotechnology* 5, 479–485.

Dehoux, P., Ribes, V., Sobczak, E. and Streeck, R.E. (1986) *Gene* 48, 155–163.

DeLisi, C. and Berzofsky, J.A. (1985) *Proc. Natl. Acad. Sci. USA* 82, 7048–7052.

Delpeyroux, F., Chenciner, N., Lim, A., Malpièce, Y., Blondel, B., Crainic, R., van der Werf, S. and Streeck, R.E. (1986) *Science* 233, 472–475.

Delpeyroux, F., Chenciner, N., Lim, A., Lambert, M., Malpièce, Y. and Streeck, R.E. (1987) *J. Mol. Biol.* 195, 343–350.

Delpeyroux, F., Peillion, N., Blondel, B., Crainic, R. and Streeck, R.E. (1988) *J. Virol.* 62, 1836–1839.

Eble, B.E., MacRae, D.R., Lingappa, V.R. and Ganem, D. (1987) *Mol. Cell Biol.* 7, 3591–3601.

Eichmann, K. and Rajewsky, K. (1975) *Eur. J. Immunol.* 5, 661–666.

Elfassi, E., Romet-Lemonne, J.-L., Essex, M., McLane, M.F. and Haseltine, W.A. (1984) *Proc. Natl. Acad. Sci. USA* 81, 3526–3528.

Elfassi, E., Haseltine, W.A. and Dienstag, J.L. (1986) *Proc. Natl. Acad. Sci. USA* 83, 2219–2222.

Emini, E.A., Larson, V., Eichberg, J., Conrad, P., Garsky, V.M., Lee, D.R., Ellis, R.W., Miller, W.J., Anderson, C.A. and Gerety, R.J. (1989) *J. Med. Virol.* 28, 7–12.

Ertl, H.C.J., Homans, E., Tournas, S. and Finberg, R.W. (1984) *J. Exp. Med.* 159, 1778–1783.

Feitelson, M.A. and Miller, R.H. (1988) *Proc. Natl. Acad. Sci. USA* 85, 6162–6166.

Feitelson, M.A., Marion, P.L. and Robinson, W.S. (1981) *J. Virol.* 39, 447–454.

Feitelson, M.A., Millman, I., Halbherr, T., Simmons, H. and Blumberg, B.S. (1986) *Proc. Natl. Acad. Sci. USA* 83, 2233–2237.

Feitelson, M.A., DeTolla, L.J. and Zhou, X.-D. (1988) *J. Virol.* 62, 1408–1415.

Ferns, R.B. and Tedder, R.S. (1984) *J. Gen. Virol.* 65, 899–908.

Ferns, R.B. and Tedder, R.S. (1986) *J. Gen. Virol.* 19, 193–203.

Ferrari, C., Penna, A., Sansoni, P., Giuberti, T., Neri, T.M., Chisari, F.V. and Fiaccodori, F. (1986) *Clin. Exp. Immunol.* 67, 497–506.

Ferrari, C., Mondelli, M.U., Penna, A., Fiaccadori, F. and Chisari, F.V. (1987a) *J. Immunol.* 139, 539–544.

Ferrari, C., Penna, A., Giuberti, T., Tong, M.J., Ribera, E., Fiaccadori, F. and Chisari, F.V. (1987b) *J. Immunol.* 139, 2050–2058.

Ferrari, C., Chisari, F.V. Ribera, E., Penna, A. and Mondelli, M.U. (1988a) *J. Immunol.* 141, 1155–1160.

Fujiyama, A., Miyanohara, A., Nozaki, C., Yoneyama, T., Ohtomo, N. and Matsubara, K. (1983) *Nucleic Acids Res.* 11, 4601–4610.

Galibert, F., Mandart, E., Fitoussi, F., Tiollais, P. and Charnay, P. (1979) *Nature* 281, 646–650.
Galibert, F., Chen, T.N. and Mandart, E. (1982) *J. Virol.* 41, 51–65.
Galle, P.R., Schlicht, H.J., Fischer, M. and Schaller, H. (1988) *J. Virol.* 62, 1736–1740.
Garcia, P.D., Ou, J.-H., Rutter, W.J. and Walter, P. (1988) *J. Cell Biol.* 106, 1093–1104.
Gerin, J.L., Purcell, R.H. and Lerner, R.A. (1985) in *Vaccines 85* (R.A. Lerner, R.M. Chanock and F. Brown, Eds.) Cold Spring Harbor Laboratory, Cold Spring Harbor NY. pp. 235–239.
Gerin, J.L., Tennant, B.C., Popper, H., Tyeryar, F.J. and Purcell, R.H. (1986) in *Vaccines 86* (F. Brown, R.M. Chanock and R.A. Lerner, Eds.) Cold Spring Harbor Laboratory, Cold Spring Harbor NY. pp. 383–386.
Gerlich, W.H., Feitelson, M.A., Marion, P.L. and Robinson, W.S. (1980) *J. Virol.* 36, 787–795.
Graham, C.M., Barnett, B.C., Hartlmayr, I., Burt, D.S., Faulkes, R., Skehel, J.J. and Thomas, D.B. (1989) *Eur. J. Immunol.* 19, 523–528.
Gripon, P., Diot, C., Thézé, N., Fourel, I., Loreal, O., Bréchot, C. and Guguen-Guillouzo, C. (1988) *J. Virol.* 62, 4136–4143.
Guerrero, E., Gavilanes, F. and Peterson, D.L. (1988) In *Viral Hepatitis and Liver Disease* (A.J. Zuckerman, Ed.) Alan R. Liss, New York. pp. 606–613.
Gust, I.D., Burrell, C.J., Coulepis, A.G., Robinson, W.S. and Zuckerman, A.J. (1986) *Intervirology* 25, 14–29.
Hanaoka, K., Takasaki, S., Kobata, A., Miyamoto, H., Nakamura, T. and Mayumi, M. (1986) *J. Biochem.* 99, 1273–1276.
Hanson, R.G., Hoofnagle, J.H., Minuk, G.Y., Purcell, R.H. and Gerin, J.L. (1984) *Clin. Exp. Immunol.* 57, 257–264.
Heermann, K.-H., Kruse, F., Seifer, M. and Gerlich W.H. (1987) *Intervirology* 28, 14–25.
Heermann, K.-H, Waldeck, F. and Gerlich, W.H. (1988) In *Viral Hepatitis and Liver Disease* (A.J. Zuckerman, Ed.) Alan R. Liss, New York. pp. 697–700.
Heijtink, R.A., de Wilde, G.A., van Hattum, J. and Schalm, S.W. (1989) *J. Med. Virol.* 27, 95–99.
Hemmerling, A.E., Fuchs, J., Steichele, G. and Thoma, H.A. (1989) *International Symposium on Progress in Hepatitis B Immunization.* INSERM/WHO, Paris, May 1989.
Hess, J., Stemler, M., Will, H., Schröder, C.H., Kühn, J. and Braun, R. (1988) *Med. Microbiol. Immunol.* 177, 195–205.
Hirsch, R., Colgrove, R. and Ganem, D. (1988) *Virology* 167, 136–142.
Huang, S.-N. and Neurath, A.R. (1979) *Lab. Invest.* 40, 1–17.
Imamura, T., Araki, M., Miyanohara, A., Nakao, J., Yonemura, H., Ohtomo, N. and Matsubara, K. (1987) *J. Virol.* 61, 3543–354.
Ionescu-Matiu, I., Kennedy, R.C., Sparrow, J.T., Culwell, A.R., Sanchez, Y., Melnick, J.L. and Dreesman, G.R. (1983) *J. Immunol.* 130, 1947–1951.
Ise, I., Tsuda, F., Aihara, S., Machida, A., Takai, E., Miyamoto, H., Akahane, Y., Miyakawa, Y. and Mayumi, M. (1988) *Hepatology* 8, 1089–1093.
Iwarson, S., Tabor, E., Thomas, H.C., Goodall, A., Waters, J., Snoy, P., Shih, J.W.K. and Gerety, R.J. (1985) *J. Med. Virol.* 16, 89–96.
Jacobs, E., Rutgers, T., Voet, P., Dewerchin, M., Cabezón, T. and De Wilde, M. (1989) *Gene,* in press.
Jean-Jean, O., Levrero, M., Will, H., Perricaudet, M. and Rossignol, J.M. (1989) *Virology* 170, 99–106.
Jilbert, A.R., Freiman, J.S., Gowans, E.J., Holmes, M., Cossart, Y.E. and Burrell, C.J. (1987) *Virology* 158, 330–338.
Jin, Y., Shih, W.-K. and Berkower, I. (1988) *J. Exp. Med.* 168, 293–306.
Junker, M., Galle, P. and Schaller, H. (1987) *Nucleic Acids Res.* 15, 10117–10132.
Kaneko, S. and Miller, R.H. (1988) *J. Virol.* 62, 3979–3984.
Kang, C.Y., Bishop, D.H.L., Seo, J.-S., Matsuura, Y. and Choe, M. (1987) *J. Gen. Virol.* 68, 2607–2613.
Kay, A., Mandart, E., Trépo, C. and Galibert, F. (1985) *EMBO J.* 4, 1287–1292.
Kennedy, R.C. and Dreesman, G.R. (1983) *J. Immunol. 130,* 385–389.

Kennedy, R.C., Dreesman, G.R., Sparrow, J.T., Culwell, A.R., Sanchez, Y., Ionescu-Matiu, I., Hollinger, F.B. and Melnick, J.L. (1983a) *J. Virol. 46,* 653–655.

Kennedy, R.C., Adler-Storthz, K., Henkel, R.D., Sanchez, Y., Melnick, J.L. and Dreesman, G.R. (1983b) *Science 221,* 853–855.

Kennedy, R.C., Melnick, J.L. and Dreesman, G.R. (1984a) *Science 223,* 930–931.

Kennedy, R.C., Sparrow, J.T., Sanchez, Y., Melnick, J.L. and Dreesman, G.R. (1984b) *Virology* 136, 247–252.

Kennedy, R.C., Eichberg, J.W., Landford, R.E. and Dreesman, G.R. (1986a) *Science* 232, 220–223.

Kennedy, R.C., Eichberg, J.W. and Dreesman, G.R. (1986b) *Virology 148,* 369–374.

Kniskern. P.J., Hagopian, A., Montgomery, D.L., Burke, P., Dunn, N.R., Hofmann, K.J., Miller, W.J. and Ellis, R.W. (1986) *Gene* 46, 135–141.

Kniskern, P.J., Hagopian, A., Burke, P., Dunn, N., Emini, E.A, Miller, W.J., Yamazaki, S. and Ellis R.W. (1988) *Hepatology* 8, 82–87.

Kobayashi, M. and Koike, K. (1984) *Gene* 30, 227–232.

Koike, K., Akatsuka, T. and Miyamura, T. (1988) *Virology* 163, 233–235.

Korba, B.E., Wells, F., Tennant, B.C., Yoakum, G.H., Purcell, R.H. and Gerin, J.L. (1986) *J. Virol.* 58, 1–8.

Korba, B.E., Wells, F., Tennant, B.C., Cote, P.J. and Gerin, J.L. (1987) *J. Virol.* 61, 1318–1324.

Korba, B.E., Cote, P.J., Wells, F.V., Baldwin, B., Popper, H., Purcell, R.H., Tennant, B.C. and Gerin, J.L. (1989) *J. Virol.* 63, 1360–1370.

Kuroda, S., Itoh, Y., Miyazaki, T., Otaka-Imai, S. and Fujisawa, Y. (1989) *Gene* 78, 297–308.

Lai, M.E., Farci, P., Figus, A., Balestrieri, A., Arnone, M. and Vyas G.N. (1989) *Blood* 73, 17–19.

Landford, R.E., Luckow, V., Kennedy, R.C., Dreesman, G.R., Notvall, L. and Summers, M.D. (1989) *J. Virol.* 63, 1549–1557.

Langley, K.E., Egan, K.M., Barendt, J.M., Parker, C.G. and Bitter, G.A. (1988) *Gene* 67, 229–245.

Leclerc, C., Charbit, A., Molla, A. and Hofnung, M. (1989) *Vaccine* 7, 242–248.

Liang, X.-H., Stemler, M., Will, H., Braun, R., Tang, Z.-Y. and Schröder, C.H. (1988) *J. Med. Virol.* 25, 329–337.

Lieberman, H.M., Tung, W.W. and Shafritz, D.A. (1987) *J. Med. Virol.* 21, 347–359.

Lo, S.J., Chien, M.-L. and Lee, Y.-H.W. (1988) *Virology* 167, 289–292.

Machida, A., Ohnuma, H., Takai, E., Tsuda, F., Tanaka, T., Naito, M., Munekata, E., Miyakawa, Y. and Mayumi, M. (1989) *Mol. Immunol.* 26, 413–421.

Mack, D.H., Bloch, W., Nath, N. and Sninsky, J.J. (1988) *J. Virol.* 62, 4786–4790.

Mandart, E., Kay, A. and Galibert, F. (1984) *J. Virol.* 49, 782–792.

Margalit, H., Spouge, J.L., Cornette. J.L., Cease, K.B., DeLisi, C. and Berzofsky, J.A. (1987) *J. Immunol.* 138, 2213–2229.

Marion, P.L., Oshiro, L.S., Regnery, D.C., Scullard, G.H. and Robinson, W.S. (1980) *Proc. Natl. Acad. Sci. USA* 77, 2941–2945.

Mason, W.S., Seal, G. and Summers, J. (1980) *J. Virol.* 36, 829–836.

McMahon, G., McCarthy, L., Dottavio, D., Nogueira, N. and Ostberg, L. (1989) Abstract, *1989 Meeting on Hepatitis B Viruses,* Sept. 1989, Cold Spring Harbor, NY.

Meyers, M.L., Trepo, L.V., Nath, N. and Sninsky. J.J. (1986) *J. Virol.* 57, 101–109.

Michel, M.-L., Pontisso, P., Sobczak, E., Malpièce, Y., Streeck, R.E. and Tiollais, P. (1984) *Proc. Natl. Acad. Sci. USA* 81, 7708–7712.

Michel, M.-L., Mancini, M., Sobczak, E., Favier, V., Guetard, D., Bahraoui, E.M. and Toillais, P. (1988) *Proc. Natl. Acad. Sci. USA* 85, 7957–7961.

Milich, D.R. (1987) *Immunol. Rev.* 99, 71–103.

Milich, D.R. and Chisari, F.V. (1982) *J. Immunol.* 129, 320–325.

Milich, D.R. and McLachlan, A. (1986) *Science* 234, 1398–1401.

Milich, D.R., Alexander, H. and Chisari, F.V. (1983) *J. Immunol.* 130, 1401–1407.

Milich, D.R., Leroux-Roels, G.G., Louie, R.E. and Chisari, F.V. (1984) *J. Exp. Med.* 159, 41–56.

Milich, D.R., Peterson, D.L., Leroux-Roels, G.G., Lerner, R.A. and Chisari, F.V. (1985a) *J. Immunol.* 134, 4203–4211.

Milich, D.R., Louie, R.E. and Chisari, F.V. (1985b) *J. Immunol.* 134, 4194–4202.

Milich, D.R., McLachlan, A., Chisari, F.V., Nakamura, T. and Thornton, G.B. (1986a) *J. Immunol.* 137, 2703–2710.

Milich, D.R., McLachlan, A., Chisari, F.V., Kent, S.B. and Thornton, G.B. (1986b) *J. Immunol.* 137, 315–322.

Milich, D.R., McLachlan, A., Chisari, F.V. and Thornton, G.R. (1986c) *J. Exp. Med. 164,* 532–547.

Milich, D.R., McLachlan, A., Moriarty, A. and Thornton, G.B. (1987a) *J. Immunol.* 138, 4457–4465.

Milich, D.R., McLachlan, A., Moriarty, A. and Thornton, G.B. (1987b) *J. Immunol.* 139, 1223–1231.

Milich, D.R., McLachlan, A., Thornton, G.B. and Hughes, J.L. (1987c) *Nature* 329, 547–549.

Milich, D.R., Hughes, J.L., McLachlan, A., Thornton, G.B. and Moriarty, A. (1988a) *Proc. Natl. Acad. Sci. USA* 85, 1610–1614.

Milich, D.R., McLachlan, A., Stahl, S., Wingfield, P., Thornton, G.B., Hughes, J.L. and Jones, J.E. (1988b) *J. Immunol.* 141, 3617–3624.

Miller, R.H. (1987) *Science* 236, 722–725.

Miller, R.H. (1988) *Virology* 164, 147–155.

Millman, I., Halbherr, T. and Simmons, H. (1982) *Infect. Immun.* 35, 752–757.

Miyanohara, A., Imamura, T., Araki, M., Sugawara, K., Ohtomo, N. and Matsubara, K. (1986) *J. Virol.* 59, 176–180.

Molnar-Kimber, K.L., Jarocki-Witek, V., Dheer, S.K., Vernon, S.K., Conley, A.J., Davis, A.R. and Hung, P.P. (1988) *J. Virol.* 62, 407–416.

Mondelli, M.U., Bortolotti, F., Pontisso, P., Rondanelli, E.G., Williams, R., Realdi, G., Alberti, A. and Eddleston, A.L.W.F. (1987) *Clin. Exp. Immunol.* 68, 242–250.

Moriarty, A.M., Alexander, H. and Lerner, R.A. (1985) *Science* 227, 429–433.

Morin, J.E., Lubeck, M.D., Barton, J.E., Conley, A.J., Davis, A.R. and Hung, P.P. (1987) *Proc. Natl. Acad. Sci. USA* 84, 4626–4630.

Murray, K., Bruce, S.A., Wingfield, P., van Eerd, P., de Reus, A. and Schellekens, H. (1987) *J. Med. Virol.* 23, 101–107.

Nassal, M. (1988) *Gene* 66, 279–294.

Nassal, M., Galle, P.R. and Schaller, H. (1989) *J. Virol.* 63, 2598–2604.

Neurath, A.R. (1989) *Immunology of Prophylactic Immunization* (A.J. Zuckerman, Ed.), MTP Press, Lancaster, U.K., pp. 210–242.

Neurath, A.R. and Kent, S.B.H. (1985) In *Immunochemistry of Viruses: The Basis for Serodiagnosis and Vaccines* (M.H.V. Van Regenmortel and A.R. Neurath, Eds.) Elsevier, Amsterdam. pp. 325–366.

Neurath, A.R. and Kent, S.B.H. (1988) *Adv. Virus Res.* 34, 65–142.

Neurath, A.R. and Strick, N. (1979) *J. Gen. Virol.* 42, 645–649.

Neurath, A.R. and Strick, N., (1989) In *Immunology of Proteins and Peptides* (Z. Atassi, Ed.), Plenum Press, New York, pp. 237–250.

Neurath, A.R., Strick, N., Szmuness, W., Stevens, C.E. and Harley, E.J. (1979) *J. Gen. Virol.* 42, 493–504.

Neurath, A.R., Kent, S.B.H., Strick, N., Stark, D. and Sproul, P. (1985) *J. Med. Virol.* 17, 119–125.

Neurath, A.R., Kent, S.B.H., Strick, N. and Parker, K. (1986) *Cell* 46, 429–436.

Neurath, A.R., Jameson, B.A. and Huima, T. (1987a) *Microbiol. Sci.* 4, 45.

Neurath, A.R., Kent, S.B.H., Strick, N., Parker, K., Courouсé, A.-M., Riottot, M.M., Petit, M.A., Budowska, A., Girard, M. and Pillot, J. (1987b) *Mol. Immunol.* 24, 975–980.

Neurath, A.R., Kent, S.B.H., Adamowicz, P., Riottot, M.M., Price, P., Strick, N., Parker, K., Petit, M.A., Budkowska, A., Girard, M. and Pillot, J. (1987c) *Mol. Immunol.* 24, 561–568.

Neurath, A.R., Kent, S.B.H., Strick, N. and Parker, K. (1988a) *Ann. Inst. Pasteur/Virol.* 139, 13–38.

Neurath, A.R., Strick, N., Kent, S.B.H., Parker, K., Chin, S.K., Girard, M., Ralph, H.E. and Valinsky, J. (1988b) In *Synthetic Peptides: Approaches to Biological Problems* (J.P. Tam and E.T. Kaiser, Eds.) Alan R. Liss, New York, in press.

Neurath, A.R., Strick, N., Parker, K. and Kent, S.B.H. (1988c) *J. Med. Virol.* 24, 137–151.

Neurath, A.R., Seto, B. and Strick, N. (1989a) *Vaccine* 7, 234–236.

Neurath, A.R., and Strick, N. and Girard, M. (1989b) *Mol. Immunol.* 26, 53–62.

Neurath, A.R., and Strick, N. and Lee, E.Y.S. (1990) *J. Gen. Virol.,* in press.

Nisonoff, A and Lamoyi, E. (1981) *Clin. Immunol. Immunopathol.* 21, 397–406.

Ogston, C.W., Schechter, E.M., Humes, C.A. and Pranikoff, M.B. (1989) *Virology* 169, 9–14.

Ohnuma, H., Machida, A., Takai, E., Tsuda, F., Miyamoto, H., Tanaka, T., Oda, K., Usuda, S., Nakamura, T., Miyakawa, Y. and Mayumi, M. (1986) *Clin Exp. Immunol.* 67, 709–715.

Ohori, H. and Matsuda, K. (1989) *Virology* 168, 40–47.

Okamoto, H., Imai, M., Shimozaki, M., Hoshi, Y., Iizuka, H., Gotanda, T., Tsuda, F., Miyakawa, Y. and Mayumi, M. (1986) *J. Gen. Virol.* 67, 2305–2314.

Okamoto, H., Imai, M., Miyakawa, Y. and Mayumi, M. (1987a) *Biochem. Biophys. Res. Commun.* 148, 500–504.

Okamoto, H., Imai, M., Tsuda, F., Tanaka, T., Miyakawa, Y. and Mayumi, M. (1987b) *J. Virol.* 61, 3030–3034.

Okamoto, H., Tsuda, F., Sakugawa, H., Sastrosoewignjo, R.I., Imai, M., Miyakawa, Y. and Mayumi, M. (1988) *J. Gen. Virol.* 69, 2575–2583.

Okamoto, H., Omi, S., Wang, Y., Itoh, Y., Tsuda, F., Tanaka, T., Akahane, Y., Miyakawa, Y. and Mayumi, M. (1989) *Mol. Immunol.* 26, 197–205.

Ono, Y., Onda, H., Sasada, R., Igarashi, K., Sugino, Y. and Nishioka, K. (1983) *Nucleic Acids Res.* 11, 1747–1757.

Onodera, S., Ohori, H., Yamaki, M. and Ishida, N. (1982) *J. Med. Virol.* 10, 147–155.

Ou, J.-H., Laub, O. and Rutter, W.J. (1986) *Proc. Natl. Acad. Sci. USA* 83, 1578–1582.

Patzer, E.J., Nakamura, G.R., Hershberg, R.D., Gregory, T.J., Crowley, C., Levinson, A.D. and Eichberg, J.W. (1986) *Biotechnology* 4, 630–636.

Persing, D.H., Varmus, H.E. and Ganem, D. (1986a) *Science* 234, 1388–1391.

Persing, D.H., Varmus, H.E. and Ganem, D. (1986b) *J. Virol.* 60, 177–184.

Persing, D.H., Varmus, H.E. and Ganem, D. (1987) *J. Virol.* 61, 1672–1677.

Peterson, L.B., Wilner, G.D. and Thomas, D.W. (1983) *J. Immunol.* 130, 2542–2545.

Petit, M.-A., Dubanchet, S. and Capel, F. (1989) *Mol. Immunol.* 26, 531–537.

Pfaff, E., Salfeld, J., Gmelin, K., Schaller, H. and Theilmann, L. (1987) *Virology* 158, 456–460.

Pohl, C.J., Cote, P.J., Purcell, R.H. and Gerin, J.L. (1986) *J. Virol.* 60, 943–949.

Ponzetto, A., Cote, P.J. Jr., Ford, E.C., Purcell, R.H. and Gerin, J.L. (1984) *J. Virol.* 52, 70–76.

Price, P.M., Mohamad, A., Zelent, A., Neurath, A.R. and Acs, G. (1988) *DNA* 7, 417–422.

Price, P.M., Reichelderfer, C.F., Johansson, B.E., Kilbourne, E.D. and Acs, G. (1989) *Proc. Natl. Acad. Sci. USA* 86, 1453–1456.

Pride, M.W., Thakur, A.N., Ogra, P.L., Evans, R.L. and Thanavala, Y. 104 (1988). In *Modern Approaches to New Vaccines including prevention of AIDS.* Cold Spring Harbor Laboratory, Cold Spring Harbor NY. (Abstr. 104)

Pugh, J.C., Weber, C., Houston, H. and Murray, K. (1986) *J. Med. Virol.* 20, 229–246.

Pugh, J., Zweidler, A. and Summers, J. (1989) *J. Virol.* 63, 1371–1376.

Radziwill, G., Zentgraf, H., Schaller, H. and Bosch, V. (1988) *Virology* 163, 123–132.

Reed, R.G. and Burrington, C.M. (1989) *J. Biol. Chem.* 264, 9867–9872.

Roisman, F.R., Mota, A.H. and Fainboim, L. (1989) *J. Clin. Immunol.* 9, 10–15.

Roitt, I.M., Cooke, A., Male, D.K., Hay, F.C., Guarnotta, G., Lydyard, P.M., DeCarvalho, L.P., Thanavala, Y.M. and Ivanyi, J. (1981) *Lancet* i, 1041–1045.

Romet-Lemonne, J.-L., McLane, M.F., Elfassi, E., Haseltine, W.A., Azocar, J. and Essex, M. (1983) *Science* 221, 667–669.

Roossinck, M.J. and Siddiqui, A. (1987) *J. Virol.* 61, 955–961.

Roossinck, M., Jameel, S., Loukin, S. and Siddiqui, A. (1986) *Mol. Cell. Biol.* 6, 1393–1400.

Rothbard, J.B. and Taylor, W.R. (1988) *EMBO J.* 7, 93.

Rutgers, T., Cabezón, T., Harford, N., Vanderbrugge, D., Descurieux, M., Van Opstal, O., Van Wijnendaele, F., Hauser, P., Voet, P., and De Wilde, M. (1988) In *Viral Hepatitis and Liver Disease* (A.J. Zuckerman, Ed.) Alan R. Liss, New York. pp. 304–308.

Rutgers, T., Cabezón, T. and De Wilde, M. (1990) In *Immunochemistry of Viruses II* (M.H.V. Van Regenmortel and A.R. Neurath, Eds.) Elsevier, Amsterdam. pp. 103–124.

Sacks, D.L., and Sher, A. (1983) *J. Immunol.* 131, 1511–1515.

Salfeld, J. (1985) Gene Products of Hepatitis B Virus. Thesis, University of Heidelberg, F.R.G.

Salfeld, J., Pfaff, E., Noah, M. and Schaller, H. (1989) *J. Virol.* 63, 798–808.

Samanta, H. and Youn, B.W. (1989) *Vaccine* 7, 69–76.

Schaeffer, E. and Sninsky, J.J. (1984) *Proc. Natl. Acad. Sci. USA* 81, 2902–2906.

Schlicht, H.J., Kuhn, C., Guhr, B., Mattaliano, R.J. and Schaller, H. (1987a) *J. Virol.* 61, 2280–2285.

Schlicht, H.-J., Salfeld, J. and Schaller, H. (1987b) *J. Virol.* 61, 3701–3709.

Schlicht, H.-J., Bartenschalager, R. and Schaller, H. (1989a) *J. Virol.* 63, 2995–3000.

Schlicht, H.-J., Radziwill, G. and Schaller, H. (1989b) *Cell* 56, 85–92.

Schödel, F. and Will, H. (1989) *Infect. Immun.* 57, 1347–1350.

Schödel, F., Sprengel, R., Weimer, T., Fernholz, D., Schneider, R. and Will, H. (1989) In *Advances in Viral Oncology*, Vol. 8 (G. Klein, Ed.) Raven Press, New York. pp. 73–101.

Seeger, C., Ganem, D., and Varmus, H.E. (1984a) *J. Virol.* 51, 367–375.

Seeger, C., Ganem, D., and Varmus, H.E. (1984b) *Proc. Natl. Acad. Sci. USA* 81, 5849–5852.

Seeger, C., Marion, P.L., Ganem, D. and Varmus, H.E. (1987) *J. Virol.* 61, 3241–3247.

Sells, M.A., Chen, M.-L. and Acs, G. (1987) *Proc. Natl. Acad. Sci. USA* 84, 1005–1009.

Sharpe, A.H., Gauton, G.N., McDade, K.K., Fields, B.N. and Greene, M.I. (1984) *J. Exp. Med.* 160, 1195–1205.

Shevach, E.M. and Rosenthal, A.S. (1973) *J. Exp. Med.* 138, 1213.

Siddiqui, A., Jameel, S. and Mapoles, J. (1987) *Proc. Natl. Acad. Sci. USA* 84, 2513–2517.

Siddiqui, A., Gaynor, R., Srinivasan, A., Mapoles, J. and Farr, R.W. (1989) *Virology* 169, 479–484.

Simon, K., Lingappa, V.R. and Ganem, D. (1988) *J. Cell Biol.* 107, 2163–2168.

Spandau, D.F. and Lee, C.-H. (1988) *J. Virol.* 62, 427–434.

Sprengel, R., Kuhn, C., Manso, C. and Will, H. (1984) *J. Virol.* 52, 932–937.

Sprengel, R., Kuhn, C., Will, H. and Schaller, H. (1985) *J. Med. Virol.* 15, 323–333.

Sprengel, R., Kaleta, E.F. and Will, H. (1988) *J. Virol.* 62, 3832–3839.

Standring, D.N., Ou, J.-H., Masiarz, F.R. and Rutter, W.J. (1988) *Proc. Natl. Acad. Sci. USA* 85, 8405–8409.

Stannard, L.M., Hantz, O. and Tréepo, C. (1983) *J. Gen. Virol.* 64, 975–980.

Stibbe, W. and Gerlich, W.H. (1982) *Virology* 123, 436–442.

Sureau, C., Romet-Lemonne, J.-L., Mullins, J.I. and Essex, M. (1986) *Cell* 47, 37–47.

Sureau, C., Eichberg, J.W., Hubbard, G.B., Romet-Lemonne, J.L. and Essex, M. (1988) *J. Virol.* 62, 3064–3067.

Sy, M.S., Bach, B.A. Dohi, Y., Nisonoff, A., Benacerraf, B. and Greene, M.I. (1980) *J. Exp. Med.* 151, 896–909.

Tachibana, K., Tanaka, T., Usuda, S., Okomoto, H., Tsuda, F., Akahane, Y., Miyakawa, Y. and Mayumi, M. (1989) *Viral Immunol.* 2, 25–29.

Takahashi, K., Machida, A., Funatsu, G., Nomura, M., Usuda, S., Aoyagi, S., Tachibana, K., Miyamoto, H., Imai, M., Nakamura, T., Miyakawa, Y., and Mayumi, M. (1983) *J. Immunol.* 130, 2903–2907.

Takehara, K., Ireland, D. and Bishop. D.H.L. (1988) *J. Gen. Virol.* 69, 2763–2777.

Thanavala, Y.M. and Roitt, I.M. (1986a) *Int. Rev. Immunol.* 1, 27–39.

Thanavala, Y.M., Bond, A., Hay, F.C. and Roitt, I.M. (1985) *J. Immunol. Methods* 83, 227–232.

Thanavala, Y.M., Brown, S.E., Howard, C.R., Roitt, I.M. and Steward, M.W. (1986b) *J. Exp. Med.* 164, 227–236.

Thanavala, Y.M., Pride, M., Evans, R. and Thakur, A. (1989) In *7th International Congress of Immunology,* Berlin (Abstr. 31–21).

Theilmann, L., Salfeld, J., Kommerell, B. and Pfaff, E. (1988) *Hepatogastroenterology* 35, 104–106.

Thomas, W.R., Morahan, G. and Miller, J.F.A.P. (1983) *J. Immunol.* 130, 2079–2082.

Thornton, G.B., Milich, D., Chisari, F., Mitamura, K., Kent, S.B.H., Neurath, A.R., Purcell, R. and Gerin, J. (1987) In *Vaccines 87* (R.M. Chanock, R.A. Lerner, F. Brown and H. Ginsberg, Eds.) Cold Spring Harbor Laboratory, Cold Spring Harbor NY. pp. 77–80.

Thornton, G.B., Moriarty, A.M., Milich, D.R., Eichberg, J.W., Purcell, R.H. and Gerin, J.L. (1989) In *Vaccines 89* (R.A. Lerner, H. Ginsberg, R.M. Chanock and F. Brown, and Eds.) Cold Spring Harbor Laboratory, Cold Spring Harbor NY. pp. 467–471.

Toh, H., Hayashida, H. and Miyata, T. (1983) *Nature* 305, 827–829.

Townsend, A., Rothbard, J., Gotch, F.M., Bahadur, G., Wraith, D. and McMichael, A.J. (1986) *Cell* 44, 959–968.

Trevisan, A., Realdi, G., Alberti, A., Ongaro, G., Pornaro, E. and Meliconi, R. (1982) *Gastroenterology* 82, 218–222.

Tsurimoto, T., Fujiyama, A. and Matsubara, K. (1987) *Proc. Natl. Acad. Sci. USA* 84, 444–448.

Tuttleman, J.S., Pugh, J.C. and Summers, J.W. (1986) *J. Virol.* 58, 17–25.

Twu, J.-S. and Robinson, W.S. (1989) *Proc. Natl. Acad. Sci. USA* 86, 2046–2050.

Uchida, T., Suzuki, K., Okuda, Y. and Shikata, T. (1988) *Hepatology* 8, 760–765.

Urban, M.K., O'Connell, A.P. and London, W.T. (1985) *J. Virol.* 55, 16–22.

Uy, A., Bruss, V., Gerlich, W.H., Köchel, H.G. and Thomssen, R. (1986) *Virology* 155, 89–96.

Valenzuela, P., Quiroga, M., Zaldivar, J., Gray, P. and Rutter, W.J. (1980) In *Animal Virus Genetics* (B.N. Fields and R. Jaenisch, Eds.) Academic Press, New York. pp. 57–70.

Valenzuela, P., Coit, D. and Kuo, C.H. (1985a) *Biotechnology* 3, 317–320.

Valenzuela, P., Coit, D., Medina-Selby, M.A., Kuo, C.H., Van Nest, G., Burke, R.L., Urdea, M.S. and Graves, P.V. (1985b) In *Vaccines* 85 (R.A. Lerner, R.M. Chanock and F. Brown, Eds.) pp. 285–290. Cold Spring Harbor Laboratory, Cold Spring Harbor, NY.

Vento, S., Hegarty, J.E., Alberti, A., O'Brien, C.J., Alexander, G.J.M., Eddleston, A.L.W.F. and Williams, R. (1985) *Hepatology* 5, 192–197.

Wampler, D.E., Lehman, E.D., Boger, J., McAleer, W.J. and Scolnick, E.M. (1985) *Proc. Natl. Acad. Sci. USA* 82, 6830–6834.

Wands, J.R., Fujita, Y.K., Isselbacher, K.J., Degott, C., Schellekens, H., Dazza, M.-C., Thiers, V., Tiollais, P. and Brèchot, C. (1986) *Proc. Natl. Acad. Sci. USA* 83, 6608–6612.

Waters, J.A., Jowett, T.P. and Thomas, H.C. (1986a) *J. Med. Virol.* 19, 79–86.

Waters, J., Pignatelli, M., Galpin, S., Ishihara, K. and Thomas, H.C. (1986b) *J. Gen. Virol.* 67, 2467–2473.

Weimer, T., Salfeld, J. and Will, H. (1987) *J. Virol.* 61, 3109–3113.

Wen, Y.-M., Xu, Y.-Y., Zhang, W. and Liu, Y.-Z. (1990) submitted for publication.

Werner, B.G., Smolec, J.M., Snyder, R. and Summers, J. (1979) *J. Virol.* 32, 314–322.

Wu, J.Y., Newton, S., Judd, A., Stocker, B. and Robinson, W.S. (1989) *Proc. Natl. Acad. Sci. USA* 86, 4726–4730.

Yaginuma, K., Shirakata, Y., Kobayashi, M. and Koike, K. (1987) *Proc. Natl. Acad. Sci. USA* 84, 2678–2682.

Yanagi, Y., Yoshikai, Y., Leggett, K., Clark, S., Aleksander, I. and Mak, T. (1984) *Nature* 308, 145–149.

Yokosuka, O., Omata, M. and Ito, Y. (1988) *Virology* 167, 82–86.

Youn, B.W. and Samanta, H. (1989) *Vaccine* 7, 60–68.

Eds. M.H.V. van Regenmortel and A.R. Neurath
Immunochemistry of viruses, II. The basis for serodiagnosis and vaccines.

Herpes simplex virus 20

M.I. MUGGERIDGE,[1,2], S.R. ROBERTS[1,2], V.J. ISOLA[1,2], G.H. COHEN[1,2] and R.J. EISENBERG[2,3]

Department of Microbiology[1] and Center for Oral Health Research[2], School of Dental Medicine, and Department of Pathobiology, School of Veterinary Medicine[3], University of Pennsylvania, Philadelphia, PA 19104, U.S.A.

1. Introduction

Herpes simplex viruses types 1 and 2 (HSV-1 and HSV-2) are enveloped DNA viruses that cause human diseases, including cold sores, eye and genital infections, and encephalitis (Corey, 1985; Whitley, 1985). The virion envelope contains seven glycoproteins, of which gB, gD and gH are essential for virus infection (Sarmiento et al., 1979; Gompels and Minson, 1986; Cai et al., 1988a; Desai et al., 1988; Ligas and Johnson, 1988; Fuller et al., 1989). The genes for gC, gE, gG and gI can be deleted or inactivated without destroying virus infectivity in cell culture (Ruyechan et al., 1979; Holland et al., 1983; Longnecker and Roizman, 1986; Longnecker and Roizman, 1987; Neidhardt et al., 1987; Weber et al., 1987; Harland and Brown, 1988); however, gC, gE and gI interact with components of the immune system and may contribute to HSV pathogenesis (Baucke and Spear, 1979; Friedman et al., 1984; Johnson and Feenstra, 1987; Kubota et al., 1987; McNearney et al., 1987; Neidhardt et al., 1987; Johnson et al., 1988; Dubin et al., 1990). Because they reside on the virion surface and infected cell plasma membranes, the glycoproteins are major targets of the immune response to HSV infection (Glorioso et al., 1985; Ross et al., 1985; Yasukawa and Zarling, 1985; Bishop et al., 1986; Zarling et al., 1986; Blacklaws et al., 1987; Marsden, 1987; Rosenthal et al., 1987). In the search for an effective vaccine, an understanding of the importance of specific immune mechanisms and delineation of the epitopes which they recognize is clearly desirable. Recent reviews have focused on the immunology of HSV infections (Nash et al., 1985; Norrild, 1985; Wildy and Gell, 1985; Dix, 1987), progress in vaccine development (Meignier, 1985; Hall and Katrak, 1986; Dix, 1987), and the structures and functions of HSV glycoproteins (Spear, 1985a; Spear 1985b; Marsden, 1987). The object of this review is to describe the antigenic structures of the three best characterized glycoproteins (gB, gC and gD), as of June 1989. Discussion of the remaining glycoproteins is premature at this time as little is known about their antigenicity; for the same reason, we have described only B-cell epitopes, as mapping of T-cell epitopes is currently limited to gD (DeFreitas et al., 1985; Heber-Katz et al., 1985; Heber-Katz et al., 1988).

Two complementary approaches to antigenic mapping have been taken. One begins with the arrangement of monoclonal antibodies (mAbs) in groups, based on their type-specificity, reaction with protein fragments, and blocking assays, and anticipates that members of a group bind to the same antigenic site; components of each site are then identified by the protein fragments to which the mAbs bind. The other starts with the selection of neutralization-resistant viruses (*mar* mutants), followed by cross-neutralization studies to identify mAbs with overlapping epitopes. Although this approach is limited to mAbs with neutralizing activity, the residues involved in antibody binding can be identified by sequencing of the mutant genes. We have described epitopes and antigenic sites as continuous or discontinuous. The former are those which remain intact after denaturation of the protein, and presumably are composed of residues adjacent to each other in the protein sequence; the latter are those which are destroyed by denaturation because they comprise noncontiguous residues, brought together by folding of the protein into its native state.

2. Glycoprotein B

2.1. Introduction

gB has an essential role in entry of HSV into the cell, which appears to occur by fusion with the plasma membrane (Para et al., 1980; DeLuca et al., 1981). Thus, viruses with temperature-sensitive (ts) mutations in gB (Sarmiento et al., 1979; Haffey and Spear, 1980; Little et al., 1981), and mutants in lacking gB entirely (Cai et al., 1988a), can bind to cells but not initiate infection. In both cases, the defect is overcome by polyethylene glycol (PEG)- induced fusion between virus and cell membranes. In addition, virus entry can be prevented by an anti-gB mAb (B4) even when the virus has been preadsorbed to the cell surface (Highlander et al., 1988).

gB-1 has 874 amino acids after removal of the signal sequence (Bzik et al., 1984; Pellet et al., 1985; Bzik et al., 1986; Claesson-Welsh and Spear, 1986; Pachl et al., 1987; Stuve et al., 1987), whereas gB-2 has 882 (Bzik et al., 1986; Stuve et al., 1987). In this review of gB, we will refer to the first amino acid after the signal as residue 1, and numbers from the literature have been adjusted where necessary.

The homology between gB-1 and gB-2 is 87% (Bzik et al., 1986). There are six potential *N*-linked glycosylation sites in the external portion of gB-1, of which five are used (Bzik et al., 1984; Pellet et al., 1985; Stuve et al., 1987; Cai et al., 1988b). gB-2 has seven sites (Bzik et al., 1986; Stuve et al., 1987). The protein can be extracted from virions and infected cells in a dimeric form (Sarmiento and Spear, 1979; Haffey and Spear, 1980; Eberle and Courtney, 1982), and also can be chemically cross-linked within the virion envelope to produce a high molecular weight complex (Zhu and Courtney, 1988). Dimerization occurs within minutes of translation (Claesson-Welsh and Spear, 1986), and appears to be necessary for the formation of several epitopes (Chapsal and Pereira, 1988; Pereira et al., 1989).

Many anti-gB mAbs have been isolated (Pereira et al., 1981; Showalter et al., 1981; Kousoulas et al., 1984; Para et al., 1985; Marlin et al., 1986; Fujinaga et al., 1987), and two laboratories have undertaken an antigenic analysis of gB. Antigenic sites have been referred to as I, II, III, and IV (Highlander et al., 1989), or domains D1a, b and c, D2a, b and c and D3a, b and c (Pereira et al., 1989) (Fig. 20.1); we have made some suggestions about where these systems overlap, but experimental reconciliation is awaited.

2.2. Continuous epitopes

Mapping of continuous epitopes is still preliminary: approximate locations have been deduced from reactions with truncated fragments, but only mAbs with epitopes at the N-terminus have reacted with synthetic peptides. Pereira et al. (1989) identified several mAbs that react with fragment 1–190, and these define the D1 domains (Fig. 20.1A). Five of the mAbs react with peptides, mapping domain D1a within residues 1–20 and D1b within 16–50 (Fig. 20.2E). Selection and sequencing of *mar* mutants has also been successful: D1b mAb H1397, which binds to peptide 31–50, selected a mutation at residue 47 (Fig. 2A); and H1392 selected a mutation at amino acid 32, placing its epitope in domain D1b (Fig. 20.2A) (Kousoulas et al., 1988; Pereira et al., 1989). Antigenic site IV may overlap domain D1b, as mAb B6 selected a mutation at amino acid 55 (Fig. 20.2A) (Highlander et al., 1989).

Domains D2c and D3c are located between residues 190 and 475 (Figs. 20.1A and 2E), but the failure of the mAbs concerned to react with a set of peptides covering residues 181–470 has frustrated more precise mapping (Pereira et al., 1989). Other mAbs do not recognize a 1–475 fragment, and presumably have epitopes in the C-terminal half of gB; two of these mAbs, which are type 1-specific, bind to a hybrid containing the N-terminal half of gB-2 and the C-terminal half of gB-1 (Kousoulas et al., 1989).

Reaction with the gB-2/gB-1 hybrid has also been used to map the continuous epitope of a type 2-specific mAb (H368) to the N-terminal half of gB-2 (Chapsal and Pereira, 1988; Kousoulas et al., 1989). The H368 epitope is missing from the cleaved form of gB-2 made in Vero cells (Pereira et al., 1981), just as some of the D1a, b and c epitopes are absent from the corresponding form of gB-1 (Chapsal and Pereira, 1988); by analogy, it seems likely that the H368 epitope is near the N-terminus of gB-2.

2.3. Discontinuous epitopes

Several discontinuous antigenic sites have been identified. Those which are type-common or type 1-specific have been partially mapped and will be discussed in detail, but almost no mapping has been reported for those which are type 2-specific.

2.3.1. Type-common epitopes

Domain D2a is defined by four mAbs which selected *mar* mutations between residues

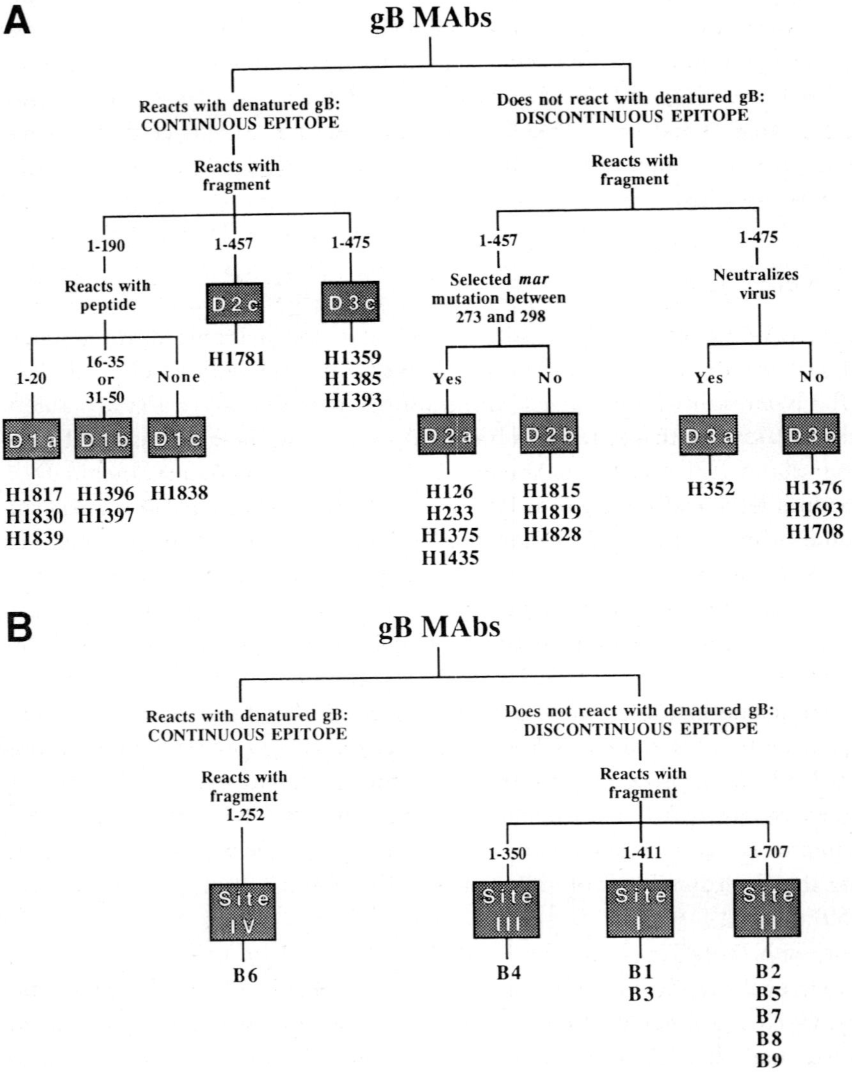

Fig. 20.1. Schemes for grouping of anti-gB mAbs: (A) according to Pereira and colleagues; (B) according to Glorioso and colleagues.

273 and 298 (Figs. 20.1 A and 2D) (Pellett et al., 1985; Kousoulas et al., 1988; Pereira et al., 1989). Cross-neutralization studies showed that three of these mAbs (H126, H1375 and H1435) have overlapping epitopes (Kousoulas et al., 1988), but there is no evidence for overlap of H233, which may therefore define a distinct antigenic site. A second mutant selected with H126 has changes at positions 305 and 443 (Pellett et al., 1985; Kousoulas et al., 1988); however, it is not known if both are required for neutralization-resistance. mAb B4, which defines site III, selected a *mar* mutation at position 275 (Figs. 20.1 B and 2D) (Highlander et al., 1989), suggesting that site III and domain D2a overlap. mAb 3S also binds to site III, since it fails to neutralize the B4 *mar* mutant (Showalter et al., 1981; Marlin et al., 1986). B4 recognizes a 1–350 fragment of gB (Highlander et al., 1988), whereas the D2a mAbs recognize a 1–457 fragment but not 1–440 (Pereira et al., 1989) (Fig. 20.2E). Domain D2a may therefore include downstream residues not required for the B4 epitope.

Domain D2b is also present within the sequence 1–457 but not 1–440 (Figs. 20.1 A and 2E) (Pereira et al., 1989), and the difference between D2a and D2b is unclear, as cross-neutralization studies with D2b mAbs and D2a *mar* mutants were not reported. Domains D3a and D3b are present within 1–475 but not 1–457 (Fig. 20.2E) (Pereira et al., 1989), the distinction apparently being that the D3a mAb has neutralizing activity, whereas the D3b mAbs do not (Chapsal and Pereira, 1988).

Site II is contained within a 1–707 fragment but not 1–565 (Fig. 20.2E) (Highlander et al., 1988). Two site II mAbs selected a *mar* mutation at residue 564 (Fig. 20.2B) (Marlin et al., 1986; Highlander et al., 1989), and cross-neutralization experiments identified 5 additional mAbs (B7, B8, B9, 40S and 61S) that recognize this site (Showalter et al., 1981; Marlin et al., 1986). The 564 mutation produces two phenotypic changes in the virus in addition to its antigenic effect: first, it slows down the rate of penetration into cells (Highlander et al., 1988); second, it makes gB maturation temperature-sensitive (Marlin et al., 1986), as does the mutation at amino acid 275 in site III. A double *mar* mutant, incorporating both amino acid changes, is ts for virus production as well as gB maturation. Several ts viruses with defective processing of gB were previously isolated without recourse to mAbs, and the gB precursor which they synthesize at non-permissive temperatures does not react with site II or site III mAbs (Marlin et al., 1986), suggesting faulty folding. The gB genes of two ts mutants have been sequenced (Bzik et al., 1984) and found to have mutations at amino acids 243 (tsJ20) and 347 (tsJ12) (Fig. 20.2D).

2.3.2. Type 1-specific epitopes

Site I is defined by mAbs B1 and B3 and is listed as discontinuous because it is not seen by mAbs on a Western blot of a denaturing gel (Highlander et al., 1989). However, the evidence that it is discontinuous is equivocal. Unlike sites II and III, it remains intact in precursor gB synthesized by ts mutants at the non-permissive temperature (Marlin et al., 1986), and is unaffected by amino acid insertions scattered

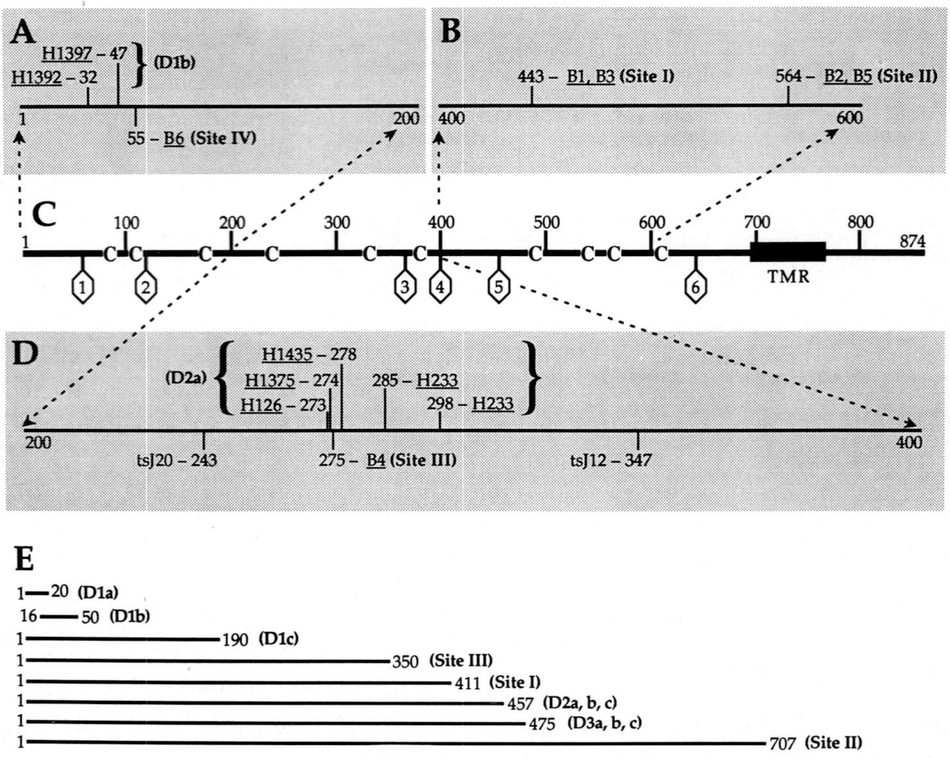

Fig. 20.2. Mapping of antigenic sites on gB-1. (A) *mar* mutations in domain D1b and site IV, with mAbs underlined; (B) *mar* mutations in sites I and II; (C) stick figure of gB-1, showing positions of cysteine residues (C), *N*-linked carbohydrates (hexagons) and transmembrane region (TMR); (D) *mar* mutations in domain D2a and site III; (E) smallest fragment containing the indicated antigenic sites.

throughout the protein (Highlander et al., 1989). If the site is actually continuous, its presence in a 1–411 fragment but not 1–350 (Fig. 20.2E) suggests that it lies between residues 350 and 411 (Highlander et al., 1988). Both mAbs selected a *mar* mutation at amino acid 443 (Fig. 20.2B) (Highlander et al., 1989); presumably, resistance to neutralization results from a conformational effect. As noted in section 2.3.1, 443 was one of two altered residues in a *mar* mutant selected with the D2a mAb H126 (Pellett et al., 1985; Kousoulas et al., 1988). Since the mutation of 443 alone has no influence on site III (Marlin et al., 1986), which we believe to be related to domain D2a, perhaps it is sufficient to alter the conformation of site I but needs an additional change to alter site III and domain D2a. This could be tested by reacting the H126 *mar* mutants with B1, B3 and B4, and the B1 or B3 mutants with H126.

2.4. Implications for gB structure

Mutagenesis of the gB gene suggests that the external part of the protein has two structural (rather than antigenic) domains, and that the N-terminal domain folds independently of the C-terminal domain. Thus, insertions between residues 159 and 407, but not between residues 433 and 704, affect binding of mAbs to site III, whereas binding to site II is reduced by insertions anywhere between positions 159 and 704 (Cai et al., 1988b; Highlander et al., 1989). Since all of these mutants have impaired processing of *N*-linked carbohydrates and are functionally defective (Cai et al., 1988b), insertions downstream of 407 may affect the structure of a C-terminal domain that includes site II, whereas insertions upstream of 407 affect the structure of the entire protein, including an N-terminal domain containing site III. In addition, a deletion mutant lacking residues 523–845 is recognized by mAbs to sites I and III, but not II (Cai et al., 1987). However, the boundary for correct folding of an N-terminal domain is unclear. Highlander et al. (1988) find that truncated fragments as short as 1–350 contain site III, whereas Pereira et al. (1989) find no discontinuous epitopes in a 1–440 fragment. Furthermore, the 1–350 fragment, and another truncated at 411, are not processed correctly or secreted, suggesting malfolding (Cai et al., 1988b).

Pereira et al. (1989) interpret their data in terms of the gB dimer. Of 30 mAbs with discontinuous epitopes, the 11 that recognize domains D2a, D2b, D3a or D3b bind a 1–475 fragment, which does not dimerize. Therefore, discontinuous epitopes can form in the N-terminal half of monomeric gB, independently of the C-terminal half. In contrast, 19 mAbs do not react with 1–475 or with full-length monomeric gB; they react only with the intact dimer (Chapsal and Pereira, 1988; Pereira et al., 1989), suggesting that epitopes in the C-terminal half may involve residues from both gB monomers. Curiously, however, the mAbs that bind the 1–475 fragment fail to bind full-length monomeric gB, which was derived from the dimer by reduction with 2-mercaptoethanol (Chapsal and Pereira, 1988). It would be interesting to test all the antibodies against the monomeric gB produced by in vitro translation (Claesson-Welsh and Spear, 1987), and also against a 1–680 fragment reported to form dimers (Pachl et al., 1987).

2.5. Biological properties of mAbs

All the B-series mAbs have neutralizing activity; however, only B4 (site III) neutralizes in the absence of complement (Marlin et al., 1986), and this occurs at the level of penetration rather than attachment (Highlander et al., 1988). B1 and B3 (site I) and B6 (site IV) reduce the rate of virus penetration, but only B6 inhibits cell-to-cell spread of virus (Highlander et al., 1988); the site II mAbs do neither. Many of the H-series mAbs have neutralizing activity, but it is not clear which ones require complement. Domains D1a, b and c, D2a, b and c, and D3a are the targets of neutralizing mAbs; domains D3b and D3c are not (Chapsal and Pereira, 1988). Of 11 anti-gB mAbs described by Para et al. (1985), only one neutralized virus, even in the presence of complement. Several were tested for inhibition of virus adsorption or cell fusion and found to have no effect (Noble et al., 1983; Fuller and Spear, 1985).

3. Glycoprotein C

3.1. Introduction

gC is not essential for infectivity in cell culture, as gC$^-$ mutants are viable (Ruyechan et al., 1979; Holland et al., 1983; Holland et al., 1984). However, its interaction with complement component C3b suggests an important role in modulation of the immune response against HSV (Friedman et al., 1984; Fries et al., 1986; Eisenberg et al., 1987; Kubota et al., 1987; McNearney et al., 1987).

gC-1 contains 511 amino acids (Frink et al., 1983; Draper et al., 1984; Homa et al., 1986), of which 25 constitute the signal sequence (Kikuchi et al., 1984) (Fig. 20.3C). The protein has nine sites for *N*-linked glycosylation (Frink et al., 1983), as well as a number of *O*-linked oligosaccharides (Olofsson et al., 1983). Its homologue in HSV-2 is smaller, containing 480 amino acids with seven sites for *N*-linked glycosylation (Dowbenko and Lasky, 1984; Swain et al., 1985). The two proteins are more homologous in the C-terminal region of the protein; five glycosylation sites and all eight cysteines are conserved. Cross-linking studies show that gC is capable of forming oligomers (Zhu and Courtney, 1988).

Several laboratories have isolated mAbs against gC, the majority of which are type 1-specific (Pereira et al., 1980; Showalter et al., 1981; Braun et al., 1983; Friedman et al., 1984; Marlin et al., 1985; Para et al., 1985; Koga et al., 1986; Seidel-Dugan et al., 1988). Most of the epitope mapping studies have been carried out in one laboratory and have focused on gC-1, with no distinction made between continuous and discontinuous epitopes (Holland et al., 1984; Marlin et al., 1985; Homa et al., 1986; Sjöblom et al., 1987).

3.2. gC-1

3.2.1. Epitope mapping

A panel of gC-1-specific mAbs selected two classes of neutralization-resistant viruses: those which contain gC (*mar* mutants) and those which do not (gC$^-$ mutants), the latter phenotype resulting from frameshift or nonsense mutations that produce truncated forms of gC (Holland et al., 1983; Holland et al., 1984; Marlin et al., 1985). Both mutant classes provided information about the location of antigenic sites. Cross-neutralization experiments with the *mar* mutants divided the mAbs into two groups, defining sites I and II (Figs. 20.3A and 20.3B) (Marlin et al., 1985), and the lack of competition between group I and II mAbs in a binding assay suggested that the sites are physically separate (Marlin et al., 1985). Analysis of the truncation

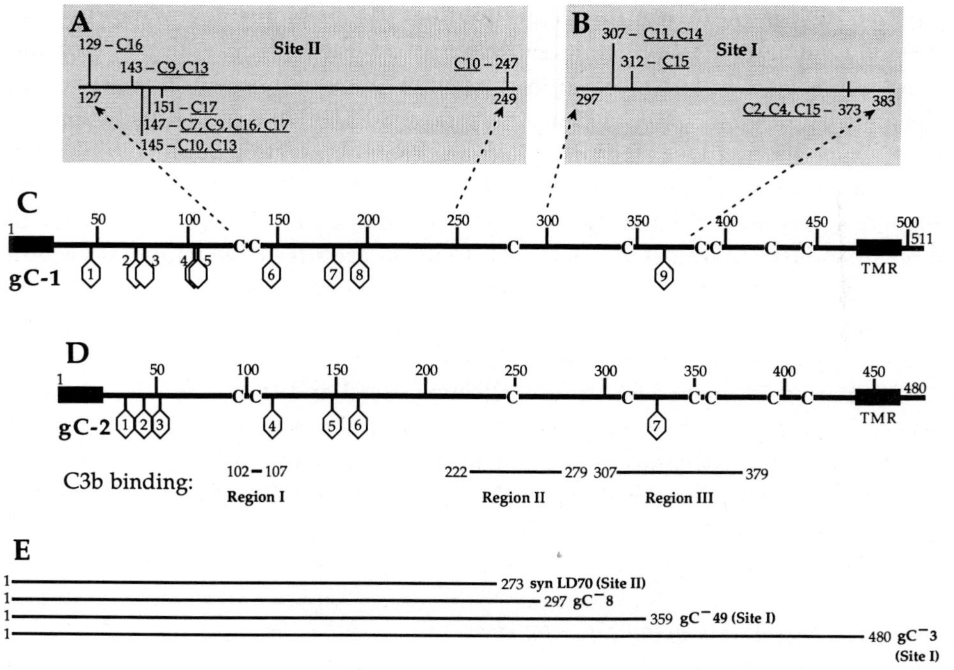

Fig. 20.3. Mapping of antigenic sites and C3b binding sites on gC. (A) *mar* mutations in site II of gC-1, with mAbs underlined; (B) *mar* mutations in site I of gC-1; (C) stick figure of gC-1, showing positions of cysteine residues (C), *N*-linked carbohydrates (hexagons) and transmembrane region (TMR); (D) stick figure of gC-2, including regions needed for C3b binding; (E) smallest fragment containing the indicated antigenic sites of gC-1.

mutants supports this conclusion (Kikuchi et al., 1984; Marlin et al., 1985; Homa et al., 1986).

The six group I mAbs recognize at least three different epitopes within site I, judging by their patterns of reactivity with the *mar* mutants (Marlin et al., 1985). Six mutants have subsequently been sequenced, and changes were found at residues 307, 312 and 373 (Fig. 20.3B) (Wu et al., 1990). Four of these mAbs were tested with truncated fragments of gC-1 (Fig. 20.3E), with the following results. They all react with the $gC^{-}3$ mutant (1–480); three also react with $gC^{-}49$ (1–359); none react with $gC^{-}8$ (1–297) (Marlin et al., 1985; Homa et al., 1986). Those reacting with 1–359 included C15, which suggests that the *mar* mutation at 373 prevents neutralization via a conformational effect on site I. The major part of site I therefore appears to lie just downstream of residue 297.

The 16 mAbs in group II, which include eight produced by Showalter et al. (1981), apparently recognize at least six different epitopes; in fact, their patterns of reactivity with the *mar* mutants are sufficiently different that site II was divided into subsites IIa, IIb and IIc (Marlin et al., 1985). The proximity of these subsites is emphasized by the observation that representative mAbs from the three subgroups are mutually competitive in a binding assay and all react with 1–273, the shortest truncated fragment (Fig. 20.3E) (Marlin et al., 1985). Sequencing of the mutants has identified seven amino acids within site II (Fig. 20.3A) (Wu et al., 1990), but the results have blurred the distinctions between subgroups. Thus, a mutation at 147 was selected by C16 (IIa), C9 and C17 (IIb), and C7 (IIc). Also, mutants with changes at positions 145 and 247 were resistant to neutralization by mAbs from all three subgroups (Marlin et al., 1985). All 16 mAbs are affected by mutations within residues 143–151, and the authors suggest that these nine amino acids form the core of site II. Interestingly, five of the seven altered amino acids in site II were originally arginines, all located between residues 129 and 151.

The spacing of amino acid changes affecting sites I and II suggests that they are discontinuous, and we have found that group I and II mAbs do not react with gC-1 electrophoresed on denaturing gels (Seidel-Dugan, unpublished). The only mAbs found to react with denatured gC on Western blots are H595 (Braun et al., 1983) and Fd8 (Koga et al., 1986), neither of which has been examined for recognition of gC mutants.

3.2.2. Importance of carbohydrate for antigenicity

Sjöblom et al.(1987) used a periodate-sialidase treatment to show that removal of sialic acid and galactose from gC-1 carbohydrate chains reduces the binding of mAbs in group II but not group I. Binding of mAbs to gC was not inhibited by high concentrations of monosaccharides, so carbohydrate does not constitute an epitope. Rather, the data suggest that carbohydrate modification affects the conformation of site II and, in this regard, it is of interest that the majority of the carbohydrates, both *N*-linked and *O*-linked, are in the N-terminal half of the protein (Fig. 20.3C) (Frink et al., 1983; Kikuchi et al., 1984; Dall'Olio et al., 1985).

3.3. gC-2

A grouping scheme for mAbs directed against gC-2 has been proposed, based on whether they recognize continuous or discontinuous epitopes, their type-specificity, and the results of blocking assays (Seidel-Dugan et al., 1988; Seidel-Dugan et al., 1990). Five groups have been identified, but none of the epitopes have been localized.

3.4. Biological properties of mAbs

Of the 22 mAbs used to define antigenic sites I and II, all have type 1-specific, complement-dependent neutralizing activity (Showalter et al., 1981; Marlin et al., 1985), as do two produced by Pereira et al. (1980) and three out of eight produced by Para et al. (1985). Only the last set was tested for the ability to block virus adsorption; four have some activity, but there is no correlation with neutralization (Para et al., 1985). In addition, 10 mAbs, including representatives from groups I and II, have no effect on virus penetration (Fuller and Spear, 1987; Highlander et al., 1987). One out of three gC-2-specific mAbs has complement-dependent neutralizing activity (Para et al., 1985).

3.5. Mapping of C36 binding sites

gC-1 in HSV-1-infected cells acts as a receptor for complement component C3b (Friedman et al., 1984). As purified gC-1 binds directly to C3b, the interaction does not require any other viral or cellular proteins (Eisenberg et al., 1987; McNearney et al., 1987). HSV-2-infected cells do not express a C3b receptor, but purified gC-2 binds C3b (Eisenberg et al., 1987; McNearney et al., 1987), as do cells transfected with a plasmid expressing gC-2 (Seidel-Dugan et al., 1988). These results suggest that C3b receptor activity may be masked on HSV-2-infected cells.

Analysis of gC-2 mutants has been used to partially map the C3b binding site, which appears to be discontinuous. Small insertions in three regions abolished C3b binding without disrupting protein folding, suggesting their importance for receptor activity (Fig. 20.3D) (Seidel-Dugan et al., 1990). Region I is located within amino acids 102–107, region II within 222–279, and region III within 307–379. Region III is the most conserved sequence among gC homologues of other herpesviruses, and also has some structural features similar to a conserved motif found in a family of proteins including human complement receptors (Reid et al., 1986; Klickstein et al., 1987). The first 48 amino acids after the signal sequence are not required, as they can be deleted with no effect on C3b binding (Seidel-Dugan et al., 1988).

Blocking experiments suggest that gC-1 and gC-2 bind to similar sites on C3b, and some mapping of the binding site on gC-1 has been carried out using gC$^-$ and *mar* mutants (Friedman et al., 1986). The results show no correlation between the C3b binding site and antigenic sites I or II, as some mutants within each site are C3b receptor positive while others are negative. However, some of the *mar* mutants which affect C3b binding are near areas of gC-1 corresponding to regions I, II and III of gC-2. Thus, residue 373 of gC-1 would be within region III, residue 247 would be just upstream of region II, and residue 129 would be just upstream of region I. Therefore, the results to date suggest that gC-1 and gC-2 have similar binding sites for C3b.

4. Glycoprotein D

4.1. Introduction

gD, together with gB and gH, is involved in penetration of virus into the cell. Ligas and Johnson (1988) showed that replication of a gD-minus mutant is blocked at a stage after attachment, and that the block is bypassed by PEG-mediated fusion. Moreover, complement-independent neutralization by anti-gD mAbs occurs by inhibition of penetration rather than attachment (Highlander et al., 1987). Cell lines expressing large amounts of gD are resistant to infection (Campadelli-Fiume et al., 1988a; Johnson and Spear, 1989), leading to speculation that gD binds to a cell receptor after initial attachment of virus to the cell, and that the receptor is sequestered in the gD cell lines. This idea is supported by the finding that infection of cells can be prevented by prior exposure to UV-inactivated virus but not gD-minus virus (Johnson and Ligas, 1988).

gD-1 has 369 amino acids (Watson et al., 1982; Lasky et al., 1984; McGeoch et al., 1985; Minson et al., 1986) and gD-2 has 368 (Watson, 1983), after removal of the 25-amino-acid signal sequence (Eisenberg et al., 1984). The proteins are 85% homologous, and both have three *N*-linked oligosaccharides (Cohen et al., 1983; Matthews et al., 1983) and two or three *O*-linked oligosaccharides (Serafini-Cessi et al., 1988). Correct folding of gD-1 requires six of its seven cysteine residues (Wilcox et al., 1988), and all six are conserved in gD-2.

There are a large number of mAbs against gD (Pereira et al., 1980; Showalter et al., 1981; Eisenberg et al., 1982; Pereira et al., 1982; Noble et al., 1983; Seigneurin et al., 1983; Friedman et al., 1984; Rawls et al., 1984; Eisenberg et al., 1985; Cohen et al., 1986; Koga et al., 1986; Minson et al., 1986; Highlander et al., 1987; Isola et al., 1989), many of which fall into a grouping scheme based on the following criteria: reaction with continuous or discontinuous epitopes; type specificity; recognition

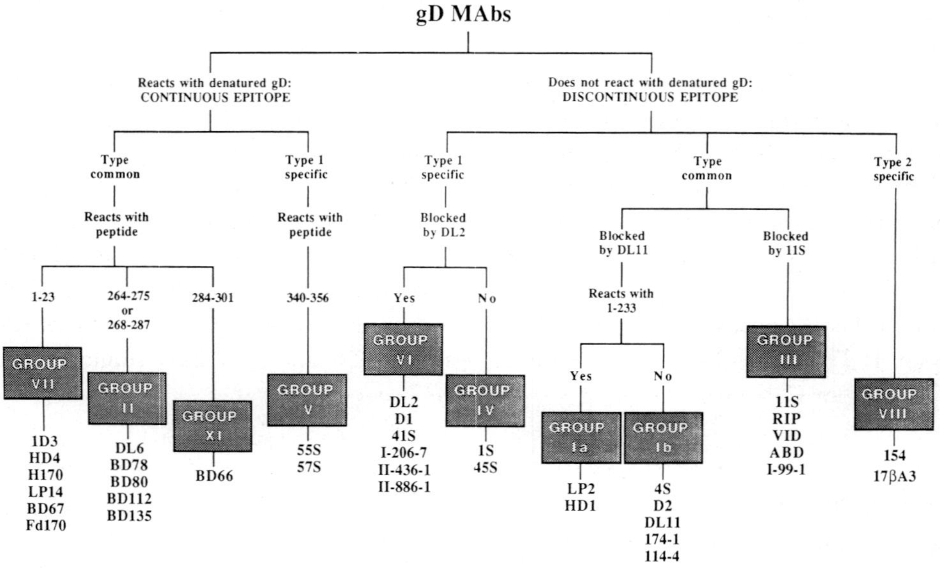

Fig. 20.4. Scheme for grouping of anti-gD mAbs.

of truncated fragments; and the ability of mAbs to block each other in a binding assay (Fig. 20.4). The resulting groups and their corresponding antigenic sites will be described in detail below.

4.2. Continuous epitopes

Recognition of synthetic peptides by rabbit sera prepared against native gD (Isola et al., 1989), and by human sera (Welling-Wester, personal communication), suggests that continuous epitopes are concentrated near the N-terminus and to either side of the transmembrane region. The clustering of continuous mAb epitopes in precisely these regions supports this idea; thus, antigenic site VII is near the N-terminus, sites II and XI lie between residues 264 and 301, and site V is near the C-terminus (Fig. 20.5C).

Six type-common mAbs recognize peptide 1–23, and their epitopes form antigenic site VII (Figs. 20.4 and 5C) (Cohen et al., 1984; Isola, 1988). Fine mapping localized the H170 and LP14 epitopes to residues 11–19 (Dietzschold et al., 1984) and 9–21 (Minson et al., 1986; Bosch et al., 1987), respectively. In addition, LP14 selected a *mar* mutation at residue 16 (Fig. 20.5B) (Minson et al., 1986). Rabbit antiserum made against a chimeric protein containing amino acids –5–55 of gD-1 reacts with a set of overlapping peptides spanning residues 1–54 (Kocken et al., 1988), re-

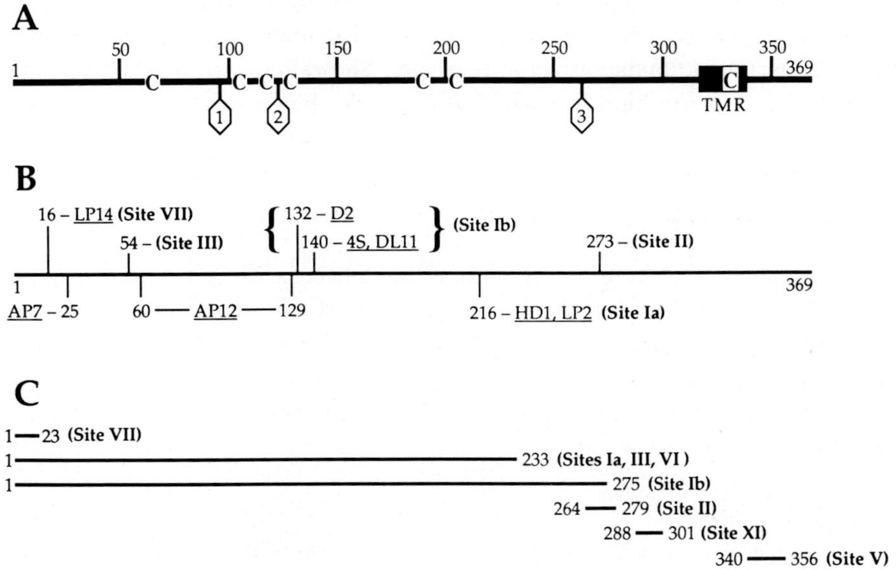

Fig. 20.5. Mapping of antigenic sites on gD. (A) Stick figure of gD-1, showing positions of cysteine residues (C), *N*-linked carbohydrates (hexagons) and transmembrane region (TMR); (B) *mar* mutations, with mAbs underlined, and other point mutations affecting antigenic sites; (C) smallest fragment containing the indicated antigenic sites.

vealing the existence of epitopes adjacent to site VII. Some of the epitopes are type-specific, as anti-gD-2 rabbit serum reacts with peptides 1–16 and 8–23 only if they have the type 2 residues at positions 7 and 21 (Dietzschold et al., 1984). However, no type-specific mAbs have been mapped to this site.

Site II is defined by mAbs that bind to peptides within residues 264–287, and which block the binding of DL6 (Fig. 20.4). It contains at least two type-common epitopes, DL6 itself recognizing residues 272–279, and four other mAbs (BD78, BD80, BD112 and BD135) reacting with peptide 264–275 (Eisenberg et al., 1985; Isola et al., 1989). A type-specific epitope is also present, as anti-gD-2 rabbit serum reacts only with the type 2 sequence for peptide 266–279 (Isola et al., 1989). Site II is predicted to include a β-turn that would be destroyed by changing amino acid 273 from Pro to Ser, and this mutation does in fact abolish the binding of DL6 and BD78 (Fig. 20.5B) (Isola et al., 1989). It is important to note that group II, as described here, does not correspond to the group II described by Highlander et al. (1987).

Group XI contains one mAb, BD66 (Fig. 20.4), which reacts strongly with peptide 284–301 and weakly with peptide 289–301 (Isola et al., 1989). Its recognition of full-length gD-1 is not diminished by a 5-amino-acid insertion between amino acids 287 and 288 (Muggeridge and Trifillis, unpublished), placing the epitope between residues 288 and 301 (Fig. 5C). Despite the proximity to the DL6 epitope, BD66 and DL6 do not block each other.

Site V lies in the cytoplasmic tail of gD-1, as shown by the reaction of group V mAbs with peptide 340–356 (Figs. 20.4 and 5C) (Eisenberg et al., 1985). Under most conditions, the antibodies are type 1-specific (Showalter et al., 1981; Eisenberg et al., 1982), but 57S can bind to native gD-2 in a dot-blot assay (Eisenberg et al., 1982).

4.3. Discontinuous epitopes

mAbs which recognize discontinuous epitopes have been arranged in four groups, one of which is subdivided (Fig. 20.4). Groups I and III are type-common, groups IV and VI are type 1-specific, and group VIII is type 2-specific. However, this may not reflect the true complexity of the antigenic sites, as some type-specific and type-common mAbs have overlapping epitopes and therefore belong to the same group. This will be discussed further.

4.3.1. Type-common epitopes

Antibodies in group I block the binding of mAbs HD1 and DL11, whereas those in group III block mAb 11S (Fig. 20.4). There is no blocking between these groups.

Seven mAbs have been assigned to group I, and their epitopes constitute antigenic site I (Fig. 20.4) (Eisenberg et al., 1982; Muggeridge et al., 1988). Cross-neutralization studies with *mar* mutants showed that site I can be split into Ia and Ib, such that binding of the Ia mAbs is not affected by mutations selected with the Ib mAbs, and vice versa (Muggeridge et al., 1988). The Ia and Ib antibodies can also be distinguished by their reactivity with truncation and deletion mutants. Thus, all the group I mAbs react with a 1–275 fragment, but only the Ia mAbs react with a 1–233 fragment and with gD lacking residues 234–244 or 243–286 (Cohen et al., 1986; Muggeridge et al., 1988), showing that part of site Ib maps to 234–275 (Fig. 20.5C). Sequencing of the *mar* mutants identified three of the amino acids within site I (Minson et al., 1986; Muggeridge et al., 1990). Both Ia mAbs selected a change at position 216 (Fig. 20.5B), consistent with the truncation mutant data. The Ib mutations were at residues 132 and 140 (Fig. 20.5B), identifying a second part of site Ib, which is presumably close to the 234–275 segment in the folded protein. The amino acid change at position 140 creates a new *N*-linked glycosylation site, which is used; deglycosylation with endo F restores binding of mAbs DL11 and D2, but not 4S, from which two conclusions can be drawn. First, the carbohydrate chain blocks site Ib, preventing neutralization by Ib mAbs; second, amino acid 140 is part of the 4S epitope.

Five mAbs, including all three human mAbs against gD, have been assigned to group III, and their epitopes constitute antigenic site III (Fig. 20.4) (Eisenberg et al., 1982; Isola, 1988). All five mAbs recognize a 1–233 fragment of gD-1 (Fig. 20.5C), and also a 30-kDa proteolytic fragment (Isola, 1988). The 30-kDa fragment was thought to represent amino acids 1–182 (Cohen et al., 1986), but tryptic peptide analysis shows that it probably contains amino acids 1–226 (Isola, 1988). The N-terminal 20 amino acids are not required for binding, as their removal by endoproteinase-LysC does not affect group III mAbs (Isola, 1988), so site III is within residues 21–226. Knowledge of individual residues in site III is sparse. The contribution of amino acid 54 (Fig. 20.5B) was discovered as a result of a spontaneous mutation within a cloned gD gene; the expressed protein failed to bind three group III mAbs (RIP, VID and I-99-1) (Sodora, unpublished). The two remaining group III mAbs were unaffected, indicating the presence of at least two epitopes within site III.

4.3.2. Type 1-specific epitopes

The two groups of type 1-specific mAbs are IV and VI (Fig. 20.4). The criteria for assigning an antibody to one of these groups have changed as new information has become available. To clarify this situation, we have defined DL2 as the prototype group VI mAb. Those which block its binding to gD-1 are in group VI, and those which do not are in group IV (Fig. 20.4). The latter is thus a group defined by exclusion, rather than inclusion, so its members may represent more than one antigenic site.

Group IV currently contains mAbs 1S and 45S (Fig. 20.4) (Showalter et al., 1981), the latter having been moved from group VI (Eisenberg et al., 1982; Isola, 1988). Previous mapping data for site IV were obtained with mAb 41S (Cohen et al., 1986), which has now been reassigned to group VI (see below), so no information is available on the location of group IV epitopes.

Group VI contains six mAbs (Fig. 20.4), II-436-1 and 41S having been moved from group IV (Isola, 1988). DL2 and 41S bind to a 1–233 fragment of gD-1 (Fig. 20.5C) (Cohen et al., 1986) and also to a 1–226 fragment produced by proteolysis (Cohen et al., 1986; Isola, 1988), but no individual amino acids in site VI have been identified. Binding of mAbs II-886-1 and I-206-7, but not of the other 4, depends on the presence of *N*-linked carbohydrate attached to residue 121 (Sodora et al., 1989).

4.3.3. Type 2-specific epitopes

Two mAbs constitute group VIII (Fig. 20.4) (Pereira et al., 1982; Rawls et al., 1984), although there is no evidence that they recognize the same antigenic site. Despite its type 2 specificity, 17βA3 reacts with gD from the type 1 isolate SP140; DNA sequencing showed that SP140 gD has the type 2 residue at position 72, identifying it as part of the 17βA3 epitope (Rawls et al., 1984).

4.3.4. Groups defined by cross-neutralization

Highlander et al. (1987) arranged 11 mAbs in four groups, based on neutralization tests with a panel of *mar* mutants. In naming these groups, some consistency was maintained with the Cohen and Eisenberg scheme. Thus, group I, containing only D2, was so named because this antibody falls into Cohen and Eisenberg's group I. However, group II (consisting of D1, D3, 1S and 50S) does not correspond to Cohen and Eisenberg's group II. Of these mAbs, the type 1-specific 1S and D1 were placed in groups IV and VI, respectively, whereas the type-common D3 and 50S could not be grouped (Isola, 1988). Given that they all have overlapping epitopes (Highlander et al., 1987), presumably they all bind to site VI, since D1 is the only one for which a site has been positively identified. This in turn suggests two things: first, that a discontinuous site can contain both type-common and type-specific epitopes; and second, that when a mAb blocks none of the other antibodies in a binding assay, as found with D3, 1S and 50S, the result should be interpreted with caution.

Groups IX and X define at least two new antigenic sites. Group IX contains mAbs D4 and D7, neither of which falls convincingly into one of the Cohen and Eisenberg groups (Isola, 1988). D5, D8, D9 and 4S were assigned to group X because they neutralize the *mar* mutants selected with D1, D2, D3 and D4. However, there is no evidence that they have overlapping epitopes, so they may recognize more than one antigenic site. In fact, 4S has been shown to bind to site I (Muggeridge et al., 1988) (see section 4.3.1).

4.3.5. Ungrouped mAbs

Additional mAbs are currently being studied. Some may represent new groups, expanding the number of antigenic sites, and possible examples already exist amongst those well-studied mAbs which could not be grouped. For example, AP7 and AP12 (Minson et al., 1986) do not behave like group I or group III antibodies, despite recognizing type-common discontinuous epitopes (Isola, 1988). Competitive binding experiments indicate some overlap between AP12 and LP2 (Minson et al., 1986), but there is no other evidence for its inclusion in group I. Sequencing of *mar* mutants has identified amino acids 60 and 129 as part of the AP12 epitope, and 25 as part of the AP7 epitope (Fig. 20.5B) (Minson et al., 1986; Minson, personal communication).

4.4. Implications for gD structure

Most continuous epitopes in gD appear to be either near the N-terminus or in the C-terminal third of the protein. In contrast, the majority of discontinuous epitopes are contained within residues 1–233. Moreover, six of the seven point mutations found to affect discontinuous epitopes are between residues 54 and 216 (Fig. 20.5B). The correlation between these findings and the locations of cysteine residues is striking (Fig. 20.5A), and suggests that disulfide bonds hold the central part of gD in the correct conformation for maintenance of discontinuous epitopes. The loss of gD conformation after site-directed mutagenesis of individual cysteines, or deletion of segments which include cysteines, supports this idea (Cohen et al., 1988; Wilcox et al., 1988).

4.5. Biological properties of mAbs

Of the mAbs described by Minson et al. (1986), LP14 (group VII) has the highest titer in a 50% plaque reduction assay, but is unable to neutralize more than 80% of the virus; LP2 (group I) has a lower 50% titer, but can neutralize essentially all of the virus. This is a feature of the group I mAbs (Muggeridge et al., 1988), although D1 (group VI) has the same property (Muggeridge, unpublished). AP7 and AP12 neutralize only in the presence of complement (Minson et al., 1986).

LP14 and AP12, but not AP7 and LP2, cause a reduction in plaque size when added to infected monolayers, and also inhibit syncytium formation (Minson et al., 1986). In the absence of complement, AP7 has the intriguing ability to increase the titer of HSV-2 but not HSV-1 (Minson et al., 1986).

All nine mAbs described by Highlander et al. (1987) neutralize virus in the presence of complement; only D1 and D2 have high complement-independent titers, though D3 and D8 have measurable activity. The fact that these four mAbs can neutralize virus preadsorbed to cell surfaces, and all nine antibodies reduce the rate of virus penetration, strongly supports neutralization by inhibition of penetration rather than attachment (Highlander et al., 1987). Most of the mAbs (including D1 and D2) block syncytium formation and D6 reduces the size of plaques when added to infected monolayers (Highlander et al., 1987).

Noble et al. (1983) found that all of their gD mAbs inhibit syncytium formation. The neutralization titers were subsequently reported by Para et al. (1985). Those with high complement-independent activity are III-114-4 and III-174-1 (both group I) and III-255-2 (ungrouped). II-436-1 (group VI) has lower activity, and I-99-1 (group III) neutralizes in the presence of complement. Two other group VI mAbs (I-206-7 and II-886-1) have little or no neutralizing activity, but are the most effective at preventing adsorption of virions to cell monolayers (Fuller and Spear, 1985). This, and the occurrence of complement-independent neutralization at antibody concentrations far lower than those required to block adsorption, implied that neutralization must occur by a different mechanism. In fact, Fuller and Spear (1987) have now used electron microscopy to show that mAbs III-114-4 and III-174-1 prevent penetration of adsorbed virions into the cell.

5. Antigenic sites and protection against infection

For the purposes of vaccine development, it would be valuable to know which antigenic sites are important for protection against infection. Two possible approaches are to immunize animals either with mAbs against defined epitopes, or with virus mutants carrying mutations in specific antigenic sites.

5.1. Passive immunization with mAbs

Unfortunately, the majority of mAbs used in passive immunization studies have not been assigned to the groups described in sections 2, 3 and 4, and there is little data with which to correlate antigenic sites and protection. The most comprehensive study was carried out by Kümel et al. (1985), who tested several mAbs against gB, gC and gD for their ability to protect mice against an i.c. infection with HSV-1. Overall, mAbs against gC were most effective, followed by gB and then gD. There was no difference between mAbs recognizing sites I and II of gC; of the gB mAbs, only B3 (site I) was as effective as the gC mAbs. mAbs to sites II, III and IV of gB had lower activity, as did the gD mAb D2 (site Ib). Dix et al. (1981) had found earlier that the anti-gD mAb HD1 (site Ia) protects mice against a lethal footpad challenge with HSV-1 or HSV-2, whereas the gC-1-specific mAb HC1 (ungrouped) protects only against HSV-1. In addition, gD mAbs LP2 (site Ia) and AP7 (ungrouped) prevented zosteriform spread of HSV-1 inoculated into the flanks of mice, as did the gH mAb LP11 (Simmons and Nash, 1985).

Ungrouped mAbs against gB, gC, gD, gE and gG of HSV-2 all gave some protection against a lethal footpad challenge with HSV-2 (Balachandran et al., 1982), and mAbs against gB, gC, gD and gE protected against lethal ocular infections with HSV-1 (Rector et al., 1982; Rector et al., 1984). In the latter study, one anti-gD mAb (55S) was ineffective, presumably because its epitope is in the cytoplasmic tail.

Another set of mAbs against gB, gC, gD and gE reduced the severity of keratitis following ocular infection, but only those against gB and gD reduced the frequency of latent infections (Metcalf et al., 1988). Two mAbs, whose targets were tentatively identified as gD and gH, protected against a lethal footpad or i.p. infection with HSV-1, whereas a gB mAb did not (Chan et al., 1985), and Roberts et al. (1985) found that neither a gC mAb nor a gB mAb prevented death following i.c. infection with HSV-1. Finally, human mAbs against gB protected mice against lethal i.p. infections with either HSV-1 or HSV-2 (Fujinaga et al., 1987).

5.2. Immunization with *mar* mutants

If it could be demonstrated that loss of a particular antigenic site reduced the protective value of immunization, that site would be implicated as an important target of the immune response. However, it seems unlikely that single mutations within a site would change its structure sufficiently to affect the polyclonal response, so the results of Kümel et al. (1985) are perhaps not surprising. They used neuroattenuated *mar* mutants to immunize mice, and demonstrated that gC-1 mutants selected by mAbs C10 (site II) and C5 (ungrouped) can protect against a lethal i.c. challenge with wild-type HSV-1, as can a mutant with amino acid changes in sites I, II, III and IV of gB-1.

6. Concluding remarks

Progress in mapping antigenic sites of gB, gC and gD has been rapid, as demonstrated by the many recent references, and we are grateful to Drs. Lenore Pereira and Joseph Glorioso for allowing us to include 'in press' data. The next review of this subject will surely include studies of the other HSV glycoproteins, together with information on T-cell epitopes. Antigenic mapping by definition identifies those parts of a protein which are recognized by components of the immune system. It has also provided valuable insights into the structures of gB, gC and gD, and will be a crucial factor in the selection of mutants for mapping of their functional sites. Antigenic and functional sites may or may not coincide; ultimately crystallization and three-dimensional structure determination will provide a framework to link the two together.

7. References

Balachandran, N., Bacchetti, S. and Rawls, W.E. (1982) *Infect. Immun.* 37, 1132–1137.
Baucke, R. and Spear, P. (1979) *J. Virol.* 32, 779–789.
Bishop, G.A., Kümel, G., Schwartz, S.A. and Glorioso, J.C. (1986) *J. Virol.* 57, 294–300.
Blacklaws, B.A., Nash, A.A. and Darby, G. (1987) *J. Gen. Virol.* 68, 1103–1114.

Bosch, D.L., Geerligs, H.J., Weyer, W.J., Feijlbrief, M., Welling, G.W. and Welling-Wester, S. (1987) *J. Virol.* 61, 3607–3611.

Braun, D.K., Pereira, L., Norrild, B. and Roizman, B. (1983) *J. Virol.* 46, 103–112.

Bzik, D.J., DebRoy, C., Fox, B.A., Pederson, N.E. and Person, S. (1986) *Virology* 155, 322–333.

Bzik, D.J., Fox, B.A., DeLuca, N.A. and Person, S. (1984) *Virology* 133, 301–314.

Cai, W., Gu, B. and Person, S. (1988) *J. Virol.* 62, 2596–2604.

Cai, W., Person, S., DebRoy, C. and Gu, B. (1988) *J. Mol. Biol.* 201, 575–588.

Cai, W., Person, S., Warner, S.C., Zhou, J. and DeLuca, N.A. (1987) *J. Virol.* 61, 714–721.

Campadelli-Fiume, G., Arsenakis, M., Farabegoli, F. and Roizman, B. (1988) *J. Virol.* 62, 159–167.

Chan, W.L., Lukig, M.L. and Liew, F.Y. (1985) *J. Exp. Med.* 162, 1304–1318.

Chapsal, J.M. and Pereira, L. (1988) *Virology* 164, 427–434.

Claesson-Welsh, L. and Spear, P.G. (1986) *J. Virol.* 60, 803–806.

Claesson-Welsh, L. and Spear, P.G. (1987) *J. Virol.* 61, 1–7.

Cohen, G., Long, D., Matthews, J., May, M. and Eisenberg, R. (1983) *J. Virol.* 46, 679–689.

Cohen, G.H., Dietzschold, B., Ponce de Leon, M., Long, D., Golub, E., Varrichio, A., Pereira, L. and Eisenberg, R.J. (1984) *J. Virol.* 49, 102–108.

Cohen, G.H., Isola, V.J., Kuhns, J., Berman, P.W. and Eisenberg, R.J. (1986) *J. Virol.* 60, 157–166.

Cohen, G.H., Wilcox, W.C., Sodora, D.L., Long, D., Levin, J.Z. and Eisenberg, R.J. (1988) *J. Virol.* 62, 1932–1940.

Corey, L. (1985) In *The Herpesviruses. Immunobiology and Prophylaxis of Human Herpesvirus Infections* (B. Roizman and C. Lopez, Eds.) Plenum Press, New York. pp. 1–35.

Dall'Olio, F., Malagolini, N., Speziali, V., Campadelli-Fiume, G. and Serafini-Cessi, F. (1985) *J. Virol.* 56, 127–134.

DeFreitas, E.C., Dietzschold, B. and Koprowski, H. (1985) *Proc. Natl. Acad. Sci. USA* 82, 3425–3429.

DeLuca, N., Bzik, D.J., Person, S. and Snipes, W. (1981) *Proc. Natl. Acad. Sci. USA* 78, 912–916.

Desai, P.J., Schaffer, P.A. and Minson, A.C. (1988) *J. Gen. Virol.* 69, 1147–1156.

Dietzschold, B., Eisenberg, R.J., Ponce de Leon, M., Golub, E., Hudecz, F., Varrichio, A. and Cohen, G.H. (1984) *J. Virol.* 52, 431–435.

Dix, R.D. (1987) *Prog. Med. Virol.* 34, 89–128.

Dix, R.D., Pereira, L. and Baringer, J.R. (1981) *Infect. Immun.* 34, 192–199.

Dowbenko, D.J. and Lasky, L.A. (1984) *J. Virol.* 52, 154–163.

Draper, K.G., Costa, R.H., Lee, G.T.-Y., Spear, P.G. and Wagner, E.K. (1984) *J. Virol.* 51, 578–585.

Dubin, G., Frank, I. and Friedman, H.M. (1990) *J. Virol.* 64, 2725–2731.

Eberle, R. and Courtney, R.J. (1982) *J. Virol.* 41, 348–351.

Eisenberg, R.J., Long, D., Hogue-Angeletti, R. and Cohen, G.H. (1984) *J. Virol.* 49, 265–268.

Eisenberg, R.J., Long, D., Pereira, L., Hampar, B., Zweig, M. and Cohen, G.H. (1982) *J. Virol.* 41, 478–488.

Eisenberg, R.J., Long, D., Ponce de Leon, M., Matthews, J.T., Spear, P.G., Gibson, M.G., Lasky, L.A., Berman, P., Golub, E. and Cohen, G.H. (1985) *J. Virol.* 53, 634–644.

Eisenberg, R.J., Ponce de Leon, M., Friedman, H.M., Fries, L.F., Frank, M.M., Hastings, J.C. and Cohen, G.H. (1987) *Microbial Pathogenesis* 3, 423–435.

Friedman, H.M., Cohen, G.H., Eisenberg, R.J., Seidel, C.A. and Cines, D.B. (1984) *Nature* 309, 633–635.

Friedman, H.M., Glorioso, J.C., Cohen, G.H., Hastings, J.C., Harris, S.L. and Eisenberg, R.J. (1986) *J. Virol.* 60, 470–475.

Fries, L.F., Friedman, H.M., Cohen, G.H., Eisenberg, R.J., Hammer, C.H. and Frank, M.M. (1986) *J. Immunol.* 137, 1636–1641.

Frink, R.J., Eisenberg, R.J., Cohen, G.H. and Wagner, E.K. (1983) *J. Virol.* 45, 634–647.

Fujinaga, S., Sugano, T., Matsumoto, Y.-I., Masuho, Y. and Mori, R. (1987) *J. Infect. Dis.* 155, 45–53.

Fuller, A.O., Santos, R.E. and Spear, P.G. (1989) *J. Virol.* 63, 3435–3443.

Fuller, A.O. and Spear, P.G. (1985) *J. Virol.* 55, 475–482.

Fuller, A.O. and Spear, P.G. (1987) *Proc. Natl. Acad. Sci. USA* 84, 5454–5458.

Glorioso, J., Kees, U., Kümel, G., Kirchner, H. and Krammer, P.H. (1985) *J. Immunol.* 135, 575–582.

Gompels, U. and Minson, A. (1986) *Virology* 153, 230–247.

Haffey, M.L. and Spear, P.G. (1980) *J. Virol.* 35, 114–128.

Hall, M.J. and Katrak, K. (1986) *Vaccine* 4, 138–150.

Harland, J. and Brown, S.M. (1988) *J. Gen. Virol.* 69, 113–124.

Heber-Katz, E., Hollosi, M., Dietzschold, B., Hudecz, F. and Fasman, G.D. (1985) *J. Immunol.* 135, 1385–1390.

Heber-Katz, E., Valentine, S., Dietzschold, B. and Burns-Purzycki, C. (1988) *J. Exp. Med.* 167, 275–287.

Highlander, S.L., Cai, W., Person, S., Levine, M. and Glorioso, J.C. (1988) *J. Virol.* 62, 1881–1888.

Highlander, S.L., Dorney, D.J., Gage, P.J., Holland, T.C., Cai, W., Person, S., Levine, M. and Glorioso, J.C. (1989) *J. Virol.* 63, 730–738.

Highlander, S.L., Sutherland, S.L., Gage, P.J., Johnson, D.C., Levine, M. and Glorioso, J.C. (1987) *J. Virol.* 61, 3356–3364.

Holland, T.C., Homa, F.L., Marlin, S.D., Levine, M. and Glorioso, J. (1984) *J. Virol.* 52, 566–574.

Holland, T.C., Marlin, S.D., Levine, M. and Glorioso, J. (1983) *J. Virol.* 45, 672–682.

Homa, F.L., Purifoy, D.J.M., Glorioso, J.C. and Levine, M. (1986) *J. Virol.* 58, 281–289.

Isola, V.J. (1988) Ph.D. Thesis. University of Pennsylvania.

Isola, V.J., Eisenberg, R.J., Siebert, G.R., Heilman, C.J., Wilcox, W.C. and Cohen, G.H. (1989) *J. Virol.* 63, 2325–2334.

Johnson, D.C. and Feenstra, V. (1987) *J. Virol.* 61, 2208–2216.

Johnson, D.C., Frame, M.C., Ligas, M.W., Cross, A.M. and Stow, N.D. (1988) *J. Virol.* 62, 1347–1354.

Johnson, D.C. and Ligas, M.W. (1988) *J. Virol.* 62, 4605–4612.

Johnson, R.M. and Spear, P.G. (1989) *J. Virol.* 63, 819–827.

Kikuchi, G.E., Coligan, J.E., Holland, T.C., Levine, M., Glorioso, J.C. and Nairn, R. (1984) *J. Virol.* 52, 806–815.

Klickstein, L.B., Wong, W.W., Smith, J.A., Weis, J.H., Wilson, J.G. and Fearon, D.T. (1987) *J. Exp. Med.* 165, 1095–1112.

Kocken, C.H.M., Geerligs, H.J., Bos, C.A., Ab, Ġ., Weijer, W.J., Drijfhout, J.W., Welling, G.W. and Welling-Wester, S. (1988) *Arch. Virol.* 103, 267–274.

Koga, J., Chatterjee, S. and Whitley, R.J. (1986) *Virology* 151, 385–389.

Kousoulas, K., Arsenakis, M. and Pereira, L. (1989) *J. Gen. Virol.* 70, 735–741.

Kousoulas, K.G., Huo, B. and Pereira, L. (1988) *Virology* 166, 423–431.

Kousoulas, K.G., Pellett, P.E., Pereira, L. and Roizman, B. (1984) *Virology* 135, 379–394.

Kubota, Y., Gaither, T.A., Cason, J., O'Shea, J.J. and Lawley, T.J. (1987) *J. Immunol.* 134, 1137–1142.

Kümel, G., Kaerner, H.C., Levine, M., Schröder, C.H. and Glorioso, J.C. (1985) *J. Virol.* 56, 930–937.

Lasky, L.A., Dowbenko, D., Simonsen, C.C. and Berman, P.W. (1984) *Bio/Technology* 2, 527–532.

Ligas, M.W. and Johnson, D.C. (1988) *J. Virol.* 62, 1486–1494.

Little, S.P., Jofre, J.T., Courtney, R.J. and Schaffer, P.S. (1981) *Virology* 115, 149–160.

Longnecker, R. and Roizman, B. (1986) *J. Virol.* 58, 583–591.

Longnecker, R. and Roizman, B. (1987) *Science* 236, 573–576.

Marlin, S.D., Highlander, S.L., Holland, T.C., Levine, M. and Glorioso, J.C. (1986) *J. Virol.* 59, 142–153.

Marlin, S.D., Holland, T.C., Levine, M. and Glorioso, J.C. (1985) *J. Virol.* 53, 128–136.

Marsden, H.S. (1987) In *Molecular basis of virus disease. Fortieth symposium of the Society for General Microbiology* (W.C. Russell and J.W. Almond, Eds.) Cambridge University Press, pp. 259–288.

Matthews, J.T., Cohen, G.H. and Eisenberg, R.J. (1983) *J. Virol.* 48, 521–533.

McGeoch, D.J., Dolan, A., Donald, S. and Rixon, F.J. (1985) *J. Mol. Biol.* 181, 1–13.

McNearney, T.A., Odell, C., Holers, V.M., Spear, P.G. and Atkinson, J.P. (1987) *J. Exp. Med.* 166, 1525–1535.

Meignier, B. (1985) In *The Herpesviruses. Immunobiology and Prophylaxis of Human Herpesvirus Infections* (B. Roizman and C. Lopez, Eds.) Plenum Press, New York. pp. 265–296.

Metcalf, J.F., Chatterjee, S., Koga, J. and Whitley, R.J. (1988) *Intervirology* 29, 39–49.

Minson, A.C., Hodgman, T.C., Digard, P., Hancock, D.C., Bell, S.E. and Buckmaster, E.A. (1986) *J. Gen. Virol.* 67, 1001–1013.

Muggeridge, M.I.., Isola, V.J., Byrn, R.A., Tucker, T.J., Minson, A.C., Glorioso, J.C., Cohen, G.H. and Eisenberg, R.J. (1988) *J. Virol.* 62, 3274–3280.

Muggeridge, M.I., Wu, T.-T., Johnson, D.C., Glorioso, J.C., Eisenberg, R.J. and Cohen, G.H. (1990) *Virology,* 174, 375–387.

Nash, A.A., Leung, K.-N. and Wildy, P. (1985) In *The Herpesviruses. Immunobiology and Prophylaxis of Human Herpesvirus Infections* (B. Roizman and C. Lopez, Eds.) Plenum Press, New York. pp. 87–102.

Neidhardt, H., Schröder, C.H. and Kaerner, H.C. (1987) *J. Virol.* 61, 600–603.

Noble, A.G., Lee, G.T.-Y., Sprague, R., Parish, M.L. and Spear, P.G. (1983) *Virology* 129, 218–224.

Norrild, B. (1985) In *The Herpesviruses. Immunobiology and Prophylaxis of Human Herpesvirus Infections* (B. Roizman and C. Lopez, Eds.) Plenum Press, New York. pp. 69–86.

Olofsson, S., Sjöblom, I., Lundström, M., Jeansson, S. and Lycke, E. (1983) *J. Gen. Virol.* 64, 2735–2747.

Pachl, C., Burke, R.L., Stuve, L.L., Sanchez-Pescador, L., Nest, G.V., Masiarz, F. and Dina, D. (1987) *J. Virol.* 61, 315–325.

Para, M.F., Baucke, R.B. and Spear, P.G. (1980) *J. Virol.* 34, 512–520.

Para, M.F., Parish, M.L., Noble, A.G. and Spear, P.G. (1985) *J. Virol.* 55, 483–488.

Pellett, P.E., Kousoulas, K.G., Pereira, L. and Roizman, B. (1985) *J. Virol.* 53, 243–253.

Pereira, L., Ali, M., Kousoulas, K., Huo, B. and Banks, T. (1989) *Virology* 172, 11–24.

Pereira, L., Dondero, D.V., Gallo, D., Devlin, V. and Woodie, J.D. (1982) *Infect. Immun.* 35, 363–367.

Pereira, L., Dondero, D., Norrild, B. and Roizman, B. (1981) *Proc. Natl. Acad. Sci. USA* 78, 5202–5206.

Pereira, L., Klassen, T. and Baringer, J.R. (1980) *Infect. Immun.* 29, 724–732.

Rawls, W.E., Balachandran, N., Sisson, G. and Watson, R.J. (1984) *J. Virol.* 51, 263–265.

Rector, J.T., Lausch, R.N. and Oakes, J.E. (1982) *Infect. Immun.* 38, 168–174.

Rector, J.T., Lausch, R.N. and Oakes, J.E. (1984) *J. Gen. Virol.* 65, 657–661.

Reid, K.B.M., Bentley, D.R., Campbell, R.D., Chyng, L.P., Sim, R.B., Kristensen, T. and Tack, B.F. (1986) *Immunology Today* 7, 230–234.

Roberts, P.L., Duncan, B.E., Raybould, T.J. and Watson, D.H. (1985) *J. Gen. Virol.* 66, 1073–1085.

Rosenthal, K.L., Smiley, J.R., South, S. and Johnson, D.C. (1987) *J. Virol.* 61, 2438–2447.

Ross, C., Glorioso, J., Sacks, S., Lavery, C. and Rawls, W.E. (1985) *J. Virol.* 54, 851–855.

Ruyechan, W.T., Morse, L.S., Knipe, D.M. and Roizman, B. (1979) *J. Virol.* 29, 677–697.

Sarmiento, M., Haffey, M. and Spear, P.G. (1979) *J. Virol.* 29, 1149–1158.

Sarmiento, M. and Spear, P.G. (1979) *J. Virol.* 29, 1159–1167.

Seidel-Dugan, C., Cohen, G.H. and Eisenberg, R.J. (1990) *J. Virol.,* in press.

Seidel-Dugan, C., Ponce de Leon, M., Friedman, H.M., Eisenberg, R.J. and Cohen, G.H. (1990) *J. Virol.,* 64, 1897–1906

Seidel-Dugan, C., Ponce de Leon, M., Friedman, H.M., Fries, L.F., Frank, M.M., Cohen, G.H. and Eisenberg, R.J. (1988) *J. Virol.* 62, 4027–4036.

Serafini-Cessi, F., Dall'Olio, F., Malagolini, N., Pereira, L. and Campadelli-Fiume, G. (1988) *J. Gen. Virol.* 69, 869–877.

Showalter, S.D., Zweig, M. and Hampar, B. (1981) *Infect. Immun.* 34, 684–692.

Simmons, A. and Nash, A.A. (1985) *J. Virol.* 53, 944–948.

Sjöblom, I., Lundström, M., Sjögren-Jansson, E., Glorioso, J.C., Jeansson, S. and Olofsson, S. (1987) *J. Gen. Virol.* 68, 545–554.

Sodora, D.L., Cohen, G.H. and Eisenberg, R.J. (1989) *J. Virol.,* 63, 5184–5193.

Spear, P.G. (1985) In *Immunochemistry of Viruses. The Basis for Serodiagnosis and Vaccines* (M.H.V. van Regenmortel and A.R. Neurath, Eds.) Elsevier, Amsterdam. pp. 425–446.

Spear, P.G. (1985) In *The Herpesviruses* (B. Roizman, Ed.) Plenum Press, New York. pp. 315–356.

Stuve, L.L., Brown-Shimer, S., Pachl, C., Najarian, R., Dina, D. and Burke, R.L. (1987) *J. Virol.* 61, 326–335.

Swain, M.A., Peet, R.W. and Galloway, D.A. (1985) *J. Virol.* 53, 561–569.

Watson, R.J. (1983) *Gene* 26, 307–312.

Watson, R.J., Weis, J.H., Salstrom, J.S. and Enquist, L.W. (1982) *Science* 218, 381–384.

Weber, P.C., Levine, M. and Glorioso, J.C. (1987) *Science* 236, 576–579.

Whitley, R.J. (1985) In *The Herpesviruses* (B. Roizman, Ed.) Plenum Press, New York. pp. 1–44.

Wilcox, W.C., Long, D., Sodora, D.L., Eisenberg, R.J. and Cohen, G.H. (1988) *J. Virol.* 62, 1941–1947.

Wildy, P. and Gell, P.G.H. (1985) *Br. Med. Bull.* 41, 86–91.

Wu, C.-T.B., Levine, M., Homa, F., Highlander, S. and Glorioso, J.C. (1989) *J. Virol.*, 64, 856–863.

Yasukawa, M. and Zarling, J.M. (1985) *J. Immunol.* 134, 2679–2682.

Zarling, J.M., Moran, P.A., Burke, R.L., Pachl, C., Berman, P.W. and Lasky, L.A. (1986) *J. Immunol.* 136, 4669–4673.

Zhu, Q. and Courtney, R.J. (1988) *Virology* 167, 377–384.

Eds. M.H.V. van Regenmortel and A.R. Neurath
Immunochemistry of viruses, II. The basis for serodiagnosis and vaccines.

Poxviruses

21

Y. ICHIHASHI

Department of Virology, Faculty of Medicine, Niigata University, Asahimachi, Niigata 951, Japan

1. Introduction

Immunity to a poxvirus was first recognized by Jenner in 1796 in the case of smallpox, and immunization with vaccinia virus, a progeny of the virus introduced by him, eradicated human smallpox virus infection in 1979 (Jenner, 1798; Fenner et al., 1988). During the intervening 180 years, the study of poxviruses led to a wealth of important findings such as the principle of vaccination, recognition of the viral host range, size of virus particles, replication in embryonated eggs, in vitro neutralization by antiserum, route of virus spreading in the body, recombination in mixedly infected cells, analysis of biochemical and morphological steps of replication, discovery of interferon, histocompatibility restriction in cell-mediated immunity and requirements for the eradication of viral disease. Together these advances led to the development of a number of basic virological and immunological concepts.

Recent advances in genetic engineering have opened a new dimension in poxvirus research. The use of vaccinia virus as an expression vector in eukaryotic cells (Panicali and Paoletti, 1982; Machett et al., 1982) made it possible to obtain knowledge about the biological activity of the introduced gene as well as that of the poxvirus genes, and holds considerable promise for the development of a new generation of vaccines.

We will discuss in this chapter the antigenic properties of orthopoxviruses with particular reference to (1) the membrane where epitopes are located, (2) the immunological basis of species specificity, (3) the immunological reactivity of poxvirus epitopes, and (4) the remaining problems that need to be solved.

2. General properties of poxviruses

Present taxonomy recognizes about 60 viral species in the poxviridae; within the subfamily chordopoxvirinae there are six genera and within the subfamily entomopoxvirinae there are three genera. Vaccinia virus of the chordopoxvirinae, genus orthopoxvirus, is the type species. The viruses of the entomopox subfamily do not resemble any known vertebrate or plant virus group, probably as a result of their long isolation in specific insect hosts. Several viruses (human molluscum contagiosum, elephantpox, Yaba monkey tumor virus) are not assigned to any taxonomic phylum due to lack of information (Matthews, 1979).

The unique structure of the virion is the major criterion used for classifying poxviruses. The brick shaped virion (about 150 × 200 × 300 nm) is covered by fine tubular material, and encases a thick-walled core and one or two proteinaceous lateral bodies. The genome in the core is a double stranded DNA of 1.6–2.2 × 10^5 base pairs with covalently bound ends. The DNA sequences of chordopoxvirinae viruses are highly homologous in their two-thirds central region. Both end regions contain long terminal repeats unique to each species. After penetration into the cell, the DNA replication process occurs within the cytoplasm. Marker rescue occurs in the cells doubly infected with different species of orthopoxviruses (Hanafusa et al., 1959; Ball, 1987; Upton et al., 1988), and the progeny sample contains large numbers of recombinant viruses.

The process of poxvirus infection and replication is characterized by a series of interdependent morphological and biochemical events. As suggested by the DNA homology and the presence of similar virus-associated enzymes in vaccinia and entomopox viruses, the basic processes are similar in all members of the family. The properties of poxviruses are described in several reviews (Dales and Pogo, 1981; Fenner et al., 1989; Moss, 1990; Fenner, 1990; Arif, 1984).

3. Membranes of poxviruses and host cells

In immunological experiments with poxviruses, one must take into account the unique characteristics of the viral membrane. The outermost membrane of the virus is generated at first as an immature viral membrane in the viroplasm. It grows by self-assembly without any contact with cellular membranes (Fig. 21.1). The intrinsic curvature leads to the immature viral membrane forming a sac which packages one copy of the viral DNA genome and a fixed amount of viroplasmic substance. The contents of the immature virus metamorphose into a core wall and lateral bodies in the subsequent maturation process (Dales and Siminovitch, 1961). At the same time, the spherical immature virus rapidly transforms into the brick-shaped mature virus particle. The membrane of the mature virus becomes associated with the surface tubule, HA and a substantial amount of phospholipids. These compo-

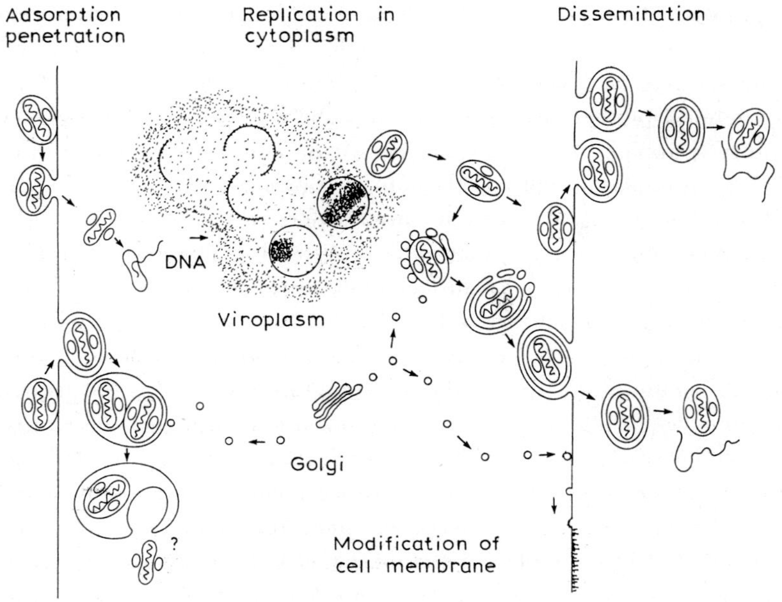

Fig. 21.1. Schematic drawing of virus-cell membrane interactions.

components seem to become integrated from the outside into the immature virus membrane.

Dimensions of the virus membrane look similar to those of the cellular lipid bilayer membrane. In contrast to the plasma membrane, the virus membrane is resistant to detergent. The amount of phospholipids in the virus seems to be insufficient to form a complete lipid bilayer membrane around the virus (Stern and Dales, 1974; Ichihashi and Oie, 1983; Oie, 1985). To disrupt the viral membrane, both a reducing agent and a detergent are necessary (Esterbrook, 1966; Ichihashi et al., 1984). The poxvirus membrane maintains its structure mainly by disulfide bridges, and thus resembles a capsid membrane rather than a lipid bilayer membrane.

Electron microscopic studies have shown that the membrane of the mature virus has affinities for at least three substances in addition to the receptor of the host cell. One is a fine tubular structure outside of the viral membrane. The protein composing the surface tubule of the vaccinia virus is P58; anti-surface tubule serum neutralizes the virus and blocks the cell-cell fusion of infected cells (Stern and Dales, 1976). The second is a virus-encoded factor named V. During the dissemination process, the mature particles are occluded into the matrix protein of an A-type inclusion body in the presence of the V factor. The A-type inclusion is commonly observed in cells infected with orthopox viruses having the ATI^+ trait (Ichihashi and Matsumoto, 1968; Funahashi et al., 1988). Vaccinia virus has ATI^- V^+ markers. Entomopox viruses produce occlusion bodies. The third is a substance lining the cytoplasmic side of the host cell membrane, which triggers budding of the virus from the plasma membrane and/or the adherence of small Golgi vesicles around the mature virus particles (Ichihashi et al., 1971; Morgan, 1976; Hiller and Weber, 1985) (Fig. 21.1). The surrounding Golgi vesicles then fuse mutually to form two layers of continuous membranes wrapping around the mature virus. The membrane-wrapped virus migrates to the protoplasmic surface of the plasma membrane where the outer wrapping membrane fuses with the plasma membrane. The inner surface of the vesicular membrane finally becomes the outer surface of the plasma membrane. The Golgi vesicles transport the virus-encoded surface proteins to the plasma membrane, and the other layer of the membrane around the disseminated virus usually dissociates extracellularly, but remains attached in several species (Nakano, 1985).

The membrane fusion and fission during the wrapping-transport follow the general rule of membrane flow (Palade, 1975); as a result, it is unclear whether this wrapping process is a cellular reaction provoked by recognition of a foreign substance in the cytoplasm, or a virus-directed dissemination process that recruits the cellular exocytosis system. The host cell seems to recognize the virus by some components residing on the inner surface of the plasma membrane and on the cytoplasmic surface of Golgi vesicles. Vaccinia virus HA may possibly have one such recognition component. It is synthesized on the rough endoplasmic reticulum, glycosylated in the Golgi apparatus and transported to the plasma membrane following the pathway of membrane proteins (Shida and Dales, 1981, 1982; Shida and Matsumoto, 1983; Shida, 1986). Accordingly, the affinity between a virus surface protein (VP37K) and HA molecules (Oie et al., 1990) could be the mechanism underlying the specific association of virus with the inner surface of the plasma membrane and the cytoplasmic surface of Golgi vesicles. However, the transmembrane domain of the HA, to which VP37K binds, may not be exposed to the cytoplasm, and mutant HAs lacking the cytoplasmic domain can associate with the virus. Moreover, the amounts of virus in the culture medium do not differ significantly between HA-free strains and viruses that possess HA. Therefore, the HA is not likely to be involved in the intracellular transportation of virus. The intracellular virus transportation seems to be a cellular expulsion reaction against the cytoplasmic virus. What makes this specific recognition possible remains to be elucidated.

Thus, the majority of particles obtained from a homogenate of infected cells is a naked-form virus. The virus harvested from the culture fluid is a mixture of naked virus and virus wrapped by the host cell membrane. The virus genome codes for the naked virus membrane proteins, and the wrapping membrane has its origin in the host cell membrane. The two morphologically distinct types of virion differ in their surface epitope compositions and in their biological activities. It has been shown that the extracellular virus is highly infectious and that its antigenicity is distinguishable from that of the naked virus (Appleyard et al., 1971; Boulter and Appleyard, 1973; Thornton, 1980; Payne and Kristensson, 1985).

In vaccinia virus infections of tissue culture cells, a few particles establish the replication site in the cytoplasm while a majority of the inoculated particles are phagocytosed into phagosomes (Kohn, 1989). Participation of the wrapping membrane in the viral penetration process has not been demonstrated by microscopic observation. On the other hand, it has been clearly demonstrated that infection occurs by fusion between the outer membrane of the naked virus and the plasma membrane of the host cell (Chang and Metz, 1976). The virus membrane has the capacity to fuse with the lipid bilayer in spite of its capsid-like nature. At the same time, the uncoating results in the release of core and lateral bodies into the cytoplasm of the host cell. No inverted wrapping process or uncoating from the phagosome has been observed in the cytoplasm at the early phase of infection. Antiserum against the host cell membrane has no neutralizing activity against purified virus preparations, and treatment of the virus with trypsin removes the wrapping membrane, and the titer of the treated virus sample increases 3–10-fold (Ichihashi and Oie, 1982a; 1982b). These findings indicate that the outer membrane of the naked virus has all the functions required for adsorption and penetration into the cytoplasm. The wrapping membrane is not indispensable for infection even though it may influence viral infec-

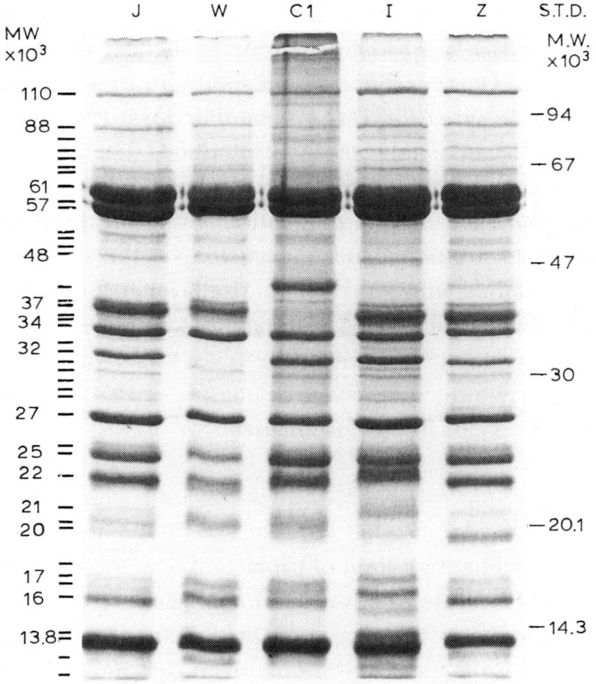

Fig. 21.2. SDS-polyacrylamide gel electrophoresis of orthopoxviruses. Purified viruses are electrophoresed in a 11% gel and stained with Coomassie brilliant blue. J, IHD-J strain vaccinia virus; W, IHD-W strain vaccinia virus; C1, CPR-C1 strain cowpox virus (derived from Brighton red pock strain); I, Ishibashi strain mousepox virus; Z, Zaire 599 strain monkeypox virus.

tivity, probably by controlling the susceptibility to phagocytosis or by helping proteolytic activation.

In view of the above facts, the purified naked-form virus is considered to be adequate material for the comparison of virus-specific epitopes among poxviruses.

4. Structural proteins of the virus

Fig. 21.2 shows profiles of the orthopoxvirus proteins separated by SDS polyacrylamide gel electrophoresis. The large poxvirus genome codes for over 100 proteins, and single dimension electrophoresis resolves 30–40 proteins as structural proteins (Oie and Ichihashi, 1981). There is no consensus on the naming of these poxvirus proteins at present. The approximate molecular size measured by SDS-polyacrylamide gel electrophoresis and that calculated from the open reading frame have been adopted by different researchers, the values usually being different. Because of differences in post-transcriptional processing, in vitro transcripts from isolated genes have different sizes and charges from those of in vivo products. In addition, among various poxvirus strains substantial differences exist in the molecular sizes of apparently related proteins. For example, a major glycoprotein of IHD-J strain of vaccinia virus has an apparent molecular size of 37 kDa while the equivalent molecule of the cowpox CPR-C1 strain is 42 kDa. Vaccinia VP17K easily forms a 25-kDa dimer, and corresponds to monkeypox virus 15-kDa protein. The VP16K of NR4 strain vaccinia virus corresponds to the 14-kDa protein of Ishibashi-111 strain mousepox virus.

These characteristics of poxvirus proteins lead to difficulties in identification. To confirm relationships, it is necessary to detect a sequence homology in the genes between a standard vaccinia strain and other virus strains. Monoclonal antibodies against continuous epitopes may give evidence for possible homology.

5. Immunological species specificity of orthopoxviruses

5.1. Studies with polyclonal antisera

Vaccination with vaccinia virus induces a high level of protection against smallpox virus infection in humans. This fact indicates a strong immunological cross-reactivity between the viruses. On the other hand, the unique character of each orthopoxvirus species suggests the existence of some species- and/or strain-specific epitopes (Downie and McCarthy, 1950; Ikuta et al., 1979; Baxby, 1982). Agar double diffusion analysis of orthopox viruses supported this by revealing species-specific antigens reactive with cross-absorbed sera (Esposito et al., 1977). Anti-monkeypox sera that have been absorbed extensively with homogenate of vaccinia-infected cells, can distinguish monkeypox from vaccinia viruses (Hutchinson et al., 1977). The serum of monkeypox-infected patients contains antibody different from that of a person immunized with vaccinia. However, all hyperimmune sera tested have been found to react with the various detectable structural proteins of all orthopoxviruses in Western blot analysis. No species-specific protein has been detected among the structural proteins of orthopoxviruses. Structural components closely resemble each other among orthopoxviruses, while species-specific epitopes may arise from slight changes in the common proteins.

5.2. Studies with monoclonal antibodies

The viral epitopes present on the surface of the virion have been compared by ELISA using 1200 monoclonal antibodies raised against vaccinia and monkeypox viruses. (Ichihashi and Oie, 1988). Almost all the clones (94.5%) were cross-reactive with the 11 viruses tested including vaccinia (IHD-J, WR, Elstree), monkeypox (Zaire 599, Copenhagen, WHO 21, 218, 240), mousepox (Ishibashi-111), cowpox (CPR C1) and a taterapox. In general, the results agree with previous observations made with polyclonal antisera. A majority of the epitopes were found to be common, while a small number of them were apparently species-specific.

No strain-specific epitope was found in the three vaccinia strains and also not in five monkeypox isolates. The vaccinia strains have been passaged independently in the laboratory for over 30 years. The monkeypox viruses were obtained at remote sites in Africa from different hosts (monkey, human, squirrel) with long intervals between isolation times. These passage conditions did not produce a strain-specific epitope detectable by any of the randomly selected 1200 monoclonal antibodies. The epitope composition of vaccinia viruses is stable under laboratory conditions and that of monkeypox viruses is stable in the natural environment of wild monkeys.

Immunological species specificity exists in two forms: (1) There are several species-specific epitopes of vaccinia or monkeypox viruses, but some of the epitopes may not be truly species-specific since smallpox virus also seems to contain them. (2) Composition of the common epitopes is species-specific. Some epitopes common to vaccinia and monkeypox viruses do not exist in mousepox or in cowpox virus. When monoclonal antibodies were raised using mousepox virus, some epitopes common between mousepox and vaccinia virus were not found in cowpox or monkeypox virus. The common epitopes appear in each species in mosaic fashion, and the epitope mosaic shows species-specific as well as strain-specific patterns. The proportion of the first group epitopes compared to the second group epitopes is less than 0.5%.

The four species of orthopoxviruses are immunologically distinct according to the composition of their epitopes. To establish these patterns of species-specific composition among orthopoxviruses, may require a hybridization phase and static phase. In the hybridization phase, a rare incidence of hybridization occurs in the host doubly infected with two species of poxviruses. It may produce a prototype of a species adapted to the specific host. Then in the static phase, a new species-specific epitope mosaic may be constructed from the pool of parental epitopes for adaptation to the changed environment by self-recombination in every replication cycle in the host species. Poxvirus DNA undergoes homologous recombination during the replication cycles, as evidenced by efficient marker rescue and high rate of recombinant virus in the progeny virus. Some species-specific epitopes may be produced by mutation and will spread rapidly, if the mutation is advantageous for the virus. An example is the taterapox virus. Its epitope composition suggests that it is derived from vaccinia and smallpox. Thus, exchange of genes by self-recombination in the replication cycle is the mechanism to maintain genetic homogeneity in the specific host. The ease of recombination secures reasonable flexibility for species specificity. On the other hand, the genetic stability of species specificity will depend mainly on the stringency of the host range. In other words, the genome composed of genes well adapted to the new host species is selected and determines the composition of epitopes. This forms the basis for the species concept in orthopoxvirus taxonomy. A typical example of such species generation may be vaccinia virus which has been passed artificially through humans, cows, rabbits, developing eggs and tissue culture cells, and has no known wild host.

A competitive binding inhibition test using monkeypox- and vaccinia-specific monoclonal antibodies showed that polyclonal anti-vaccinia serum did not compete with monkeypox specific monoclonal antibody and vice versa in the case of anti-monkeypox serum. When these monoclonal antibodies were used as probes, the serum of monkeypox-infected patients, who had been vaccinated by vaccinia virus, interfered with both vaccinia- and monkeypox-specific monoclonal antibodies in competitive binding inhibition assay. Patients' sera contained specific antibody reactive to the vaccinia epitope which is not detectable in the challenging monkeypox virus. Since the anti-vaccinia titer in the serum of vaccinated persons drop to a normal level within 12 months after immunization, the immunological memory for vaccinia virus may have been recalled. As the vaccinia-specific epitope is not present in the monkeypox virus, the recall of memory that was triggered by the monkeypox infection does not depend on stimulation with a precisely corresponding epitope but on some higher order regulation in antigen presentation or on specific stimulation of memory clones. This original antigenic sin (cross-activation) phenomenon should be taken into account when using vaccinia virus as a recombinant vaccine.

5.3. Antigens reactive with monoclonal antibodies

The screening of monoclonal antibodies by ELISA using purified virus antigen selects clones reactive with surface-exposed components of virions (Wilton et al., 1986). The intensity of the reaction is sometimes different among closely related viruses, and the reactive epitope may reside in antigens of different sizes. Fig. 21.3 shows an example of such cross-reactivity between the NR4 strain of vaccinia virus and the Ishibashi strain of mousepox virus revealed with anti-monkeypox monoclonal antibodies. All monoclonal antibodies in the figure are cross-reactive, except for anti-VP32K monoclonal antibody (J2B2). NR4 and its parent IHD-W strain are known to lack the VP32K trait. Reactive antigens of vaccinia virus are VP88K, VP54K, VP34K, VP32K, VP25K, VP21-20K and a cluster of VP17K-13K (VP17K, 16K, 14K, 13.8K, and VP13K). The monoclonal antibody producers reactive to these antigens are repeatedly established in every hybridization, and

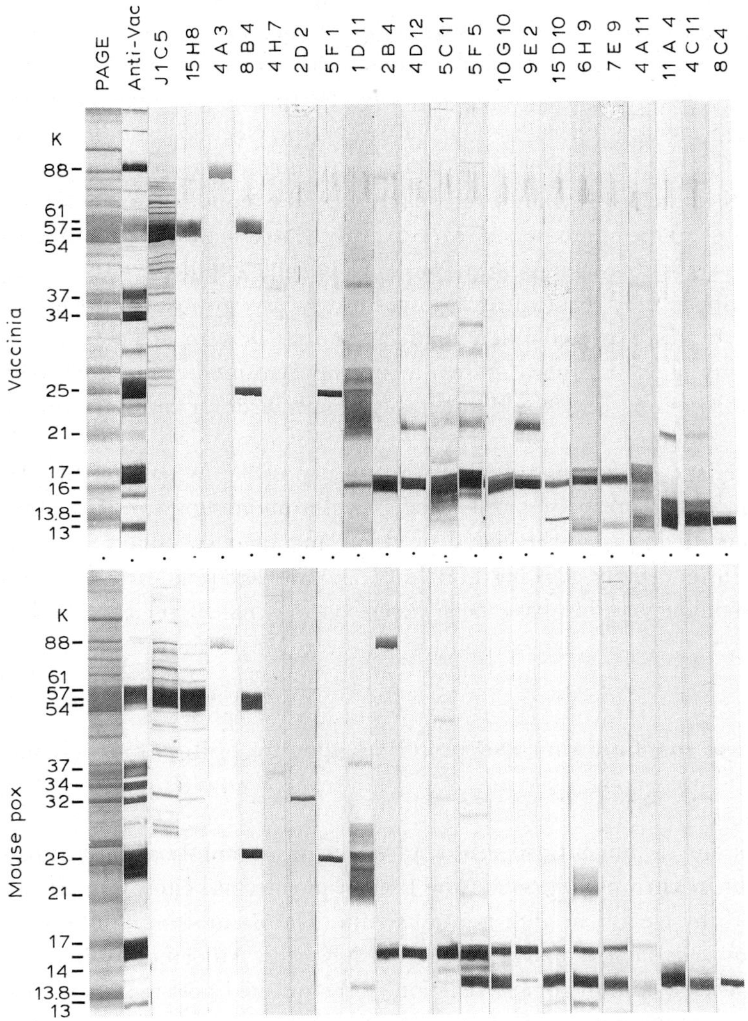

Fig. 21.3. Western blotting profiles of vaccinia and mousepox viruses. 0NR4 strain vaccinia virus (rifampicin-resistant mutant derived from IHD-W strain) and Ishibashi strain mousepox virus are electrophoresed and subjected to Western blotting analysis using anti-mousepox monoclonal antibodies. PAGE, profiles in polyacrylamide gel electrophoresis. Coomassie blue staining. Anti-Vac, anti-vaccinia virus polyclonal rabbit serum.

about half of the established monoclonal antibody clones are reactive with antigens of 17–13 kDa range (Oie and Ichihashi, 1987).

The monoclonal antibodies show various kinds of specificities. The group of anti-VP17K-13K monoclonal antibodies in Fig. 21.3 did not compete with each other against mousepox virus, indicating differences in their precise recognition of epitopes on the same antigen. The high antigenicity of VP17K-13K suggests its abundance on the virus surface, or high antigenicity of the epitopes. Some monoclonal antibodies are reactive to two proteins. This is not due to a mixture of antibody clones. Probably these proteins are constituting a single unit on the virus, and the junction sites of the components may produce discontinuous epitopes (Van Regenmortel, 1986; Barlow et al., 1986). Some monoclonal antibodies react with different molecules of vaccinia and mousepox viruses. The VP16K of the NR4 strain of vaccinia may correspond to the 14-kDa protein of the Ishibashi strain of mousepox virus. The VP88K and VP21K of the NR4 strain react weakly with the monoclonal antibodies (J2B4 and J6H9, while those of mousepox react strongly. Some monoclonal antibodies (J1C5, J1D11) react with a wide range of proteins. Possible explanations are that the monoclonal antibody may react with virus-specific glycosyl-groups or with some other virus-specific but partially common residues, or that it is an artifact due to non-specific reactivity of the antibody. When monoclonal antibodies are used as a tool in protein identification, they should preferably be specific for a known continuous epitope of the protein.

Application of genetic engineering methods makes it possible to isolate genes of vaccinia virus, monoclonal antibodies and specific antiserum being commonly used for isolating and identifying the gene-coded products and their functions. The list of such viral proteins is rapidly growing (Table 21.1). Foreign gene products obtained from recombinant vaccinia have been described in a recent review (Piccini and Paoletti, 1988).

6. Immune defense mechanisms and poxvirus specific epitopes on virus-infected cells

Normal serum contains an unidentified substance acting as an initial barrier against inoculated virus. In in vitro plaque assay, the titer of purified vaccinia virus is reduced to about 1/10 by treatment with normal serum. The natural immunity may also be mediated by cells such as macrophages and NK cells. An orthopoxvirus infection causes a full-scale immunological reaction in the infected host including humoral (lymphokines, interferons, antibodies) and cellular immune reactions (macrophages, natural killer, delayed hypersensitivity, cytotoxic T lymphocytes). When enough vaccinia virus (10^7 pfu) is injected into a mouse, the titer of the neutralizing antibody begins to increase at the 6th day, reaches a plateau after one month, then declines to a normal level after 4–6 months. Immunological protection occurs before

TABLE 21.1

Virus-encoded proteins

*Hind*III fragment	Virus-induced components	Reference
C	Telomeres	Bertholet et al., 1986 Parsons and Pickup, 1990
C	19K, Vaccinia growth factor	Venkeatesan et al., 1982 Blomquist et al., 1984 Twardzik et al., 1985 Stroobant et al., 1985 Reisner, 1985 Brown et al., 1985 Eppstein et al., 1985 King et al., 1986 Chang et al., 1987 Porter and Archard, 1987 Shultz et al., 1987 Upton et al., 1987 Mark et al., 1988 Buller et al., 1988a, b
C-N	37K, Complement control factor	Kotwal and Moss, 1988a Kotwal and Moss, 1988b
N	60K, α-Amanitin resistance	Villarreal and Hruby, 1986 Tamin et al., 1988
M-K	29K, Host range	Gillard et al., 1986.
K 1 K 3	42.2K, Serine protease inhibitor 48.8K	Boursnell et al., 1988.
F	37K, Ribonucleotide reductase	Slabaugh and Mathews, 1984 Slabaugh et al., 1988 Tengelsen, et al., 1988
F	37K, Vaccinia envelope protein	Hiller and Weber, 1985 Hirt et al., 1986
F-E	11K, VP11, DNA binding, phosphorylated basic	Kao et al., 1981 Hiller and Weber, 1982 Wittek et al., 1984 Tsao et al., 1986 Kao and Bauer, 1987

TABLE 21.1 (continued)

*Hind*III fragment	Virus-induced components	Reference
E	100K, DNA polymerase	Jones and Moss, 1984 Traktman et al., 1984 Earl et al., 1986 De Filippes, 1989 Traktman et al., 1989
O	32.5K Host range	Gillard et al., 1989
I	M1 subunit of ribonucleotide reductase	Hull et al., 1986 Child et al., 1990
L	25K, VP8	Yang and Bauer, 1988 Yang et al., 1988
L–J	20K, Thymidine kinase	Zaslavski and Yakobson, 1975 Weir et al., 1982 Hruby and Ball, 1982 Hruby and Guarino, 1982 Hruby et al., 1983 Weir and Moss, 1983 Franke et al., 1985 Buller et al., 1985 Boyle and Couper, 1986 Weir and Moss, 1987
J–H	147K, 22K, RNA polymerase Subunit genes are distributed in fragments E,J,H,D, and A.	Baroudy and Moss, 1980 Morrison et al., 1985 Broyles and Moss, 1986 Thompson et al., 1989
H	35K, Ag35,	Gordon et al., 1988 Rosel et al., 1986
D	ORF14, L65 D8, 35K	Niles et al., 1986 Weinrich et al., 1985 Niles and Seto, 1988 Lee-Chen and Niles, 1988
D	63K, Rifampicin-sensitive component, core protein	Tartaglia and Paoletti, 1985 Tartaglia et al., 1986 Baldick and Moss, 1987 Miner and Hruby, 1989
D	Nucleoside triphosphate phosphohydrolase I	Kahn and Esteban, 1990

→

TABLE 21.1 (continued)

*Hind*III fragment	Virus-induced components	Reference
D	33K, mRNA capping enzyme	Niles et al., 1989
D	DNA dependent ATPase I	Kunzi and Traktman, 1989
A	132K, RNA polymerase subunit	Morrison and Moyer, 1986 Jones et al., 1987 Enzinger M.J., 1987 Patel and Pickup, 1989 Hooda-Dhingra et al., 1989 Hooda-Dhingra et al., 1990
A	62K, Major core protein	Rosel and Moss, 1985 Weir and Moss, 1985
A	39K	Maa and Esteban, 1987
A	85K, Hemagglutinin	Shida and Matsumoto, 1983 Shida, 1986 Seki et al., 1990 Oie et al., 1990
A	160K, A-type inclusion	Patel et al., 1986 Funahashi et al., 1988
A	14K, Virion surface protein	Rodriguez et al., 1985 Dallo et al., 1987 Dallo and Esteban, 1987 Gong et al., 1989
B	40K, 38K, Serine protease inhibitor	Kotwal and Moss, 1989 Palumbo et al., 1989
?	130K, Early promotor binding factor	Yuen et al., 1987
?	11K, VP11b, Protein synthesis inhibitor, Virion component	Person-Fernandez and Beaud, 1986
?	Transcription termination factor. RNA capping enzyme?	Rohrmann et al., 1986 Shuman et al., 1987
?	Surface tubule, 58K	Stern and Dales, 1976
?	Alkaline protease	Arzoglou et al., 1979

neutralizing antibody can be detected. An infection with highly toxic mousepox virus (LD_{50} = 0.1 pfu assayed in BALB/c mice/Vero cells) will kill mice within seven days, whereas mice will survive 100 pfu of the virus when challenged with vaccinia virus 2–3 days after immunization. Mice were found to survive a similar challenge with mousepox virus 12 months after immunization, and showed clear secondary reactions; however, the virus may persist in the vaccinated mouse (Buller and Wallace, 1985). The ability of virus to establish infection and the ability of host effectors to inhibit viruses in vivo is much more complex than the cell culture system can reveal.

6.1. Immune response against virions

The binding of an antibody molecule to a viral epitope affects the function of the antigen molecule. Blockade of adsorption or inhibition of virus-cell fusion obviously will result in neutralization. However, in the case of orthopoxviruses, the antigenic structures critical for infection and the host receptors recognized by the virus have not been identified. It has been suggested that both the VP54K surface tubule and the 14-kDa protein may correspond to the viral ligand (Stern and Dales, 1976; Oie and Ichihashi, 1981; Dallo et al., 1987). The receptor of epidermal growth factor has been shown to be the viral receptor (Eppstein et al., 1985), although the virus can infect NR-G cells lacking the EGF receptor (Stroobant et al., 1985). Vaccinia virus is able to infect various cell species, and neutralizing monoclonal antibodies have been shown to react with several virion surface proteins. It is possible, therefore, that the virus may have multiple receptors, that the adsorption ligand is composed by several proteins, or that the adsorption and penetration phases may be multi-step processes (Oie and Ichihashi, 1987). Since our understanding of the virus infection process is fragmentary, the mechanism of neutralization is still completely obscure.

The neutralizing efficiency of monoclonal antibody against purified virus has been found in a plaque reduction test to be lower than that of polyclonal antibody. So far, we have not been able to find any monoclonal antibody which requires complement for its neutralizing activity. Addition of normal serum as a source of complement consistently resulted in elevated neutralizing activity of the monoclonal antibody, but this enhancement was attributable to the normal serum itself. The higher functional effectiveness of a polyclonal serum may be brought about by cooperative interactions between antibody populations and epitopes. Even when each antibody clone in a polyclonal serum does not have enough neutralizing activity, the binding of non-neutralizing antibodies may have a cumulative effect or it could expose new antigenic sites. Chain reactions with the antibodies in serum that react with new antigenic sites may finally lead to neutralization.

A majority of antibody clones in serum may have low neutralizing ability, and this may be the reason for the low neutralizing activity of the majority of monoclonal antibodies. There are several other possible explanations: (1) The blockage of one step of the poxvirus infection by a monoclonal antibody may be compensated by an alternative process. For instance, if the poxvirus has several kinds of ligands for adsorption to cells, association of a monoclonal antibody to one of them may result in incomplete neutralization. (2) Proteolytic processing may reactivate some of the neutralized virus. Treatment of the neutralized virus with trypsin can abrogate the neutralizing activity of the associated antibody in vitro. Since the proteolytic cleavage of virus surface proteins is essential for infection (Ichihashi and Oie, 1983), the host protease may reactivate some of the neutralized virus, even when the monoclonal antibody is reactive to the critical site for infection. (3) Critical sites for infection may be masked from the antibody molecule. The phenomenon of proteolytic activation of infectivity suggests that certain indispensable penetration components are masked and not accessible to the critical sites until the end of the adsorption phase (Ichihashi and Oie, 1982a). If this were the case, the monoclonal antibody clone reactive to the essential epitope for infection may possibly be missed during the two-step selection of neutralizing monoclonal antibody clones. It is indeed possible that the majority of clones positive in the first step ELISA are reactive to epitopes on the intact virus surface, but not to those on the activated virus.

6.2. Immune response against cell surface antigens

The surface of poxvirus-infected cells expresses virus-specific antigens in three different ways. The first is mediated by virus-cell fusion that occurs at infection. Evidence is that the penetration step of poxvirus infection occurs by virus-cell membrane fusion (Chang and Metz, 1976). The virus membrane is integrated into the host cell membrane and leaves in it the structural components of the viral outer membrane (Mallon and Holowczak, 1985). UV-irradiated vaccinia virus can proceed to this step, and the host cell becomes susceptible to lysis by vaccinia-specific cytotoxic T lymphocytes under the restriction of the major histocompatibility complex (Hapel et al., 1980; Oie and Ichihashi, 1981; Domber and Holowczak, 1986).

The second is an antigenic change of the cellular membrane apparently introduced by the attachment of mature virus particles. Ferritin-conjugated anti-vaccinia antibody is able to label the external surface of the cell membrane where the mature virus becomes attached from the inside (Ichihashi et al., 1971). The wrapped virus is also able to react with the ferritin-conjugated antibody through what looks like an extrusion or diffusion of viral surface antigen through the membrane. Another more plausible explanation is that the virus-infected cells modify their surface membrane by a virus-encoded transmembrane protein whose cytoplasmic domain has specific affinity for the surface of the mature virion. Vaccinia hemagglutinin may play such a role since it has specific affinity for the virus surface major glycoprotein VP37K (Oie and Ichihashi, 1990).

The third is the synthesis of transmembrane or membrane-associated proteins as evidenced by an analysis of vaccinia hemagglutinin (Shida and Matsumoto, 1983; Shida, 1986). Vaccinia hemagglutinin is synthesized on membrane-bound ribosomes of the rough endoplasmic reticulum, extruded co-translationaly into endoplasmic reticulum lumen mediated by the N-terminal signal sequence, *N*-glycosylated and anchored to the lipid bilayer by its hydrophobic amino acid sequence near the C-terminal. The synthesized hemagglutinin is sent from endoplasmic reticulum to the Golgi system where the high mannose oligosaccharides are converted to the complex type. The glycosyl epitopes may be produced by host glycosyltransferases. The *O*-glycosylated protein is transported to the plasma membrane through the membrane flow system. The wrapping transport of mature virus to plasma membrane depends on this Golgi vesicle transportation. Early surface antigen (S antigen, Ueda, personal communication), which appears on infected cell membrane before viral DNA synthesis, and 37-kDa envelope antigen (Hiller and Weber, 1982; Hirt et al., 1986) may be expressed on the infected cell membrane following a synthetic process similar to that found with hemagglutinin.

Possibly all the viral antigens on the infected cell membrane can be target molecules for specific cytolysis mediated by the antibody-complement system, cytotoxic T cells, cells which cause antibody-dependent cell-mediated cytolysis or delayed type hypersensitivity. Cytotoxic T cells (CTL) play a central role (Doherty et al., 1976). The effector T cell carries the $CD8^+$ ($Lyt\text{-}2^+$) phenotype, and the function is triggered by an interaction of the T-cell receptor with the antigen under the restriction of class I components of the major histocompatibility complex (MHC).

Which viral constituents become targets of specific CTL is not clear. Since recognition of CTL antigen requires compatibility of class I MHC, the viral antigen seems to exist in association with the MHC molecules ('self + viral') or in MHC modified by the virus (altered self). The formation of CTL target antigen may require processing of viral antigen, and maybe different from the epitopes on the viral antigen. No poxvirus-specific monoclonal antibody or T cell clone has been demonstrated so far to be reactive with the CTL antigen. However, it is known that viral structural components in the inoculated UV-inactivated virus induce CTL target antigens on mouse L 929 cells. UV-inactivated virus can fuse with the cell membrane and induce the early functions, although it does not induce the late function of the virus. In contrast to UV-inactivated virus, heat-inactivated virus has lost its capacity to fuse with the cell membrane, and it does not induce detectable levels of the CTL-recognizable antigens in L fibroblast cells. Therefore, the viral proteins in the inoculated virus may associate directly with MHC class I molecules after the incorporation into L cell membrane by virus-cell fusion. These facts indicate that certain viral antigens are recognized to be not self by the MHC complex when they appear integrated in the cell membrane, and form CTL target. The endogenous antigens of vaccinia virus may become the target antigen of the effector CTL in the same way. As far as the virus surface antigens are concerned, the antigen processing mechanism working in antigen presenting cells may not participate in the CTL target formation. However, it cannot be excluded at present that membrane protease or MHC molecules of the L cell have the capacity to process the viral antigens.

The virus-induced endogenous antigens, such as early surface antigen or hemagglutinin, appear on the infected cell membrane. In the case of influenza virus, MHC class I- and class II-restricted CTL are generated, and the target cell of the former CTL is induced by live virus whereas that of the latter is induced by inactivated virus or isolated hemagglutinin (Morrison et al., 1986). It is not clear yet in the vaccinia system whether the MHC on L929 cells recognizes differentially the exogenous and endogenous poxvirus proteins or not.

The $CD8^+$ cytotoxic T cell is induced by the helper function of the $CD4^+$ T cell ($L3T4^+$) which responds to antigens in the context of the class II MHC determinant (Cantor and Boyse, 1975). Specific $CD8^+$ CTL can also be induced without help from the $CD4^+$ T cell (Widmer and Bach, 1981; Sprent and Schaefer, 1985). Mousepox virus infection develops both types of virus-specific stimulation to generate MHC-restricted $CD8^+$ CTL in mice where the $CD4^+$ T cell-dependent helper function is long-lasting (Mark et al., 1987; Mizuochi et al., 1989). The helper T cell secretes interleukin-2 which strongly affects the induction of CTL (Ramshaw et al., 1987; Flexner et al., 1987; Mizuochi et al., 1986). It remains to be elucidated how lymphocyte differentiation is controlled by the message antigens of helper cells, and what relationships exist between the mechanism of message and target antigen formation.

6.3. Soluble factors

The ability of a virus to take over the cellular biosynthetic or regulation system of the host is well documented. In this regard, the finding of vaccinia-coded α -TGF like growth factor (Blomquist et al., 1984), serine protease inhibitors (Pickup et al., 1986; Boursnell et al., 1988) and of a factor resembling human complement 4b binding protein (Kotwal and Moss, 1988a,b) suggests that the virus affects a wide range of host activities beyond the cellular level. Vaccinia growth factor stimulates the proliferation of neighboring cells before their infection (Buller et al., 1988a,b). Tumorigenicity of Yaba monkeypox virus and molluscum contagiosum virus may be related to the poxvirus growth factor (Porter and Archard, 1987). Vaccinia C4b binding protein inhibits the classical pathway of complement-mediated hemolysis (Kotwal and Moss, 1988a). The higher effectiveness of cell-mediated immunity over humoral immunity for the elimination of virus from an infected body may cause interference by the viral C4b binding protein. Additional soluble factors effective in controlling host reactions to virus infection are likely to exist. A combination of intracellular biosynthetic controls and of extracellular effects determine the characteristics of adaptation to the host, and hence the pathogenicity of the virus. The corresponding genes are located near the long terminal repeat (LTR) region where sequences differ widely among orthopoxvirus species. The soluble factors seem to play a role in viral adaptation to the environment. The sequence homology of the genes coding for the soluble factors suggests that they were acquired from the host genome or from another co-infected virus rather than through independent evolution. This may point to the existence of a mechanism for the incorporation of foreign genes.

7. Conclusions

The immunological species-specificity of orthopoxviruses is determined in two ways; the large number of common epitopes existing in a species-specific composition and a few species-specific epitopes. The membrane of the infected cell expresses viral antigens by virus-cell fusion and by synthesized transmembrane proteins. In addition, there are soluble nonstructural antigens. All of the virus-encoded components may become targets of humoral or cell-mediated immunological reactions. Certain biological functions have been found to be associated with a number of viral epitopes or antigens but the information is still insufficient for a systematic analysis of viral functions in terms of immunological reactions of the host against virus infection.

8. References

Appleyard, G., Hapel, A.J., and Boulter, E.A. (1971) *J. Gen. Virol.* 13, 9–17.
Arif, B.M. (1984) *Adv. Virus Res.* 29, 195–213.
Arzoglou, P., Drillien, R., and Kirn, A. (1979) *Virology* 95, 211–214.
Ball, L.A.. (1987) *J. Virol.* 61, 1788–1795.
Baroudy, B.M. and Moss, B. (1980) *J. Biol. Chem.* 255. 4372–4395.
Baldick, C.J. Jr. and Moss, B. (1987) *Virology* 156, 138–145.
Barlow, D.J., Edwards, M.S., and Thornton, J.M. (1986) *Nature* 322, 747–748.
Baxby, D. (1982) *J. Gen. Virol.* 58. 251–262.
Bertholet, C., Stocco, P., Meir, E.V. and Wittek, R. (1986) *EMBO J.* 5. 1951–1957.
Blomquist, M.C., Hunt, L.T. and Barker, W.C. (1984) *Proc. Natl. Acad. Sci.* 81. 7363–7367.
Boulter, E.A. and Appleyard, G. (1973) *Proc. Med. Virol.* 16. 86–108.
Boursnell, M.E.G., Foulds, I.J., Campbell, J.I. and Binns, M.M. (1988) *J. Gen. Virol.* 69, 2995–3003.
Boyle, D.B. and Couper, B.E.H. (1986) *J. Gen. Virol.* 67. 1591–1600.
Brown, J. P., Twardzik, D.R., Marquardt, H. and Todaro, G. (1985) *Nature* 313. 491–493.
Broyles, S.S. and Moss, B. (1986) *Proc. Natl. Acad. Sci. USA* 83. 3141–3145.
Buller, R.M.L., and Wallace, G.D. (1985) *Lab. Anim. Sci.* 35. 473–476.
Buller, R.M.L., Chakrabarti, S., Moss, B., and Fredrickson, T. (1988a) *Virology* 164, 182–192.
Buller, R.M.L., Chakrabarti, S., Cooper, J.A., Twardzik, D.R., and Moss, B. (1988b) *J. Virol.* 62. 866–874.
Buller, R.M.L., Smith, G.L., Cremer, K., Notkins, A., and Moss, B. (1985) *Nature.* 317. 813–815.
Cantor, H., and Boyse, E.A. (1975) *J. Exp. Med.* 141. 1390–1399.
Chang, A., and Metz, D.H. (1976) *J. Gen. Virol.* 32. 275–282.
Chang, W., Upton, C., Hu, S.-L., Purchio, A.F., and McFadden, G. (1987) *Mol. Cell. Biol.* 7. 535–540.
Child, S.J., Palumbo, G.J., Buller, M.L. and Hruby, D.E. (1990) *Virology* 174, 625–629.
Dallo, S., and Esteban, M. (1987) *Virology* 159. 408–422.
Dallo, S., Rodriguez, J.F., and Esteban, M. (1987) *Virology* 159. 423–432.

Dales, S. and Pogo, B.G.T. (1981) *Biology of Poxviruses*. Virology Monographs Vol. 18. Springer-Verlag, New York.

Dales, S., and Siminovitch, L. (1961) *J. Biophys. Biochem. Cytol.* 10. 475–503.

De Filippes, F.M. (1989) *J. Virol.* 63, 4060–4063.

Doherty, P.C., Blanden, R.V., and Zinkernagel, R.M. (1976) *Transplant. Rev.* 29. 89–124.

Domber E., and Holowczak, J.A. (1986) *Virology* 152. 331–342.

Downie, A.W., and McCarthy, K. (1950) *Br. J. Exp. Pathol.* 31. 789–796.

Earl, P.L., Jones, E.V., Moss, B. (1986) *Proc. Natl. Acad. Sci. USA* 83. 3659–3663.

Ensinger, M.J. (1987) *J. Virol.* 61, 1842–1850.

Eppstein, D.A., Marsh, Y.V., Schreiber, A.B., Newman, S.R., Todaro, G.J., and Nestor, J.J., Jr. (1985) *Nature* 318. 663–665.

Esposito, J.J., Obijeski, J.F., and Nakano, J.H. (1977) *J. Med. Virol.* 1. 35–47.

Esterbrook, K.B. (1966) *J. Ultrastruct. Res.* 14. 484–496.

Fenner, F., Henderson, D.A., Arita, I., Jezek, Z. and Landnyi, D. (1988) *Smallpox and its Eradication.* WHO Geneva.

Fenner, F., Wittek, R. and Dumbell, K.R. (1989) *The Orthopoxviruses*, Academic Press, San Diego.

Fenner, F. (1990) *Virology* (Fields, B.N. et al., eds.) Raven Press, New York. pp. 2113–2136.

Flanke, C.A., Roseman, N.A., and Hruby, D.E. (1985) *Virus Res.* 3. 13–17.

Flexner, C., Hugin, A. and Moss, B. (1987) *Nature* 330, 259–262.

Funahashi, S., Sato, T., and Shida, H. (1988) *J. Gen. Virol.* 69. 35–47.

Gillard, S., Spehner, D., Drillien, R. and Kirn, A. (1986) *Proc. Nat. Acad. Sci. USA* 83. 5573–5577.

Gillard, S., Spehner, D., Drillien, R., and Kirn, A. (1989) *J. Virol.* 63, 1814–1817.

Gong, S., Lai, C., Dallo, S. and Esteban, M. (1989) *J. Virol.* 63, 4507–4514.

Gordon, J., Kovala, T., and Dales, S. (1988) *Virology* 167. 361–369.

Hanafusa, T., Hanafusa, H., and Kamahora, J. (1959) *Biken's J.* 2. 77–84.

Hapel, A.J., Bablanian, R., and Cole, G.A. (1980) *J. Immunol.* 124. 1990–1996.

Hiller, G., and Weber, K. (1982) *J. Virol.* 44. 647–657.

Hiller, G., and Weber, K. (1985) *J. Virol.* 55. 651–659.

Hirt, P., Hiller, G., and Wittek, R. (1986) *J. Virol.* 58. 757–764.

Hooda-Dhingra, U., Patel, D.D., Pickup, D.J. and Condit, R.C. (1990) *Virology* 174, 60–69.

Hooda-Dhingra, U., Thompson, C.L. and Condit, R.C. (1989) *J. Virol.* 63, 714–729.

Hruby, D.E., and Ball, L.A. (1982) *J. Virol.* 43. 403–409.

Hruby, D.E., Maki, R.A., Miller, D.B., and Ball, L.A. (1983) *Proc. Natl. Acad. Sci. USA* 80. 3411–3415.

Hruby, D.E., and Guarino, L.A. (1984) *Virus Res.* 1. 315–320.

Hull, R., Sadler, J. and Longstaff, M. (1986) *EMBO J.* 5, 3083–3090.

Hutchinson H.D., Ziegler, D.W., Wells, D.E., and Nakano, J.H. (1977) *Bull. WHO* 55. 613–623.

Ichihashi, Y. (1981) *Virology* 113. 277–284.

Ichihashi, Y. and Matsumoto, S. (1968) *Virology* 36, 262–270.

Ichihashi, Y., Matsumoto, S. and Dales, S. (1971) *Virology* 46. 507–532.

Ichihashi, Y. and Oie, M. (1982a) *Virology* 116. 297–305.

Ichihashi, Y. and Oie, M. (1982b) *Virology* 122. 290–296.

Ichihashi, Y. and Oie, M. (1983) *Virology* 130. 306–317.

Ichihashi, Y. Oie, M., and Tsuruhara, T. (1984) *J. Virol.* 50. 929–938.

Ichihashi, Y. and Oie, M. (1988) *Virology* 163. 133–144.

Ikuta, K., Miyamoto, H. and Kato, S. (1979) *J. Gen. Virol.,* 44. 557–563.

Jenner, E. (1798) *Classics of Medicine and Surgery.* Dover, New York. pp. 213–240.

Jones, E.V., and Moss, B. (1984) *J. Virol.* 49. 72–77.

Jones, E.V., Puckett, C., and Moss, B. (1987) *J. Virol.* 61. 1765–1771.

Kahn, J.S. and Esteban, M. (1990) *Virology* 174, 459–471.

Kao, S-Y., Ressner, E., Kates, J., and Bauer, W.R. (1981) *Virol.* 111. 500–508.

Kao, S-Y., and Bauer, W.R. (1987) *Virol.* 159. 399–407.
King, C.S., Cooper, J.A., Moss, B. (1986) *Mol. Cell. Biol.* 6. 332–336.
Kohn, A. (1979) *Adv. Virus Res.* 24. 223–276.
Kotwal, G.J., and Moss, B. (1988a) *Nature* 335. 176–178.
Kotwal, G.J., and Moss, B. (1988b) *Virology* 167. 524–537.
Kotwal, G.J. and Moss, B. (1989) *J. Virol.* 63, 600–606.
Kunzi, M.S. and Traktman, P. (1989) *J. Virol.* 63, 3999–4010.
Lee-Chen, G.J., and Niles, E.G. (1988) 163. 80–92.
Maa, J.S,, and Esteban, M. (1987) *J. Virol.* 61. 3910–3919.
Machett, M., Smith, G.L., Moss, B. (1982) *Proc. Natl. Acad. Sci. USA* 79. 7415–7419.
Mallon, V. (R), and Holowczak, J.A. (1985) *Virol.* 141. 201–220.
Matthews, R.E.F. (1979) *Intervirology.* 12. 129–296.
Mark, R., Buller, L., Holmes, K.L., Hugin, A., Frederickson, T.N. and Morse III, H.C. (1987) *Nature.* 328. 77–79.
Mark, R., Buller, L., Chakrabarti, S., and Moss, B. (1988) *Virology* 164. 182–192.
Mizuochi, T., Ono, S., Malek, T.R., and Singer, A. (1986) *J. Exp. Med.* 163. 603–619.
Mizuochi, T., Hugin, A.W., Morse III, H.C., Singer, A., and Buller, R.M. (1989) *J. Immunol.* 142. 270–273.
Morgan, C. (1976) *Virology* 73. 43–58.
Morrison, D.K., Carter, J.K., and Moyer, R.W. (1985) *J. Virol.* 55. 670–680.
Morrison, L.A., Lukacher, A.E., Braciale, V.L., Fan, D.P., and Braciale, T.J. (1986) *J. Exp. Med.* 163. 903–921.
Moss, B. (1990) *Virology* (Fields, B.N. et al., eds.) Raven Press, New York. pp. 2079–2112.
Nakano, J. H. (1985) *Laboratory Diagnoses of Viral Infections.* Marcel Dekker, New York. pp. 401–423.
Niles, E.G., Condit, R.C., Caro, P., Davidson, K., Matusick, L., and Seto, J. (1986) *Virology* 153. 96–112.
Niles, E.G., and Seto, J. (1988) *J. Virol.* 62. 3772–3778.
Niles, E.G., Lee-Chen, G.-J., Shuman, S., Moss, B. and Broyles, S.S. (1989) *Virology* 172, 513–522.
Oie, M. and Ichihashi, Y. (1981a) *Virology* 113. 263–276.
Oie, M. and Ichihashi, Y. (1981b) *Microbiol. Immunol.* 25. 361–375.
Oie, M. (1985) *Virology* 142. 299–306.
Oie, M., and Ichihashi, Y. (1987) *Virol.* 157. 449–459.
Oie, M., Shida, H. and Ichihashi, Y. (1990) *Virol.* in press.
Palade, G. (1975) *Science.* 189. 347–358.
Palumbo, G.J., Pickup, D.J., Fredrickson, T.N., McIntyre, L.J. and Byller, R.M.L. (1989) *Virol.* 172. 262–273.
Panicali, D. and Paoletti, E. (1982) *Proc. Natl. Acad. Sci. USA* 79, 4927–4931.
Parsons, B. and Pickup, D. (1990) *Virol.* 175. 69–80.
Patel, D.D., Pickup, D.J. and Joklik, W.K. (1986) *Virology* 149, 174–189.
Patel, D. and Pickup, D.J. (1989) *J. Virol.* 63. 1076–1086.
Payne, L.G., and Kristensson, K. (1985) *J. Gen. Virol.* 66. 643–646.
Perkus, M.E., Piccini, A., Lipindskas, B.R., and Paoletti, E. (1985) *Science.* 229. 981–984.
Perkus, M.E., Panicali, D., Mercer, S., and Paoletti, E. (1986) *Virology* 152. 285–297.
Person-Fernandez, A., and Beaud, G. (1986) *J. Biol. Chem.* 261. 8283–8289.
Piccini, A., and Paoletti, E. (1988) *Adv. Virus Res.* 34. 43–64.
Pickup, D., Ink, B.S., Hu, W., Ray, C.A., and Joklik, W.K. (1986) *Proc. Natl. Acad. Sci. USA* 83. 7698–7702.
Porter, C.D., and Archard, L.C. (1987) *J. Gen. Virol.* 68. 673–682.
Ramshaw, I.A., Andrew, M.E., Phillips, S.M., Boyle, D.B., and Coupar, B.E. (1987) *Nature.* 329. 545–546.
Reisner, A.H. (1985) *Nature.* 313. 801–803.

Rodriguez, J.F., Janeczko, R., and Esteban, M. (1985) *J. Virol.* 56. 482–488.
Rohrmann, G., Yuen, L., and Moss, B. (1986) *Cell.* 46. 1029–1035.
Rosel, J., and Moss, B. (1985) *J. Virol.* 56. 830–838.
Rosel, J., Earl, P.L., Weir, J.P., and Moss, B. (1986) *J. Virol.* 60. 436–449.
Seki, M., Oie, M., Ichihashi, Y. and Shida, H. (1990) *Virology* 175, 372–384.
Shida, H. (1986) *Virology* 150. 451–462.
Shida, H., and Matsumoto, S. (1983) *Cell* 33. 423–434.
Shultz, G.S., White, M., Mitchell, R., Brown, G., Lynch, J., Twardzik, D.R., and Todaro, J. (1987) *Science.* 235. 350–352.
Shuman, S., Broyles, S.S., Moss, B. (1987) *J. Biol. Chem.* 262. 12372–12380.
Shuman, S., Golder, M., Moss, B. (1989) *Virology* 170. 302–306.
Slabaugh, M.B., and Mathews, C.K. (1984) *J. Virol.* 52. 501–506.
Slabaugh, M.B., Roseman, N., Davis, R., and Mathews, C.K. (1988) *J. Virol.* 62. 519–527.
Sprent, J., and Schaefer, M. (1985) *J. Exp. Res.* 162. 2068–2088.
Stern, W., and Dales, S. (1974) *Virology* 62. 293–306.
Stern, W., and Dales, S. (1976) *Virology* 75. 232–241.
Stroobant, P., Rice, A.P., Gullick, W.J., Cheng, D.J., Kerr, I.M., and Waterfield, M.D. (1985) *Cell* 42. 383–393.
Tamin, A., Villarreal, E.C., Weinrich, S.L. and Hruby, D.E. (1988) *Virol.* 165. 141–150.
Tartaglia, J., and Paoletti, E. (1985) *Virology* 147. 394–404.
Tartaglia, J., Piccini, A., and Paoletti, E. (1986) *Virology* 150. 45–54.
Tengelsen, L.A., Slabaugh, M.B., Bibler, J.K., and Hruby, D.E. (1988) *Virology* 164. 121–131.
Thompson, C.L. and Condit, R.C. (1986) *Virology* 150. 10–20.
Thompson, C.L., Hooda-Dhingra, U. and Condit, R.C. (1989) *J. Virology* 63. 705–713.
Thornton, B. (1980) *Br. J. Pathol.* 61. 444–450.
Traktman, P., Sridhar, P., Condit, R.C., and Robert, B.E. (1984) *J. Virol.* 49. 125–131.
Traktman, P., Kelvin, M. and Pacheco, S. (1989) *J. Virol.* 63. 841–846.
Tsao, H., Ren, G.-F., and Chu, C.-M. (1986) *J. Virol.* 57. 693–696.
Twardzik, D.R., Brown, J.P., Ranchalis, J.E., Todaro, G.J., and Moss, B. (1985) *Proc. Natl. Acad. Sci. USA* 82. 5300–5304.
Upton, C., Macen, J.L., and McFadden, G. (1987) *J. Virol.* 61. 1271–1275.
Upton, C., Delange, A.M., and McFadden, G. (1987) *Virology* 160. 20–30.
Upton, C., Macen, J.L., Maranchuk, R.A., DeLange, A.M. and McFadden, G. (1988) *Virology* 166. 229–239.
Van Regenmortel, M.H.V. (1986) *Trends Biochem. Sci.* 11, 36–39.
Venkeatesan, S., Gershowitz, A., and Moss, B. (1982) *J. Virol.* 44. 637–646.
Villarreal, E.C., and Hruby, D.E. (1986) *J. Virol.* 57. 65–70.
Weinrich, S.L., Niles, E.G., and Hruby, D.E. (1985) *J. Virol.* 55. 450–457.
Weir, J.P., Bajszar, G., and Moss, B. (1982) *Proc. Natl. Acad. Sci. USA* 79. 1210–1214.
Weir, J.P., and Moss, B. (1983) *J. Virol.* 46. 530–537.
Weir, J.P., and Moss, B. (1985) *J. Virol.* 56. 534–540.
Weir, J.P., and Moss, B. (1987) *Virology* 158. 206–210.
Widmer, M.B., and Bach, F.H. (1981) *Nature.* 294. 750–752.
Wilton, S., Gordon, J., and Dales, S. (1986) *Virology* 148. 84–96.
Wittek, R., Hanggi, M., and Hiller, G. (1984) *J. Virol.* 49. 371–378.
Yang, W.-P., Kao, S.-Y., and Bauer, W.R. (1988) *Virology* 167. 585–590.
Yang, W.-P., and Bauer, W.R. (1988) *Virology* 167. 578–584.
Yuen, L., Davison, A.J., and Moss, B. (1987) *Proc. Natl. Acad. Sci. USA* 84. 6069–6073.
Yuen, L., Dionne, J., Arif, B. and Richardson, C. (1990) *Virology* 175, 427–433.
Zaslavski, V., and Yakobson, E. (1975) *J. Virol.* 16. 210–213.

Eds. M.H.V. van Regenmortel and A.R. Neurath
Immunochemistry of viruses, II. The basis for serodiagnosis and vaccines.

Plant viruses 22

M.H.V. VAN REGENMORTEL

Institut de Biologie Moléculaire et Cellulaire du CNRS, 15 rue Descartes, 67084 Strasbourg, France

The antigenic properties of plant viruses are of interest mainly because serological techniques are extremely useful for routine diagnosis of infected plants as well as for assessing the degree of similarity between different viruses. The present chapter will summarize the advances that have occurred in the serological identification and antigenic analysis of plant viruses since an earlier review of this topic appeared five years ago (Van Regenmortel 1985).

1. New methodological advances

1.1. Monoclonal antibodies

The advantages of mAb reagents over polyclonal antisera for the detection of plant viruses are well-known and need not be reiterated here (Halk 1986; Halk and De Boer 1985; Van Regenmortel 1986a). By 1985, mAbs had been prepared against 33 plant viruses (Van Regenmortel 1986a) and since then, reports have appeared concerning the preparation of mAbs to tobacco etch (Dougherty et al., 1985), beet necrotic yellow vein (Merten et al., 1985), African cassava mosaic (Thomas et al., 1986), citrus tristeza (Vela et al., 1986), odontoglossum ringspot (Dore et al., 1987a), tomato mosaic (Dekker et al., 1987), peanut mottle (Sherwood et al., 1987), maize streak (Dekker et al., 1988), wheat soil-borne mosaic (Bahrani et al. 1988), plum pox (Himmler et al., 1988), lettuce necrotic yellows (Dietzgen and Francki 1988), cucumber mosaic (Porta et al., 1989), peanut clump (Huguenot et al., 1989) and tomato spotted wilt viruses (Sherwood et al., 1989).

In addition to providing a reliable means for detecting the viruses in infected plants, mAb reagents make it possible to differentiate between related viral strains in a consistent and reproducible manner (Diaco et al., 1986; Thomas et al., 1986; Massalski and Harrison 1987; Huguenot et al. 1989). In contrast, the inherent qualitative and quantitative variability of polyclonal antisera often makes it difficult to differentiate closely related viral serotypes (Porta et al., 1989). Since mAbs are epitope-specific rather than virus-specific or strain-specific, it is possible either to select mAbs that do not recognize an antigenically related virus (because the epitope is not shared between the two viruses) or on the contrary, to select mAbs that do not differentiate between the two viruses because they recognize an identical epitope present in both. A single amino acid substitution may suffice to abolish the reactivity of a mAb that recognizes the modified epitope (Altschuh et al., 1985), while two viruses that differ in the sequence of their coat proteins by as much as 18% may not be distinguished by some mAbs (Briand et al., 1982). In general, it is observed that mAbs fail to cross-react with two viruses when the serological differentiation index (SDI) between them is equal to or larger than 3–4 (Dore et al., 1987a). The SDI corresponds to the average number of two-fold dilution steps separating homologous from heterologous precipitation titers (Van Regenmortel 1975). Reliable SDI values corresponding to those measured in precipitation tests can also be obtained by ELISA (Jaegle and Van Regenmortel 1985; Clark and Barbara 1987).

It is now widely recognized that many mAbs raised to a viral antigen do not react equally well with the antigen in different ELISA formats (Van Regenmortel 1986a; McCullough 1986; Dekker et al., 1989). This property of mAbs is due to the fact that they are able to recognize specific conformational features of the antigen and are often specific for either the polymerized or dissociated form of the viral coat protein (Dekker et al., 1987).

The common ELISA procedure of coating the virus directly to the microtiter plate in the presence of carbonate buffer pH 9.6 tends to dissociate the particles into protein subunits which then become preferentially bound to the solid phase (Halk and De Boer 1985; Diaco et al., 1986; Massalski and Harrison 1987). This was clearly demonstrated in a recent study using gold-labeled antibodies in which the ELISA was carried out on electron microscope grids (Dore et al., 1988). This approach made it possible to visualize the viral antigen to which the mAb bound in the solid-phase immunoassay. At the virus concentration (1 μ/ml) used for coating plates in ELISA, no intact tobacco mosaic virus particles became adsorbed to the solid phase at pH 9.6; only dissociated protein subunits became adsorbed and were detected by mAbs specific for cryptotopes. Many mAbs raised against plant viruses are specific for neotopes (Van Regenmortel 1986a), and these reagents will not detect the viral antigen present on so-called virus-coated plates when the antigen is actually dissociated subunits. However, anti-neotope mAbs will detect intact virus particles trapped on plates coated with a first layer of antibodies (Dore et al., 1988). Since subunits from different viruses usually are antigenically more closely related than the corresponding intact virions (the newly exposed surfaces in the dissociated subunits tend to be more conserved among related proteins), the extent of cross-reactivity measured with antigen-coated plates tends to be greater than when antibody-coated plates are used (Koenig and Paul 1982; Jaeglé and Van Regenmortel 1985).

It has been claimed that when double antibody sandwich (DAS)-ELISA is carried out solely with mAbs instead of with polyclonal antisera, it is imperative to use, as the coating and labeled antibody, mAbs specific for different epitopes of the virus (Diaco et al., 1986). Results obtained with many different viruses, however, demonstrate that this is not the case and that there are always enough accessible epitopes on an intact virus particle for the same mAbs to bind first as the coating antibody and subsequently as the detecting biotinylated antibody (Dekker et al., 1987; Dore et al., 1987a; Porta et al., 1989; Huguenot et al., 1989). Since a mAb may lose its activity after adsorption to plastic or after conjugation to a label, tests involving the use of the same antibody in both positions in DAS-ELISA may appear to be inferior because of antibody instability (Van Regenmortel 1986a; Torrance and Pead 1986).

When mAbs are used to investigate the presence of cross-reactions between plant viruses, it is essential not to use excessively high reagent concentrations that may give rise to nonspecific reactions. In ELISA, antigen concentrations of 1–10 μg/ml and culture supernatant concentrations of 0.1–1.0 mg/ml protein tend to produce spurious nonspecific reactions, at least when the usual blocking agent, BSA, is used (Zimmermann and Van Regenmortel 1989). Using such conditions, Dietzgen (1986) claimed that 10 mAbs raised against arabis mosaic virus which recognized seven distinct epitopes of this virus, all cross-reacted with the unrelated tobacco mosaic and potato leaf roll viruses. However, when defatted milk was used as blocking agent, these spurious cross-reactions between unrelated viruses were abolished (Zimmermann and Van Regenmortel 1989). Nonspecific reactions can also result from electrostatic binding of antibodies to the basic domain of viral subunits, a phenomenon that can be inhibited by the polyanion heparin (Dietzgen and Francki 1987).

1.2. Improved immunoassays

ELISA methods continue to be the most popular assay for detecting plant viruses (Clark and Bar Joseph 1984). The most commonly used conjugate in these assays is alkaline phosphatase-labeled antibody prepared by conjugation with glutaraldehyde. However, because of the small size of biotin and the gentle conditions required for coupling this compound to antibody, biotinylated antibodies followed by enzyme-labeled avidin are becoming increasingly used as the detecting system (Diaco et al., 1985; Hewish et al., 1986). Recently, it was shown that the use of biotinylated viral antibody instead of enzyme-labeled antibody (Zrein et al., 1986) could overcome one of the disadvantages of DAS-ELISA (compared to indirect ELISA using anti-immunoglobulin reagents), namely its excessive serotype specificity (Koenig 1981; Van Regenmortel 1982). It seems that biotinylation of viral antibody is less detrimental to its capacity to recognize a broad range of serologically related viruses than is conjugation to an enzyme. In addition, assay sensitivity is also increased (Kendall et al., 1983; Hewish et al., 1986; Zrein et al., 1986).

It has been shown that when Fab′ fragments are conjugated to enzyme by means of maleimide compounds, the yield and binding activity of the conjugated antibody is higher than when other coupling agents are used (Ishikawa et al., 1983). A single thio group, remote from the antigen binding site, is present on the Fab′ fragment and interacts with the maleimide compound bound to a single, fully active peroxidase molecule. When such Fab′ conjugates were used for detecting tobamoviruses, it was found that they increased the sensitivity of homologous virus detection by direct or indirect DAS-ELISA, although they did not overcome the narrow serotype specificity characteristic of enzyme-labeled antibodies (Weiss et al., 1989). This shortcoming appeared to be linked to the monomeric nature of the Fab′ conjugates and not to the conjugation procedure.

Various other improvements of ELISA for the detection of plant viruses have been described which increase the sensitivity, specificity or ease of antigen detection (Edwards and Cooper 1985; Torrance 1987; Mowat and Dawson 1987; Hughes and Thomas 1988).

The sensitivity of virus detection has been further improved by the introduction of a time-resolved fluoroimmunoassay (TR-FIA). This assay, which uses a Europium label, avoids the non-specific background fluorescence normally associated with fluorescence measurements (Hemmilä et al., 1984). When used for detecting potato viruses, TR-FIA was found to be about 5–100 times more sensitive than DAS-ELISA (Siitari and Kurpa 1987; Sinijärv et al., 1988; Saarma et al., 1989).

Two immunoblotting techniques have become increasingly used for detecting plant viruses and their coat proteins: the electroblot and dot-blot immunoassays. The electroblot immunoassay is useful for measuring cross-reactions between viral proteins (Burgermeister and Koenig 1984; Koenig and Burgermeister 1986) and for demonstrating proteolytic breakdown of proteins (Ehlers and Paul 1986). The dot-blot immunoassay, in which the antigen is applied directly to a nitrocellulose membrane (Banttari and Goodwin 1985; Hibi and Saito 1985) is about five times more sensitive than ELISA (Berger et al., 1985).

Synthetic peptides corresponding to regions of viral-encoded proteins are commonly used for raising anti-peptide antibodies able to bind the viral proteins (see Chapter 1). This approach has been used successfully for example for studying the appearance in infected leaves of non-structural proteins of alfalfa mosaic virus (Berna et al., 1986; Van Pelt-Heerschap et al., 1987), cauliflower mosaic virus (Ziegler et al., 1985) and cowpea mosaic virus (Wellink et al., 1987). Antibodies to synthetic peptides have been used also for studying the conformational changes that occur when tomato bushy stunt virus (TBSV) particles swell under slightly alkaline conditions. It was found that residues 138–154 of the coat protein of TBSV contain an epitope of the dissociated protein which is accessible to antibody binding in the virion only when the particles are in the swollen state (Jaeglé et al., 1988).

2. Antigenic structure of plant viruses

2.1. Tobamoviruses

Since the antigenic structure of tobacco mosaic viruses (TMV) has been described in detail in a recent review (Van Regenmortel 1986b), the present discussion will be limited to data obtained during the period 1985–1989.

The specificity of 18 mAbs directed to TMV has been analyzed by measuring their ability to bind to viral mutants, to other tobamoviruses, to dissociated viral subunits and to peptide fragments of the coat protein (Altschuh et al., 1985). By means of computer-generated images of the surface residues of the protein subunit, it was possible to identify certain clusters of residues that formed discontinuous epitopes recognized by some of the mAbs (Dore et al., 1987b). About half of the 18 mAbs were so specific for the quaternary structure of TMV that they were unable to react with monomeric viral subunits. Furthermore, only three of the 18 mAbs were able to bind to some tryptic peptides of TMV protein (Altschuh et al., 1985).

In a separate study, 30 mAbs raised against the isolated TMV protein were examined for their capacity to bind intact virus, TMV protein and a series of synthetic peptides representing virtually the entire sequence of TMV protein (Al Moudallal et al., 1985). Five mAbs which did not react with intact virions reacted with the isolated coat protein as well as with four synthetic peptides (residues 76–88, 90–117, 115–134 and 134–146) corresponding to regions that are hidden when the subunit is assembled into the viral capsid. These five antibodies were therefore referred to as anti-cryptotope antibodies. In addition, the region 19–32 was recognized by several anti-TMV protein antisera. Both the regions 19–32 and 115–134 correspond to α-helices in TMV protein, which may explain why shorter peptides within these regions (peptides 18–25, 22–29, 27–33, 129–134) were inactive; presumably peptides of a length of about 15–20 residues are required for producing the helical conformation necessary for recognition by antibody (Al Moudallal et al., 1985).

In a subsequent study, a reduced binding was observed when some of the anti-cryptotope mAbs reacted with TMV mutants possessing an amino acid substitution outside the peptide sequences recognized by these mAbs (Dore et al., 1987b). Furthermore, by comparing the common accessible surface between TMV and other tobamoviruses, it was possible to narrow down the region recognized by certain anti-crytotope mAbs. These results led to the conclusion that the linear sequence of the peptides recognized by these mAbs did not represent the entire epitope. Just as in the case of neotope-specific antibodies, anti-cryptotope antibodies can thus be specific for epitopes that comprise residues that are adjacent in space but not directly linked in the sequence.

Of the 30 mAbs studied by Al Moudallal et al.(1985), 22 reacted both with the virus and the dissociated viral subunits and were thus specific for epitopes known as metatopes (Van Regenmortel 1982). Since none of these antibodies reacted with synthetic peptides, the precise location of the metatopes on the viral subunit was not established (Dore et al., 1987b). In an attempt to visualize the anti-metatope binding sites on TMV particles, immunoelectron microscopy experiments were carried out with the different mAbs (Dore et al., 1988). It was found that the anti-metatope antibodies reacted only with one extremity of the viral rods as well as with disks while the anti-neotope antibodies bound to the entire longitudinal surface of the virus particles. When several mixtures of two different anti-metatope mAbs were tested by immunoelectron microscopy, it was found that in all cases only one of the rod extremities was recognized by gold-labeled antibodies. In order to establish if these antibodies were binding to the extemity containing the 3′ or the 5′ end of the RNA, virus particles were partially uncoated by the action of 6 M urea, a procedure which uncovers RNA tails at the 5′ terminus (Blowers and Wilson 1982). By using an immunoelectron microscopy technique which permits the visualization of both the RNA tails and the gold-labeled mAbs, it could be shown that the anti-metatope antibodies recognized the extremity of the particle containing the 5′ end of the RNA (Dore et al., 1989). In view of the orientation of the RNA molecule inside the helical virus rod (Namba and Stubbs 1986), this means that the mAbs bound to the so-called bottom face of the subunit containing the two α helices made up of residues 73–89 and 111–135.

The anti-metatope mAbs were also used to try to establish the location of the binding sites on the stacked disk aggregates of TMV protein. It is believed that the two-layer disks are capable of switching to 'lock-washer' intermediates which then assemble into helical rods (Durham et al., 1971; Bloomer and Butler 1986). Such a transformation implies that opposite subunit surfaces are exposed at the two ends of short rods of stacked disks and thus that the mAbs should be able to bind to only one end of the stacked disks. Contrary to this expectation, it was found that the anti-metatope antibodies bound to both ends of the stacked disks and thus that the same face of the subunit was exposed at both ends (Dore et al., 1989). These results indicate that the stacked disks in solution have the same geometry as the crystalline four-layer aggregate which is a symmetric dimer of a polar pair of rings (Gilbert and Klug 1974) and thus that the disks cannot be converted directly into helical rods (Raghavendra et al., 1985, 1988).

The reason why 22 out of 30 mAbs obtained from a single fusion experiment with mouse spleen cells (Al Moudallal et al., 1985) recognized the same TMV subunit surface had remained a puzzle for several years. However, since the mice used for hybridoma production had been immunized with a concentrated TMV protein preparation which probably contained stacked disks, the presence of the same subunit surface on both ends of the disks explains the predominance of anti-metatope mAbs recognizing this particular surface.

2.2. Potyviruses

The potyvirus group is the largest group of plant viruses and comprises more than

150 members. mAbs have been produced against various potyviruses for diagnostic purposes (Gugerli and Fries 1983; Hill et al., 1984) as well as for epitope mapping. In a study of the antigenic structure of tobacco etch virus (TEV), ten mAbs raised against this virus were grouped into two classes (Dougherty et al., 1985). The first group (3 mAbs) recognized epitopes at the surface of virus particles that were specific for TEV while the second group (7 mAbs) reacted with epitopes that were not exposed on the surface and were present in a number of different potyviruses. This observation is consistent with the known conservation of internal regions of the viral protein subunit that are involved in protein-RNA or protein-protein interactions.

Proteolytic treatment of intact TEV particles was found to remove 29 amino acid residues from the N-terminus of the protein and to abolish the binding of TEV-specific mAbs (Allison et al., 1985). In a subsequent study of the proteolytic degradation of six potyviruses, it was shown that trypsin treatment removed the N-terminal region (corresponding to a length of 30–67 residues, depending on the virus) as well as 18–20 residues from the C-terminus of the protein (Shukla et al., 1988). After the removal of both termini, fully assembled virus particles were left which contained subunit cores of 216–218 residues and which were indistinguishable from untreated native particles in the electron microscope. The core particles lacked the surface epitopes that were recognized by the majority of the polyclonal antibodies raised against intact virus particles and that were specific for individual potyviruses. The trypsin-resistant core protein region contained epitopes shared among all potyviruses tested (Shukla et al., 1988; Shukla and Ward 1988) and is a useful immunogen for raising antibodies able to recognize most potyviruses (Shukla et al., 1989).

2.3. Potexviruses

In common with the tobamoviruses and potyviruses, potexviruses also have the N- and C-termini of their coat proteins exposed on the external surface of the virus particles (Koenig et al., 1978; Sawyer et al., 1987). When the mAbs to potato virus X (PVX) obtained by Torrance et al.(1986) were used to analyze the epitopes of the virus, three groups of determinants could be distinguished (Koenig and Torrance 1986). The first group of epitopes was located on the protruding N-terminus of the capsid protein; the second group corresponded to epitopes that became exposed only after denaturation or disruption of the particles caused by direct adsorption of virions to the ELISA plates. These epitopes which correspond to cryptotopes present on dissociated protein subunits are not destroyed by extensive denaturation, e.g. by SDS treatment. The third group corresponded to epitopes that are exposed on intact virus particles but are not located in the N-terminal region of the protein.

In a subsequent study, Sober et al.(1988) identified two epitopes on PVX particles by means of mAbs. One of the epitopes was located on the 68 amino acid N-terminal fragment of the coat protein isolated by CNBr cleavage. The second epitope was shown by competition ELISA to be distinct from the first one and was sensitive to denaturation of the viral protein. By means of polyclonal antibodies, an epitope has also been identified on the 48-amino-acid C-terminal region of PVX protein (Radavsky et al., 1988).

2.4. Isometric plant viruses

The antigenic structure of southern bean mosaic virus (SBMV) has been analyzed with mAbs raised against the virus. SBMV particles swell in EDTA at pH 7.5 and the swollen virions can be contracted by the addition of divalent cations or by adjusting the virus preparation to pH 5.0. Nine mAbs were tested for their ability to react with native virus, swollen virus, swollen virus contracted by calcium ions or by pH adjustment, and with purified coat protein (Tremaine et al., 1985a, b). Six mAbs were found to react with epitopes present on the dissociated protein and on the swollen particles, presumably on the subunit interfaces which became exposed upon swelling. These cryptotope-specific mAbs reacted differently in various immunoassays, suggesting that they recognized different conformations of the viral protein (Tremaine et al., 1985b). Three mAbs reacted with native virus and two of these reacted also with swollen virus contracted by calcium ions or by pH adjustment (Tremaine et al., 1985a). In a subsequent study with a mAb raised against the N-terminal region of SBMV protein, it was found that the corresponding epitope, located in the N-terminal 30 residues, was inaccessible in native virus particles but became exposed when the virus was swollen or disrupted following adsorption to an ELISA plate (MacKenzie and Tremaine 1986).

The particles of another isometric plant virus, tomato bushy stunt virus (TBSV), also swell when the pH is raised from 6 to 8.0 (Krüse et al., 1982). This structural phase transition of TBSV particles has been studied by means of antibodies raised against three synthetic peptides corresponding to residues 28–40, 138–154, and 380–387 of TBSV coat protein (Jaeglé et al., 1988). Antibodies to peptides 28–40 and 380–387 reacted with both compact and swollen particles, as well as with dissociated subunits. In contrast, antibodies to peptide 138–154 reacted only with swollen particles and with dissociated subunits. The epitope in the region 138–154 is a cryptotope located at the interface of subunits whose relative position is altered during the swelling process.

The results obtained in these studies of SBMV and TBSV illustrate the usefulness of mAbs and antipeptide antibodies for studying conformational changes in virus particles such as those associated with the swelling of virus particles.

3. References

Al Moudallal, Z., Briand, J.P. and Van Regenmortel, M.H.V. (1985) *EMBO J.* 4, 1231–1235.

Allison, R.F., Dougherty, W.G., Parks, T.D., Willis, L., Johnston, R.E., Kelly, M.E. and Armstrong, F.B. (1985) *Virology* 147, 309–316.

Altschuh, D., Al Moudallal, Z., Briand, J.P. and Van Regenmortel, M.H.V. (1985) *Mol. Immunol.* 22, 329–337.

Bahrani, Z., Sherwood, J.L., Sanborn, M.R. and Keyser, G.C. (1988) *J. Gen. Virol.* 69, 1317–1322.

Banttari, E.E. and Goodwin, P.H. (1985) *Plant Dis.* 69, 202–205.

Berger, P.H., Thornbury, D.W. and Pirone, T.P. (1985) *J. Virol. Methods* 12, 31–39.

Berna, A., Briand, J.P., Stussi-Garaud, C. and Godefroy-Colburn, T. (1986) *J. Gen. Virol.* 67, 1135–1147.

Bloomer, A.C. and Butler, P.J.G. (1986) in *The Plant Viruses, Vol. 2. The Rod-Shaped Plant Viruses* (M.H.V. Van Regenmortel and H. Fraenkel-Conrat, Eds.) Plenum Press, New York. pp. 19–57.

Blowers, J.E. and Wilson, T.M.A. (1982) *J. Gen. Virol.* 61, 137–141

Briand, J.P., Al Moudallal, Z. and Van Regenmortel, M.H.V. (1982) *J. Virol. Methods* 5, 293–300.

Burgermeister, W. and Koenig, R. (1984) *Phytopathol. Z.* 111, 15–25.

Clark, M.F. and Bar-Joseph, M. (1984) *Methods Virol.* 7, 51–85.

Clark, M.F. and Barbara, D.J. (1987) *J. Virol. Methods* 15, 213–222

Dekker, E.L., Dore, I., Porta, C. and Van Regenmortel, M.H.V. (1987) *Arch. Virol.* 94, 191–203

Dekker, E.L., Pinner, M.S., Markham, P.G. and Van Regenmortel, M.H.V. (1988) *J. Gen. Virol.* 69, 983–990

Dekker, E.L., Porta, C. and Van Regenmortel, M.H.V. (1989) *Arch. Virol.* 105, 269–286

Diaco, R., Hill, J.H., Hill, E.K., Tachibana, H. and Durand, D.P. (1985) *J. Gen. Virol.* 66, 2089–2094

Diaco, R., Lister, R.M., Hill, J.H. and Durand, D.P. (1986) *J. Gen. Virol.* 67, 353–362

Dietzgen, R.G. (1986) *Arch. Virol.* 91, 163–173

Dietzgen, R.G. and Francki, R.I.B. (1987) *J. Virol. Methods* 15, 159–164

Dietzgen, R.G. and Francki, R.I.B. (1988) *Virology* 166, 486–494

Dore, I., Dekker, E.L., Porta, C. and Van Regenmortel, M.H.V. (1987a) *J. Phytopathol.* 20, 317–326

Dore, I., Altschuh, D., Al Moudallal, Z. and Van Regenmortel, M.H.V. (1987b) *Mol. Immunol.* 24, 1351–1358

Dore, I., Weiss, E., Altschuh, D. and Van Regenmortel, M.H.V. (1988) *Virology* 162, 279–289

Dore, I., Ruhlmann, C., Oudet, P., Cahoon, M., Caspar, D.L.D., Van Regenmortel, M.H.V. (1989) *Virology,* 162, 279–289.

Dougherty, W.G., Willis, L. and Johnston, R.E. (1985) *Virology* 144, 66–72

Durham, A.C.H., Finch, J.T. and Klug, A. (1971) *Nature New Biol.* 229, 37–42.

Edwards, M.L. and Cooper, J.I. (1985) *J. Virol. Methods* 11, 309–319

Ehlers, U. and Paul, H.L. (1986) *J. Phytopathol.* 115, 294–304

Gilbert, P.F.C. and Klug, A. (1974) *J. Mol. Biol.* 86, 193–207

Gugerli, P. and Fries, P. (1983) *J. Gen. Virol.* 64, 2471–2477

Halk, E.L. (1986) *Methods Enzymol.* 118, 766–780

Halk, E.L. and De Boer, S.H. (1985) *Annu. Rev. Phytopathol.* 23, 321–350

Hemmilä, I., Dakubu, S., Mukkala, V.M., Siitari, H. and Lövgren, T. (1984) *Anal. Biochem.* 137, 335–343

Hewish, D.R., Shukla, D.D. and Gough, K.H. (1986) *J. Virol. Methods* 13, 79–85

Hibi, T. and Saito, Y. (1985) *J. Gen. Virol.* 66, 1191–1194

Hill, S.A. (1984) *Methods in Plant Virology.* Blackwell, Oxford

Himmler, G., Brix, U., Steinkellner, H., Laimer, M., Mattanovich, D. and Katinger, H.W.D. (1988) *J. Virol. Methods* 22, 351–358

Hughes, D.A. and Thomas, B.J. (1988) *Ann. Appl. Biol.* 112, 117–126

Huguenot, C., Givord, L., Sommermeyer, G. and Van Regenmortel, M.H.V. (1989) *Res. Virol.* 140, 87–102

Ishikawa, E., Imagawa, M., Hashida, S., Yoshitake, S., Hamaguchi, Y. and Ueno, T. (1983) *J. Immunoassay* 4, 209–327

Jaeglé, M., Briand, J.P., Burckard, J. and Van Regenmortel, M.H.V. (1988) *Ann. Inst. Pasteur Virol.* 139, 39–50

Jaeglé, M. and Van Regenmortel, M.H.V. (1985) *J. Virol. Methods* 11, 189–198

Kendall, C., Ionescu-Matiu, I. and Dreesman, G.R. (1983) *J. Immunol. Methods* 56, 329–339

Koenig, R. (1981) *J. Gen. Virol.* 55, 53–62

Koenig, R. and Burgermeister, W. (1986) In *Developments and Applications in Virus Testing* (R.A.C. Jones and L. Torrance, Eds.) Association of Applied Biologists, Wellesbourne. pp. 121–137.

Koenig, R. and Paul, H.L. (1982) *J. Virol. Methods* 5, 113–125.

Koenig, R. and Torrance, L. (1986) *J. Gen. Virol.* 67, 2145–2151.

Koenig, R., Tremaine, J.H. and Shephard, J.F. (1978) *J. Gen. Virol.* 38, 329–337.

Krüse, J., Krüse, K.M. and Witz, J. (1982) *J.Mol.Biol.* 162, 393–417

MacKenzie, D.J. and Tremaine, J.H. (1986) *J. Gen. Virol.* 67, 727–735

Massalski, P.R. and Harrison, B.D. (1987) *J. Gen. Virol.* 68, 1813–1821

McCullough, K.C. (1986) *Arch. Virol.* 87, 1–36

Merten, O.W., Heberle-Bors, E., Himmler, G., Reiter, S., Messner, P. and Katinger, H. (1985) *Dev. Biol. Stand.* 60, 451–456.

Mowat, W.P. and Dawson, S. (1987) *J. Virol. Methods* 15, 233–247

Namba, K. and Stubbs, G. (1986) *Science* 231, 1401–1406

Porta, C., Devergne, J.C., Cardin, L., Briand, J.P. and Van Regenmortel, M.H.V. (1989) *Arch. Virol.* 104, 271–285

Radavsky, J.L., Grama, D.L., Zaikin, A.A., Witer, S.S., Turova, I.P., Bobkova, A.F., Goldstein, M.I., Järvekülg, L.V., Sober, J.P. and Saarma, M.J. (1988) *Bioorg. Khim.* 14, 20–26

Raghavendra, K., Adams, M.L. and Schuster, T.M. (1985) *Biochemistry* 24, 3298–3304

Raghavendra, K., Kelley, J.A., Khairallah, L. and Schuster, T.M. (1988) *Biochemistry* 27, 7583–7588

Saarma, M., Järvekülg, L., Hemmilä, I., Siitari, H. and Sinijärv, R. (1989) *J. Virol. Methods* 23, 47–54

Sawyer, L., Tollin, P. and Wilson, H.R. (1987) *J. Gen. Virol.* 68, 1229–1232

Sherwood, J.L., Sanborn, M.R. and Keyser, G.C. (1987) *Phytopathology* 77, 1158–1161

Sherwood, J.L., Sanborn, M.R., Keyser, G.C. and Myers, L.D. (1989) *Phytopathology* 79, 61–64.

Shukla, D.D., Strike, P.M., Tracy, S.L., Gough, K.H. and Ward, C.W. (1988) *J. Gen. Virol.* 69, 1497–1508

Shukla, D.D. and Ward, C.W. (1988) *J. Gen. Virol.* 69, 2703–2710

Shukla, D.D., Jilka, D., Tosic, M. and Ford, R.E. (1989) *J. Gen. Virol.* 70, 13–23

Siitari, H. and Kurppa, A. (1987) *J. Gen. Virol.* 68, 1423–1428

Sinijärv, R., Järvekülg, L., Andreeva, E. and Saarma, M. (1988) *J. Gen. Virol.* 69, 991–998

Sober, J., Järvekülg, L., Toots, I., Radavsky, J., Villems, R. and Saarma, M. (1988) *J. Gen. Virol.* 69, 1799–1807

Thomas, J.E., Massalski, P.R. and Harrison, B.D. (1986) *J. Gen. Virol.* 67, 2739–2748

Torrance, L. (1987) *J. Virol. Methods* 15, 131–138

Torrance, L., Larkins, A.P. and Butcher, G.W. (1986) *J. Gen. Virol.* 67, 57–67

Torrance, L. and Pead, M.T. (1986) In *Developments and Applications in Virus Testing* (R.A.C. Jones and L. Torrance, Eds) Association of Applied Biologists, Wellesbourne. pp. 103–120.

Tremaine, J.H., MacKenzie, D.J. and Ronald, W.P. (1985a) *Virology* 144, 80–87

Tremaine, J.H., Ronald, W.P. and MacKenzie, D.J. (1985b) *Phytopathology* 75, 1208–1212

Van Pelt-Heerschap, H., Verbeek, H., Huisman, M.J., Loesch-Fries, L.S. and Van Vloten-Doting, L. (1987) *Virology* 161, 190–197

Van Regenmortel, M.H.V. (1975) *Virology* 64, 415–420

Van Regenmortel, M.H.V. (1982) *Serology and Immunochemistry of Plant Viruses.* Academic Press, New York, pp. 1–302.

Van Regenmortel, M.H.V. (1985) In *Immunochemistry of Viruses. The Basis for Serodiagnosis and Vaccines* (M.H.V. Van Regenmortel and A.R. Neurath, eds.) Elsevier, Amsterdam, pp. 467–478.

Van Regenmortel, M.H.V. (1986a) In *Developments and Applications in Virus Testing* (R.A.C. Jones and L. Torrance, Eds) Association of Applied Biologists, Wellesbourne, pp. 89–101.

Van Regenmortel, M.H.V. (1986b) In *The Plant Viruses, Vol. 2. The Rod-Shaped Plant Viruses* (M.H.V. Van Regenmortel and H. Fraenkel-Conrat, Eds) Plenum Press, New York, pp. 79–104.

Vela, C., Cambra, M., Cortes, E., Moreno, P., Miguet, J.G., Perez de San Roman, C. and Sanz, A. (1986) *J. Gen. Virol.* 67, 91–96

Weiss, E. and Van Regenmortel, M.H.V. (1989) *J. Virol. Methods* 24, 11–26

Wellink, J., Jaeglé, M. and Goldbach, R. (1987) *J. Gen. Virol.* 61, 236–238

Ziegler, V., Laquel, P., Guilley, H., Richards, K. and Jonard, G. (1985) *Gene* 36, 271–279

Zimmermann, D. and Van Regenmortel, M.H.V. (1989) *Arch. Virol.* 106, 15–22

Zrein, M., Burckard, J. and Van Regenmortel, M.H.V. (1986) *J. Virol. Methods* 13, 121–128

Eds. M.H.V. van Regenmortel and A.R. Neurath
Immunochemistry of viruses, II. The basis for serodiagnosis and vaccines.

Bacteriophages 23

J.L. CARRASCOSA[1] and E. KELLENBERGER[2]

[1]*Centro de Biologia Molecular, Facultad de Ciencias, Universidad Autonóma de Madrid, Madrid, Spain, and*
[2]*Department of Microbiology, Biozentrum, University of Basel, Klingelbergstrasse 70, 4056 Basel, Switzerland*

1. Introduction

Although bacteriophages played an important role in early work aimed at understanding the underlaying mechanisms of virus neutralization (for references see Mandel, 1979), their immunochemistry has not been a subject of strong interest, except for basic questions related to their life cycle. Since bacteriophage are only rarely associated with diseases (for instance in certain well-documented enteric infections; Flewett et al., 1976), they have not been studied by the powerful immunological methods used in clinical virology. Nevertheless, the development of new experimental techniques, both on the biochemical side and in electron microscopy, has made it possible to combine these with immunological methods in the study of detailed aspects of the structure and function of bacteriophage.

The techniques used fall into four categories: (a) immunocytochemistry by immunolabeling of thin-sectioned infected cells. This allows the localization of antigens in situ in the context of the intracellular process of particle assembly (section 2); (b) immunolabeling of isolated particles for direct observation in the electron microscope. This technique can localize different antigens on a particle (section 4), or even different epitopes on a protein (section 5); (c) the widely applied technique of immunoblots from gel electrophoretic samples (Towbin et al., 1979) which we will not discuss in this chapter; (d) Immuno-coprecipitation of complexes combined with an analysis by gel electrophoresis and immunoblotting of the constituent proteins (section 3).

In this review we have chosen examples related to a well-studied system, the bacteriophage T-even, to show various ways of combining biochemical methods with immunology and electron microscopy in the study of different aspects related to the structure and morphogenesis of viral structures.

The structure of the mature bacteriophage T4 studied by electron microscopy had been refined by the use of immunolabels (Yanagida and Ahmad Zadeh, 1970). This technique provided new information on the tail fibers and produced confirmatory evidence for data obtained earlier by indirect methods. In section 4 we discuss more recent applications for studying phage-related particles.

The morphogenesis of bacteriophage has been a subject of interest for a long time and has offered insights into the basic mechanisms of assembly of biological macromolecules (for reviews, see Murialdo and Becker, 1978; Wood and King, 1979; Casjens and King, 1988). Nevertheless, although some general rules of assembly have been established, there are particular aspects of the process that demand further clarification. We will focus on some of these aspects to show that a combination of approaches can be very informative.

Bacteriophage T4 morphogenesis is a complex process that comprises assembly of proheads and their successive maturation to particles that have become packaged with DNA; this renders the viral head ready to interact with the tail that was previously independently assembled through a sequential pathway of intermediates. This process involves not only phage-encoded proteins but also cellular components that play important transient functions during the intermediate assemblies. The classical approach for studying viral morphogenetic pathways has been the use of conditional-lethal mutants. These mutants produce — under restrictive conditions — an altered, mostly nonfunctional protein, thus leading to the accumulation of intermediates (or aberrant products). The study of these intermediates indicates which aborted function was carried out by the particular gene product (Epstein et al., 1965).

During the very early steps of the T4 morphogenetic pathway, a particular phenotype was observed, caused by one of three different types of mutations: one in a phage gene (gp31) (Laemmli et al., 1970) and the other two in host genes (Georgopoulos et al., 1972; Simon et al., 1975). The altered phenotype was characterized by the presence of protein aggregates called 'lumps'. These lumps were variable in size and shape and were attached to the membranes (Kellenberger, 1966). Since they are difficult to isolate, the biochemical analysis of such aggregates is not easy (Laemmli et al., 1970) and their immunochemical characterization is a good example for the power of immunologically based analytical methods.

A further application (section 4) is the use of Fab fragments for localizing protein domains and their relative movements during conformational changes. For such applications modern methods of information processing involving the use of multiple, identical molecules are necessary in order to increase resolution.

Presently, receptors are of central interest in biology. The phage receptors on the bacterial surface again provide an excellent experimental model system which we will shortly discuss in section 6.

2. Immunocytochemical localization of viral components in cells.

To understand the role of the different components involved in the formation of lumps and their possible interactions, Dürrenberger et al. (1990) studied thin sections of different mutants of *Escherichia coli* infected with different T4 mutants. Three systems were particularly informative: *E. coli* wild type infected with T4 mutant in gene 31 (the viral gene involved in the lump formation) and *E. coli* hd B3-1 and *E. coli* B178 groE$_{L44}$ infected with T4 wild type. Both *E. coli* mutants presented the lump phenotype after phage T4 infection. The proteins involved were, respectively, fatA (which causes an alteration of the bacterial membrane), and groE, a heat shock protein that, when mutated, reduces the phage yield by a factor up to 10^{-6} (Georgopoulos et al., 1972; Simon et al., 1975).

The lumps were specifically labeled with sera against the major T4 head protein (gp23) but not with gp22 and gp68 antibodies (Fig. 1a,c,d). The distribution of protein p22 (the major scaffolding protein) as revealed by immunolabeling, falls into two subclasses: the infection with a mutant in gene 31 shows the p22 label concentrated on a relatively dense, ribosome-free area adjacent to the membrane (Fig. 1b). When *E. coli* mutants (either in gene fatA or groE) become infected with wild-type T4, the label of protein p22 is found dispersed over the whole cytoplasm (Fig. 1f,d). These results suggest that the two host mutations lead to an abortive phenotype characteristic of an early step in head formation, while the mutation of phage gene 31 arrests the process at a later step.

3. Immunocoprecipitation of interacting protein complexes

During the first steps of phage head formation, the bacterial protein groE and the connector protein of bacteriophage λ has been shown to form a complex (Murialdo, 1979). Also, a similar complex was observed in *Bacillus subtilis* infected with phage φ29 (Carrascosa et al., 1982). These complexes are transient and difficult to isolate by standard biochemical methods; in order to determine which T4 proteins interact with groE, Dürrenberger et al. (1989) precipitated T4 infected extracts with sera against groE. The analysis by polyacrylamide gel electrophoresis revealed (Fig. 2) that two viral proteins co-precipitated with groE: gp20 (the connector protein) and gp23 (the major head protein). These complexes were not found when a mutated groE$_{L44}$ was provided. A further test to characterize the interactions between the T4 connector protein (gp20) and groE was done by immunoblot (Fig. 3). That the complex consisting of both proteins was coprecipitated by sera against p20 was unambiguously confirmed by blotting the polypeptides, respectively, with anti-groE and anti-gp20 sera.

Based on these data, an interesting hypothesis was proposed for the interaction of these two proteins. It has been described that groE and other related proteins can

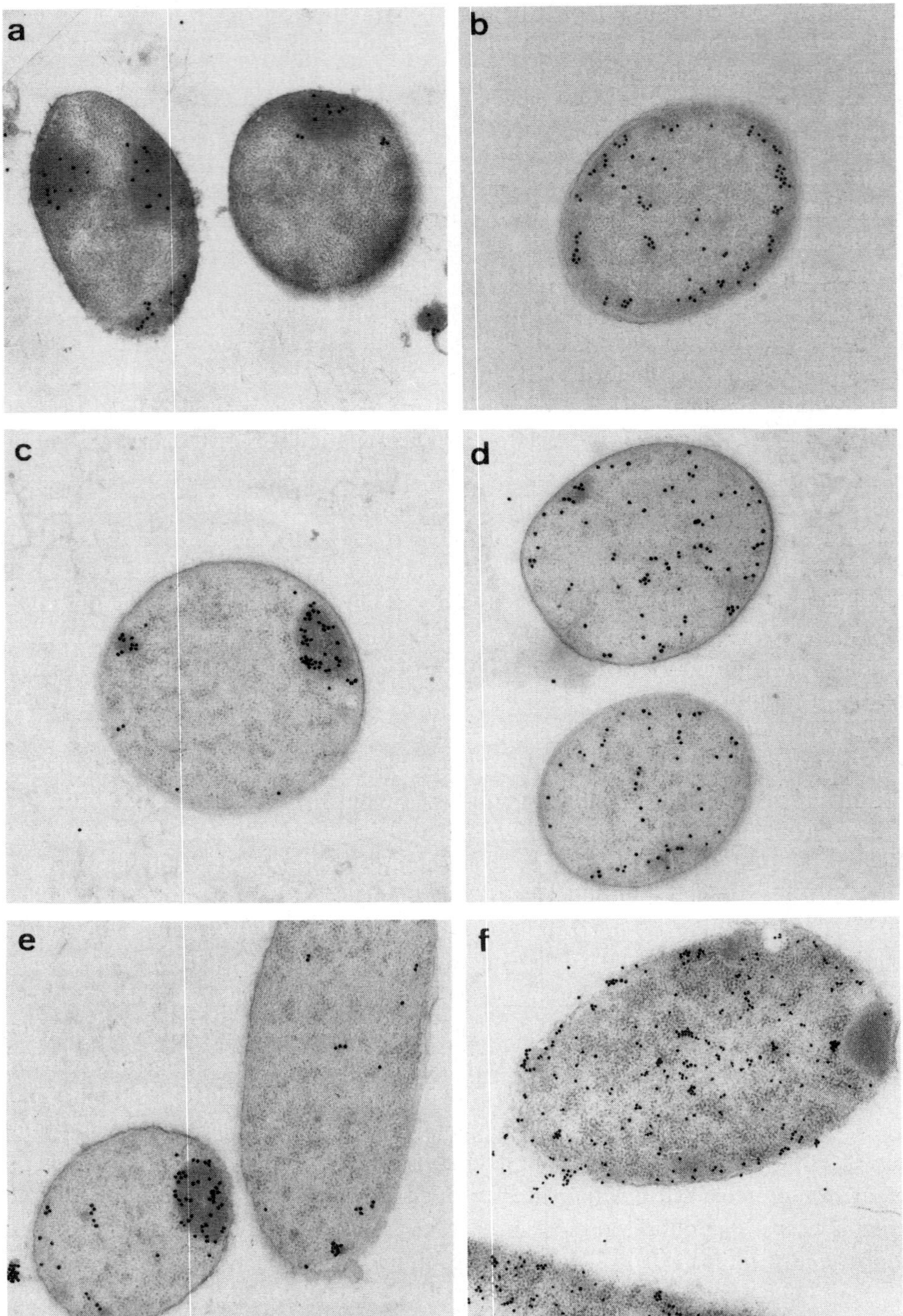
a
b
c
d
e
f
1 μm

act as a 'chaperone' for the assembly of proteins (Hemmingsen et al., 1988; Bochkareva et al., 1988; Golubinoff et al., 1989). In the particular cases of gp20 and gp23, the correct assembly for the connector and capsomer respectively could be assisted by the groE and the resulting complex could then migrate to the membrane where the further steps of assembly would proceed.

4. Immunomicroscopical localization of viral components.

The *connector* is a viral component that plays a fundamental role in phage head morphogenesis. For this reason it has received much attention and attempts have been made to follow its function along the head assembly pathway. We have already discussed its possible role as an initiator of the head assembly of many phage. Phage P22 seems to be an exception because both in vivo (Bazinet and King, 1988) and in vitro (Prevelidge et al., 1988) normal prohead assembly proceeds without the portal protein or connector, while with T-even phage the connector is needed in vivo and in vitro and with other phage at least in vivo (for further discussions see Kellenberger, 1990). The connector-directed assembly of scaffold and shell seems to be simultaneous. Assembly of the scaffold and shell proteins directed by the connector leads to the prohead. If this hypothesis is correct, then the connector must be located in a peculiar vertex of the phage head. Probably, the most sensitive method to study this localization is immunoelectron microscopy. Antibodies raised against gp20 were incubated with proheads and were visualized by labeling either with ferritin-conjugated goat anti-rabbit antibodies or with protein A-gold complex (Driedonks and Caldentey, 1983). The label was found at one specific vertex of the prohead, associated with a morphological component visible by negative staining (Figure 4). The figure shows that it is also possible to detect the gp20 connector in expanded heads (which represent a later stage of the head formation), thus confirming the presence of gp20 as portal protein in the proximal vertex of the capsid during the whole head assembly process.

The potential for *immunodetection of specific structural components* in viral particles, under conditions where the presence of different species of components renders their

---

←

Fig. 23.1. The label for gp23 (a,c,e) and for gp22 (b,d,f) on *E. coli B* infected with T4.31 (am N54) at 37°C (a,b) and on *E. coli B178* groE$_{L44}$ infected with T4D (c) are shown. Lumps occur in all three infections and are composed of gp23 (a.c.e). The three infections differ in the localization of gp22. In the 31$^-$ infection the gp22 is located over the viroplasm situated along the inner membrane (b). In this viroplasm naked scaffolding cores can be distinguished at higher magnifications. In the other two infections, the gp22 label is homogeneously distributed over the ribosomal area (d,f). These infections do not build up head-related particles. The label is made by the precoupling method on Lowicryl HM20 thin sections; the background is in all preparations smaller than 0.1 counts per μm^2 (from Dürrenberger 1989 with permission).

host	E.Coli B	E.Coli B	B178 groEL44	B178 groEL44	E.Coli B
infection	none	T4D	none	T4D	23(amH11)
pulse time	2 min	12 min p.i. to 14 min p.i.	2 min	12 min p.i. to 14 min p.i.	12 min p.i. to 14 min p.i.

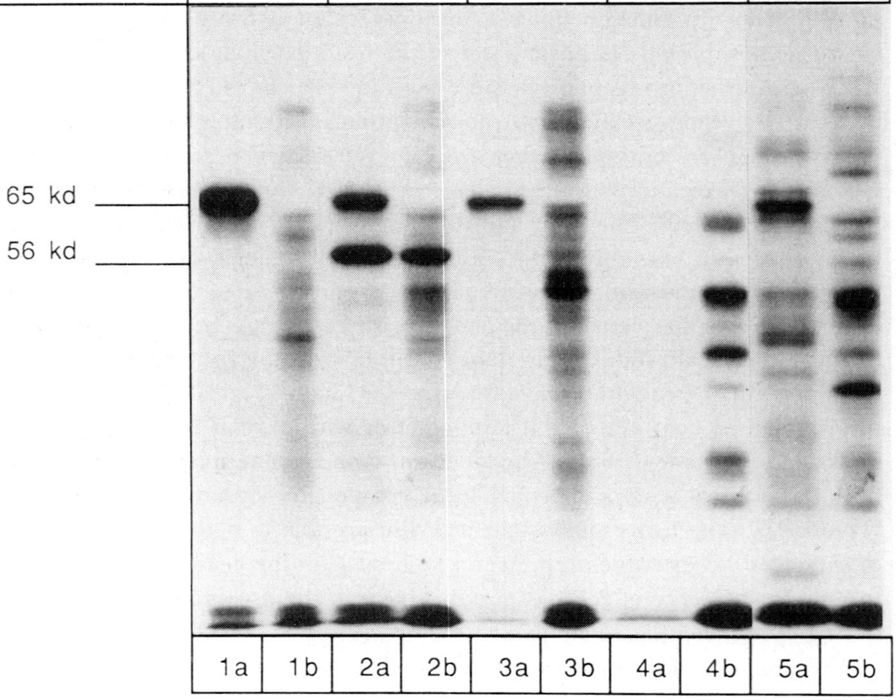

Fig. 23.2. Shows the analysis of the immunocoprecipitates of T4 gene product in complex with groE protein. The lanes marked with a are precipitates and coprecipitates and the lanes marked with b are the supernatants from which the a-lanes have been precipitated. The lanes 1a/1b and 3a/3b are control-experiments where *E. coli* proteins have been radioactively pulse-labeled to show that the serum is monospecific. The lanes 2a/2b, 4a/4b and 5a/5b are from experiments where only T4 proteins are pulse-labeled to show coprecipitates with groE, because shortly after infection the host DNA is broken down and no host proteins can anymore be newly synthesized. Some of the groE protein is found in complex with gp20 and gp23 after infection (lane 2a). In a 23^- infection most gp20 is found in a complex with groE (lane 5a). The mutated $groE_{L44}$ protein does not interact with any of the T4 proteins at all temperatures (lane 4a) from (Dürrenberger 1989, with permission).

precipi-tation	anti 20	anti 20	anti 20	anti 20	none	anti 20	anti 20	anti 20	anti 20	none
blot anti	groE	groE	groE	groE	groE	gp 20	gp 20	gp 20	gp 20	gp 20
host	groEL44	groEL44	B	B	B	groEL44	groEL44	B	B	B
infec-tion	T4	23-	T4	23-	T4	T4	23-	T4	23-	T4
lane	1	2	3	4	5	6	7	8	9	10

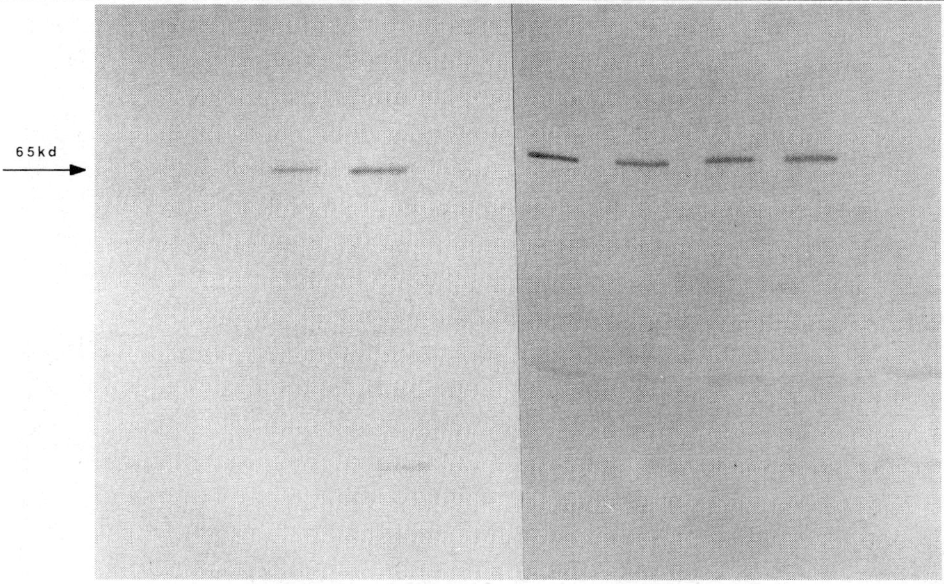

Fig. 23.3. An immunoblot analysis of gels of coprecipitates, as in Fig. 2, is shown. The single lanes are specified on top of the figure. The figure demonstrates that groE can also be coprecipitated in complex with gp20 (from Dürrenberger 1989, with permission).

analysis by conventional biochemical methods more difficult, has been made to dissect certain steps of the phage head morphogenetic pathways.

T4 mutants in gene 17 produced, under restrictive conditions, two types of head-related, empty small (ESP) and empty large (ELP) particles that differed slightly in their size. Protein analysis confirmed that the major head protein of both particles was in the cleaved state (characteristic of late steps of head maturation), but the ESPs presented this protein arranged in an unexpanded state, while the ELPs were expanded (Carrascosa and Kellenberger, 1978; Steven and Carrascosa, 1979).

The characteristics of ESPs (small size, roughness, etc.) suggested that they were more similar to the T4 proheads, whereas ELPs were almost like empty final capsids. To further establish this point, a mixture of ESPs and ELPs were incubated with serum against hoc protein. This protein is a structural component of the normal, mature T4 heads assembled in the latest steps of head maturation (Carrascosa and Kellenberger, 1978). Fig. 5 shows that only the expanded ELPs bind the antibody while the ESPs gave no reaction. This result agrees with the idea that these two head related particles represent successive steps of the morphogenesis and demonstrates that hoc protein only binds to the head after the expansion of the lattice into which the proteolyzed major head protein rearranges.

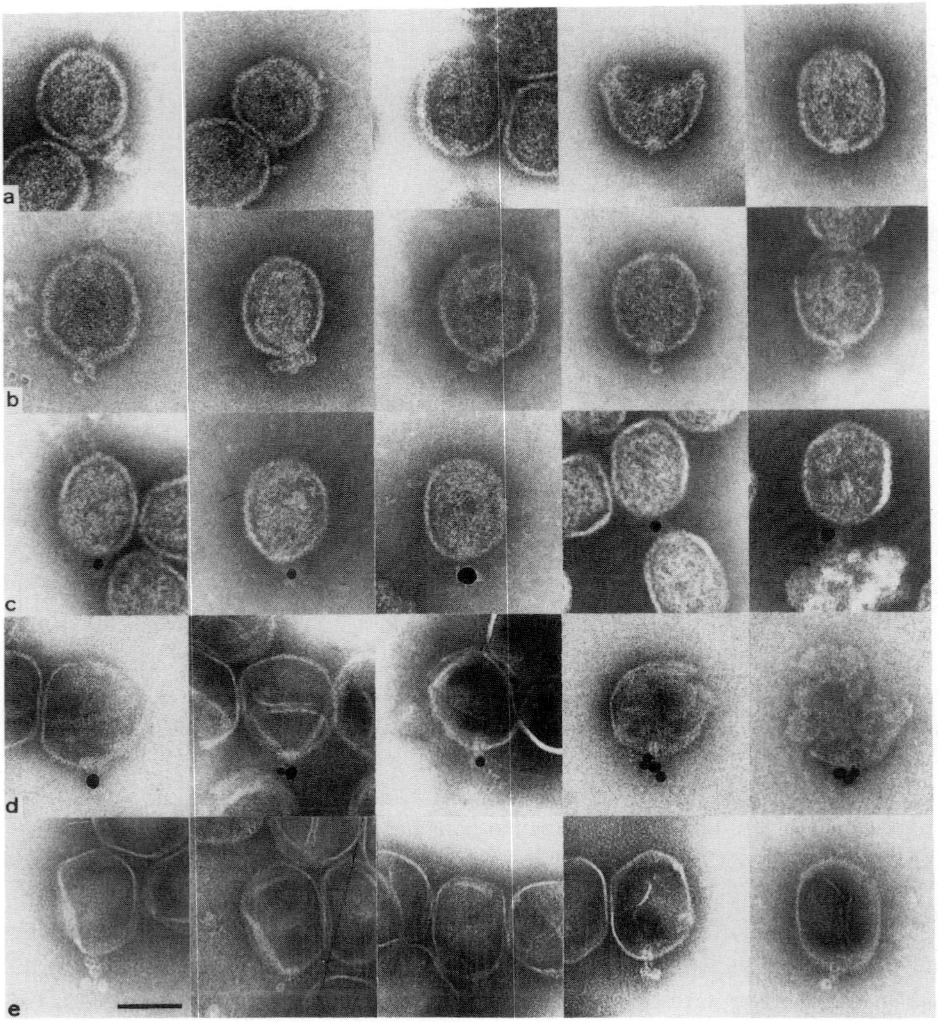

Fig. 23.4. A gallery of electron micrographs of head-related structures showing the location of gp20. (a) Proheads that show substructure in the 'plug' at the proximal vertex. The central hole and the radial protrusions of the gp20 oligomer can be recognized. (b) Proheads incubated with anti-gp20 antibodies and ferritin label. (c) Proheads labeled with anti-gp20 antibodies and protein A-gold. The gold label appears as an intense black spot and has a 'hole-like' appearance. However, a real hole would be white on the micrograph. (d) Capsids labeled as in (c). (e) Capsids labeled as in (b). Preparations were negatively stained with 2% sodium phosphotungstate (pH 7.0). The bar represents 50 nm (from Driedonks and Caldentey, 1983, with permission).

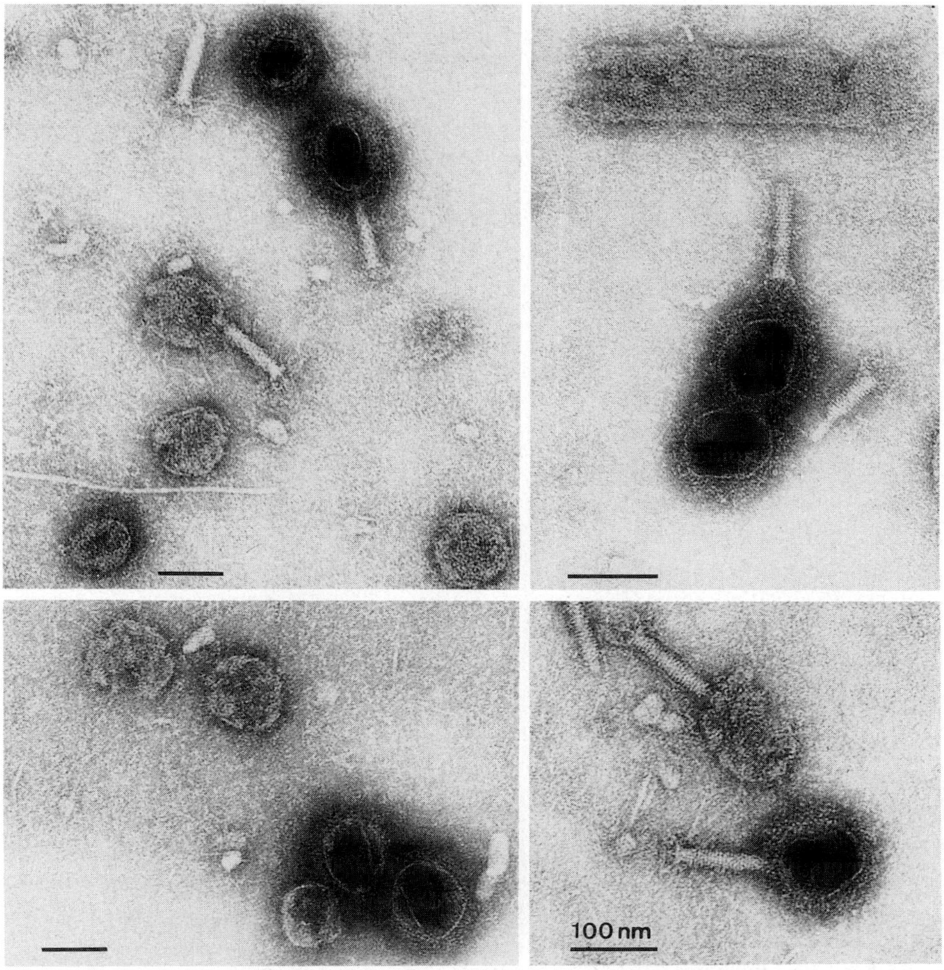

Fig. 23.5. Expansion of the shell of the prohead into the capsid is accompanied by binding of hoc (and soc). A mixture of 17⁻ particles (ESPs and ELPs) was reacted against antiHOC serum, washed and negatively stained. While the rough ESPs do not react, the ELPs show clear labeling all over the capsid, thus revealing the presence of HOC in these particles.

The immunochemical study of the *tail fiber* of various strains of T-even phage has produced very interesting results. The pioneering approach of Yanagida and Amad Zadeh (1970) for localizing the different surface antigens of bacteriophage T4 was based on the fact that the antibodies were visible even without a heavy metal label when the antigens form a multimeric structure (Anderson et al., 1961). The antibody-coating then shows up like a sort of fur (Figs.5 and 6a,b,c). Yanagida and Amad Zadeh (1970) used a complete anti-T4 serum which they depleted with lysates of am-mutants in different genes. The am-fragments of the considered gene

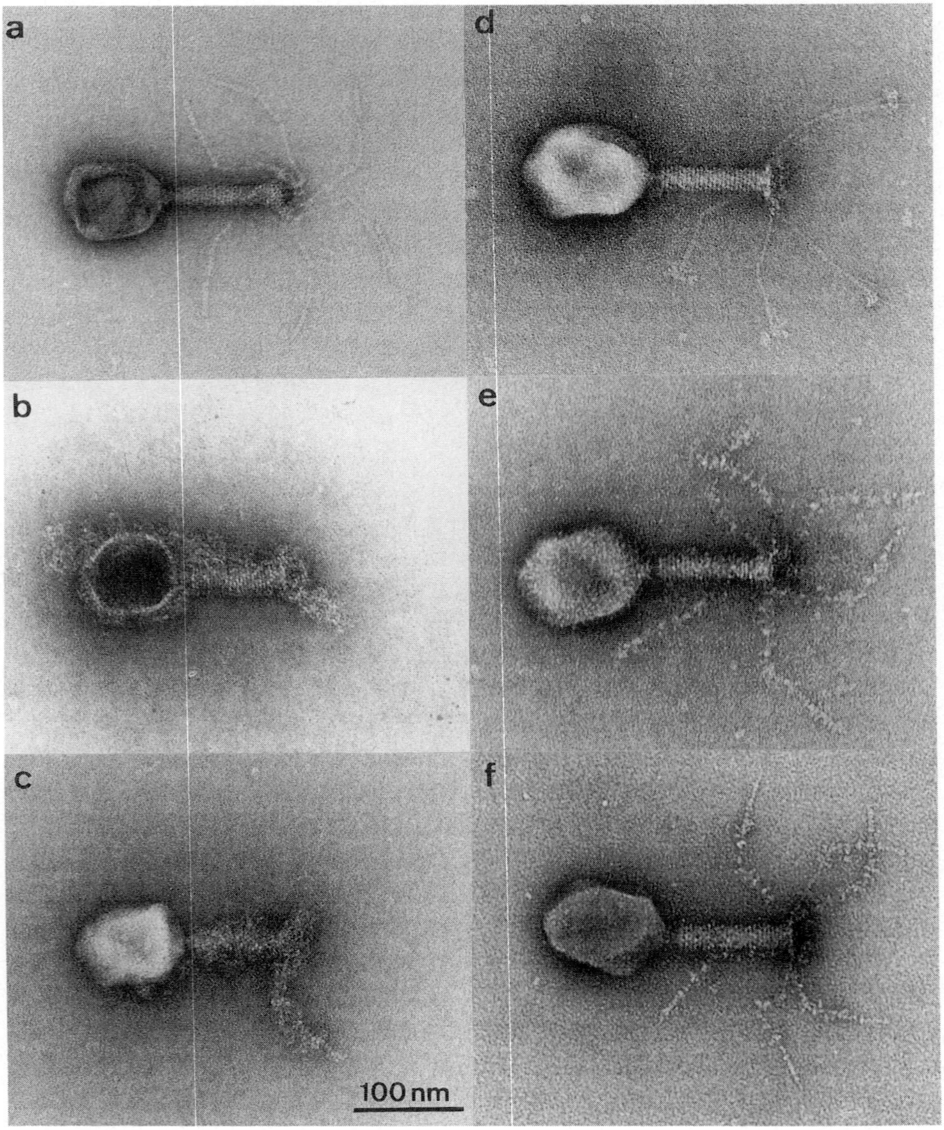

Fig. 23.6. Analysis of T-even tail fibers by direct immunolabeling and negative staining. (a) Phage TuII*-46 without label. (b) Phage TuII*-6 labeled with anti-TuII*-6-serum. (c) Phage T2 labeled with anti-T4 serum; the head is not labeled. (d) Phage T4 labeled with anti-gp314 serum. ORF 314 is a DNA sequence in phage λ which shows homology with the sequence of the distal part of the gene 37 of T4 for the distal half fiber. (e) Phage TuII*-6 labeled with Fab fragments of the homologous antiserum. (f) Phage TuII*-46 labeled with Fab fragments of the serum of Fig. 6e. The tail tips show no epitopes against this serum. (Unpublished micrographs, courtesy of H. Schwarz)

product did not adsorb the serum and therefore the remaining activity was directed specifically against the considered gene product. With this technique and immunoelectron microscopy they confirmed the presence of gp23 of the capsid and gp18 of the tail sheath and produced for the first time the sequence of antigens in the tail fiber: starting from the base plate as a proximal half fiber with gp34, followed by a short piece of gp36 after the kink and then by the distal half fiber with gp37.

The group of Henning refined this technique by using antibodies produced against isolated gene products (Figure 6d). The main result of their work was the demonstration that gp38 — or its analogue in other strains — is contributing to the host specificity of the strain. By immunoelectron microscopy they could show that this protein was located at the tip region of the tail fiber (Riede et al., 1987). By these immunological techniques they could also demonstrate the curious fact that a corresponding protein of phage λ could functionally replace the gp38 of the T-even phage. The hybrid phage now needs OmpC as receptor (Montag et al., 1989).

5. Antibody labeling of protein domains

Although it has been possible to extend the resolution of electron microscopy to the molecular level, mainly by the application of image processing techniques which extract averaged information in two and three dimensions, it is often difficult to map protein domains by direct imaging (for a review, see Stewart, 1988a,b; Radermacher, 1988). Immunomicroscopy offers the possibility to use antibodies either monoclonal or monospecific, raised against specific domains of proteins to label them in their assembled complexes. On the bacteriophage T-even tail fiber, individual epitopes labeled by Fab can be distinguished clearly by direct electron microscopy (Schwarz et al., 1983). The presence and position of these epitopes can now be compared among the different relatives (Fig.6 e,f).

By applying image processing of many identical proteins, it is possible to compare the high resolution averaged information of the images with and without antibodies, which leads to a tentative mapping of the corresponding epitopes. The method is limited by the actual size of the immunological probe which is either the entire immunoglobulin G molecule or the Fab fragment. (Baschong and Wrigley, 1990). Fragmented Fab (e.g., Fv) has also given promising results (U. Aebi, personal communication). Uncertainty concerning the location of the non-interacting (non-specific) area of the antibody or antibody fragment still limits the attainable resolution.

Aebi and colleagues (Aebi et al., 1977; Kistler et al., 1978, 1982; Buhle and Aebi, 1984) studied the localization of different polypeptides or polypeptide domains of a protein assembled in an ordered supramolecular lattice. Although such a crystalline order is not an absolute requirement of the method, it makes it possible to apply the straight forward methods of image processing to enhance the resolution. These studies have been made with the shells of either in vitro matured polyheads that are constituted of the cleaved major shell protein (gp23) or of the in vivo obtained giant phage of the same constitution. The lattices of these two structures are virtually identical.

The averaged images of the gp23* assembled into giants are shown in Fig. 7. Both the unprocessed and the processed images are shown.

Buhle and Aebi (1984) used tubular variants of bacteriophage T4 (matured polyheads). Two different Fab fragments of (quasi monovalent) antibodies directed against two different epitopes of gp23*, were incubated with the polyheads and the samples were then either negatively stained or freeze-dried.

By using a method known as surface reconstruction (Smith and Kistler, 1977), it became apparent that Fab labeling changed the appearance of the gp23* lattice

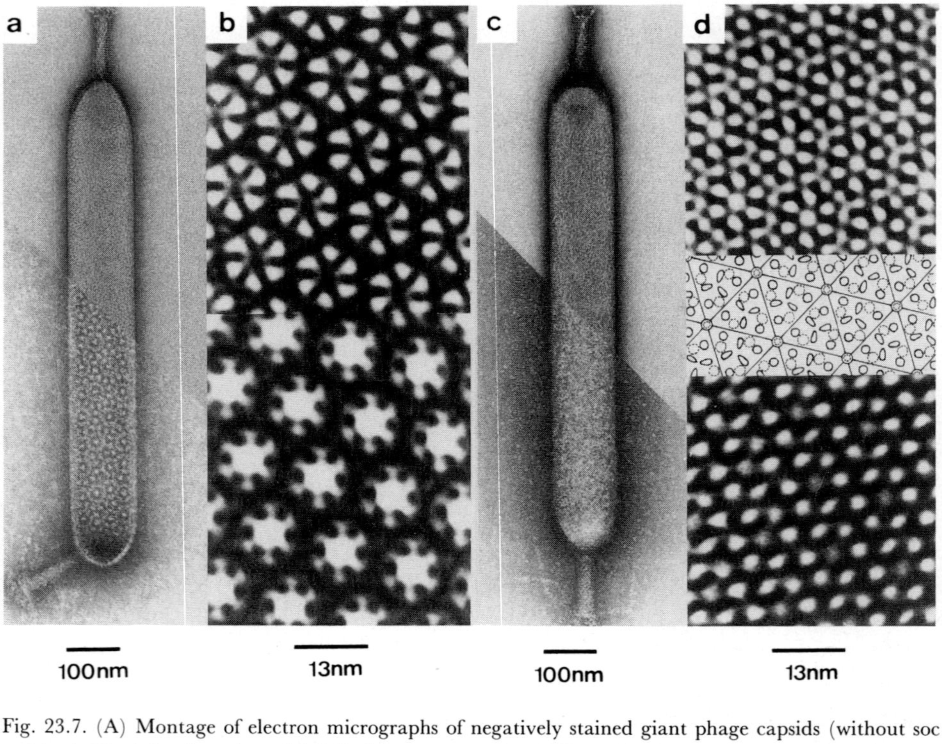

Fig. 23.7. (A) Montage of electron micrographs of negatively stained giant phage capsids (without soc and hoc) (Top) that have been labeled with an excess of anti-P23 Fab fragments (Bottom) prepared from a serum raised against P23-containing coarse polyheads (without soc and hoc). (B) Optical filtered areas of the micrographs shown to the left in A. (C) Montage of electron micrographs of negatively stained giant phage capsids (with soc and hoc), (Top) that have been labeled with an excess of anti-P23 Fab fragments (Bottom) prepared from a serum raised against P23-containing coarse polyheads. (D) Optically filtered areas of the micrographs shown to the left in C. In the middle, the unlabeled (solid lines) and the labeled (broken lines) stain-excluding patterns are schematically superimposed. (Aebi et al., 1977, with permission)

(Fig. 8). There were clear differences depending on the type of Fab used: one of them (label 1), seemed to bind (i) peripherally to the six 'petals' of the gp23* capsomer and (ii) near the 'center' of the capsomer (i.e. a centro-symmetric arrangement of 6 protein subunits that builds the capsid shell). The other type of Fab fragments (label 2), bound specifically to a site near the 'inner ring' of the capsomer. The labeling of the center of the capsomer with each Fab was distinct in the sense that label 1 bound more closely to the center, while label 2 was localized more to an inner ring surrounding the central area of the capsomer. This precise localization of specific areas by antibody labeling opens up the possibility of using antibodies produced against chosen synthetic peptides to map a particular part of the primary sequence within three-dimensional structures of the folded peptide. In addition it makes it possible to follow the movement of protein domains during conformational changes. An application of this approach will be reported now.

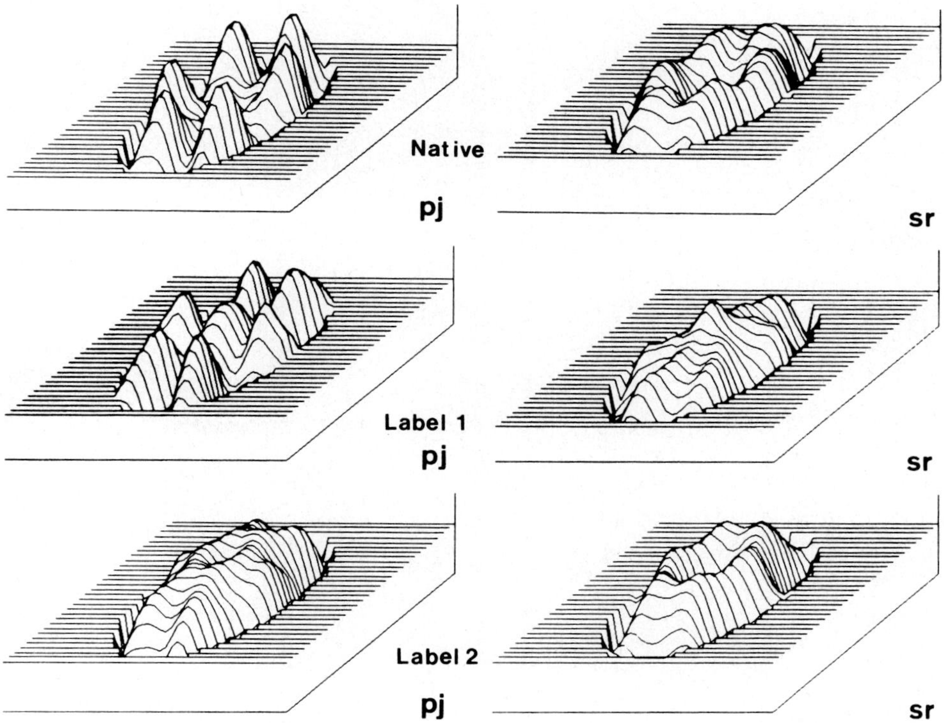

Fig. 23.8. Projection plots of averaged capsomers from negatively stained (left) and of surface reconstructions from freeze-dried and metal-shadowed (right) native (top), label 1 (middle), and label 2 (bottom) Fab-labeled polyheads. It should be noted that in the case of the negatively stained capsomers, the projection plots do not display actual three-dimensional information about the samples. The z-coordinate merely represents relative grey levels of the averaged projection data. In the case of the unidirectionally metal-shadowed capsomers (right), however, the z coordinate represents surface elevations such that the projection plots display actual surface reliefs of the unlabeled (top) and Fab fragment labeled (middle and bottom) averaged gp23* capsomers. (Buhle and Aebi, 1984, with permission).

The tubular variant of the prohead of phage T4 is produced intracellularly and is therefore constituted of the still uncleaved major head protein gp23. It is not further processed intracellularly despite the presence of both the T4ppase (T4 prohead protease) and its zymogen (gp21) (for a summary and references see Black and Showe, 1983). Isolated polyheads, however, can be processed in vitro by addition of crude preparations (lysates) containing the protease. In such a system the progressive transformation of the lattice has been studied extensively (for summaries

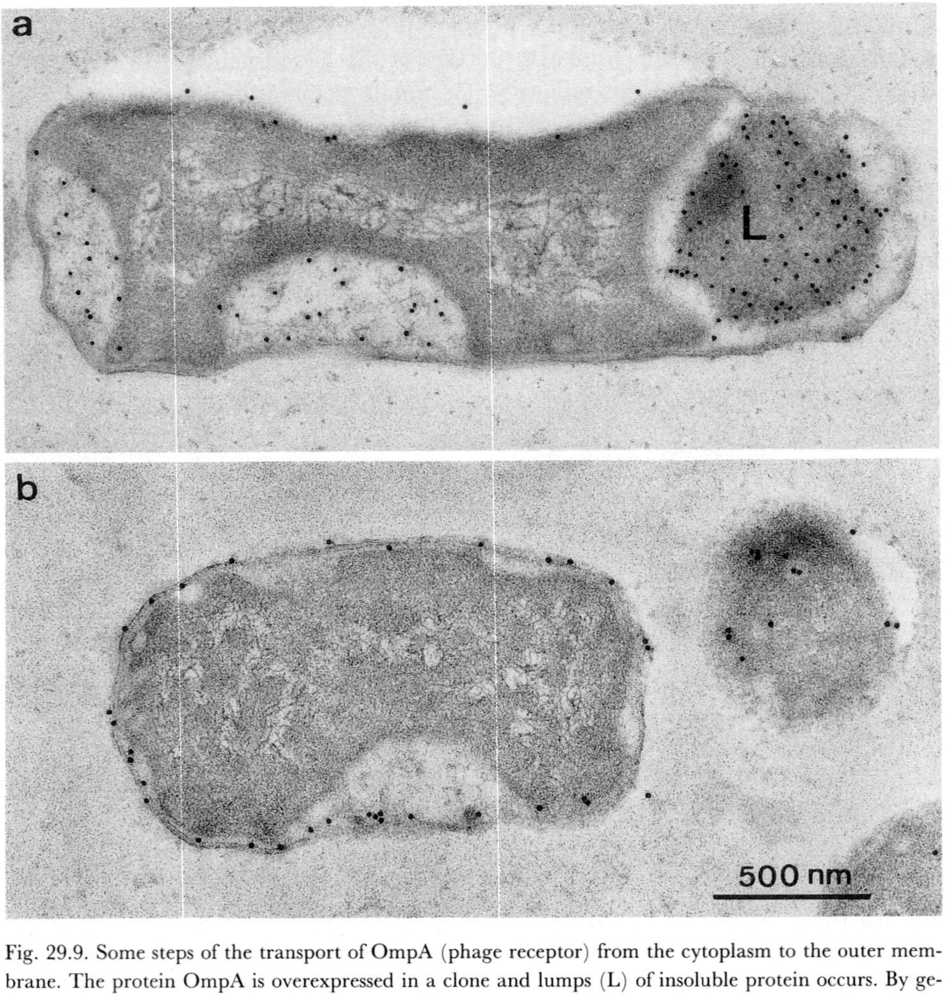

Fig. 29.9. Some steps of the transport of OmpA (phage receptor) from the cytoplasm to the outer membrane. The protein OmpA is overexpressed in a clone and lumps (L) of insoluble protein occurs. By genetic engineering different deletions were introduced into the gene, of which the consequences are arrests at specific steps of the transport. Thin sections of bacteria embedded in Lowicryl K4M are labeled with OmpA antibodies. In (a) (del. 1–229) the protein enters the periplasm (of which the volume is increased by plasmolysis). In (b), (del. 43–84) the protein is still able to reach its destination in the outer membrane (unpublished micrographs, courtesy of H. Schwarz).

and references see Kellenberger, 1978 and Black and Showe, 1983). By using a monospecific anti-gp23 Fab fragment, Kistler et al. (1978, 1982) were able to show that a given epitope moved from the inside of the polyhead to the outside as a consequence of protein cleavage and lattice expansion.

6. Immunocytochemistry of a phage T-even receptor

The tip of the distal tail fiber of T-even phage (see section 4) interacts specifically with an outer membrane protein of its host. OmpA, abundantly present in *E. coli* K12 is the receptor for T-even phage K3 (Schweizer et al., 1978). The adhesin-receptor interaction of phage is an excellent model system for studying protein-protein interactions, which goes back to Max Delbruck who very much emphasized this feature. The group around Henning in Tübingen has caused this field to flourish anew. The *E. coli* host system, has the pecularity of a complex cellular envelope since the inner (plasma) membrane is separated from the outer membrane by the periplasm. The receptor protein of OmpA is synthesized in the cytoplasm as a precursor which is processed during export (Freundl et al., 1987, 1989; Close et al., 1988). This export was studied by constructing various deletion mutants. Immunocytochemistry provided the means to study whether the protein either stayed in the cytoplasm (MacIntyre et al., 1988) or became stuck in the inner membrane or reached the periplasm without going further or — finally — by integrating into the outer membrane. These studies were made possible by the ingenious device of producing a plasmolysis, by which the periplasm was strongly enlarged so as to allow a rapid and significant count of the label that eventually accumulated in the periplasm (Freudl et al., 1987, 1985). These situations are shown in Fig. 9.

7. Acknowledgements

Our thanks go to the many colleagues and friends who allowed us to use their published and unpublished documents, and to Uli Aebi for reading and commenting the manuscript. We are grateful to Regula Niederhauser for patiently typing the manuscript. We are particularly indebted to Marlies Zoller, Hedi Frefel and Margrit Steiner for the excellent photographic work and the arrangement of micrographs. J.L. Carrascosa acknowledges a grant from the Commision Interministerial de Ciencia y Technologia (PB87-0365) and the institutional help from Fundacion Ramón Areces. Experimental work in the laboratory of E. Kellenberger has been subsidized by the Kanton Basel-Stadt and the Swiss National Science Foundation.

8. References

Aebi U., ten Heggeler, B., Onorato, L., Kistler, J. and Showe, M.K. (1977) *Proc. Natl. Acad. Sci. USA* 74, 5514–5518.

Anderson, T.F., Yamamoto, N., Hummeler, K. (1961) *J. appl. Phys.* 32, 1639.

Baschong, W., and Wrigley, N.G. (1990) *J. Electron Microsc Techn.* 14, 313.

Bazinet, Ch. and King, J. (1988) *J. Mol. Biol.* 202, 77–86.

Black, L.W. and Showe, M.K. (1983) In *Bacteriophage T4,* (C.K. Mathews, E. Kutter, G. Mosig and P.B. Berget, Eds.) American Society for Microbiology, Washington, D.C.

Bochkareva, E.S., Lissin, N.M. and Gishovich, A.S. (1988) *Nature* 336, 254–257.

Buhle, E.L., Jr. and Aebi, U. (1984) *J. Ultrastruct. Res.* 89, 165–178.

Carrascosa, J.L. and Kellenberger, E. (1979) *J. Virol.* 25, 931–844.

Casjens, S. and Hendrix, R. (1988) In *The Bacteriophage* (R. Calendar, Ed.) Vol. 1. Plenum Press, New York. pp. 15–91.

Dürrenberger, M. (1989) Thesis. Dept. Microbiol., University of Basel.

Dürrenberger, M., Villiger, W. and Kellenberger, E. (1990) submitted for publication.

Driedonks, R.A. and Caldentey J. (1983) *J. Mol. Biol.* 166, 341–360.

Epstein, R.H., Bolle, A., Steinberg, C.M., Kellenberger, E. Boy de la Tour, E., Chevalley, R., Edgar, R.S., Susman, M., Denhardt, G.H. and Lielausis, A. (1963) *Cold Spring Harbor Symp. Quant. Biol.* 28, 375–392.

Flewett, T.H., Bryden, A.S. and Davies, H. (1974) *J. Clin. Pathol.* 27, 603–614.

Freudl, R., Schwarz, H. Degen, M. and Henning, U. (1987) *J. Bacteriol.,* 169, 66–71.

Freudl, R., Schwarz, H. Degen, M. and Henning, U. (1989) *J. Mol. Biol.* 205, 771–775.

Georgopoulos, C.P., Hendrix, R.W., Kaiser, A.D. and Wood, W.B. (1972) *Nature* 239, 38–41.

Goloubinoff, P., Gatenby, A.A. and Lorimer, G.H. (1989) *Nature* 337, 44–47.

Hemmingsen, S.M., Woolford, C., van der Vies, S.M. Tilly, K., Dennis, D.T., Georgopoulos, C.P., Hendrix, R.W. and Ellis, R.J. (1988) *Nature* 333, 330–334.

Kellenberger, E. (1966) In *Principles of Biomolecular Organization* (G.E.W. Wolstenholme and H. O'Connor, Eds.) Little, Brown and Co., Boston. pp. 192–228.

Kellenberger, E. (1978) In *Electron Microscopy* Vol. III (J.M. Sturgess, Ed.) pp. 441–449.

Kellenberger, E. (1990) *Eur. J. Biochem.* 190, 233–248.

Kistler, J., Aebi, U., Onorato, L., ten Heggeler, B. and Showe, M.K. (1978) *J. Mol. Biol.* 126, 571–589.

Kistler, J., Showe, M.K., Carrascosa, J.L., Stirmer, B. and Onorato-Showe, L. (1982) *J. Mol. Biol.* 162, 607–622.

Klose, M., Schwarz, H., MacIntyre, S., Freudl, R., Eschbach, M.-L., Henning, U. (1988) *J. Biol. Chem.* 263, 13291–13296.

Laemmli, U.K., Beguin, F. and Gujer-Kellenberger, G. (1970) *J. Mol. Biol.* 47, 69–85.

MacIntyre, S., Freudl, R., Eschbach, M.-L. and Henning, U. (1988) *J. Biol. Chem.* 263, 19053–19059.

Mandel, B. (1979) In *Comprehensive Virology* H. Fraenkel-Conrat and R.R. Wagner, Eds.) Vol. 15, Plenum Press, New York and London. pp. 37–121.

Montag, D., Schwarz, H., and Henning, U. (1989) *J. Bacteriol.* 171, 4378–4384.

Murialdo, H. and Becker, A. (1978) *Microbiol. Rev.* 42, 529–576.

Murialdo, H. (1979) *Virology* 96, 341–367.

Prevelidge, P.E. Thomas, D. and King, J. (1988) *J. Mol. Biol.* 202, 743–757.

Riede, I., Drexler, K., Schwarz, H., and Henning, U. (1987) *J. Mol. Biol.* 194, 23–30.

Radermacher, M. (1988) *J. Electron Microsc. Techn.* 9, 359–394.

Schwarz, H., Riede, J., Sonntag, I., Henning, U. (1983) *EMBO J.* 2, 375–380.

Schweizer, M., Hindennach I., Garten, W. Henning, U. (1978) *Eur. J. Biochem.* 82, 211–217.

Simon, L.D., McLaughlin, T.J.M., Snover, D., Ou, J., Grisham, C. and Loeb, M. (1975) *Nature* 256, 379–383.

Smith, R.P. and Kistler, J. (1977) *J. Ultrastruct. Res.* 61, 124–133.
Steven, A.C. and Carrascosa, J.L. (1979) *J. Supramol. Struct.* 10, 1–11.
Stewart, M. (1988a) *J. Electron Microsc. Techn.* 9, 301–324.
Stewart, M. (1988b) *J. Electron Microsc. Techn.* 9, 325–358.
Towbin, H., Staehelin, T., Gordon, J. (1979) *Proc. Natl. Acad. Sci. USA* 76, 4350–4354.
Wood, W.B. and King, J. (1979) In *Comprehensive Virology* (H. Fraenkel-Conrat and R.R. Wagner, Eds.) Vol. 13. Plenum Press, New York and London.
Yanagida, M., Ahmad-Zadeh, C. (1970) *J. Mol. Biol.* 51, 411–421.

Subject index